# ENCYCLOPEDIA OF COMMON NATURAL INGREDIENTS

## USED IN FOOD, DRUGS, AND COSMETICS

**Second Edition**

# ENCYCLOPEDIA OF COMMON NATURAL INGREDIENTS

## USED IN FOOD, DRUGS, AND COSMETICS

**Second Edition**

ALBERT Y. LEUNG
**Consultant, Natural Products**
**Glen Rock, New Jersey**

STEVEN FOSTER
**Consultant, Medicinal and Aromatic Plants**
**Fayetteville, Arkansas**

**A Wiley-Interscience Publication**

**JOHN WILEY & SONS, INC.**

**New York • Chichester • Brisbane • Toronto • Singapore**

This text is printed on acid-free paper.



***Library of Congress Cataloging in Publication Data***

Leung, Albert Y.
Encyclopedia of common natural ingredients used in food, drugs, and cosmetics / by Albert Y. Leung and Steven Foster. -- 2nd ed.
p. cm.
"A Wiley-Interscience publication."
Includes index.
ISBN 0-471-50826-8 (acid-free paper)
1. Natural products--Encyclopedias. I. Foster, Steven, 1957– II. Title.
QD415.A25L48 1996
660'63--dc20 94-49668

Printed in the United States of America

10 9 8 7 6 5 4 3

**To Barbara and Our Children**

**Amy and Camille**

# Foreword to the First Edition

The publication of *Encyclopedia of Common Natural Ingredients Used in Food, Drugs, and Cosmetics* is a welcome addition to the libraries of those of us interested in natural products. The reasons for publishing this unique encyclopedia are aptly dealt with by the author in the Preface, and the principal audience has been identified as practicing technologists in the food, drug, and cosmetic industries and their purchasing agents and marketers. But, as well, it should prove to be an important reference for teaching and research in economic botany, food technology, natural products chemistry, and pharmacognosy, for it brings together information about a variety of substances which, for various reasons, are not included in recent compendia dealing with one or another of these disciplines. Yet, as the author points out, these are materials which find significant usage in our society.

Dr. Albert Y. Leung's education as a pharmacist and pharmacognosist, coupled with his extensive experience in natural products industries, provide him with a unique background that accounts for his successful synthesis of this information into a practical compendium. The material is accurately and succinctly presented, the individual monographs are selectively supplemented with a current bibliography that allows for further reading on a particular product, and the selection of products included has been skillful.

Dr. Leung is to be commended for his efforts in bringing us this most worthy publication.

ARA G. PAUL, PH.D.

*Professor of Pharmacognosy*
*and Dean*
*College of Pharmacy*
*The University of Michigan*
*Ann Arbor, Michigan*

# Foreword to the First Edition

By some peculiar irony, the rapid technological advances made by the chemical industry since World War II have worked to obscure the solid basic knowledge the industry once had of some of its natural building blocks, the botanicals that were (and still are) the prime ingredients in so many drugs, cosmetics, flavors, industrial reodorants, and so on. The recently trained chemist, pharmacologist, or food flavorist (or, for that matter, the person involved in sales, marketing, or purchasing of these materials) in all likelihood has missed the fact that these materials have considerable historical significance, that they still have application in so diverse a list of products, and even what specific role they play in familiar products. After all, these older, possibly no-longer-glamorous natural substances may seem unsophisticated and awkward to handle to those trained in the glories of what might be called synthetic chemistry—the molecular juggling of carbons, hydrocarbons, acids, and alcohols to evolve pristine crystals and powders.

Many of the veteran bench chemists with experience in natural materials have retired or passed on to their ultimate reward (hopefully, a golf cart or a fishing boat in some warmer clime), so the time is coming when there will be less use of such fascinating ingredients as bloodroot, horehound, or ylang ylang oil. The veteran chemists used these materials to make cough remedies or perfume oils before there were synthetics, and when they are gone, the individual little pockets of knowledge have been in danger of dying out. They appreciated that these unique materials provide special product attributes, in the same way that classic spices do for a good chef. Then too, much of the chemical and biological information has been buried in foreign scientific literature, thus making it unavailable to the average technologist.

Dr. A. Y. Leung has been observing this widening information gap for several years, perceiving that one logical way to bridge it was to put together a compendium of materials of natural origin. He has gone about the task with logic and a sense of order, selecting the cardinal facts without deluging the reader or peruser of the book with a veritable mountain of biological data. As befits the only reference book that covers food, drug, and cosmetic aspects of common natural ingredients, Dr. Leung has identified each entry according to biological name, its alternative or slang description, a general description of the plant from which it is derived, chemical composition, pharmacological or biological activity, and uses and commercial preparations. And for those needing more information, he has included a comprehensive list of references.

Such handy organization of material makes this book especially useful to the working chemist or technologist, to the purchasing director, and to the person in sales or product development or marketing, for in one fell swoop he or she is given clear, comprehensive information with no unnecessary embellishment. Exotica become less exotic, the strange becomes more familiar.

Because of the ongoing work of the *Cosmetic Ingredient Review* and the Research Institute for Fragrance Materials, it's a safe bet that in the not-so-distant future there will be a demand for a second edition of this monumental work. These efforts will produce much information about the toxicity or safety of these materials, information that will give better clues as to whether it may be opportune or diplomatic or safe from a regulatory standpoint to persist in using a material that may be allergenic or sensitizing. Also, Dr. Leung has tried to emphasize the quality of commercial preparations, with an eye toward the purchasers and end users who will ultimately find the volume so useful.

Anyone with a sense of romance will cherish the names of these materials, and anyone with an appreciation for order and thorough documentation will regard this book as useful and to-the-point.

DONALD A. DAVIS, EDITOR
*Drug and Cosmetic Industry*
*New York*

# Preface to the Second Edition

Fifteen years have passed since the publication of the First Edition. During this period, basic information on most of the traditional ingredients in the Encyclopedia has remained essentially the same. However, usages and use trends of many of these ingredients have changed. A whole new field of food/drug products, loosely categorized as "health foods," has established itself in North America; so has the use of herbal teas. Although a few conventional food and drug companies have tried to capitalize on the market generated by these new fields, most members of the medical, pharmaceutical, and food establishments have so far chosen to ignore them. However, these rapidly expanding fields of health foods and herbal teas have made it very difficult for one to ignore. For this reason, up-to-date and accurate information on their ingredients should be made readily available, not only to provide useful data for technologists and consumers interested in these ingredients, but also to counterbalance the proliferation of promotional literature from marketers that is often grossly inaccurate and misleading as well as negative information from opposing interest groups that is based on bias and self-interest and not on relevant traditional and scientific data available. Consequently, I have included information to support the "new" uses in this Edition. As Chinese herbs constitute a majority of all natural products used in the world, which have increasingly found their way into American cosmetic, health food, and herbal tea products in recent years, new ingredients described in this Second Edition reflect this trend.

Due to other commitments that limited my availability for this revision, I enlisted the capable assistance of Steven Foster who has earned a reputation for his writings. He has been instrumental in updating much of the information in the original entries of the Encyclopedia as well as introducing most of the new non-Chinese ingredients in this revision.

Concomitant with the development of health foods and herbal teas, many books on natural products have appeared since the First Edition was published. Some of these books contain well-researched information, while others are simply indiscriminate compilations of data, which only help to perpetuate the confusion relating to information in the herbal products field. In order to help minimize the spread of dubious data, books containing excessive outdated, secondary and/or misleading data as well as those consisting primarily of indiscriminate compilations of data, including some English titles on Chinese medicinal plants (even though aggressively marketed), are not included in the **general references.**

An unusual, but positive, alliance emerged during the past decade. Numerous scientists and practitioners from traditional fields (pharmacognosy, pharmacology, chemistry, medicine, botany, etc.) have joined forces with herbalists and manufacturers and developers of health foods and herbal teas to promote research and information dissemination in the field of medicinal plants and herbal products. Thus, the Herb Research Foundation, established in 1983 with an advisory board of respected scientists in various fields, has been engaged in promoting research in the various aspects of herbs. Along with the more recently founded American Botanical Council (1988), it publishes *HerbalGram*, a quarterly journal, which provides accurate information on many commonly used natural products. In this new Edition, we have selected as ***general references*** some of the books that we find useful and which we believe had an impact in the industry in recent years. Also, the extensive use of information from the Chinese literature in this new Edition reflects the greatly increased availability of data on natural products from China during the past decade. As there is no standard translation of Chinese pharmaceutical and biomedical titles, I have used the transliterated titles of such references whenever there is a possibility of confusion. The *pin-yin* system of transliteration has been selected over the Wade-Giles system because the former is now standard in Chinese literature originating in the People's Republic of China, which is by far the more abundant than that originating elsewhere.

Along with the greatly increased availability of information and books on natural products, I have observed a tendency in both the professional and lay press in the overly simplified interpretation of this information. It is tempting to assign the biological activity of a compound present in a natural product to the product itself prematurely, irrespective of the amount present. For example, taking this approach, the common spice, ginger, could easily be turned into a panacea as it contains dozens of active compounds each of which by itself has been shown to have various biological activities. These activities include: antimicrobial (essential oil components such as linalool, geraniol, chavicol, 1,8-cineole, etc.); narcotic (cumene); spasmolytic (borneol, myrcene); analgesic (borneol, gingerols, shogaols); diuretic (asparagine); antihistaminic (citral); lipotropic (lecithins); antiinflammatory ($\alpha$-curcumene, borneol); sedative (gingerols, shogaols); hypotensive (1,8-cineole, gingerols); hypertensive (shogaols); liver protectant (borneol); cardiotonic (gingerols); antipyretic (borneol, gingerols, shogaols); insect repellant (*p*-cymene, geraniol, myrcene); antibronchitic, antitussive and expectorant (1,8-cineole); nutrient (vitamins, minerals, amino acids); and others.[1] Under certain conditions or in specific dosage forms, ginger could indeed exert some of these effects. But to say that the spice ginger is narcotic or hypotensive is an oversimplification in interpretation.

This brings up one of the major challenges in natural products research, especially in Chinese herbs, which is to make sense out of their myriad of traditional uses. Some of the answers seem to lie in the complex chemical nature of these products. The bioavailability of these chemicals in an herbal formula or in an ingested herb is most likely very selective and dependent upon the physiological state of the individual consumer. This may be one of the major reasons why ginseng and other tonics have been used for so many centuries in China for so many different conditions, and yet despite extensive research over the past thirty years, generating thousands of research publications, ginseng has still not been "proven effective" by modern science. In our current state of specialization and advanced instrumentation and analytical and biological technology, it is very easy for a chemist to discover new chemicals or find known active chemicals in trace amounts in any plant material or for a pharmacologist to test the pharmacological activities of chemicals that are isolated only in traces from plant drugs, which would invariably result in publications that in turn would boost the political and financial status of the researchers involved. There is nothing wrong about such research. However, the challenge is to refrain from over-interpreting the results which are often blown out of context by proponents or opponents of the herbal drug as "preliminary evidence" to promote or restrict use of this particular herb.

There is also a general tendency to consider biomedical publications from Chinese sources as of inferior quality, which consequently should not be taken seriously. However, in my opinion, the most common flaw in publications on natural products, a good amount from "advanced" countries, is the failure of the investigators to identify correctly and quality control the material they are studying. This leads to results that cannot be duplicated and contributes further to the overflow of misinformation or useless information in this field.

While use of natural ingredients in processed foods and cosmetics was at its peak when the First Edition was published, use of these ingredients in drugs was on the decline. Now, the trend is reversed. More and more natural ingredients are being used in "herbal formulations" for the prevention and often the treatment of illnesses, most of which are related to side effects of our modern lifestyle or are common diseases that normally will resolve themselves with adjustment of lifestyle and without drug treatment. The former include obesity, hyperlipemia, and stress-related conditions while the latter include some digestive problems, minor aches and pains, and the common cold and its related symptoms. Based on traditional consumption patterns and use history, many of these formulations contain ingredients that can truly be considered as food ingredients while others fall under the category of drugs; and still others can be considered as either food or drug, depending on usage. Most of the original entries in the First Edition serve as ingredients in both foods and drugs in conventional usage, that is, processed foods and over-the-counter (OTC) drugs. In the Second Edition, I have added over seventy new entries and included a new category of usage called "**Health Food/Herb Teas**." All food and drug uses of commercial products that do not fall under conventional processed foods or cosmetic or OTC drug categories are grouped under "**Health Food/Herb Teas**." In this section we simply report on perceived uses of individual ingredients in the health food/herb tea category. This information is not intended to confirm efficacy or safety for a given indication. Rather, it is meant only to indicate for what purposes consumers may be using these products.

The debate whether health foods or herbal teas should be classified legally as genuine foods/teas or as drugs still goes on. While health food and herbal tea companies consider their products as composed of food ingredients, the medical and pharmaceutical industries generally view them or prefer to classify them as drugs requiring strict federal control. Although these opposing views are obviously dictated by economic and political considerations, the truth, in reality, lies somewhere in between. I expect this ongoing debate to continue for a long time. In the meantime, the new "**Health Food/Herb Teas**" category should be adequate in covering reported uses in these areas. However, the information reported here should in no way be construed to be an endorsement of the reported usages.

Also, the "**Folk Medicine**" category has been changed to "**Traditional Medicine**" to accommodate Chinese traditional medical usages of the new Chinese drug and cosmetic entries, as well as to recognize the role traditional medicine now plays in primary health care delivery, particularly in developing countries. Since 1978 the World Health Organization (WHO) and dozens of collaborating institutions worldwide have sought to assess the value and extent of the use of plants in health care systems. WHO has estimated that as much as 80% of the world's population rely chiefly on traditional medical systems, primarily in the form of plants, plant extracts, and active principles. Observing traditional, historic, folkloric, or ethnobotanical uses of plants is regarded as a useful approach for targeting research leads in the development of new drugs from plants. A recent survey of medicinal plants used in therapy worldwide found that 119 distinct chemical substances derived from 91 species are used as drugs in one or more countries. Of these plant-derived substances, 74% were discovered following chemical studies to determine the active compounds responsible for the use of the plant in traditional medicine.[2] While many traditional uses may not be validated as safe or efficacious by current scientific methodology, they can provide valuable leads for new or expanded utilization in the future.

Under "regulatory status," information has been included on German regulatory monographs. German health authorities have established a separate expert commission ("Commission E") to develop standardized therapeutic monographs on herbal medicines. It has produced nearly 300 "Therapeutic Monographs on Medicinal Products for Human Use." Each monograph, published in the German Federal Gazette (*Bundesanzeiger*), includes details on the name of the drug, constituents, indications (including those for the crude drug or preparations), contraindications (if any), side effects (if known), interactions with other drugs or agents (if known), details on dosage of the crude drug or preparations, the method of administration, and the general properties or therapeutic value of the herb or herb product. The German monograph system is considered to be the best governmental information source on medicinal plant usage produced by a Western industrialized nation.[3] It also serves as the model for the development of a European phytomedicine monograph system produced by the European Scientific Cooperative on Phytotherapy (ESCOP) for use by European Union member countries.

Since the publication of the First Edition, use of natural ingredients in cosmetics had been slowly declining until more recently, when a new surge of interest in Chinese cosmetic ingredients prompted the introduction of a number of Chinese natural products into American cosmetics. Although used for centuries in China, these ingredients are new to most American cosmetic formulators. Some of these new ingredients can be found among the more than twenty-four main entries that I have included in this revision. Others (more than twenty-two) can be found under the new section titled "**Chinese Cosmetic Ingredients**." This section describes in brief some of the more commonly used natural ingredients in Asia, which may now be found in new cosmetic products on the domestic market.

Despite renewed talks in the herbal/botanical industry to standardize quality and to assure purity of herbal ingredients, trade practices in this industry have not changed significantly during the past decade. And irrespective of claims by individual suppliers, manufacturers, and associated trade groups on quality, no *meaningful* assay standards or quality assurance methods have been introduced to guarantee purity and quality of many natural ingredients. Thus, for example, the most commonly used ingredients, such as aloe vera and ginseng, still lack meaningful assay standards and are frequently adulterated. The practice of this intentional adulteration is implicitly encouraged by manufacturers who purchase only low-priced ingredients and who will simply accept dubious "certificates of purity" from suppliers as the sole proof of quality and by the common practice of employing "label claims" in the cosmetic industry. Only a very small number of companies have their own programs to standardize and control the identity and purity of the herbal ingredients used in their products. In addition, due to ignorance, even some well-known herbs, especially in their powdered forms, are misidentified yet distributed as genuine in the industry. These include echinacea, eleuthero, ginseng, and numerous Chinese herbs such as fo-ti. Thus, it is obvious that much remains to be done in assuring the identity and quality of natural ingredients in the health foods/herb teas field.

This adulteration/misidentification has caused a major problem in the research on commercial natural products. Due to the failure of researchers to recognize the importance of identifying the correct source of test materials, results of studies on unidentifiable commercial herbal products (e.g., "ginseng capsules" or "aloe vera") are irreproducible and mostly worthless. Because of this problem, one should exercise extreme caution when quoting results of these studies. A well-publicized example is an uncontrolled study on ginseng(?) resulting in the so-called "ginseng abuse syndrome" which was published in a reputable journal.[4] This study has been repeatedly quoted worldwide for the past 16 years both in scientific journals and in the lay press. None of the people who quoted this study seemed to have read the original publication, noticed, realized, or cared that the results of that study were based on uncontrolled test materials that included not only Asian ginseng (identity and purity doubtful), but also American ginseng (?), Siberian ginseng (?), desert ginseng (canaigre)(?), caffeinated drinks, other drugs the subjects happened to be taking, as well as other unidentified materials (could be anything) in commercial "ginseng" products! Unfortunately, this is not the only incidence of such publications or research by researchers and editors who lack expertise in the natural products area and who otherwise are eminent in their own fields. If such papers were submitted to journals of natural products such as the *HerbalGram*, *Planta Medica*, *and Journal of Natural Products*, they would be rejected outright. This clearly demonstrates the need for experts of other disciplines to be aware of the intricacies of natural products when investigating, reporting, and evaluating these products.

Another important point to remember when studying natural products is that it is sometimes not enough just to identify correctly the botanical source of the natural product to be studied, especially where Chinese herbs are concerned. While in most cases with Western medicinal plants it is sufficient to simply assure their botanical identity, it is not so with Chinese herbal materials. In addition to their correct botanical sources, Chinese herbal materials require further clarification, including plant parts used and whether or not the materials are simply cleaned and dried or are specially treated with other herbs and/or boiled in water or wine. Thus, simply identifying an herbal drug as *Ephedra sinica* Stapf can mean one of at least two different drugs with distinctly different medicinal properties: *mahuang* (stem) is diaphoretic, among other properties, while *mahuanggen* (root) is antiperspirant. Another example is *Polygonum multiflorum* Thunb., from which at least three different herbal products are derived, each with distinct medicinal characteristics: stem, raw root tuber (*heshouwu*), and cured root tuber (*zhiheshouwu*). It is obvious the Western term for it, fo-ti, is meaningless. A voucher specimen of *Polygonum multiflorum* to go with fo-ti would further add to the confusion and would not determine whether the fo-ti shipment in question is the laxative (raw root) or the tonic (cured root).

A recent trend in the herbal industry is to market so-called standardized extracts, such as ginseng extract standardized to "ginsenosides" content or Siberian ginseng extract to "eleutherosides" content. However, as there are normally more than one (or one type of) active component in a natural product, standardization based on one particular type of chemical component is not representative of the total activity of the product. Consequently, these arbitrarily selected components can only be useful as a "marker" of product quality. And these "markers" are only valid for extracts that are total extractions of the herbs concerned. Extraction processes designed to extract these "markers" selectively would produce extracts that are not representative of the original herbs. Thus, a "standardized" ginseng or Siberian ginseng extract may be devoid of polysaccharides that are also biologically active. Also, the ginsenosides in a ginseng extract may not be from ginseng itself but rather from another much cheaper, non-ginseng source (see ***ginseng***). To be fair to both tradition and modern science, one should not be overzealous in trying to equate a chemical constituent to a traditional herbal drug.

As more and more biological and toxicological research is performed on commercial natural products, it is increasingly apparent that scientific evaluation of individual purified components from these natural products has rarely produced results that are consistent with the property of the products *in toto*. Consequently, one should not be prematurely alarmed if one of numerous components in a long-used natural product is shown to have toxic effects in the laboratory, unless further research on the product in its complete form produces the same effects. Conversely, one should not be over-optimistic in claiming a particular herb or natural product as "cure" for a certain disease after studies have indicated that one of its numerous chemical components exhibits a positive effect on the disease. This is especially true if this component is present only in minute quantities whose effect may be overshadowed by those of other compounds present, thus making the herb (in its original form) inactive as a cure for the disease. Oleanolic acid is a typical example. It is widely distributed in nature. A recent double-blind study involving 152 cancer patients demonstrated it to have immunomodulating effects (enhanced phagocytosis, E-rosette formation, and delayed hypersensitivity), improving the general condition in two-thirds of the patients. Preliminary studies have also indicated it to be effective against hepatitis and HIV.[5] However, all this does not mean eating cloves and olives, both containing oleanolic acid, will necessarily produce such an effect.

We are currently being literally choked by an overabundance of data on natural products, much of which either has not been evaluated or is of dubious value. In this Second Edition, as in the First Edition, every effort has been made to evaluate all original publications available to assure that the research methods and findings are of decent quality. And I have paid particular attention to the identity and quality control of the test materials. Papers reporting on results of studies based on unidentified or unidentifiable test materials are not cited under the respective entries (e.g., the so-called "ginseng abuse syndrome" not cited under ginseng) for the same reason that results of research on an unidentified "yellow powder" as due to riboflavin or curcumin would not be reported in a medical or pharmaceutical journal.

Appearing in the English literature mostly for the first time, the information on the new Chinese natural products in this Second Edition has been gathered from dozens of major Chinese classical and modern works and from over fifty Chinese journals on traditional and herbal medicine. I have tried my best to present a balanced view of the traditional and modern aspects of Chinese herb use. The new ingredients selected for this new edition generally reflect the trend in current commercial use of natural products in America. I hope this new edition will provide the readers with an accurate update on the original entries of the First Edition as well as an overview of the huge resources in Chinese herbal ingredients.

1. J.A. Duke, *HerbalGram*, **17**, 20 (1988).

2. N. R. Farnsworth et al., *Bulletin of the World Health Organization*, **63**, 6, 965 (1985).
3. V. E. Tyler, *HerbalGram*, **30**, 24 (1994).
4. R.K. Siegel, *JAMA*, **241**, 1614 (1979).
5. Y. Sun et al., *Chin. J Clin. Pharmacol.*, **6**, 72 (1990).

ALBERT Y. LEUNG
*Glen Rock, New Jersey*
*July 1995*

# Preface to the First Edition

About 500 natural ingredients are currently used in commercial food, drug, and cosmetic products. These do not include antibiotics, vitamins, and many other natural substances that constitute prescription drugs nor medicinal herbs that are not readily available in commerce. Some of these ingredients are pure chemicals isolated from natural sources while others are extracts of botanicals. Our daily food, drug, and cosmetic items often contain these ingredients. Many of the substances used in foods are also used in drugs and cosmetics, where higher concentrations are involved.

Three major reasons have prompted me to compile this encyclopedia. First, no reference books are presently available that specifically and simultaneously deal with commonly used natural ingredients in processed foods, over-the-counter drugs, and cosmetics. Since many natural flavor ingredients and food additives are also drug and cosmetic ingredients when used in higher concentrations, there has been an acute need for a compact reference book that provides condensed and accurate information on these substances, saving the reader much time and effort that otherwise would have to be spent in consulting various handbooks and journals.

Second, most of the currently available technical reference books in the English language on food, drug, or cosmetic ingredients contain limited and out-of-date information regarding naturally derived substances. Many formerly official botanical drugs that are no longer official in the *United States Pharmacopoeia* (U.S.P. XIX) or the *National Formulary* (N.F. XIV) are still widely used in nonprescription pharmaceutical preparations and in food products. Yet they are largely neglected or ignored by editors or authors of readily available handbooks. Presumably, when a botanical drug is deleted from a currently official compendium, there should no longer be any interest in it. Formerly official drugs such as arnica, chamomile, rhubarb, valerian, white pine, and witch hazel are still widely used today in foods, drugs, and cosmetics; so are many plants that have never been admitted as official drugs, examples of which are alfalfa herb, annatto seed, chicory root, fenugreek seed, ginseng root, and rose hips. There is still ongoing, active research on many of these natural products, particularly outside the United States. Since these botanicals are very much a part of our culture and daily life, information on them should be readily available. This encyclopedia is intended to furnish correct, up-to-date information on these materials.

Third, there is a general information gap regarding natural products between technologists of the botanical industry and those of the food, drug, and cosmetic industries, between members of the academic and research communities and those in industry, as well as between the consumer and the industry concerned. Information readily available to one group is often not available to the others. One of the objectives of this book is to try to bridge this gap by supplying information that would make different groups more aware of the practices and happenings outside of their own circle regarding the use of natural ingredients.

In this encyclopedia, each natural product is presented in alphabetical order according to its most common name, with each natural ingredient being cross-referenced with its scientific name (Latin binomial) in the Index. As a natural ingredient often has several common names (synonyms), the reader is advised to use the Index if an ingredient cannot be found in the text under a particular synonym. Data on about 310 natural ingredients are furnished. Information included in each item includes plant or other sources, habitats, parts used or derived from, method of preparation, brief physical description, chemical composition, pharmacology or biological activities, common commercially available forms in the United States and their qualities, uses, and regulatory status, whenever applicable.

Data on chemical compositions of natural ingredients are constantly increasing as analytical techniques keep improving. Often an ingredient contains hundreds of chemical constituents, yet only a few (occasionally arbitrarily selected) are listed in this encyclopedia. For further information on other compounds the reader is referred to the original references cited. Incidentally, the absence of a particular compound in a natural ingredient does not necessarily mean that it is actually absent; it may simply mean that nobody has analyzed for it in this particular ingredient. On the other hand, its reported presence in a natural ingredient means only that someone has investigated it in this particular ingredient using a particular analytical technique for whatever reason. Also, the mere presence of a toxic chemical in a natural ingredient does not necessarily make this ingredient toxic. Its concentration and biological availability should be taken into account when the toxicity of the ingredient is considered.

The data on pharmacology or biological activities (be they favorable or unfavorable) reported in this book should be viewed with caution as often they were single reports or reports from a single laboratory or research group that have not been substantiated by other studies. Furthermore, it should be kept in mind that results from animal studies are not necessarily applicable to humans. Purity of the test material (which is often not sufficiently stressed) should also be taken into account when evaluating such data.

Uses are categorized into four major areas: (1) pharmaceutical and/or cosmetic, (2) food, (3) folk medicine, and (4) others. Pharmaceutical and cosmetic uses refer to current uses in commercially available products mainly in the United States. No attempt has been made to identify the function of each ingredient in a product, as often there are over a dozen botanical components present in a single preparation, making it an impossible task. The same situation applies to the food area where the majority (200–250) of the ingredients used in food products are broadly identified only as flavor ingredients. The specific function and use level of a particular ingredient in a flavor formulation are often proprietary information, which is seldom publicly available. Consequently, food uses are reported in this encyclopedia by food categories, as in the report on "Average Maximum Use Levels" published by the Flavor and Extracts Manufacturers' Association of the United States (FEMA). Only in cases where the functions of the ingredients have become widely known in the trade or otherwise in open literature (e.g., fenugreek extract as a major flavoring agent in artificial maple syrup, yucca extracts as foaming agents in rootbeer, and absinthium as a flavor ingredient in vermouths, etc.) are they specifically mentioned in this book. Sometimes an ingredient is reported used in various types of food products yet federal regulations have approved its used in only one particular type of product. This appears to be a typical case of information dissemination lag. Under folk medicinal uses are listed only those traditional uses that are reported in reliable sources available to me, primarily in the English, German, and Chinese languages; they are by no means complete and they should not be regarded as endorsement of such uses. They are included in this volume because of their popular interest. Under the fourth category ("others") are listed potential or unusual uses that do not fall in above categories.

Use levels in foods reported in this encyclopedia are based on the FEMA report; a manufacturer may foreseeably use an ingredient in an amount five times the average maximum use and still be considered within good manufacturing practice. Use levels reported for cosmetics are based on values reported in the Monographs on "Fragrance Raw Materials" prepared by Opdyke of The Research Institute for Fragrance Materials, Inc. and published in *Food and Cosmetic Toxicology*.

Under regulatory status, GRAS means generally recognized as safe as sanctioned by the Food and Drug Administration (FDA); an ingredient described as having been approved for food use is not necessarily GRAS. For more precise and up-to-date information, the reader is referred to §182 and its appropriate sections under Title 21 of the *Code of Federal Regulations* (formerly §121.101), to §172.510 (formerly §121.1163), and to other appropriate sections, to the FDA, and to the latest notices and rulings published in the *Federal Register*.

A glossary of terminology commonly used in the botanical industry is found in the Introduction. Since the primary purpose of this encyclopedia is to serve as a practical reference guide for practicing technologists in the food, drug, and cosmetic industries and their purchasing agents and marketers, theoretical considerations and basic principles in the fields concerned are omitted. For these topics, the reader is referred to standard texts on these subjects such as BALSAM AND SAGARIN, FURIA, HARBORNE, LEWIS AND ELVIN-LEWIS, REMINGTON, and TYLER, listed in the General References.

In the General References are listed textbooks and handbooks from which general and sometimes specific information was obtained. They are identified in the text by the names of the authors in small capital letters, and if there are more than two authors, by the name of the first author. If an author has more than one book, it is identified by a number such as 1 or 2 immediately following the author's name (e.g., BAILEY 2); the number refers to the order of appearance of this author's books in the list.

Specific references are cited under References immediately following each entry, numbered according to their order of citation in the text.

It is hoped that this encyclopedia will serve as a handy and useful reference to technical and nontechnical members of the food, drug, and cosmetic industries, to teachers and students of corresponding sciences and related fields, and to the general public who want to know more about natural ingredients.

ALBERT Y. LEUNG
*Glen Rock, New Jersey*
*January 1980*

# Acknowledgments

I have spent two years and thousands of working hours of intensive efforts in compiling this encyclopedia. A book of this scope could not have been written without the assistance or participation of others. I am indebted to all who have been part of my heritage and background and to those who have contributed to my scientific training, especially to the University of Michigan and to the Lilly Endowment Foundation.

My indebtedness to Dr. Madis Laboratories, Inc. of South Hackensack, New Jersey is acknowledged, for it was during my years as Director of Research and Development there that the need of a book like this was first realized. I wish to thank my clients and friends in the food, drug, and cosmetic industries from whom much of the knowledge and information I have included in this Encyclopedia was learned. In this regard, special thanks go to Dr. James Chan of Botanicals International, Long Beach, California, and to one of my clients, the Penick Corporation, Lyndhurst, New Jersey, for its patronage.

Numerous other individuals have also contributed to this Encyclopedia by supplying me with specific information and/or reviewing part of the manuscript. I wish to acknowledge the assistance of these colleagues, especially that of Dr. N. R. Farnsworth, College of Pharmacy, University of Illinois at the Medical Center, Chicago, Illinois; Dr. E. E. Terrell, The Plant Taxonomy Laboratory, USDA Beltsville Agricultural Research Center, Beltsville, Maryland; Dr. G. E. Inglett, USDA Northern Regional Research Center, Peoria, Illinois; and Dr. K. J. Harkiss, University of Bradford, Postgraduate School of Studies in Pharmacy, Bradford, England. Also, all the authors who have furnished me with reprints of their original publications are hereby gratefully acknowledged.

During the research and preparation of the manuscript I used the facilities of numerous libraries. I wish to express my appreciation to these libraries and their librarians, including in alphabetical order: College of Medicine and Dentistry of New Jersey (CMDNJ) Library; the Fair Lawn Public Library; Glen Rock Public Library; Montclair College Library; National Library of Medicine; New York Public Library and its branches; Ramapo College Library; Ridgewood Public Library; Rutgers University Libraries; and William Paterson College Library. Special thanks are due my local library, the Glen Rock Public Library, for obtaining copies of numerous patents.

I would like to thank those manufacturers who had the courtesy to reply to my inquiries for product information.

I am specially indebted to Dean Ara G. Paul and Mr. Donald A. Davis for providing forewords to this volume.

Special appreciation is hereby extended to Dr. and Mrs. Peter S. Chan, Mr. and Mrs. Joseph B. Gorini, Mr. Thomas J. Hanrahan, Dr. Raul I. and Mrs. Alba Ludmer, Mr. Don and Mrs. Barbara Scott, and Mr. Leonard E. and Mrs. Ruth Willett for their genuine interest and concern in my project and for their encouragements and support.

Last, but not the least, I wish to express my gratitude to my beloved wife, Barbara, without whose assistance, understanding, and encouragements this encyclopedia would not have been completed. She did much of the literature search and occasionally provided helpful criticisms on my manuscript, as well as efficiently organizing my materials so I could concentrate on reading and writing. When no technical typists could handle the task and erratic schedule, Barbara also typed the manuscript from my mostly handwritten drafts. Furthermore, during these two years, she maintained a household as normal as possible for our two young daughters, who at times for days on end could say only brief hellos to their father. In this regard I wish also to thank my older daughter, Amy, for her patience and understanding and for her occasional care of Camille during this period.

ALBERT Y. LEUNG
*January 1980*

Since the first edition of Dr. Leung's Encyclopedia was published in 1980, this book has been my first reference of choice when looking for detailed, succinct information on natural products. It was also the first reference I would recommend to colleagues seeking information on natural ingredients in commerce. The opportunity to co-author the second edition was largely serendipitous, sparked by a colleague's query on the availability of the first edition. Dr. Dennis Awang, then Head of the Natural Products Section, Bureau of Drug Research, Health and Welfare Canada, called me looking for information on how to obtain a copy of the first edition, which was no longer available from the publisher. I immediately called Dr. Leung to ask if he had any extra copies he would be willing to part with. He indicated that no copies were available, and furthermore he was working on a second edition; a daunting task, given the proliferation of new research on natural products in the past decade. During that phone call, he enlisted my help in completing the second edition. It was with honor and excitement that I accepted the assignment. I would like to express my deep gratitude to Albert Y. Leung for giving me the opportunity to work with him on this exciting project.

During a period of twenty-four months this book consumed most of my time. I am grateful for the generous assistance of numerous colleagues, friends and associates who provided data and new research materials for this book. In particular I would like to thank Dr. Dennis Awang, Consultant, MediPlant, Ottawa, for sending numerous papers, as well as his exceptionally detailed and useful comments on the manuscript; Prof. Rudolf Bauer, Institute for Pharmaceutical Biology, Heinrich Heine University, Düsseldorf for data on the genus *Echinacea*; Mark Blumenthal, Executive Director, American Botanical Council, Austin, Texas, for supplying volumes of information from his files, as well as English translations of the German Commission E Monographs on human therapeutic uses of medicinal plants; Dr. Lyle Craker, Plant and Soil Sciences, University of Massachusetts, Amherst, for details on diversity of aromatic plants; Dr. James A. Duke, Germplasm Resources Laboratory, USDA, Beltsville, for keeping me abreast of research developments; Dr. Norman Farnsworth, Programme for Collaborative Research in the Pharmaceutical Sciences, University of Illinois, Chicago, for access to the NAPRALERT database; and Dr. Varro Tyler, Lilly Distinguished Professor of Pharmacognosy, Purdue University, for his rapid responses to specific problems.

I would also like to thank a number of librarians, without whose help, I could not have completed the work. Thanks go to Rebecca Perry (past director), and Michael Flannery, Director, Lloyd Library and Museum, Cincinnati, for promptly fulfilling requests for obscure papers; Dr. Reinhard Liersch, Head of the Library and Documentation Department, Madaus, AG, Cologne, for providing up-to-date information on European regulatory developments and difficult-to-find European literature; and Bonnie Shipper, Shaman Pharmaceuticals, Inc., South San Francisco, for exchanging information on specific medicinal plant groups.

I am grateful to my able assistant, Mary Pat Boian, who spent endless hours on the thankless task of retyping, reformatting, and cross-checking the first edition with new entries. Thanks, too, to Jessica Crandall for helping to correct page proofs.

Finally, I would like to express my deepest thanks to Jude Farar, and my children, Colin and Abbey, for allowing me the freedom, quiet, and time necessary to concentrate on completing my work on the manuscript.

STEVEN FOSTER

# Contents

# Introduction

People have been using natural products since the dawn of human history. Only during the last century, however, have we started to know something about the chemistry of some of these products. With our increasing knowledge of chemistry and related sciences, we have begun to duplicate some of the natural chemicals and at the same time make modifications in these compounds, or sometimes produce completely new ones. Consequently, since the advent of the Synthetic Era several decades ago, many natural drugs have been replaced by synthetic ones; natural flavors and fragrances have been duplicated or simulated by manufactured chemicals. However, the number of natural products used in drugs is still sizable, comprising almost half of all prescriptions written. This number has not changed appreciably for the last two decades, especially with reference to botanicals.[1] At least 250 plants or their extracts are currently used in commercial food products broadly classified as flavoring ingredients (FEMA). The present trend points to an increasing interest in the use of natural products, particularly in foods and cosmetics.

To define a natural product is not a straight-forward task; for, strictly speaking, everything is derived from nature. Nevertheless, by natural products it is generally meant that products are not made by chemical synthesis. Theoretically a natural chemical is the same as its synthetic counterpart in every respect. However it must be pointed out that unless this chemical is absolutely pure (which it seldom is) it would contain different impurities, depending on its sources. The impurities present in a naturally derived food, drug, or cosmetic ingredient are bound to be different from those of its synthetic counterpart, and if there is more than one way to synthesize this compound, then the impurities would be different from one synthetic process to another. The relative toxicities or merits of these small differences have not been determined. If an impurity, whether it is a natural or synthetic chemical, has unusually high latent biological activity, a minute quantity of it present in a chemical would produce physiological effects besides those elicited by the pure chemical itself. These effects may not be immediately apparent. Most if not all of existing standards for food, drug, and cosmetic ingredients do not have provisions for pinpointing small amounts of impurities, as it is impractical to set absolute purity standards for these ingredients. Consequently in practice most of these materials are permitted to have a range of error built into their purity assays. This range of error can be due either to the assay methods themselves or to actual impurities present in the chemical. In some cases, as analytical methodology advances, this range has become progressively narrower. However before this range becomes negligible, one should not equate a naturally derived chemical with its synthetic counterpart, and their sources should be indicated, as is the case with certain flavor chemicals.

There are several definitions of a natural product. In the case of flavoring substances, some definitions of a natural product (flavor) limit the product to be one obtained from natural sources by physical processes only. Other definitions allow hydrolysis and fermentation as permissible processes. For all practical purposes in this book, a natural product is defined as a product that is derived from plant, animal, or microbial sources, primarily through physical processing, sometimes facilitated by simple chemical reactions such as acidification, basification, ion exchange, hydrolysis, and salt formation as well as microbial fermentation. These chemical reactions do not drastically alter the chemical structure of the natural product to be isolated.

Ingredients used in foods, drugs, and cosmetics can be divided into two main categories, namely, active and inactive. Active ingredients can be considered as those that supply energy to the body or serve as its nutrients (foods and some food additives), or cause physiological changes in or on the body (drugs and cosmetics) when taken internally or applied externally. Inactive ingredients are substances which, based on prevalent data, do not exert physiological actions when ingested or applied to the body. Their primary function is to facilitate the ultimate intake or utilization of the active ingredients. Among food products, basic foodstuffs such as flour, starch, and milk are not included in this book, although they are considered active ingredients. Only food additives are considered. However in drug and cosmetic products, both active and inactive substances are included.

Food additives are a large group of substances that are added to foods either directly or indirectly during the growing, storage, or processing of foods for one or more of the following purpose:[2]

1. Improve or maintain nutritional value
2. Enhance quality
3. Reduce wastage
4. Enhance consumer acceptability
5. Improve keeping quality
6. Make the food more readily available
7. Facilitate preparation of the food

There are about 2500 direct food additives currently used by the food industry. Out of this number perhaps 12 to 15% are natural products. Many of these food additives are also drugs when used in larger quantities. Some of these are also used in cosmetics. The total number of the more commonly used natural food, drug and cosmetic ingredients in this Encyclopedia is about 310 (first edition).

In spite of the fact that plants have been used for therapeutic purposes for millennia, only a relatively few plants or plant derivatives are currently officially recognized in the United States as effective drugs. This is largely due to the difficulties encountered in plant drug research and the limitations of scientific methodology employed. Quite often, premature publicity on unconfirmed research data has tainted the reputation of many botanical drugs. Since many drug plants have rather complicated chemical compositions and analytical technology has not been adequate in determining their identities and qualities once extracts are made from them, adulteration, sophistication, or substitution has been common. This has led to inconsistencies in drug potency, and many natural drugs have probably been removed from officially recognized status as a result. Many natural drugs formerly recognized by the *United States Pharmacopeia* (U.S.P.) and *National Formulary* (N.F.) are no longer official in these compendia; yet many of these continue to be used in pharmaceutical preparations.

Some of the food, drug, and cosmetic ingredients are pure chemicals isolated from plants, animals or microbes. However most are in the form of extracts, oleoresins, fixed oils, and volatile oils, among others. The following glossary contains most of the commonly encountered terms used in the botanical industry.

# Glossary

ABSOLUTES. Absolutes are entirely alcohol-soluble extracts prepared by alcohol extraction of ***concrètes*** or related fat-soluble or waxy materials, alcohol-insoluble substances being removed before evaporation of the solvent.

ALKALOIDS. Alkaloids can be broadly defined as natural amines (nitrogen containing compounds) which have pharmacological properties and which are generally of plant origin. They are widely distributed throughout the plant kingdom. They usually exhibit basic properties, though there are exceptions. Most alkaloids are insoluble or only slightly soluble in water, but their salts are water soluble. Many naturally derived drugs are alkaloids; well-known examples are morphine, codeine, cocaine, caffeine, nicotine, emetine, atropine, and quinine.

BALSAMS. By conventional definition, balsams are mixtures of ***resins*** that contain relatively large amounts of cinnamic or benzoic acid or their esters. Typical balsams are balsam Peru, balsam Tolu, styrax, and benzoin. Canada balsam, Oregon balsam, and copaiba balsam are not true balsams since they do not contain benzoic or cinnamic acid or their esters. Balsams are insoluble in water but soluble in alcohol.

CONCRETES. Concrètes are water-insoluble but hydrocarbon-soluble extracts prepared from natural materials by using hydrocarbon-type solvents. They are primarily used in perfumery and in the preparation of ***absolutes***.

DECOCTIONS. Decoctions are dilute aqueous ***extracts*** prepared by boiling the botanicals with water for a specific period of time, followed by straining or filtering. These are normally not commercially available in the United States.

DRY EXTRACTS. Dry extracts are the same as ***powdered extracts.*** This term is used mainly in the United Kingdom and in Commonwealth countries.

ELIXIRS. According to the U.S.P. definition, elixirs are clear, sweetened, hydroalcoholic liquids intended for oral use. They contain flavoring substances and, in the case of medicated elixirs, active medicinal agents. Their primary solvents are alcohol and water, with glycerin, sorbitol, and syrup sometimes used as additional solvents and/or sweetening agents. They are prepared by simple solution or admixture of the several ingredients.

ENZYMES. Enzymes are proteins produced by living organisms which can bring about specific changes in other compounds (called substrates). Enzymes are also called organic catalysts; they are not consumed in the reactions they catalyze but are regenerated at the end of such reactions. The most commonly used enzymes in the food and drug industries are ***proteases***, amylases, ***lipases,*** and pectinases.

ESSENTIAL OILS. Essential oils are also known as volatile oils, ethereal oils, or essences. When exposed to the air they evaporate at room temperature. They are usually complex mixtures of a wide variety of organic compounds (e.g., hydrocarbons, alcohols, ketones, phenols, acids, ethers, aldehydes, esters, oxides, sulfur compounds, etc.). They generally represent the odoriferous principles of the plants from which they are obtained. Most of these compounds are derived from isoprene and are terpenes at different stages of oxidation. Essential oils are generally isolated by distillation (most commonly steam distillation), solvent extraction, or expression. Clove, cinnamon, and peppermint oils are obtained by steam distillation, whereas bergamot and lemon oils and sweet and bitter orange oils are obtained by expression.

EXTRACTS. Extracts are generally but not necessarily concentrated forms of natural substances obtained by treating crude materials containing these substances with a solvent and then removing the solvent completely or partially from the preparations. Most commonly used extracts are ***fluid extracts*** (***liquid extracts***), ***solid extracts***, ***powdered extracts*** (***dry extracts***), ***tinctures***, and ***native extracts***.

FATS. Fats are glycerol esters (glycerides) of ***fatty acids***. They are semisolids or solids at room temperature and are generally produced from botanicals by expression and from animal materials by extraction or rendering.

FATTY ACIDS. Fatty acids are carboxylic acids obtained from natural sources, mostly from ***fats***. They can be both saturated and unsaturated. Examples of saturated fatty acids are palmitic and stearic acids; unsaturated ones are oleic, linoleic, and linolenic acids.

FIXATIVES. Fixatives are materials, usually high boiling and of high molecular weight, that retard the evaporation of the more volatile components in perfume formulations.

FIXED (FATTY) OILS. Fixed oils are chemically the same as ***fats***. They differ only physically from fats in that they are generally liquids at room temperature.

FLUID EXTRACTS. These ***extracts*** are commonly hydroalcoholic solutions with ***strengths*** of 1:1. The alcohol content varies with each product. Fluid extracts are prepared either from ***native extracts*** or ***solid extracts*** by adjusting to the prescribed strength with alcohol and water or by direct extraction of the botanicals with alcohol–water mixtures as directed in the official compendia. The latter method usually produces more desirable products due to the fewer steps involved in processing. Fluid extracts are also known as ***liquid extracts***.

GLYCOSIDES. Glycosides are sugar-containing compounds which on hydrolysis yield one or more sugars. They contain two components in their molecules, glycone and aglycone. The glycone is the sugar component, which can be glucose, rhamnose, xylose, arabinose, or other sugars. When the glycone is glucose, the glycoside is commonly known as a glucoside. The aglycone is the nonsugar component of the glycoside; it can be any type of compound such as sterols, triterpenes, anthraquinones, hydroquinones, tannins, carotenoids, and anthocyanidins. They are a very important group of natural products, are widely present in plants, and constitute major classes of drugs. Well-known drug examples are digitalis glycosides, sennosides, cascarosides, ginseng glycosides, rutin, and arbutin. Glycoside-containing materials that are used in foods include grape skin color (betanin), soapbark, fenugreek, alfalfa (saponins), and licorice (glycyrrhizin).

GRANULAR EXTRACTS. Granular extracts are produced in the same way as ***powdered extracts***. They are also of the same potency as powdered extracts. The only difference between them is that granular extracts have larger particle sizes.

GUMS. Gums are hydrocolloids. They are polysaccharides of high molecular weight and can be dissolved or dispersed in water to form a viscous colloidal solution. The most commonly used natural gums are seaweed extracts (agar, algin, carrageenan, furcellaran), tree exudates (acacia, ghatti, karaya, tragacanth), tree extracts (larch gum), seed gums (guar, locust bean, quince seed) and microbial gums (dextran, xanthan).

GUM RESIN. Gum resins are ***resins*** occurring admixed with ***gums***. They usually also contain small amounts of ***volatile oils*** and sometimes are also called ***oleogum resins.*** Common examples are myrrh, gamboge, asafetida, galbanum, and olibanum.

INFUSIONS. Infusions are sometimes the same as ***decoctions***. They are generally dilute aqueous ***extracts*** containing the water-soluble ingredients of the botanicals. They are prepared by extracting the botanicals with boiling water. The resulting extracts are not concentrated further. Because of the dilute and aqueous nature of infusions and decoctions, they are very susceptible to microbial deterioration. Infusions are not normally available commercially in the United States.

LIPASES. Lipases are lipolytic ***enzymes*** that hydrolyze ***fats*** or ***fixed oils*** into their glycerol and ***fatty acid*** components. They are chiefly used in the dairy industry as flavor producers or modifiers and in medicine as digestive aids.

LIPIDS. Lipids are fatty materials that are soluble in fat solvents (ether, chloroform, alcohol, etc.) They include ***fatty acids, fats, waxes, fixed oils,*** steroids, lecithins, and fat-soluble vitamins (vitamins A, D, and K).

LIQUID EXTRACTS. Liquid extracts are British equivalents of ***fluid extracts.*** They are used in the United Kingdom and the Commonwealth countries.

MENSTRUUMS. Menstruums (or menstrua) are solvents used for extraction, for example, alcohol, acetone, and water.

NATIVE EXTRACTS. In the commercial manufacture of extracts, a botanical is first extracted with an appropriate solvent such as denatured alcohol, alcohol, methanol, water, or mixtures of these solvents. The extract is then concentrated under reduced pressure at low temperatures until all solvent is removed. The viscous, semisolid concentrated ***extract*** at this state is called a native extract by some manufacturers. The native extracts are usually of high potency from which ***solid***, ***fluid***, and ***powdered extracts*** of various ***strengths*** can be prepared by diluting with suitable diluents. If the botanical has ***resins*** and ***volatile oils*** as its active principles and the solvent used is a fat solvent, the resulting native extract is equivalent to a prepared ***oleoresin***.

OLEOGUM RESINS. See gum resin.

OLEORESINS. Oleoresins are mixtures of mostly ***resins*** and ***volatile oils.*** They either occur naturally or are prepared by solvent extraction of botanicals. Prepared oleoresins are made by extracting the oily and resinous materials from botanicals with fat solvents (hexane, acetone, ether, alcohol). The solvent is then removed under vacuum, leaving behind a viscous, semisolid extract which is an oleoresin. Examples of prepared oleoresins are paprika, ginger, and capsicum (see also ***native extracts***). Examples of natural oleoresins are gum turpentine, Oregon balsam, and Canada balsam.

POWDERED EXTRACTS. Powdered extracts are prepared from ***native extracts*** by diluting the native extracts to the specified ***strengths*** with appropriate diluents (lactose, dextrose, sucrose, starch, etc.) and/or anticaking agents (calcium phosphate, magnesium carbonate, magnesium oxide, etc.), followed by drying, usually under vacuum, to yield dry solids. These are then ground into fine powders to form powdered extracts or into coarse granules to produce ***granular extracts***.

PROTEASES. Proteases, or proteinases, are proteolytic enzymes which act on proteins by attacking specific peptide linkages in the proteins and hydrolyzing them. Depending on their specific applications, commonly used proteases can be of plant, animal, or microbial origin. They find uses in tenderizing meat, modifying dough in baking, chillproofing beer, cheese making, in wound debridement, as digestive aids, in relieving inflammations, bruises, and blood clots, as well as in other industries (leather, textile, dry cleaning, waste control). Examples of widely used plant proteases are bromelain, ficin, and papain; common animal proteases are pepsin and rennin. Proteases are usually divided into two types. Endopeptidases break up internal peptide bonds of the protein chain, producing peptides. Exopeptidases, on the other hand, cleave terminal peptide linkages, producing amino acids. Most commercial proteases are mixtures of different protease fractions and usually have both endopeptidase and exopeptidase activities. Commercial proteases come in many different grades which vary widely in proteolytic strengths. Few published studies on proteases, particularly commercial plant proteases, specify activity of enzymes used, and hence results are generally quantitatively irreproducible.

RESINS. Resins are natural products that either occur naturally as plant exudates or are prepared by alcohol extraction of botanicals that contain resinous principles. Naturally occurring resins are solids or semisolids at room temperature. They are soluble in alcohol and alkali solutions but are insoluble in water. They are usually noncrystalline, transparent or translucent, and soften or melt on heating. Chemically they are complex oxidation products of terpenes. They rarely occur in nature without being mixed with ***gums*** and/or ***volatile oils***, forming ***gum resins***, ***oleoresins***, and ***oleogum resins***. Hence in commerce the term "resins" is often used to include all above resinous materials. During preparation of a resin, the alcoholic extract is poured into an excess of water or acidified water and the precipitated resin is collected, washed, and dried. Typical examples of prepared resins are podophyllum and jalap resins. A prepared resin may also be derived from a natural oleoresin by removing the volatile oil by heat or from a natural gum resin by extracting its resin with alcohol followed by removal of the solvent. Resins prepared by alcohol extraction of natural resinous materials are sometimes referred to as resinoids. Resinoids may be considered as purified forms of certain resins; they are usually prepared from resins by extraction with hydrocarbons.

SAPONINS. Saponins are ***glycosides*** generally with sterols or triterpenes as their aglycones, although there are exceptions. They have the ability of forming foams when their aqueous solutions are shaken. The aglycone portions are called sapogenins. Many saponins are hemolytic. However their foam-forming properties are utilized in beverages. Common examples are extracts of soapbark, yucca, and sarsaparilla. Other saponin-containing natural ingredients include alfalfa, fenugreek, senega, ginseng, and licorice.

SOFT EXTRACTS. This is a British term which is equivalent to our ***native*** or ***solid extracts***. It is used in the United Kingdom and the Commonwealth countries.

SOLID EXTRACTS. Solid extracts are also known as pilular extracts. They are usually thin to thick, viscous liquids or semisolids prepared from ***native extracts*** by adjusting the latter to the correct ***strength*** with suitable diluents (liquid glucose, corn syrup, glycerol, propylene glycol, etc.). Solid extracts are generally of the same ***strength*** as their corresponding ***powdered extracts***.

STRENGTH OF EXTRACTS. The potencies or strengths of botanical drug ***extracts*** are generally expressed in two ways. If they contain known active principles, for example, alkaloids in belladonna and ipecac, their strengths are commonly expressed in terms of their content of active compounds. Otherwise they are expressed in terms of their total extractives in relation to the crude drug. Thus a strength of 1:4 means one part of extract is equivalent to or derived from four parts of crude drug. This method of expressing drug strength is not accurate since in commerce an extract of a certain strength, for example, 1:3, may be different in actual potency depending on the manufacturer and the process or equipment used in the production. Examples of such variations can be found in extracts of senna and cascara, where the active principles are well known; yet, the strengths are sometimes expressed in terms of total extractives. Consequently a ***solid extract*** of cascara (1:3) or a ***fluid extract*** of senna (1:1) from one supplier may contain several times more anthraglycosides than officially equivalent products from another supplier. Even though all these products may pass current or past pharmacopoeial specifications, some of them may actually contain little or no active ingredients. Furthermore, in recent years, strengths of extracts are also expressed in reverse numerical order (e.g., 4:1 instead of 1:4), causing much confusion. Hence one should not rely solely on numerical designations in reporting or interpreting strengths.

TINCTURES. Tinctures are alcoholic or hydroalcoholic solutions usually containing the active principles of botanicals in comparatively low concentrations. They are generally prepared either by maceration or percolation or by dilution of their corresponding ***fluid extracts*** or ***native extracts***. The ***strengths*** of tinctures are generally 1:0.1 or 1:0.2.

VOLATILE OILS. See ***essential oils.***

WAXES. Technically, waxes are esters of ***fatty acids*** with alcohols, both of high molecular weight and straight-chained. But in reality waxes also contain free fatty acids, free fatty alcohols, and hydrocarbons. They are extensively used in pharmaceutical and cosmetic ointments, creams, and lotions.

## REFERENCES


See the General References for ARCTANDER; FEMA; FURIA; FURIA AND BELLANCA; REMINGTON; U.S.P. XIX; U.S.P. XXI.

1. N. R. Farnsworth and R. W. Morris, *Am. J. Pharm.*, **148,** 46 (1976).
2. National Academy of Sciences, "The Use of Chemicals in Food Production, Processing, Storage and Distribution," Washington, D.C, 1973.

# ENCYCLOPEDIA OF COMMON NATURAL INGREDIENTS

## USED IN FOOD, DRUGS, AND COSMETICS

**Second Edition**

# Natural Ingredients

## ABSINTHIUM

**Source:** ***Artemisia absinthium*** L (Family Compositae or Asteraceae).

*Synonyms.* Wormwood, common wormwood, absinthe, armoise, Wermut, absinthe grande, and Wermutkraut.

### GENERAL DESCRIPTION

Shrubby perennial herb with grayish white stems covered with fine silky hairs, 30–90 cm high; leaves also silky, hairy, and glandular, 2- to 3-pinnatisect, petiolate lobes, mostly obtuse; odor aromatic, spicy; taste bitter; native to Europe, northern Africa, and western Asia, naturalized in North America; extensively cultivated. Parts used are the leaves and flowering tops (fresh and dried), harvested just before or during flowering; from these a volatile oil is obtained by steam distillation (FERNALD, YOUNGKEN).

### CHEMICAL COMPOSITION

Contains up to 1.7% volatile oil that is composed mainly of thujone ($\alpha$- and $\beta$-) and $\beta$-caryophyllene. Thujone constitutes a major part of the oil.[1] Bitter principles absinthin (a dimeric guaianolide), anabsinthin (formed from absinthin), and artabsin (analogous monomer of absinthin);[2,3] artemetin (5-hydroxy-3,6,7,3′,4′-pentamethoxyflavone);[4] other recently isolated lactones include arabsin, artabin, ketopelenolide a (a germacranolide),[5,6] artanolide, artemoline, and deacetylglobicin (monomeric guaianolides),[7] artenolide, anabsin, isoabsinthin, and absintholide (dimeric guaianolides).[5,6,8,9] Other constituents present in lesser amounts in the oil include pinene, sabinene, phellandrene, cadinene, camphene, bisabolene, thujyl alcohol, myrcene, 1-8 cineole, and azulenes (chamazulene; 3,6-dihydrochamazulene; 5,6-dihydrochamazulene), among others.[10,11] Chamazulene concentration up to 0.29% has been reported in flowers at the beginning of flowering (or about 17% of the oil).[12] Also *cis*- and *trans*-epoxyocimenes have been isolated from an Italian absinthium oil of which they made up 16 to 57%.[13]

A recent study found that varying geographical origin, altitude, and exposures affects qualitative and quantitative differences in the essential oil. Plant material collected in Argentina (Patagonia) was composed of 59.9% $\beta$-thujone (2.34% $\alpha$-thujone), sabinyl acetate (18.11%), *cis*-epoxyocimene (1.48%), caryophyllene (1.92%), linalool (1.15%), and sabinene (1.09%), with trace amounts of $\alpha$-pinene, $\alpha$-terpineol, germacrene D, neryl acetate, neryl propionate, nerol, geranyl propionate, and geraniol (<1% each).[11]

Miscellaneous constituents include inulobiose (an oligofructoside),[14] coumarins, phenolic acids, flavonoids,[15,16] carotene (ca. 0.05%), ascorbic acid (0.12–0.26%), tannins (4.0–7.7%),[17] 3,7-dioxabicyclo[3,3,0]-octanes ($6\alpha$ and $6\beta$),[6] pipecolic acid[18] and sterols, including an antipyretic sterol ($24\zeta$-ethylcholesta-7,22-dien-$3\beta$-ol).[19,20]

### PHARMACOLOGY OR BIOLOGICAL ACTIVITIES

It has been reported that the psychological actions of absinthe liqueur and marijuana are similar. This has led to the postulation that thujone and tetrahydrocannabinol, active principles of absinthe and marijuana, respectively, interact with a common receptor in the central nervous system.[21] Thujone is toxic; ingestion of large doses causes convulsions.

Absinthium has bitter tonic properties, but habitual use or large doses cause restlessness, insomnia, nightmares, vomiting, vertigo, tremors, and convulsions.[22] Antitumor and cytotoxic activities have also been reported.[23] The popular 19th-century alcoholic beverage absinthe—made by macerating absinthium and other aromatic herbs in alcohol, distilling the spirit, then adding flavorings or coloring—is recognized for its adverse pharmacological effects, including hallucinations and addiction. Absinthe was banned in many countries soon after the turn of the 19th century (1907 in Switzerland; 1912 in the United States). Its sale persisted in France until 1915.[24] Vincent van Gogh (1853–1890) was believed to have been addicted to absinthe. His well-known erratic behavior, resulting in self mutilation of his left ear and, finally, suicide, has been attributed to congenital psychosis exacerbated by consumption of high levels of absinthe near the end of his life.[25]

Azulenes (also sometimes major components in absinthium oil) have antiinflammatory and antipyretic properties. The essential oil is antibacterial against *Staphylococcus aureus*, the penicillin-resistant strain *S. aureus* (H57), *Klebsiella pneumoniae*, and *Pseudomonas aeruginosa*.[26]

When applied externally, the essential oil is nontoxic.[27]

Recently, an antimalarial compound, qinghaosu (artemisinin, arteannuin), has been isolated from the Chinese drug, *qinghao* (*Artemisia annua* L.), that is effective against chloroquinine-resistant malaria.[28] It is also a potent phytotoxin,[29] and recent studies showed it to have both immunosuppressive and immunostimulant activities in mice.[30]

## USES

**Medicinal, Pharmaceutical, and Cosmetic.** Oil is used as an ingredient in certain rubefacient preparations; extracts now rarely used internally, except in some bitter tonics for anorexia and dyspeptic symptoms (bitter value of at least 15,000).[31] The oil is used as a fragrance component in soaps, detergents, creams, lotions, and perfumes, with maximum use levels of 0.01% in detergents and 0.25% in perfumes.[27]

**Food.** Absinthium is widely used in flavoring alcoholic bitters and in vermouth formulations; average maximum use level of 0.024% reported. The oil and extracts are also used in alcoholic beverages as well as in other categories of foods such as nonalcoholic beverages, frozen dairy desserts, candy, baked goods, and gelatins and puddings. Reported average maximum use levels for the oil is about 0.006% in the last four food categories.

**Health Food/Herb Teas.** Not commonly used; cut and sifted herb as tea (infusion or decoction) reportedly used as a bitter digestive stimulant (HOFFMAN).

**Traditional Medicine.** Reportedly used as aromatic bitter for anorexia, as a digestive tonic, and as choleretic for liver and gallbladder disorders, usually in the form of a dilute extract; also as emmenagogue, febrifuge, and vermifuge; anthelmintic activity probably the result of lactones related to santonin found in wormseed (*A. cina* Berg.) and other *Artemisia* species.

**Others.** Absinthium has been studied in Europe as a potential source of azulenes.[32–34]

## COMMERCIAL PREPARATIONS

Crude, dilute extracts (tincture and fluidextract) and the essential oil. Absinthium was formerly official in U.S.P. (1830–1890) and N.F. (1916).

*Regulatory status.* Has been approved for food use by the FDA, provided the finished food is thujone free (§172.510). In the UK, it is on the GSL (General Sales List). The herb is subject of a German Commission E monograph in aqueous extract, at doses equivalent to 2–3 g of herb, for treatment of anorexia, dyspeptic symptoms, etc.[31].

## REFERENCES


See the General References for APhA; BIANCHI AND CORBETTA; BLUMENTHAL; FEMA; FERNALD; GUENTHER; HOFFMAN; KARRER; MASADA; MERCK; SAX; STAHL; TUTIN 5; YOUNGKEN.

1. J. Slepetys, *Polez. Rast. Priblat. Respub. Beloruss., Mater., Nauch, Konf.*, **2nd**, 289 (1973); through *Chem. Abstr.*, **81**, 60957z (1974).
2. T. Nozoe and S. Ito in L. Zechmeister, ed., *Fortschritte der Chemie Organischer Naturstoffe*, Vol. 19, Springer-Verlag, Vienna, Austria, 1959, p. 1.
3. J. Beauhaire et al., *Tetrahedron Lett.*, **21**, 3191 (1980).
4. K. Venkataraman in L. Zechmeister, ed., in *Fortschritte der Chemie Organischer Naturstoffe*, Vol. 17, Springer-Verlag, Vienna, Austria, 1959, p. 1.
5. S. K. Zakirov et al., *Khim. Prir. Soedin.*, **4**, 548 (1976); through *Chem. Abstr.*, **86**, 16805c (1977).
6. M. Dermanovic et al., *Glas. Hem. Drus., Beogard*, **41**, 287 (1976); through *Chem. Abstr.*, **87**, 98796h (1977).
7. S. Z. Kasymov et al., *Khim. Prir. Soedin.*, **5**, 667 (1987); through *Chem. Abstr.*, *102,* 146143v (1985).
8. A. Ovezdurdyev et al., *Khim. Prir. Soedin.*, **5**, 667 (1987); through *Chem. Abstr.* **108**, 164684a (1988).
9. J. Beauhaire et al., *Tetrahedron Lett.*, **25**, 2751 (1984).
10. D. J. Bertelli and J. H. Crabtree, *Tetrahedron*, **24**, 2079 (1968).
11. T. Sacco and F. Chialva, *Planta Med.*, **54**, 93 (1988).
12. J. Slepetys, *Liet. TSR Mokslu Akad. Darb., Ser. C*, **4**, 29 (1974); through *Chem. Abstr.*, **83**,111201z (1975).
13. F. Chialva et al., *Riv. Ital. Essenze, Profumi, Piante Off, Aromi, Saponi, Cosmet., Aerosol*, **58**, 522 (1976); through *Chem. Abstr.*, **86**, 161105d (1977).
14. M. L. Tourn and A. Lombard, *Atti Acad. Sci. Torino, Cl. Sci. Fis., Mat. Nat.*, **5–6,** 941 (1974); through *Chem. Abstr.*, **84**, 176678a (1976).
15. L. Swiatek and E. Dombrowicz, *Farm. Pol.*, **40**, 729 (1984); through *Chem. Abstr.*, **103**, 34884b (1985).
16. G. A. Zhukov and V. V. Timofeev, *Khim. Prir. Soedin.*, **3**, 447 (1987); through *Chem. Abstr.*, **107**, 140977b (1987).
17. J. Slepetys, *Liet. TSR Mokslu Akad. Darb., Ser. C*, **1**, 43 (1975); through *Chem. Abstr.* **83**, 160852w (1975).
18. V. Rosetti and A. Garrone, *Phytochemistry*, **14**, 1467 (1975).
19. M. Ikram et al., *Planta Med.*, **53**, 389 (1987).
20. M. D. Sayed et al., *Egypt J. Pharm. Sci.*, **19**, 323 (1980).
21. J. del Castillo et al., *Nature*, **253**, 365 (1975).
22. J. D. Kinloch. *Practitioner*, **206**, 44 (1971).
23. R. Truhaut, paper presented at Atti Società Italiana Fitochimica, St. Vincent, Apr. 26–27, 1984.

24. D. D. Vogt. *J. Ethnopharmacol.*, **4**, 337 (1981).
25. W. Arnold. *J.A.M.A.*, **3042** (Nov. 25, 1988).
26. V. K. Kaul et al., *Indian J. Pharm.*, **38**, 1, 21 (1975).
27. D. L. J. Opdyke, *Food Cosmet. Toxicol.*, **13**(Suppl.), 721 (1975).
28. Q. C. Xu, *Yaoxue Tongbao*, **17**(9), 32 (1982).
29. S. O. Duke et al., *ACS Symp. Ser.*, **380**, 318 (1988).
30. P. Y. Lin et al., *Asia Pac. J. Pharmacol.*, **3**, 197 (1988).
31. Monograph *Absinthii herba*, *Bundesanzeiger*, no. 228 (Dec. 5, 1984).
32. L. Sommer et al., *Farmacia (Bucharest)*, **15**, 689 (1967); through *Chem. Abstr.*, **68**, 24524d (1968).
33. J. Slepetys, *Polez. Rast. Priblat. Respub. Beloruss., Mater. Nauch. Konf.*, **2**, 294 (1973); through *Chem. Abstr.*, **81**, 62501v (1974).
34. L. Sommer et al., *Farmacia (Bucharest)*, **22**, 277 (1974); through *Chem. Abstr.*, **81**, 148527v (1974).

# ACACIA

**Source:** ***Acacia senegal*** (L.) Willd. and other ***Acacia*** spp. (Family Leguminosae or Fabaceae).

*Synonyms*. Gum acacia, gum arabic, gum Senegal, gummae mimosae, gomme arabique, gomme de Senegal, and kher.

## GENERAL DESCRIPTION

The dried gummy exudate from stems and branches of *Acacia senegal* (L.) Willd. (syn. *A. verek* Guill. et Perr.) or other related African *Acacia* species. *Acacia senegal* has triple spines at the base of its branchlets, which distinguishes it from many other *Acacia* spp. in its range. The thorny trees, which grow to a height of about 6 m, are tapped by making transverse incisions in the bark and peeling off a thin strip of the bark. The gummy exudates form as tears on the surface of the wounds and are collected after they have hardened, usually in 2 or more weeks.

The Republic of Sudan supplies most of the world's acacia and produces the best quality product. *A. senegal* ranges from Senegal to northeastern Africa, south to Mozambique. Other suppliers include Senegal, Mauritania, Chad, Nigeria, Tanzania, and Ethiopia.

Acacia is one of the most water soluble plant gums; 1 part acacia can dissolve in 2 parts water, forming a weakly acidic solution with pH 4.5–5.5. Its solutions have lower viscosities than those of other natural gums. It is insoluble in alcohol, chloroform, ether, and oils and very slightly soluble in glycerol and propylene glycol. It is almost odorless and has a bland mucilaginous taste.

Acacia contains a peroxidase that, unless destroyed by heating briefly at 100°C, forms colored compounds with certain amines and phenols (e.g., aminopyrine, antipyrine, epinephrine, cresol, eugenol, guaiacol, phenol, tannins, thymol, vanillin, etc.) It also causes partial destruction of many alkaloids, including atropine, hyoscyamine, scopolamine, homatropine, morphine, apomorphine, cocaine, and physostigmine.

Acacia is incompatible with heavy metals, which destroy the gum by precipitation. Borax and alcohol also precipitate it, but the process can be prevented or reversed.

## CHEMICAL COMPOSITION

Acacia contains neutral sugars (rhamnose, arabinose, and galactose), acids (glucuronic acid and 4-methoxyglucuronic acid), calcium, magnesium, potassium, and sodium.[1] Its structure is quite complex and still not completely known. Its backbone chain consists of D-galactose units, and its side chains are composed of D-glucuronic acid units with L-rhamnose or L-arabinose as end units.[2] The molecular weight of acacia is still unsettled; it has been reported to be between 200,000 and 300,000 and as high as 600,000 (FURIA, KENNEDY).[2,3] This wide discrepancy appears to be explained by a recent enzymatic study the findings of which indicate the structure of crude *A. senegal* gum to consist of varying numbers of polysaccharide units of molecular weight 200,000 linked to a protein core,[4] forming an arabinogalactan-protein complex.[5]

## PHARMACOLOGY OR BIOLOGICAL ACTIVITIES

Ingested orally, acacia is nontoxic. However, some people are allergic to its dust and develop skin lesions and severe asthmatic attacks when in contact with it. Acacia can be digested by rats to an extent of 71%; guinea pigs and rabbits also seem to use it for energy, as do humans to a certain extent.[6]

Acacia, along with other plant polysaccharides, has been studied as a potential hypocholesterolemic agent.[7,8] Data so far reported have not indicated it to be hypocholesterolemic; on the contrary, it has been suggested that acacia may actually elevate serum or tissue cholesterol levels in rats.[8]

## USES

Acacia gum has been in use since ancient times.

**Medicinal, Pharmaceutical, and Cosmetic.** Mainly in the manufacture of emulsions and in making pills and troches (as an excipient); as a demulcent for inflammations of the throat or stomach and as a masking agent for acrid-tasting substances such as capsicum (MARTINALDE); also as a film-forming agent in peel-off facial masks.

**Food.** Currently, its major use is in foods, where it performs many functions, e.g., as a suspending or emulsifying agent, stabilizer, adhesive, and flavor fixative and to prevent crystallization of sugar, among others. It is used in practically all categories of processed foods, including candy, snack foods, alcoholic and nonalcoholic beverages, baked goods, frozen dairy desserts, gelatins and puddings, imitation dairy products, breakfast cereals, and fats and oils, among others. Its use levels range from <0.004% (40 ppm) in soups and milk products to 0.7–2.9% in nonalcoholic beverages, imitation dairy, and snack foods to as high as 45% in candy products.

## COMMERCIAL PREPARATIONS

Available in crude, flake, powdered, granular, and spray-dried forms. It is official in N.F. and F.C.C.

*Regulatory Status.* Has been affirmed as GRAS (§184.1330).

## REFERENCES

See the General References for DAVIDSON; FEMA; FURIA; GLICKSMAN; GOSSELIN; KEAY; KENNEDY; LAWRENCE; MARTINALDE; REMINGTON; SAX; TERRELL; WHISTLER AND BEMILLER; YOUNGKEN.

1. J. Artaud et al., *Ann. Falsif. Expert. Chim.*, **70**, 511 (1977); through *Chem. Abstr.*, **88**, 75s59q (1978).
2. R. L. Whistler in H. W. Schultz et al., eds., *Symposium on Foods: Carbohydrates and Their Roles*, AVI, Westport, Conn., 1969, p. 73.
3. F. Aprile et al., *Ind. Carta*, **16**, 3 (1978); through *Chem. Abstr.*, **88**, 122893f (1978).
4. S. Connolly et al., *Carbohydr. Polym.*, **8**, 23 (1988).
5. Y. Akiyama et al., *Agri. Biol. Chem.*, **48**, 235 (1984).
6. A. Jeanes, *ACS Symp. Ser.*, **15**, 336 (1975).
7. S. Kiriyama et al., *J. Nutr.*, **97**, 382 (1969).
8. A. C. Tsai et al., *J. Nutr.*, **106**, 118 (1976).

# ACEROLA

**Source:** ***Malpighia glabra*** L. and ***M. punicifolia*** L. (Family Malpighiaceae).

*Synonyms*. Barbados cherry; Puerto Rican cherry, West Indian cherry (*M. punicifolia*); huesito (*M. glabra*).

## GENERAL DESCRIPTION

Acerola is the fruit of a shrub or small tree that grows to a height of 5 m. Fruits (drupes) are globose, ovoid, or subglobose, 1 to 2 cm in diameter, bright red, slightly resembling cherries; hence the name *cherry*. Mature fruits are juicy and soft, with a pleasant tart flavor.[1]

Both *M. glabra* and *M. punicifolia* have been reported in the literature as a source of acerola with high vitamin C content.[1–3] However *M. punicifolia* or its hybrid with *M. glabra* appears to be the correct source (MORTON 2, WATT AND MERRILL).[1]

*Malpighia punicifolia* is native to the West Indies and is also found in northern South America, Central America, Florida, and Texas.[1] Its fruit, acerola, is the richest known source of natural vitamin C.

## CHEMICAL COMPOSITION

Contains 1–4.5% vitamin C (ascorbic acid and dehydroascorbic acid, mainly the former) in edible portion of fruit (cf. 0.05% in peeled orange), which makes up about 80% of the fruit. Vitamin C content varies with ripeness of the fruit (highest in green and lowest in fully ripe fruit), seasons, climate, and localities.[1,3,4]

Other vitamins present include 4,300–12,500 IU/100 g vitamin A (cf. 11,000 IU/100 g for raw carrots); thiamine, riboflavin, and niacin in concentrations comparable to those in other fruits.[1]

Miscellaneous constituents include calcium, iron and phosphorus in comparable concentrations to those of apple; *l*-malic acid; dextrose, fructose, and sucrose; evidence of a heat-resistant enzyme (not completely deactivated at 103°C) that breaks down ascorbic acid during storage of pasteurized juice, resulting in carbon dioxide buildup, causing swelling of cans or explosion of bottles.[1,2]

## USES

**Food.** As a source of natural vitamin C, in the form of juice, tablet, or capsule. However, as most of the vitamin C is destroyed during processing, acerola products are usually supplemented with the synthetic vitamin (see ***rose hips***).

**Health Food/Herb Teas.** Tablets, capsules, or other products, often combined with other herbs.

**Traditional Medicine.** The fruits have reportedly been used for the treatment of dysentery, diarrhea, and liver disorders (CSIR VI).

**Others.** The bark, which contains 20–25% tannin has been used in the manufacture of leathers (MORTON 4).

### COMMERCIAL PREPARATIONS

Available as fresh fruit for home consumption in certain East Coast supermarkets and ethnic stores; also in juice and spray-dried form. Canned juice of the fruits has been used to enhance ascorbic acid content of other juices, such as pear, apricot, and grape juice.

### REFERENCES


See the General References for CSIR VI; MORTON 2; MORTON 4; TERRELL; WATT AND MERRILL.

1. C. G. Moscoso, *Econ. Bot.*, **10**, 280 (1956).
2. R. E. Berry et al., *Food Prod. Dev.*, **14**, 109 (1977).
3. H. Y. Nakasone et al., *Proc. Am. Soc. Hort. Sci.* **89**, 161 (1966).
4. A. Schillinger, *Z. Lebensm, Unters. Forsch.*, **131**, 89 (1966)


# ACONITE

**Source:** ***Aconitum napellus*** L. and other ***Aconitum*** spp. (Family Ranunculaceae).

*Synonyms.* Aconitum, monkshood, and wolfsbane.

### GENERAL DESCRIPTION

Perennial herbs consisting of many subspecies, varieties, clones, and forms; up to 1.5 m high with tuberous roots that resemble turnips; native to mountainous regions of central Europe; naturalized in Asia, Africa, and North America; cultivated in Russia, Germany, Spain, and France. Part used is the dried tuberous root. Of the 100 northern temperate species in the genus, 35 species in China have been investigated chemically.

### CHEMICAL COMPOSITION

Total alkaloids 0.2–2%, consisting mainly of aconitine (acetylbenzoylaconine), picraconitine (benzoylaconine), aconine, and napelline (isoaconitine, pseudoaconitine); others include 12-epidehydronapelline, 12-epiacetyldehydronapelline, 1,14-diacetylneoline, N-deethylaconitine, aconosine, 14-acetylneoline, hokbusine A, senbusine A, senbusine C, mesaconitine, neoline and songoramine.[1–3] Alkaloid content decreases with altitude from 0.82% of fresh root of plants grown at 1750 m to 0.29% at 2500 m. Aconitine content is greatest in winter-dormant tubers (FROHNE AND PFÄNDER)[4].

On hydrolysis, aconitine yields picraconitine, which in turn yields aconine on further hydrolysis.

Compounds identified from raw (dried) *A. carmichaelii* Debx. include aconitines, coryneine chloride, and higenamine, all of which have been implicated in cardioactivity of the tubers.[5]

Other constituents include aconitic acid, itaconic acid, succinic acid, malonic acid, fructose, maltose, melibiose, mannitol, starch, fat, and resin.

## PHARMACOLOGY OR BIOLOGICAL ACTIVITIES

Aconite is a strong and fast-acting poison, affecting both the heart and the central nervous system. Its active principles are aconitine and its related alkaloids. As little as 2 mg aconitine may cause death from paralysis of the heart or respiratory center. The lethal dose for adults generally ranges only from 3 to 6 mg of aconitine, readily contained in a few grams of plant material (FROHNE AND PFÄNDER).

When applied to the skin, aconite produces tingling and then numbness; poisoning may result from percutaneous absorption.

Extracts of *A. carmichaelii* have shown cardiotonic activity, including ionotropic and chronotropic activity, leading to hypotension and/or hypertension. Analgesic and anesthetic activities have been reported. Hyoscine potentiates the action of aconitine.[5] A recent study confirms that aconitine and related compounds exhibit antiinflammatory and analgesic properties in experimental animals.[6] A component of prepared (processed) lateral rootlets of *A. carmichaelii* (considered a separate drug in Chinese tradition), *dl*-demethylcoclaurine has been shown to raise the heart rate in sinus arrhythmia patients.[7]

Certain *Aconitum* species are reported to have antitumor activity in laboratory animals, and others show antibacterial, antifungal, and antiviral activities. Extracts of various species also have antipyretic properties (FARNSWORTH 3).[8]

During the past decade, extensive studies have been carried out on the chemistry and pharmacology of aconite in general. While hypaconitine is found to be the active neuromuscular blocking agent in Asian aconite,[9] higenamine (*dl*-demethylcoclaurine) and other chemical components (including a nonalkaloidal fraction) are the cardiotonic principles. These cardiotonic substances are heat resistant, and their activities are realized after prolonged decocting, whereby the deadly aconitine is hydrolyzed to the much less toxic aconine (WANG).

## USES

**Medicinal, Pharmaceutical, and Cosmetic.** Now rarely used internally in the United States; its current use is mainly in liniments (rubefacients), often with belladonna, for external applications only.

**Traditional Medicine.** Used internally as a cardiac depressant and mild diaphoretic; externally as local analgesic in facial neuralgia, rheumatism, and sciatica. Related species such as *A. chinense* Paxt. and *A. kusnezoffii* Reichb. are widely used in Chinese medicine for rheumatoid arthritis, chronic nephritis, sciatica, and other ailments. These and other Asian *Aconitum* spp. have been valued for their analgesic, antiinflammatory (antirheumatic); antibiotic (antiseptic), antipyretic, and cardiotonic activities. The methods used for treating these crude drugs (roots) are numerous and quite different from that for *A. napellus*. One method involves soaking and washing in clear water for several days and treating with licorice, ginger, black beans, and other drugs, followed by boiling or steaming and then drying. The resulting product therefore cannot be compared directly with the American or European product or any other unprocessed product (JIANGSU, NANJING, WANG).

## COMMERCIAL PREPARATIONS

Crude and extracts. Strengths (see ***glossary***) of extracts are expressed in weight-to-weight ratios. Crude was formerly official in U.S.P. (1850–1936) and N.F. (1942).

*Regulatory Status.* Subject of a negative

German therapeutic monograph, due to toxicity that can occur within the therapeutic dose (including vomiting, dizziness, muscle spasms, hypothermia, paralysis of respiratory system, and rhythmic heart disorders).[10]


**REFERENCES**

See the General References for BLUMENTHAL; CLAUS; FARNSWORTH 3; FERNALD; FOGARTY; FROHNE AND PFÄNDER; GOSSELIN; JIANGSU; KARRER; MARTINDALE; MERCK; NANJING; SAX; WANG.

1. G. de la Fuente et al., *Heterocycles*, **27**, 1109 (1980).
2. E. Arlandini et al., *J. Nat. Prod.*, **50**. 937 (1987).
3. H. Hikino et al., *J. Nat. Prod.*, **47**, 190 (1984).
4. A. Crema, *Arch. Ital. Sci. Farmacol.*, **7**, 119 (1957); through *Chem. Abstr.*, **51**, 16748e (1957).
5. N. G. Bisset, *J. Ethnopharmacol.*, **4**, 247 (1981).
6. M. Murayama et al., *J. Pharm. Pharmacol.*, **35**, 135 (1991).
7. P. G. Xiao and K. J. Chen. *Phytotherapy Res.*,**1**, 2, 5 (1987).
8. J. L. Hartwell, *Lloydia*, **34**, 103 (1971).
9. M. Kimura et al., *Jpn. J. Pharmacol.*, **48**, 290 (1988).
10. Monograph *Aconitum napellus*, *Bundesanzeiger*, no. 193 (Oct. 15, 1987).


# AGAR

**Source:** Red algae, including: ***Gelidium cartilagineum*** (L.) Gaill., ***Gelidium amansii*** Lamour., ***Gracilaria confervoides*** (L.) Grev., other ***Gelidium*** and ***Gracilaria*** species as well as species of the genera ***Pterocladia, Ahnfeltia, Acanthopeltis,*** and ***Suhria***.

*Synonyms*. Agar-agar, Japanese isinglass, vegetable gelatin, Chinese gelatin, Japanese gelatin, gelose, gelosa, layor carang, and colle du Japon.

## GENERAL DESCRIPTION

Agar is the dried hydrophilic, colloidal extract of various red algae (Class Rhodophyceae); the more commonly used red algae are *Gelidium cartilagineum* (L.) Gaill., *Gelidium amansii* Lamour., *Gracilaria confervoides* (L.) Grev., other *Gelidium* and *Gracilaria* species as well as species of the genera *Pterocladia*, *Ahnfeltia*; *Acanthopeltis*, and *Suhria*. Agar is extracted from the algae by boiling them in water at a neutral or slightly acidic pH. The hot liquor is filtered and on cooling forms a gel, which is purified by freezing and thawing followed by drying.

The major agar producer has been and still is Japan. Other producing countries include U.S., Spain, Portugal, Chile, Taiwan, Korea, Morocco, New Zealand, Australia, Argentina, and Mexico.

Agar is insoluble in cold water but readily soluble up to 5% in boiling water. The solution (sol) on cooling to 35 to 40°C forms a firm, resilient gel that does not melt below 85°C. This ability to gel at a much lower temperature than the melting temperature of the gel, commonly called hysteresis lag, is uniquely long in agar, and many of its uses depend on this property. Agar gels also have the property of shrinking and exuding water from their surface (syneresis), particularly when broken. The gel strength of agar can be increased by addition of dextrose, sucrose, and locust bean gum, while it tends to weaken with gelatin, algin, starch, and karaya gum. The colorless, tasteless powder can absorb up to 200 times its volume of water when forming a gel.

Agar solutions have low viscosity; their degree of clarity and color (yellowish to colorless) depend on the quality and source of the agar, as do their gel strength, gelling temperature, and the degree of syneresis. Quality is largely affected by extraction procedures. Physical and rheological properties of agar that are the greatest determiners of quality are the average molecular weight and molecular weight distribution.[1]

Agar is insoluble in organic solvents; it is precipitated from aqueous solution by alcohol and tannin.

## CHEMICAL COMPOSITION

The structure of agar is still not fully determined, the problem being complicated by the large number of commercial sources of agar. Nevertheless, it has been generally believed that all agars consist of two major polysaccharides (neutral agarose and charged agaropectin), although several studies have indicated a much more complicated structure.[2–6] Agarose is the gelling fraction and agaropectin is the nongelling fraction. Both are composed of a linear chain of alternating $\beta$-D-galactopyranose and 3,6-anhydro-$\alpha$-L-galactopyranose residues, with agaropectin having a higher proportion of uronic acid, sulfate, and pyruvic acid residues.[6] Commercial agar may contain free amino acids (arginine, glutamic acid, aspartic acid, and threonine) and free sugars (galactose and gluconic acid).[7] It may also contain other sugar residues including 4-*O*-methyl-L-galactose, 6-*O*-methyl-D-galactose,[8–10] D-xylose, and *O*-methylpentose as well as boric acid (approx. 0.1%)[11] and various inorganic cations ($Na^+$, $K^+$, $Ca^{2+}$, $Mg^{2+}$, etc.).[12]

## PHARMACOLOGY OR BIOLOGICAL ACTIVITIES

Agar is nontoxic and can be ingested in large doses without much distress. It passes through the intestinal tract mostly unabsorbed. However, a 1981 study found that mice fed agar had significantly more colon tumors per animal (twice as many) than those fed diets without agar, despite the fact that agar-fed animals had decreased levels of fecal neutral sterol and bile acid concentrations.[13] One study has demonstrated agar to have antipeptic activities *in vitro*.[14] Another study has suggested that it may elevate serum or tissue cholesterol levels in rats.[15] When fed to rats at 5 and 15% levels in their diet, agar impaired protein utilization.[16]

## USES

**Medicinal, Pharmaceutical, and Cosmetic.** As a bulk laxative, particularly in chronic constipation; in the manufacture of emulsions, suspensions, gels, and hydrophilic suppositories; in dentistry as basic constituent of reversible impression and duplicating materials.

**Food.** Used in canned meat and fish products as gel filler or gel binder; in baked goods (icings and glazes); and in confectionery, dairy products, processed fruits, sweet sauces, and reconstituted vegetables, among others. Highest average maximum use level usually about 0.4% in baked goods.

Agar has been used as a food in the Far East for centuries.

**Others.** A major use of agar is in culture media for microorganisms. It is one of the most widely used media for biotechnology purposes.

**COMMERCIAL PREPARATIONS**

Available in flakes, strips, and powders; grades and quality vary, with the bacteriological grades demanding the most stringent quality. Some high-quality agars from certain commercial sources have higher congealing temperatures than that required by the F.C.C. and N.F. This is due to the source of algae used.

*Regulatory Status*. GRAS.[17]

**REFERENCES**

See the General References for FEMA; FURIA; GLICKSMAN; GOSSELIN; LAWRENCE; MARTINDALE; PHILLIPS; REMINGTON; TYLER 2; UPHOF; WHISTLER AND BEMILLER; WREN.


1. F. Zanetti et al., *Planta Med.*, **58**(S1), A696 (1992).
2. K. B. Guiseley, in A. Standen, ed., *Kirk-Othmer Encyclopedia of Chemical Technology*, Vol. 17, 2nd ed., Wiley-Interscience, New York, 1968, p. 763.
3. M. Duckworth and W. Yaphe, *Carbohydr. Res.*, **16**, 189 (1971).
4. M. Duckworth and W. Yaphe, *Carbohydr. Res.*, **16**, 435 (1971).
5. K. Young et al., *Carbohydr. Res.*, **16**, 446 (1971).
6. K. Izumi, *Carbohydr. Res.*, **17**, 227 (1971).
7. K. Hayashi and K. Nonaka, *Nippon Shokuhin Kogyo Gakkaishi*, **14**, 66 (1967); through *Chem. Abstr.*, **78**, 10216t (1968).
8. C. Araki et al., *Bull. Chem. Soc. Jpn.*, **40**, 959 (1967); through *Chem. Abstr.*, **67**, 91046s (1967).
9. M. Duckworth et al., *Carbohydr. Res.*, **18**, 1 (1971).
10. Y. Karamanos et al., *Carbohydr. Res.*, **187**, 93 (1989).
11. H. Hayashi et al., *Shokuhin Eiseigaku Zasshi*, **29**, 390 (1988); through *Chem. Abstr.*, **110**, 171863c (1989).
12. H. A. Kordan, *Biochem. Physiol. Pflanz.*, **183**, 355 (1988).
13. H. P. Glauert et al., *Food. Cosmet. Toxicol.*, **19**, 281 (1981).
14. M. W. Gouda and G. S. Godhka, *Can. J. Pharm. Sci.*, **12**, 4 (1977).
15. A. C. Tsai et al., *J. Nutr.*, **106**, 118 (1976).
16. S. Y. Shiau et al., *Nutr. Rep. Int.*, **39**, 281 (1989).
17. Anon., *Fed. Regist.* **42**(161), 41876 (1977).


# ALETRIS

**Source:** ***Aletris farinosa*** L. (Family Liliaceae).

*Synonyms*. Stargrass, whitetube stargrass, unicorn root, true unicorn root, colic root, ague root, ague grass, starwort, and blazing star.

**GENERAL DESCRIPTION**

Perennial herb with grasslike leaves up to 20 cm long, formed as a rosette around a slender, naked flowering stem that grows up to almost 1 m high; flowers white, tubular, mealy at base; native to North America from southern Maine south to Florida and west to Wisconsin and Texas. Parts used are the dried rhizome and roots, which are collected in the fall (FOSTER AND DUKE, YOUNGKEN).

## CHEMICAL COMPOSITION

Not much chemical data are available on this drug. Diosgenin is the only significant compound that has been isolated from it.[1] Diosgenin and gentrogenin have also been isolated from two Japanese *Aletris* species: *A. foliata* and *A. formosana*.[2]

Other constituents reported to be present in alteris include an amber volatile oil said to be pharmacologically active, a resinous material, and a saponinlike glycoside that yields diosgenin on hydrolysis.

## PHARMACOLOGY OR BIOLOGICAL ACTIVITIES

Aletris was found to be estrogenic, though the estrogen has not been identified and no other study has since been made to confirm this activity.[3,4]

Diosgenin is one of the starting materials for the manufacture of steroid hormones. It is likely that the alleged estrogenic properties of aletris are the result of some diosgenin-derived steroid that has escaped characterization.

## USES

**Medicinal, Pharmaceutical, and Cosmetic.** It has been and is still used in proprietary preparations for the treatment of female disorders such as dysmenorrhea and other menstrual discomforts, in laxatives, and also as an antiflatulent.

**Health Food/Herb Teas.** Crude root, powdered, or cut and sifted used in tablets, capsules, tinctures, teas, often in combination with other herbs for menstrual disorders and as a bitter digestive tonic (CRELLIN AND PHILPOTT).

**Traditional Medicine.** Reported to be used by the Catawba of North and South Carolina to relieve colic and as an antidiarrheal tea; in Appalachia as a treatment for rheumatism and stomach pains; as a tonic and sedative and as a diuretic. A root decoction has been used as a bitter tonic for indigestion and to promote appetite. *Aletris spicata* root has been used as an antitussive in coughs and wheezing, as a tonic in infantile emaciation, and as a vermifuge for round worms.[5]

## COMMERCIAL PREPARATIONS

Crude and extracts; strengths (see ***glossary***) of extracts are expressed in weight-to-weight ratios. Crude was formerly official in U.S.P. (1820–1860) and N.F. (1916–1942).

## REFERENCES

See the General References for CRELLIN AND PHILPOTT; FERNALD; FOSTER AND DUKE; KROCHMAL AND KROCHMAL; LEWIS AND ELVIN-LEWIS; MERCK; YOUNGKEN.

1. R. E. Marker et al., *J. Am. Chem. Soc.*, **62**, 2620 1940.
2. T. Okanishi et al., *Chem. Pharm. Bull.*, **23**, 575 (1975).
3. C. L. Butler and C. H. Costello, *J. Am. Pharm. Assoc.*, **33**, 177 (1944).
4. C. H. Costello and E. V. Lynn, *J. Am. Pharm. Assoc.*, **39**, 177 (1950).
5. S. Foster. *East-West Botanicals: Comparisons of Medicinal Plants Disjunct Between Eastern Asia and Eastern North America*, Ozark Beneficial Plant Project, Brixey, Mo., 1986.

# ALFALFA

**Source:** ***Medicago sativa*** L. (Family Leguminosae or Fabaceae).

*Synonym.* Lucerne.

## GENERAL DESCRIPTION

Perennial herb with a deep taproot; leaves resemble those of clover; grows to a height of 1 m with mostly bluish purple flowers in the typical subspecies. Native to the Near East (western Asia and east Mediterranean regions); now cultivated extensively throughout the world. Parts used are the aerial parts. The species has several distinct variants including *M. sativa* (*sensu stricto*), and subsp. *falcata* (L.) Arcangeli (syn. *M. falcata* L.). The former is a purple-flowered form with strongly coiled legumes, originating from an arid continental climate in alkaline soils, principally from Turkey. Wild and cultivated *M. sativa* subsp. *sativa* and their progeny are relatively low in hemolytic saponins. *M. sativa* subsp. *falcata* has yellow flowers and uncoiled fruits, originating from cool, upland, humid climates in acidic soils and is comparatively higher in hemolytic saponins. Both taxa are involved in the parentage of numerous commercial alfalfa cultivars.[1] Modern western European and North American cultivars have intermediate levels of hemolytic alfalfa saponins due to hybridization and introgressions involving *M. sativa* subsp. *falcata*.[2]

## CHEMICAL COMPOSITION

Alfalfa has been one of the most studied plants. The chemical constituents reported to be present in this plant include the following:

Saponins (2–3%) which on hydrolysis yield the aglycones medicagenic acid, soyasapogenols A, B, C, D, and E, and hederagenin and the glycones glucose, arabinose, xylose, rhamnose, galactose, and glucuronic acid;[3–8] sterols (β-sitosterol, α-spinasterol, stigmasterol, cycloartenol, and campesterol, with β-sitosterol as the major component);[9–11] high molecular weight alcohols (octacosanol, triacontanol); and paraffins (nonacosane, triacontane, hentriacontane).[12] β-Sitosterol also occurs as esters with fatty acids (mainly palmitic, lauric, and myristic). Triacontanol has been shown to be a plant growth regulator that increases the growth of rice, corn, and barley as well as the yield of tomato, cucumber, and lettuce.[13]

Flavones and isoflavones (tricin, genistein, daidzein, biochanin A, formononetin, and (−)-5′-methoxysativan); coumarin derivatives (coumestrol, medicagol, sativol, trifoliol, lucernol, and daphnoretin); and pectin methylesterase (an enzyme present in significant quantities believed to be one of the causes for bloating in cattle by releasing pectic acids that combine with calcium in the rumen to form a resinous material, trapping gases produced during digestion).[4,14–16]

Alkaloids (trigonelline, which is in seeds only; stachydrine; and homostachydrine); plant acids (malic, oxalic, malonic, maleic and quinic etc.); vitamins and growth factors (vitamins A, $B_1$, $B_6$, $B_{12}$, C, E, and $K_1$; niacin; pantothenic acid; biotin; folic acid; etc.); amino acids (valine, lysine, arginine, leucine, isoleucine, tryptophan, phenylalanine, methionine, and threonine; asparagine in high concentrations in seeds); sugars (sucrose, fructose, arabinose, xylose, galactose, ribose, manno-heptulose, and D-glycero-D-manno-octulose); plant pigments (chlorophyll, xanthophyll, β-carotene, anthocyanins); crude fibers (17–25%); proteins (15–25% in dehydrated alfalfa meal); minerals; and trace elements (Ca, P, K, Mn, Fe, Zn, Cu) (KARRER, LIST AND HÖRHAMMER).

Medicarpin-β-D-glucoside (in roots); cerebrosides (sphingosines); plastocyanins and ferredoxins; benzoylmesotartaric acid, and benzoyl-(*S*),(−)-malic acid;[17–19] three phytoalexins;[20] medicosides A, C, G, I, J and L (triterpene glycosides) in roots;[21] and a new amino acid, medicanine, ((*S*)-*N*-(3-hydroxypropyl)-azetidine-2-carboxylic acid) from seedlings.[22]

Traces of cannabinol, caffeine, scopolamine, isocoumarin, phenylpentadienal, phenylhexadiene, and nepetalactone have been reported in one study to be present in commercial solid extracts of alfalfa and red clover.[23] Whether or not these compounds were results of contamination or adulteration remains to be confirmed.

## PHARMACOLOGY OR BIOLOGICAL ACTIVITIES

Medicagol has been reported to have antifungal properties.[24] Coumestrol, genistein, biochanin A, and daidzein have estrogenic activities on ruminants.[14,15] Studies also have demonstrated that coumestrol when fed to pullets increases the age of maturity and depresses egg production.[25]

Alfalfa saponins are hemolytic;[26] they also interfere with the use of vitamin E and are believed to be one of the causes of ruminant bloat.[4,27] Hemolytic saponins have been found in concentrations of up to 8% in commercial alfalfa sprouts. Alfalfa saponins are reported to be fungitoxic, antimicrobial, insecticidal, piscicidal and taste repellent to rats, swine, and poultry, while attractive to rabbits. Oral toxicity in humans is considered low, because they are not absorbed by the gut and then enter the bloodstream. Intravenous-administered alfalfa saponins are highly toxic to mammals.[2]

Medicagenic acid and its glycoside (but not soyasapogenol and glycoside) are toxic to L-cells in culture, lowering mitotic index, viability, and growth of the cells as well as inducing cell death.[28] When administered intramuscularly to Wistar rats, they caused pathological changes in internal organs, especially the kidneys and liver.[29] However, when alfalfa saponins were incorporated in the diets of male rats for up to 6 months, no evidence of toxicity was found after analyzing numerous biochemical and histological parameters.[26]

Root saponins have been shown to prevent the expected increase in plasma cholesterol associated with the ingestion of a high-cholesterol diet in monkeys, suggesting the potential of using alfalfa saponins in treating hypercholesterolemia in humans (also see ***quillaia***).[30] Male rats fed a complex of alfalfa top saponins (1% of diet for 6 months) had reduced levels of serum cholesterol and triglycerides, with no evidence of toxicity. Alfalfa top saponins have been shown to lower plasma cholesterol, decrease intestinal absorption of cholesterol, increase excretion of neutral steroids and bile acids in fecal matter, and prevent atherosclerosis.[31] More recent studies have shown that ingesting large amounts of alfalfa seeds can produce reversible pancytopenia with splenomegaly in humans, probably due to the activity of canavanine.[32] The seeds or sprouts may induce systemic lupus erythematosus (SLE).[33–35] Persons with or predisposed to SLE are cautioned to curtail or eliminate alfalfa product intake (TYLER 1).

Quercetin isolated from alfalfa seed has been shown to have antioxidant activity.[26,36]

## USES

**Medicinal, Pharmaceutical, and Cosmetic.** The unsaponifiable extract has been claimed to be beneficial in treating skin conditions, including damage caused by radiotherapy and in the healing of gums after orthodontic operations.[9] Alfalfa is also reportedly used in peelable facial masks. (deNAVARRE).

**Food.** Extract used as a flavor ingredient in most major categories of food products, including nonalcoholic and alcoholic beverages, frozen desserts, candy, baked goods, gelatins and puddings, and meat and meat products, with highest average maximum use level of 0.05% in the last category.

**Health Food/Herb Teas.** Alfalfa sprouts are a favorite salad ingredient among health food enthusiasts. Dried leaves used in tablets, capsules, teas, tinctures, etc. reported as a source of chlorophyll, vitamins, minerals, and protein, with unsubstantiated benefit in conditions such as rheumatoid arthritis, to prevent absorption of cholesterol, treating diabetes, stimulating appetite, and as a general tonic (TYLER 1).

**Traditional Medicine.** Reportedly used as a nutrient to increase vitality, appetite, and weight in humans; also as a diuretic, galactogogue, and to increase peristaltic action of the stomach and bowels, resulting in increased appetite. More recently for the treatment of asthma and hay fever (JIANGSU).

**Others.** Alfalfa meal is used extensively as a poultry and cattle feed and as a source of raw material for the manufacture of leaf protein intended for human consumption. Alfalfa is also a source of chlorophyll manufacture, with silkworm droppings being now the major source.

**COMMERCIAL PREPARATIONS**

Crude and extracts.

*Regulatory Status.* GRAS (§182.10 and §182.20).

**REFERENCES**

See the General References for BAILEY 2; FEMA; JIANGSU; KARRER; LIST AND HÖRHAMMER; TYLER 1; WILLAMAN AND SCHUBERT.


1. E. Small and B. S. Brookes, *Econ. Bot.*, **38**, 83 (1984).
2. E. Small et al., *Econ. Bot.*, **44**, 226 (1990).
3. M. Jurzysta, *Pamiet. Pulawski*, **62**, 99 (1975); through *Chem. Abstr.*, **84**, 56500q (1976).
4. J. F. Morton, *Morris Arboretum Bull.*, **26**, 24 (1975).
5. E. Nowacki et al., *Biochem. Physiol. Pfanz.*, **169**, 183 (1976); through *Chem. Abstr.* **84**, 102400g (1976).
6. B. Gestetner, *Phytochemistry* **10**, 2221 (1971).
7. B. Berrang et al., *Phytochemistry*, **13**, 2253 (1974).
8. R. L. Baxter, *J. Nat. Prod.*, **53**, 298 (1990).
9. P. de Froment, Fr. Demande, 2,187,328 (1974); through *Chem. Abstr.*, **81**, 68543g (1974).
10. S. Ito and Y. Fujino, *Nippon Nogei Kagaku Kaisha*, **47**, 229 (1973); through *Chem. Abstr.*, **80**, 68380f (1974).
11. S. Ito and Y. Fujino, *Obihiro Chikusan Daigaku Gakujutsu Kenkyu Hokoku, Dai-I-Bu*, **9**, 817 (1976); through *Chem. Abstr.*, **85**, 189227y (1976).
12. H. Choichiro et al., *Nippon Kagaku Zasshi*, **77**, 1247 (1956); through *Chem. Abstr.*, **53**, 20316i (1959).
13. S. K. Ries et al., *Science*, **195**, 1339 (1977).
14. P. J. Schaible, *Poultry: Feeds and Nutrition*, AVI, Westport, Conn., 1970, p. 358.
15. R. F. Keeler, *Lloydia*, **38**, 56 (1975).
16. R. W. Miller et al., *J. Nat. Prod.*, **52**, 634 (1989).
17. Y. Sakagami et al., *Agri. Biol. Chem.*, **38**, 1031 (1974).
18. E. G. Sarukhanyan et al., *Dokl. Akad. Nauk. Arm. SSR*, **64**, 112 (1977);

through Chem. Abstr., **87**, 65369t (1977).

19. T. Yoshihara and S. Sakamura, *Agri. Biol. Chem.*, **41**, 2427 (1977).
20. P. M. Dewick and M. Martin, *J. Chem. Soc. Chem. Comm.*, 637 (1976).
21. A. E. Timbekova and N. K. Abubakirov, *Khim. Prir. Soedin.*, **5**, 607 (1986); through *Chem. Abstr.*, **106**, 116476w (1987).
22. S. Fushiya et al., *Heterocycles*, **22**, 1039 (1984).
23. S. R. Srinivas, *Dev. Food Sci.*, **18**, 343 (1988).
24. P. G. Stecher et al., eds., *The Merck Index*, 8th ed., Merck & Co., Rahway, N.J., 1968, p. 648.
25. M. Mohsin and A. K. Pal, *Indian J. Exp. Biol.*, **15**, 76 (1977).
26. B. J. Hudson and S. E. Mahgoub, *J. Sci. Food Agric.*, **31**, 646 (1980).
27. A. J. George, *Food Cosmet. Toxicol.*, **3**, 85 (1965).
28. M. Slotwinska, *Ann. Univ. Mariae Curie-Sklodowska, Sect. C*, **38**, 177 (1986); through *Chem. Abstr.*, **107**, 213108f (1987).
29. M. Gorski et al., *Ann. Univ. Marie Curie-Sklodowska, Sect. C*, **35**, 167 (1980); through *Chem. Abstr.*, **97**, 34467a (1982).
30. M. R. Malinow et al., *Steroids*, **29**, 105 (1977).
31. M. R. Malinow et al., *Food Cosmet. Toxicol.*, **19**, 443, (1981).
32. M. R. Malinow et al., *Lancet* **1**, 615 (1981).
33. M. R. Malinow et al., *Science*, **216**, 415 (1984).
34. J. L. Roberts and J. A. Hayashi, *N. Engl. J. Med.*, **308**, 1361 (1983).
35. P. E. Prete, *Arthritis Rheum.*, **28**, 1198 (1985).
36. J. M. Araufo and D. E. Pratt, *Cienc. Tecnol. Aliment.*, **5**, 22 (1985).

# ALGIN

**Source: Brown algae**, commonly including members of the following genera: ***Macrocystis***, ***Laminaria***, and ***Ascophyllum***.

*Synonyms*. Salts of alginic acid (alginates), particularly sodium alginate.

## GENERAL DESCRIPTION

Algin is a collective term for the hydrophilic colloidal substance isolated from certain brown algae (class Phaeophyceae). The most commonly used algae include members of the following genera: *Macrocystis*, *Laminaria*, and *Ascophyllum*.

*Macrocystis pyrifera* (L.) C.A. Agardh., commonly known as the giant kelp, which grows along the West Coast, is the major source in the United States, while *Ascophyllum nodosum* (L.) LeJolis and *Laminaria digitata* (L.) Edmonson as well as other related species are used by countries bordering the Atlantic Ocean. *Laminaria* species are also used by Japanese producers.

The process for algin manufacture basically involves a prewash of the seaweed whereby undesirable salts are leached out and removed, followed by extraction with a dilute alkaline solution that solubilizes the alginic acid present in the seaweed. The resulting thick and viscous mass is clarified, and the algin is obtained as free alginic acid on treatment with mineral acids. The alginic acid can then be converted to sodium alginate. Sodium alginate is the major form of algin currently used.[1,2]

The major producing countries include the United States, UK, Norway, France, and Japan.

Alginic acid and its calcium salt are insoluble in water, but its ammonium, sodium, potassium, and magnesium salts as well as its propylene glycol ester are readily soluble in cold and hot water to form viscous solutions. The viscosity of algin solutions depends on various factors, including concentration, pH, degree of polymerization (DP), temperature, and presence of polyvalent metal ions. Viscosity increases with DP; it decreases with increase in temperature but will regain its original value on cooling to its initial temperature, provided the solutions are not held above 50°C for long periods. Between pH 4 and 10, the viscosity of algin solutions is generally stable.

Algin solutions form gels with calcium ions, due to the formation of insoluble calcium alginate; these gels are not thermally reversible but may be liquefied by calcium sequestrants.

Propylene glycol alginate is more acid tolerant than the other alginates. Its solutions are stable below pH 4 (down to pH 2.6).[1,2]

## CHEMICAL COMPOSITION

Alginic acid is a linear polymer consisting of (1→4)-linked residues of $\beta$-D-mannopyranosyluronic acid and $\alpha$-L-gulopyranosyluronic acid. These D-mannuronic acid and L-guluronic acid residues are arranged in the polymer chain in blocks. Blocks of mannuronic acid are separated from those of guluronic acid by blocks that are made up of random or alternating units of mannuronic and guluronic acids.[2–4] The homogeneous blocks (those composed of either acid residues alone) are less readily hydrolyzed than the interconnecting heterogeneous blocks.[5] Alginates from different sources vary in their proportions of blocks of mannuronic and guluronic acid residues; values of mannuronic acid to guluronic acid ratios range from 0.3 to 2.3 for some alginate samples.[6–8] These values can be readily determined by infrared spectroscopy.[6]

Molecular weights of alginates range from 10,000 to 1,870,000 depending on algal sources or methods of analysis.[8–11]

## PHARMACOLOGY OR BIOLOGICAL ACTIVITIES

Current available data based on experiments on animals show algin (alginic acid and its sodium and calcium salts, and propylene glycol alginate) to be generally nontoxic.[12] It is apparently not digested, though this remains to be confirmed.[12,13]

Sodium alginate has the ability to reduce strontium absorption, and the sodium alginate with the highest proportion of guluronic acid is the most effective.[14] Sodium alginate can also decrease the retention of other radioactive divalent metallic ions in rats in the following order: Ba > Sr > Sn > Cd > Mn > Zn > Hg >, with Ba levels being reduced to 3% of control values and Cd and Mn levels, to about 50% in three weeks.[15]

Studies have shown that orally fed alginic acid and sodium alginate depress plasma and/or liver cholesterol levels in rats;[16,17] only algin with a high DP is active. Its hypocholesterolemic activity was attributed to its ability to inhibit cholesterol absorption from the gut.[16]

Laminine dioxalate (from *Laminaria* sp. has been shown to have hypotensive activity (DER MARDEROSIAN AND LIBERTI).

## USES

Algin has been available commercially for several decades and currently is widely used. Its applications generally depend on its thickening, gel-forming, and stabilizing properties.

**Medicinal, Pharmaceutical, and Cosmetic.** Sodium alginate is used as a binding and disintegrating agent in tablets and as a binding agent and demulcent in lozenges;

as film former in peel-off facial masks; as suspending and thickening agent in water-miscible gels, lotions, and creams; and as a stabilizer for oil-in-water emulsions. Calcium alginate is used as absorbable hemostatic; potassium alginate (in conjunction with calcium sulfate and sodium phosphate) is used as an irreversible dental impression material.

**Food.** Used in virtually every category of food products, with average maximum use level of about 1% in such products as candy, gelatins and puddings, condiments, relishes, processed vegetables, fish products, and imitation dairy products. Other products in which it is used in lower levels include alcoholic and nonalcoholic beverages, frozen dairy desserts, baked goods, meat and meat products, milk products, fats and oils, cheese, egg products, soups, snack foods, and others.

**Traditional Medicine.** Several *Laminaria* species have been used as folk medicines in northeastern Japan for the prevention and treatment of hypertension (DER MARDEROSIAN AND LIBERTI).

**Others.** A 0.2% sodium alginate spray as an effective fungicide against fungal infection of rice by *Pyricularia orysae* Cav. was claimed by a Japanese patent.[18] Alginic acid is used as a sizing agent for textiles and in adhesive formulations.

## COMMERCIAL PREPARATIONS

Sodium, potassium, ammonium, and calcium salts of alginic acid, and propylene glycol alginate. Alginic acid is official in N.F. and F.C.C.; alginates (potassium, propylene glycol, and sodium) are official in F.C.C.

*Regulatory Status.* Approved for use in foods (§182.7133, §182.7187, etc.).

## REFERENCES

See the General References for DER MARDEROSIAN AND LIBERTI; FEMA; FURIA; MARTINDALE; PHILLIPS; UPHOF; WHISTLER AND BEMILLER.


1. K. B. Guiseley, in A. Standen, ed., *Kirk-Othmer Encyclopedia of Chemical Technology*, Vol. 17, 2nd ed., Wiley-Interscience, New York, 1968, p. 768.
2. A. Wylie, *R. Soc. Health J.*, **93**, 309 (1973).
3. A. Haug et al., *Carbohydr. Res.*, **32**, 217 (1974).
4. O. Smidsrød, *Carbohydr. Res.*, **27**, 107 (1973).
5. D. A. Rees, *Adv. Carbohydr. Chem. Biochem.*, **24**, 296 (1969).
6. W. Mackie, *Carbohydr. Res.*, **20**, 413 (1971).
7. A. Penman and G. R. Sanderson, *Carbohydr. Res.*, **25**, 273 (1972).
8. M. H. Ji et al., *Proc. Joint China-U.S. Phycol. Symp.*, 393 (1983).
9. M. Fujihara and T. Nagumo, *J. Chromatogr.*, **465**, 386 (1989).
10. A. Ball et al., *Int. J. Biol. Macromol.*, **10**, 259 (1988).
11. R. S. Doubet and R. S. Quatrano, *J. Chromatogr.*, **264**, 479 (1983).
12. W. H. McNeely and P. Kovacs, *ACS Symp. Ser.*, **15**, 269 (1975).
13. S. Viola et al., *Nutr. Rep. Int.*, **1**, 367 (1970).
14. E. R. Humphreys and G. R. Howells, *Carbohydr. Res.*, **16**, 65 (1971).
15. A. J. Silva et al., *Health Phys.*, **19**, 245 (1970).
16. K. Ito and Y. Tsuchiya, *Proc. Int. Seaweed Symp.*, **7**, 558 (1972); through *Chem. Abstr.*, **82**, 149970s (1975).
17. E. Tsuji et al., *Eiyogaku Zasshi*, **33**,

273 (1975); through *Chem. Abstr.*, **85**, 31888e (1976).
18. T. Misato et al., Jpn. Kokai, 76,110,022 (1976); through *Chem. Abstr.*, **86**, 101922h (1977).

# ALKANET

**Source:** ***Alkanna tinctoria*** (L.) Tausch (Family Boraginaceae).

*Synonyms*. Alkanna, anchusa, orcanette, dyer's alkanet, and Spanish bugloss.

## GENERAL DESCRIPTION

Biennial or perennial herb about 0.5 m high with hairy leaves and blue or purple trumpet-shaped flowers; indigenous to southeastern Europe, particularly Hungary, Greece, and the Mediterranean region. Part used is the dried root.

## CHEMICAL COMPOSITION

Contains up to 5% alkannin (coloring principle) that has been called anchusin, anchusic acid, and alkanna red; it occurs mainly in the cortex. Other constituents present include tannin, alkannin isovalerate, alkannin angelate,[1] and alkannan;[2] wax substances consisting of esters of saturated, straight-chain monocarboxylic acids having even numbers of carbon atoms ($C_{16}$ to $C_{26}$); and straight-chain alcohols ($C_{24}$ to $C_{28}$) with one double bond.[3,4]

Alkannin is soluble in organic solvents but almost insoluble in water. Its buffered aqueous solutions are red at pH 6.1, purple at pH 8.8, and blue at pH 10.0.

Flavones from *A. orientalis* include kaempferol, kaempferol-3,6-dimethylether, kaempferol 3-glucoside, kaempferol 7-glucoside, kaempferol 3-rutinoside, quercetin, quercetin 3-glucoside, and quercetin 3-rutinoside.[5]

## PHARMACOLOGY OR BIOLOGICAL ACTIVITIES

Alkannin when fed to mice for 15 weeks at 1% of the diet has been shown to be nontoxic; it was not deposited in abdominal fat and was excreted in urine.[6] Alkannins have been shown to have antimicrobial and wound-healing properties and have been used externally for the treatment of ulcers.[7] Shikonin, the 1′*R*-isomer of alkannin, has immunomodulatory effects at low dosage, and is immunosuppressive in higher doses (HARBOURNE AND BAXTER). Alkannin angelate and alkannin isovalerate are claimed to have 80 and 85% healing effects respectively, in patients with leg ulcers.[8]

## USES

**Medicinal, Pharmaceutical, and Cosmetic.** In lipsticks and hair dyes.[9]

**Food.** Current use mainly as dye for sausage casings; oleomargarine, and shortening; also ink to mark food products.

**Traditional Medicine.** Used as an astringent in diarrhea and abscesses.

**Others.** Alkannin is used as a pH indicator. Its alcoholic solution is used in microscopy for the detection of oils and fats.

## COMMERCIAL PREPARATION

Not widely available either in crude or extract forms; crude was formerly official in U.S.P.

*Regulatory Status*. Approved by the USDA Meat Inspection Division as a food dye with specific limitations.

## REFERENCES

See the General References for FURIA; GLASBY 2; HARBOURNE AND BAXTER; HOCKING; MARTINDALE; MERCK; POUCHER; TERRELL; UPHOF; USD 23rd; WREN; YOUNGKEN.

1. V. P. Papageorgiou and G. A. Digenis, *Planta Med.*, **39**, 81 (1980).
2. R. D. Gibbs, *Chemotaxonomy of Flowering Plants*, Vol. 2, McGill-Queens University Press, Montreal, 1974, p. 700.
3. A. G. Varvoglis, *Chem. Chron.*, **1**, 156 (1972); through *Chem. Abstr.*, **77**, 141697j (1972).
4. B. Papageorgiou, *Chem. Chron.*, **6**, 365 (1977); through *Chem. Abstr.*, **86**, 185918x (1977).
5. Mansour et al. *J. Nat. Prod.*, **49**, 355 (1981).
6. L. Majlathova, *Nahrung*, **15**, 505 (1971).
7. V. P. Papageorgiou, *Planta Med.*, **39**, 193 (1980).
8. V. P. Papageorgiou, Ger. Offen., 2,829,744 (1979); through *Chem. Abstr.*, **91**, 1814415 (1979).
9. P. Hatinguais and R. Belle, Fr. Demande FR 2,477,872 (1981); through *Chem. Abstr.*, **96**, 11513c (1982).


# ALLSPICE

**Source:** ***Pimenta dioica*** (L.) Merr. (syn. *P. officinalis* Lindl.; *Eugenia Pimenta* DC.) (Family Myrtaceae).

*Synonyms*. Pimenta, Jamaica pepper, and pimento.

## GENERAL DESCRIPTION

Tree 8–20 m high, with opposite, leathery, oblong leaves 5–15 cm long; fruit globose, about 6 mm in diameter; native to the West Indies, Central America, and Mexico. Part used is the dried, full-grown but unripe fruit; leaves are also used. Major producers include Jamaica and Cuba.

West Indian allspice berries are smaller than Central American and Mexican berries, but they have stronger and smoother flavor. The relatively harsher flavor and aroma of Central American and Mexican berries are due to their relatively high abundance of monoterpene hydrocarbons, especially myrcene, in their essential oil.[1]

## CHEMICAL COMPOSITION

Allspice contains about 4% volatile oil, which is rather stable compared with those of tarragon and black pepper.[2] However there is evidence that storage of the undried berries under conditions that prevent rapid removal of moisture can increase the volatile oil content by up to 50%; it appears that enzymes released in the fruit after harvest are responsible for producing volatile components from their precursors.[3] The major component of the volatile oil (known as pimenta, pimento, or allspice oil) is eugenol, present in 60–80%. Other constituents include methyleugenol, 1,8-cineole, *l*-$\alpha$-phellandrene, caryophyllene, and two epimeric 10-cadinols (2%).[4] Total identified constituents number more than three dozen.[4,5]

Besides volatile oil, other constituents include quercetin glycosides,[6] catechins, proanthocyanidins,[7] protein, lipids, carbohydrates, vitamins (A, C, thiamine, riboflavin, niacin), and minerals.[8]

The leaf oil (pimenta leaf oil) contains more eugenol (up to 96%) than the berry oil and is similar in composition to clove leaf oil.[9,10] Annual leaf oil production exceeds that of the oil of the berries.

## PHARMACOLOGY OR BIOLOGICAL ACTIVITIES

Eugenol, the major component of both allspice berry and leaf oils, has local antiseptic and anesthetic properties. It is considered anticonvulsant, antimitotic, antioxidant, and spasmolytic. Eugenol has been shown to have central nervous system depressant activity and inhibits prostaglandin synthesis in human colonic muscoa (HARBOURNE AND BAXTER).

When pimento oil and eugenol were applied on intact shaved abdominal skin of the mouse, no percutaneous absorption was observed.[11]

Eugenol, aqueous extracts of allspice, and allspice oil, along with numerous other spices and their volatile oils, have been demonstrated to enhance trypsin activity;[12] they also exhibit larvicidal properties.[13]

## USES

**Medicinal, Pharmaceutical, and Cosmetic.** Allspice oil has been used medicinally as an aromatic carminative at dose of 0.05 to 0.2 mL. It is also used in cosmetics as an ingredient in fragrance formulations, for spicy, clovelike notes. Eugenol is used as a dental antiseptic and anesthetic.

**Food.** Allspice, its oil, and its oleoresin (less so) are currently extensively used in many food products, including alcoholic and nonalcoholic beverages, frozen dairy desserts, candy, baked goods, gelatins and puddings, meat and meat products, condiments and relishes, and others. The leaf oil is also used for flavoring in all these food products. The highest average maximum use level of the berry oil is in candy (ca. 0.025%).

**Traditional Medicine.** Formerly, berries used as an appetite stimulant; stomachache, painful menstruation; leaves used for pain, fever, cold remedy; toothache; anodyne; astringent; carminative. In Jamaica, the fruit is used to treat influenza and stomachache. Used in Guatemala to treat rheumatism. In the Dominican Republic, the fruits, decocted with salt, are given to treat vomiting (WENIGER AND ROBINEAU).

**Others.** Like certain other spices such as rosemary and sage, allspice has antioxidative properties[14] and is a potential source of new natural antioxidants.

## COMMERCIAL PREPARATIONS

Crude, oleoresin, berry and leaf oils. Allspice and allspice oil were formerly official in N.F.; allspice oil and pimenta leaf oil are official in F.C.C.

*Regulatory Status.* GRAS (§182.10 and §182.20).

## REFERENCES

See the General References for ARCTANDER; AYENSU; BAILEY 1; BAUER, FEMA; FURIA; GOSSELIN; GUENTHER; HARBOURNE AND BAXTER; KARRER; MARSH; MARTINDALE; ROSENGARTEN; TERRELL, WENIGER AND ROBINEAU.

1. C. L. Green and F. Espinosa, *Dev. Food Sci.*, **18**, 3 (1988).
2. E. G. Chinenova et al., *Konserv, Ovoshchesush. Prom.*, **24**, 31 (1969); through *Chem. Abstr.*, **71**, 48402a (1969).
3. P. R. Ashurst, *An. Acad. Bras. Cienc.*, **44**(Suppl.),198 (1972); through *Chem.*

*Abstr.*, **83**, 95022j (1975).

4. J. W. Hogg et al., *Am. Perfum. Cosmet.*, **86**, 33 (1971).
5. J. Nabney and F. V. Robinson, *Flavour Ind.*, **3**, 50 (1972).
6. B. Voesgen and K. Herrmann, *Z. Lebensm. Unters. Forsch.*, **170**, 204 (1980).
7. J. M. Schulz and K. Herrmann, *Z. Lebensm. Unters. Forsch.*, **171**, 278 (1980).
8. M. Teotia et al., *Indian Food Packer*, **41**(5), 49 (1987).
9. E. Calderon Gomez et al., *Rev. Colomb. Cienc. Quim. Farm.*, **2**, 37 (1974); through *Chem. Abstr.*, **83**, 103122j (1975).
10. M. E. Veek and G. F. Russell, *J. Food Sci.*, **38**, 1028 (1973).
11. F. Meyer and E. Meyer, *Arzeim. Forsch.*, **9**, 516 (1959).
12. Y. Kato, *Koryo*, **113**, 17, 24 (1975); through *Chem. Abstr.*, **84**, 149393x, 178343m (1976).
13. K. Oishi et al., *Nippon Suisan Gakkaishi*, **40**, 1241 (1974); through *Chem. Abstr.*, **82**, 84722r (1975).
14. Y. Saito et al., *Eiyo To Shokuryo*, **29**, 505 (1976); through *Chem. Abstr.*, **87**, 150314r (1977).

# ALMONDS

**Source:** Sweet almond ***Prunus dulcis*** (Mill.) D. A. Webb (syn. *Prunus amygdalus* Batsch var. *dulcis* (DC.) Koehne).
Bitter almond ***Prunus dulcis*** (Mill.) D. A. Webb var. ***amara*** (DC.) H. E. Moore (syn. *Prunus amygdalus* Batsch var. *amara* (DC.) Focke) (Family Rosaceae).

## GENERAL DESCRIPTION

The almond tree, *Prunus dulcis*, is also known as *P. communis* (L.) Arcang, *Amygdalus dulcis* Mill., and *A. communis* L., in addition to the above synonyms. It grows to a height of about 7 m and has several varieties; two of them, var. *dulcis* and *amara*, yield sweet and bitter almonds, respectively. The tree is native to western Asia and is now extensively cultivated in the Mediterranean countries and in California. The fruit is botanically classified as a drupe (same as peach or plum), except that its outer portion is leathery, not fleshy and edible like the peach; the almond is its seed.

Sweet almonds are used as food, but bitter almonds are not; this is due to the presence of amygdalin in bitter almonds that can be hydrolyzed to yield deadly hydrocyanic acid (HCN).

Two major types of products are derived from the almond, namely a fixed oil and a volatile oil.

The fixed oil is commonly called almond oil, expressed almond oil, or sweet almond oil; it is made from both sweet and bitter almonds by pressing the kernels. It does not contain benzaldehyde or HCN.

The volatile oil is called bitter almond oil. It is obtained by water maceration and subsequent steam distillation of the expressed and partially deoleated bitter almonds or kernels of other *Prunus* species that contain amygdalin; these species include apricot (*P. armeniaca* L.), peach (*P. persica* (L.) Batsch.), and plum (*P. domestica* L.). During maceration, the enzyme (emulsin) present hydrolyzes the amygdalin into sugar, benzaldehyde, and HCN, the last two being distilled by steam. Sweet almond does not yield a volatile oil.

## CHEMICAL COMPOSITION

Both sweet and bitter almonds have similar chemical composition and contain 35–55% fixed oil (MERCK).[1,2] The only difference appears to be the presence of amygdalin

(3–4%) in bitter almond and its absence or presence in trace amounts in sweet almonds.[3,4] Other constituents reported to be present in sweet and/or bitter almonds include protein (18–25%); emulsin; prunasin (0.005% in bitter almond);[5] daucosterol and other sterols (e.g., sitosterol, citrostadienol, 24-methylene-cycloartanol);[2,6] calcium oxalate;[7] zinc; copper;[8] tocopherols (mostly $\alpha$);[9,10] trace amounts of vitamins A, B complex, and E; and amino acids, including glutamic acid, aspartic acid, and arginine.[11]

Expressed almond oil has been reported to contain 53 individual triglycerides of which triolein and dioleolinolein make up 32 and 33%, respectively;[12] fatty acids present include oleic (66–72%), linoleic (18–22%), palmitic (5.7–7.9%), stearic, lauric, myristic, and palmitoleic acids (MERCK).[7,13,14]

Bitter almond oil contains mostly benzaldehyde (95%) and HCN (2–4%). For food and flavor uses the HCN is removed, and the resulting oil is almost pure benzaldehyde.

A recently issued Japanese patent claims to have isolated from almonds low molecular weight peptides (mol. wt. 1000–10,000) that have analgesic and antiinflammatory activities.[15] Two antiinflammatory peptides (mol. wt. 257,000 and 19,000) have also been isolated from Chinese almonds (peach kernels, *Prunus persica*).[16]

Almonds also contain varying amounts (3.11–5.25%) of soluble nonreducing sugars (sucrose, raffinose, and stachyose), depending on the variety.[16–18]

## PHARMACOLOGY OR BIOLOGICAL ACTIVITIES

Expressed almond oil has emollient, demulcent, and mildly laxative properties. It is a weak antibacterial. While easily absorbed and digested orally, it is slowly absorbed through intact skin.[11] Based on prevalent data, almond oil and almond meal are nonirritating and nonsensitizing to the skin and are considered safe for cosmetic use.[11]

Bitter almond oil, containing 2–4% HCN, is poisonous; and fatal poisoning of an adult after taking 7.5 mL has been reported.

Bitter almond oil, FFPA (free from prussic acid, outdated term for HCN), can be regarded as pure benzaldehyde; it has antipeptic, local anesthetic, and antispasmodic properties; it also has narcotic properties at high doses; ingestion of 50–60 mL can be fatal due to central nervous depression with respiratory failure (GOSSELIN).[19]

## USES

**Medicinal, Pharmaceutical, and Cosmetic.** Sweet almond oil is used as a laxative in doses up to 30 mL as well as a solvent for parenterally administered drugs and a solvent for hemorrhoid injectable solutions.[11] It is also used as an emollient and emulsifier for chapped hands, in lotions (both moisturizing and night skin care preparations), suntan gels, blushers, makeup bases, skin cleansing preparations, creams, and as an ointment base. It is used in cosmetic formulations in concentrations up to 50%; 25% in lipstick formulations. Almond meal is used as a skin cleanser and in medicated soaps.[11]

Bitter almond oil was formerly used in the United States as cough sedative and as an antipruritic. It is no longer used commercially for these purposes.

**Food.** Bitter almond oil (FFPA) is widely used as a flavor ingredient in most categories of food products, including alcoholic and nonalcoholic beverages, frozen dairy desserts, candy, baked goods, gelatins and puddings, and others, generally at use levels below 0.05%. Synthetic benzaldehyde is even more widely used for the same purposes.

Sweet almonds have been used as food for thousands of years. They are a good source of protein, fats, calcium, iron, potassium, phosphorus, and trace minerals such as zinc and copper.

**Health Food/Herb Teas.** The controversial unofficial anticancer drug laetrile is mainly amygdalin isolated from kernels of apricot, peach, and other related fruits (TYLER 2).[20] Almond oil is often used as a base for massage oil products.

**Traditional Medicine.** In traditional Chinese medicine, apricot (*P. armeniaca*) kernels are used as an antitussive and antiasthmatic and in treating tumors. Apricot tree inner bark in the form of a decoction is used in treating apricot kernel poisoning, reportedly with great success (JIANGSU). Sweet almond seed or seed oil has been used as a folk cancer remedy for bladder, breast, mouth, spleen, and uterine cancers, among others.

## COMMERCIAL PREPARATIONS

Crude and oils (sweet and bitter, FFPA); sweet almond oil is official in N.F., and bitter almond oil (FFPA) is official in F.C.C.

*Regulatory Status.* Bitter almond oil (FFPA) is GRAS (§182.20).

## REFERENCES

See the General References for ARCTANDER; BAILEY 2; CLAUS; DUKE 3; FEMA; GUENTHER; JIANGSU; MARTINDALE; NANJING; SAX; TERRELL; TYLER 2; WATT AND MERRILL.


1. E. Y. Babekova and V. F. Shcheglova, *Sb. Rab. Aspirantov, Tadzh. Univ. Ser Biol. Nauk*, **3**, 24 (1969); through *Chem. Abstr.*, **75**, 128632y (1971).
2. T. M. Jeong et al., *Lipids*, **9**, 921 (1974).
3. I. Karkocha, *Rocz. Panstev. Zakl. Higi.*, **24**, 703 (1973); through *Chem. Abstr.*, **81**, 11904w (1974).
4. M. Nishijima et al., *Tokyo Toritsu Eisei Kenkyusho Kenkyu Nempo*, **26**, 183 (1975); through *Chem. Abstr.*, **84**, 178441s (1976).
5. U. Schwarzmaier, *Phytochemistry*, **11**, 2358 (1972).
6. T. M. Jeong et al., *Lipids*, **10**, 634 (1975).
7. J. Seidemann, *Seifen, Öle, Fette, Wachse*, **99**, 302 (1973); through *Chem. Abstr.*, **79**, 64789c (1973).
8. K. G. D. Allen et al., *Nutr. Rep. Int.*, **16**, 227 (1977).
9. G. Lambersten et al., *J. Sci. Food Agri.*, **13**, 617 (1962); through *Chem. Abstr.*, **58**, 5986e (1963).
10. F. Hotellier and P. Delaveau, *Ann. Pharm. Fr.*, **30**, 495 (1972); through *Chem. Abstr.*, **78**, 62044m (1973).
11. K. T. Fisher, *J. Am. Coll. Toxicol.* **2**, 85 (1983).
12. N. S. Geiko et al., *Khlebopek. Konditer. Prom-st.*, **8**, 25 (1975); through *Chem. Abstr.*, **83**, 204934f (1975).
13. K. Aitzetmueller and M. Ihrig, *Fette Wiss. Technol.*, **90**, 464 (1988).
14. M. Farines et al., *Rev. Fr. Corps Gras*, **33**, 115 (1986); through *Chem. Abstr.*, **105**, 759266 (1986).
15. M. Kubo, Jpn. Kokai Tokkyo Koho JP 62,198,399 [87,198,399] (1987); through *Chem. Abstr.*, **108**, 82088v (1988).
16. F. J. Lopez Andreu et al., *An. Edafol. Agribiol.*, **44**, 207 (1985); through *Chem. Abstr.*, **104**, 128578c (1986).
17. X. D. Fang et al., *Zhongyao Tongbao*, **11**, 37 (1986).
18. F. S. Calicto et al., *J. Agri. Food Chem.*, **29**, 509 (1981).
19. D. L. J. Opdyke, *Food Cosmet. Toxicol.*, **14**, 693 (1976).
20. J. Jee et al., *J. Pharm. Sci.*, **67**, 438 (1978).

# ALOE (AND ALOE VERA)

**Source:** ***Aloe vera*** (L.) N. L. Burm. (syn. *Aloe barbadensis* Mill.; *A. perfoliata* L. var. *vera* L.); ***A. arborescens*** Miller var. ***natalensis*** Berger; ***A. ferox*** Mill. and its hybrids with ***A. africana*** Mill. and ***A. spicata*** Baker; ***A. perryi*** Baker (Family Liliaceae).

## GENERAL DESCRIPTION

There are two major products derived from the leaves of *Aloe* spp. The yellow bitter juice present in specialized cells beneath the thick epidermis yields the drug aloe, which is obtained from all above species. The parenchymatous tissue in the center of the leaf contains a mucilaginous gel that yields aloe gel or aloe vera gel; it is currently obtained from *A. vera*.[1]

All above species are perennial succulents native to Africa, and later spread to other parts of the world. They are not cacti, and they also should not be confused with the American aloe, or century plant, which is an *Agave* species. Considerable confusion has arisen over the nomenclature of *Aloe* species. Currently, there are more than 360 accepted species of *Aloe*. N. L. Burman's *Aloe vera* binomial was published in 1768, thus has priority over Miller's *A. barbadensis*.[2]

*Aloe vera* yields Curaçao aloe, or Barbados aloe, which is produced in the West Indies (Curaçao, Aruba, Bonaire). *Aloe ferox* and its hybrids yield Cape aloe, produced in South Africa. Other *Aloe* species yield aloes of lesser importance, though *A. arborescens* var. *natalensis* Berger has recently emerged as an important research subject in Japan, where it is called Kidachi aloe.

Aloe is obtained by cutting the leaves at their base and letting the yellow bitter juice drain out. The water is evaporated off from the juice by heat, and the resulting light to dark brown mass is the drug aloe.

Commercial aloin is a concentrated form of aloe, containing high concentrations of anthraglycosides (mostly barbaloin). This commercial product is not pure aloin.

Aloe vera gel is prepared by numerous methods, some patented and others proprietary.[3–6] These methods essentially involve expression and/or solvent extraction often with harsh physical and chemical treatments; the resulting gel products vary considerably in properties and generally are not representative of the fresh gel.

## CHEMICAL COMPOSITION

Aloe contains cathartic anthraglycosides as its active principles; these are mostly C-glucosides, notably barbaloin, which is a glucoside of aloe-emodin. Aloes from most species contain concentrations between 10 and 20%, though some contain levels of 30% barbaloin. A single *A. vera* (*A. barbadensis*) plant from Mannar (Sri Lanka) was found to contain 57% barbaloin in its exudate. The highest concentrations of barbaloin are found in young mature leaf exudates, decreasing in older leaves toward the base of the plant.[7] The concentrations of anthraglycosides vary with the types of aloe ranging from 4.5 to 25% of aloin. Other constituents present include aloesin and its aglycone aloesone (a chromone),[8] free anthraquinones (e.g., aloe-emodin), and resins.

The composition of aloe vera gel is still not clear. Studies to date have indicated the gel to consist of more than one type of polysaccharide. While one study showed it to contain at least four different partially acetylated linear glucomannans with $(1\rightarrow4)$-glycosidic linkages,[9] others revealed an acidic galactan, mannan, glucomannan, arabinan, and/or glucogalactomannan. The ratios of hexoses in each polysaccharide differ widely among the studies; so do their molecular weights.[10–12] Other constituents reported or otherwise claimed to be present include other polysaccharides (containing galactose, xylose, and arabinose), steroids, organic acids, enzymes, antibiotic principles, amino acids, "biogenic stimulators," "wound-healing hormones" saponins and minerals.[1,3–5,13–18]

The polysaccharides constitute 0.2–0.3% of the fresh gel and 0.8–1.2% of dry matter. Postproduction autodegradation of the glucomannan polysaccharides produces mainly mannans. The gel polysaccharides, consisting mainly of mannose and glucose in a 1:3 ratio, can degrade in 48 h at room temperature, with a decrease in glucose content and an increase in mannose:glucose ratio to >10. Addition of an algal sulfated polysaccharide (extracted from a species of red microalgae) has been found to inhibit the degradation and browning of aloe polysaccharide.[19]

Carboxypeptidase, a serine carboxypeptidase enzyme, in *A. arborescens* and other species has been suggested to be a primary antithermic agent from aloe species.

## PHARMACOLOGY OR BIOLOGICAL ACTIVITIES

Drug aloe and its purified form aloin have cathartic properties, acting on the colon. They are extremely bitter and are considered the least desirable among the plant purgative drugs (see ***cascara*** and ***senna***) due to their bitterness and their tendency to produce more griping and irritation. Overdosage causes abdominal pain, bloody diarrhea, hemorrhagic gastritis, and sometimes nephritis (GOSSELIN, MARTINDALE). Aloin induces secretion of electrolytes and water in the intestinal lumen, inhibiting reabsorption of electrolytes and water from the large intestine, thereby increasing intestinal volume via increased filling pressure, which stimulates peristalsis.[20]

Aloe-emodin (see ***buckthorn***) has been reported to have anticancer and antiviral (herpes simplex types 1 and 2) activities. A methanolic extract of the whole leaf is also claimed to be active against the viruses.[21,22]

An alcoholic extract of drug aloe is also reported to have anticancer activity (JIANGSU).

Recent studies have focused on the pharmacology of aloe gel and its mechanisms of action. A number of studies have explored the immunoreactive or immunomodulatory activity of various *Aloe* species.[23–25] Lectins isolated from the gel portion of *A. vera* were found to have strong hemagglutination and mitogenic activities.[26] Aloctin A isolated from the leaves of *A. arborescens*, in cultures with human T cells and accessory cells, resulted in the production of different amounts of lymphokines such as interleukin-2 (IL-2), IL-3 and interferon-$\gamma$ (INF-$\gamma$). It has recently been reported that aloctin A could augment the levels of natural killer cell activity and could induce cells cytotoxic to syngeneic and allogenic tumor cells *in vitro*; it also exhibits antiinflammatory, antiulcer (inhibiting gastric secretion and gastric lesions), and antitumor activity. The lectin was also found to be effective against edema and adjuvant arthritis in rats.[24,25]

Carboxypeptidase (from *A. arborescens*) was found to have a significant analgesic effect (comparable with bromelain) and inhibited the acceleration of vascular permeability in mice with abdominal acetic acid inflammation. It also exhibited a significant healing effect in thermal burns of rats; however, it was more effective when used as a prophylactic rather than in postburn applications. It has been suggested that carboxypeptidase may be a main antiinflammatory agent of aloe, though other compounds could contribute to the effect.[27,28]

A recent review of pharmacological and clinical studies concluded that *Aloe* promotes wound healing and is of therapeutic value in thermal injuries and a wide variety of soft tissue injuries, preventing progressive dermal ischemia following thermal injury, frostbite, electrical injury; in flap physiology, intraarterial drug abuse, and other injuries irrespective of trauma etiology. Aloe penetrates injured tissue, relieves pain, is antiinflammatory, and dilates capillaries, increasing blood supply to the injury. Aloe has an antithromboxane activity, while maintaining prostaglandin ratio, without causing injured blood vessel collapse.[29]

Oil-in-water aloe extracts significantly increase soluble collagen levels, suggesting a topical antiaging effect.[30] An excellent review of aloe therapeutic uses is available.[2]

## USES

**Medicinal, Pharmaceutical, and Cosmetic.** Currently the only official recognized use of aloe is as an ingredient in compound benzoin tincture, presumably for its beneficial properties on the skin.

Aloe and aloin are extensively used as active ingredients in laxative preparations, often with other cathartics such as buckthorn, cascara, and senna; belladonna extracts are often included to lessen griping. Aloin is also used in antiobesity preparations.

In Germany, concentrated dried aloe leaf juice is used for conditions in which ease of defecation and soft stool are desired, e.g., anal fissures, hemorrhoids, postanorectal surgery, and refractory constipation.[20]

Aloe gel and sometimes drug aloe are used as moisturizer, emollient, or wound healer in various cosmetic and pharmaceutical formulations.

Extracts of aloe or aloin are used in sunscreen and other cosmetic preparations.

**Food.** Aloe extracts are used as a flavor ingredient primarily in alcoholic and nonalcoholic beverages and in candy to impart a bitter note. Based on the reported average maximum use levels of about 0.02% in alcoholic (186 ppm) and nonalcoholic (190 ppm) beverages and 0.05% in candy, the extracts used must be tinctures or greatly diluted extracts, as standard extracts (e.g., solid extract or fluid extract) would contain too much active anthraglycosides to be safely used.

**Health Food/Herb Teas.** Aloe vera gel is used in nonalcoholic beverages that are commonly known as aloe vera juice. It is normally produced from aloe vera gel by diluting with water and mixing with citric acid and preservatives. It is also sometimes mixed with fruit juices and/or herbal extracts. Despite label claims, "pure" aloe vera juice is rarely pure; instead, it contains only a minor percentage of aloe vera gel.[1]

Aloe gel products are available in liquid and solid forms. The most popular liquid products are the 10×, 20×, and 40× concentrates, while spray-dried aloe vera is the most popular solid product. Although commercial liquid concentrates are usually genuine, the more highly concentrated they are, the more degradation they have undergone, as evidenced by their lack of viscosity.[1]

Despite claims to be 200× concentrated pure aloe gel, solid products normally contain high proportions of carriers such as gums (acacia, guar, locust bean), lactose, mannitol, hydrolyzed starch, and/or others.[1]

**Traditional Medicine.** Fresh aloe vera gel is well known as a domestic medicine.[1,31-33] For this reason aloe vera is also called the "burn", "first aid", or "medicine" plant. When freshly obtained, the gel has the property of relieving thermal burn and sunburn as well as promoting wound healing; it also has moisturizing and emollient properties. The plant is widely used as a home remedy for these purposes.[1] It is the most widely used herbal folk remedy among the general population in the United States (LUST).

## COMMERCIAL PREPARATIONS

Crude, aloin, and extracts in various forms. Crude Barbados and Cape aloes are official in U.S.P.

Aloin was official in N.F. XI; its current quality is generally governed by standards set forth in this compendium, which do not require specific assays for anthraglycosides.

Aloe gel products are available in liquid and solid forms; qualities vary greatly depending on suppliers. Because the principal component in the gel is a glucomannan similar to guar and locust bean gums, these gums are frequently mixed with aloe gel to increase its viscosity and yield. Currently there is no meaningful assay method that distinguishes genuine liquid gel products from adulterant mixtures, though dried concentrated aloe gel can be evaluated by infared spectroscopy.[1]

*Regulatory Status*. Aloe has been approved for food use (§172.510). Dried Aloe vera leaf juice (and preparations), calculated to contain at least 28% hydroxyanthracene derivatives (as anhydrous barbaloin), and *A. ferox* dried leaf juice (and preparations), calculated to contain at least 18% hydroxyanthracene derivatives (as anhydrous barbaloin), are the subject of a German BGA monograph.[20]

## REFERENCES

See the General References for AHPA; APhA; BLUMENTHAL; FEMA; JIANGSU; LUST; MERCK; TERRELL; USD 26th; YOUNGKEN.


1. A. Y. Leung, *Drug Cosmet. Ind.*, **120**(6), 34 (1977).
2. D. Grindlay and T. Reynolds, *J. Ethnopharmacol.*, **16**, 117 (1986).
3. A. Farkas, U.S. Pat. 3,103,466 (1963).
4. A. Farkas and R. A. Mayer, U.S. Pat. 3,362,951 (1968).
5. H. H. Cobble, U.S. Pat. 3,892,853 (1975).
6. H. Matsui and T. Matsukura, Jpn. Kokai 75 155,664 (1975).
7. Q. J. Groom and T. Reynolds, *Planta Med.*, **53**, 345 (1987).
8. D. K. Holdsworth, *Planta Med.*, **22**, 54 (1972).
9. D. C. Gowda et al., *Carbohydr. Res.*, **72**, 210 (1979).
10. G. Mandal and A. Das, *Carbohydr. Res.*, **87**, 249 (1980).
11. Q. N. Haq and A. Hannan, *Bangladesh J. Sci. Ind. Res.*, **16**, 68 (1981); through *Chem. Abstr.*, **98**, 14377p (1983).
12. G. Mandal et al., *Indian J. Chem., Sect. B*, **22B**, 890 (1983).
13. T. D. Rowe and L. M. Parks, *J. Am. Pharm. Assoc.*, **30**, 262 (1941).
14. E. Roboz and A. J. Haagen-Smit, *J. Am. Chem. Soc.*, **70**, 3248 (1948).
15. L. J. Lorenzetti et al., *J. Pharm. Sci.*, **53**, 1287 (1964).
16. L. B. Fly and I. Kiem, *Econ. Bot.*, **17**, 46 (1963).
17. G. D. Bouchey and G. Gjerstad, *Q. J. Crude Drug Res.*, **9**, 1445 (1969).
18. G. Gjerstad, *Adv. Frontiers Plant Sci.*, **28**, 311 (1971).
19. A. Yaron, *Phytotherapy Res.*, **7**, S11 (1993).
20. Monograph *Aloe*, *Bundesanzeiger*, no. 154 (Aug. 21, 1985).
21. S. M. Kupchan and A. Karim, *Lloydia*, **39**, 223 (1976).
22. R. J. Sydiskis and D. G. Owen, U.S. Pat. 4,670,265 (1987).
23. L. A.'t Hart et al., *J. Ethnopharmacol.* **23**, 61 (1988).
24. K. Imanishi, *Phytotherapy Res.*, **7**, S20 (1993).
25. H. Saito, *Phytotherapy Res.*, **7**, S14 (1993).
26. W. D. Winters, *Phytotherapy Res.*, **7**, S23 (1993).
27. S. Ito et al., *Phytotherapy Res.*, **7**, S26 (1993).

28. M. Obata et al., *Phytotherapy Res.*, **7**, S30 (1993).
29. J. P. Heggers et al., *Phytotherapy Res.*, **7**, S48 (1993).
30. I. E. Danhof et al., *Phytotherapy Res.*, **7**, S53 (1993).
31. J. F. Morton, *Econ. Bot.*, **15**, 311 (1961).
32. G. Gjerstad and T. D. Riner, *Am. J. Pharm.*, **140**, 58 (1968).
33. J. F. Nieberding, *Am. Bee J.*, **114**, 15 (1974).

# ALTHEA ROOT

**Source:** ***Althaea officinalis*** L. (Family Malvaceae).

*Synonyms.* Althaea and marshmallow root.

## GENERAL DESCRIPTION

Perennial herb, velvety hairy, up to 1.5 m high with three-lobed coarsely serrate leaves, pink 3-cm-wide flowers in peduncled clusters. Grows in marshes and moist places; under cultivation, adaptable to drier soils; native to Europe and naturalized in the United States in salt marshes from Massachusetts to Virginia; locally elsewhere. Preferred part is the peeled root collected in the fall; whole dried root and dried leaves enter commerce.

## CHEMICAL COMPOSITION

Has been reported to contain starch, pectin, mucilage, sugar, fats, tannin, asparagine, and calcium oxalate. The mucilage content is generally considered to be 25–35%, but that of the homogeneous mucilaginous polysaccharides is much lower. The mucilage content has been found to change considerably with season (6.2–11.6%), being highest in winter.[1] A purified, homogeneous mucilage, Althaea-mucilage O, has been demonstrated to be composed of L-rhamnose:D-galactose:D-galacturonic acid:D-glucuronic acid in the molar ratio of 3:2:3:3, with a molecular weight of about 34,000 (as the ammonium salt).[2] The sequence of the component sugars and the configurations of the glycoside linkages have also been examined.[3]

Scopoletin, quercetin, kaempferol, chlorogenic acid, caffeic acid and *p*-coumaric acids have recently been found in the roots.[4]

## PHARMACOLOGY OR BIOLOGICAL ACTIVITIES

Althaea root extracts reportedly have demulcent, soothing properties on the mucous membranes, and an antitussive effect. It is believed that these properties are mainly caused by the mucilaginous substances present. Althaea-mucilage O has been shown to have strong hypoglycemic activity.[5] Crude root may delay absorption of other drugs taken at the same time.[6,7]

## USES

**Medicinal, Pharmaceutical, and Cosmetic.** Mainly as a demulcent in various pharmaceutical preparations, particularly cough medicines. In Europe, root and leaf preparations used for irritation of the oral or pharyngeal mucosa and associated dry irritable cough; root also for mild inflammation of the gastric mucosa.[6,7] Root is used crude or in formulations at a daily dose of 6 g; leaf used in 5-g daily dose or equivalent in formulations.[6,7]

**Food.** Used to a limited extent in alcoholic and nonalcoholic beverages, frozen desserts, candy, baked goods, and gelatins and puddings. Extracts of the root are used in confectionaries. Reported use levels are very low, usually below 0.002% (20 ppm).

**Traditional Medicine.** Has been reportedly used for more than 2000 years in Europe both internally and externally as a wound healer, remedy for coughs, sore throat, and stomach troubles, among other ailments; in ointments to relieve chapped hands and chilblains (BIANCHINI AND CORBETTA, FOSTER).

### COMMERCIAL PREPARATIONS

Crude and extracts. Crude was formerly official in N.F. and U.S.P.; extracts come in various nonstandard forms, depending on users' requirements.

*Regulatory Status.* Approved for food use (§172.510). In Germany leaf and root subject of therapeutic monographs, approved for irritation of the oral or pharyngeal mucosa and associated dry irritable cough; root also for mild inflammation of the gastric mucosa.[6,7]

### REFERENCES

See the General References for BAILEY 2; BIANCHINI AND CORBETTA; BLUMENTHAL; FEMA; FOSTER; GLEASON AND CRONQUIST; GOSSELIN; KARRER; MARTINDALE; MERCK; TUTIN 2; USD 23rd; YOUNGKEN.

1. G. Franz, *Planta Med.*, **14**, 90 (1966).
2. M. Tomoda et al., *Chem. Pharm. Bull.*, **25**, 1357 (1977).
3. M. Tomoda et al., *Chem. Pharm. Bull.*, **28**, 824 (1980).
4. St. N. I. Ionkova and D. Kolev, *Fitoterapia* **63**, 474 (1992).
5. M. Tomoda et al., *Planta Med.*, **53**, 824 (1987).
6. Monograph *Althaea folium*, *Bundesanzeiger*, no. 43 (Mar. 2, 1989).
7. Monograph *Althaea radix*, *Bundesanzeiger*, no. 43 (Mar. 2, 1989).

# AMBRETTE SEED

**Source:** ***Abelmoschus moschatus*** Medic. (syn. *Hibiscus abelmoschus* L.) (Family Malvaceae).

*Synonyms.* Musk seed, and muskmallow.

### GENERAL DESCRIPTION

Annual or biennial herb with bristly hairs, up to 2 m high; showy flowers yellow, crimson centered, about 10 cm across; seeds musk fragrant, flat, and kidney shaped; indigenous to India; widely cultivated in tropical countries, including the West Indies, Java, Indonesia, and Africa. Part used is the seed. The kidney-shaped, grayish brown seeds, about 3 mm in diameter, are traded (as musk grains, musk pods), and their aromatic oil is obtained via steam distillation of the crushed, dried seeds (CSIR 1).

### CHEMICAL COMPOSITION

Compounds isolated from the seed oil include ambrettolide ((*Z*)-7-hexadecen-16-olide), ambrettolic acid, and farnesol as well as 12,13-epoxyoleic acid, malvalic acid, sterculic acid, and $C_{10}$–$C_{18}$ acids (oleic, palmitic, $C_{10}$, $C_{12}$, $C_{14}$, $C_{16}$, $C_{18}$).[1]

Recently (*Z*)-5-tetradecen-14-olide, (*Z*)-5-dodecenyl acetate, and (*Z*)-5-tetradecenyl acetate have also been isolated from the absolute in yields of 0.5, 0.01, and 0.4%, respectively. Also, 2-*trans*,6-*trans*-farnesylacetate; 2-*cis*,6-*trans*-farnesylacetate; and oxacyclononadec-10-en-2-one (an ambrettolide homolog) have recently been reported in the seed coat. The floral musky odor of the oil is primarily the result of ambrettolide, and (*Z*)-5-tetradecen-14-olide.[2,3]

Other compounds reported in seeds include methionine sulfoxide, phospholipids ($\alpha$-cephalin, phosphatidylserine, phosphatidylserine plasmalogen, and phosphatidylcholine plasmalogen),[4,5] and sterols, including campesterol, sitosterol, stigmasterol, ergosterol, and cholesterol.[6] Sizable amounts of palmitic and myristic acids may also occur in ambrette seed oil or concrete, depending on the method of manufacture (ARCTANDER). The long-chain fatty acids result in a crude product of a waxy nature (Ambrette beurre); the fatty acids, removed with alkali, dilute alcohol, calcium, or lithium salts, produce a yellow clear to amber liquid that possesses the musky fragrance of ambrettolide (BAUER). Yield of the oil is from 0.2 to 0.6%.

## PHARMACOLOGY OR BIOLOGICAL ACTIVITIES

Based on available data, ambrette seed oil and its major odor principle, ambrettolide, are nontoxic.[7]

## USES

**Medicinal, Pharmaceutical, and Cosmetic.** Oil and absolute are used in sophisticated types of perfumes and in soaps, detergents, creams, and lotions; maximum use level reported is 0.12% for the oil in perfumes.[7] The oil is valued for its sweet, rich wine or brandylike floral musky scent with a unique bouquet and roundness. While resembling animal-derived musk scents, ambrette seed oil lacks the fecal note, sometimes found in the former (CSIR I). This fine fragrance is one of the most expensive of essential oils.

**Food.** Ambrette seed and its tincture are used in preparing vermouths and bitters. The oil and absolute are also used in vermouths and bitters, but they are more commonly used in flavoring other types of food products, including nonalcoholic beverages, frozen dairy desserts, candy, baked goods, and gelatins and puddings. The use levels are usually very low, <0.001% (10 ppm).

**Traditional Medicine.** Reportedly use as a stimulant and an antispasmodic; in Chinese medicine, to treat headache. In Western traditions; a folk medicine with stimulant, aromatic, antispasmodic, and insecticidal activity for protecting woolen garments from moths (CSIR I).

**Others.** The stem bark has been used as a fiber (78% cellulose); the root mucilage used as a paper sizing material (CSIR I).

## COMMERCIAL PREPARATIONS

Seeds, oil, absolute, concrete, and tincture; oil is official in F.C.C.

*Regulatory Status.* GRAS (§182.10 and §182.20).

## REFERENCES

See the General Reference for ARCTANDER; BAILEY 1; BAUER; CSIR I; FEMA; GUENTHER; KARRER; JIANGSU; TERRELL; USD 23rd; WREN.

1. M. Hashmi et al., *J. Oil Technol. Assoc. India*, **14**, 64 (1982); through *Chem. Abstr.*, **98**, 86284d (1983).
2. B. Maurer and A. Grieder, *Helv. Chim. Acta*, **60**, 1155 (1977).
3. T. Y. Nee et al., *Phytochemistry*, **25**, 2157 (1986).
4. L. Peyron, *Bull. Soc. Fr. Physiol. Veg.* **7**, 46 (1961); through *Chem. Abstr.*, **56**, 12014b (1962).
5. K. C. Srivastava and S. C. Rastogi, *Planta Med.*, **17**, 189 (1969).
6. U. K. Chauhan, *Proc. Natl. Acad. Sci., India, Sect. B*, **54**, 236 (1984); through *Chem. Abstr.*, **103**, 157419m (1985).
7. D. L. J. Opdyke, *Food Cosmet. Toxicol.*, **13**, 705, 707 (1975).

# ANGELICA

**Source:** ***Angelica archangelica*** L. (syn. *Archangelica officinalis* Hoffm.) (Family Umbelliferae or Apiaceae).

*Synonyms.* Garden angelica and European angelica.

### GENERAL DESCRIPTION

Stout biennial or perennial herb, up to 2 m high with a large rhizome; fruit with thick corky wings; native to northern and eastern Europe, to The Netherlands, and Iceland; eastward to Siberia; cultivated in Belgium, Hungary, Germany, and other countries; naturalized elsewhere. Parts used are the rhizome and roots, fruits, and stem, with the stem less extensively used; currently the roots and rhizome are the most frequently used.

Other *Angelica* spp. are also used, but infrequently.

### CHEMICAL COMPOSITION

Angelica is very rich in coumarins, which occur throughout the plant.

The root (root and rhizome) contains 0.3–1% volatile oil that is composed mainly of *d*-$\alpha$-phellandrene, $\alpha$-pinene, limonene, $\beta$-caryophyllene, linalool, borneol, acetaldehyde, and four macrocyclic lactones ($\omega$-tridecanolide, 12-methyl-$\omega$-tridecanolide, $\omega$-pentadecanolide, and $\omega$-heptadecanolide), among others (MASADA);[1,2] coumarins, including osthol, angelicin, osthenol, umbelliferone, archangelicin, bergapten, ostruthol, imperatorin, umbelliprenine, xanthotoxol, xanthotoxin, oxypeucedanin, oreoselone, phellopterin, marmesin, byakangelicol, and 2′-angeloyl-3′-isovaleryl vaginate, with osthol in major concentration (ca. 0.2% of root);[3–7] plant acids (angelic, aconitic, citric, malic, oxalic, malonic, fumaric, succinic, caffeic, chlorogenic, quinic, lauric, tridecanoic, myristic, pentadecanoic, palmitic, palmitoleic, stearic, oleic, linoleic, linolenic, petroselinic, behenic acids, etc.).[4,8,9] Other constituents include resin, starch, sugars (sucrose, fructose, glucose, umbelliferose), archangelenone, (a flavonone), $\beta$-sitosteryl palminate, and arachinate (KARRER).[4,9,10]

The fruits (commonly known as seeds) contain about 1% volatile oil, consisting mainly of $\beta$-phellandrene and other terpenes similar to those in root oil (MASADA),[11] and coumarins, including imperatorin, bergapten, iso-imperatorin, isopimpinellin, 8-hydroxy-5-methoxypsoralen, 4-methoxy-7-hydroxypsoralen, phellopterin, xanthotoxol, and xanthotoxin, with imperatorin and bergapten in major concentration (0.5% and 0.1%, respectively) (KARRER).[12,13] The seed oil of a Pakistani variety, *A. archangelica* L. var. *Himalaica* (CLARKE) E. Nasir, contains hexylmethylphthalate as its major component (36%).[14]

## PHARMACOLOGY OR BIOLOGICAL ACTIVITIES

Angelica is considered to have diaphoretic and expectorant properties. Certain coumarins (e.g., bergapten, xanthotoxin) are known to be phototoxic (see ***bergamot oil***). Angelica root and seed oils are obtained by steam distillation and are not expected to contain these coumarins. However, extracts (e.g., absolute, solid extract, fluidextract) may contain such coumarins. Root oil (but not seed oil) has been reported to be phototoxic.[1,11] Angelica root oil has also been reported to exhibit antibacterial and antifungal properties.[1,15]

Calcium-antagonist-like effects have been observed *in vitro* in a number of *Angelica* species. Calcium antagonists have been of recent interest in cardiovascular disease research where relaxation of vascular smooth muscle is desirable. A recent study investigated the calcium-antagonist activity of angelica root extracts prepared with twenty different solvents. It was determined that chloroform is the most suitable solvent for extraction of nonpolar biologically active *A. archangelica* root coumarin fractions.[16]

Cardiovascular effects have also been reported for *Angelica sinensis* (Oliv.) Diels (Chinese angelica or *danggui*). An injectable aqueous extract of the root was rated at 91% effective in the treatment of 111 patients with acute ischemic cerebrovascular disease.[17]

Angelicin has relaxant activity on a wide variety of smooth muscles of various animal species, being almost as active as papaverine in some isolated muscle preparations.[18]

## USES

**Medicinal, Pharmaceutical, and Cosmetic.** Now rarely used in pharmaceutical preparations. Its major use currently is as a fragrance ingredient in soaps, detergent, creams, lotions, and perfumes. Reported maximum use levels for both root and seed oils are usually very low, the highest being about 0.1% for both in the perfume category.[1,11]

**Food.** Used as a flavor ingredient in most major categories of food products, including alcoholic (bitters, liqueurs, vermouths) and nonalcoholic beverages, frozen dairy desserts, candy, baked goods, and gelatins and puddings. The seed and root oils and the root extract are more commonly used; average maximum use levels are low, usually below 0.01%, except for the seed extract, which is reported to be 0.2% in alcoholic beverages.

**Health Food/Herb Teas.** Both dried seeds and cut and sifted or powdered root occasionally used as tea flavoring; also in tinctures or oral formulations, primarily for menstrual regulation and as an expectorant (FOSTER).

**Traditional Medicine.** Reportedly used in Europe for centuries to treat bronchial ailments, colds, coughs, and stomach troubles caused by indigestion; also used in cosmetics for its allegedly quieting and soothing effect on the nerves of the skin (deNAVARRE). The roots and seeds have been used in the treatment of arthritic disease, nervous conditions, insomnia, hyperacidity, and intestinal disturbances as well as for antiinflammatory, diuretic, and diaphoretic activity.[16]

In Chinese medicine, at least 10 *Angelica* species are used, including *A. dahurica* (Fisch.) Benth. et Hook., *A. anomala* Lalem., *A. formosana* Boiss., and *A. sinensis* (Oliv.) Diels. The last one is known as "*danggui*"; it is a well-known drug for treating female ailments and has been used for several thousand years, with a reputation in China perhaps second only to ginseng (FARNSWORTH, FOGARTY, JIANGSU, NANJING).

**Others.** Could serve as a source of bergapten, xanthotoxin, and other coumarins, which have been shown to be effective in treating psoriasis and vitiligo (see ***bergamot***).

## COMMERCIAL PREPARATIONS

Crude, extracts, and oils. Root and seed were both formerly official in U.S.P. and N.F.; both root oil and seed oil are official in F.C.C.

*Regulatory Status*. GRAS (§182.10 and §182.20). In Germany the fruit and roots are subjects of official monographs.[19,20] The crude root at a daily dose of 4.5 g and galenical preparations are indicated for internal use for appetite loss and digestive ailments, including mild gastrointestinal tract spasms, and flatulence.[20] Crude fruit (seed) and preparations are not recommended for reported diuretic and diaphoretic activity, because efficacy and safety have not been established.[19]

## REFERENCES

See the General Reference for BIANCHINI AND CORBETTA; BLUMENTHAL; deNAVARRE; FEMA; FOSTER; GUENTHER; JIANGSU; MARTINDALE; deNAVARRE AND JOHNSON; TERRELL; TUTIN 2; UPHOF.

1. D. L. J. Opdyke, *Food Cosmet. Toxicol.*, **13**(Suppl.), 713 (1975).
2. J. Taskinen, *Acta Chem. Scand.*, *B*, **29**, 637 (1975).
3. W. Steck and B. K. Bailey, *Can. J. Chem.*, **47**, 2425 (1969).
4. A. B. Svendsen, *Blyttia*, **11**, 96 (1953); through *Chem. Abstr.*, **50**, 7963d (1956).
5. A. Chatterjee and S. Dutta, *Indian J. Chem.*, **6**, 415 (1968).
6. P. Härmälä et al., *Planta Med.*, **58**, 288 (1992).
7. J. Carbonnier and D. Molho, *Planta Med.*, **44**, 162 (1982).
8. G. I. Kas'yanov et al., *Khim. Prir. Soedin.*, **1**, 108 (1977); through *Chem. Abstr.*, **87**, 50189k (1977).
9. B. E. Nielsen and H. Kofod, *Acta Chem. Scand.*, **17**, 1161 (1963).
10. S. C. Basa et al., *Chem. Ind. (London)*, **13**, 355 (1971).
11. D. L. J. Opdyke, *Food Cosmet. Toxicol.*, **12**(Suppl.), 821 (1974).
12. A. Patra et al., *Indian J. Chem.*, **14B**, 816 (1976); through *Chem. Abstr.*, **86**, 136337y (1977).
13. T. Beyrich, *Arch. Pharm.*, **298**, 672 (1965); through *Chem. Abstr.*, **64**, 3956b (1966).
14. M. Ashraf et al., *Pak. J. Sci. Ind. Res.*, **23**, 73 (1980); through *Chem. Abstr.*, **94**, 109075v (1981).
15. N. Saksena and H. H. S. Tripathi, *Fitoterapia*, **56**, 243 (1985).
16. P. Härmälä et al., *Planta Med.*, **58**, 176 (1992).
17. P. G. Xiao and K. J. Chen. *Phytotherapy Res.*, **1**, 53 (1987).
18. G. K. Patnaik et al., *Planta Med.*, **53**, 517 (1987).
19. Monograph *Angelicae fructus*, *Bundesanzeiger*, no. 101 (June 1, 1990).
20. Monograph *Angelicae radix*, *Bundesanzeiger*, no. 101 (June 1, 1990).

# ANGOSTURA BARK

**Source:** ***Galipea officinalis*** Hancock (Family Rutaceae).

*Synonyms*. Angustura, cusparia, and carony bark.

## GENERAL DESCRIPTION

*Galipea officinalis* is a tree native to northern South America, reportedly growing on mountains in Venezuela at an altitude between 200 and 300 m above sea level. It is now generally believed to be the source of angostura bark (USD 23rd).

Formerly, angostura bark was believed to be the bark of *Cusparia trifoliata* Engl. (syn. *C. febrifuga* Humb.), a tree related to *G. officinalis* reported to grow in Brazil. Much of the scientific literature before 1960 was based on this species.

There does not appear to be any authoritative work that definitively establishes the source and identity of angostura bark. The irreproducible results on certain alkaloids (e.g., cuspareine, cusparidine, galipidine) may have been due to different sources of bark employed (GLASBY 1).

## CHEMICAL COMPOSITION

Contains two unstable bitter principles: angostura bitters 1 and 2 (3,5-dihydroxy-5-ethoxy-2-syringoyl-1-methyl-4-*O*-β-D-glucopyranosylcyclopentane and 3,5-dihydroxy-5-ethoxy-2-vanilloyl-1-methyl-4-*O*-β-D-glucopyranosylcyclopentane);[1] alkaloids (cusparine, galipine, cuspareine, galipoline, galipoidine, galipidine, quinaldine, 4-methoxyquinaldine, quinoline, 2-*n*-amylquinoline, etc.) (GLASBY 1);[1,2] and a volatile oil (1–2%) containing cadinene and galipene (KARRER; WREN). The stem and root bark of a related species, *Galipea trifoliata* Aublet, contains phebalosin, ramosin, and galipein (coumarins).[3]

## PHARMACOLOGY OR BIOLOGICAL ACTIVITIES

Has been reported to have antipyretic, antiperiodic, antidysenteric, and bitter tonic properties. Large doses may produce nausea.

The alkaloids, particularly cusparine and galipine, have been demonstrated to have antispasmodic properties.[4] They also exhibited respiration-excitation effects in dogs.[5]

## USES

**Medicinal, Pharmaceutical, and Cosmetic.** Used in bitter tonics.

**Food.** Extracts used in most categories of food products such as alcoholic (bitters) and nonalcoholic beverages, frozen dairy desserts, candy, baked goods, gelatins and puddings, and gravies. Average maximum use level in alcoholic beverages is reported to be about 0.3%.

The well-known "angostura bitters" does not contain angostura bark at all but is made from a mixture of gentian root and other botanicals.

**Traditional Medicine.** Used as a febrifuge, antidiarrheal, and bitter tonic in dose of 0.3–1 g.; large doses cathartic and emetic.

## COMMERCIAL PREPARATIONS

Crude and extracts.

*Regulatory Status.* GRAS (§182.10 and §182.20).

## REFERENCES


See the General References for ARCTANDER; FEMA; GLASBY 2; GUENTHER; MARTINDALE; UPHOF; WREN.

1. C. H. Brieskorn and V. Beck, *Phytochemistry*, **10**, 3205 (1971).
2. C. H. Brieskorn and V. Beck. *Präp. Pharm.*, **6**, 177 (1970); through *Chem. Abstr.*, **74**, 79530g (1971).
3. K. R. Wirasutisna et al., *Phytochemistry*, **26**, 3372 (1987).
4. C. E. T. Kraukau, *Kgl. Fysiograf. Saellskap. Lund, Foerh.*, **15**, 289 (1945); through *Chem. Abstr.*, **43**, 3527i (1949).
5. L. Binet and M. V. Strumza, *Therapie*, **8**, 669 (1953).


# ANISE (AND STAR ANISE)

**Source:** ***Anise*** *Pimpinella anisum* L. (syn. *Anisum vulgare* Gaertn.; *A. officinarum* Moench.) (Family Umbelliferae or Apiaceae); ***Star anise*** *Illicium verum* Hook. f. (Family Illiciaceae).

*Synonyms*. Anise seed, aniseed, sweet cumin (*P. anisum*); illicium, Chinese anise, and Chinese star anise (*I. verum*).

### GENERAL DESCRIPTION

Anise is an annual herb, usually less than 0.6 m high; leaves alternate below, opposite above; native to Greece and Egypt, now widely cultivated. Part used is the dried ripe fruit; anise oil is obtained from it by steam distillation.

Star anise is an evergreen tree usually 4–6 m high but may reach 12 m; indigenous to southeastern Asia; extensively cultivated in southern China, also in Vietnam, India, Japan. Part used is the dried, ripe fruit that consists of 5–13, usually 8, seed-bearing woody follicles (one seed per follicle) attached to a central axis in the shape of a star, hence the name star anise. In Chinese also called "eight-horned anise" or simply "eight horns," referring to the usually eight-follicled fruit. China is the major producer of star anise. Star anise oil is obtained by steam distillation.

Japanese star anise should not be confused with true star anise (Chinese star anise). Japanese star anise is obtained from a related species, *Illicium lanceolatum* A.C. Smith (formerly believed to be *Illicium anisatum* L. or *I. religiosum* Sieb. et Zucc.), which grows in southern China, Taiwan, and Japan; it looks like a smaller, deformed version of Chinese star anise and is highly poisonous. A 10–15% aqueous extract is used in China as agricultural insecticide.

Due to the traditional use of anise oils with licorice in licorice candy, the flavor of anise is often confused with that of licorice, particularly among the public, and is erroneously described as licorice-like.

### CHEMICAL COMPOSITION

Anise contains 1–4% volatile oil (FURIA AND BELLANCA);[1] coumarins (bergapten, umbelliprenine, umbelliferone, scopoletin);[2,3] lipids (ca. 16%), including fatty acids ($C_{16}$, $C_{18}$, $C_{20}$, $C_{22}$, $C_{24}$, $C_{26}$, $C_{30}$, etc.), $\beta$-amyrin, and stigmasterol and its salts (palmitate and stearate) (MARSH);[2,4] flavonoid glycosides (quercetin-3-glucuronide, rutin, luteolin-7-glucoside, isoorientin, isovitexin, apigenin-7-glucoside (apigetrin), etc.);[5] myristicin;[6] protein (ca. 18%); carbohydrate (ca. 50%); and others (MARSH).

Anise oil contains 75–90% *trans*-anethole; estragole (methylchavicol); anise ketone (*p*-methoxyphenylacetone); and $\beta$-

caryophyllene. Other compounds in minor concentrations include anisaldehyde, anisic acid (oxidation products of anethole), linalool, limonene, $\alpha$-pinene, acetaldehyde, *p*-cresol, creosol, hydroquinine, $\beta$-farnesene, $\gamma$-himachalene, and *ar*-curcumene (KARRER).[1,7]

Star anise contains about 5% volatile oil (ca. 10% in follicles and 2.5% in seeds) (NANJING), catechins and proanthocyanidins, among others.[8]

Star anise oil contains *trans*-anethole (80–90%) as its major component; other constituents include estragole, 1,4-cineole, $\beta$-bisabolene, $\beta$-farnesene, $\alpha$-copaene, *cis*- and *trans*-$\alpha$-bergamotene, caryophyllene, nerolidol, methylanisoate, *trans*-methylisoeugenol, cadinene, foeniculin, 3-carene, *d*-$\alpha$-pinene, phellandrene, $\alpha$-terpineol, hydroquinine, traces of *cis*-anethole, and safrole (KARRER).[9–12] Presence of safrole disputed.[10,11]

## PHARMACOLOGY OR BIOLOGICAL ACTIVITIES

Oil of anise or star anise has carminative and expectorant properties. Its major component, anethole, has been reported to be the cause of dermatitis (erythema, scaling, and vesiculation) in some people.[13] Anethole has two isomers (*trans* and *cis*), with the *cis* isomer being 15–38 times more toxic to animals than the *trans* isomer, depending on the species (MERCK).[12] Current U.S.P. and F.C.C. specifications for anethole do not require differentiation between the isomers. Anethole (no isomer given) has been reported to be mutagenic in Ames *Salmonella* reversion assay.[14]

Anethole, anisaldehyde, and myristicin (in aniseed), along with *d*-carvone (present in *P. anisum* plant), have been found to have mild insecticidal properties,[15] Anethole also inhibits growth of mycotoxin-producing *Aspergillus* species in culture.[16]

Anethole was formerly considered an active estrogenic agent of the essential oil of anise. However, further research suggests that the active estrogenic compounds are polymers of anethole, such as dianethole and photoanethole.[17]

## USES

Anise oil and star anise oil are used interchangeably in the United States, both being officially recognized as anise oil in the U.S.P. and F.C.C.

**Medicinal, Pharmaceutical, and Cosmetic.** Both anise and star anise oils are used as carminative, stimulant, mild spasmolytic, weak antibacterial, and expectorant in cough mixtures and lozenges, among other preparations. Internally used for dyspeptic complaints; externally as an inhalant for congestion of the respiratory tract.

Both oils are used to mask undesirable odors in drug and cosmetic products and as fragrance components in toothpastes, perfumes, soaps, detergents, creams, and lotions, with maximum use levels of 0.25% anise oil and 0.4% star-anise oil in perfumes.[12,13]

**Food.** Anise, star anise (to a lesser extent), anise oil, and star anise oil are widely used as flavoring ingredients in all major categories of foods, including alcoholic (bitters, brandies, and liqueurs, e.g., anisette) and nonalcoholic beverages, frozen dairy desserts, candy (e.g., licorice candies), baked goods, gelatins and puddings, and meat and meat products. Highest average maximum use levels for anise oil are about 0.06% (570 ppm) in alcoholic beverages and 0.07% (681 ppm) in candy.

Both anise and star anise are widely used as domestic spices; the former is mainly used by Westerners while the latter is used primarily by Asians, especially in Chinese foods.

**Health Food/Herb Teas.** Whole, crushed, or ground crude drug for infusion, and other galenical preparations.[18]

**Traditional Medicine.** Anise and star anise are reportedly used as aromatic carminative, stimulant, and expectorant; also as estrogenic agents to increase milk secretion, promote menstruation, facilitate child birth, increase libido, and alleviate symptoms of male climacteric;[17] star anise has been used in Chinese medicine for similar purposes for 1,300 years (JIANGSU).

## COMMERCIAL PREPARATIONS

Crude and essential oils. Both oils are official in N.F. and F.C.C. Preparations containing 5–10% essential oil are used as a respiratory inhalant in Germany.[18]

*Regulatory Status.* GRAS (§182.10 and §182.20). Anise seed and star anise seed are subjects of German official monographs; 3.0 g of seed or 0.3 g essential oil (mean daily dose) allowed as a bronchial expectorant for upper respiratory tract congestion and as gastrointestinal spasmolytic.[18,19]

## REFERENCES

See General References for ARCTANDER; BAILEY 2; BLUMENTHAL; GUENTHER; JIANGSU; LUST; MARTINDALE; MASADA; NANJING; REMINGTON; TERRELL; UPHOF; YOUNGKEN.

1. M. B. Embong et al., *Can. J. Plant. Sci.*, **57**, 681 (1977).
2. T. Kartnig and G. Scholz, *Fette, Seifen, Anstrichmit.*, **71**, 276 (1969); through *Chem. Abstr.*, **71**, 57561d (1969).
3. T. Kartnig et al., *Planta Med.*, **27**, 1 (1975).
4. A. Szegfu et al., *Acta Pharm. Hung.*, **42**, 162 (1972); through *Chem. Abstr.*, **77**, 111449n (1972).
5. J. Kunzemann and K. Hermann, *Z. Lebebsm. Unters. Forsch.*, **164**, 194 (1977).
6. J. B. Harborne et al., *Phytochemistry*, 8, 1729 (1969).
7. R. Tabacchi et al., *Helv. Chem. Acta*, **57**, 849 (1974).
8. J. M. Schulz and K. Herrmann, *Z. Lebensm. Unters. Forsch.*, **171**, 278, (1980).
9. H. M. Okely and M. F. Grundon, *J. Chem. Soc., D*, **19**, 1157 (1971).
10. R. Kaempf and E. Steinegger, *Pharm. Acta Helv.*, **49**, 87 (1974).
11. J. Bricout, *Bull. Soc. Chim. Fr.*, **9–10**, 1901 (1974); through *Chem. Abstr.*, **83**, 15470r (1975).
12. D. L. J. Opdyke, *Food Cosmet. Toxicol.*, **13**(Suppl.), 715 (1975).
13. D. L. J. Opdyke, *Food Cosmet. Toxicol.*, **11**, 865 (1973).
14. J. Sekizawa and T. Shibamoto, *Mutat. Res.*, **101**, 127 (1982).
15. G. T. Carter, *Diss. Abstr. Int., B*, **37**, 766 (1976).
16. H. Hitokoto et al., *Appl. Environ. Microbiol.*, **39**, 818 (1980).
17. M. Albert-Puleo, *J. Ethnopharmacol.*, **2**, 337 (1980).
18. Monograph *Anisi fructus Bundesanzeiger*, no. 122 (July 6, 1988).
19. Monograph *Anisi stellati fructus Bundesanzeiger*, no. 122 (July 6, 1988).

# ANNATTO

**Source:** ***Bixa orellana*** L. (Family Bixaceae).

*Synonyms.* Arnotta, annotta, achiote, and achiotillo.

## GENERAL DESCRIPTION

Shrub or small tree, up to 10 m high; native to tropical America; extensively cultivated in tropical areas. Part used is the seed, which contains the coloring principles in an orange-red waxy covering.

Major producers of annatto seeds include India, Jamaica, Kenya, Mexico, and Peru; qualities differ considerably depending on sources and seasons.

## CHEMICAL COMPOSITION

The coloring principles are carotenoids, mostly bixin and norbixin, with bixin (especially the *cis* isomer) in major concentration. *cis*-Bixin is unstable and during extraction is usually converted to *trans*-bixin, the stable isomer, which is also known as isobixin. Bixin is a monomethyl ester of norbixin (a dicarboxylic acid); it is readily hydrolyzed by alkalis during alkali extraction of annatto to the dicarboxylic acid (MERCK).[1,2] Annatto also contains $\beta$-carotene (6.8–11.3 mg/100 g), an essential oil (with ishwarane as the major component), pentosans, pectin, 13–17% protein, and tannins, among others.[3,5]

The coloring principle in oil-soluble annatto preparations is free bixin, while that of water-soluble (usually alkaline) annatto extracts is an alkali salt of norbixin, commonly the potassium salt.

Both bixin and norbixin in the free acid state are insoluble in water but soluble in organic solvents (e.g., acetone, alcohols) and aqueous alkaline solutions.

Bixin is one of the more stable natural yellow colors. However, it loses much of its tinctorial power gradually on storage, the process being accelerated by light and heat. Hence, for manufacturing purposes, fresh seeds are preferred.

Tinctorial strength of bixin is comparable to that of $\beta$-carotene, but bixin is the more stable.

Annatto is reported to be most stable at pH 8, with decreased stability in the pH 4 to 8 regions.[6]

## PHARMACOLOGY OR BIOLOGICAL ACTIVITIES

Although bixin and norbixin are carotenoids, they do not have vitamin A activity.

No pharmacological or toxicity studies on annatto are available.

## USES

Currently the major commercial uses of annatto colors are in foods, though they are being tested in drugs and cosmetics and are slowly being incorporated in such products, e.g., hair dyes.

Aqueous alkaline extracts are extensively used in coloring cheeses (especially cheddar) and to a lesser extent ice creams and other dairy products. Oil-soluble extracts are used in oily food products such as salad oils, popcorn oil, margarine, and sausage casings. Major food categories in which annatto color is used include alcoholic and nonalcoholic beverages, frozen dairy desserts, baked goods, meat and meat products, condiments and relishes, fats and oils, snack foods, and gravies, with highest average maximum use level reported for the extract in baked goods (ca. 0.24%).

## COMMERCIAL PREPARATIONS

Crude, oil- and water-soluble extracts and spray-dried powders.

Also available for domestic use as ground or whole seeds in supermarkets or ethnic stores in metropolitan areas in the United States.

*Regulatory Status*. Approved for use as a general colorant (§73.30, §73.1030, and §73.2030).

## REFERENCES


See the General References for FEMA; FURIA; MORTON 2; TERRELL; UPHOF.

1. P. Karrer and E. Jucker, *Carotenoids*, Elsevier Applied Science Publishers, Barking, UK, 1950.
2. F. Mayer and A. H. Cook, *The Chemistry of Natural Coloring Matters*, ACS Monograph Series **89**, Rheinhold, New York, 1943.
3. Z. Angelucci et al., *Colet. Inst. Tecnol. Aliment (Campinas Braz.)*, **11**, 89 (1980); through *Chem. Abstr.*, **95**, 148861p (1981).
4. R. Bressani et al., *Arch. Latinoam. Nutr.*, **33**, 356 (1983); through *Chem. Abstr.*, **100**, 84410c (1984).
5. C. Srinivasulu and S. N. Mahapatra, *Indian Perfum.*, **26**, 132 (1982).
6. D. E. Auslander et al., *Drug Cosmet. Ind.*, **121**(6), 55 (1977).


# ARNICA

**Source:** ***Arnica montana*** L. (Family Compositae or Asteraceae).

*Synonyms*. European arnica, mountain tobacco, leopard's bane, and wolf's bane.

### GENERAL DESCRIPTION

Perennial herb, up to about 0.6 m high; native to mountainous regions of Europe; cultivated in northern India. Parts used are the dried flower heads.

Several American *Arnica* species have also been used: *A. fulgens* Pursh, *A. sororia* Green, and *A. cordifolia* Hook. They are called American arnica and are native to the western Rocky Mountains regions.

### CHEMICAL COMPOSITION

Contains up to 1% (normally about 0.3%) of a viscous volatile oil about half of which is composed of fatty acids, with palmitic, linoleic, myristic, and linolenic acids in predominant concentrations. Aromatic constituents present include terpenes, thymol, thymol methyl ether, 4-hydroxythymol dimethyl ether, and isobutyric acid thymyl ether.[1–5] Other constituents are resins, bitter principle (arnicin), sesquiterpene lactones (pseudoguaianolides) such as helenalin, 11,13-dihydrohelenalin, helenalin and 11,13-dihydrohelenalin esters, 2$\beta$-ethoxy-6-*O*-isobutyryl 1-2,3-dihydrohelenalin and 6-*O*-isobutyryltetrahydrohelenalin;[6–8] tannin, arnisterin (a sterol), carotenoids ($\alpha$- and $\beta$-carotene, cryptoxanthin, lutein (xanthophyll), etc.), (KARRER, MERCK);[9] flavonoids (betuletol, 6-methoxykaempferol, hispidulin, jaceosidin, pectolinarigenin, etc.);[10,11] phenolic acids (*p*-hydroxybenzoic, *p*-coumaric, gentisic, ferulic, caffeic, vanillic, etc.).[12]

Several pseudoguaianolides have been reported in leaves of *A. montana*, such as arnifolin; arnicolides A, B, C, and D; and loliolide.[13,14]

## PHARMACOLOGY OR BIOLOGICAL ACTIVITIES

Arnica extract has been reported to increase the resistance of animals to bacterial infections by stimulating phagocytosis of the bacteria involved, particularly *Listeria monocytogenes* and *Salmonella typhimurium*.[15]

Arnica is believed to have wound-healing and stimulant properties. However it is an irritant to mucous membranes, and ingestion may cause burning pain in the stomach, diarrhea, vomiting, giddiness, intense muscular weakness, decrease or increase of the pulse rate, and collapse. One ounce (ca. 30 mL) of the tincture (1:0.2, or 20%) has been reported to produce serious but not fatal symptoms. (USD 23rd).

Arnica has also been reported to cause allergic contact dermatitis in humans.[16] Sesquiterpenes, especially helenalin and its derivatives are believed to be the sensitizers.[6,17]

The sesquiterpene lactones helenalin acetate and 11,13-dihydrohelenalin have antibacterial and antifungal activities *in vitro*.[18] Numerous esters of helenalin have been shown to have antiinflammatory activity in mice and rats.[19]

Acidic polysaccharide fractions from Arnica extract have strong immunostimulating activities *in vitro*.[18] Helenalin also had immunostimulating activity in a preliminary screening.[20]

## USES

**Medicinal, Pharmaceutical, and Cosmetic.** No longer (or rarely) used in preparations intended for internal use. Current use is mostly as a local antiinflammatory and immunostimulant in the form of a tincture or salve for external application to sprains and bruises.[20–22] Also used in hair tonics and anti-dandruff preparations. Oil is occasionally used in perfumes and other cosmetic preparations.[23]

**Food.** Used (though not widely) as a flavor ingredient in alcoholic (ca. 0.03%) and nonalcoholic (0.02%) beverages, frozen dairy desserts (0.03%), candy (0.04%), baked goods (0.08%), and gelatins and puddings (0.04%), with reported average maximum use levels in parentheses. These figures apparently cannot apply to the crude flowers as they are too high to be safe in all categories, except perhaps alcoholic beverages because of their limited volume of intake.

**Health Food/Herb Teas.** Various ointments, salves, lotions, tincture, primarily homeopathic products, for external use only (FOSTER AND CARAS, LUST, WREN).

**Traditional Medicine.** Reported used as a diaphoretic, diuretic, stimulant; externally antiphlogistic, analgesic, antiseptic, vulnerary for hematomas, dislocations, contusions, fracture-induced edema, and insect bites.[20]

## COMMERCIAL PREPARATIONS

Crude, tincture, and extracts. Crude, tincture, and fluid extract were formerly official in N.F.; crude flowers and root formerly in U.S.P.

*Regulatory Status*. Has been approved for food use in alcoholic beverages only (§172.510). External preparations subject of a positive German monograph.[22]

## REFERENCES

See the General Reference for ARCTANDER; BLUMENTHAL; FEMA; FOSTER AND CARAS; LUST; MARTINDALE; WREN; YOUNGKEN.

1. G. Willuhn, *Planta Med.*, **21**, 221 (1972).
2. H. Kating et al., *Planta Med.*, **18**, 130 (1970).
3. G. Willuhn, *Planta Med.*, **22**, 1 (1972).
4. G. Willuhn, *Planta Med.*, **21**, 329 (1972).
5. R. Schmitz and H. Kating, *Planta Med.*, **31**, 310 (1977).
6. H. D. Herrmann et al., *Planta Med.*, **34**, 299 (1978).
7. G. Willuhn et al., *Planta Med.*, **50**, 35 (1984).
8. G. Willuhn et al., *Planta Med.*, **49**, 226 (1983).
9. M. Vanhaelen, *Planta Med.*, **23**, 308 (1973).
10. I. Merfort, *Planta Med.*, **50**, 107 (1984).
11. I. Merfort and D. Wendisch, *Planta Med.*, **53**, 434 (1987); **54**, 247 (1988).
12. L. Swiatek and J. Gora, *Herba Pol.*, **24**, 187 (1978); through *Chem. Abstr.*, **91**, 52764a (1979).
13. V. Herout, in H. Wagner and L. Hörhammer, eds., *Pharmacognosy* and *Phytochemistry*, 1st Int. Congress, Munich, 1970, Springer-Verlag, Berlin, Germany, 1971, p. 93.
14. M. Holub et al., *Phytochemistry*, **14**, 1659 (1975).
15. H. Buschmann, *Fortschr. Veterinärmed.*, **20**, 98 (1974); through *Chem. Abstr.*, **81**, 130910e (1974).
16. J. C. Mitchell, in V. C. Runeckles, ed., *Recent Advances in Phytochemistry*, Vol. 9, Plenum, New York, 1975, p. 119.
17. G. Willuhn, *Deut. Apoth. Ztg.*, **126**, 2038 (1986).
18. G. Willuhn et al., *Pharm. Ztg.*, **127**, 2183 (1982); through *Chem. Abstr.*, **98**, 14238u (1983).
19. I. H. Hall et al., *Planta Med.*, **53**, 153 (1987).
20. H. Wagner et al., *Planta Med.*, **50**, 139 (1985).
21. H. Wagner et al., *Arzneim. Forsch.*, **35**, 1069 (1985).
22. Monograph *Arnicae flos*, *Bundesanzeiger*, no. 228 (Dec. 5, 1984).
23. H. B. Heath, *Cosmet. Toilet.*, **92**, 19 (1977).

# ARTICHOKE

**Source:** ***Cynara scolymus*** L. (Family Compositae or Asteraceae).

*Synonyms*. Globe artichoke.

### GENERAL DESCRIPTION

Large thistle-like perennial herb, up to about 1 m high; native to southern Europe, North Africa, and the Canary Islands; widely cultivated. Parts used are the leaves; the immature flowerheads with fleshy bracts are eaten as a vegetable. It should not be confused with Jerusalem artichoke, which is the tuber of *Helianthus tuberosus* L.

### CHEMICAL COMPOSITION

Contains up to 2% *o*-diphenolic derivatives such as caffeic acid, 1-, 3-, 4-, and 5-caffeoylquinic acids, 1,5-di-*O*-caffeoylquinic acid (cynarin), and 1,3-di-*O*-caffeoylquinic acid; flavonoids (0.1–1.0%), including glycosides luteolin-7-$\beta$-rutinoside (scolymoside), luteolin-7-$\beta$-D-glucoside, and 4-$\beta$-D-glucoside; glycolic and glyceric acids; taraxasterol, $\Psi$-taraxasterol; inulin; guaianolides (grosheimin, cynaropicrin, 8-epigrosheimin); cynaratriol;[1,2] sugars; enzymes (KARRER);[3–10] and a volatile oil consisting of β-selinene and caryophyllene as its major components, with α-cadrene, oct-1-en-3-one, hex-1-en-3-one, decanal, non-*trans*-2-enal, phenylacetaldehyde, and eugenol as the major aromatic principles.[11,12]

Cynarin, 1,3-dicaffeoylquinic acid, 3-caffeoylquinic acid (chlorogenic acid), and scolymoside are believed to be the active constituents.[7,8,13] The maximum content of cynarin is obtained by aqueous ebullition of the drug.[14]

## PHARMACOLOGY OR BIOLOGICAL ACTIVITIES

Artichoke is reported to have stimulant, choleretic (increasing bile secretion), and diuretic properties. Cynarin and scolymoside have choleretic properties.[8] Cynarin also has been reported to be hypolipidemic in humans, reducing blood cholesterol and triglyceride levels (MARTINDALE)[15] though one study showed negative results in rats.[16]

Artichoke has been reported to cause allergic contact dermatitis.[17]

Hydroalcoholic extracts of artichoke have been demonstrated to have choleretic and cholesterol-lowering properties in rats.[18]

## USES

**Medicinal, Pharmaceutical, and Cosmetic.** Cynarin, one of the active ingredients, is used in the treatment of hyperlipidemia and various dyspeptic symptoms for choleretic effect, primarily in Europe. Use is contraindicated in known allergies to composite family members, in bile duct obstructions, and in gallstones (except under medical advice).[19]

**Food.** Leaves and their extracts used mainly as a flavor ingredient in alcoholic beverages (bitters, liqueurs, etc.). Reported average maximum use level for leaves is 0.0016% (16 ppm).

Cynarin and chlorogenic acid can be used as sweeteners (LEWIS AND ELVIN-LEWIS).

**Health Food/Herb Teas.** Dried or ground leaves, or expressed juice of fresh plant, in capsules, tablets, tincture, or other products for oral use (WREN).

**Traditional Medicine.** Has reportedly been used in Europe since Roman times as a choleretic and diuretic, among others (BIANCHINI AND CORBETTA).

## COMMERCIAL PREPARATIONS

Crude and extracts; there are no uniform standards for extracts.

*Regulatory Status.* Approved for food use in alcoholic beverages only (§172.510). Leaves subject of a German therapeutic monograph indicated as a choleretic for dyspeptic problems.[19]

## REFERENCES

See the General References for BAILEY 2; BIANCHINI AND CORBETTA; BLUMENTHAL; FEMA; LIST AND HÖRHAMMER; MARTINDALE; WREN; UPHOF.


1. H. O. Bernhard et al., *Helv. Chim. Acta*, **62**, 1288 (1979).
2. P. Barbetti et al., *Stud. Carciofo Congr. Int.*, **3**, (1979), **77** (1981); through *Chem. Abstr.*, **97**, 195778t (1982).
3. M. Jaruzelski et al., *Herba Pol.*, **22**, 144 (1976); through *Chem. Abstr.*, **86**, 127146h (1977).
4. L. I. Dranik, *Fenolnye Soedin. Ikh Biol. Funkts., Mater. Vses. Simp.*, **1** (1966), **53** (1968); through *Chem. Abstr.*, **71**, 19515j (1969).

5. E. Bombardelli et al., *Fitoterapia*, **48**, 143 (1977).
6. H. Wagner et al., *Chem. Ber.*, **104**, 2118 (1971).
7. J. Michaud, *Bull. Soc. Pharm. Bordeaux*, **106**(4), 181 (1967); through *Chem. Abstr.*, **69**, 12905m (1968).
8. P. Massacci, *Corriere Farm.*, **22**, 69 (1967); through *Chem. Abstr.*; **67**, 89667b (1967).
9. D. G. Constatninescu et al., *Pharmazie*, **22**(3), 176 (1967).
10. L. Panizzi and M. L. Scarpati, *Nature*, **174**, 1062 (1954).
11. R. G. Buttery et al., paper presented by the ACS 175th National Meeting, Division of Agriculture and Food, paper no. 35, 1978; through *Chem Abstr.*
12. A. J. MacLeod et al., *Phytochemistry*, **21**, 1647 (1982).
13. M. Scemama and P. Garde, *Plant. Med. Phytother.*, **5**, 39 (1971).
14. J. Ingelsias, et al., *Plant. Med. Phytother.*, **19**(3), 202 (1985).
15. W. H. Hammerl et al., *Wien. Med. Wochenschr.*, **123**, 601 (1973).
16. P. E. Altman Jr. and I. L. Honigberg, *J. Pharm. Sci.*, **61**, 610 (1972).
17. J. C. Mitchell, in V. C. Runeckles, ed., *Recent Advances in Phytochemistry*, Vol. 9, Plenum, New York, 1975, p. 119.
18. A. Lietti, *Fitoterapia*, **48**, 153 (1977).
19. Monograph *Cynarae folium*, *Bundesanzeiger*, no. 122 (July 6, 1988); with correction (Sept. 1, 1990).

# ASAFETIDA

**Source:** ***Ferula assa-foetida*** L. or other ***Ferula*** spp. (Family Umbelliferae or Apiaceae).

*Synonyms.* Asafoetida, gum asafetida, devil's dung, and food of the gods.

## GENERAL DESCRIPTION

A large branching perennial herb, up to 3 m high; native to southwestern Asia (eastern Iran and western Afghanistan). Asafetida is the oleogum resin obtained by incising or cutting the living rhizomes and roots. It is a solid or semisolid with a persistent alliaceous (garliclike) odor and bitter acrid taste. An essential oil (asafetida oil) is obtained by steam distillation.

## CHEMICAL COMPOSITION

Contains 40–64% resinous material composed of ferulic acid, asaresinotannols, umbelliferone, and umbelliferone ethers (e.g., farnesiferols A, B, and C, kamolonol) among others; about 25% gum, which is composed of glucose, galactose, L-arabinose, rhamnose, and glucuronic acid; volatile oil (3–21%; usual range 7–9%), consisting of disulfides and polysulfides as its major components, notably 2-butyl propenyl disulfide (*E*- and *Z*-isomers), with monoterpenes ($\alpha$- and $\beta$-pinene, etc.), free ferulic acid, valeric acid, and traces of vanillin, among others, also present.[1–8]

The sulfides are responsible for the characteristic flavor of asafetida.[9]

## PHARMACOLOGY OR BIOLOGICAL ACTIVITIES

Asafetida has been reported to have hypotensive activities in animals.[10] It has also been demonstrated to increase blood coagulation time.[11] More recently ferulic acid and its sodium salt (sodium ferulate) have been shown to inhibit platelet aggregation and platelet malondialdehyde production in rats.[12]

Asafetida is believed to have carminative, expectorant, and antispasmodic properties.

Animals (dogs, cats, rabbits, and deer) are repelled by its odor.

Available data indicate asafetida to be relatively nontoxic, and ingestion of 0.5 oz. (ca.15 g) has been reportedly produced no untoward effects (GOSSELIN).

## USES

**Medicinal, Pharmaceutical, and Cosmetic.** Now rarely used in pharmaceutical preparations. Main use in cosmetics is as a fixative or fragrance component in perfumes.

**Food.** Asafetida is reported to be an ingredient in Worcestershire sauce. Together with the oil and fluidextract, it is reportedly used in nonalcoholic beverages, frozen dairy desserts, candy, baked goods, gelatins and puddings, meat and meat products, and condiments and relishes, among others. Use levels are usually very low (<0.004%).

**Traditional Medicine.** Used for various ailments, including flatulence, cough, asthma, bronchitis, hysteria, and convulsions, among others. Also used in Chinese medicine (since the 7th century) as a nerve stimulant in treating neurasthenia; in chronic bronchitis; and as an expectorant, antiflatulent, and laxative.

## COMMERCIAL PREPARATIONS

Crude, oil, tincture, and extracts. Crude and tincture were formerly official in N.F. Commercial crude asafetida differs widely in quality, depending on sources and is not necessarily derived from *Ferula assafoetida*.[13]

*Regulatory Status.* Approved for use in foods. (§182.20).

## REFERENCES

See the General References for ARCTANDER; BAILEY 2; CLAUS; FEMA; GUENTHER; JIANGSU; LUST; MERCK; MORTON 1; NANJING; USD 23rd.


1. G. H. Mahran et al., *Bull. Fac. Pharm., Cairo Univ.*, **12**, 119 (1975); through *Chem. Abstr.*, **85**, 74896y (1976).
2. L. Caglioti et al., *Helv. Chim. Acta*, **41**, 2278 (1958).
3. L. Caglioti et al., *Helv. Chim. Acta*, **41**, 2557 (1959).
4. A. Kjaer et al., *Acta Chem. Scand., Ser. B*, **30**, 137 (1976).
5. H. Naimie et al., *Collect. Czech. Chem. Commun.*, **37**, 1166 (1972).
6. B. Rajanikanth et al., *Phytochemistry*, **23**, 899 (1984).
7. O. Hofer et al., *Monatsh. Chem.*, **115**, 1207 (1984).
8. M. Ashraf et al., *Pak. J. Sci. Ind. Res.*, **23**, 68 (1980); through *Chem. Abstr.*, **94**, 127135u (1981).
9. K. O. Abraham et al., *Indian Food Packer*, **36**(5), 67 (1982); through *Chem. Abstr.*, **98**, 159253e (1982).
10. R. G. Sarkis'yan, *Med. Zh. Uzb.*, **9**, 23 (1969); through *Chem. Abstr.*, **72**, 119891t (1970).
11. M. M. Mansurov, *Med. Zh. Uzb.*, **6**, 46 (1967); through *Chem. Abstr.*, **68**, 20794u (1968).
12. Z. Z. Yin et al., *Zhongguo Yaoli Xuebao*, **7**, 336 (1986).
13. M. M. Samimi and W. Unger, *Planta Med.*, **36**, 128 (1979).

# ASH, PRICKLY

**Source:** ***Zanthoxylum americanum*** Mill. or ***Z. clava-herculis*** L. (sometimes erroneously spelled ***Xanthoxylum***) (Family Rutaceae).

*Synonyms*. Northern prickly ash, angelica tree, pepper wood, and yellow wood (*Z. americanum*); southern prickly ash, sea ash, Herculis' club, and prickly yellow wood (*Z. clava-herculis*); toothache tree and xanthoxylum.

## GENERAL DESCRIPTION

Both species are shrubs or small trees, up to about 3 m high, with prickly stems and petioles; native to North America. Northern prickly ash (*Z. americanum*) grows from Quebec south to Mississippi and west to Oklahoma, while southern prickly ash (*Z. clava-herculis*) grows farther to the south, from southern Virginia to Florida, Texas, and perhaps Mexico. Part used is the dried bark.

## CHEMICAL COMPOSITION

Northern prickly ash contains coumarins (xanthyletin, xanthoxyletin, alloxanthoxyletin), alkaloids (laurifoline, nitidine, chelerythrine, tembetarine, magnoflorine, and candicine, with first two in major amounts),[1,2] resins, tannins, and an acrid volatile oil. Other compounds isolated include 8-(3,3-dimethylallyl)-alloxanthoxyletin (a coumarin) from root bark, and two furoquinoline alkaloids ($\gamma$-fagarine and skimmianine) from leaves.[1]

Southern prickly ash contains alkaloids (laurifoline, magnofoline, tembetarine, and candicine in root bark; chelerythrine, nitidine, and tembetarine in stem bark), amides (herculin, neoherculin, and a cinnamamide), pluviatilol $\gamma,\gamma$-dimethylallyl ether, *N*-acetylanonaine, lignans (asarinin and sesamin) (KARRER)[2–4] tannins, resins, and an acrid volatile oil.

Previously thought to be present in prickly ash bark, berberine was not detected in later studies,[1,3] though it has been reported as a major alkaloid in another *Zanthoxylum* species, *Z. monophyllum* Lam.[5]

## PHARMACOLOGY OR BIOLOGICAL ACTIVITIES

Has irritant, diaphoretic, carminative, and antidiarrheal properties (MERCK, USD 23rd).

Asarinin has been reported to have antitubercular activities (MERCK).

Asarinin, sesamin, neoherculin, pluviatilol $\gamma,\gamma$-dimethylallyl ether, and *N*-acetylanonaine are fish poisons.[4]

## USES

**Food.** Extract is reportedly used as a flavor component in major food categories such as alcoholic and nonalcoholic beverages, frozen dairy desserts, candy, baked goods, and gelatins and puddings. Highest average maximum use level is approximately 0.01%.

**Health Food/Herb Teas.** Used in combination with other herbs in capsules, tablets, tinctures, fluid extract, or decoction for increased circulation, antispasmodic in colds, rheumatism, poor digestion, arthritis, and as "blood purifier" (FOSTER AND DUKE; KROCHMAL AND KROCHMAL).

**Traditional Medicine.** Used to treat toothache; as a tonic and a stimulant; both internally and externally to treat rheumatism; as a diaphoretic in fever; to treat sores, ulcers, and cancer (as an ingredient in Hoxsey "cure" in the 1950s); and others.[6,7]

**COMMERCIAL PREPARATIONS**

Crude and extracts; crude was formerly official in N.F. and U.S.P.

*Regulatory Status.* GRAS (§182.20).

**REFERENCES**

See the General References for CLAUS; FEMA; FERNALD; FOSTER AND DUKE; KROCHMAL AND KROCHMAL; LEWIS AND ELVIN-LEWIS; MERCK; USD 23rd.

1. G. C. Sun, *Diss. Anstr. Int.*, *B*, **35**, 5826 (1975).
2. F. Fish et al., *Lloydia*, **38**, 268 (1975).
3. F. Fish and P. G. Waterman, *J. Pharm. Pharmacol.*, **25**(Suppl.), 115P (1973).
4. K. V. Rao and R. Davies, *J. Nat. Prod.*, **49**, 340 (1986).
5. F. R. Stermitz and I. A. Sharifi, *Phytochemistry*, **16**, 2003 (1977).
6. J. F. Morton, *Bull. Med. Libr. Assoc.*, **56**, 161 (1968).
7. J. L. Hartwell, *Lloydia*, **34**, 103 (1971).

# ASPARAGUS

**Source:** ***Asparagus officinalis*** L. (Family Liliaceae).

*Synonym.* Garden asparagus.

**GENERAL DESCRIPTION**

A dioecious perennial herb with erect and much-branched stem, up to 3 m high; leaves scalelike; native to Europe and western Asia, widely cultivated. Aerial stems (asparagus spears) arising from rhizomes are used as vegetable. Fleshy fibrous roots and to a lesser extent seeds are used for medicinal purposes.

**CHEMICAL COMPOSITION**

Roots contain inulin and eight fructo-oligosaccharides;[1] two glycosidic bitter principles (officinalisnin-I and officinalisnin-II, isolated from dried root in 0.12 and 0.075% yields, respectively;[2] $\beta$-sitosterol, sarsasapogenin, and nine steroidal glycosides (named asparagosides A to I, in order of their increasing polarity);[3,4] asparagusic acid;[5] and others.

Shoots (spears and tips) contain sulfur-containing acids, including asparagusic, dihydroasparagusic, and *S*-acetyldihydroasparagusic acids;[5,6] $\alpha$-amino-dimethyl-$\gamma$-butyrothetin (an *S*-methylmethionine derivative), among others;[7] a glycosidic bitter principle that is different from the two found in roots;[8] flavonoids (rutin, hyperoside, isoquercitrin, cosmosiin, kaempferol-3-*O*-L-rhamno-D-glucoside, kaempferol, quercetin, etc.);[9,10] asparagine, arginine, tyrosine, sarsasapogenin, $\beta$-sitosterol, succinic acid, sugars, and others (KARRER, JIANGSU; MERCK).[11]

Asparagusic acid and its derivatives are plant growth inhibitors, inhibiting the growth of lettuce;[6] it also has nematicidal properties, thus being responsible for the resistance of asparagus to several plant parasitic nematodes.[5]

Methylmercaptan (a hydrolysis product of the *S*-containing compounds) or asparagine-aspartic acid monoamide is believed to be present in urine after eating asparagus, causing its peculiar odor (MERCK).

Seeds have been reported to contain large amounts of NaOH-soluble polysaccharides, which are linear molecules consisting of β-glucose and β-mannose 1→4 linked with α-galactose as a terminal group, with a glucose:mannose ratio of 1:1.[12] They also contain 15.3% oil that is composed of 43.47% arachidic, 22.16% oleic, 11.52% palmitic, 11.34% linoleic, 5.78% behenic, 3.59% stearic, and 2.14% linolenic acids as well as 1.43% unsaponifiable matter consisted mostly of β-sitosterol.[13]

## PHARMACOLOGY OR BIOLOGICAL ACTIVITIES

The roots are reported to have diuretic and hypotensive properties (JIANGSU). Animal experiments indicate slight diuretic activity of herb preparations.[14]

Fibers isolated from the vegetable are claimed to have mutagen-adsorbing (cancer-preventing) properties.[15]

## USES

**Medicinal, Pharmaceutical, and Cosmetic.** Roots are used in diuretic preparations. Herb is used in diuretic galenical preparations in Germany, though claimed efficacy is not substantially documented.[14]

**Food.** Seed and root extracts reportedly used in alcoholic beverages at an average maximum use level of 0.0016% (16 ppm). Seeds have been used as coffee substitutes (UPHOF).

**Traditional Medicine.** Roots are used as diuretic, laxative, and in neuritis and rheumatism; also reportedly used in treating cancer.[16] In Chinese medicine, it is used to treat various parasitic diseases among other ailments.

Fruits (berries) have been reported used as contraceptives.[17,18]

Shoots are used in homemade preparations to cleanse face and to dry up pimples and sores (ROSE).

## COMMERCIAL PREPARATION

Crude and extracts.

*Regulatory Status.* Undetermined in the United States. Herb subject of a German therapeutic monograph, though therapeutic use is not recommended (as diuretic) due to insufficient scientific evidence.[14] A positive therapeutic German monograph exists for use of root preparation in irrigation therapy for inflammatory disease of the urinary tract and prevention of renal gravel (contraindicated in inflammatory kidney disease or edema caused by cardiac or renal disease).[19]

## REFERENCES

See the General References for BAILEY 1; BAILEY 2; BIANCHINI AND CORBETTA; BLUMENTHAL; FEMA; GOSSELIN; KARRER; JIANGSU; LEWIS AND ELVIN-LEWIS; UPHOF; WREN.

1. N. Shiomi et al., *Agr. Biol. Chem.*, **40**, 567 (1976).
2. K. Kawano et al., *Agr. Biol. Chem.*, **39**, 1999 (1975).
3. G. M. Goryanu et al., *Khim. Prir. Soedin.*, **3**, 400 (1976); through *Chem. Abstr.*, **85**, 90178w (1976).
4. G. M. Goryanu and P. K. Kintya, *Khim. Prir. Soedin.*, **6**, 762 (1976); through *Chem. Abstr.*, **87**, 53516a (1977).
5. M. Takasugi et al., *Chem. Lett.*, **1**, 43 (1975); through *Chem. Abstr.*, **82**, 121683b (1975).
6. H. Yanagawa et al., *Tetrahedron Lett.*, **25**, 2549 (1972).
7. R. Tressl et al., *J. Agr. Food Chem.*, **25**, 455 (1977).

8. S. Sakamura et al., *Nippon Shokuhin Kogyo Gakkaishi*, **14**, 491 (1967); through *Chem. Abstr.*, **69**, 74477g (1968).
9. M. Woeldecke and K. Herrmann, *Z. Lebensm. Unters. Forsch.*, **155**, 151 (1974).
10. T. Kartnig et al., *Planta Med.*, **51**, 288 (1985).
11. R. Tressl et al., *J. Agr. Food Chem.*, **25**, 459 (1977).
12. R. Goldberg, *Phytochemistry*, **8**, 1783 (1969).
13. Y. R. Prasad and S. S. Nigam, *Proc. Natl. Acad. Sci., India Sect. A*, **52**, 396 (1982).
14. Monograph *Asparagi herba*, *Bundesanzeiger*, no. 127 (July 12, 1991).
15. Y. Sasaki et al., Jpn Kokai Tokkyo Koho JP 6140764 (1986); through *Chem. Abstr.*, **104**, 220607j (1986).
16. J. L. Hartwell, *Lloydia*, **33**, 97 (1970).
17. C. S. Barnes et al., *Lloydia*, **38**, 135 (1975).
18. V. J. Brondegaard, *Planta Med.*, **23**, 167 (1973).
19. Monograph *Asparagi rhizoma*, *Bundesanzeiger*, no. 127 (July 12, 1991).

# ASPIDIUM

**Source:** ***Dryopteris filix-mas*** (L.) Schott or ***D. marginalis*** (L.) Gray (Family Polypodiaceae).

*Synonyms*. European aspidium and male fern (*D. filix-mas*); American aspidium and marginal fern (*D. marginalis*).

## GENERAL DESCRIPTION

Both are stout perennial ferns with fronds up to about 1 m long; *D. filix-mas* is the larger of the two. Sori in *D. filix-mas* are large and near the midvein, while those of *D. marginalis* are small and near the margin. *Dryopteris filix-mas* is found in Europe, Asia, North America, South America, and northern Africa; *D. marginalis* grows in eastern and central North America, from Nova Scotia to Georgia and Kansas. Parts used are the dried rhizomes and stipes.

## CHEMICAL COMPOSITION

Most of the studies were done on *D. filix-mas*, which contains 6.5–15% oleoresin. The oleoresin contains not less than 24% of the active principles filicin which is a collective term for several ether-soluble phloroglucides (filixic acids, flavaspidic acids, para-aspadin, albaspidin, desaspidin, etc.);[1,2] aspidinol; *n*-alkanes from $C_{27}$ to $C_{33}$, with $C_{29}$ and $C_{31}$ in major concentrations; triterpene hydrocarbons (12-hopene, 11,13 (18)-hopadiene, 9(11)-fernene);[3] lignins;[4] volatile oil, resins, and others. Presence of aspidinol and desaspidin was not observed in a more recent study.[2]

*Dryopteris marginalis* contains margaspidin, flavaspidic acids, paraaspidin, phloraspin, and others.[5,6]

## PHARMACOLOGY OR BIOLOGICAL ACTIVITIES

Filicin is anthelmintic (especially against tapeworms), but it is a violent poison if absorbed. Castor oil promotes its absorption in the gut and should never be used with it as a laxative. Poisoning symptoms include nausea, vomiting, bloody diarrhea, dizziness, delirium, tremors, convulsions, coma, respiratory or cardiac failure, visual disturbances leading to temporary or permanent blindness and others.

Margaspidin has been reported to have antiinflammatory activity in rats.[5]

Extracts of *D. filix-mas* are active against vesicular stomatitis virus in monkey cell cultures.[7]

Extracts containing phloroglucinols from *Dryopteris crassirhizoma* Nakai have antimicrobial activities *in vitro*[8] and antitumor activities in transplanted tumors (ARS, $S_{180}$, $U_{14}$, $B_{22}$, etc.) in rats and mice.[9]

### USES

**Medicinal, Pharmaceutical, and Cosmetic.** Oleoresin is used for expulsion of tapeworms along with a saline laxative such as magnesium or sodium sulfate.

**Traditional Medicine.** Has been reported to be used in treating tumors.[10]

### COMMERCIAL PREPARATIONS

Crude and oleoresin. Was formerly official in U.S.P.

### REFERENCES

See the General References for BAILEY 1; CLAUS; GOSSELIN; KARRER; MARTINDALE; MERCK; USD 26th; YOUNGKEN.


1. C. J. Widen, *Helv. Chim. Acta*, **54**, 2824 (1971).
2. M. Guley and T. Soylemezoglu, *Ankara Univ. Eczacilik Fak. Mecm.*, **6**, 214 (1976); through *Chem. Abstr.*, **87**, 193871n (1977).
3. F. Bottari et al., *Phytochemistry*, **11**, 2519 (1972).
4. O. Faix et al., *Holzforschung*, **31**(5), 137 (1977); through *Chem. Abstr.*, **87**, 197273s (1977).
5. H. Otsuka et al., *Takeka Kenkyusho Ho*, **30**, 225 (1971); through *Chem. Abstr.*, **76**, 126q (1972).
6. H. S. Puri et al., *Planta Med.*, **33**, 177 (1978).
7. G. P. Husson et al., *Ann. Pharm. Fr.*, **44**, 41 (1986); through *Chem. Abstr.*, **105**, 108018n (1986).
8. S. A. Vichkanova et al., *Rast. Resur.*, **18**, 93 (1982).
9. D. H. Li et al., *Zhongcaoyao*, **17**(6), 14 (1986).
10. J. L. Hartwell, *Lloydia*, **33**, 288 (1970).


# ASTRAGALUS

**Source:** ***Astragalus membranaceus*** (Fisch.) Bge. (syn. *A. propinguus* B. Schischk.) ***A. mongholicus*** Bge. (syn. *A. membranaceus* (Fish.) Bge. var. *mongholicus* (Bge.) Hsiao), and other Chinese ***Astragalus*** spp. (Family Leguminosae or Fabaceae).

*Synonyms*. Membranous milk vetch (*A. membranaceus*), Mongolian milk vetch (*A. membranaceus* var. *mongholicus*), milk vetch (*Astragalus* spp.), and *huangqi*.

### GENERAL DESCRIPTION

Perennial herbs, up to about 1 m high (normally 0.5–0.8 m). Most are native to northern China and some to high regions such as in Sichuan, Yunnan, and Tibet;

extensively cultivated. Although several *Astragalus* species serve as source of astragalus root, *A. membranaceus* and *A. mongholicus* yield most of the root in commerce and on which most the chemical and pharmacological research has been performed.[1]

Part used is the dried root from 4- to 7-year-old plants collected in the spring before leaves appear or in autumn after they have fallen. After the root is dug up, the crown and rootlets are removed along with dirt and then usually sun dried. The most commonly used forms are raw astragalus (dried root) and cured (honey-treated) astragalus; the former usually comes in slices (size and shape like tongue depressors) which is produced by thoroughly moistening the raw root, cutting into thick slices, and drying; and the latter is produced by frying the sliced root with honey (25–30 parts to 100 parts of root) over medium heat until no longer sticky to the touch.

## CHEMICAL COMPOSITION

Astragalus root contains numerous types of active components that include saponins, polysaccharides, flavonoids, free amino acids and trace minerals.[2] Others include coumarin, folic acid, nicotinic acid, choline, betaine, phenolic acids (ferulic, isoferulic, caffeic and chlorogenic acids, etc.), sitosterol, sucrose, and linoleic and linolenic acids (HU, JIANGSU).[2,3]

There are more than 40 triterpene glycosides (saponins) isolated from roots of *A. membranaceus*, *A. mongholicus*, and other *Astragalus* species, including astragalus saponins I, II, and III; astragalosides I, II, III, IV, V, VI, VII, and VIII; acetylastragaloside I; isoastragalosides I and II; astrasieversianin I, II, III, IV (astragaloside I), V, VI, VII (isoastragaloside II), VIII (astragaloside II), IX, X, XI, XII, XIII, XIV, XV, and XVI; cyclogaleginosides A and B; astramembrannin I (astragaloside IV, astrasieversianin XIV) and astramembrannin II (cyclogaleginoside B); and soyasaponin I. Soyasapogenol B is the aglycone of soyasaponin I and astragaloside VIII; and cycloastragenol (cyclogalegigenin, astramembrangenin) is the aglycone of the other saponins.[3–5]

Polysaccharides (from *A. mongholicus* root) include astragalan I (mol. wt. 36,300; D-glucose:D-galactose:L-arabinose = 1.75 : 1.63 : 1; with a trace of pentose), astragalan II (mol. wt. 12,300; $\alpha$-(1→4)(1→6)-glucan), and astragalan III (mol. wt. 34,600; $\alpha$-(1→4)(1→6)-glucan);[2,6] and AG-1 [$\alpha$-glucan with $\alpha$-(1→4):$\alpha$-(1→6) ratio of 5:2], AG-2 [water-insoluble $\alpha$-(1→4)-glucan], AH-1 (acidic; galacturonic acid/glucuronic acid:glucose:rhamnose:arabinose = 1:0.04:0.02:0.01), and AH-2 (glucose:arabinose = 1:0.15).[2,6,7]

Flavonoids include kaempferol, quercetin, isorhamnetin, calycosin, formononetin, rhamnocitrin, kumatakenin, (3*R*)-2′,3′-dihydroxy-7,4-dimethoxy-isoflavone, L-3-hydroxy-9-methoxypterocarpan, (6a*R*, 11a*R*)-10-hydroxy-3,9-dimethoxypterocarpan, calycosin glucoside, 9,10-dimethoxypterocarpan-3-*O*-$\beta$-D-glucoside and 2′-hydroxy-3′,4′-dimethoxy-isoflavone-7-*O*-$\beta$-D-glucoside, among others.[2,8]

Of over 20 free amino acids identified, asparagine, glutamic acid, canavanine, proline, arginine, $\beta$-aminobutyric acid, aspartic acid, and alanine are present in the highest concentrations; they make up 0.50–1.26% of astragalus root, depending on sources.[2,9,10]

Among more than 20 trace minerals found in astragalus, magnesium (1108–1761 ppm), iron (94–694 ppm), manganese (8–52 ppm), zinc (11–23 ppm), copper (5–9 ppm), rubidium (11–13 ppm), molybdenum (0.1–10 ppm) and chromium (0.3–0.8 ppm) are the major ones, while europium, tantalum, hafnium and thorium are only present in traces, with concentrations not yet determined.[2,11–13]

## PHARMACOLOGY OR BIOLOGICAL ACTIVITIES

Astragalus root is a highly valued Chinese herbal tonic; like ginseng, its pharmacological properties are varied. Its decoctions, alcoholic extracts and/or powder have been reported to have numerous biological activities in humans and experimental animals, including immunopotentiating effects;[14–20] antibacterial and antiviral; promoting nucleic acid synthesis in liver and spleen; elevating and/or reducing cAMP and cGMP levels in blood, liver and spleen of mice;[16] antiinflammatory; cardiovascular effects (hypotensive, vasodilating, etc.);[21] inhibiting experimentally induced hypoglycemia and hyperglycemia in mice;[22] contracting smooth muscles; prolonging life span of silkworm and of cells *in vitro*;[14,23] antioxidant effects (e.g., increasing superoxide dismutase activity); improving learning and memory; promoting cartilage growth *in vitro*; liver protection in experimental hepatitis by preventing liver glycogen reduction caused by carbon tetrachloride treatment; reducing urinary protein in chronic and in experimental nephritis; diuretic; improving stamina; and others (JIANGSU, WANG, ZHOU AND WANG).[24,25] Above effects are not due to a single compound or one single class of compounds but rather to different types of components, with the saponins and polysaccharides playing a major role.[14–26]

Toxicity of astragalus root is very low: 75 g/kg and 100 g/kg per os did not cause adverse effects in mice within 48 h; and the $LD_{50}$ in mice (i.p.) was 40 g/kg. The former oral doses are 375 and 500 times the usual effective diuretic dose in humans (WANG, ZHOU AND WANG).

## USES

**Medicinal, Pharmaceutical, and Cosmetic.** Extracts of astragalus root are used in skin care cosmetics (e.g., hand and facial creams and lotions) for its traditional healing and nourishing as well as vasodilating properties; also used in hair tonics for similar effects (ZHOU).

**Health Food/Herb Teas.** Powdered crude and/or extracts are used singly or in combination with other herbs in capsule, tablet or liquid (syrup or drink) form primarily as a general (*qi*) tonic to improve body resistance (immunity); also used in sliced or tea bag–cut form in tea or soup mix packets (FOSTER AND YUE).

**Traditional Medicine.** One of the major Chinese *qi* (energy) tonics, with a recorded use history of 2000 years. Raw root is traditionally considered to benefit the body's resistance (*yiwei gubiao*), promote diuresis, reduce swelling, promote suppuration (drains pus, *tuo du*), and regenerate tissue or promote muscle growth (*sheng ji*). Cured root is said to reinforce the Middle Burner and replenish the vital energy (*buzhong yiqi*).

Raw astragalus used mainly in spontaneous and night sweating, edema, chronic sores and abscesses, unhealing wounds and ulcers, and painful joints; cured astragalus primarily as an energy (*qi*) tonic to treat general weakness, fatigue, lack of appetite, diarrhea caused by spleen deficiency (*pi xu xie xie*), rectal prolapse, and uterine bleeding. Uses of the two occasionally overlap.

Modern/recent uses include prevention and treatment of the common cold and influenza; stomach ulcer;[27] neurodermatitis; and diabetes, for which high doses (>60g) are sometimes used (CHP, JIANGSU). Astragalus root is also one of the *fuzheng guben* (strengthening body defense therapy) herbs currently being tried in AIDS treatment.[28]

Usual daily oral dose for adults is 9–30 g.

## COMMERCIAL PREPARATIONS

Raw astragalus (readily available as sticks or slices in several grades); powdered crude; and extracts (aqueous, hydroalcoholic, glycolic). Most extracts come without standardized strengths and powdered astragalus may contain adulterants such as starches and preextracted plant materials.

*Regulatory Status.* U.S. regulatory status not clear; sold as dietary supplement.

## REFERENCES

See the General References for CHP; FOSTER AND YUE; HU; JIANGSU; WANG; ZHOU; ZHOU AND WANG.


1. S. S. Sun et al., *Zhongguo Yaoxue Zazhi*, **25**, 643 (1990).
2. Z. S. Qi, *Zhongcaoyao*, **18**(5), 41 (1987).
3. K. He and H. K. Wang, *Yaoxue Xuebao*, **23**, 873 (1988).
4. I. Kitagawa et al., *Chem. Pharm. Bull.*, **31**, 716 (1983).
5. I. Kitagawa et al., *Chem. Pharm. Bull.*, **31**, 689 (1983).
6. S. D. Fang et al., *Youji Huaxue* **1**, 26 (1982); through *Chem. Abstr.*, **96**, 177941r (1982).
7. Q. S. Huang et al., *Yaoxue Xuebao*, **17**, 200 (1982).
8. G. B. Lu et al., *Zhongcaoyao*, **15**(10), 20 (1984).
9. Y. X. Gong, *Zhongcaoyao*, **18**(11), 37 (1987).
10. R. Xiao et al., *Zhongyao Tongbao*, **9**(6), 30 (1984).
11. X. Z. Liu et al., *Chin, J. Integr. Trad. West. Med.*, **5**, 235 (1985).
12. X. Zhao et al., *Chin. J. Integr. Trad. West. Med.*, **8**, 419 (1988).
13. C. L. Ma and S. K. Wang, *Zhongcaoyao*, **16**(6), 4 (1985).
14. Z. D. Du and Y. M. Lai, *Zhongcaoyao*, **19**(9), 40 (1988).
15. B. Y. Du and J. Y. Zhang, *Zhongyao Tongbao*, **12**(6), 53 (1987).
16. C. S. Geng, *Chin, J. Integr. Trad. West. Med.*, **6**, 62 (1986).
17. B. H. S. Lau et al., *Phytother. Res.*, **3**(4), 148 (1989).
18. D. C. Wang et al., *Chin. J. Oncol.*, **11**, 180 (1989).
19. D. T. Chu et al., *J. Clin. Lab. Immunol.*, **25**, 119 (1988).
20. D. T. Chu et al., *J. Clin. Lab. Immunol.*, **25**, 125 (1988).
21. Y. D. Zhang et al., *Yaoxue Xuebao*, **19**, 333 (1984).
22. X. R. Li et al., *Zhongchengyao*, **11**(9), 32 (1989).
23. M. She, *Zhongyao Tongbao*, **10**(6), 41 (1985).
24. Y. Kang et al., *Zhongcaoyao*, **20**(11), 21 (1989).
25. X. R. Li et al., *Zhongchengyao*, **11**(3), 27 (1989).
26. H. Wagner, *Naturheilpraxis*, **3**, 256 (1983).
27. J. Grujic-Vasic et al., *Planta Med.*, **55**, 649 (1989).
28. S. M. Bao, *Zhongguo Zhongyao Zazhi*, **14**(5), 59 (1989).

# ATRACTYLODES (*BAIZHU* AND *CANGZHU*)

*Atractylodes* yields two widely used herbs, *BAIZHU* and *CANGZHU* (see individual entries).

# AVOCADO

**Source:** ***Persea americana*** Mill. (syn. *P. gratissima* Gaertn.; *Laurus persea* L.) (Family Lauraceae).

*Synonyms*. Alligator pear, avocato, and ahuacate.

## GENERAL DESCRIPTION

Large tree up to about 20 m high; fruit large (5–20 cm long) and fleshy, pyriform, ovate, or spherical, with a thin to thick sometimes woody skin; native to tropical America (Mexico, Central America); widely cultivated. Parts used are the fruit and seed.

There are several commercial varieties in the United States, including the Mexican avocado and the West Indian avocado. The Mexican avocado (*P. americana* var. *drymifolia*) is grown in California, while the larger West Indian avocado (*P. americana* var. *americana*) is produced in Florida.[1] Avocado oil (a fixed oil) is produced by expressing the dried pulp of the fruit.

## CHEMICAL COMPOSITION

Pulp contains a fatty oil (4–40%, depending on season, location, climate, etc., but usually about 16%); about 2% protein; 6–9% carbohydrates and sugars (glucose, fructose, D-mannoheptulose, a taloheptulose, and an alloheptulose); two bitter substances (1-acetoxy-2,4-dihydroxyheptadeca-16-ene and 1,2,4-trihydroxyheptadeca-16-ene); carnitine; and others.[2–10]

The pulp oil (avocado oil) consists mainly of glycerides of oleic acid; it also contains highly variable amounts of unsaponifiable matter (1.6–11.3%),[2,3] consisting of sterols (β-sitosterol, campesterol, 24-methylenecycloartanol, citrostadienol, etc.) and hydrocarbons, volatile acids (propionic, butyric, valeric, etc.) amino acids, and vitamin D (higher than in butter or eggs), among others.[11,12]

Seeds contain polyhydric alcohols (volemitol, perseitol, arabinitol, galactitol, myoinositol, D-erythro-D-galacto-octitol, glycerol), rare sugars (D-mannoheptulose, etc.);[13] avocatins (avocadene, avocadyne, avocadenone acetate, avocadynone acetate, avocadenofuran, avocadynofuran, avocadienofuran, isoavocadienofuran, etc.);[14] 4,8″-biscatechin (a condensed flavanol);[15] $C_{17}$ oxygenated aliphatic unsaturated compounds (1,2,4-trihydroxyheptadeca-16-ene, etc.),[6,7,16] some of which have an unpleasant bitter flavor,[6,7] and other usual constituents, including protein and fats.

Mexican avocado leaves have been reported to yield by steam distillation 3.1% of an essential oil that consists of 95% estragole (see ***sweet basil*** for toxicity) and 5% anethole.[17]

The peel contains *l*-epicatechin (an antioxidant);[18] and proanthocyanins isolated from avocado fruit are claimed to be deodorants for kitchens and other rooms.[19]

## PHARMACOLOGY OR BIOLOGICAL ACTIVITIES

Avocado oil is believed to have healing and soothing properties to the skin.

4,8″-Biscatechin, a condensed flavanol isolated from avocado seeds, has been reported to have antitumor activity against Sarcoma 180 in mice and Walker 256 in rats.[15]

$C_{17}$ oxygenated unsaturated aliphatics (especially 1,2,4-trihydroxyheptadeca-16-

ene) isolated from avocado (pulp and seeds) have been shown to be highly bactericidal against Gram-positive bacteria, especially *Staphylococcus aureus*.[16,20]

Poisoning of cattle, horses, goats, rabbits, canaries, and fish by avocado (leaves, fruit, bark, seeds) have been reported (LEWIS AND ELVIN-LEWIS).

## USES

**Medicinal, Pharmaceutical, and Cosmetic.** Pulp oil is used in massage creams, muscle oils, hair products, and others. A pharmaceutical preparation containing the seed oil (nonsaponifiable fraction) has been patented for use in the treatment of sclerosis of the skin, pyorrhea, arthritis, and others.[21] Pulp is used in face creams.

**Food.** Pulp has been used as a food for thousands of years in tropical America; is a good source of vitamin D and potassium.

**Traditional Medicine.** Pulp used as a hair pomade to stimulate hair growth, to hasten suppuration of wounds, and as an aphrodisiac and emmenagogue. Seeds used for dysentery and diarrhea by American Indians and to relieve toothache in the Philippines. Leaves and bark are also similarly used.

## COMMERCIAL PREPARATIONS

Fruit and pulp oil.

## REFERENCES

See the General References for BAILEY 2; MORTON 2; POUCHER; UPHOF; WATT AND MERRILL.

1. L. O. Williams, *Econ. Bot.*, **31**, 315 (1977).
2. T. Itoh et al., *Fruits*, **30**, 687, (1975); through *Chem. Abstr.*, **84**, 86758x (1976).
3. M. H. Bertoni et al., *Ann. Asoc. Quim. Argent.*, **55**, 257, (1967); through *Chem. Abstr.*, **69**, 85524w (1968).
4. G. G. Slater et al., *J. Agr. Food Chem.*, **23**, 468 (1975).
5. D. Pearson, *J. Sci. Food Agr.*, **26**, 207 (1975).
6. G. Ben-Et et al., *J. Food Sci.* **38**, 546 (1973).
7. B. I. Brown, *J. Agr. Food Chem.*, **20**, 753 (1972).
8. I. Johansson and N. K. Richtmyer, *Carbohydr. Res.*, **13**, 461 (1970).
9. J. N. Ogata et al., *J. Agr. Food Chem.*, **20**, 113 (1972).
10. M. Tada et al., *Nippon Eiyo Shokuryo Gakkaishi*, **37**, 13 (1984); through *Chem. Abstr.*, **101**, 71364m (1984).
11. M. A. Joslyn and W. Stepka, *Food Res.*, **14**, 459 (1949).
12. A. Zanobini et al., *Boll. Soc. Ital. Biol. Sper.*, **50**, 887, (1974); through *Chem. Abstr.*, **83**, 7489p (1975).
13. N. K. Richtmyer, *Carbohydr. Res.*, **12**, 135 (1970).
14. H. M. Alves et al., *Ann. Acad. Brasil Cienc.*, **42**(Suppl.), 45 (1970); through *Chem. Abstr.*, **75**, 115910n (1971).
15. M. M. De Oliveira et al., *Ann. Acad. Brasil Cienc.* **44**, 41 (1972); through *Chem. Abstr.*, **78**, 119383c (1973).
16. I. Néeman et al., *Appl. Microbiol.*, **19**, 470 (1970).
17. D. I. Acosta de Iglesias et al., *Riv. Ital. Essenze, Profumi, Piante Offic., Aromi, Saponi, Cosmet., Aerosol*, **58**, 158 (1976); through *Chem. Abstr.*, **85**, 25263f (1976).
18. M. Nose and N. Fujino, *Nippon Shokuhin Kogyo Gakkaishi*, **29**, 507 (1982); through *Chem. Abstr.*, **98**, 3719a (1983).

19. H. Date et al., Jpn. Kokai Tokkyo Koho JP 60160962 (1985); through *Chem. Abstr.*, **104**, 39535n (1986).
20. I. Néeman et al., Fr. Demande 2,075,994 (1971); through *Chem. Abstr.*, **77**, 70928x (1972).
21. H. Thiers, Neth. Appl. 6,601,888 (1966); through *Chem. Abstr.*, **66**, 5777r (1967).

# BALM, LEMON

**Source:** *Melissa officinalis* L. (Family Labiatae or Lamiaceae).

*Synonyms*. Melissa, common balm, bee balm, and balm.

## GENERAL DESCRIPTION

An aromatic (lemony) perennial herb, up to about 1 m high, growing in the Mediterranean region, western Asia, southwestern Siberia, and northern Africa; widely cultivated. Parts used are the dried leaves often with flowering tops; an essential oil is obtained from these by steam distillation.

## CHEMICAL COMPOSITION

Contains about 0.1–0.2% volatile oil composed mainly of oxygenated compounds such as citral (a and b), caryophyllene oxide, citronellal, eugenol acetate, and geraniol, plus smaller amounts of terpene hydrocarbons, including *trans*- and *cis*-$\beta$-ocimene, caryophyllene, $\alpha$-cubebene, copaene, and $\beta$-bourbonene;[1–5] polyphenols (caffeic acid, protocatechuic acid, etc.) and a tannin composed of caffeic acid units;[6–9] flavonoids (luteolin-7-glucoside and rhamnazin), rosmarinic acid,[10] triterpenoids (ursolic acid, etc.), and glucosides of geraniol, nerol, eugenol, benzyl alcohol, $\beta$-phenylethyl alcohol, neric acid and geranic acid, among others (KARRER).[6,9,11,12]

## PHARMACOLOGY OR BIOLOGICAL ACTIVITIES

Hot-water extracts of balm have strong antiviral properties (in egg- and cell-culture systems) against Newcastle disease, mumps, herpes simplex, vaccinia, and other viruses. Polyphenols (other than caffeic acid) and the tannin present have been shown to be responsible for these antiviral properties.[7,8]

Freeze-dried aqueous extracts of balm have antithyrotropic and antigonadotropic activities.[13]

Balm oil has been reported to have antibacterial activities, especially against *Mycobacterium phlei* and *Streptococcus hemolytica*,[5] as well as antifungal activities.[14] It also has antihistaminic and antispasmodic activities in isolated guinea pig ileum and other preparations, with eugenol acetate in the oil being the most effective antispasmodic agent.[5,15]

## USES

**Medicinal, Pharmaceutical, and Cosmetic.** Occasionally used in certain pharmaceutical preparations, as a carminative and mild tranquilizer. Oil more often used as a component in perfumes.

**Food.** Balm extract and oil are used in major categories of food products such as alcoholic (bitters, vermouths, etc.) and nonalcoholic beverages, frozen dairy desserts, candy, baked goods, and gelatins and puddings. Highest average maximum use level reported is 0.5% of extract in baked goods.

**Health Food/Herb Tea.** Cut and sifted herb, powdered herb, liquid and dried extracts for infusions, tinctures, etc., used as mild sleep aid as well as a stomachic. Dried leaves used for tea in doses of 1.5–4.5 g of the herb in infusion. Often used in combination with other herbs.[16]

**Traditional Medicine.** Reportedly used as a carminative, diaphoretic, antispasmodic, stomachic, and emmenagogue in all sorts of ailments, including female discomforts, nervous problems, insomnia, cramps, headache, toothache, sores, tumors, and insect bites. Considered a cure-all; used primarily in an infusion or poultice form (FOSTER).

**COMMERCIAL PREPARATIONS**

Crude; extracts, and oil; oil is seldom unadulterated, (ARCTANDER). Crude formerly official in U.S.P.

*Regulatory Status*. GRAS (§182.10 and §182.20). Formerly official in the U.S.P. from 1840 to 1890. The leaves and preparations thereof are the subject of a positive German therapeutic monograph, indicated for difficulty in falling asleep caused by nervous conditions, and functional gastrointestinal symptoms.[16]

**REFERENCES**


See the General References for ARCTANDER; BIANCHINI AND CORBETTA; BLUMENTHAL; FEMA; FOSTER; GOSSELIN; GUENTHER; LUST; ROSE; UPHOF; YOUNGKEN.

1. F. Enjalbert et al., *Fitoterapia*, **54**, 59 (1983).
2. N. Stankeviciene et al., *Polez. Rast. Priblat. Respub. Beloruss., Mater. Nauch. Konf.*, **2** (1973), 264 (1973); through *Chem. Abstr.*, **81**, 111399d (1974).
3. S. Kapetanovic and S. Dugumovic, *Acta Pharm. Jugosl.*, **18**(304),127 (1968); through *Chem. Abstr.*, **73**, 63152d (1970).
4. F. W. Hefendehl, *Arch. Pharm. (Weinheim)*, **303**, 345 (1970).
5. H. Wagner and L. Sprinkmeyer, *Dtsch. Apoth. Ztg.*, **113**, 1159 (1973).
6. H. Thieme and C. Kitze, *Pharmazie*, **28**, 69 (1973).
7. L. S. Kucera and E. C. Herrmann Jr., *Proc. Soc. Exp. Biol. Med.*, **124**, 865 (1967).
8. E. C. Herrmann, Jr., and L. S. Kucera, *Proc. Soc. Exp. Biol. Med.*, **124**, 869 (1967).
9. I. Morelli, *Boll. Chim. Farm.*, **116**, 334 (1977).
10. U. Gerhardt and A. Schroeter, *Fleischwirtschaft*, **63**, 1628 (1983).
11. A. Mulkens et al., *Pharm. Acta Helv.*, **60**(9–10), 276 (1985); through *Chem. Abstr.*, **103**, 211224j (1985).
12. M. Burgett, *Bee World*, **61**(2), 44 (1980); through *Chem. Abstr.*, **93**, 66820u (1980).
13. M. Auf'mkolk et al., *Endocrinology (Baltimore)*, **115**, 527 (1984).
14. A. Mulkens et al., *Pharm. Acta Helv.*, **60**(9–10), 276 (1985); through *Chem. Abstr.*, **103**, 211224j (1985).
15. A. M. Debelmas and J. Rochat, *Plant. Med. Phytother.*, **1**, 23 (1967).
16. Monograph *Melissae folium*, *Bundesanzeiger*, no. 228 (Dec. 5, 1984; with revision (Mar. 13, 1990).


# BALM OF GILEAD BUDS

**Source:** ***Populus tacamahacca*** Mill, (syn. *P. balsamifera* Du Roi) or ***P. candicans*** Ait. (Family Salicaceae).

*Synonyms*. Poplar buds and balsam poplar buds.

**GENERAL DESCRIPTION**

There has been much confusion regarding the sources of this botanical. In N.F. XI, it was officially described as derived from *P. tacamahacca* Mill. (syn. *P. balsamifera* L.) or from *P. candicans* Ait. However according to SARGENT, *P. balsamifera* L. is not the same as *P. tacamahacca* Mill. but is a different species; he listed *P. balsamifera* Du Roi as the synonym for *P. tacamahacca*.

Balm of Gilead has also been used as a synonym for Canada balsam (*Abies balsamea*) and mecca balsam (*Balsamodendron opobalsamum* Engl.), among others.

*Populus tacamahacca* (balsam poplar) is a tree often up to 33 m high, trunk about 3 m in diameter, with stout, erect branches.

*Populus candicans* (true balm of gilead) has been considered as a variety of the balsam poplar; only the pistillate tree is known.

*Populus balsamifera* L. is described as a large tree with massive spreading branches and stout yellow-brown, often angular branchlets.

All above three species are native to North America (SARGENT).

Balm of Gilead buds are the leaf buds collected in the spring before they open. Precise source is not certain; all above species as well as other *Populus* species are probably used.

## CHEMICAL COMPOSITION

Contains about 2% volatile oil; resins; $C_{25}$, $C_{27}$ and $C_{29}$ *n*-alkanes; salicin and populin; phenolic acids (e.g. caffeic); chalcones; and others. Compounds reportedly present in the volatile oil include *d*-cadinene, cineole, *ar*-curcumene, bisabolene, farnesene, *d*-$\alpha$-bisabolol, $\beta$-phenethyl alcohol, acetophenone (KARRER)[1-3] and humulene ($\alpha$-caryophyllene) (CLAUS, FURIA AND BELLANCA, YOUNGKEN).

## PHARMACOLOGY OR BIOLOGICAL ACTIVITIES

Is considered to have stimulant and expectorant properties.

Salicin (a glucoside of salicyl alcohol) has antipyretic, antirheumatic, analgesic and other properties as well as toxicity of salicylates (GOODMAN AND GILMAN, MARTINDALE).

## USES

**Medicinal, Pharmaceutical, and Cosmetic.** Its major use is in cough preparations often together with white pine and wild cherry barks, bloodroot, and spikenard root, as in white pine compound or its variations.

**Traditional Medicine.** Used for relieving minor aches and pains; in colds and coughs; locally for sores, bruises and cuts, and for healing pimples. Bark and leaves are also similarly used.

## COMMERCIAL PREPARATIONS

Crude and extracts. Crude was formerly official in N.F. and U.S.P.

*Regulatory Status.* Has been approved for food use in alcoholic beverages only (§172.510). Subject of a German therapeutic monograph; allowed in external preparations for superficial skin injuries, hemorrhoids, frost bite, and sunburn.[4]

## REFERENCES

See the General References for ARCTANDER; CLAUS; GOSSELIN; KROCHMAL AND KROCHMAL; LUST; MERCK; POUCHER; ROSE; SARGENT; UPHOF; YOUNGKEN.

1. E. Wollenweber and W. Weber, *Z. Pflanzenphysiol.*, **69**, 125 (1973).
2. S. Frantisek et al., *Collect. Czech. Chem. Commun.*, **18**, 364 (1953).
3. O. Isaac et al., *Dtsch. Apoth. Ztg.*, **108**, 293 (1968).
4. Monograph *Populi gemma*. *Bundesanzeiger*, no. 22 (Feb. 1, 1990).

# BALSAM CANADA

**Source:** ***Abies balsamea*** (L.) Mill. (Family Pinaceae).

*Synonyms*. Canada balsam, Canadian balsam, Canada turpentine, balsam fir, balsam of fir, balsam fir Canada, and balm of Gilead.

## GENERAL DESCRIPTION

Evergreen tree up to 20 m high with trunk usually 30–45 cm in diameter; native to eastern North America, reaching Minnesota and Wisconsin. Canada balsam is not a true balsam, because it does not contain benzoic or cinnamic acid or their esters (see ***glossary***). It is an oleoresin occurring normally in the bark, collected by puncturing the vesicles on the bark. It is a light greenish yellow, viscous liquid that solidifies on exposure to air.

## CHEMICAL COMPOSITION

Contains a volatile oil (up to 30%) and an odorless resin. The volatile oil is composed entirely of monoterpenes (37.4% $\beta$-phellandrene, 36.4% $\beta$-pinene, 23.5% $\alpha$-pinene, and 2.7% $\alpha$-phellandrene) according to one report.[1] The resin makes up the rest of the oleoresin and contains neutral and acidic materials (abietic and neoabietic acids).

## PHARMACOLOGY OR BIOLOGICAL ACTIVITIES

Considered as nontoxic when applied externally.[2] Internal toxicity data not available.

## USES

**Medicinal, Pharmaceutical, and Cosmetic.** Used in certain ointments and creams as antiseptic and as treatment for hemorrhoids. As a fixative or fragrance ingredient in soaps, detergents, creams, lotions, and perfumes. Maximum use levels reported are 0.15% in soaps and 0.2% in perfumes.[2] Used in dentistry as an ingredient in root canal sealers and dentifrices (ADA)[3]

**Food.** Both oleoresin and oil have limited use in major categories of foods, including alcoholic and nonalcoholic beverages, frozen dairy desserts, candy, and gelatins and puddings. Use levels are low, generally below 0.001% (10 ppm).

**Traditional Medicine.** Used externally by American Indians for burns, sores, and cuts and to relieve heart and chest pains. Also reportedly used in treating tumors.[4]

**Others.** Due to its ability to dry to a brittle, clear glasslike residue, Canada balsam (usually freed from volatile oil and dissolved in xylene) is extensively used as a cement for lenses and prepared microscopic slides. Its use in some "balsam" hair grooming products probably takes advantage of this property to stiffen hair and give it "body."

## COMMERCIAL PREPARATIONS

Oleoresin and oil. Oleoresin was formerly official in U.S.P.

*Regulatory Status*. Only needles and twigs of *A. balsamea* and their appropriate derivatives have been approved for food use (§172.510).

**REFERENCES**

See the General References for ARCTANDER; CLAUS; FEMA; GUENTHER; KROCHMAL AND KROCHMAL; MERCK; SARGENT; UPHOF.

1. H. J. Petrowitz et al., *Riechst., Aromen Körperpflegem.*, **12**, 1 (1962).
2. D. L. J. Opdyke, *Food Cosmet. Toxicol.*, **13**, 449 (1975).
3. K. M. Kosti, U.S. Pat 4,348,378 (1982).
4. J. L. Hartwell, *Lloydia*, **33**, 288 (1970).

# BALSAM COPAIBA

**Source:** ***Copaifera officinalis*** (Jacq.) L. or other South American ***Copaifera*** species (Family Leguminosae or Fabaceae).

*Synonyms.* Copaiba, copaiva, and Jesuit's balsam.

## GENERAL DESCRIPTION

Trees considerably branched, up to 18 m high, growing in tropical South America (particularly Brazil, Colombia, and Venezuela). Copaiba balsam is an oleoresin that accumulates in cavities within the tree trunk and is tapped by drilling holes into the wood of the trunk; it is not a true balsam (see ***glossary***).

Copaiba oil is obtained by direct vacuum distillation of the oleoresin containing large amounts of the volatile oil (60–90%) (ARCTANDER).

## CHEMICAL COMPOSITION

Copaiba balsam contains 30–90% volatile oil; the rest being resins and acids (ARCTANDER, YOUNGKEN).[1] Compounds reported in the oleoresin include caryophyllene (major component of the volatile oil), copaene, $\beta$-bisabolene, $\gamma$-humulene, caryophyllene oxide, $\alpha$-ylangene, $\alpha$-multijugenol;[2,3] and terpenic acids such as copalic, copaiferic, copaiferolic, Hardwick, 7-hydroxy-Hardwick, enantioagathic, and eperu-8(20)-en-15,18-dioic acids, among others.[3–6]

## PHARMACOLOGY OR BIOLOGICAL ACTIVITIES

Copaiba is believed to have diuretic, expectorant, disinfectant, and stimulant properties.

The oil has been reported to exhibit antibacterial activity.[1,7]

The oleoresin from Brazilian *Copaifera* species showed marked antiinflammatory activity using various animal experimental models.[8]

Available data indicate it to be relatively nontoxic,[1,9] though large doses are reported to cause vomiting and diarrhea as well as measles-like rash (MARTINDALE, SAX).

## USES

**Medicinal, Pharmaceutical, and Cosmetic.** Oleoresin and oil occasionally used in pharmaceutical preparations (diuretics, cough medicines, etc.). Both copaiba balsam and oil are widely used in cosmetic preparations (soaps, bubble baths, detergents, creams, lotions, and perfumes), the former primarily as a fixative while the latter as a fragrance component. Maximum use level reported for both is 0.8% in perfumes.[1,9,10]

**Food.** Occasionally used as a flavor component in most major categories of foods, including alcoholic and nonalcoholic beverages, frozen dairy desserts, candy, baked goods, gelatins and puddings, and meat and meat products. Average maximum use levels are usually very low, less than 0.002% (16 ppm).

**Traditional Medicine.** Has been used for several centuries in Europe in the treatment of chronic cystitis and bronchitis; and in treating hemorrhoids, chronic diarrhea, and others. Also used in Latin America for same types of conditions (MORTON 1).

### COMMERCIAL PREPARATIONS

Oleoresin ("balsam") and oil; copaiba balsam was formerly official in N.F., and the oil is official in F.C.C.

*Regulatory Status.* Both oleoresin and oil have been approved for food use (§172.510).

### REFERENCES

See the General References for ARCTANDER; CLAUS; FEMA; GUENTHER; LEWIS AND ELVIN-LEWIS; MARTINDALE; MORTON 1; TERRELL; YOUNGKEN.

1. D. L. J. Opdyke, *Food Cosmet. Toxicol.*, **14**(Suppl.), 687 (1976).
2. G. Delle Monache et al., *Tetrahedron Lett.*, **8**, 659 (1971).
3. M. Ferrari et al., *Phytochemistry*, **10**, 905 (1971).
4. F. Delle Monache et al., *Ann. Chim. (Rome),* **59**, 539 (1969); through *Chem. Abstr.*, **71**, 124707w (1969).
5. F. Delle Monache et al., *Ann. Chim. (Rome),* **60**, 233 (1970); through *Chem. Abstr.*, **73**, 25690v (1970).
6. J. R. Mahajan and G. A. L. Ferreira, *Ann. Acad. Brasil. Cienc.*, **43**, 611 (1971); through *Chem. Abstr.*, **77**, 140339g (1972).
7. J. C. Maruzzella and N. A. Sicurella, *J. Am. Pharm. Assoc.*, **49**, 692 (1960).
8. A. C. Basile, et al., *J. Ethnopharmacol.*, **22**, 101 (1988).
9. D. L. J. Opdyke, *Food Cosmet. Toxicol.*, **11**, 1075 (1973).
10. M. J. Del Nunzio, *Aerosol Cosmet.*, **7**(41), 7 (1985); through *Chem. Abstr.*, **104**, 56164q (1986).

# BALSAM FIR OREGON

**Source:** ***Pseudotsuga taxifolia*** (Lam.) Britt. (syn. P. *douglasii* Carr.; *P. mucronata* (Raf.) Sudw.) (Family Pinaceae).

*Synonyms.* Oregon balsam and balsam Oregon.

### GENERAL DESCRIPTION

Tree commonly known as Douglas fir, Douglas spruce, or red fir, varying greatly in size, depending on localities, often up to 60 m high with a trunk of about 1 m in diameter; native to western North America, particularly the Pacific Coast (California, Oregon, Washington, and British Columbia); cultivated in Europe.

Oregon balsam is an oleoresin that occurs in the tree trunk and is usually collected from felled trees. It is a light amber or yellow, viscous liquid with a piney odor. It is not a true balsam (see ***glossary***).

**CHEMICAL COMPOSITION**

Although considerable chemical data are available on the needles, wood, and bark (KARRER)[1–6] chemical information on the "balsam" (oleoresin) itself is practically nonexistent, except that it has been reported to be an oleoresin of the turpentine type, yielding a volatile oil on steam distillation and has properties and uses similar to those of Canada balsam (ARCTANDER, CLAUS).

**PHARMACOLOGY OR BIOLOGICAL ACTIVITIES**

Data on its pharmacological or toxicological properties are not available.

**USES**

Similar to those of Canada balsam, and it has been used as an adulterant of the latter oleoresin. However, Oregon balsam is not suited for use in microscopy, because it is less viscous and slower drying than Canada balsam; it does not dry to a glassy and brittle film as Canada balsam.

It is not reportedly used in foods.

**COMMERCIAL PREPARATION**

Oleoresin ("balsam").

**REFERENCES**

See the General References for ARCTANDER; GOSSELIN; GUENTHER; SARGENT; UPHOF; YOUNGKEN.

1. E. Von Rudloff, *Can. J. Bot.*, **50**, 1025 (1972).
2. E. Von Rudloff, *Pure Appl. Chem.* **34**, 401 (1973).
3. H. M. Graham and E. F. Kurth, *Ind. Eng. Chem.* **41**, 409 (1949); through *Chem. Abstr.*, **43**, 3195a (1949).
4. E. F. Kurth and H. J. Klefer, *Tappi*, **33**, 183 (1950); through *Chem. Abstr.*, **44**, 8681h (1950).
5. H. L. Hergert and E. F. Kurth, *Tappi*, **35**, 59 (1952); through *Chem. Abstr.*, **46**, 6823h (1952).
6. R. D. Kolesnikova et al., *Khim. Prir. Soedin.*, **5**, 613 (1976); through *Chem. Abstr.*, **86**, 127087q (1977).

# BALSAM PERU

**Source:** ***Myroxylon pereirae*** (Royle) Klotzsch (syn. *M. balsamum* var. *pereirae* (Royle) Harms) (Family Leguminosae or Fabaceae).

*Synonyms*. Peru balsam, Peruvian balsam, Indian balsam, and black balsam.

**GENERAL DESCRIPTION**

*Myroxylon pereirae* is a large tree, up to about 25 m high, native to Central America. Balsam Peru is a pathological product obtained from the exposed wood after strips of bark are removed from the tree trunk. The exudation is soaked up by rags wrapped around the trunk that are then boiled with water. The balsam sinks to the bottom and is separated.

In its commercial crude form, balsam Peru is a dark-brown, viscous liquid, reddish brown and transparent in thin layers, with an aromatic vanilla-like odor and a bitter acrid taste. It is a true balsam (see ***glossary***).

## CHEMICAL COMPOSITION

Contains 50–64% of a high-boiling volatile oil, referred to as cinnamein; and 20–28% resin. The volatile oil consists mainly of benzoic and cinnamic acid esters such as benzyl benzoate, benzyl cinnamate, and cinnamyl cinnamate (styracin), with small amounts of nerolidol, free benzyl alcohol, and free benzoic and cinnamic acids also present. Other constituents include traces of styrene, vanillin, and coumarin (KARRER, REMINGTON).

## PHARMACOLOGY OR BIOLOGICAL ACTIVITIES

Balsam Peru has mild antiseptic and antibacterial properties and is believed to promote the growth of epithelial cells; also antiparasitic (especially for scabies).[1]

It is one of the most common contact allergens, and dermatitis as a result of contact with this balsam has been well documented.[2–4]

## USES

**Medicinal, Pharmaceutical, and Cosmetic.** Balsam Peru is used extensively in topical preparations for the treatment of wounds, indolent ulcers, scabies, diaper rash, hemorrhoids, anal pruritus, bedsores, intertrigo, and others; in hair tonic and antidandruff preparations, feminine hygiene sprays, and as a fixative or fragrance ingredient in soaps, detergents, creams, lotions, and perfumes, with maximum use level up to 0.8% in perfumes. Balsam Peru oil and resinoid, obtained by high-vacuum distillation and/or solvent extraction of the balsam, are also used in cosmetics.[2,5]

Balsam Peru is also used in dental preparations, especially for the treatment of dry socket (postextraction alveolitis) and as a component in certain dental impression materials, and dentifrices.[6]

**Food.** Balsam and oil are extensively used as a flavor ingredient in major categories of foods, including alcoholic and nonalcoholic beverages, frozen dairy desserts, candy, baked goods, and gelatins and puddings, with the highest reported average maximum use level of about 0.0015% (15.33 ppm) for the balsam in candy.

**Traditional Medicine.** Reportedly used in treating cancer.[7]

## COMMERCIAL PREPARATIONS

Balsam, resinoid, and oil. Balsam Peru was formerly official in N.F. XII.

*Regulatory Status.* GRAS (§182.20). Subject of a German therapeutic monograph; externally preparations allowed for infected or poorly healing wounds, burns, ulcers, frost-bite, ulcus cruris, bruises, hemorrhoids.[1]

## REFERENCES

See the General References for ADA; ARCTANDER; BLUMENTHAL; FEMA; GOSSELIN; GUENTHER; JIANGSU; MARTINDALE; PHILLIPS; TERRELL; YOUNGKEN.

1. Monograph *Balsamum peruvianum*, *Bundesanzeiger*, no. 173 (Sept. 18, 1986).
2. D. L. J. Opdyke, *Food Cosmet. Toxicol.* **12**(Suppl.), 951 (1974).
3. H. Ebner, *Hautarzt*, **25**, 123 (1974); through *Food Cosmet. Toxicol.*, **13**, 484 (1975).
4. K. E. Malten et al., *Dermatologica*, **147**, 241 (1973); through *Chem. Abstr.*, **80**, 94133x (1974).
5. D. L. J. Opdyke, *Food Cosmet. Toxicol.*, **12** (Suppl.), 953 (1974).
6. K. M. Kosti, U.S. Pat. 4,348,378 (1982).
7. J. L. Hartwell, *Lloydia*, **33**, 97 (1970).

# BALSAM TOLU

**Source:** ***Myroxylon balsamum*** (L.) Harms (syn. *M. toluiferum* H.B.K.; *Toluiferum balsamum* L.) (Family Leguminosae or Fabaceae).

*Synonyms*. Tolu balsam, Thomas balsam, resin tolu, and opobalsam.

## GENERAL DESCRIPTION

*Myroxylon balsamum* is a tall tree native to northern South America (Colombia, Peru, Venezuela), cultivated in the West Indies. Balsam tolu is a pathological product obtained from the tree trunk by making V-shaped incisions through the bark and sap wood. The liquid balsam is collected in gourds and solidifies on aging. It is a plastic solid that darkens and hardens on aging. Like balsam Peru, it is a true balsam.

Tolu balsam has an aromatic vanillalike odor and an aromatic, mildly pungent taste; it is insoluble in water but soluble in alcohol, acetone, benzene, and chlorinated hydrocarbons.

## CHEMICAL COMPOSITION

Reported to contain resin, free cinnamic and benzoic acids, and volatile oil composed mainly of esters of these acids with small amounts of terpenes. Concentrations of these constituents vary greatly with each report, probably due to the great differences in quality of commercial products and the lack of requirements for determining specific components in official compendia (e.g., U.S.P.). Other constituents present include triterpene acids (oleanolic acid, sumaresinolic acid, etc.), other triterpenoids (e.g., 20R, 24$\xi_2$-ocotillone), and traces of eugenol and vanillin, among many other compounds (CLAUS, KARRER, MERCK, AND REMINGTON).[1–4] None of the recent reports identified with certainty the sources of the samples investigated, with the exception of some museum specimens that were shown to contain benzyl cinnamate not found in modern commercial samples.[2,5]

## PHARMACOLOGY OR BIOLOGICAL ACTIVITIES

It has mild antiseptic and expectorant properties. Allergic reactions to tolu balsam occur in some individuals.[3]

## USES

**Medicinal, Pharmaceutical, and Cosmetic.** Balsam Tolu is extensively used as a flavor and mild expectorant in cough medicines (e.g., syrups, lozenges, etc.); also used as an ingredient in Compound Benzoin Tincture or similar formulations for treatment of bedsores, cracked nipples, lips, and minor cuts on the skin and for inhalation to treat laryngitis and croup. Both balsam and its oil (obtained by dry or steam distillation) are used as a fixative or fragrance ingredient in cosmetics including soaps, detergents, creams, lotions, and perfumes, with maximum use levels of 0.1% in soaps and 0.2% in perfumes reported.[3]

**Food.** Used as a flavor ingredient in chewing gum and other major categories of food products, such as alcoholic and nonalcoholic beverages, frozen dairy desserts, candy, baked goods, and gelatins and puddings. Use levels are usually rather low.

**Traditional Medicine.** Reportedly used in treating cancer.[6]

## COMMERCIAL PREPARATIONS

Crude, resinoid, extracts, and tincture. Tincture is official in N.F.

*Regulatory Status*. Approved for food use (§172.510). Subject of a German therapeutic monograph;[7] allowed in preparations with a mean daily dose of 0.6 g for catarrhs of the respiratory tract.

## REFERENCES

See the General References for ARCTANDER; BLUMENTHAL; CLAUS; FEMA; GUENTHER; UPHOF; USD 26th; YOUNGKEN.

1. I. Wahlberg et al., *Acta Chem. Scand.*, **25**, 3285 (1971).
2. K. J. Harkiss and P. A. Linley, *J. Pharm. Pharmacol.*, **25**(Suppl.), 146P (1973).
3. D. L. J. Opdyke, *Food Cosmet. Toxicol.*, **14**(Suppl.), 689 (1976).
4. I. Wahlberg and C. R. Enzell, *Acta Chem. Scand.*, **25**, 70 (1971).
5. K. J. Harkiss, personal communication.
6. J. L. Hartwell, *Lloydia*, **33**, 97 (1970).
7. Monograph *Balsamum tolutanum*, *Bundesanzeiger*, no. 173 (Sept. 18, 1986).

# BARBERRY

**Source:** ***Berberis vulgaris*** L., ***B. aquifolium*** Pursh (syn. *Mahonia aquifolium* Nutt.). or other ***Berberis*** species (Family Berberidaceae).

*Synonyms*. Common barberry, European barberry (B. vulgaris); Oregon grape, trailing mahonia (*B. aquifolium*), berberis.

### GENERAL DESCRIPTION

*Berberis vulgaris* is a deciduous spiny shrub that may reach 5 m in height; native to Europe, naturalized in eastern North America.

*Berberis aquifolium* is an evergreen spineless shrub, 1–2 m high; native to the Rocky Mountains, extending to British Columbia and California.

Parts used are the dried rhizome and roots (*B. aquifolium*) and barks of stem and root (*B. vulgaris*).

### CHEMICAL COMPOSITION

*Berberis* spp. are rich in isoquinoline alkaloids; those reportedly present in *B. vulgaris* include berberine, berbamine, oxyacanthine, jatrorrhizine, columbamine, palmatine, isotetrandine (berbamine methyl ether), bervulcine, and magnoflorine;[1] those present in *B. aquifolium* include aromoline, obamegine, oxyberberine, berbamine, and oxyacanthine.[2] Other constituents in *Berberis* spp. include tannin, resin, fat, and starch.

Berberine and other *Berberis* alkaloids were reported to be toxic to seedlings of dog rose and horsechestnut, causing atrophy.[3]

### PHARMACOLOGY OR BIOLOGICAL ACTIVITIES

Certain *Berberis* alkaloidal salts, particularly berberine, oxyacanthine, and columbamine, have bactericidal activities.[4,5] In one study berberine chloride was demonstrated to have higher activity than chloramphenicol against *Staphylococcus epidermidis*, *Neisseria meningitidis*, *Escherichia coli*, and other bacteria. Oxyacanthine chloride at 0.01% and columbamine chloride at 1.0% killed *Bacillus subtilis* and *Colpidium colpoda*.[5]

Some reports have stated berberine sulfate to be amebicidal and trypanocidal and berbamine, isotetrandine, and hydroxyacanthine to have hypotensive properties (GLASBY 1, MARTINDALE).[6]

Berberine has antifibrillatory activity, elevating the ventricular fibrillation threshold to electrical stimulation in anesthetized cats.[7]

Berberine also has anticonvulsant, sedative, and uterine stimulant activities (see ***goldenseal***).

A fraction of a root extract of *B. vulgaris* containing 80% berbamine and three unidentified isoquinoline alkaloids has been reported to have spasmolytic effects on smooth muscles.[8]

A recent study of berbamine in mice infected with influenza viruses yielded results that indicate it to be an immunostimulating agent.[9] It also exhibits various cardiovascular effects, including hypotensive and antiarrhythmic.[10]

## USES

**Medicinal, Pharmaceutical, and Cosmetic.** Berberis is used as an ingredient in certain tonic preparations. Berberine salts are used in ophthalmic products, usually in eyedrops and eyewashes.

**Health Food/Herb Teas.** Crude root, cut and sifted and powdered used in capsules, teas, and other products, primarily as a bitter tonic (FOSTER AND DUKE).

**Traditional Medicine.** In Europe and the United States, *Berberis* reportedly used as a bitter tonic, antipyretic, and antihemorrhagic, usually referring to *B. vulgaris*. In China, various *Berberis* species are also used for similar purposes, and *B. vulgaris* has actually been listed as a related drug, sometimes used as an adulterant, to *Dichroa febrifuga* Lour., which is widely used in China for its antimalarial and antipyretic activities (NANJING).

## COMMERCIAL PREPARATIONS

Barberry (*B. vulgaris*) is available as crude and as extracts. Qualities of extracts may vary, because the only standards for potency are strengths (see ***glossary***) based on a weight to weight ratio of extract to crude drug. Crude was formerly official in U.S.P.

Berberine salts are readily available.

*Regulatory Status.* Root of *B. vulgaris* is the subject of a German therapeutic monograph.[11]

## REFERENCES

See the General References for BAILEY 2; BLUMENTHAL; CLAUS; FARNSWORTH 3; FOGARTY; FOSTER AND DUKE; GOSSELIN; JIANGSU; MERCK; NANJING.

1. M. Ikram, *Planta Med.*, **28**, 353 (1975).
2. D. Kostalova et al., *Chem. Pap.*, **40**, 389 (1986).
3. V. M. Oleksevich, *Introd. Eksp. Ekol. Rosl.* **1**, 224 (1972); through *Chem. Abstr.*, **82**, 52393t (1975).
4. Z. Kowalewski et al., *Arch. Immunol. Ther. Exp.*, **20**, 353 (1972); through *Chem. Abstr.*, **77**, 135606n (1972).
5. E. Andronescu et al., *Clujul Med.*, **46**, 627 (1973); through *Chem. Abstr.*, **81**, 100062n (1974).
6. L. P. Naidovich et al., *Farmatsiya (Moscow)*, **25**, 33 (1976); through *Chem. Abstr.*, **86**, 114974b (1977).
7. D. C. Fang et al., *Zhongguo Yaoli Xuebao*, **7**, 321 (1986).
8. P. Manalov et al., *Eksp. Med. Morfol.*, **24**, 41 (1985); through *Chem. Abstr.*, **103**, 189503a (1985).
9. L. Jin and W. Z. Sui, *Zhongguo Yaoli Xuebao.*, **7**, 475 (1986).
10. F. L. Li, *Yaoxue Xuebao*, **20**, 859 (1985).
11. Monograph *Berberis vulgaris*, *Bundesanzeiger*, no. 43 (March 2, 1989).

# BASIL, SWEET

**Source:** ***Ocimum basilicum*** L. (Family Labiatae or Lamiaceae).

*Synonyms*. Basil and common basil.

## GENERAL DESCRIPTION

Annual herb, about 0.5 m high, thought to be native to Africa and tropical Asia; cultivated worldwide (e.g., Europe, India, and the United States). There are many varieties, some of which have different compositions and flavoring characteristics. The plant is also strongly affected by environmental factors such as temperature, geographic location, soil, and amount of rainfall.[1,2] Parts used are the dried leaves and flowering tops.

An essential oil is obtained by steam distillation. There are two major types of commercial basil oils, namely, the true sweet basil oil and the so-called exotic, or Reunion, basil oil. True sweet basil oil is distilled in Europe and the United States; exotic basil oil is produced in the Comoro Islands, the Seychelles, and the Malagasy Republic. The two differ mainly in their contents of *d*-camphor, linalool, and methyl chavicol (estragole). Generally, the former does not contain camphor and the latter contains little or no linalool; also the former is levorotatory and the latter is dextrorotatory.

## CHEMICAL COMPOSITION

The volatile oil (ca. 0.08%) contains *d*-linalool and methyl chavicol as the major components, with the former up to 55% and the latter about 70%, depending on the sources (MASADA).[3,4] Other components include methyl cinnamate (which has been reported to be the major component (ca. 28%) of a variety of sweet basil), 1,8-cineole, eugenol, borneol, ocimene, geraniol, anethole; 10-cadinols, $\beta$-caryophyllene, $\alpha$-terpineol, camphor, 3-octanone, methyleugenol, safrole, sesquithujene, and 1-epibicyclosesquiphellandrene as well as juvocimene 1 and juvocimene 2, which are potent juvenile hormone mimics (JIANGSU).[4–8] There are great variations in concentrations of these components in the volatile oils from different sources.

Other constituents present in sweet basil include protein (14%), carbohydrates (61%), and relatively high concentrations of vitamins A and C (MARSH), rosmarinic acid (an antioxidant);[9,10] and thymol and xanthomicrol (a flavone).[11]

## PHARMACOLOGY OR BIOLOGICAL ACTIVITIES

The volatile oil of a variety of sweet basil was shown to have antiwormal activities. This study was prompted by the reported use of the fresh juice of this plant to treat a maggots-infested nasal disease in India.[12]

Methyl cinnamate, methyl chavicol and, to a lesser extent, ocimene, cineole, and linalool have insecticidal activities.[13]

Sweet basil oil is reported to be nontoxic.[14]

Estragole (methyl chavicol), a major component in some sweet basil oils, has been shown to produce tumors (hepatocellular carcinomas) in mice.[15]

Xanthomicrol has been reported to have cytotoxic and antineoplastic activities.[11]

## USES

**Medicinal, Pharmaceutical, and Cosmetic.** Used as a fragrance ingredient in perfumes, soaps, hair dressings, dental creams, and mouth washes.

**Food.** Used as a spice and in chartreuse liqueur.

The oil and oleoresin are extensively used as a flavor ingredient in all major food products, usually in rather low use levels (mostly below 0.005%).

**Health Food/Herb Teas.** Fresh herb now widely available in United States; considered by some a source of vitamin C (FOSTER).

**Traditional Medicine.** Reportedly used for head colds and as a cure for warts and worms, as an appetite stimulant, carminative, and diuretic, among others.[16]

More widely used as a medicinal herb in the Far East, especially in China and India. It was first described in a major Chinese herbal around A.D. 1060 and has since been used in China for spasms of the stomach and kidney ailments, among others; it is especially recommended for use before and after parturition to promote blood circulation. The whole herb is also used to treat snakebite and insect bites (JIANGSU, NANJING).

## COMMERCIAL PREPARATIONS

Crude, essential oil, and oleoresin.

*Regulatory Status*. GRAS (§182.10 and §182.20). Subject of a German therapeutic monograph; claimed efficacies not well substantiated; allowed as flavor corrigent at 5% or less.

## REFERENCES

See the General References for ARCTANDER; BLUMENTHAL; FEMA; FOSTER; GUENTHER; JIANGSU; MASADA; MORTON 1; NANJING; ROSENGARTEN; TERRELL; YOUNGKEN.


1. D. Pogany, *Diss. Abstr., B*, **28**, 1871 (1967).
2. S. N. Sobti et al., *Lloydia*, **41**, 50 (1978).
3. B. C. Gulati et al., *Parfüm. Kosmet.*, **58**, 165 (1977).
4. S. S. Nigam and A. K. Rao, *Riechst., Aromen, Körperpflegem.*, **18**, 169 (1968).
5. S. J. Terhune et al., paper presented at the 6th International Congress of Essential Oils, 1974, p. 153.
6. M. S. Karawya et al., *J. Agr. Food Chem.*, **22**, 520 (1974).
7. J. W. Hogg et al., *Am. Perfum. Cosmet.*, **86**, 33 (1971).
8. W. S. Bowers and R. Nishida, *Science*, **209**(4460), 1030 (1980).
9. U. Gerhardt and A. Schroeter, *Fleischwirtschaft*, **63**, 1628 (1983).
10. A. Reschke, *Z. Lebensm. Unters. Forsch.*, **176**, 116 (1983).
11. M. O. Fatope and Y. Takeda, *Planta Med.*, **54**, 190 (1988).
12. M. L. Jain and S.R. Jain, *Planta Med.*, **22**, 66 (1972).
13. R. S. Deshpande and H. P. Tipnis, *Pesticides*, **11**(5), 11 (1977).
14. D. L. J. Opdyke, *Food Cosmet. Toxicol.*, **11**, 867 (1973).
15. N. R. Drinkwater et al., *J. Natl. Cancer Inst.*, **57**, 1323 (1976).
16. Monograph *Basicilici herba*, *Bundesanzeiger*, (March 18, 1992).


# BAY, SWEET

**Source:** ***Laurus nobilis*** L. (Family Lauraceae).

*Synonyms*. Bay, laurel, bay laurel, Grecian laurel, true bay, and Mediterranean bay.

## GENERAL DESCRIPTION

*Laurus nobilis* is an evergreen tree, up to 20 m high, native to the Mediterranean region; extensively cultivated. Part used is the dried leaf.

An essential oil, commonly known as

laurel leaf oil, is produced by steam distillation of the leaves and branchlets.

Its leafy branchlets were used in wreaths by the ancient Greeks and Romans to crown their victors.

There are several botanicals known under the name of *bay*. For example, West Indian bay is *Pimenta racemosa* (Mill.) J. W. Moore, and California bay is *Umbellularia californica* Nutt. The word *bay* in the literature may mean any one of these botanicals, among others.

**CHEMICAL COMPOSITION**

Contains 0.3–3.1% volatile oil that is composed mainly of cineole (30–50%), $\alpha$-pinene (ca. 12%), linalool (ca. 11%), $\alpha$-terpineol acetate (ca. 10%), $\alpha$-terpineol, $\beta$-pinene, sabinene, limonene, methyl eugenol (3,4-dimethoxyallylbenzene), eugenol, *p*-cymene, camphene, and dehydro-1,8-cineole as well as phenylhydrazine, piperidine, and geraniol.[1] Oil content is highest in autumn and lowest in spring, with old leaves containing the most oil.[2] Other constituents reported include costunolide, laurenobiolide (germacranolides), catechins, proanthocyanidins,[3] alkaloids (reticuline, boldine, launobine, isodomesticine, neolitsine, nandigerine, etc.)[4] and plant acids (e.g., butyric, caproic, enanthic acids, etc.).[5–11]

**PHARMACOLOGY OR BIOLOGICAL ACTIVITIES**

Methyl eugenol, a constituent of sweet bay and California bay oils (at 4 and 5.4%, respectively) as well as a variety of West Indian bay and other species, has been reported to have sedative and narcotic properties in mice, producing sedation at low doses and reversible narcosis at higher doses; it prevented the death of mice treated with lethal convulsant doses of strychnine.[8]

The essential oil has been reported to have bactericidal and fungicidal properties; it also depressed the heart rate and lowered blood pressure in animals. Formulations containing sweet bay leaf and its volatile oils have been claimed to have antidandruff activities.

Allergic reactions (contact dermatitis) to sweet bay have been documented.[12,13]

Bay leaf and some of its volatile compounds (esp. cineole, phenylhydrazine, geraniol and piperidine) have been shown to repel cockroaches.[14]

**USES**

**Medicinal, Pharmaceutical, and Cosmetic.** The oil is used mainly as a fragrance ingredient in creams, lotions, perfumes, soaps, and detergents. Maximum use level reported is 0.2% in perfumes.[12]

**Food.** Sweet bay is a common household spice known as bay leaf.

Both the spice and oil are extensively used in processed foods, including alcoholic (oil only) and nonalcoholic beverages, frozen dairy desserts, baked goods, meat and meat products, condiments and relishes, and others. Use levels are generally low; highest reported are in condiments and relishes, which are 0.1% for the spice and 0.02% for the oil.

**Traditional Medicine.** Has been reportedly used in cancer[15] and as a cholagogic, general stimulant, carminative, and diaphoretic.

**COMMERCIAL PREPARATIONS**

Crude and oil.

*Regulatory Status.* GRAS (§182.10 and §182.20).

## REFERENCES

See the General References for ARCTANDER; BAILEY 1; BIANCHINI AND CORBETTA; FEMA; GUENTHER; JIANGSU; MASADA; ROSENGARTEN; TERRELL; UPHOF; YOUNGKEN.

1. H. Hokwerda et al., *Planta Med.*, **44**, 116 (1982).
2. Z. Putievsky et al., *Isr. J. Bot.*, **33**, 47 (1984).
3. J. M. Schulz and K. Hermann, *Z. Lebensm. Unters. Forsch.*, **171**, 278 (1980).
4. B. Pech and J. Bruneton, *J. Nat. Prod.*, **45**, 560 (1982).
5. U. Asllani, *Bull. Univ. Shteteror Tiranes, Ser. Shkencat Natyr.*, **23**, 93 (1969).
6. N. A. Gugunava, *Subtrop. Kul't.*, **3**, 84 (1971); through *Chem. Abstr.*, **76**, 138336q (1972).
7. M. G. Pertoldi and B. Stancher, *Atti Congr. Qual.*, **6th**, 303 (1967); through *Chem. Abstr.*, **73**, 91168g (1970).
8. J. T. MacGregor et al., *J. Agr. Food Chem.*, **22**, 777 (1974).
9. J. W. Hogg et al., *Phytochemistry*, **13**, 868 (1974).
10. H. Tada and K. Takeda, *Chem. Pharm. Bull.*, **24**, 667 (1976); through *Chem. Abstr.*, **85** 171296 (1976).
11. K. Tori et al., *Tetrahedron Lett.*, **5**, 387 (1976).
12. D. L. J. Opdyke, *Food Cosmet. Toxicol.*, **14**, 337 (1976).
13. J. C. Mitchell in V. C. Runeckles, ed., *Recent Advances in Phytochemistry*, Vol. 9, Plenum, New York, 1975, p. 119.
14. M. M. Verma, *Diss. Abstr. Int. B*, **41**(12, Pt. 1), 4514 (1981).
15. J. L. Hartwell, *Lloydia*, **32**, 247 (1969).

# BAY, WEST INDIAN

**Source:** ***Pimenta racemosa*** (Mill.) J. W. Moore (syn. *P. acris* Kostel) (Family Myrtaceae).

*Synonyms.* Myrcia, bay, and bay rum tree.

### GENERAL DESCRIPTION

Tree with leathery leaves, up to about 8 m high; native to the West Indies; cultivated in Venezuela, Puerto Rico, and the Caribbean Islands. Part used is the leaf from which a volatile oil (commonly called bay oil or Myrcia oil) is obtained by steam distillation, with yield of up to 3.9% reported.[1]

Although sometimes also referred to as bay leaf in the literature, the commonly used domestic spice is sweet bay (*Laurus nobilis*), not West Indian Bay.

### CHEMICAL COMPOSITION

The main components of bay oil are eugenol (up to 56%), chavicol (up to 22%), and myrcene (up to 21%). Those present in lesser amounts include 1,8-cineole, limonene, isoeugenol, linalool, methyl eugenol (3,4-dimethoxyallylbenzene), estragole (methyl chavicol), $\alpha$-terpineol, and others.[1–3]

Two varieties, an anise-scented and a lemon-scented variety, have been reported to yield volatile oils with quite different proportions of the above components. Thus the anise-scented variety contains methyl eugenol (43%) and methyl chavicol (32%) as the major components, and the lemon-scented variety contains mostly citral (>80%).[1]

## PHARMACOLOGY OR BIOLOGICAL ACTIVITIES

The volatile oil has antiseptic and astringent properties; it is considered moderately toxic on oral administration because of its relatively high content of phenols (GOSSELIN). However, no allergic reactions in humans have been reported.[4]

## USES

**Medicinal, Pharmaceutical, and Cosmetic.** Volatile oil used extensively as a fragrance ingredient in bay rum; also in creams, lotions (particularly aftershave and hair lotions), soaps, detergents, and perfumes, with maximum use level of 1.5% in certain perfumes.[4]

**Food.** Volatile oil, oleoresin, and extract (less extensively) all used as a flavor ingredient in major categories of food products, including alcoholic and nonalcoholic beverages, frozen dairy desserts, candy, baked goods, gelatins and puddings, meat and meat products, and condiments and relishes at very low levels, usually below 0.01%.

**Traditional Medicine.** Reportedly used in cancer therapy.[5]

**Others.** Volatile oils from anise-scented and lemon-scented varieties could serve as sources of methyl eugenol and citral, respectively. Methyl eugenol has been reported to have sedative and anticonvulsant activities. (see ***sweet bay***).

## COMMERCIAL PREPARATIONS

Volatile oil (regular and terpeneless), oleoresin, and extracts. The volatile oil was formerly official in N.F.; it is currently official in F.C.C.

*Regulatory Status.* Myrcia oil is GRAS (§182.20).

## REFERENCES

See the General References for ARCTANDER; BAILEY 1; FEMA; GUENTHER; MASADA.

1. D. McHale et al., *Food Chem.*, **2**, 19 (1977).
2. R. G. Buttery et al., *J. Agr. Food Chem.*, **22**, 773 (1974).
3. Analytical Methods Committee, *Analyst*, **100**, 593 (1975).
4. D. L. J. Opdyke, *Food Cosmet. Toxicol.*, **11**, 869 (1973).
5. J. L. Hartwell, *Lloydia*, **33**, 288 (1970).

# BAYBERRY BARK

**Source:** ***Myrica cerifera*** L. (Family Myricaceae).

*Synonyms.* Southern bayberry, southern wax myrtle, and wax myrtle bark.

## GENERAL DESCRIPTION

Evergreen shrub or small tree up to 13 m high; branchlets waxy; fruits round, grayish green, and coated with bluish wax, which can be removed by boiling in water. Native to eastern United States from New Jersey to South Florida and west to Texas; also grows in the Bahamas, the West Indies, and Bermuda. Part used is the dried root bark.

**CHEMICAL COMPOSITION**

Contains tannins, triterpenes (myricadiol, taraxerol, and taraxerone), and myricitrin (a flavonoid glycoside).[1] Other constituents reported to be present include an acrid astringent resin, gum, and starch.

**PHARMACOLOGY OR BIOLOGICAL ACTIVITIES**

Dried root bark has astringent, emetic, and antipyretic properties.

Myricitrin has been reported to exhibit choleretic, bactericidal, spermatocidal, and paramecicidal activities; myricadiol has mineralcorticoid activity.[1]

Tannins and phenols isolated from bayberry bark have been reported to be carcinogenic in NIH black rats per subcutaneous administration.[2]

An extract from the "sap-containing portions" of *M. cerifera* is claimed to have eliminated all pain and symptoms of a severe case of arthritis.[3]

**USES**

**Medicinal, Pharmaceutical, and Cosmetic.** Root bark has been and probably still is used as astringent, tonic, and stimulant to indolent ulcers and as an ingredient in Composition Powder used for colds and chills.

**Health Food/Herb Teas.** Powdered root bark still seen as an ingredient in Composition Powders, for colds and fevers (FOSTER AND DUKE).

**Traditional Medicine.** Used in Puerto Rico to treat stubborn ulcers (MORTON 2). Root bark used historically in the United States as an astringent and in larger doses an emetic, for chronic gastritis, diarrhea, dysentery, leucorrhea, jaundice, fevers; externally for hard-to-heal ulcers (FOSTER AND DUKE).

**Other.** Fruit is source of bayberry wax for candles.

**COMMERCIAL PREPARATIONS**

Available generally as crude botanical.

*Regulatory Status*. Bayberry root bark was included in N.F. IV-V (1916–1926).

**REFERENCES**

See the General References for BAILEY 1; FOSTER AND DUKE; KROCHMAL AND KROCHMAL; LUST; MARTINDALE; MERCK; MORTON 2; YOUNGKEN.

1. B. D. Paul et al., *J. Pharm. Sci.*, **63**, 958 (1974).
2. G. J. Kapadia et al., *J. Natl. Cancer Inst.*, **57**, 207 (1976).
3. H. L. Burnett, U.S. Pat. 4,321,262 (1982).

# BEE POLLEN

**Source:** Pollen collected by bees or harvested directly from flowers for commercial use.

*Synonyms*. Pollen, typha pollen, pollen typhae, *puhuang*, buckwheat pollen, rape pollen, maize pollen, pine pollen, Pollen Pini, *songhuafen*, etc.

**GENERAL DESCRIPTION**

Pollen is composed of microspores (male

reproductive elements) of seed-bearing plants. Bee pollen refers to pollen collected by bees that is in turn harvested for commercial distribution.

The sources and types of bee pollen are extremely variable. Known species that yield commercial bee pollen include buckwheat, rape, maize, and pine, among others. Typha pollen (*puhuang*) is collected from *Typha* species (*T. angustata* Bory et Chaub., *T. angustifolia* L., *T. latifolia* L., etc.) and has probably the longest and most extensively documented use history, dating back to the *Shen Nong Ben Cao Jing* (circa 200 B.C.–A.D. 100). Pine pollen is collected from numerous *Pinus* species (including *P. thunbergii* Parl., *P. massoniana* Lamb. and *P. tabulaeformis* Carr.) and has been in traditional Chinese medical records since the 7th century when it was first described in the *Tang Ben Cao*.

Commercial bee pollen is collected by means of netlike pollen traps, set up next to the beehives, that remove some of the pollen from the hind legs of worker bees as they return to their hives. The collected pollen is manually rid of impurities (dirt, floral parts, insect fragments, etc.) and dried. Major bee pollen–producing countries include China and Spain. Bee pollen from China is mostly derived from buckwheat (*Fagopyrum esculentum* Moench) and rape (*Brassica campestris* L).

To collect typha and pine pollen, the male inflorescence or flowerhead is picked in spring or summer when the flowers just start to bloom. It is sun dried; the pollen is then mechanically separated from the floral parts and other impurities. Major producers are northeastern provinces of China.

## CHEMICAL COMPOSITION

Pollen is very rich in nutrients. However, its chemistry varies greatly depending on its botanical source and contains: 3–16% water; 5.9–28.3% crude protein; 14.6–21.9% amino acids, with some in free form; 1–20% lipids; up to 44% carbohydrates; 4–10% simple sugars; 2–2.5% flavonoids; vitamins (A, $B_1$, $B_2$, C, $D_2$, E, $K_1$, $K_3$, folic acid, nicotinic acid, etc.); 19–24 trace elements; sterols; and others.[1–5]

The following are some examples of specific chemical constituents reported to be present in certain types of pollen but not necessarily in others: pentacosane, isorhamnetin glycoside, narcissin, free palmitic and stearic acids, 6-aminopurine, turanose and an oligosaccharide, sitosterol, and α-typhasterol in *puhuang* (ZHOU AND WANG);[6] β-sitosterol and cholesterol, ursolic acid, rutin (0–17%), C-3/C-8″-biapigenin, palmitic acid, nonacosane, luteolin, tricetin, kaempferol-3-*O*-sophoroside and kaempferol-3-*O*-β-D-gluco-7-*O*-β-D-glucoside in buckwheat pollen.[7]

## PHARMACOLOGY OR BIOLOGICAL ACTIVITIES

Many biological studies have been performed on bee pollen. However, results of these are extremely difficult to evaluate or duplicate due to the highly variable nature of this food/drug. The following are some of the biological activities of pollen and its extracts: hypolipemic in humans and experimental animals (*puhuang* and pollen mixture);[8,9] antiatherosclerotic in experimental animals (*puhuang* and pollen mixture);[10,11] protecting liver from experimental injury (rape pollen and pollen mixture);[12,13] inhibiting prostatic hypertrophy in aged dogs (rape pollen and pollen mixture);[14–16] immunoregulating and antioxidant (rape pollen and *puhuang*);[11,17,18] antiulcer in humans and experimental animals (*puhuang* and rape pollen);[19,20] antifatigue in mice (maize pollen);[21] laxative in humans (maize pollen and unspecified product);[22,23] antiinflammatory (*puhuang*); and uterine stimulant (*puhuang*) effects, among others (WANG, ZHOU AND WANG).[24]

Bee pollen could cause allergic reactions. However, no serious toxic side effects as a result of its ingestion have been reported. As it is a uterine stimulant, pregnant women are advised not to use it (*puhuang*) (CHP).

## USES

**Medicinal, Pharmaceutical, and Cosmetic.** Extracts (hydroalcoholic and lipoid) of pollen are used in skin-care products (facial and hand creams and lotions) for its nutritional and traditional healing and skin-softening properties (ETIC).

**Health Food/Herb Teas.** Used extensively as a food supplement in tablet, capsule, or liquid (drink or syrup) form (TYLER 1).

**Traditional Medicine.** Used for centuries by different cultures as a nutrient.
In China, typha pollen (*puhuang*) was first described 2000 years ago as sweet tasting, neutral, and having diuretic, hemostatic and stasis-dispersing properties. It has since been used to treat bleeding of different kinds (nosebleed, vomiting blood, coughing blood, metrorrhagia, bloody diarrhea, traumatic injuries, etc.), amenorrhea, dysmenorrhea, abdominal pain, painful urination and mouth sores; recently also successfully used to treat constipation and externally, eczema.

Pine pollen (*songhuafen*) also has a long use history, dating back to the 7th century A.D. Traditionally regarded as sweet tasting, warming, benefiting vital energy, removing wetness (*zao shi*), astringent and hemostatic; used topically in treating eczema, pustular eruptions, diaper rash, bleeding caused by traumatic injuries and other skin conditions; also used internally to treat alcohol intoxication, chronic diarrhea and rheumatism (CHP, JIANGSU).

## COMMERCIAL PREPARATIONS

Bee pollen comes in powdered or granular form, usually with color ranging from yellow to orange, depending on sources. Typha pollen comes in two types, one mixed with anthers and filaments while the other is pure pollen. Extracts (water, hydroalcoholic and lipoid) are also available.

*Regulatory Status.* U.S. regulatory status not determined. Subject of a German therapeutic monograph; allowed as an appetite stimulant.

## REFERENCES

See General References for BLUMENTHAL; CHP; JIANGSU; LU AND LI; NATIONAL; TYLER 1; WANG; ZHOU AND WANG; ZHU.

1. C. Y. Guo et al., *Jilin Zhongyiyao*, (4), 35 (1990).
2. L. Z. Mao et al., *Yingyang Xuebao*, **12**(1), 121 (1990).
3. S. S. Jia and S. M. Yang, *Zhongcaoyao*, **19**(1), 47 (1988).
4. V. E. Tyler, *The New Honest Herbal*, George F. Stickley Co., Philadelphia, 1987, p. 184.
5. T, Seppanen et al., *Phytother. Res.*, **3**(3), 115 (1989).
6. L. F. Chen et al., *Zhongguo Yaoli Xuebao*, **8**(2), 123 (1987).
7. J. X. Wei et al., *Zhongguo Zhongyao Zazhi*, **15**(5), 37 (1990).
8. B. Z. Zhang et al., *Chin. J. Integr. Med.*, **5**, 141 (1985).
9. P. G. Xiao and K. J. Chen, *Phytother. Res.*, **1**(2), 53 (1987).
10. J. Wojcicki et al., *Atherosclerosis*, **62**(1), 39 (1986).
11. Y. L. Yin et al., *Zhongguo Zhongyao Zazhi*, **17**, 374 (1992).
12. J. Wojcicki et al., *Acta Pharmacol. Toxicol. Suppl.*, **59**(7), 233 (1986).

13. M. S. Wang et al., *Zhongcaoyao*, **18**(5), 25 (1987).
14. B. C. Qian et al., *Zhonghua Laonian Yixue Zazhi*, **6**, 177 (1987).
15. X. L. Liu et al., *Zhongcaoyao*, **21**(4), 20 (1990).
16. M. Kimura et al., *Planta Med.*, **52**, 148 (1986).
17. F. Zhao and E. Y. Zhan, *Zhongyao Tongbao*, **11**(8), 45 (1986).
18. B. C. Qian et al., *Zhongguo Zhongyao Zazhi*, **15**(5), 45 (1990).
19. J. Chen et al., *Zhongcaoyao*, **20**(1), 27 (1989).
20. X. H. Yang et al., *Zhongyao Tongbao*, **12**(8), 48 (1987).
21. S. T. Yu et al., *Zhongyao Tongbao*, **13**(12), 44 (1988).
22. S. T. Yu et al., *Zhongcaoyao*, **19**(2), 26 (1988).
23. J. Q. Zhou et al., *Chin. J. Integr. Med.*, **8**, 357 (1988).
24. B. C. Qian, *Chin. J. Integr. Med.*, **9**, 125 (1989).
25. Monograph *Pollen*, *Bundesanzeiger* (Jan. 17, 1991).

# BEESWAX

**Source:** Honeycomb of the honeybee (***Apis*** spp.).

*Synonyms*. White beeswax, yellow beeswax, bleached beeswax, white wax, and yellow wax.

## GENERAL DESCRIPTION

Beeswax is the wax obtained from the honeycomb of the honeybee, *Apis mellifera* L., as well as other *Apis* species, including *A. cerana* Fabricius (Family Apidae).

After the honey is removed from the honeycombs the combs are washed rapidly and thoroughly with water. They are then melted with hot water or steam, strained, and run into molds to cool and harden.[1]

There are three major beeswax products: yellow beeswax, white beeswax (bleached beeswax), and beeswax absolute (*absolute cire d'abeille*). Yellow beeswax is the crude beeswax first obtained from the honeycombs. White beeswax and beeswax absolute are derived from yellow beeswax, the former from bleaching with the combined action of air, sunlight, and moisture (or with peroxides) and the latter by extraction with alcohol.

Beeswax is produced worldwide.

Yellow beeswax is a yellow to brownish yellow or grayish brown solid with an agreeable honeylike odor and faint but characteristic taste; it melts between 62° and 65°C.

White beeswax is a yellowish-white solid with a faint, characteristic odor, less pronounced than yellow beeswax; it is almost tasteless and translucent in thin layers; melts between 62° and 65°C.

Both yellow wax and white wax are insoluble in water, slightly soluble in cold alcohol, partly soluble in cold benzene, and completely soluble in chloroform, ether, and fixed and volatile oils.

Beeswax absolute is a pale yellow solid with a mild, sweet, and oily odor reminiscent of good linseed oil with a trace of honey notes, depending on sources.

## CHEMICAL COMPOSITION

Beeswax (yellow and white) contains about 71% esters of fatty acids (mostly palmitic and 15-hydroxypalmitic acid) and $C_{24}$ to $C_{34}$ straight-chained mono- and sometimes di-alcohols, composed of 35% $C_{46}$ to $C_{48}$ monoesters and 12% free acids (cerolein). Up to 23% of the monoesters is myricyl palmitate, which together with myricyl alcohol has been referred to as myricin (ARCTANDER, JIANGSU, REMINGTON).[1,2]

Beeswax absolute contains mostly cerolein; also aromatic volatile compounds.[3]

Cerolein is soluble in cold alcohol; myricin is insoluble in cold alcohol but sparingly soluble in boiling alcohol.

Myricyl alcohol, also known as triacontanol, has been shown to be a plant growth regulator, increasing yields of tomato, cucumber, and lettuce (see ***alfalfa***).

## PHARMACOLOGY OR BIOLOGICAL ACTIVITIES

Although beeswax is generally regarded as inert and nontoxic, allergic reactions have been reported (MARTINDALE).[4]

## USES

**Medicinal, Pharmaceutical, and Cosmetic.** Both yellow beeswax and white beeswax are used as thickener, emulsifier, or stiffening agents in ointments, baby products, bath preparations, cold creams, emollient creams, eye and facial makeups, lotions, lipsticks, hair dressings, hair conditioners, shaving products, suntan products, suppositories, and others; also used as a tablet polishing component.[5,6]

Beeswax absolute is used as a fragrance ingredient in soaps, lotions, creams, and perfumes in levels up to 0.4% in perfumes.[4]

**Food.** White beeswax and beeswax absolute are used as thickener, emulsifier, or flavor ingredients in all major categories of foods, including nonalcoholic and alcoholic beverages, frozen dairy desserts, baked goods, gelatins and puddings, confectioner frosting, and sweet sauces. White wax is also used as a candy glaze or polish. Use levels are usually low, the highest being in candy (ca. 0.05%).

**Traditional Medicine.** In Chinese medicine, beeswax is used to treat diarrhea and hiccups and to relieve pain, among others. For internal use, it is usually dissolved in hot alcohol or wine.

**Others.** As a source of triacontanol for increasing crop yield.[7,8]

## COMMERCIAL PREPARATIONS

Yellow beeswax, white beeswax, and beeswax absolute; first two are official in F.C.C. and N.F.

*Regulatory Status.* GRAS (§182.1973 and §182.1975).[9]

## REFERENCES

See the General References for ARCTANDER; FEMA; JIANGSU; REMINGTON; SAX; YOUNGKEN.

1. E. S. McLoud in A. Standen, ed., *Kirk-Othmer Encyclopedia of Chemical Technology*, Vol. 22, 2nd ed., Wiley-Interscience, New York, 1970, p. 156.
2. A. P. Tulloch, *Lipids*, **5**, 247 (1970).
3. C. E. M. Ferber and H. E. Nursten, *J. Sci. Food Agr.*, **28**, 511 (1977).
4. D. L. J. Opdyke, *Food Cosmet. Toxicol.*, **14**(Suppl.) 691 (1976).
5. S. L. Pulco, *Cosmet. Toilet.*, **102**(6) 57 (1987).
6. Anon., *J. Am. Coll. Toxicol.*, **3**(3), 1 (1984).
7. Z. X. Huang et al., *Huaxue Shiji*, **9**, 299 (1987); through *Chem. Abstr.*, **108**, 145317s (1988).
8. C. Devakumar et al., *Indian J. Agr. Sci.*, **56**, 744 (1986); through *Chem. Abstr.*, **106** 1768a (1987).
9. *Fed. Regist.* **42**(146), 38609 (1977).

# BEET COLOR, RED

**Source:** ***Beta vulgaris*** L. (Family Chenopodiaceae).

## GENERAL DESCRIPTION

Red beet color is the coloring material derived from the red beet root, *Beta vulgaris* L. *Beta vulgaris* has several varieties with roots ranging in size from small to thick and in color from whitish or yellowish (sugar beets) to deep bloodred (certain garden beets).[1]

## CHEMICAL COMPOSITION

The coloring principles present in red beet juice are known as betalains (quaternary ammonium amino acids). They consist mostly of betacyanins (red), with a small amount of betaxanthins (yellow). Betanin and to a lesser extent isobetanin account for most of the betacyanins present, while vulgaxantin-I and vulgaxanthin-II are the major betaxanthins. Betanin is a glucoside of betanidin, and isobetanin is its $C_{15}$ epimer. There is evidence that betanin occurs in red beet root as a sulfate linked through the sugar moiety at the 3- or 6-position.[2,3] Cyclodopa glucoside has recently been found in red beet juice, strengthening its role as intermediate in the biosynthesis of betanin.[4]

Red beet color is most stable at pH 4.5–5.5; it is rather unstable outside this range. There is evidence of an enzyme present in red beet that specifically destroys the betanin chromophore at an optimal pH of 3.35 and a temperature optimum of 42°C.[5] Betacyanin and betaxanthin decolorizing enzyme with an optimal pH of 3.4 have recently been reported in beet root tissue.[6]

Red beet pigments are heat labile, especially in the presence of metals (e.g., Cu, Mn, Fe, Zn); and at temperatures above 121°C, betanin is rapidly destroyed. Copper is the most efficient catalyst for the breakdown. The color can be stabilized by sequestrants and/or antioxidants such as citric acid, sorbic acid, and ascorbic acid;[7,8] though ascorbic acid has also been shown to decrease the color stability of betanin in aqueous solutions.[9]

Although the tinctorial power of betanin is quite high, its concentration in most commercial beet colors is only 1–2%, making it necessary to use these colors at relatively high levels to achieve the desired color effects. At these high levels, the characteristic beet flavor usually is perceptible. According to a patented process, the pigment content can be considerably increased, at the same time eliminating the beet flavor and aroma.[10,11]

Red beet color has been reported to have weakly mutagenic activities per Ames test,[12] though these results are not substantiated by others.[13] It did not initiate or promote hepatocarcinogenesis in rats during a short-term study.[14]

## USES

Used in coloring various food products.[15]

## COMMERCIAL PREPARATIONS

Powdered beet root, juice, concentrated juice, and spray-dried powder are available in different coloring strengths.

*Regulatory Status*. Has been approved for food use (§73.40 and §73.260); exempt from certification.

## REFERENCES

See the General References for BAILEY 2.

1. B. V. Ford-Lloyd and J. T. Williams, *Bot. J. Linn. Soc.*, **71**, 89 (1975).
2. T. J. Mabry and A. S. Dreiding in T. J. Mabry et al., eds., *Recent Advances in Phytochemistry*, Vol. 1, Appleton-Century-Crofts, New York, 1968, p. 145.
3. J. B. Harborne in L. Reinhold et al., eds., *Progress in Phytochemistry*, Vol. 4, Pergamon, Oxford, UK, 1977, p. 189.
4. H. Wyler et al., *Herb. Chim. Acta*, **67**, 1348 (1984); through *Chem. Abstr.*, **101**, 167164x (1984).
5. M. S. Ul'yanova et al., *Dokl. Akad. Nauk SSSR*, **200**, 990 (1971); through *Chem. Abstr.*, **76**, 43395u (1972).
6. C. C. Shih and R. C. Wiley, *J. Food Sci.*, **47**, 164 (1981).
7. A. F. Fang-Yung and A. V. Khotivari, *Isv. Vyssh. Uchebn. Zaved., Pishch. Tekhnol.*, **6**, 152 (1975); through *Chem. Abstr.*, **84**, 104045u (1976).
8. N. I. Oragvelidze et al., USSR 565,049 (1977); through *Chem. Abstr.*, **87**, 116653r (1977).
9. G. Muschiolik and H. Schmandke, *Nahrung*, **22**, 637 (1978).
10. J. P. Adams et al., *J. Food Sci.*, **41**, 78 (1976).
11. J. Von Elbe and C. H. Ammundson, *Ger. Offen.*, 2,545,975 (1976).
12. M. Ishidate Jr. et al., *Food Chem. Toxicol.*, **22**, 623 (1984).
13. K. Kawana et al., *Kanagawa-ken Eisei Kenkyusho Kenkyu Hokoku*, **13**, 27 (1983); through *Chem. Abstr.*, **100**, 208041e (1984).
14. J. H. Von Elbe and S. J. Schwartz, *Arch. Toxicol.*, **49**, 93 (1981).
15. S. J. Schwartz et al., *Food Chem. Toxicol.*, **21**, 531 (1983).

# BELLADONNA

**Source:** ***Atropa belladonna*** L. or its variety ***acuminata*** Royle ex Lindl. (Family Solanaceae).

*Synonym.* Deadly nightshade.

### GENERAL DESCRIPTION

Perennial herb up to 1 m high with black fruit (a berry); native to central and southern Europe and Asia Minor; now cultivated worldwide, including the United States, UK, China, and India. Parts used are the dried leaves, (including flowering and fruiting tops) and roots.

The specific epithet *belladonna* is of Italian origin, meaning "beautiful lady." This refers to the former practice of Italian women in using the juice of the berry on the eyes to dilate the pupils, giving them a striking appearance.

### CHEMICAL COMPOSITION

Leaves and roots contain tropane alkaloids (0.3–0.5%) that are composed mainly of *l*-hyoscyamine (95–98%) and traces of *l*-scopolamine (hyoscine) and atropine (*dl*-hyoscyamine); their concentration and proportions vary greatly with age of the plant. On extraction most of the *l*-hyoscyamine is racemized to atropine. Other alkaloids isolated include *l*-hyoscyamine *N*-oxide (equatorial and axial) and *l*-hyoscine *N*-oxide (equatorial isomer); roots contain cuscohygrine, which is absent in the leaves. A total of at least 14 alkaloids have been found in the root.[1,2] Scopolin, scopoletin, 7-methylquercetin, and a methylkaempferol are flavonoids among others present in leaves (STAHL).[3,4]

## PHARMACOLOGY OR BIOLOGICAL ACTIVITIES

The activity of belladonna is due to its alkaloids, primarily atropine. Atropine is anticholinergic, both central and peripheral. Its effect on the central nervous system is first stimulation and then depression. Its peripheral anticholinergic effects include reducing secretions (e.g., sweat, tears, saliva, nasal, gastric, and intestinal), decreasing gastric and intestinal motility, and increasing heart rate. Other activities include dilatation of the pupil, increase of intraocular pressure, and photophobia. The activities of *l*-hyoscyamine and *l*-scopolamine are essentially the same as those of atropine, except that scopolamine is a powerful hypnotic and usually slows the heart rate rather than increasing it.

Toxicity symptoms caused by overdose include dryness of mouth, intense thirst, difficulty in swallowing, burning pain in the throat, dilatation of the pupils with blurred vision and photophobia, flushing with hot and dry skin, high fever, fast heart rate with palpitations and elevated blood pressure, urge to urinate but inability to, constipation, restlessness, confusion, excitement, hallucinations, and delirium. Death may result from respiratory failure. Above symptoms have been described as "blind as a bat, dry as a bone, red as a beet, hot as a hare, and mad as a hatter" (GOSSELIN).

## USES

**Medicinal, Pharmaceutical, and Cosmetic.** Its extracts and isolated alkaloids are widely used in both over-the-counter and prescription drugs, including sedatives, antispasmodics in bronchial asthma and whooping cough, cold and hay fever remedies, ophthalmic preparations, laxatives (to lessen griping), suppositories for hemorrhoids, liniments for treatment of muscular rheumatism, sciatica, and neuralgia (often with aconite extract); also in the treatment of Parkinson's disease and intestinal and biliary colic.

Hyoscine is used in antimotion sickness preparations (MERCK).

Hyoscine-containing plants have been used for centuries in traditional Chinese medicine as anesthetics. Recently hyoscine from *Flos daturae* (flowers from *Datura* species) has been used as a general anesthetic in China reportedly with considerable success (JIANGSU).[5]

## COMMERCIAL PREPARATIONS

Available as crude and as various extracts. Belladonna and its fluid, solid, and powdered extracts are official in U.S.P.

*Regulatory Status*. Leaves and roots, calculated to specified levels of tropane alkaloids are the subject of a German therapeutic monograph, indicated for treatment of spasms and colic pains in the gastrointestinal tract and bile ducts.[6]

## REFERENCES

See the General References for APhA; BLUMENTHAL; CLAUS; GOODMAN AND GILMAN; JIANGSU; MARTINDALE; NANJING; USD 26th.

1. T. Hartmann et al., *Planta Med.*, **52**, 390 (1986).
2. F. Oprach et al., *Planta Med.*, **52**, 513 (1986).
3. J. D. Phillipson and S. S. Handa, *Phytochemistry*, **14**, 999 (1975).
4. G. Clair et al., *C.R. Hebd. Seances Acad. Sci. Ser. D.*, **282**, 53 (1976); through *Chem. Abstr.* **84**, 132617a (1976).
5. Anon., *Am. J. Chin. Med.*, **3**, 91 (1975).
6. Monograph *Atropa belladonna*, *Bundesanzeiger*, no. 223 (Nov. 30, 1985).

# BENZOIN

**Source:** ***Styrax*** spp. (Family Styraceae).

*Synonyms*. Gum benzoin, gum benjamin, Siam benzoin, Sumatra benzoin.

## GENERAL DESCRIPTION

Benzoin is the balsamic resin obtained from various *Styrax* spp. *Styrax benzoin* Dry. and *S. paralleloneurus* Perkins yield Sumatra benzoin; *S. tonkinensis* (Pierre) Craib ex Hartwich and other related *Styrax* species (Section *Anthostyrax*, Family Styraceae) yield Siam benzoin.

Benzoin-producing *Styrax* species are mostly small to medium trees (up to 20 m high) growing in tropical Asia. Sumatra benzoin is produced from trees growing in Malay Peninsula, Sumatra, and Java; Siam benzoin is from trees growing in Laos, Vietnam, Cambodia, China, and Thailand (formerly Siam).

Benzoin is a pathological product formed when the tree trunk is injured. It is produced by incising the bark; the exuded balsamic resin hardens on exposure to air and sunlight, and is collected.

Benzoin resinoid is prepared from crude benzoin by extraction with solvents such as benzene and alcohol, followed by their subsequent removal.

## CHEMICAL COMPOSITION

Benzoin contains chiefly esters of cinnamic and benzoic acids together with free acids. Amounts and types of esters and acids vary widely with its source.[1]

Sumatra benzoin contains 70–80% coniferyl cinnamate, cinnamyl cinnamate (styracin), and coniferyl benzoate; about 10% free cinnamic acid; small amounts of benzoic acid; and traces of benzaldehyde, vanillin, and styrene. Siam benzoin contains 60–80% coniferyl benzoate and cinnamyl benzoate; about 12% free benzoic acid; *d*-siaresinolic acid; and traces of cinnamic acid and vanillin (TYLER 3; YOUNGKEN).

According to a more recent source, Sumatra benzoin contains about 90% resinous matter composed mainly of sumaresinolic acid and coniferyl cinnamate, with 10–20% benzoic acid and 10–30% cinnamic acid; other constituents present include 2–3% phenylpropyl cinnamate, 1% vanillin, and traces of cinnamyl cinnamate, styrene, and benzaldehyde. Siam benzoin contains 70–80% resinous matter composed primarily of siaresinolic acid and coniferyl benzoate, with 11.7% benzoic acid, 2.3% cinnamyl benzoate, and 0.3% vanillin also present; cinnamic acid is reported absent (JIANGSU).

Benzoins from three *Styrax* species of Chinese origin are reported to be similar to Siam benzoin in balsamic acids content, consisting only of benzoic acid (JIANGSU).

#### PHARMACOLOGY OR BIOLOGICAL ACTIVITIES

Benzoin vapor (with steam) has expectorant properties; its solutions (e.g., tincture) have local antiseptic properties.

It is regarded as moderately toxic, probably due to occasional contact dermatitis developed in some individuals when using Compound Benzoin Tincture, which contains, in addition to benzoin, aloe, storax, balsam tolu, and others (GOSSELIN).[2]

#### USES

Both Siam benzoin and Sumatra benzoin are official in pharmacopoeias of many countries. Both types are official in U.S.P., but in the United States, Sumatra benzoin is more customarily used in pharmaceutical preparations, while Siam benzoin is used in flavors and fragrances.

**Medicinal, Pharmaceutical, and Cosmetic.** As antiseptic, astringent, and expectorant; in vaporizer fluids for inhalation to relieve respiratory discomforts; in Compound Benzoin Tincture, which is widely used as a skin protectant; and as an antiseptic and styptic on small cuts. Tincture also used in dentistry to treat inflammation of gums and oral herpetic lesions.

Benzoin, especially Siam benzoin, has antioxidative and preservative properties and is used in cosmetics for these properties. The resinoid is extensively used as a fixative in perfumes, soaps, detergents, creams, and lotions, in amounts up to 0.8% in perfumes (ARCTANDER).[2,3]

**Food.** Classified as a natural flavor; used in most categories of foods, including alcoholic and nonalcoholic beverages, frozen dairy desserts, candy (e.g., chocolate glaze), baked goods, and gelatins and puddings. Use levels usually quite low, with highest average maximum level of 0.014% reported in candy and baked goods.

#### COMMERCIAL PREPARATIONS

Crude benzoin, benzoin tincture, fluidextract, and resinoid are all readily available. Benzoin is currently official in U.S.P.

*Regulatory Status.* Has been approved for food use (§172.510).

#### REFERENCES

See the General References for ADA; ARCTANDER; CLAUS; FEMA; GUENTHER; JIANGSU; TYLER 3; YOUNGKEN.

1. A. Nitta et al., *Yakugaku Zasshi*, **104**, 592 (1984).
2. D. L. J. Opdyke, *Food Cosmet. Toxicol.*, **11**, 871 (1973).
3. D. O. Gyane, *Drug Cosmet. Ind.*, **118**, 36 (1976).

## BERGAMOT OIL

**Source:** ***Citrus bergamia*** Risso et Poit. (syn. *C. aurantium* L. subsp. *bergamia* Wright et Arn.) (Family Rutaceae).

#### GENERAL DESCRIPTION

Small tree native to tropical Asia; extensively cultivated in the Calabrian coast in southern Italy. Part used is the peel of the fresh, nearly ripe fruit. Bergamot oil is obtained by cold expression of the peel; it is also known as expressed bergamot oil from which rectified or terpeneless bergamot oil is produced by vacuum distillation or by selective solvent extraction, or by chromatography (ARCTANDER).

## CHEMICAL COMPOSITION

About 300 compounds have been reported to be present in the expressed oil, including 30–60% linalyl acetate, 11–22% linalool and other alcohols, sesquiterpenes ($\alpha$-*trans*-bergamotene, caryophyllene, $\beta$-farnesene, humulene, $\beta$-bisabolene), terpenes (limonene, *p*-cymene, $\gamma$-terpinene, phellandrene, $\alpha$- and $\beta$-pinene), $C_{20}$ to $C_{33}$ *n*-alkanes, and furocoumarins (bergaptene, bergamottin, citroptene, 7-methoxy-5-geranoxycoumarin, bergaptol, isopimpinellin, and xanthotoxin, with bergapten at 0.30–0.39%).[1–6]

Rectified (terpeneless) oil contains lower concentration of terpene components than the expressed oil and no coumarins (ARCTANDER).[7]

## PHARMACOLOGY OR BIOLOGICAL ACTIVITIES

Certain furocoumarins (particularly bergapten and xanthotoxin, also known as 5-methoxypsoralen and 8-methoxypsoralen, respectively) present in the expressed oil have been shown to be phototoxic when tested on human skin. Hyperpigmentation of the face and neck in some cases is thought to be due to photosensitization with cosmetics containing these coumarins.[5,6] When used with longwave ultraviolet light, they have been shown to be effective in the treatment of psoriasis, vitiligo, and mycosis fungoides.[8,9]

## USES

**Medicinal, Pharmaceutical, and Cosmetic.** Extensively used in perfumes (especially eau de cologne), creams, lotions, and soaps, with use levels up to 0.25% in creams and lotions and 3% in perfumes.[6] Could serve as source of furocoumarins.

**Food.** Widely used as an ingredient in flavor formulations with fruity citrus notes in most major food categories, including alcoholic and nonalcoholic beverages, frozen dairy desserts, candy, baked goods, gelatins and puddings, and meat and meat products. Highest average maximum use level is 0.02% in gelatins and puddings.

## COMMERCIAL PREPARATIONS

Bergamot oil expressed and bergamot oil rectified; the former is official in F.C.C.; formerly officinal in N.F.

*Regulatory Status.* GRAS (§182.20).

## REFERENCES

See the General References for FEMA; GUENTHER; MASADA.


1. M. Mammi de Leo, *Essenze Deriv. Agrum.*, **46**, 181 (1976); through *Chem. Abstr.*, **86**, 127088r (1977).
2. G. Calabro and P. Curro, *Ann. Fac. Econ. Commer., Univ. Studi Messina*, **10**, 67 (1972); through *Chem. Abstr.* **79**, 64737j (1973).
3. A. Liberti and G. Goretti, *Atti Conv. Naz. Olii Essenz. Sui Deriv. Agrum.*, **1–2**, 69 (1974); through *Chem. Abstr.*, **83**, 103125n (1975).
4. U. R. Cieri, *J. Assoc. Offic. Anal. Chem.*, **52**, 719 (1969).
5. S. T. Zaynoun et al., *Br. J. Dermatol.*, **96**, 475 (1977).

6. D. L. J. Opdyke, *Food Cosmet. Toxicol.*, **11**, 1031 (1973).
7. D. L. J. Opdyke, *Food Cosmet. Toxicol.*, **11**, 1035 (1973).
8. B. B. Mandula et al., *Science*, **193**, 1131 (1976).
9. T. Lakshmipathi et al., *Br. J. Dermatol.* **96**, 587 (1977).

# BILBERRY

**Source:** ***Vaccinium myrtillus*** L. (Family Ericaceae).

*Synonyms*. Whortleberry, huckleberry, dwarf bilberry.

## GENERAL DESCRIPTION

Deciduous freely branched shrub up to about 35(–60) cm high; arising from a creeping rhizome. Found in heaths, moors and woods from most of Europe (mountains in southern Europe), and montane and subalpine western North America from British Columbia to Alberta, south to Arizona and New Mexico. The parts used are the fruits and leaves.

## CHEMICAL COMPOSITION

Fruits or fruit juice contain at least 3% anthocyanosides such as procyanidins $B_1$, $B_2$, $B_3$, $B_4$; myrtillin; flavonoids include quercitrin, hyperoside, isoquercitrin, astragalin; flavan-3-ols including (+)-catechin and (−)-epicatechin; phenolic acids including caffeic, chlorogenic, *p*-coumaric, ferulic, syringic, gallic, protocatechuic, *p*-hydroxybenzoic, *m*-hydroxybenzoic, vanillic, *m*-coumaric, and *o*-coumaric acids, and a hydroxybenzoic acid derivative; vitamin C; quinolizidine alkaloids myrtine and epimyrtine in aerial parts; arbutin and other hydroquinone derivatives, ubiquitous in other *Vaccinum* spp., are absent in bilberry.[1–4]

## PHARMACOLOGY OR BIOLOGICAL ACTIVITIES

Bilberry fruits and leaves exhibit astringent and diuretic activity. Recent studies have supported capillary-strengthening activity, resulting in clinical use of fruit extracts, particularly in Italy, for various microcirculation diseases, including venous insufficiency of the lower limbs, varicose veins, atherosclerosis, and degenerative retinal conditions. Anthocyanosides extracted from the fruit have shown a protective effect on the liberation of lactate dehydrogenase in heart and plasma and cardiac isoenzymes, indicating an angina-protecting effect;[5] inhibiting retinal phosphoglucomutase and glucose-6-phosphatase;[6] a hypoglycemic effect (due to neomyrtillin content in leaves);[7] vasoprotective (twofold as active in protecting capillary permeability as rutin); and antiedema activity.[8] Long-lasting capillary resistance activity of the anthocyanins results from a greater affinity of the compounds to skin and kidney tissue, rather than plasma.[9] The anthocyanins decrease collagen hydrolysis, significantly reducing permeability of the blood-brain barrier;[10] have a vasodilating effect, stimulating local synthesis of vasodilator prostaglandins;[11,12] also inhibit platelet aggregation and thrombus formation, via stimulation of $PGI_2$-like substances in vascular tissue.[13] Orally administered leaf extracts lower plasma cholesterol and triglycerides.[14]

## USES

**Medicinal, Pharmaceutical, and Cosmetic.** In Europe, leaf preparations are used for the supportive treatment of diabetes mellitus; prevention and treatment of gastrointestinal, kidney, and urinary tract disorders as well as arthritis, dermatitis, functional heart problems, gout, hemorrhoids, poor circulation, and for metabolic stimulation of circulation. Efficacy poorly established; therapeutic use of the leaves not recommended.[15] Fruit preparations, calculated at a daily dose of 20–60 g., are used for the treatment of acute diarrhea, and for localized mild inflammation of the mucous membranes of the mouth and throat.[16]

**Food.** The fruit is best known for its food value; used in alcoholic and nonalcoholic beverages; conserves, pastries, compote, syrups, or eaten raw; fruit extracts also as red coloring in wine.

**Health Food/Herb Teas.** Dried fruits, a nutritive in capsulated products; teas; primarily for improved vision.[17]

**Traditional Medicine.** Historically, in Europe, the fruits and to a lesser extent the leaves, have been used for astringent and antiseptic activity in diarrhea, dysentery, dyspepsia, intestinal dyspepsia in infants; leaf tea as antidiabetic.

## COMMERCIAL PREPARATIONS

Crude, or extract (typically standardized to 36% anthocyanosides), in doses of 160 mg, 1–2 times daily.[17]

*Regulatory Status.* Undetermined in the United States. Both the leaves and the fruits are subjects of German therapeutic monographs.[15,16] Efficacy of leaves is not documented;[15] fruits allowed for acute diarrhea and mild inflammation of mouth and throat.[16]

## REFERENCES

See the General References for AHPA; BLUMMENTHAL; STEINMETZ; TUTIN 2; UPHOF; WEISS; WREN.

1. P. Slosse and C. Hootelé, *Tetrahedron Lett.*, **34**, 397 (1978).
2. P. Slosse and C. Hootelé, *Tetrahedron Lett.*, **37**, 4287 (1981).
3. M. Azar et al., *J. Food Sci.*, **52**, 1255 (1987).
4. H. Friedrich and J. Schönert, *Planta Med.*, **24**, 90 (1973).
5. M. Marcollet et al., *C. R. Soc. Biol.*, **163**, 8 (1969); through *Chem. Abstr.*, **72** 130053s (1970).
6. C. Cluzel et al., *C. R. Soc. Biol.*, **163**, 147 (1969); through *Chem. Abstr.*, **72** 130823t (1970).
7. R. N. Zozulya et al., *Rast. Resur.*, **11**, 87 (1975).
8. A. Lietti et al., *Arzneim. Forsch.*, **26**, 829 (1976).
9. A. Lietti and G. Forni, *Arzneim. Forsch.*, **26**, 832 (1976).
10. A. M. Robert et al., *J. Med.*, **8**, 321 (1977).
11. V. Bettini et al., *Fitoterapia*, **55**, 265 (1984).
12. V. Bettini et al., *Fitoterapia*, **56**, 3 (1985).
13. P. Morazzoni and M. J. Magistretti, *Fitoterapia*, **57**, 11 (1986).
14. A. Cignarella et al., *Planta Med.*, **58S**, A581 (1992).
15. Monograph *Myrtilli folium*, *Bundesanzeiger*, no. 76 (April 23, 1987).
16. Monograph *Myrtilli fructus*, *Bundesanzeiger*, no. 76 (April 23, 1987; correction (March 13, 1990).
17. R. McCaleb, *Better Nutr.*, 28 (July 1992).

# BIRCH OIL, SWEET

**Source: *Betula lenta*** L. (syn. *B. carpinefolia* Ehrh.) (Family Betulaceae).

*Synonyms.* Black birch and cherry birch oil.

## GENERAL DESCRIPTION

Tree up to about 25 m high with dark reddish brown bark, which is not peeling but broken into plates; native to southern Canada and northern United States, from Maine to Ohio and south to Florida and Alabama.

There are numerous species of birch with habitats spanning several continents.

Sweet birch oil is produced by steam distillation of the warm water-macerated bark. During maceration the enzyme system present hydrolyzes gaultherin, setting free methyl salicylate, which is the major component of the oil. The yield is about 0.6%.[1]

Sweet birch oil should not be confused with other birch oils such as birch bud and birch tar oils, which are produced from different species of birch and have different physical and chemical characteristics; they are used for quite different purposes. For example, birch tar oil is obtained by destructive distillation of the wood and bark of the European white birch (*Betula pendula* Roth; syn. *B. alba* L.) and is used in psoriasis, eczema, and other chronic skin diseases.

## CHEMICAL COMPOSITION

Almost entirely of methyl salicylate (98%).

## PHARMACOLOGY OR BIOLOGICAL ACTIVITIES

Like salicylates in general, methyl salicylate has antipyretic, antiinflammatory, and analgesic properties but is much more toxic. It can be absorbed through the skin, and fatal poisoning via this route has been reported. As little as 4 mL (4.7 g) methyl salicylate may be fatal in children (GOODMAN AND GILMAN).

## USES

**Medicinal, Pharmaceutical, and Cosmetic.** Limited use as counterirritant in antiarthritic and anti-neuralgic preparations such as ointments, liniments, and analgesic balms; as an antiseptic; and as a fragrance ingredient in perfumes and other cosmetic preparations. Presently, synthetic methyl salicylate is mostly used.

*Betula pendula* leaves reportedly diuretic, used in irrigation therapy for bacterial and inflammatory disease of the urinary tract.[2]

**Food.** Extensively used for its wintergreen (or root beer) flavor in most major categories of foods, especially nonalcoholic and alcoholic beverages, frozen dairy desserts, candy, chewing gum, gelatins and puddings, and baked goods. It is a common flavor ingredient in root beers. The highest average maximum use level reported is in candy (ca. 0.1%), but most of it is probably lost during processing.

**Traditional uses.** American Indians reportedly used bark tea for fevers, stomachache, lung ailments; twig tea for fever. Essential oil reportedly used for rheumatism, gout, scrofula, bladder infection, and neuralgia (FOSTER AND DUKE).

## COMMERCIAL PREPARATIONS

Volatile oil; it was formerly official in U.S.P. and is currently official in F.C.C. monographed under methyl salicylate together with wintergreen oil. *Betula pendula* leaves subject of a German therapeutic monograph, used in irrigation therapy as a diuretic.[2]

## REFERENCES

See the General References for ARCTANDER; BAILEY 2; BLUMENTHAL; FEMA; FOSTER AND DUKE; GUENTHER; MERCK; UPHOF; USD 26th.

1. G. A. Nowak, *Am. Perfum. Cosmet.*, **81**, 37 (1966).
2. Monograph *Betulae folium*, *Bundesanzeiger*, no. 50 (March 13, 1986).

# BLACKBERRY BARK

**Source:** Section ***Eubatus*** Focke of the genus ***Rubus*** L. (Family Rosaceae).

*Synonym.* Dewberry bark.

## GENERAL DESCRIPTION

There are at least 50 species in the Section *Eubatus*, which includes both blackberries and dewberries, all native to North America. The species reportedly used as a source of blackberry bark include a dewberry species (*Rubus villosus* Ait.) and several blackberry species (*Rubus allegheniensis* Porter, *R. nigrobaccus* Bailey, *R. sativus* Brain., R. *cuneifolius* Pursch, etc.) Part used is the dried bark of the rhizome and roots collected in the spring and fall.

## CHEMICAL COMPOSITION

Active constituents believed to be tannins; other constituents reported to be present include gallic acid, villosin, starch, and calcium oxalate.

## PHARMACOLOGY OR BIOLOGICAL ACTIVITIES

Has astringent properties due to tannins present.

A study on a *Rubus* species (*R. odoratus* L.), not of Section *Eubatus*, has demonstrated its tannins to have antitumor activity against the Walker 256 carcinosarcoma.[1]

Alcoholic extracts of the leaves of *Rubus ellipticus* Sm. have shown significant uterotropic activity at 300 mg/kg p.o. (1/4 $LD_{50}$), compared with controls, and potentiated estrogenic activity.[2]

## USES

**Food.** In flavor formulations in all major categories of foods, including alcoholic and nonalcoholic beverages, frozen dairy desserts, candy, baked goods, gelatins and puddings, and sweet sauces, with highest average maximum use level of 0.08% in frozen dairy desserts.

**Health Food/Herb Teas.** Blackberry bark in capsules, tablets, teas and other preparations, primarily as an astringent (FOSTER AND DUKE).

**Traditional Medicine.** Used as an astringent and antidiarrheal (KROCHMAL AND KROCHMAL).

## COMMERCIAL PREPARATIONS

Crude and extracts.; formerly official in U.S.P. and N.F.

*Regulatory Status.* Has been approved for food use (§172.510). Root bark and leaves subject of German therapeutic monographs. Leaf preparations allowed for treatment of acute, unspecific diarrhea and mild inflammation of the oral cavity and throat. Claimed efficacy of the root bark in "dropsy" is not substantiated; therefore, use is unjustified.[3]

## REFERENCES

See the General References for BAILEY 2; FEMA; FERNALD; FOSTER AND DUKE; KROCHMAL AND KROCHMAL; MERCK; YOUNGKEN.

1. H. H. S. Fong et al., *J. Pharm. Sci.*, **61**(11), 1818 (1972).
2. A. C. Rana et al., *Fitoterapia*, **65**, 3, 222 (1994).
3. Monograph *Rubi fruticosi folium* and *Rubi fruticosi radix Bundesanzeiger*, no. 22a (Feb. 1, 1990).


# BLACK COHOSH

**Source:** ***Cimicifuga racemosa*** (L.) Nutt. (syn. *Actaea racemosa* L.) (Family Ranunculaceae).

*Synonyms*. Black snakeroot, cohosh bugbane, and cimicifuga.

## GENERAL DESCRIPTION

Perennial herb, up to 3 m high, with knotted rhizome; leaves are three divided; terminal leaflet three lobed; middle lobe is largest; flowers, white, in tall raceme; native to rich woods, eastern North America, from Maine west to Ontario and Wisconsin, and south to Georgia. Parts used are the dried rhizome and roots.

## CHEMICAL COMPOSITION

Contains triterpene glycosides, actein, racemoside, 27-deoxyactein, and cimicifugoside (cimigoside);[1,2] isoflavones including formononetin. Other constituents reported to be present include isoferulic and salicylic acids, cimigonite, tannin, and volatile oil (WREN).

## PHARMACOLOGY OR BIOLOGICAL ACTIVITIES

A water-insoluble but chloroform-soluble resinous fraction has been reported to have hypotensive activities on animals and to have peripheral vasodilatory effects on humans.[3] The plant has also exhibited hypoglycemic and antiinflammatory activities on animals.[4,5]

Racemoside is claimed to have antiulcer activity in mice.[2]

Rhizome extract components have endocrine activities, one of which has been identified as formononetin, which binds to estrogen receptors in the rat uterus.[6–8]

Cimicifugoside is believed to affect the hypothalamus-pituitary system, producing effects on the reproductive and nervous systems.[9]

Numerous clinical studies confirm efficacy as alternative to hormone therapy in neurovegetative and psychic problems associated with menopause.[10]

## USES

**Medicinal, Pharmaceutical, and Cosmetic.** Used in certain analgesic and tonic preparations, among others. In European phytomedicine, ethanol extracts corresponding to 40 mg of dried root, used for estrogen-like action and luteinizing-hormone suppression in menstrual disorders, including premenstrual discomfort, dysmenorrhea, and uterine spasms.[11]

**Health Food/Herb Teas.** Reportedly used as a sedative in herbal formulas to relieve problems related to menopause; capsules, tablets, tinctures, fluid extract, crude root in infusion or decoction, primarily for functional uterine disorders (BRADLEY, FOSTER, WREN).

**Traditional Medicine.** Reportedly used in treating sore throat, rheumatism, bronchitis, uterine disorders, and other ailments; also as a sedative. Tincture traditionally used for bronchitis, chorea, fevers, nervous disorders, lumbago, rheumatism, snakebite, menstrual disorders; aid in childbirth (FOSTER AND DUKE)

In traditional Chinese medicine, other *Cimicifuga* species such as *C. foetida* L., *C. dahurica* (Turcz.) Maxim., and *C. simplex* Womsk. have been used for centuries for similar purposes.

**COMMERCIAL PREPARATIONS**

Available as crude and extracts (fluid, solid, and powdered). Formerly official in N.F. and U.S.P. Potencies of extracts are expressed only in strength (see ***glossary***) based on weight-to-weight ratio of crude and extracts.

*Regulatory Status.* The root is the subject of a German therapeutic monograph, indicated for premenstrual discomfort and dysmenorrhea.[11]

**REFERENCES**

See the General References for BAILEY 1; BLUMENTHAL; BRADLEY; FOGARTY; FOSTER; FOSTER AND DUKE; JIANGSU; KROCHMAL AND KROCHMAL; MERCK; UPHOF; WREN; YOUNGKEN.

1. L. Radics et al., *Tetrahedron Lett.*, **48**, 4287 (1975).
2. Suntry Ltd., *Jpn. Kokai Tokyo Koho JP* **59**, 20, 298 (1984); through *Chem. Abstr.*, **101**, 43572j (1984).
3. E. Genazzani and L. Sorrentino, *Nature*, **194**, 544 (1962).
4. N. R. Farnsworth and A.B. Segelman, *Tile Till*, **57**, 52 (1971).
5. P. S. Benoit et al., *Lloydia*, **39**, 160 (1976).
6. H. Jarry and G. Harnischfeger, *Planta Med.*, **51**, 46 (1985).
7. H. Jarry et al., *Planta Med.*, **51**, 316 (1985).
8. S. Berger et al., *Planta Med.*, **54**, 579 (1988).
9. A. Petho, *Ärztl. Praxis*, **35**, 146 (1983).
10. R. Jaspersen-Schib, *Apoth. Ztg.*, **127**, 348 (1990).
11. Monograph *Cimicifugae racemosae rhizoma*, *Bundesanzeiger*, no. 43 (March 2, 1989).

# BLACK HAW BARK

**Source:** ***Viburnum prunifolium*** L. or ***V. prunifolium*** L. var. ***ferrugineum*** Torr. and Gray (syn. *V. rufidulum* Raf.) (Family Caprifoliaceae).

*Synonyms.* Stag bush, viburnum, and southern black haw (*V. rufidulum*).

**GENERAL DESCRIPTION**

*Viburnum prunifolium* is a spreading deciduous shrub or small tree, up to 5 m high; native to North America, from Connecticut to Florida and west to Michigan and Texas.

*Viburnum rufidulum* is a large shrub or small tree, up to 10 m high, with stout branches; habitat from Virginia to Florida, west to Illinois and Texas.

Parts used are the root and stem barks.

**CHEMICAL COMPOSITION**

*Viburnum prunifolium* root bark is reported to contain several active constituents (uterine relaxants), one of which has been identified as scopoletin (7-hydroxy-6-methoxy coumarin).[1] Other constituents reported to be present include plant acids (malic, citric, oxalic, and valeric acids), tannin, bitter resin, and others (MERCK). 1-Methyl-2,3-dibutyl hemimellitate has been isolated from stem bark.[2]

### PHARMACOLOGY OR BIOLOGICAL ACTIVITIES

Has been demonstrated to have uterine antispasmodic properties *in vitro*.[3]

### USES

**Medicinal, Pharmaceutical, and Cosmetic.** Root bark and its extracts are used as tonics and in uterine sedative, antidiarrheal, diuretic, and general antispasmodic preparations.

**Food.** Stem bark extract is used as a flavor ingredient primarily in alcoholic and nonalcoholic beverages in very low concentrations, with maximum use level at less than 0.001%.

**Health Food/Herb Teas.** Root bark used in capsules, tablets, tinctures, and infusion, primarily for uterine sedative and antidiarrheal activity (WREN).

**Traditional Medicine.** American Indians used the root and/or stem bark for the treatment of painful menses, to prevent miscarriage, as a postpartum antispasmodic; also for asthma (FOSTER AND DUKE).

### COMMERCIAL PREPARATIONS

Crude, fluid extract, solid extract, and powdered extract are readily available; crude and fluid extract were formerly official in N.F. and U.S.P. Strengths (see ***glossary***) of extracts are expressed in weight-to-weight ratios between crude and extracts.

*Regulatory Status.* Has been approved for food use (§172.510).

### REFERENCES

See the General References for BAILEY 1; FEMA; FOSTER AND DUKE; KROCHMAL AND KROCHMAL; UPHOF; WREN; YOUNGKEN.

1. C. H. Jarboe et al., *J. Med. Chem.*, **10**, 488 (1967).
2. C. H. Jarboe et al., *J. Org. Chem.*, **34**, 4202 (1969).
3. C. H. Jarboe et al., *Nature*, **212**, 837 (1966).

# BLESSED THISTLE

**Source:** ***Cnicus benedictus*** L. (Family Compositae or Asteraceae).

*Synonym.* Holy thistle.

### GENERAL DESCRIPTION

Rough annual, 10–60 cm high; leaves oblong, minute spinose-dentate, prominent white veins beneath; flowers yellow to 4 cm, involucral scales bristly; indigenous to waste places and fields of Mediterranean region; naturalized in central and southeastern Europe; casually established as weed in eastern United States (FOSTER AND DUKE). Part used is the herb (flowering tops).

### CHEMICAL COMPOSITION

Leaves contain lignans, 2-acetylnortracheloside, arctigenin, nortracheloside, salonitenolide, trachelogenin;[1] arctiin in fruit; sesquiterpene lactone cnicin (bitter index = 1:1,800);[2] lithospermic acid; minute amounts of volatile oil; plus tannins, mucilage, potassium and manganese salts (WREN).

**PHARMACOLOGY OR BIOLOGICAL ACTIVITIES**

Essential oil bacteriostatic against *Staphylococcus aureus*; inactive in *Bacillus coli*.[3] The extremely bitter sesquiterpene lactone cnicin has antibacterial and antitumor activity; antifeedant activity against certain insects (HARBOURNE AND BAXTER). Following oral ingestion, the lignans arctiin and tracheloside are metabolized to their genins arctigenin and trachelogenin in the intestinal tract, which have inhibitory effects on cyclic-AMP phosphodiesterase and on histamine release in rat mast cells; plus $Ca^{2+}$ and platelet-activating factor antagonist activities (see also ***burdock*** and ***safflower***).[4]

**USES**

**Food.** Extract an ingredient in alcoholic beverages (Benedictine); also in bitters.

**Health Food/Herb Teas.** Leaf capsules, tablets; tea, extract, tincture, primarily as bitter digestive,[5] antiflatulent and in gallbladder disease.

**Traditional Medicine.** Leaves a stomachic, diuretic, appetizer; folk medicine for anorexia, dyspepsia, colds, fever; externally poulticed for boils, wounds, and ulcers (FOSTER AND DUKE, WREN).

**COMMERCIAL PREPARATIONS**

Crude leaves, leaf extract.

*Regulatory Status.* Undetermined in the United States; subject of German therapeutic monograph; allowed as a bitter digestive to treat loss of appetite and dyspeptic discomfort.[5]

## REFERENCES

See the General References for AHPA; BLUMENTHAL; DUKE 2; FOSTER AND DUKE; GLEASON AND CRONQUIST; HARBOURNE AND BAXTER; NIKITAKIS; STEINMETZ; TUTIN 4; TYLER 1; UPHOF; WREN.

1. M. Vanhaelen and R. Vanhaelen-Fastré, *Phytochemistry*, **14**, 2709 (1975).
2. R. Vanhaelen-Fastré and M. Vanhaelen, *Planta Med.*, **29**, 179 (1976).
3. R. Vanhaelen-Fastré, *Planta Med.*, **24**, 165 (1973).
4. M. Nose et al., *Planta Med.*, **59**, 131 (1993).
5. Monograph *Cnici benedicti herba*, *Bundesanzeiger*, no. 193 (Oct. 15, 1987).

# BLOODROOT

**Source:** ***Sanguinaria canadensis*** L. (Family Papaveraceae).

*Synonyms*. Sanguinaria, red puccoon, red root, and Indian red paint.

## GENERAL DESCRIPTION

A low perennial herb with horizontal, branching rhizome bearing slender roots; up to about 35 cm high; both rhizome and roots contain an orange-red latex. Grows in eastern United States from New England south to Florida and west to Wisconsin and Texas. Part used is the dried rhizome, sometimes referred to as "root" in the literature.

## CHEMICAL COMPOSITION

Both rhizome and root contain numerous alkaloids of which the major ones are sanguinarine (ca. 1%), sanguidimerine, chelerythrine, and protopine. Other alkaloids reported to be present include oxysanguinarine, $\alpha$- and $\beta$-allocryptopine, sanguilutine, dihydrosanguilutine, berberine, coptisine, and homochelidonine. Also contains a reddish resin, plant acids (e.g., citric, malic), and starch.[1-3]

## PHARMACOLOGY OR BIOLOGICAL ACTIVITIES

Sanguinarine, which is the major alkaloid, has various pharmacological activities; among them are its broad antimicrobial activities, its local anesthetic property, and its implication in the pathogenesis of glaucoma in epidemic dropsy in India.[4,5] It has also been reported to have anticancer activity, though confirmatory studies are required.[6] Sanguinarine and chelerythrine have been reported to have antiplaque properties and to be effective in controlling the volatile sulfur compounds in the oral cavity, which are responsible for bad breath.[7]

Protopine has been shown to have inhibitory effects on blood platelet aggregation in experimental animals.[8]

## USES

**Medicinal, Pharmaceutical, and Cosmetic.** Used in cough remedies, almost always in combination with other drug ingredients such as spikenard root, balm of gilead bud, white pine, and wild cherry barks, as in Compound White Pine Syrup and other formulations. Also used in cosmetics reportedly for its alleged healing properties.

Sanguinarine is used as an antiplaque agent in toothpaste and mouthwash preparations.

**Traditional Medicine.** Used by American Indians to treat fever, rheumatism, skin burns, and sore throat and as a tonic.

## COMMERCIAL PREPARATIONS

Crude, tincture, fluid extract, solid extract, and powdered extract are all readily available. Crude and fluid extract were formerly official in N.F. and U.S.P. Strengths (see ***glossary***) of extracts are expressed in weight-to-weight ratios between crude and extracts.

## REFERENCES

See the General References CLAUS; KROCHMAL AND KROCHMAL; MERCK; deNAVARRE; YOUNGKEN.

1. F. Santavy in R. H. F. Manske, ed., *The Alkaloids*, Vol. 12, Academic Press, New York, 1970, p. 333.
2. M. Tin-Wa et al., *J. Pharm. Sci.*, **61**, 1846 (1972).
3. D. K. Kim and F. R Stermitz, *Phytochemistry*, **14**, 834 (1975).
4. V. Preininger in R. H. F. Manske, ed., *The Alkaloids*, Vol. 15, Academic Press, New York, 1975, p. 207.
5. I. S. Shenolikar et al., *Food Cosmet. Toxicol.*, **12**,, 699 (1974).
6. M. Tin-Wa et al., *Lloydia*, **33**, 267 (1970).
7. R. T. Boulqare et al., *J. Soc. Cosmet. Chem.*, **36**, 297 (1985).
8. H. Matsuda, *Planta Med.*, **54**, 498 (1988).

# BLUE COHOSH

**Source:** ***Caulophyllum thalictroides*** (L.) Michx. (Family Berberidaceae).

*Synonyms*. Squaw root, papoose root, and caulophyllum.

## GENERAL DESCRIPTION

Perennial herb with a thick, crooked, and horizontal rhizome, up to 1 m high; grows in eastern North America. Parts used are the dried rhizome and roots.

## CHEMICAL COMPOSITION

Alkaloids isolated from rhizome and root include methylcytisine (caulophylline), baptifoline and anagyrine (lupine alkaloids), and magnoflorine. Other constituents present include caulosaponin and resins.[1–3]

Another *Caulophyllum* species (*C. robustum* Maxim.) had been extensively studied in the former USSR and found to be rich in triterpene glycosides (caulosides A, B, C, D, E, F, and G), most of which have hederagenin as their aglycone; they possess fungicidal activities.[4–7]

## PHARMACOLOGY OR BIOLOGICAL ACTIVITIES

An alcoholic extract of its aerial parts after treatment with petroleum ether has been demonstrated to have antiinflammatory properties in rats.[8] A hot water extract and the saponin-containing fraction of the roots and rhizomes exhibited a uterine stimulant effect in isolated rat uterine muscle preparations.[3]

Methylcytisine, with similar pharmacological activity to nicotine, increases blood pressure, stimulates the small intestine, and produces hyperglycemia (FOSTER AND DUKE).

## USES

**Medicinal, Pharmaceutical, and Cosmetic.** Used in diuretic, uterine, antispasmodic, and emmenagogue as well as laxative preparations.

**Health Food/Herb Teas.** Tea, tincture, capsules, tablets, and other products used primarily for menstrual difficulties or as an aid in childbirth (FOSTER).

**Traditional Medicine.** Root infusion or decoction used by American Indians as an aid in labor and for menstruation, abdominal cramps, urinary tract infections, lung ailments, and fevers. A folk remedy for rheumatism and inflammation of the uterus. Historically prescribed for chronic uterine conditions (FOSTER AND DUKE).

### COMMERCIAL PREPARATIONS

Available as crude and extracts; crude was formerly official in N.F. (1916–1942); U.S.P (1880–1890). Strengths (see ***glossary***) of extracts are expressed as weight-to-weight ratios between crude and extract.

*Regulatory Status*. Undetermined in the United States.

### REFERENCES

See the General References for FOSTER; FOSTER AND DUKE; KROCHMAL AND KROCHMAL; MERCK; YOUNGKEN.

1. M. S. Flom et al., *J. Pharm. Sci.*, **56**, 1515 (1967).
2. M. S. Flom, *Diss. Abstr. Int. B*, **32**, 2312 (1971).
3. C. T. Che, *Diss. Abstr. Int. B*, **43**, 1049 (1982).
4. L. I. Strigina et al., *Khim. Prir. Soedin.*, **6**, 552 (1970); through *Chem. Abstr.*, **74**, 50520t (1971).
5. L. I. Strigina et al., *Phytochemistry*, **14**, 1583 (1975).
6. M. M. Anisimov, *Antibiotiki (Moscow)*, **17**, 834 (1972); through *Chem. Abstr.*, **78**, 628w (1973).
7. L. I. Strigina et al., *Khim. Prir. Soedin.*, **5**, 619 (1976); through *Chem. Abstr.*, **86**, 152616u (1977).
8. P. S. Benoit et al., *Lloydia*, **39**, 160 (1976).

# BOIS DE ROSE OIL

**Source: *Aniba rosaeodora*** Ducke (Family Lauraceae).

*Synonyms*. Rosewood oil and cayenne rosewood oil.

### GENERAL DESCRIPTION

Medium-size evergreen tree, growing wild in the Amazon region, notably Brazil, French Guiana, Peru, and East Surinam. The essential oil is obtained from chipped wood by steam distillation and occasionally water distillation. Brazil and Peru are the major producers of the oil; French Guiana is also a producer, producing cayenne bois de rose oil that is considered the best quality among the bois de rose oils (ARCTANDER, MASADA).

### CHEMICAL COMPOSITION

The major component is linalool, which is present in 90–97% in cayenne bois de rose oil and 80–90% in Brazilian oil (MASADA).[1,2] The balance is made up of cineole (up to 10%), $\alpha$-terpineol, geraniol, citronellal, limonene, $\alpha$-pinene, $\beta$-pinene, $\beta$-elemene, *cis*- and *trans*-linalool oxides, sesquiterpenes, and others (MASADA).[3,4]

### PHARMACOLOGY OR BIOLOGICAL ACTIVITIES

Available data indicate acetylated bois de rose oil to be nontoxic when applied externally.[5]

Linalool, the major component of bois de rose oil, has weak tumor-promoting properties in mice.[6,7] It also has been reported to have anticonvulsant activity in mice and rats, spasmolytic activity on isolated guinea pig ileum, antimicrobial properties, and others.[7]

USES

**Medicinal, Pharmaceutical, and Cosmetic.** Used as a source of natural linalool or linalool acetate, which are extensively used in perfumery. Acetylated bois de rose oil is reportedly used in soaps, detergents, creams, lotions, and perfumes, with maximum use level of 1.2% in perfumes.[5]

**Food.** Used extensively as a flavor ingredient in most major categories of foods, including alcoholic and nonalcoholic beverages, frozen dairy desserts, candy, baked goods, gelatins and puddings, meat and meat products, and gravies. Average maximum use level are generally below 0.003% (24.9 ppm).

COMMERCIAL PREPARATIONS

Oil and acetylated oil. Bois de rose is official in. F.C.C.

*Regulatory Status.* GRAS (§182.20).

REFERENCES

See the General References for ARCTANDER; FEMA; GUENTHER; LIST AND HÖRHAMMER; UPHOF.

1. A. Alpande de Morais et al., *Acta Amazon.*, **2**, 41 (1972); through *Chem. Abstr.*, **80**, 149023u (1974).
2. A. Alpande de Morais et al., *An. Acad. Bras. Cienc.*, **44**(Suppl.) 303 (1972); through *Chem. Abstr.*, **83**, 136703c (1975).
3. G. Chiurdoglu et al., *Ind. Chim. Belge*, **28**, 636 (1963); through *Chem. Abstr.*, **59**, 15534f (1963).
4. I. C. Nigam and L. Levi, *Perfum. Essent. Oil Record*, **54**, 814 (1963).
5. D. L. J. Opdyke, *Food Cosmet. Toxicol.* **11**, 1039 (1973).
6. F. Homburger and E. Boger, *Cancer Res.*, **28**, 2372 (1968).
7. D. L. J. Opdyke, *Food Cosmet. Toxicol.*, **13**, 827 (1975).

# BOLDO LEAVES

**Source:** ***Peumus boldus*** Mol. (syn. ***Boldu boldus*** (Mol.) Lyons) (Family Monimiaceae).

*Synonym.* Boldus.

GENERAL DESCRIPTION

Dioecious evergreen shrub or small tree, up to about 6 m high; native to mountainous regions of Chile and naturalized in Europe (Mediterranean region). Parts used are the dried leaves.

CHEMICAL COMPOSITION

Contains 0.25–0.7% alkaloids, including laurotetanine, *N*-methyllaurotetanine, boldine, isoboldine, laurolitsine, norisocorydine, isocorydine, isocorydine-*N*-oxide, and reticuline, of which boldine (0.06%) is not the major alkaloid as previously reported but may actually be absent;[1–4] about 2.5% volatile oil composed mainly of *p*-cymene (28.6%), ascaridole (16.1%), 1,8-cineole (16.0%), and linalool (9.1%), among 38 identified compounds;[4,5] flavonol glycosides (e.g., isorhamnetin-3-α-L-arabinopyranoside-7-α-L-rhamnopyranoside);[6] resin; and tannins.

PHARMACOLOGY OR BIOLOGICAL ACTIVITIES

It is reported to have choleretic, diuretic, stomachic, and cholagogic properties.[4,6]

One study indicated that its total alcoholic extract, as opposed to its partial extracts, had the highest choleretic activity in rats.[7]

Its ethanolic and ether extracts exhibited strong antioxidative properties.[8]

**USES**

**Medicinal, Pharmaceutical, and Cosmetic.** In tonic and diuretic preparations; in Europe for gastrointestinal spasms.[9] Rarely, if at all, used in cosmetics.

**Food.** Only used in alcoholic beverages (liqueurs, bitters, etc.). Average maximum use level reported is about 0.002% (16 ppm).

**Health Food/Herb Teas.** Leaves used in combination products, capsules, tablets, and infusions as antioxidant (WREN).

**Traditional Medicine.** Reportedly used as a diuretic and biliary stimulant in hepatic illnesses and cholelithiasis (gallstones) (MARTINDALE).

**COMMERCIAL PREPARATIONS**

Crude and extracts; crude and fluid extract were formerly official in N.F. Strengths (see ***glossary***) of extracts are expressed in weight-to-weight ratios.

*Regulatory Status.* Approved for food use in alcoholic beverages only (§172.510). Leaves subject of a German therapeutic monograph; allowed for mild gastrointestinal spasms and dyspeptic disorders.[9]

**REFERENCES**

See the General References for BIANCHINI AND CORBETTA; BLUMENTHAL; FEMA; GOSSELIN; MARTINDALE; RAFFAUF; TERRELL; WILLAMAN AND SCHUBERT; WREN; YOUNGKEN.

1. M. Sobiczewska and B. Borkowski, *Acta Pol. Pharm.*, **29**, 271 (1972); through *Chem. Abstr.*, **77**, 137008t (1972).
2. M. Vanhaelen, *J. Pharm. Belg.*, **28**, 291 (1973).
3. N. Didry, *Bull Soc. Pharm. Lille*, **31**, 51 (1977); through *Chem. Abstr.*, **87**, 50251z (1977).
4. H. Schindler, *Arzneim. Forsch.*, **7**, 747 (1957).
5. K. Bruns and M. Köhler, *Parfüm. Kosmet.*, **55**(8), 225 (1974).
6. E. Bombardelli et al., *Fitoterapia*, **47**, 3 (1976).
7. M. C. Levy-Appert-Collin and J. Levy, *J. Pharm. Belg.*, **32**, 13 (1977).
8. T. Hirosue et al., *Nippon Shokuhin Kohyo Gakkaishi*, **35**, 630 (1988); through *Chem. Abstr.*, **109**, 229018d (1988).
9. Monograph *Boldo folium*, *Bundesanzeiger*, no. 76 (April 23, 1987).

# BONESET

**Source:** ***Eupatorium perfoliatum*** L. (Family Compositae or Asteraceae).

*Synonyms.* Eupatorium, common boneset, feverwort, agueweed, and thoroughwort.

**GENERAL DESCRIPTION**

Perennial herb with opposite sessile clasping leaves; up to about 1.5 m high; native to eastern and central North America, from Quebec south to Florida, Alabama, and Louisiana, and west to Texas and the

Dakotas. Parts used are the dried leaves and flowering tops collected in late summer.

## CHEMICAL COMPOSITION

Much chemical information on various *Eupatorium* species has been published (KARRER).[1-8] *E. perfoliatum* is reported to contain flavonoids, including quercetin, kaempferol, quercetin-3-$\beta$-galactoside (hyperoside), kaempferol-3-$\beta$-glucoside (astragalin), quercetin-3-rutinoside (rutin), kaempferol-3-rutinoside, and eupatorin (MERCK);[1] terpenoids, including chromenes, sesquiterpenes, sesquiterpene lactones (euperfolin, euperfolitin, eufoliatin, and euperfolide, among others), and diterpenes (dendroidinic acid and hebeclinolide);[2-4] triterpenes (e.g., $\alpha$-amyrin); sterols (sitosterol and stigmasterol); dotriacontane;[5] a volatile oil; a polysaccharide (4-*O*-methylglucuronoxylan);[9,10] and resin, among others.

## PHARMACOLOGY OR BIOLOGICAL ACTIVITIES

Boneset is reported to have stimulant and diaphoretic properties; in large doses it is both emetic and cathartic (CLAUS).

Numerous sesquiterpene lactones (e.g., eupatilin, eupafolin, eupatorin acetate, and eupaformosanin) and flavones (especially eupatorin) isolated from *Eupatorium* species have been demonstrated to have cytotoxic and/or antineoplastic activities.[6-8,11]

An ethanol extract of the whole plant after treatment with petroleum ether has exhibited weak antiinflammatory property in rats.[12]

The polysaccharide has immunostimulating activities.[9,10] Preliminary screening has also indicated eufoliatin to have immunostimulating activity.[13]

## USES

**Medicinal, Pharmaceutical, and Cosmetic.** Used in certain antipyretic and urinary antiseptic preparations, among others.

**Health Food/Herb Teas.** Crude herb in infusion, also tinctures and extracts as immunostimulant and for fevers; use relatively uncommon. Use discouraged due to presence of hepatotoxic pyrrolizidine alkaloids, ubiquitous in *Eupatorium* species (FOSTER AND DUKE).

**Traditional Medicine.** Reportedly used in infusion as tonic, febrifuge, diaphoretic, emetic, and cathartic; also used to treat skin rashes (KROCHMAL AND KROCHMAL).

## COMMERCIAL PREPARATIONS

Crude; extracts not readily available. Crude was formerly official in N.F. and U.S.P. (1820–1900).

*Regulatory Status*. Undetermined in United States.

## REFERENCES

See the General References for CLAUS; FERNALD; FOSTER AND DUKE; GOSSELIN; KROCHMAL AND KROCHMAL; LIST AND HÖRHAMMER; LUST; deNAVARRE; TYLER 1; YOUNGKEN.

1. H. Wagner et al., *Phytochemistry*, **11**, 1504 (1972).
2. F. Bohlmann and M. Grenz, *Chem. Ber.*, **110**, 1321 (1977).
3. W. Herz et al., *J. Org. Chem.*, **42**, 2264, (1977).
4. F. Bohlmann et al., *Phytochemistry*, **16**, 1973 (1977).
5. X. A. Dominguez et al., *Phytochemistry*, **13**, 673 (1974).
6. M. D. Midge and A. V. R. Rao, *Indian J. Chem.*, **13**, 541 (1975).

7. K. H. Lee et al., *Phytochemistry*, **16**, 1068 (1977).
8. E. O. Arene et al., *Lloydia*, **41**, 186 (1978).
9. H. Wagner et al., *Arzneim. Forsch.*, **35**, 1069 (1985).
10. A. Vollmar et al., *Phytochemistry*, **25**, 377 (1986).
11. E. Rodriguez et al., *Phytochemistry*, **15**, 1573 (1976).
12. P. S. Benoit et al, *Lloydia*, **39**, 160 (1976).
13. H. Wagner et al., *Planta Med.*, **51**, 139 (1985).

# BORAGE

**Source:** ***Borago officinalis*** L. (Family Boraginaceae).

## GENERAL DESCRIPTION

Coarse, hispid annual, 15–100 cm high; leaves rough, wrinkled; flowers blue, star shaped with protruding cone; indigenous to dry, waste places of south Europe; grown as an ornamental or pot herb; naturalized in central, eastern, and western Europe; established as a casual weed in the eastern United States. The parts used are the nutlets (seeds), leaves, and flowers.

## CHEMICAL COMPOSITION

Leaves contain small amounts of pyrrolizidine alkaloids including lycopsamine, intermedine, their acetyl derivatives, supinine, supinidine, plus amabiline (trace amounts), choline (WREN); 9.1% fatty acids, including $\alpha$-linolenic acid (55%) and $\gamma$-linolenic acid (>4%); silicic acid (1.5%–2.2%); potassium, calcium, potassium nitrate (3%), acetic, lactic, and malic acids; $\delta$-bornesitol, cyanogens; fresh leaves contain up to 30% mucilage hydrolyzing to glucose, galactose, arabinose, and allantoin (especially in seedlings). Seed oil (28–38% lipids) of recent interest as the richest plant source of $\gamma$-linolenic acid. (GLA) 17–25%.[1,6] Other seed fatty acids include linoleic (38%), oleic (14.5–23%), palmitic (11%), and stearic (4.7%).[3,4] Seeds contain the rare pyrrolizidine thesinine (absent in flowers) and amabiline; seed oil without thesinine.[4]

## PHARMACOLOGY OR BIOLOGICAL ACTIVITIES

Seed oil of interest for GLA content as a prostaglandin precursor, especially for $PGE_1$; prostaglandins help regulate metabolic functions. Normal synthesis of GLA from linoleic acid via $\delta$-6-desaturase may be blocked or diminished in mammalian systems as the result of aging, diabetes, excessive carbohydrate intake, or fasting. GLA sources such as seeds of *Oenothera biennis* (see ***evening primrose***), various *Ribes* species, and borage serve as GLA sources for nutritional supplementation.[2–4] GLA is purported therapeutically promising in treatment of atopic eczema, premenstrual syndrome, diabetes, alcoholism, inflammation; preventive in heart disease and stroke.[5]

## USES

**Medicinal, Pharmaceutical, and Cosmetic.** Borage extract used in skin care products (NIKITAKIS).

**Food.** Seldom used in flavoring; fresh flowers with saline, cucumber-like flavor (free of toxic pyrrolizidine alkaloids) used in salads.

**Health Food/Herb Teas.** Capsulated seed oil products available as dietary supplement; dried tops sometimes used in teas.[4]

**Traditional Medicine.** Leaves reportedly used as diuretic, demulcent, emollient, expectorant, and refrigerant; in fevers, lung disease, colds; externally a poultice; a folk cancer remedy in breast or facial cancers (DUKE 2, FOSTER, WREN).

### COMMERCIAL PREPARATIONS

Crude leaves, leaf extract, and seed oil.

*Regulatory Status.* Undetermined in the United States. The seed oil has been awarded a DIN in Canada as a GLA dietary supplement for essential fatty acid deficiency.[4] Borage leaf subject of a German therapeutic monograph; use suspended due to pyrrolizidine alkaloid content.[7]

### REFERENCES

See the General References for AHPA; BLUMENTHAL; DUKE 2; FOSTER; GLEASON AND CRONQUIST; HARBOURNE AND BAXTER; NIKITAKIS; STEINMETZ; TUTIN 3; TYLER 1; UPHOF; WREN.

1. P. Quinn et al., *J. Amer. Soc. Hort. Sci.*, **114**, 511 (1989).
2. N. A. Beaubaire and J. E. Simon, *Acta Horticult.*, **208**, 101 (1987).
3. A. Whipkey et al., *JACOS*, **65**, 979 (1988).
4. D. Awang, *Can. Pharm. J.*, 121 (Mar. 1990).
5. J. Janick et al. in L. E. Craker and J. E. Simon, eds., *Herbs, Spices, and Medicinal Plants: Recent Advances in Botany, Horticulture, and Pharmacology*, Vol. 4, Oryx Press, Phoenix, Ariz., 1989, p. 145.
6. K. M. Laron et al., *J. Nat. Prod.*, **47**, 747 (984).
7. *Borago*, *Bundesanzeiger*, no. 127 (July 12, 1991).

## BORONIA ABSOLUTE

**Source:** ***Boronia megastigma*** Nees ex. Bartl. (Family Rutaceae).

### GENERAL DECISION

Shrub to about 3 m high; native to southwestern Australia. Part used is the flower, from which a concrète is produced in 0.4–0.8% yield by extraction with petroleum ether; the absolute is obtained by alcohol washing of the concrète in a yield of about 60%.[1]

### CHEMICAL COMPOSITION

Contains ionones as principal constituents, which are composed mainly of $\beta$-ionone (95%), with the rest being $d$-$\alpha$-ionone (KARRER),[2] eugenol, hydrocarbons (mostly heptacosane), and others.[3]

### PHARMACOLOGY OR BIOLOGICAL ACTIVITIES

Its major constituents, ionones, are reportedly allergenic (MERCK).

### USES

**Medicinal, Pharmaceutical, and Cosmetic.** Used mainly in expensive perfumes.

**Food.** Quite extensively used in fruit-type flavors (e.g., strawberry, raspberry, plum, and peach) in most major categories of food products, including alcoholic and nonalcoholic beverages, frozen dairy desserts, candy, baked goods, gelatins and puddings, meat and meat products, and condiments and relishes. Use levels are very low, with highest average maximum of about 0.001% (11.6 ppm) reported in baked goods.

### COMMERCIAL PREPARATION

Absolute.

*Regulatory Status*. Has been approved for food use (§172.510).

### REFERENCES

See the General References for ARCTANDER; FEMA; GUENTHER; TERRELL.

1. A. R. Penfold and J. L. Willis, *Econ. Bot.*, **8**, 316 (1954).
2. F. V. Wells, *Indian Perfum.*, **2**(Pt. 2), 27 (1958).
3. Y. R. Naves and G. R. Parry, *Helv. Chim. Acta*, **30**, 419 (1947).

# BROMELAIN

**Source:** ***Ananas comosus*** (L.) Merr. (syn. *A. sativus* Schult. f.) and varieties. (Family Bromeliaceae).

*Synonyms*. Bromelains, bromelin, and plant protease concentrate.

### GENERAL DESCRIPTION

Bromelains are sulfhydryl proteolytic enzymes obtained from the pineapple plant, a perennial herb with many varieties native to tropical America. Two kinds of bromelain are known: stem bromelain and fruit bromelain; they are derived from juices of the stem and fruit, respectively.

Commercial bromelain is usually stem bromelain and is generally prepared from the juice of pineapple wastes (mainly stems) by precipitation with organic solvents (e.g., acetone and methanol) or by ultrafiltration.[1] It is a mixture of several proteases (including carboxypeptidase) and small amounts of nonproteolytic enzymes such as acid phosphatase, peroxidase, and cellulase as well as several protease inhibitors that are polypeptides.[2–6] It usually comes in the form of a buff to light brown powder and ranges from odorless to slightly putrid, depending on the source or quality. Currently the principal sources are Hawaii, Japan, and Taiwan.

Commercial bromelain is slightly soluble in water and glycercol but insoluble in most organic solvents.

Commercial bromelain exhibits activity over the pH range of 3–10, with optima between pH 5 and 8, depending on the substrates.[6] Thus with gelatin and hemoglobin the optimal pH is 5–6; with casein it is 7–8.[6–8] Its optimal temperature range is generally 50–60°C and can withstand a temperature of up to about 70°C before inactivation occurs.[9] These temperatures are lower than those for papain.

### CHEMICAL COMPOSITION AND PROPERTIES

Stem bromelain (in its purified forms) is a basic glycoprotein with one oligosaccharide moiety and one reactive sulfhydryl group per molecule. The carbohydrate portion consists of three moles D-mannose, one mole each D-xylose and L-fucose, and two moles *N*-acetyl-D-glucosamine; this portion is covalently linked to an asparagine residue of the peptide chain through one of the *N*-acetyl-D-glucosamine components. This carbohydrate moiety does not seem to be essential for proteolytic activity.[10–12]

Stem bromelain has a molecular weight of 28,000 according to the most recent data;[13,14] previously reported values range from 18,000 to 36,000.[6,15–18] Its isoelectric point is pH 9.55.[15,18] Its principal amino terminal residue is valine, while its carboxy terminal residue is glycine.[14,17] Its complete structure has not been resolved.[19]

Fruit bromelain is an acidic protease with isoelectric point at pH 4.6.[7] Its molecular weight has been reported to be 18,000 by one group of investigators and 31,000 by others.[7,14,17] Its status as a glycoprotein is still in dispute.[7] It has alanine as its major amino terminal residue and glycine as the major carboxy terminal residue.[14,17]

Being sulfhydryl proteases like papain and ficin, both fruit and stem bromelains are inhibited by oxidizing agents such as hydrogen peroxide, methyl bromide, and iodoacetate, and certain metallic ions (e.g., $Zn^{2+}$). They are activated by reducing agents (e.g., cysteine, sulfide, bisulfite, and cyanide). Their lost activities due to inactivation by heavy metals can often be completely restored by EDTA and cysteine.[15] Other inactivators or inhibitors of stem bromelain include citric acid and gluconic acid[20] and seven polypepides present in crude bromelain.[5] Other stem bromelain activators include $Mg^{2+}$.[9] It has been claimed that commercial bromelain is activated and stabilized by certain metallic salts ($Pb(OAc)_2$, $PbCl_2$, $BaCl_2$, $SrCl_2$, and $MgSO_4$), the effect being enhanced by cysteine or thiomalic acid, and that benzoate ions have stabilizing effects on bromelain during its preparation.[21,22] It has also been claimed that commercial bromelain as well as ficin and papain can be stabilized in solution with excess amounts of glycerol and small amounts of partially hydrolyzed and solubilized collagen.[23]

Like papain, stem bromelain has broad specificity, hydrolyzing various proteinaceous substrates (e.g., proteins, amides, esters, and small peptides).

## PHARMACOLOGY OR BIOLOGICAL ACTIVITIES

Stem bromelain has been reported to exert a wide variety of pharmacological effects, including burn debridement, antiinflammatory action, prevention of epinephrine-induced pulmonary edema, smooth muscle relaxation, stimulation of muscle contractions, inhibition of blood platelet aggregation, enhanced antibiotic absorption, cancer prevention and remission, ulcer prevention, sinusitis relief, appetite inhibition, shortening of labor, and enhanced excretion of fat.[24–32] The precise nature of these effects (some of which are not produced by other proteases such as ficin, papain, and trypsin) are not clear.[26,27] It has been postulated that a minor enzymatic component present in bromelain is responsible for the release of a kinin, which stimulates the production of prostaglandin $E_1$-like compounds; these $PGE_1$-like compounds are then responsible for the physiological activities of bromelain.[27]

Bromelin was found to stimulate digestion and absorption in healthy guinea pigs, and significantly prevented the morphological, enzymatic, and functional impairment of the jejunal mucosa in experimental methotrexate-induced cytostatic enteropathy.[33]

Bromelain has been reported to be lethal toward certain plant parasitic nematodes *in vitro* and in the soil.[34]

Bromelain is inactivated in the blood *in vivo* and *in vitro*.[28] In therapeutic doses, it may cause nausea, vomiting, diarrhea, skin rash, and menorrhagia.

Preliminary studies have shown that bromelain-induced rat paw edema may serve as a useful model for evaluating the effects of nonsteroidal nonacidic antiinflammatory agents.[35]

As with papain and other proteases, bromelain may cause contact allergenic reactions in certain individuals.

## USES

**Medicinal, Pharmaceutical, and Cosmetic.** Used primarily in preparations to treat inflammation and edema associated with surgical or accidental trauma, infections, or allergies. The rationale for such uses has been challenged.[30] A patent has been issued for its use in a debriding tape for eschar and necrotic tissues.[31]

Bromelain is used in certain cosmetics, such as facial cleansers and bath preparations (see also ***papain***).

**Food.** Due to the high cost of papain, bromelain is increasingly used to replace or supplement papain usage. Current major uses of bromelain are in meat tenderizing, manufacturing precooked cereals, modifying dough (bread, wafers, pizza, etc.), and in chill-proofing beer.

Other uses or potential uses include preparation of protein hydrolyzates, liquefying fish protein to facilitate fish oil extraction, clarifying fruit juices, and manufacture of sausage casings as well as their removal from sausages.

**Health Food/Herb Teas.** Used in some vitamin and herbal formulations, mainly as a digestive aid (MARTINDALE).

**Others.** Used in bating hide and in desizing fabrics.

## COMMERCIAL PREPARATIONS

Available in numerous grades with different activities that are expressed in different enzyme units, depending on the suppliers (see ***papain***). Bromelain is official in F.C.C.

## REFERENCES

See the General References for BAILEY 1; MARTINDALE; MERCK; TERRELL; USD 26th.

1. R. M. Heinicke, U.S. Pat. 3,002,891 (1961).
2. M. El-Gharbawi and J. R. Whitaker, *Biochemistry*, **2**, 476 (1963).
3. K. R. Lynn, *Anal. Biochem.*, **77**, 33 (1977).
4. E. Doi et al., *J. Biochem.*, **75**, 1063 (1974).
5. R. L. Heinrikson and F. J. Kézdy in L. Lorand, ed., *Methods in Enzymology*, Vol. 45, Part B, Academic Press, New York, 1976, p. 740.
6. W. Berndt et al., *Z Gaströnterol.*, **6**, 185 (1968); through *Chem. Abstr.*, **70**, 54276h (1969).
7. F. Yamada et al., *J. Biochem.* **79**, 1223 (1976).
8. A. Yamamoto in G. Reed, ed., *Enzymes in Food Processing*, 2nd ed., Academic Press, New York, 1975, p. 123.
9. Y. M. Chen and H. Y. Lin, *Taiwania*, **17**, 266 (1972); through *Chem. Abstr.*, **78**, 145173d (1973).
10. Y. Yasuda et al., *Biochemistry.*, **9**, 25 (1970).
11. N. Takahashi and T. Murachi in R. L. Whistler and J. N. BeMiller, eds., *Methods in Carbohydrate Chemistry*, Vol. 7, Academic Press, New York, 1976, p. 175.
12. Y. Yasuda et al., *Biochemistry*, **10**, 2624 (1971).
13. N. Takahashi et al., *J. Biochem.*, **74**, 355 (1973).
14. T. Murachi in L. Lorand, ed., *Methods in Enzymology*, Vol. 45, Part B, Academic Press, New York, 1976, p. 475.
15. I. E. Liener, *Adv. Chem. Ser.*, **136**, 202 (1974).
16. G. Feinstein and J. R. Whitaker, *Biochemistry*, **3**, 1050 (1964).
17. S. Ota et al., *Biochemistry*, **3**, 180 (1964).
18. T. Murachi et al., *Biochemistry*, **3**, 48 (1964).
19. K. Goto, *FEBS Lett.*, **62**, 93 (1976).
20. L. A. Marin et al., *Rev. Colegio Quim. Puerto Rico*, **16**, 39 (1959); through *Chem. Abstr.*, **53**, 22119h (1959).
21. N. Hazama, Jpn. Kokai 7137,866

(1971); through *Chem. Abstr.*, **76**, 43476w (1972).
22. R. M. Heinicke, U.S. Pat. 3,293,143 (1966).
23. T. Cayle, U.S. Pat. 3,296,094 (1967).
24. T. Enomoto et al., *Jpn. J. Pharmacol.*, **17**, 331 (1967).
25. S. Mineshita and Y. Nagai, *Jpn. J. Pharmacol.*, **27**, 170 (1977).
26. W. M. Cooreman et al., *Pharm. Acta. Helv.*, **51**, 73 (1976).
27. G. E. Felton, *Hawaii Med. J.*, **36**(2) 39 (1977).
28. International Commission on Pharmaceutical Enzymes, *Farm. Tijdschr. Belg.*, **54**, 85 (1977).
29. S. J. Taussig and S. Batkin, *J. Ethnopharmacol*, **22**, 191 (1988).
30. S. Kumakura et al., *Eur. J. Pharmacol.*, **150**, 295 (1988).
31. L. L. Bolton and B. E. Constantine, Eur. Pat. Appl. EP 194, 647 (1986); through *Chem. Abstr.*, **106**, 201776p (1987).
32. H. R. Maurer et al., *Planta Med.*, **54**, 377 (1988).
33. F. Barbarine et al., *J. Nucl. Med. Allied Sci.*, **26**, 2, 97 (1982).
34. P. M. Miller and D. C. Sands, *J. Nematol.*, **9**, 192 (1977); through *Chem. Abstr.*, **87**, 162156d (1977).
35. C. S. Morimoto et al., *Gnsho* **7**, 563 (1987); through *Chem. Abstr.*, **108**, 1239676 (1988).
36. Anon., *Fed. Reg.*, **40**(143), 30995 (1975).

# BROOM TOPS

**Source:** ***Cytisus scoparius*** (L.) Link (syn. *Sarothamnus scoparius* (L.) Wimm.; *S. vulgaris* Wimm.; *Spartium scoparium* L.) (Family Leguminosae or Fabaceae).

*Synonyms.* Scotch broom, scoparius, Irish broom, hogweed, and bannal.

## GENERAL DESCRIPTION

Deciduous shrub with erect slender branches, up to about 3 m high; native to central and southern Europe; naturalized in North America; also grows in Asia and South Africa. Parts used are the dried flowering tops collected just before blooming.

Scotch broom should not be confused with its relative Spanish broom, which is *Spartium junceum* L. (see ***genet***).

## CHEMICAL COMPOSITION

Contains alkaloids, including sparteine (ca. 0.3%), genisteine, and sarothamnine; simple amines (tyramine, hydroxytyramine, epinine, etc.); scoparin (scoparoside); pigments (e.g., taraxanthin and flavoxanthin); amino acids; volatile oil; tannin; wax; fat; and sugars (KARRER).[1]

## PHARMACOLOGY OR BIOLOGICAL ACTIVITIES

The activity of broom tops is the result of its alkaloids (principally sparteine), which have cardiac depressant and curarelike properties and are highly toxic (MARTINDALE, SAX).

Sparteine also has oxytocic properties (MERCK).

Broom tops is regarded to have diuretic,, emetic, and cathartic properties. Due to tyramine, use may cause blood pressure crisis with simultaneous administrations of MAO-inhibitors.[2]

## USES

**Medicinal, Pharmaceutical, and Cosmetic.** Used in certain laxative, diuretic, and tonic preparations.[2]

**Health Food/Herb Teas.** Teas, capsules, etc., primarily as a diuretic.

**Traditional Medicine.** Reportedly used internally as a diuretic and externally to treat sore muscles, abscesses, and swellings; flowers used in hair rinses for their lightening and brightening effects (LUST, ROSE, UPHOF).

Broom flowers, seeds, and root as well as the whole herb have been reportedly used in treating tumors.[3]

#### COMMERCIAL PREPARATIONS

Crude and extracts; crude was formerly official in N.F. and U.S.P. Strengths (see ***glossary***) of extracts are based on weight-to-weight ratios.

*Regulatory Status.* Not determined in the United States. Subject of a German therapeutic monograph, with preparations not to contain more than 1 mg/mL of sparteine for treatment of functional heart and circulatory disorders.[2]

#### REFERENCES

See the General References for ARCTANDER; BAILEY 1; BIANCHINI AND CORBETTA; BLUMENTHAL; GOSSELIN; LIST AND HÖRHAMMER; LUST; MERCK; ROSE; TERRELL; TYLER 1; UPHOF.

1. K. Egger, *Planta*, **80**, 65 (1968).
2. Monograph *Cytisi scoparii herba*, *Bundesanzeiger*, no. 11 (Jan. 17, 1991).
3. J. L. Hartwell, *Lloydia*, **33**, 97 (1970).

# BUCHU

**Source:** ***Agathosma betulina*** (Berg.) Pillans (syn. *Barosma betulina* (Berg.) Bartl. et Wendl.) or ***A. crenulata*** (L.) Pillans (syn. *B. crenulata* (L.) Hook. and *B. serratifolia* (Curt.) Willd.) (Family Rutaceae).

*Synonyms.* Bookoo, buku, diosma, bucku, and bucco; round buchu (*A. betulina*) and long buchu (*A. crenulata*).

#### GENERAL DESCRIPTION

Low shrubs usually less than 2 m high with opposite and/or alternate leaves that are finely toothed at the margin and bear oil glands beneath and at the base of the teeth; native to South Africa. Parts used are the dried leaves, from which an essential oil is obtained by steam distillation. There are two types of buchu leaves. Round or short buchu is from *A. betulina*, and long buchu is from *A. crenulata*; they differ in their relative compositions of their volatile oils. Major supplies of the leaves (from both wild and cultivated plants) come from the Cape Province of South Africa, which also produces some of the world's supply of the oil; the rest of the oil is distilled in Europe and the United States.

#### CHEMICAL COMPOSITION

Contains 1.0–3.5% (usually 1.5–2.5%) volatile oil that is composed mainly of *l*-pulegone, isopulegone, diosphenol (buchu camphor), *ψ*-diosphenol, *l*-isomenthone, *d*-menthone, *d*-limonene, 8-mercapto-*p*-menthan-3-one, 8-acetylthio-*p*-menthan-3-one, and piperitone epoxide, among more than 100 other identified minor compounds;[1–6] presence of piperitone epoxide disputed.[6] Relative proportions of pulegone and diosphenols vary considerably in commercial oils, but in general round buchu (*A.*

*betulina*) leaf yields oils with high proportions of diosphenols, while long buchu (*A. crenulata*) yields oils containing high proportions of pulegone with little or no diosphenols present.[1,4,6] In addition, round buchu leaf contains higher concentrations of volatile oil than long buchu leaf.

Other constituents present in buchu leaf include flavonoids (diosmin, rutin, quercetin-3,7-glucoside, etc.), resin, mucilage, and others (KARRER, LIST AND HÖRHAMMER, MERCK).

## PHARMACOLOGY OR BIOLOGICAL ACTIVITIES

Reported to have urinary antiseptic, diuretic, and carminative properties.

## USES

**Medicinal, Pharmaceutical, and Cosmetic.** Extensively used in diuretic preparations; also in laxative, stomachic, and carminative formulas.

**Food.** Oil is used as a component in artificial fruit flavors, especially black currant flavor. Round buchu oil is preferred because of its higher contents of diosphenols and 8-mercapto-*p*-menthan-3-one, which are considered to be the more desirable flavor components.[6] Major categories of food products in which the oil is used include alcoholic and nonalcoholic beverages, frozen dairy desserts, candy, baked goods, gelatins and puddings, and condiments and relishes. Use levels reported are rather low, with average maximum highest in gelatins and puddings at about 0.002% (15.4 ppm).

**Health Food/Herb Teas.** Capsules, tablets, crude herb in teas reportedly as a diuretic and urinary antiseptic (WREN).

**Traditional Medicine.** Reportedly used as urinary tract disinfectant in disorders such as cystitis, urethritis, and others; also as a diuretic, tonic, and stimulant.[7]

## COMMERCIAL PREPARATIONS

Crude, extracts, and oil; crude and fluid extract were formerly official in N.F. Strengths (see ***glossary***) of extracts are expressed in weight-to-weight ratios. Buchu leaf is the subject of a German therapeutic monograph; however use as a urinary tract antiinflammatory and diuretic is not recommended since effectiveness is not well documented. Leaf is allowed as an aroma or flavor corrigent in teas.[7]

*Regulatory Status.* Has been approved for food use (§172.510).

## REFERENCES

See the General References for APhA; ARCTANDER; BLUMENTHAL; CLAUS; GUENTHER; LIST AND HÖRHAMMER; LUST; MARTINDALE; TERRELL; WREN; YOUNGKEN.

1. A. A. J. Fluck et al., *J. Sci. Food Agr.*, **12**, 290 (1961).
2. E. Klein and W. Rojahn, *Dragoco Rep.*, **14**, 183 (1967).
3. E. Sundt et al., *Helv. Chim. Acta*, **54**, 1801 (1971).
4. K. L. J. Blommaert and E. Bartel, *J. S. Afr. Bot.*, **42**, 121 (1976); through *Chem. Abstr.*, **87**, 58401u (1977).
5. D. Lamparsky and P. Schudel, *Tetrahedron Lett.*, **36**, 3323 (1971).
6. R. Kaiser et al., *J. Agr. Food Chem.*, **23**, 943 (1975).
7. Monograph *Barosmae folium*, *Bundesanzeiger* (Feb. 1, 1990).

# BUCKTHORN, ALDER

**Source:** ***Frangula alnus*** Mill. (syn. *Rhamnus frangula* L.) (Family Rhamnaceae).

*Synonyms*. Frangula, buckthorn, glossy buckthorn, arrow wood, and black dogwood.

## GENERAL DESCRIPTION

Shrub or small tree with shiny, dark green, short-oblong to obovate leaves 3–7 cm long; up to 6 m high; native to Europe, western Asia, and northern Africa; naturalized in North America. Part used is the dried bark aged for 1 year to rid it of an emetic principle (see ***cascara***).

## CHEMICAL COMPOSITION

Contains 3–7% anthraquinone glycosides as active principles, which include glucofrangulin A and B, frangulin A and B, emodin-1-glucoside, emodin-8-glucoside, emodin-8-*O*-$\beta$-gentiobioside, and others.[1–9] Other constituents present include an alkaloid, armepavine, which is present in fresh bark but not in dried bark[10]; tannins; flavonoids;[11] and free anthraquinones, among others (STAHL). The fresh bark also contains anthrones and anthrone glycosides which are believed to constitute the emetic principle[12]; they are oxidized to anthraquinones or their glycosides on storage.

## PHARMACOLOGICAL OR BIOLOGICAL ACTIVITIES

The active principles of buckthorn bark (anthraglycosides) are cathartic; they act on the large intestine (colon), with the diglycosides being more active than the monoglycosides (see also ***cascara*** and ***senna***).[12,13]

Aloe emodin isolated from the seed of buckthorn has been reported to have significant inhibitory activity against P-388 leukemia in mice; it exhibited such activity only when administered as a suspension in acetone-Tween 80.[14]

## USES

**Medicinal, Pharmaceutical, and Cosmetic.** Used in certain laxative preparations, more commonly in Europe. Extracts are used in sunscreen preparations.[15,16]

**Traditional Medicine.** Used as a laxative and tonic; also reportedly used in treating cancers and as a component in Hoxsey cancer "cure."[17]

## COMMERCIAL PREPARATIONS

Crude and extracts; crude and fluid extract were formerly official in N.F. and U.S.P. Strengths (see ***glossary***) of extracts are expressed in weight-to-weight ratios.

## REFERENCES

See the General References for APhA; BAILEY 2; BIANCHINI AND CORBETTA; FERNALD; LUST; MERCK; STAHL; TERRELL; YOUNGKEN.

1. B. Kaminski and W. Grzesiuk, *Farm. Pol.*, **33**, 157 (1977); through *Chem. Abstr.*, **87**, 73290a (1977).
2. A. V. Gotsiridze and E. P. Kemertelidze, *Rast. Resur.*, **13**, 64 (1977); through *Chem. Abstr.*, **86**, 161173z (1977).
3. H. Auterhoff and E. Eujen, *Deut. Apoth. Ztg.*, **112**, 1533 (1972).
4. H. Wagner and G. Demuth, *Tetrahedron Lett.*, **49**, 5013 (1972).
5. M. Rosca and V. Cucu, *Planta Med.*, **28**, 343 (1975).
6. K. Savonius, *Farm. Aikak.*, **82**(9–10), 136 (1973); through *Chem. Abstr.*, **81**, 41319b (1974).

7. M. Rosca and V. Cucu, *Planta Med.*, **28**, 178 (1975).
8. A. Bonati and G. Forni, *Fitoterapia*, **48**, 159 (1977).
9. G. Dermuth et al., *Planta Med.*, **33**, 53 (1978).
10. M. Pailer and E. Haslinger, *Monatsh. Chem.*, **103**, 1399 (1972).
11. J. C. Dauguet and R. R. Paris, *C. R. Acad. Sci. Ser. D*, **285**, 519 (1977); through *Chem. Abstr.* **87**, 197383c (1977).
12. F. H. L. van Os, *Pharmacology*, **14**(Suppl. 1), 7, 18 (1976).
13. F. A. Nelemans, *Pharmacology*, **14**(Suppl. 1), 73 (1976).
14. S. M. Kupchan and A. Karim, *Lloydia*, **39**, 223 (1976).
15. G. Prosperio, *Cosmet. Toilet.*, **91**, 34 (1976).
16. S. Bader et al., *Cosmet. Toilet.*, **96**(10), 67 (1981).
17. J. L. Hartwell, *Lloydia*, **34**, 103 (1971).

# BURDOCK

**Source:** ***Arctium lappa*** L. (syn. *A. majus* Bernh.); ***Arctium minus*** Bernh. (Family Compositae or Asteraceae).

*Synonyms*. Great burdock, edible burdock, great bur, bardana, clotbur, lappa, and beggar's buttons.

## GENERAL DESCRIPTION

*A. lappa* is a biennial or perennial herb up to about 3 m high in its second year of growth; native to Asia and Europe; naturalized in North America. Part used is the dried first-year root collected in the fall; fruits and leaves are also used.

Common burdock (*A. minus*) is also used; it resembles *A. lappa* but is smaller.

## CHEMICAL COMPOSITION

Root contains inulin (up to about 50%); polyacetylenes (0.001–0.002%, dry-weight basis) consisting mainly of 1,11-tridecadiene-3,5,7,9-tetrayne and 1,3,11-tridecatriene-5,7,9-triyne;[1] arctic acid (an acetylenic acid containing S); volatile acids (acetic, propionic, butyric, isovaleric, 3-hexenoic, 3-octenoic, costic, etc.) and nonhydroxy acids (lauric, myristic, stearic, palmitic, etc.);[2,3] a crystalline plant hormone;[4] $\gamma$-guanidino-*n*-butyric acid; tannin; polyphenolic acids (e.g., caffeic and chlorogenic), and others (KARRER, JIANGSU).[5]

Seeds contain 15–30% fixed oils; a bitter glucoside (arctiin); chlorogenic acid; two lignans (lappaols A and B); a germacranolide, and others (KARRER).[5,6]

Leaves have been reported to contain arctiol (8$\alpha$-hydroxyeudesmol), $\Delta^{9(10)}$-fukinone (dehydrofukinone), fukinone, fukinanolide, $\beta$-eudesmol, petasitolone, eremophilene, and taraxasterol, among others.[7]

## PHARMACOLOGY OR BIOLOGICAL ACTIVITIES

Burdock (fruit and root) is reported to have diuretic, diaphoretic, antipyretic, antimicrobial, and antitumor activities (FARNSWORTH, JIANGSU, LIST AND HÖRHAMMER). The antimicrobial activity was attributed to the polyacetylenes present.[1]

Burdock fruit extracts have exhibited distinct hypoglycemic activity in rats.[8]

Fresh root juice has antimutagenic properties.[9] The active agent is reported to be polyanionic with a molecular weight of >300,000; prevalent data suggest it to be a ligninlike compound containing 10% sugar.[10,11]

## USES

**Medicinal, Pharmaceutical, and Cosmetic.** Used in some diuretic, laxative, and other preparations. Also reportedly useful in cosmetic and toiletry preparations for its alleged skin-cleansing properties (deNAVARRE);[12] used in hair tonic and antidandruff preparations.

**Food.** Root is used as a food in Asia.

**Health Food/Herb Teas.** Seeds used in cold remedies. Leaves used in teas, combination products, primarily as "blood purifier" for skin ailments (acne, psoriasis, etc.). Root used as nutritive food (FOSTER AND DUKE).

**Traditional Medicine.** Root, leaves, and seeds (fruits) of both species have been used in treating cancers;[13] decoctions or tea of root have been used in rheumatism catarrh, gout, and stomach ailments. Also used as a diuretic, diaphoretic, and mild laxative, among others. Decoctions and teas of roots and leaves are also used both externally and internally for skin problems (e.g., eczema and scaly skin) (FOSTER AND DUKE, LUST, TYLER 1).

In Chinese medicine, roots collected are from plants that are at least two years old.

Dried or roasted fruits of *A. lappa* are widely used in Chinese medicine to treat colds, sore throat, tonsillitis, coughs, measles, sores and abscesses, among others, usually in combination with other drugs (JIANGSU).

## COMMERCIAL PREPARATIONS

Crude and extracts; crude root was formerly official in N.F. and U.S.P. Strengths (see ***glossary***) of extracts are expressed in weight-to-weight ratios.

*Regulatory Status.* Not determined in the United States. A German therapeutic monograph on burdock root does not recommend use since efficacy is not confirmed.[14]

## REFERENCES

See the General References for BAILEY 2; FARNSWORTH; FERNALD; FOGARTY; FOSTER AND DUKE; JIANGSU; KROCHMAL AND KROCHMAL; LUST; NANJING; ROSE; TERRELL; TYLER 1; YOUNGKEN.


1. K. E. Schulte et al., *Arzneim.-Forsch.*, **17**, 829 (1967).
2. S. Obata et al., *Nippon Nogei Kagaku Kaishi*, **44**, 437 (1970); through *Chem. Abstr.*, **74**, 108136r (1971).
3. T. Washino et al., *Nippon Nogei Kagaku Kaishi*, **59**, 389 (1985); through *Chem. Abstr.*, **103**, 52880b (1985).
4. T. Kimura, Jpn. Kokai 76115,914 (1976); through *Chem. Abstr.*, **86**, 66856r (1977).
5. Y. Yamada et al., *Phytochemistry*, **14**, 582 (1975).
6. A. Ichihara et al., *Tetrahedron Lett.*, **44**, 3961 (1976).
7. K. Naya et al., *Chem. Lett.*, **3**, 235 (1972).
8. L. O. Lapinina and T. F. Sisoeva, *Farm., Zh. (Kiev)*, **19**, 52 (1964); through *Chem. Abstr.*, **66**, 1451e (1967).
9. K. Morita et al., *Agr. Biol. Chem.*, **42**, 1235 (1978).
10. K. Morita et al., *Agr, Biol. Chem.*, **49**, 925 (1985); through *Chem. Abstr.*, **103**, 2475c (1985).
11. K. Morita et al., *Mutat. Res.*, **129**, 25 (1984); through *Chem. Abstr.*, **102**, 19347q (1985).
12. H. B. Heath, *Cosmet. Toilet.*, **92**, 19 (1977).
13. J. L. Hartwell, *Lloydia*, **31**, 71 (1968).
14. Monograph *Bardanae radix*, *Bundesanzeiger* (Feb. 1, 1990).

# CADE OIL

**Source:** ***Juniperus oxycedrus*** L. (Family Pinaceae).

*Synonyms.* Oil of cade, juniper tar, and oil of juniper tar.

## GENERAL DESCRIPTION

*Juniperus oxycedrus* is a shrub or small tree known as prickly cedar, up to about 4 m high; native to the Mediterranean region. Cade oil is obtained by destructive distillation of the branches and wood, usually in the form of shavings or chips; the resultant distillate separates into three layers of which the uppermost dark brown viscous layer is cade oil. Rectified cade oil is obtained by steam or vacuum distillation of crude cade oil.

## CHEMICAL COMPOSITION

Volatile oil contains $\delta$-cadinene as the major component (27.3%), $\beta$-caryophyllene isomer (2.0%), $\alpha$-cedrene (1.6%), $\gamma_1$-cadinene (1.3%), $\gamma_1$-muurolene (5.7%), dihydrocurcumene (3.6%), $\gamma$-gurjunene (4.9%), $\alpha$-muurolene (3.9%), $\gamma_2$-cadinene (5.3%), cubebene (2.0%), calamenene (0.3%), calacacorne (15.6%), T-muurolol (1.2%), 1,6,-dimethyl-4-isopropylnapththalene (2.7%), $\alpha$-cadinol (0.7)%,[1,2] *p*-cresol, guaiacol, viridiflorol, and others (LIST AND HÖRHAMMER; MARTINDALE).

## PHARMACOLOGY OR BIOLOGICAL ACTIVITIES

Reported to have keratolytic and antipruritic properties; also exhibited antimicrobial activities *in vitro*.[3,4]

Has been considered as an allergen (SAX), though other available data show it to be relatively nontoxic.[5]

## USES

**Medicinal, Pharmaceutical, and Cosmetic.** Cade oil is widely used in topical preparations for the treatment of parasitic skin diseases and eczema; in antiseptic wound dressings, analgesic and antipruritic preparations, and dermatologic creams and ointments; also in antidandruff shampoos, among others. Rectified cade oil is used as a fragrance component in soaps, detergents, creams, lotions, and perfumes. Maximum use level reported is 0.2% in perfumes.[5]

**Traditional Medicine.** Used in treating various skin disorders and problems of the scalp and hair loss; also used in cancers.[6]

## COMMERCIAL PREPARATIONS

Crude and rectified. Cade oil (juniper tar oil) is official in U.S.P.

## REFERENCES

See the General References for ARCTANDER; BAILEY 2; CLAUS; GUENTHER; LUST; MARTINDALE; ROSE.

1. J. C. Chalchat et al., *Flav. Frag. J.*, **3**, 19 (1988).
2. B. M. Lawrence, *Perfum. Flavor*, **14**(3), 71 (1989).
3. J. C. Maruzzella and L. Liguori, *J. Am. Pharm. Assoc.*, **47**, 250 (1958).
4. J. C. Maruzzella and P. A. Henry, *J. Am. Pharm. Assoc.*, **47**, 294 (1958).
5. D. L. J. Opdyke, *Food Cosmet. Toxicol.*, **13**(Suppl.), 733 (1975).
6. J. L. Hartwell, *Lloydia*, **33**, 288 (1970).

# CAJEPUT OIL

**Source:** ***Melaleuca leucadendra*** (L.) L.; ***M. quinquenervia*** S. T. Blake; ***M. alternifolia*** (Maiden & Betche) Cheel; or other ***Melaleuca*** species (syn: *M. leucadendra* = *M. leucadendron* L.) (Family Myrtaceae).

*Synonyms*. Cajuput, punk tree, and paperbark tree oils; tea tree oil (*M. alternifolia*).

## GENERAL DESCRIPTION

Large evergreen trees with whitish spongy bark; up to about 30 m high; native to Australia and southeastern Asia. Introduced into south Florida in the early 20th century, *M. quinquenervia* has invaded three of the four major ecosystem types of south Florida, including sawgrass prairies and mangrove and cypress swamps. It is spread by fire, resprouting quickly, flowering prolifically within 1 week of a fire. A burned tree can release millions of seeds dispersed by wind and water. The tree is considered a major ecological threat to the Everglades National Park, Big Cypress National Preserve, and Loxahatchee National Wildlife Refuge. Consequently, it has been proposed as a "noxious weed" pursuant to the Federal Noxious Weed Act of 1974.[1]

Cajeput oil is obtained by steam distillation of the fresh leaves and twigs.

## CHEMICAL COMPOSITION

Contains 14–65% cineole (depending on the ages and sources of the trees used) or terpinen-4-ol (up to 47%) as major components (JIANGSU, MERCK)[2,3] Other compounds reportedly present include 3,5-dimethyl-4,6-di-*O*-methylphloroacetophenone (present at 10% concentration in a sample of oil obtained from *M. cajeputi*),[4] pinene, terpineol, nerolidol, and traces of benzaldehyde and valeraldehyde, among others (KARRER).

In 1985 a quality standard for tea tree oil (*M. alternifolia*) was established in Australia (Australian Standard, *Oil of Melaleuca*, "Terpinen-4-ol Type," AS 2782). The standard calls for terpinen-4-ol to be 30% or more of oil composition with less than 15% cineole; higher quality oils containing 40–47% terpinen-4-ol, and 2.5% cineole. Oils high in terpinen-4-ol, and low in cineole are considered best for predictable clinical results.[3]

## PHARMACOLOGY OR BIOLOGICAL ACTIVITIES

Considered to have carminative, stimulant, and diaphoretic properties. Reported to have antimicrobial and antiseptic properties.[4–7]

Available data indicate it to be nontoxic.[5]

A 1990 clinical trial involving 124 patients provided evidence of the effectiveness of *M. alternifolia* oil in the treatment of acne vulgaris. A 5% tea tree oil in a water-based gel was less effective than a 5% benzoyl peroxide in a water-based lotion because of slower onset of action. However, clinical assessment and self-reporting of side effects indicated that the tea tree oil preparation was better tolerated by facial skin with less skin scaling, dryness, pruritus and irritation than those experienced with the benzoyl peroxide preparation (TYLER).[8]

Nerolidol has antifeeding activity against gypsy moth larvae.[9]

## USES

**Medicinal, Pharmaceutical, and Cosmetic.** Cajeput oil is used in expectorant and tonic formulations; also in antiseptic liniments, among others. In dentistry, it is used in relieving discomfort due to dry sockets. Used as a fragrance component in soaps, detergents, creams, lotions, and perfumes, with maximum use level of 0.4% in the last category.[5]

**Food.** Cajeput oil is used as a flavor component in nonalcoholic beverages, frozen dairy desserts, candy, baked goods, meat and meat products, and condiments and relishes. Average maximum use level reported is very low, less than 0.001% (9.9 ppm).

**Health Food/Herb Teas.** Tea tree oil is found in numerous product forms, either singly or with other ingredients, with a broad range of health claims, including treatments for burns, sunburn, pimples, boils, stings, ringworm, sore throat, oral infections, bronchial congestion, lice, scabies, cuts, abrasions, and vaginal infections. While antiseptic activity is well established products are sold with a wide range of hyperbolic claims (TYLER).[3]

**Traditional Medicine.** Cajeput oil is used in colds, headaches, toothache, sore and aching muscles, rheumatism, and various skin diseases; also reportedly used in treating indolent tumors.[10]

**Others.** In natural disinfectants.

### COMMERCIAL PREPARATION

Oil; formerly officinal in U.S.P.

*Regulatory Status.* Has been approved for food use (§172.510). Oil subject of a German therapeutic monograph allowed in dose of 0.2 g or less in catarrhs of the upper respiratory tract.[11]

### REFERENCES


See the General References for ADA; ARCTANDER; FEMA; JIANGSU; GUENTHER; LEWIS AND ELVIN-LEWIS; LIST AND HÖRHAMMER; MORTON 2; ROSE; TYLER 1; UPHOF; YOUNGKEN.

1. *Fed. Reg.*, **56**, 201, 52005 (Oct. 17, 1991).
2. V. K. Sood, *Perfum. Essent. Oil Rec.*, **57**, 362 (1966).
3. A. Shemesh and W. L. May. *Int. J. Alt. Compliment. Med.* (Dec. 11, 1991).
4. J. B. Lowry, *Nature (London)*, **241**, 61 (1973).
5. D. L. J. Opdyke, *Food Cosmet. Toxicol.*, **14**, 701 (1976).
6. J. C. Maruzzella and L. Liguori, *J. Am. Pharm. Assoc.*, **47**, 250 (1958).
7. J. C. Maruzzella and P. A. Henry, *J. Am. Pharm. Assoc.*, **47**, 294 (1958).
8. I. B. Bassett et al., *Med. J. Aust.*, **153**, 455 (1990).
9. R. W. Doskotch et al., *J. Chem. Ecol.*, **6**, 845 (1980).
10. J. L. Hartwell, *Lloydia*, **33**, 288 (1970).
11. Monograph *Niauli aetheroleum*, *Bundesanzeiger* (Aug. 29, 1992).


# CALAMUS

**Source:** ***Acorus calamus*** L. (Family Araceae).

*Synonyms.* Sweet flag, sweet myrtle, sweet sedge, sweet root, and sweet cinnamon.

### GENERAL DESCRIPTION

Perennial herb growing in wet or swampy areas with stiff, sword-shaped leaves; up to about 2 m high; native to Northern Hemisphere (North America, Europe, and Asia). Part used is the stout aromatic rhizome after it is peeled and dried. The essential oil is obtained by steam distillation of both the fresh and the dried unpeeled rhizome. Roots are also reportedly used.

## CHEMICAL COMPOSITION

The rhizome contains highly variable amounts of volatile oil (0.5–10%, but usually 1.5–3.5%), depending on sources. Asian (Pakistani, Japanese, Indian) plants yield more oil than the European plants, but the European oil is considered superior in flavor and fragrance qualities (JIANGSU).[1–4] Constituents present in the oil include $\beta$-asarone, *cis*-methyl isoeugenol, asarone, asarylaldehyde, calamene, linalool, calamol, calameone, eugenol, methyl eugenol, azulene, pinene, cineole, camphor, and others, with $\beta$-asarone being the major component (up to 76% of the oil; 85% in Chinese oil) and the European type containing larger numbers of aromatic compounds (JIANGSU, MASADA).[4,5] The essential oil from the rhizome of the American variety (*A. calamus* L. var. *americanus* Wolff) does not contain the carcinogenic $\beta$-asarone.[6]

Other constituents present in the rhizome include acoragermacrone, acolamone, and isoacolamone (all sesquiterpenes),[7–10] acoradin, 2,4,5-trimethoxy benzaldehyde, 2,5-dimethoxybenzoquinone, galangin, sitosterol,[11] acoric acid, tannin, resin, and others.[12]

## PHARMACOLOGY OR BIOLOGICAL ACTIVITIES

The oil and extracts have been reported to have numerous pharmacological activities among which are the spasmolytic activities on isolated animal organs (smooth muscle) and hypotensive activities in cats and rabbits as well as anticonvulsant and CNS-depressant activities (FARNSWORTH 3, JIANGSU).[13–16]

Calamus oil is toxic, and the Jammu (Indian) variety has been reported to be carcinogenic in rats.[16–18]

$\beta$-asarone has antigonadal activity against insects.[19] It is also mutagenic per Ames test; so are commercial samples of the drug containing various concentrations of it.[20]

Indian calamus root oil repels houseflies.[21]

## USES

**Medicinal, Pharmaceutical, and Cosmetic.** Oil used as a fragrance component in soaps, detergents, creams, lotions, and perfumes. Maximum use level reported is 0.4% in perfumes.[16]

Root extract used in hair tonic and antidandruff preparations.

**Traditional Medicine.** Has been used for more than 2000 years in China to treat numerous disorders, including rheumatoid arthritis, strokes, epilepsy, gastritis, and lack of appetite; also externally in skin diseases (JIANGSU). Used in Western cultures for centuries as stomachic, carminative, sedative, febrifuge, and others. When chewed it is said to kill the taste for tobacco and to clear phlegm from the throat.

## COMMERCIAL PREPARATION

Oil; crude formerly official in U.S.P. and N.F.

*Regulatory Status*. Calamus and its derivatives (oil, extracts, etc.) are prohibited from use in human food (§189.110).

## REFERENCES

See the General References for ARCTANDER; BAILEY 2; CLAUS; FARNSWORTH 3; FOGARTY; FOSTER; GUENTHER; JIANGSU; KROCHMAL AND KROCHMAL; LUST; MARTINDALE; MERCK; MORTON 1; NANJING; ROSE; UPHOF.

1. G. Jukneviciene, *Liet. TSR Mokslu Akad. Darb. Ser. C*, **4**, 69 (1972); through *Chem. Abstr.*, **79**, 83389p (1973).
2. G. Pamakstyte-Jukneviciene, *Bot. Sady Pribaltiki*, **445** (1971); through *Chem. Abstr.*, **76**, 138157g (1972).
3. M. Raquibuddowla et al., *Sci. Res. (Dacca)*, **4**, 234 (1967); through *Chem. Abstr.*, **69**, 702e (1968).
4. G. Cavazza, *Ann. Falsif. Expert. Chim.*, **69**, 833 (1976); through *Chem. Abstr.*, **87**, 19017d (1977).
5. M. Jacobson et al., *Lloydia*, **39**, 412 (1976).
6. K. Keller and E. Stahl, *Planta Med.*, **47**, 71 (1983).
7. S. Yamamura et al., *Tetrahedron*, **27**, 5419 (1971).
8. M. Niwa et al., *Chem. Lett.* **9**, 823 (1972).
9. M. Iguchi et al., *Tetrahedron Lett.*, **29**, 2759 (1973).
10. M. Niwa et al., *Bull. Chem. Soc. Jpn.*, **48**, 2930 (1975); through *Chem. Abstr.*, **84**, 40737m (1976).
11. A. Patra and A. K. Mitra, *Indian J. Chem., Sect. B*, **17B**, 412 (1979).
12. E. G. El'yashevich et al., *Khim. Prir. Soedin.* **10**, 94 (1974); through *Chem. Abstr.*, **80**, 121271z (1974).
13. T. Shipochliev, *Vet. Med. Nauki*, **5**(6), 63 (1968); through *Chem. Abstr.*, **70**, 86144e (1969).
14. J. Maj et al., *Acta Pol. Pharm.*, **23**, 477 (1966); through *Chem. Abstr.*, **66**, 93653r (1967).
15. N. S. Dhalla and I. C. Bhattacharya, *Arch. Int. Pharmacodyn. Ther.*, **172**, 356 (1968).
16. D. L. J. Opdyke, *Food Cosmet. Toxicol.*, **15**, 623 (1977).
17. Anon. *Fed. Reg.*, **33**, 6967 (1968).
18. P. M. Jenner et al., *Food Cosmet. Toxicol.*, **2**, 327 (1964).
19. B. P. Saxena et al., *Nature*, **270**, 512 (1977).
20. W. Gogglemann and O. Schimmer, *Mutat. Res.*, **121**, 191 (1983); through *Chem. Abstr.*, **99**, 187, 715w (1983).
21. V. E. Adler and M. Jacobson, *J. Environ. Sci. Health*, **A17**, 667 (1982).

# CALENDULA

**Source:** ***Calendula officinalis*** L. (Family Compositae or Asteraceae).

*Synonyms.* Pot marigold and marigold.

## GENERAL DESCRIPTION

Hairy annual to perennial (absent freezing), 20–50 cm high; leaves oblanceolate to oblong, 7–17 × 1–4 cm; flowers yellow or orange; 4–7 cm in dia; widely cultivated ornamental in Europe and North America; naturalized in south and west Europe; origin undetermined. Parts used are the flower and herb; not to be confused with *Tagetes* species, also commonly known as marigolds (FOSTER).

## CHEMICAL COMPOSITION

Six saponins (glycosides A–D, $D_2$, and F) have been isolated and their structures determined; with a basic unit of oleanolic acid 3-*O*-β-D-glucuronide;[1] triterpenoids helinatriol C and F, ursadiol, 12-ursene-3,16,21-triol;[2] faradol, brein, arnidiol, erthrodiol, calenduladiol, longispoinogenine; calendulosides A, B, C, D, G and H; α- and β-amyrin, taraxasterol, τ-taraxasterol and lupeol; flavonoids include narcissin, rutin; trace amounts of essential oil; chlorogenic acid (GLASBY 2; WREN); polysaccharides including a rhamnoarabinogalactan and arabinogalactans;[3] flower yellow pigment is a mixture of $\beta$-carotene, lycopene, violaxanthin, and other xanthophylls (CSIR II).

## PHARMACOLOGY OR BIOLOGICAL ACTIVITIES

Flower, flower/herb preparations, and extracts antiinflammatory, immunomodulating, wound healing; stimulates granulation at wound site (and increasing glycoprotein, nucleoprotein and collagen metabolism at site); antibacterial; antifungal, antiviral; antiparasitic (trichomonacidal); stimulates phagocytosis *in vitro* and in the carbon clearance test in mice; choleretic; polysaccharides with *in vitro* and *in vivo* antitumor activity (ESCOP 3).

## USES

**Medicinal, Pharmaceutical, and Cosmetic.** In Europe preparations used for skin and mucous membrane inflammations, hard-to-heal wounds, mild burns, sunburn; internally for inflammatory lesions of the oral and pharyngeal mucosa; immunostimulant for skin inflammations and *Herpes zoster* infections (ESCOP 3).[4]

A biological additive to skin care products, including face, body, and hand creams and lotions; and night creams; ointments; shampoos (NIKITAKIS).

**Food.** Flowers primarily used as mildly saline flavoring and coloring; saffron substitute.

**Health Food/Herb Teas.** Flowers in tincture ("lotion") for external/internal use; teas (FOSTER).

**Traditional Medicine.** Flower historically considered vulnerary, antiseptic, styptic; externally used as lotion or ointment for burns and scalds (1st degree), bruises, cuts, rashes, sore nipples; internally for stomach ailments; gastric and duodenal ulcers, jaundice (FOSTER; WREN). Herb and its preparations reportedly used to stimulate circulation, promote healing; for gastric hemorrhage, ulcers, spasms, glandular swelling, jaundice, anemia; externally for abscesses, wounds, bleeding, eczema.[5]

## COMMERCIAL PREPARATIONS

Crude ligulate florets; flower heads; flower and herb; tincture; ointment, etc.; crude formerly official in U.S.P. and N.F.

*Regulatory Status.* GRAS as spice, flavoring and seasoning (§182.10). Flowers are subject of a positive German therapeutic monograph.[5] Herb subject of a German therapeutic monograph; however, therapeutic use is not recommended since claimed effectiveness is not demonstrated.[6]

## REFERENCES

See the General References for BLUMENTHAL; CSIR II; ESCOP 3; FOSTER; HARBOURNE AND BAXTER; NIKITAKIS; STEINMETZ; TUTIN 4; TYLER 1; UPHOF; WREN.


1. E. Vidal-Olliveier and G. Balansard, *J. Nat. Prod.*, **52**, 1156 (1989).
2. J. Pyrek et. al., *Pol. J. Chem.*, **53**, 1071 (1979).
3. J. Varljen et al., *Phytochemistry*, **28**, 2379 (1989).
4. ESCOP, Vol. 3. Proposals for European Monographs on *Calendulae flos/Flos cum Herba*.
5. Monograph *Calendula flos*, *Bundesanzeiger*, no. 50 (Mar. 13, 1986).
6. Monograph *Calendula herba*, *Bundesanzeiger* (July 14, 1993).

# CANANGA OIL

**Source:** ***Cananga odorata*** Hook. f. et Thoms. (syn. *Canangium odoratum* Baill. forma *macrophylla*) (Family Annonaceae).

### GENERAL DESCRIPTION

Large tree with fragrant flowers; native to islands of tropical Asia (Java, Malaysia, the Philippines, the Moluccas, etc.). The oil is obtained by water distillation of the flowers. A similar essential oil, ***ylang ylang oil*** (which see), is obtained in a similar manner from *Canangium odoratum* Baill. forma *genuina*.

### CHEMICAL COMPOSITION

Contains mainly $\beta$-caryophyllene, benzyl acetate, benzyl alcohol, farnesol, $\alpha$-terpineol, borneol, geranyl acetate, methyl salicylate, benzaldehyde, safrole, linalool, eugenol, isoeugenol, limonene, and other minor components totaling over 100 compounds (MASADA).[1]

### PHARMACOLOGY OR BIOLOGICAL ACTIVITIES

Cananga oil is nontoxic, except for causing irritation when applied full strength to rabbit skin.[2]

### USES

**Medicinal, Pharmaceutical, and Cosmetic.** Used as a fragrance component in soaps, detergents, creams, lotions, and perfumes (especially men's fragrances). Maximum use level reported is 0.8% in perfumes.[2]

**Food.** Used as a flavor ingredient in alcoholic and nonalcoholic beverages, frozen dairy desserts, candy, baked goods, and gelatins and puddings, with highest average maximum use level of about 0.003% (32.3 ppm) in the last category.

### COMMERCIAL PREPARATION

Oil; official in F.C.C.

*Regulatory Status.* GRAS (§182.20).

### REFERENCES

See the General References for ARCTANDER; BAILEY 2; FEMA; FURIA AND BELLANCA; GUENTHER; MASADA.

1. R. N. Duve et al., *Int. Flavours Food Addit.*, **6,** 341 (1975).
2. D. L. J. Opdyke, *Food Cosmet. Toxicol.*, **11**, 1049 (1973).

# CAPSICUM

**Source:** ***Capsicum frutescens*** L.; ***C. annuum*** L. and its varieties; ***C. chinense*** Jacq.; ***C. baccatum*** L. var. ***pendulum*** (Willd.) Eshbaugh (syn. *C. pendulum* Willd.); ***C. pubescens*** Ruiz et Pavon. (Family Solanaceae).

*Synonyms.* Cayenne pepper, red pepper, Tabasco pepper, hot pepper, and chili pepper.

### GENERAL DESCRIPTION

There has been much dispute and confusion regarding the classification of *Capsicum.* All peppers, hot and mild (not to be confused with black and white pepper), have been at one time or another considered as fruits of a single species, *C. annuum* and its varieties, or of two species, *C. annuum* and *C. frutescens*, and their varieties (ARCTANDER, BAILEY 2, UPHOF). Currently five species and their varieties are recognized, *C. frutescens*, *C. chinense*, *C. baccatum*, *C. pubescens*, and *C. annuum* (ROSENGARTEN, TERRELL).

*Capsicum annuum* is an annual herb up to 1 m high, while the other species are usually perennial woody shrubs, all native to tropical America and now widely cultivated. All five species yield pungent fruits commonly called red pepper or simply capsicum. Mild fruits commonly known as paprika, bell pepper, sweet pepper, or green pepper are usually produced by varieties of *C. annuum*.

Capsicum oleoresin is obtained by extracting red pepper with a suitable organic solvent; extraction of sweet pepper (paprika) with similar solvents yields paprika oleoresin, which contains high concentrations of carotenoids but little or no pungent principles, depending on the process.

## CHEMICAL COMPOSITION

Capsicum contains up to 1.5% (usually 0.1–1.0%) pungent principles, which are composed mainly of capsaicin, dihydrocapsaicin, nordihydrocapsaicin, homocapsaicin, and homodihydrocapsaicin, among others, with the last two in minor concentrations.[1–3] Other constituents present include carotenoids (capsanthin, capsorubin, carotene, lutein, etc.);[4,5] fats (9–17%), proteins (12–15%), vitamins A, C, and others; and a small amount of a volatile oil made up of more than 125 components of which 24 were identified, including 4-methyl-1-pentyl-2-methyl butyrate, 3-methyl-1-pentyl-3-methyl butyrate, and isohexyl isocaproate (JIANGSU, MARSH).[6]

Mild peppers (e.g., paprika and bell pepper) contain similar constituents as *Capsicum* but with little or no pungent principles. However, a variety of sweet pepper when allowed to ripen after harvest in continuous light has been reported to produce pungent principles.[7]

## PHARMACOLOGY OR BIOLOGICAL ACTIVITIES

Capsicum is a powerful local stimulant; its oleoresin or active principles (capsaicin) are strongly irritant to the eyes and tender skin, producing an intense burning sensation, but no vesication on the skin (MARTINDALE). Capsicum tincture and alcoholic extracts are considered as moderately to highly toxic and irritating to mucous membranes (GOSSELIN, SAX).

While a single dose of capsaicin activates pain, inflammation and hypersensitivity, repeated (long-term) application in appropriately formulated product forms, leads to desensitization, analgesic, and antiinflammatory activity. Capsiacin-induced analgesia and desensitization has been explained on the basis of neuropeptide release and depletion, but the specific mechanism of action has yet to be clearly established. Investigation of the effects of capsaicin treatment on different irritant-induced lipid peroxidation changes on free radical induced membrane damage has been given importance.[8]

A recent study on antiinflammatory action of capsaicin analogs suggested that the antioxidant nature of the methoxyphenol ring of capsaicin may interfere with the oxygen radical transfer mechanism common to lipoxygenase and cyclooxygenase pathways.[8,9]

Capsaicin causes a dose-related (1–100 nM) hemolysis of human red blood cells, is associated with significant changes in erythrocyte membrane lipid components (decreasing phospholipid and cholesterol content), and an acetylcholinesterase activity. In addition, reported alteration in membranes include calcium homeostasis, lysosomal leakage, and alterations in antioxidant enzyme defense systems.[10]

Capsicum, when fed at 10% of the diet, has been reported to produce liver tumors in rats.[11] Crude juice of fresh capsicum has exhibited antibacterial properties *in vitro*.[12]

## USES

**Medicinal, Pharmaceutical, and Cosmetic.** Capsicum tincture and oleoresin are used in counterirritant preparations to treat arthritis, rheumatism, neuralgia, lumbago, and chilblains; also used in certain preparations for stopping thumb sucking or nail biting in children.

**Food.** Capsicum, in whole and ground forms, is widely used as a spice.

Capsicum and its extracts and oleoresin are widely used in food products, including alcoholic and nonalcoholic beverages, frozen dairy desserts, candy, baked goods, gelatins and pudding, meat and meat products, and condiments and relishes, among others. Highest average maximum use levels are reported in alcoholic beverages for the oleoresin and extract, 0.09 and 0.12%, respectively.

Paprika and its oleoresin are primarily used as a colorant in all above food categories to impart a yellow to orange color to the foods.

**Health Food/Herb Teas.** Used as a synergistic ingredient in various herbal formulas, including general tonics, laxatives, sedatives, and hay fever remedies (FOSTER; LUST).

**Traditional Medicine.** Capsicum is used internally as stomachic, carminative, and stimulant and to treat diarrhea, cramps, colic, toothache, and other ailments; externally as counterirritant in rheumatism, arthritis, among others

## COMMERCIAL PREPARATIONS

Crude, capsicum oleoresin, and extracts (e.g., tincture), and paprika oleoresin. Both capsicum oleoresin and tincture were formerly official in N.F. and U.S.P.; pungency is determined by a taste test and is generally expressed in Scoville units. Paprika oleoresin comes in various color strengths.

*Regulatory Status.* Capsicum (red pepper, cayenne pepper) and paprika and their oleoresins and extractives are GRAS (§182.10 and §182.20); paprika and paprika oleoresin are also approved as color additives for food use exempt from certification (§73.340 and §73.345). Capsaicin-containing topical products are approved in over-the-counter and prescription drug form in the United States.

Standardized capsaicin products are approved in Germany for therapeutic use in painful muscle spasms in the shoulder, arms, and spine.[13] Low capsaicin-containing *Capsicum* products are the subject of a negative German monograph. Efficacy for digestive disturbances, and supportive treatment of heart and circulatory functions has not been scientifically established.[14]

## REFERENCES

See the General References for BLUMENTHAL; FEMA; FOSTER; JIANGSU; LIST AND HÖRHAMMER; LUST; MARSH; NANJING; ROSENGARTEN; STAHL; USD 26th.

1. J. A. Maga, *CRC Crit. Rev. Food Sci. Nutr.*, **6**, 177 (1975).
2. S. I. Balbaa et al., *Lloydia*, **31**, 272 (1968).
3. J. Jurenitsch and W. Kubelka, *Planta Med.*, **33**, 285 (1978).
4. C. E. C. Lord and A. S. L. Tirimanna, *Mikrochim. Acta*, **1**(4–5), 469 (1976).
5. B. Camara and R Monéger, *Phytochemistry*, **17**, 91 (1978).
6. L. W. Haymon and L. W. Aurand, *J. Agr. Food Chem.*, **19**, 1131 (1971).

7. K. Iwai et *al.*, *Agr. Biol. Chem.*, **41**, 1873, 1877 (1977).
8. A. K. De and J. J. Ghosh, *Phytother. Res.*, **7**(4), 273 (1993).
9. L. Brand et al. *Agents Actions*, **31**, 329 (1990).
10. A. K. De et al., *Phytother. Res.*, **7**(4), 310 (1993).
11. F. Homburger and E. Boger, *Cancer Res.*, **28**, 2372 (1968).
12. I. A. Abdou et al., *Qual. Plant. Mater. Veg.*, **22**, 29 (1972); through *Chem. Abstr.*, **78**, 80226b (1973).
13. Monograph *Capsicum (Paprika), Bundesanzeiger*, no. 22 (Feb. 1, 1990).
14. Monograph *Capsicum (Low Capsaicin Levels), Bundesanzeiger*, no. 80 (Apr. 27, 1989).

# CARAMEL COLOR

*Synonyms.* Burnt sugar coloring and caramel.

### GENERAL DESCRIPTION

Caramel color is produced by heating sugars (corn syrup, malt syrup, molasses, invert sugar, etc.) with small amounts of ammonia or ammonium salts under controlled temperature and pressure until the sweet taste is destroyed and the desired color is obtained. Small quantities of mineral acids, bases, or salts may also be added during heating. The temperature is usually between 100° and 140°C, depending on the pressure and the process used. The resulting product is usually a thick dark brown to black liquid.

Caramel color comes in various types with different characteristics, tinctorial strengths and varying acidic pHs. Thus acid-fast beverage color has pH 2.7–3.3, foaming-beverage color has pH 3.2–3.6, while bakers' and brewers' colors have pH 4.0–4.5. It usually has a slightly bitter taste and is soluble in water and dilute alcohol.[1]

Caramel for flavoring is prepared by heating milk and sugar; it is heated to a much lesser degree than caramel color and has a characteristic pleasant flavor but little tinctorial power.

### CHEMICAL COMPOSITION

Most of the constituents in caramel colors are still unknown. Types of compounds reported to be present in certain caramels include about 50% digestible carbohydrate, 25% nondigestible carbohydrate, and 25% melanoidins. Minor components include ammonia (0.2–2.0%), iron, and copper.[1,2] Available data also show the presence of substituted imidazoles such as 4-methylimidazole (0.005–0.1%) and other nitrogenous compounds.[2–5]

### PHARMACOLOGY OR BIOLOGICAL ACTIVITIES

Caramel color is one of the most commonly used food colors, yet its pharmacology and toxicology are practically unknown except for a long-term study on one type of caramel on rats, which indicates it to be noncarcinogenic at up to 6% of the diet.[6] However lymphocytopenia (reduction of blood lymphocytes) was present in both the long-term and short-term studies.[2,6]

Caramel was recently found to be mutagenic by both the Ames test and the chromosome aberration test.[7]

### USES

**Pharmaceutical, Cosmetic, and Food.** Used to produce pale yellow to dark brown colors in all types of plant extracts intended for drug, cosmetic, and food uses. Formerly, extracts produced by the botanical industry were mostly dark due to the antiquated technology used; it has been the tradition since to add caramel color to lighter extracts that are produced by modern methods to obtain the same color effects, as it is generally believed that a light extract means low strength. Furthermore, caramel color is used directly in practically every category of food product. Its ubiquitous presence in foods is matched perhaps only by salt or sugar. Its highest average maximum use levels are reported in gravies (5.4%) and reconstituted vegetables (4.8%).

### COMMERCIAL PREPARATIONS

Various types (acid proof, beer, spirit, bakers, and confectioners, etc.); also powdered form.[8] Official in N.F. and F.C.C.

*Regulatory Status.* GRAS as a multiple purpose food substance (§182.1235); also approved as a color additive exempt from certification, to be used in foods and drugs (§73.85 and §73.1085).

### REFERENCES

See the General References for FEMA; FURIA; MARTINDALE; MERCK.

1. W. R. Fetzer in A. Standen, ed., *Kirk-Othmer Encyclopedia of Chemical Technology*, Vol. 4, 2nd ed., Wiley-Interscience, New York, 1964, p. 63.
2. I. F. Gaunt et al., *Food Cosmet. Toxicol.*, **15**, 509 (1977).
3. G. Fuchs and S. Sundell, *J. Agr. Food Chem.*, **23**, 120 (1975).
4. R. A. Wilks et al., *J. Chromatogr.*, **87**, 411 (1973).
5. M. Komoto et al., *Seito Gijutsu Kenkyukaishi*, **25**, 25 (1975); through *Chem. Abstr.*, **83**, 162259g (1975).
6. J. G. Evans et al., *Food Cosmet. Toxicol.*, **15**, 523 (1977).
7. M. Ishidate Jr. et al., *Food Chem. Toxicol.*, **22**, 623 (1984).
8. M. Berdick, *Cosmet. Toilet.*, **92**, 26 (1977).

# CARAWAY

**Source:** ***Carum carvi*** L. (syn. *Apium carvi* Crantz) (Family Umbelliferae or Apiaceae).

*Synonyms.* Caraway fruit, caraway seed, and carum.

### GENERAL DESCRIPTION

Biennial herb with second-year stem up to about 0.75 m high, widely branching; native to Europe and western Asia and naturalized in North America; widely cultivated. Part used is the dried ripe fruit. An essential oil is obtained from the fruit by steam distillation (GUENTHER).

### CHEMICAL COMPOSITION

Contains 2–8% (usually 4–7%) volatile oil, about 15% lipids, 20% protein, a $\beta(1\rightarrow4)$ mannan, and flavonoids (quercetin-3-glucuronide, isoquercitrin, etc.), among others (JIANGSU, MARSH).[1–4]

The volatile oil is composed mainly of carvone (50–60%) and limonene (ca. 40%), with minor amounts of carveol, dihydrocarveol, dihydrocarvone, thujone, pinene, phellandrene, $\alpha$-thujene, $\beta$-fenchene, and others (MASADA).[2,5–7] Concentrations of the components vary, depending on the degree of ripeness of the fruit; contents of carvone and other oxygenated components increase as fruit ripens.[6] Thus oils obtained from fully mature seeds contain more carvone and less limonene (and other terpenes) and are considered to be of a better quality (ARCTANDER).[2]

## PHARMACOLOGY OR BIOLOGICAL ACTIVITIES

Caraway is generally considered to have carminative and stomachic properties.

Caraway oil has been reported to exhibit antibacterial activities *in vitro* as well as larvicidal properties.[8,9] It also has antispasmodic and antihistaminic activities on isolated animal organs (JIANGSU).[10]

Carvone has been implicated in chemopreventative activity. Carvone induces the detoxifying enzyme glutathione *S*-transferase (GST) in several mouse target tissues. Compounds that induce an increase in the activity of GST detoxification are considered potential inhibitors of carcinogenesis.[11]

Available data indicate caraway oil to be nontoxic.[12]

## USES

**Medicinal, Pharmaceutical, and Cosmetic.** Used in some carminative, stomachic, and laxative preparations. Oil is used as a flavor in pharmaceuticals; it is also used as a fragrance component in cosmetic preparations including toothpaste, mouthwash, soaps, creams, lotions, and perfumes, with maximum use level of 0.4% reported in perfumes.

**Food.** Caraway is widely used as a domestic spice. It is also extensively used in commercial food products particularly baked goods (rye bread, etc.) and meat and meat products, among others. Caraway oil is used in all major categories of foods, including alcoholic and nonalcoholic beverages, frozen dairy desserts, candy, baked goods, gelatins and puddings, meat and meat products, condiments and relishes, and others. Highest average maximum use level is reported to be about 0.02% (225 ppm) in baked goods.

**Health Food/Herb Teas.** Used in some products as a spasmolytic aid to digestion (FOSTER; WREN).

**Traditional Medicine.** Reportedly used as an antispasmodic, carminative, expectorant, and stomachic for dyspepsia, incontinence, and indigestion;[13] also in relieving menstrual discomforts, promoting milk secretion, and others (WREN).

## COMMERCIAL PREPARATIONS

Fruit and oil; official in N.F. and F.C.C.; formerly official in U.S.P.

*Regulatory Status.* GRAS (§182.10 and §182.20). Both essential oil and fruits subjects of German therapeutic monographs.[14,15] Essential oil (in daily dose of 3–6 drops); spasmolytic, antimicrobial; for dyspeptic complaints such as mild gastrointestinal spasm, bloating and fullness.[14] Seed used similarly, though therapeutic use not recommended since efficacy is not well documented.[15]

## REFERENCES

See the General References for BAILEY 1; BLUMENTHAL; CLAUS; FEMA; FOSTER; GUENTHER; JIANGSU; LIST AND HÖRHAMMER; LUST; USD 26th; WREN.

1. I. Bochenska and J. Kozlowski, *Herba Pol.*, **15**, 251 (1969); through *Chem. Abstr.*, **73**, 73815m (1970).
2. M. B. Embong et al., *Can. J. Plant Sci.*, **57**, 543 (1977).

3. H. Hopf and O. Kandler, *Phytochemistry*, **16**, 1715 (1977).
4. J. Kunzemann and K. Herrmann, *Z. Lebensm. Unters. Forsch.*, **164**, 194 (1977).
5. H. Rothbaecher and F. Suteu, *Planta Med.*, **28**, 112 (1975).
6. D. Razinskaite, *Nauji Laimejimai Biol. Biochem., Liet. TSR Jaunuju Mokslininku Biol. Biochem. Moksline Konf.*, **35** (1967); through *Chem. Abstr.*, **70**, 94011z (1969).
7. A. Salveson and A. B. Svendsen, *Planta Med.*, **30**, 93 (1976).
8. F. M. Ramadan et al., *Chem. Mikrobiol. Technol. Lebensm.*, **2**, 51 (1972).
9. K. Oishi et al., *Nippon Suisan Gakkaishi*, **40**, 1241 (1974); through *Chem. Abstr.*, **82**, 84722r (1975).
10. A. M. Debelmas and J. Rochat, *Plant. Med. Phytother.*, **1**, 23 (1967).
11. G. Q. Zheng et al., *Planta Med*, **58**, 338 (1992).
12. D. L. J. Opdyke, *Food Cosmet. Toxicol.*, **11**, 1051 (1973).
13. J. A. Duke and C. F. Reed., *Q. J. Crude Drug Res.*, **16**, 3, 116 (1978).
14. Monograph *Carvi aetheroleum*, *Bundesanzeiger*, no. 22 (Feb. 1, 1990).
15. Monograph *Carvi fructus*, *Bundesanzeiger*, no. 22 (Feb. 1, 1990).

# CARDAMOM

**Source:** ***Elettaria cardamomum*** (L.) Maton (Family Zingiberaceae).

*Synonyms*. Cardamom seed and cardamon.

### GENERAL DESCRIPTION

Perennial herb with lance-shaped leaves borne on long sheathing stems, up to about 4 m high; native to tropical Asia; now cultivated extensively in tropical regions, particularly India (Malabar coast), Sri Lanka (Ceylon), Laos, Guatemala, and El Salvador. Parts used are the dried ripe fruits and seeds from which an essential oil is obtained by steam distillation.

### CHEMICAL COMPOSITION

Contains up to 10% (usually 3–8%) volatile oil, about 10% protein, 10% fixed oil, 20–40% starch, manganese, and iron, among others (LIST AND HÖRHAMMER, MARSH).

The volatile oil is composed mainly of α-terpinyl acetate and 1,8-cineole, each of which may be present up to 50% or more; lesser components include limonene, sabinene, linalool, linalyl acetate, α-pinene, α-terpineol, camphene, myrcene, 1,4-cineole, borneol, and others (MASADA).[1–4]

Acid constituents of the oil include acetic, butyric, decanoic, dodecanoic, citronellic, geranic, hexanoic, heptanoic, nerylic, and perillic acids.[3,4]

Compositions of oils vary, depending on types (e.g., Mysore and Malabar). Oils containing low content of cineole but high content of terpinyl acetate are considered to be of superior quality for flavor applications.[5]

### PHARMACOLOGY OR BIOLOGICAL ACTIVITIES

Cardamom is considered to have carminative, stimulant, and stomachic properties.

Cardamom oil has been reported to have antispasmodic activity on excised mouse intestine.[6]

Cardamom aqueous extract has been

demonstrated to increase trypsin activity in buffer solution.[7]

Available data indicate cardamom oil to be nontoxic.[8]

## USES

**Medicinal, Pharmaceutical, and Cosmetic.** Cardamom is used in some carminative, stomachic, and laxative preparations. Its oil is mainly used as a flavor ingredient in Compound Cardamom Spirit to flavor pharmaceuticals; it is also used as a fragrance component in soaps, detergents, creams, lotions, and perfumes, with maximum use level of 0.4% reported in perfumes.[8]

**Food.** Cardamom is used extensively as a domestic spice in curry, coffee, cakes, and bread, especially in India, Europe, the Middle East, and Latin America. Both cardamom seed and its oil are widely used as flavor components in most categories of food products, including alcoholic and nonalcoholic beverages, frozen desserts, candy, baked goods, gelatins and puddings, meat and meat products, condiments and relishes, and gravies, among others. Highest average maximum use level reported for the seed is 0.5% in gravies and about 0.01% (117 ppm) for the oil in alcoholic beverages.

**Health Food/Herb Teas.** Whole or ground cardamom used as a flavoring ingredient in India-inspired popular tea known as chai.

**Traditional Medicine.** Cardamom has been used in medicine for centuries in India and China as a carminative, stimulant, and to treat urinary problems, among others.

The cardamom used in China for these purposes is the fruit of *Amomum cardamomum* L., which is considered in Chinese medicine to be superior to that of *Elettaria cardamomum* (JIANGSU).

## COMMERCIAL PREPARATIONS

Seed and oil; official in N.F. and F.C.C.

*Regulatory Status.* GRAS (§182.10 and §182.20). Fruits subject of a German therapeutic monograph in medium daily dose of 1.5 g for treatment of dyspeptic disorders.[9]

## REFERENCES

See the General References for ARCTANDER; BIANCHINI AND CORBETTA; FEMA; GUENTHER; JIANGSU; LUST; MASADA; NANJING; ROSENGARTEN; STAHL; TERRELL; UPHOF.

1. M. Miyazawa and H. Kameoka, *Yukagaku*, **24**, 22 (1975); through *Chem. Abstr.*, **82**, 103020d (1975).
2. J. S. T. Chou, *Koryo*, **106**, 55 (1974); through *Chem. Abstr.*, **81**, 82235q (1974).
3. B. M. Lawrence, *Perfum. Flavor.*, **14**(6), 87 (1989).
4. B. M. Lawrence, *Perfum. Flavor.*, **16**(1), 39 (1991).
5. Y. S. Lewis et al., *Sixth Int. Cong. Essent. Oils (Pap.)*, 65 (1974).
6. J. Haginiwa et al., *Yakugaku Zasshi*, **83**, 624 (1963); through *Chem. Abstr.*, **60**, 999a (1964).
7. Y. Kato, *Koryo*, **113**, 17 (1975); through *Chem. Abstr.*, **84**, 149393x (1976).
8. D. L. J. Opdyke, *Food Cosmet. Toxicol.*, **12**(Suppl.), 837 (1974).
9. Monograph *Cardamomi fructus*, *Bundesanzeiger*, no. 223 (Nov. 30, 1985); and revision (Sept. 1, 1990).

# CAROB

**Source:** ***Ceratonia siliqua*** L. (Family Leguminosae or Fabaceae).

*Synonyms*. St. John's bread and locust bean.

## GENERAL DESCRIPTION

Dome-shaped evergreen tree with dark green compound leaves, consisting of two to five pairs of large, rounded glossy leaflets; fruits (pods) up to 30 cm long, indehiscent and sometimes borne on the tree trunk; tree up to 15 m high; native to southeastern Europe and western Asia; widely cultivated in the Mediterranean region. Part used is the dried ripe fruit, from which three major commercial products are obtained: carob extract of the dried pod, either roasted or unroasted; carob flour from the pulp or the whole pod; and carob gum or ***locust bean gum*** (which see) from the endosperm of the seed. Mediterranean countries (Greece, Spain, Italy, etc.) are the major producers of carob.

Carob pods are believed to be the locusts consumed by St. John the Baptist, hence the name St. John's bread. Seeds were used in ancient times as weight units for gold from which the term *carat* is reportedly derived.

## CHEMICAL COMPOSITION

Pods contain 30–70% (usually 40–50%) sugars composed of sucrose (up to 26% in pulp), fructose (13%), xylose, maltose, and dextrose, among others; proteins; amino acids (alanine, proline, valine, etc.); gallic acid; fats; starch; abscisic acid (a plant growth inhibitor); and others.[1–4]

Seeds contain protein, fats (mostly oleic, linoleic, and palmitic acids), tannins, gum (a galactomannan), and others.[2,5,6] The protein is localized in the embryo and cotyledons while the gum is present mainly in the endosperm (LIST AND HÖRHAMMER).

## PHARMACOLOGY OR BIOLOGICAL ACTIVITIES

Carob tannins have been reported to be strongly inhibitory to digestive enzymes.[7]

Carob flour is regarded to have antidiarrheal properties, while carob flowers have laxative properties.

A recent study found that rats fed 15% carob gums for 2–6 weeks lost weight compared with control animals. Maximum effect was achieved in 6 consecutive weeks of feeding. In addition, a decreased blood glucose level, cholesterol plasma level, and insulin levels, and an increase in glucose tolerance were found. The hypoglycemic and hypoinsulinemic effects as well as blood cholesterol lowering activities suggest the gum may prove a useful adjuvant in the treatment of diabetes mellitus.[8]

## USES

**Food.** Carob flour and carob extracts (carob syrup, etc.) have been used as food for centuries. Currently, the flour is popular in health foods and as a cocoa substitute, for which roasted kibbles are used, while carob extracts are widely used as flavor ingredients (e.g., butterscotch, imitation chocolate, and vanilla) in all kinds of food products, including alcoholic and nonalcoholic beverages, frozen dairy desserts, candy, baked goods, gelatins and puddings, meat and meat products, condiments and relishes, fruit and ices, sweet sauces, gravies, imitation dairy, and many others. Highest average maximum use levels reported are in imitation dairy (0.50%), fruit and ices (0.50%), gravies (0.46%), sweet sauces (0.46%), and condiments and relishes (0.42%).

**Health Food/Herb Teas.** Carob flour is widely used in health food products, including weight-loss formulations, "energy" bars, tea formulations, and other products, primarily as a chocolate substitute.

Use of carob in the United States is entirely as a chocolate substitute for "organophiles" who distrust chocolate; imitation chocolate products containing carob include brownies, carob chip cookies, candy bars, bits, creams, fudge, carob-flavored milk, etc.[9]

**Traditional Medicine.** Carob flour reportedly used as an antidiarrheal for babies; decoction of pods used for catarrhal infections (UPHOF).

### COMMERCIAL PREPARATIONS

Roasted and unroasted crude (kibbles), syrup, and extracts; extracts usually come in specific flavor strengths, depending on users' requirements.

*Regulatory Status.* GRAS (listed as carob bean and St. John's bread both under (§182.20)).

### REFERENCES

See the General References for ARCTANDER; BIANCHINI AND CORBETTA; FEMA; UPHOF.

1. T. G. Loo, *Public R. Trop. Inst. Amsterdam*, **288** (1969); through *Chem. Abstr.*, **72**, 80601w (1970).
2. Y. Vardar et al., *Qual. Plant. Mater. Veg.*, **21**, 367 (1972); through *Chem. Abstr.*, **78**, 41779q (1973).
3. M. A. Joslyn et al., *J. Sci. Food Agr.*, **19**, 543 (1968).
4. B. H. Most et al., *Planta*, **92**, 41 (1970).
5. J. Artaud et al., *Ann. Falsif. Expert. Chim.*, **70**(749), 39 (1977); through *Chem. Abstr.*, **86**, 185961f (1977).
6. J. Artaud et al., *Ann. Falsif. Expert. Chim.*, **69**(737), 23 (1976); through *Chem. Abstr.*, **87**, 54828r (1977).
7. M. Tamir and E. Alumot, *J. Sci. Food Agr.*, **20**, 199 (1969).
8. A. M. Forestieri, *Phytotherapy Res.*, **3**(1), 1, (1989).
9. J. Ott, *The Cacahutal Eater: Ruminations of an Unabashed Chocolate Addict*, Natural Products, Co., Vashon, Wash., 1985.

# CARRAGEENAN

**Source:** ***Chondrus crispus*** (L.) Stackh., ***Euchema***, and ***Gigartina*** species or related red algae of the order Gigartinales.

*Synonyms.* Carragheenan, carrageenin, chondrus extract, and Irish moss extract.

### GENERAL DESCRIPTION

Carrageenan is a seaweed gum (hydrocolloid) obtained from various red algae growing along the Atlantic coast of Europe and North America, with *C. crispus* (Irish moss) as its major source. It occurs in the intercellular matrix and cell walls of the algae and constitutes 60–80% of their salt-free dry weight.[1–3]

In the manufacture of carrageenan, the dried seaweed is first cleaned with cold water and mechanical devices to remove salt and other extraneous materials. It is then extracted with hot water containing calcium or sodium hydroxide. The extract is clarified by filtration, its pH adjusted to slightly basic, and carrageenan is obtained either by direct drum or roll drying of the filtrate or by precipitation with alcohol (e.g., ethyl or isopropyl), depending on the type or purity desired.[4,5] The United States and European countries (e.g., Denmark, France, and Spain) are the major carrageenan producers, with the United States being by far the largest.

Carrageenan comes in many types with different solubilities and gel characteristics, depending on the process and types of algae used for its manufacture. It readily dissolves in water to form viscous solutions or gels, depending on its chemical composition. Its reaction to heating and shearing forces also depends on its chemical nature.

Carrageenan has high reactivity with certain proteins, particularly milk protein, to form weak to strong gels.

## CHEMICAL COMPOSITION

Carrageenan is a sulfated, straight-chain galactan composed of residues of D-galactose and 3,6-anhydro-D-galactose with a molecular weight usually of 100,000–500,000. It contains a high content of sulfate (20–40% dry weight basis). The number and position of the sulfate groups and the ratio of galactose to 3,6-anhydrogalactose vary greatly. Carrageenan generally contains two major fractions, a gelling fraction called κ-carrageenan and a nongelling fraction called λ-carrageenan; κ-carrageenan contains D-galactose, 3,6-anhydro-D-galactose and ester sulfate groups, while λ-carrageenan contains D-galactose and its monosulfate and disulfate esters. Other types of carrageenan include ι-carrageenan, which is composed mainly of monosulfates of D-galactose and 3,6-anhydro-D-galactose (FURIA).[2,6]

λ-Carrageenan is readily soluble in cold water to form a viscous solution regardless of the cations present, whereas κ-carrageenan is precipitated by potassium ions. The potassium salt of carrageenan, however, is soluble in water on heating and forms an elastic gel on cooling; the elasticity or rigidity of the gel depends on the amounts of potassium ions present. ι-Carrageenan (obtained mostly from *Euchema spinosum*) forms thermally reversible elastic gelatinlike gels with calcium ions.

Solutions and gels of carrageenan are degraded rapidly by low pH and high temperatures. Degraded carrageenans (molecular weight ≤20,000) do not have the viscosity or gelling properties of food-grade carrageenans (molecular weight 100,000–500,000).

## PHARMACOLOGY OR BIOLOGICAL ACTIVITIES

Carrageenan has been reported to exhibit many pharmacological activities in animals, including lowering of blood cholesterol level, reducing gastric secretions and food absorption, and increasing water content of the gut when large doses are ingested. When administered parenterally, it has shown anticoagulant, hypotensive, and immunosuppressive activities. Furthermore, when injected into a rodent's paw, carrageenan produces a reproducible inflammatory condition, which is extensively used as a model for screening potential antiinflammatory drugs.[2,7]

One study demonstrated λ-carrageenan to be much more toxic than ι-carrageenan and κ-carrageenan; it produced acronecrosis in mice after a single intraperitoneal injection and caused death in 90% of the animals within a month. ι-Carrageenan and κ-carrageenan did not produce acronecrosis.[8]

Carrageenan, both in the degraded (molecular weight ≤20,000) and undegraded forms, has been reported to alleviate peptic and duodenal ulcers in humans. The degraded form is partly absorbed and is reported to cause ulceration of the cecum and colon of certain experimental animals (e.g., guinea pigs, rabbits and Rhesus monkeys), but no adverse effects on the colon have been observed in humans in up to 2 years of treatment (MARTINDALE). In experimental animal models carrageenan induced a greater colorectal tumor incidence, or at least caused them to appear earlier.[9] Food-grade carrageenans (molecular weight 100,000–500,000) and carrageenans with a molecular weight of more than 50,000 are generally not absorbed through the gut; food-grade carrageenan is believed to be nontoxic.[2]

Carrageenan has inhibitory effects on pepsin activity *in vitro*. Its degraded form (no viscosity) and forms with low and high viscosities all exhibit antiproteolytic activities *in vitro* against papain.[10]

## USES

**Medicinal, Pharmaceutical, and Cosmetic.** Used extensively as binder, emulsifier, or stabilizer in toothpastes; also in hand lotions, creams, tablets, and others.

The degraded form is used in preparations for treating peptic ulcers, primarily in Europe (France).

**Food.** Carrageenan (or its salts) is extensively used in milk products such as chocolate milk, ice cream, sherberts, cottage cheese, cream cheese, evaporated milk, milk desserts, puddings, yogurts, infant formulas, and others. It is also used on other foods such as gravies, thickening sauces, bread doughs, meat products, jams, and jellies, among others. Its major functions are as thickening, gelling, emulsifying, stabilizing, and suspending agents, preventing the settling of solids, and in the case of ice cream, to prevent ice crystal formation, among others. For use in gel products such as jams and jellies, its use level is usually 0.5–1.1%; ***locust bean gum*** (which see) is often used with carrageenan to improve its gel strength and elasticity.

**Health Food/Herb Teas.** Carrageenan is used in various weight-loss formulations; also in drinks, especially aloe vera, fruit juice, and herbal drinks.

**Traditional Medicine.** Reported as a demulcent and a nutrient; used in tuberculosis, coughs, bronchitis, and intestinal problems, usually in the form of a decoction of the seaweed (*C. crispus*) (WREN).

## COMMERCIAL PREPARATIONS

Carrageenan and its ammonium, calcium, potassium, and sodium salts come in many grades and types to meet specific end use requirements. Official in F.C.C. and N.F.

*Regulatory Status.* Carrageenan and its salts have been approved for food use (§172.620, §172.623, §172.626, and §182.7255).

## REFERENCES

See the General References for FEMA; FURIA; GLICKSMAN; LAWRENCE; LEWIS AND ELVIN-LEWIS; LUST; MARTINDALE; UPHOF; WHISTLER AND BEMILLER; WREN; YOUNGKEN.

1. E. L. McCandless et al., *Can. J. Bot.*, **55**, 2053 (1977).
2. D. J. Stancioff and D. W. Renn, *ACS Symp. Ser.*, **15**, 282 (1975).
3. R. L. Whistler and C. L. Smart, *Polysaccharide Chemistry*, Academic Press, New York, 1953, p. 218.
4. C. T. Blood in L. W. Codd et al., eds., *Chemical Technology: An Encyclopedic Treatment*, Vol. 5, Barnes & Noble, New York, 1972, p. 27.
5. K. B. Guiseley in A. Standen, ed., *Kirk-Othmer Encyclopedia of Chemical Technology*, Vol. 17, 2nd ed., Wiley-Interscience, New York, 1968, p. 763.
6. R. L. Whistler in W. H. Schultz et al.,

eds., *Symposium on Foods: Carbohydrates and Their Roles*, AVI, Westport, Conn., 1969, p. 73.
7. J. E. Sawicki and P. J. Catanzaro, *Int. Arch. Allergy Appl. Immunol.*, **49**, 709 (1975).
8. A. W. Thomson and C. H. W. Horne, *Br. J. Exp. Pathol.*, **57**, 455 (1976).
9. E. E. Deschner et al., *Clin. Gastroenterol.*, **10**(3), 755 (1981).
10. F. Arnal-Peyrot and J. Adrian, *Med. Nutr.*, **13**, 49 (1977); through *Chem. Abstr.*, **86**, 185066t (1977).

# CARROT OILS

**Source:** ***Daucus carota*** L. (Family Umbelliferae or Apiaceae).

*Synonyms*. Oil of carrot, wild carrot, and Queen Anne's lace.

## GENERAL DESCRIPTION

Annual or biennial herb with erect, much-branched stem; up to about 1.5 m high. The common cultivated carrot, *D. carota* L. subsp. *sativus* (Hoffm.) Arcang., has an edible fleshy, orange-red taproot, while the wild carrot, or Queen Anne's lace, *D. carota* L. subsp. *carota*, has an inedible, tough whitish root; wild carrot is native to Europe, Asia, and North America; naturalized in North America. Part used is the dried fruit, from which carrot seed oil is obtained by steam distillation. Carrot root oil is obtained by solvent extraction of the red carrot (root); it contains high concentrations of carotenes ($\alpha$, $\beta$, etc.).

## CHEMICAL COMPOSITION

Carrot seed oil contains $\alpha$-pinene (up to 13.3%), $\beta$-pinene, carotol (up to 18.29%), daucol, limonene, $\beta$-bisabolene, $\beta$-elemene, *cis*-$\beta$-bergamotene, $\gamma$-decalactone, $\beta$-farnesene, geraniol, geranyl acetate (up to 10.39%), caryophyllene, caryophyllene oxide, methyl eugenol, nerolidol, eugenol, *trans*-asarone, vanillin, asarone, $\alpha$-terpineol, terpinen-4-ol, $\gamma$-decanolactone, coumarin, and $\beta$-selinene, among others. Other constituents present include palmitic acid, butyric acid, and others (JIANGSU).[1-10] Oil content is highly variable, from 0.05 to 7.15%.[10]

## PHARMACOLOGY OR BIOLOGICAL ACTIVITIES

Carrot seed oil has been reported to exhibit vasodilatory and smooth-muscle relaxant activities on isolated animal organs. It also depressed cardiac action in frog and dog hearts, among other activities.

Available data indicate it to be nontoxic.[5]

## USES

**Medicinal, Pharmaceutical, and Cosmetic.** Carrot seed oil is used primarily as a fragrance component in soaps, detergents, creams, lotions, and perfumes. Highest use level reported is 0.4% in perfumes.[5]

Carrot root oil is used in certain sunscreen preparations and as a source of $\beta$-carotene and vitamin A.

**Food.** Carrot seed oil is used as a flavor ingredient in most major categories of food products, including alcoholic (particularly liqueurs) and nonalcoholic beverages, frozen dairy desserts, candy, baked goods, gelatins and puddings, meat and meat products; condiments and relishes, and soups, usually in rather low use levels. ($<$0.003%).

Carrot root oil is used mainly as a yellow food color because of its carotene content.

**Traditional Medicine.** Seeds used as a diuretic and emmenagogue and for flatulence in the form of a decoction or infusion; in Chinese medicine to treat chronic dysentery and as an anthelmintic.

**Others.** Carrot seed oil can serve as sources of carotol and daucol, which have potential as starting materials for the synthesis of new fragrance compounds.[4,11]

**COMMERCIAL PREPARATIONS**

Available as carrot oil (which can be either seed oil or root oil); seed oil is official in F.C.C.

*Regulatory Status.* Carrot (seed) oil is GRAS (§182.20), and carrot (root) oil has been approved for use as a food color (§73.300); no distinction between root oil and seed oil is given in §182.20.

**REFERENCES**

See the General References for ARCTANDER; BAILEY 2; CLAUS; FEMA; GUENTHER; JIANGSU; LUST; MARTINDALE; NANJING; TERRELL.

1. R. M. Siefert et al., *J. Sci. Food Agr.*, **19**, 383 (1968).
2. L. N. Chelovskaya and A. G. Nikolaev, *Mezhdunar. Kongr. Efirnym Maslam (Mater.)*, **4th**(Pub. 1972); **2**, 227 (1968); through *Chem. Abstr.*, **81**, 29465p (1974).
3. H. Strzelecka and T. Soroczynska, *Farm. Pol.*, **30**, 13 (1974); through *Chem. Abstr.*, **81**, 68331m (1974).
4. J. Kuleska et al., *Riechst. Aromen, Körperpflegem.*, **23**, 34 (1973).
5. D. L. J. Opdyke, *Food Cosmet. Toxicol.* **14**, 705 (1976).
6. J. B. Harborne et al., *Phytochemistry*, **8**, 1729 (1969).
7. G. V. Pigulevskii and V. I. Kovaleva, *Rast. Resur.*, **2**, 527 (1966); through *Chem. Abstr.*, **66**, 79489k (1966).
8. A. S. Cheema et al., *Riechst. Aromen, Körperpflegem.*, **25**, 138 (1875).
9. J. W. Hogg et al., *Cosmet. Perfum.*, **89**, 64 (1974).
10. B. M. Lawrence, *Perfum. Flavor.*, **15**, 4, 63 (1990).
11. J. Kulesza et al., *An. Acad. Bras. Cienc.*, **44**(Suppl.), 412 (1972); through *Chem. Abstr.*, **83**, 136708h (1975).

# CASCARA SAGRADA

**Source:** ***Rhamnus purshiana*** DC. (syn. *Frangula purshiana* (D.C.) A. Gray, J. C. Cooper) (Family Rhamnaceae).

*Synonyms.* Cascara, sacred bark, and chittem bark.

**GENERAL DESCRIPTION**

Small to medium-size, deciduous tree with reddish brown bark and hairy twigs; up to about 13 m high; native to the Pacific Coast of North America (northern California, Oregon, Washington, British Columbia, Idaho, and Montana). Part used is the bark, which is removed from trees with trunk diameter of about 10 cm or more; it is then allowed to dry and age for 1 year before use, as the fresh bark has an emetic principle that is destroyed on prolonged storage or by heating. This emetic principle is now generally considered to be composed of monoanthrones and their *O*-glycosides that are oxidized to nonemetic anthraquinones or anthrone *C*-glycosides on storage or heat treatment.[1]

## CHEMICAL COMPOSITION

Contains 6–10% (usually ca. 8%) anthraglycosides as its active principles which consist primarily of *C*-glucosides (cascarosides A, B, C, and D; and barbaloin and chrysaloin), with minor concentrations of *O*-glycosides (e.g., frangulin) also present. Other constituents include free anthraquinones (e.g., emodin, aloe-emodin, isoemodin, and chrysophanol, also known as chrysophanic acid), resins, tannins, and lipids, among others (STAHL).[1–7]

Cascarosides account for two-thirds or more of the total anthraglycosides present; the remaining consists mostly of aloins, with the other glycosides in only minor amounts. Cascarosides A and B are *O*-glucosides of barbaloin, while cascarosides C and D are *O*-glucosides of chrysaloin. Mild acid hydrolysis of cascarosides yields aloins, which, however, can be broken down to their aglycones only by strong oxidative hydrolysis.

Cascarosides are bitterless, but aloins are extremely bitter. These aloins are probably the so-called bitter principles of cascara in the old literature.

## PHARMACOLOGY OR BIOLOGICAL ACTIVITIES

The anthraglycosides have cathartic properties. Cascarosides A and B are responsible for most of these properties in cascara. They act on the large intestine by inducing increased peristalsis and are generally considered to be safe drugs with a minimum of side effects (see ***senna***).[8–10] Cascarosides are more active than their hydrolyzed products (aloins and free anthraquinones).[11,12]

Aloe-emodin isolated from ***buckthorn*** (which see) had antileukemic activity against the P-388 lymphocytic leukemia in mice.

## USES

**Medicinal, Pharmaceutical, and Cosmetic.** Used extensively in laxative preparations; also used in sunscreens.

**Food.** Only the bitterless extract is reportedly used as a flavor component in foods, including nonalcoholic beverages, frozen dairy desserts, and candy and baked goods. Average maximum use level is below 0.008% (75 ppm).

**Health Food/Herb Teas.** Crude aged bark used in laxative and detoxicant teas; extracts in herbal formulas (capsules, tablets, drinks, etc.) for the laxative and alleged detoxicant effects.

**Traditional Medicine.** Used as a laxative and bitter tonic and in gallstones and liver ailments. Also reportedly used in cancer.[13]

## COMMERCIAL PREPARATIONS

Crude and fluid extract, aromatic fluid extract, solid extract (bitter, bitterless, aromatic), powder extract, granular extract, and cascarosides concentrates; crude, aromatic fluid extract, and bitter fluid and powder extracts are official in U.S.P. Strengths of extracts (see ***glossary***) are expressed in weight-to-weight ratios as well as total cascarosides content.

*Regulatory Status.* Has been approved for food use (§172.510). Subject of a German therapeutic monograph; allowed in laxative formulations.[14]

## REFERENCES

See the General References fox APhA; BAILEY 1; BLUMENTHAL; ESCOP 1; FEMA; KROCHMAL AND KROCHMAL; LUST; MARTINDALE; STAHL; YOUNGKEN.

1. F. H. L. van Os, *Pharmacology*, **14**(Suppl. 1), 7 (1976).
2. F. J. Evans et al., *J. Pharm. Pharmacol.*, **27**(Suppl.), 91P (1975).
3. J. W. Fairbairn et al., *J. Pharm. Sci.*, **66**, 1300 (1977).
4. H. Wagner and G. Demuth, *Z. Naturforsch., C.*, **29**(7–8); 444 (1974).
5. H. Wagner and G. Demuth, *Z. Naturforsch., B.*, **31b**, 267 (1976).
6. G. Demuth, *On the Chemical Structure of Cascarosides*, doctoral dissertation, University of Munich, Munich, 1972.
7. A. Y. Leung, *Drug Cosmet. Ind.*, **121**(6), 42 (1977).
8. F. A. Nelemans, *Pharmacology*, **14**(Suppl. 1), 73 (1976).
9. E. W. Godding, *Pharmacology*, **14**(Suppl. 1), 78 (1976).
10. F. H. L. van Os, *Pharmacology*, **14**(Suppl. 1), 18 (1976).
11. J. W. Fairbairn and G. E. D. H. Mahran, *J. Pharm. Pharmacol.*, **5**, 827 (1953).
12. J. W. Fairbairn and S. Simic, *J. Pharm. Pharmacol.*, **16**, 450 (1964).
13. J. L. Hartwell, *Lloydia*, **34**, 103 (1971).
14. Monograph *Rhamni purshianae cortex*, *Bundesanzeiger*, no. 228 (Dec. 5, 1984); replaced (July 21, 1993).

# CASCARILLA BARK

**Source: *Croton eluteria*** (L.) Sw. (Family Euphorbiaceae).

*Synonyms*. Cascarilla, sweetwood bark, and sweet bark.

### GENERAL DESCRIPTION

Large shrub to small tree, flowering and fruiting year round; up to 12 m high; native to the West Indies (Bahamas, Jamaica, Cuba, etc.); also grows in tropical America (Mexico, Colombia, and Ecuador).[1] Part used is the dried bark from which an essential oil is obtained by steam distillation; mainly from the Bahamas.

### CHEMICAL COMPOSITION

Contains 1.5–3.0% volatile oil, a bitter principle (cascarillin A), resins, tannin, starch, and lipids, among others (LIST AND HÖRHAMMER).

The volatile oil consists primarily of *p*-cymene, camphene, dipentene, *d*-limonene, β-caryophyllene, α-terpineol, α- and β-pinene, α-thujene, borneol, terpinen-4-ol, eugenol, and others.[2] Additional constituents include cascarillic acid, cineole, methylthymol, cuparophenol, cascarilladiene, and casarillone (MASADA).[3–6]

### PHARMACOLOGY OR BIOLOGICAL ACTIVITIES

The bark is regarded to have bitter tonic properties.

The essential oil has been reported to have antimicrobial activities.[7,8] Available data indicate it to be nontoxic.[9]

### USES

**Medicinal, Pharmaceutical, and Cosmetic.** Tinctures and extracts are used in certain bitter tonic preparations; essential oil as a fragrance component in soaps, detergents, creams, lotions, and perfumes (particularly oriental types and men's fragrances), with maximum use level of 0.4% reported in perfumes.[9] Oil respected by compounders, especially for men's fragrances, because of its power and tenacity.[2]

**Food.** Extract is reportedly used in alcoholic (bitters) and nonalcoholic beverages, with average maximum use levels of 0.01% and about 0.08% (775 ppm), respectively. Essential oil is used as a flavor ingredient in most major categories of food products, including alcoholic and nonalcoholic beverages, frozen dairy desserts, candy, baked goods, and condiments and relishes. Highest average maximum use level is 0.007% (72.1 ppm) reported for the last category.

**Traditional Medicine.** Used as aromatic bitter, aromatic stimulant, and body tonic (AYENSU).

**Others.** Used in flavoring smoking tobacco.

### COMMERCIAL PREPARATIONS

Crude, extracts, and essential oil; oil is official in F.C.C.

*Regulatory Status.* GRAS (§182.20).

### REFERENCES

See the General References for ARCTANDER; AYENSU; BAILEY 2; FEMA; GUENTHER; MERCK; TERRELL; UPHOF; YOUNGKEN.

1. C. D. Adams, *Flowering Plants of Jamaica*, University of the West Indies, Mona, Jamaica, 1972, p. 414.
2. B. M. Lawrence, *Perfum. Flavor.*, **2**(1), 3 (1977).
3. A. Claude-LaFontaine et al., *Bull. Soc. Chim. Fr.*, **9–10**, 2866 (1973).
4. A. Claude-LaFontaine et al., *Bull. Soc. Chim. Fr.*, **1–2**, 88 (1976).
5. O. Motl and A. Trka, *Parfüm. Kosmet.*, **54**, 5 (1973).
6. O. Motl et al., *Phytochemistry*, **11**, 407 (1972).
7. J. C. Maruzzella and N. A. Sicurella, *J. Am. Pharm. Ass.*, **49**, 692 (1960).
8. J. C. Maruzzella and L. Liguori, *J. Am. Pharm. Ass.*, **47**, 250 (1958).
9. D. L. J. Opdyke, *Food Cosmet. Toxicol.*, **14**, 707 (1976).

# CASSIE ABSOLUTE

**Source:** ***Acacia farnesiana*** (L.) Willd. (Family Leguminosae or Fabaceae).

*Synonyms.* Sweet acacia, huisache, and popinac absolute.

### GENERAL DESCRIPTION

Thorny shrub to small tree, 3–9 m high, with very fragrant flowers; believed to be native of the Old World, now widespread and cultivated in subtropical and tropical regions of the world.[1] Parts used are the flowers from which a concrète is first prepared by extraction with petroleum ether; the absolute is then obtained by alcohol extraction of the concrète. Produced primarily in Cannes, France; also in India.

### CHEMICAL COMPOSITION

The absolute contains approximately 25% of volatile constituents, which are composed mainly of benzyl alcohol, methyl salicylate, farnesol, and geraniol, with more than 40 other minor compounds, including $\alpha$-ionone, geranyl acetate, linalyl acetate, nerolidol, dihydroactinidiolide, *cis*-3-methyl-dec-3-en-1-ol, *cis*-3-methyl-dec-3-enoic acid, and *trans*-3-methyl-dec-4-enoic acid. The last three compounds are responsible for much of the characteristic fragrance of cassie oil.[2–5]

The nonfragrant material present accounts for about 75% of the absolute and consists mostly of high molecular weight lipids (e.g., hydrocarbons and waxes).[2]

## PHARMACOLOGY OR BIOLOGICAL ACTIVITIES

Reportedly antispasmodic, aphrodisiac, astringent, demulcent, antidiarrheal, febrifuge, antirheumatic, and stimulant (DUKE 1, DUKE 2). No additional pharmacological or toxicological data are available.

## USES

**Medicinal, Pharmaceutical, and Cosmetic.** The absolute is used as a fragrance component in some high-cost perfumes.

**Food.** The absolute is used as a flavor ingredient (fruit flavors) in most major categories of food products, including alcoholic and nonalcoholic beverages, frozen dairy desserts, candy, baked goods, and gelatins and puddings, with average maximum use levels generally below 0.002%.

**Traditional Medicine.** Flowers used as antispasmodic, aphrodisiac, and insecticide in the form of an infusion; also used in baths for dry skins. In India, leaves brewed in water used for gonorrhea; root chewed for sore throat. Dried gum ground to powder for diarrhea.[6] Root has been reportedly used in treating stomach cancer in Venezuela.[7] It is also used in China to treat rheumatoid arthritis and pulmonary tuberculosis.

**Others.** Bark and pods (23% tannin) used for tanning; gum marketed with other acacia gums; used in confectionery.

## COMMERCIAL PREPARATIONS

Absolute and oil.

*Regulatory Status.* Has been approved for food use (§172.510).

## REFERENCES

See the General References for ARCTANDER; BAILEY 1; CSIR 1; DUKE 1; DUKE 2; GUENTHER; JIANGSU; LIST AND HÖRHAMMER; MORTON 2; ROSE; UPHOF.

1. C. D. Adams, *Flowering Plants of Jamaica*, University of the West Indies, Mona, Jamaica, 1972, p. 336.
2. E. Demole et al., *Helv. Chim. Acta*, **52**, 24 (1969).
3. M. S. Karawaya et al., *Bull. Fac. Pharm. Cairo Univ.*, **13** 183 (1974); through *Chem. Abstr.*, **86**, 60396u (1977).
4. A. El-Hamidi and I. Sidrak, *Planta Med.*, **18**, 98 (1970).
5. A. M. El-Gamassy and I. S. Rofaeel, *Egypt J. Hortic.*, **2**, 39, 53 (1975); through *Chem. Abstr.*, **84**, 35180v, 35181w (1976).
6. M. B. Siddiqui and W. Husain, *Fitoterapia*, **62**(4), 325 (1991).
7. J. L. Hartwell, *Lloydia*, **33**, 97 (1970).

# CASTOR OIL

**Source:** ***Ricinus communis*** L. (Family Euphorbiaceae).

*Synonyms.* Ricinus, palma christi, castor bean, and castor seed oil.

## GENERAL DESCRIPTION

Plant is an annual herb (up to 5 m high) when grown in temperate zones, but is a perennial shrub or tree (up to about 15 m high) in warmer climates; it has many varieties. It is generally believed to be native of Africa or India and is extensively cultivated worldwide. Parts used are the ripe seeds, from which colorless to pale castor oil is obtained by cold pressing in 25–35% yield; hot pressing and solvent extraction yield darker grades that are of lower quality. Castor oil is remarkably stable and does not turn rancid easily. Major castor oil-producing countries include Brazil, China, and India.[1,2]

## CHEMICAL COMPOSITION

Castor oil contains mostly a triglyceride of ricinoleic acid, which is present up to 90%; other acids present include linoleic, oleic, stearic, and dihydroxystearic acids. Ricinoleic acid is a hydroxy acid, and as a result of hydrogen bonding of its hydroxyl groups, castor oil has a characteristically high viscosity.[1–3]

Castor bean (seed) contains a highly poisonous protein (ricin), which remains in the seed cake (pomace) after castor oil is expressed from it. Ricin has been reported to contain 18 different amino acids and to have a molecular weight of 53,000–54,000. Steam or moist cooking of the pomace destroys the ricin. The seed also contains ricinine (an alkaloid), lectins, and a very powerful heat-stable allergen (MARTINDALE).[1,4]

## PHARMACOLOGY OR BIOLOGICAL ACTIVITIES

Castor oil has cathartic properties, acting on the small intestine and producing purgation 2–8 h after ingestion; its usual dose is about 15 mL. Large doses may produce, besides purgation, nausea, vomiting, and colic. It also has emollient properties on the skin and is soothing to the eyes.

Castor oil facilitates the absorption of oil-soluble anthelmintics and should not be used with them (see ***aspidium***).

Castor seed is extremely toxic due to its content of ricin, which is not present in the oil; chewing a single seed may be fatal to a child. The allergen present in the seed can cause serious symptoms (asthma, eye irritations, hay fever, skin rashes, etc.) in certain individuals (MARTINDALE).

A recent feeding study involving chicks found that at concentrations of 0.5%–5% of diet castor seed produced toxicity characterized by locomotor disturbances, impaired vision, abnormal posture, growth depression, anemia with significant increases in serum sorbitol dehydrogenase, glutamic dehydrogenase, glutamic oxaloacetic transaminase, potassium, and total hepatic and cardiac lipids. Decreases in hepatic vitamin A, serum protein totals, and manganese were also observed.[5]

## USES

**Medicinal, Pharmaceutical, and Cosmetic.** Used as a cathartic, particularly in the treatment of food poisoning and in evacuation of the bowel before X-ray examination; as a solvent or vehicle in some parenteral and ophthalmic preparations; as an ingredient in lipsticks, hair-grooming products, ointments, creams, lotions, transparent soaps, suppository bases, and others.

**Food.** Castor oil is used as an antisticking and release agent in hard candy production and as a component of protective coatings in tablets (vitamins, minerals, bioflavonoids, etc.); as a flavor component (e.g., butter and nut flavors) in major categories of foods such as nonalcoholic beverages, frozen dairy desserts, candy, baked goods, and meat and meat products, with highest average maximum use level of 0.055% reported for frozen dairy desserts.

**Traditional Medicine.** Castor oil has been used for centuries in India, Egypt, and China as a cathartic and externally for sores and abscesses, among others; seeds also reportedly used as an oral contraceptive in Algiers.[6]

**Others.** By far the largest use of castor oil is as its dehydrated or partially dehydrated form in industrial lubricants, coatings, paints, varnishes, and others; also in synthesis of urethanes, foams, plastics, and certain perfume chemicals, among others.[1,2,7]

**COMMERCIAL PREPARATIONS**

No. 1 and no. 3 quality oils. Castor oil is official in U.S.P. and F.C.C.

*Regulatory Status.* Has been approved for food use (§172.510, etc.).

**REFERENCES**

See the General References for BAILEY 2; BIANCHINI AND CORBETTA; FEMA; GRIEVE; JIANGSU; LEWIS AND ELVIN-LEWIS; MARTINDALE; MERCK; NANJING; UPHOF; USD 26th; YOUNGKEN.

1. G. J. Hutzler in A. Standen, ed., *Kirk-Othmer Encyclopedia of Chemical Technology*, Vol. 4, 2nd ed., Wiley-Interscience, New York, 1967, p. 524.
2. L. A. O'Neill in L. W. Codd et al., eds., *Chemical Technology: An Encyclopedic Treatment*, Vol. 5, Barnes & Noble, New York, 1972, p. 187.
3. T. Khadzhiiski and M. Kalichkov, *Nauch. Tr., Vissh Inst. Khranit. Vkusova Prom., Plovdiv*, **21**, 61 (1974); through *Chem. Abstr.*, **85**, 149023x (1976).
4. N. Koja and K. Mochida, *Eisei Kagaku*, **20**, 204 (1974); through *Chem. Abstr.*, **82**, 26883h (1975).
5. S. M. A. El Badwi et al., *Phytother. Res.*, **6**, 4, 205 (1992).
6. V. J. Brondegaard, *Planta Med.* **23**, 167 (1973).
7. C. N. Subramanian, *Indian Chem. J. Annu.*, **107** (1972); through *Chem. Abstr.*, **81**, 68308j (1974).

# CASTOREUM

**Source:** ***Castor fiber*** L. or ***C. canadensis*** Kuhl (Family Castoridae).

*Synonyms.* Secretion of Canadian beaver (*C. canadensis*) and Siberian or European beaver (*C. fiber*).

**GENERAL DESCRIPTION**

Beavers are large pale brown to chestnut-brown rodents. The Canadian beaver inhabits lakes and rivers of Canada and northern United States, while the Siberian beaver is found in Europe and Siberia. Castoreum is the secretion accumulated in glands located near the pubis (between anus and sex organs) of these animals. These scent glands with their secretion (castoreum) are collected and dried, from which extracts (absolute, tincture, etc.) are prepared by solvent extraction. Canadian castoreum is considered superior in quality to the Siberian castoreum.[1]

**CHEMICAL COMPOSITION**

Contains 1–2% volatile oil; 0.33–2.5% castorin (a waxy crystalline substance separated from the hot alcoholic extract on cooling); up to 80% of an alcohol-soluble resinoid material; acids (benzoic, salicylic, cinnamic acids, etc.); phenols (phenol, *o*-ethylphenol, *p*-ethylphenol, *p*-propylphenol, chavicol, betuligenol, etc.); ketones (acetophenone and its derivatives, an ionone derivative, etc.); castoramine; cholesterol and other alcohols (benzylalcohol, *cis*-1,2-cyclohexanediol, etc.); 1.4% calcium phosphate; and others (ARCTANDER, LIST AND HÖRHAMMER, POUCHER).[2]

Canadian castoreum and Siberian castoreum differ considerably in their relative concentrations of certain of these constituents, with the Siberian material generally higher in volatile oil, castorin, and resinoid matter (LIST AND HÖRHAMMER).

## PHARMACOLOGY OR BIOLOGICAL ACTIVITIES

Believed to have sedative, nervine, and other properties. No pharmacological data are available.

Recent tests (primarily dermatological) using castoreum tincture have indicated it to be nontoxic.[3]

## USES

**Medicinal, Pharmaceutical, and Cosmetic.** Rarely used in pharmaceuticals. Main use (generally as a tincture) is in cosmetics as a fragrance component or fixative in perfumes (particularly men's fragrances and Oriental types), soaps, creams, and lotions, with maximum use level of 0.4% reported in perfumes.[3]

**Food.** Extracts used as flavor components (particularly in vanilla flavors) in most major categories of foods such as alcoholic and nonalcoholic beverages, frozen dairy desserts, candy, baked goods, gelatins and puddings, meat and meat products, and gravies. Average maximum use levels reported are usually below 0.009% (93.7 ppm).

**Traditional Medicine.** Used in amenorrhea, dysmenorrhea, hysteria, restless sleep, and as analeptic and nervine, among others.

## COMMERCIAL PREPARATIONS

Mainly crude.

*Regulatory Status.* GRAS (§182.50).

## REFERENCES

See the General References for ARCTANDER; FEMA; LIST AND HÖRHAMMER; MARTINDALE; POUCHER.

1. E. Shiftan in A. Standen, ed., *Kirk-Othmer Encyclopedia of Chemical Technology*, 2nd ed., Vol. 14, Interscience, New York, 1967, p. 717.
2. Z. Valenta et al., *Experientia*, **17**, 130 (1961).
3. D. L. J. Opdyke, *Food Cosmet. Toxicol.*, **11**, 1061 (1973).

# CATECHU (BLACK AND PALE)

**Source:** ***Black catechu:*** *Acacia catechu* (L.f.) Willd. (Family Leguminosae or Fabaceae). ***Pale catechu:*** *Uncaria gambir* (Hunter) Roxb. (Family Rubiaceae).

*Synonyms.* Dark catechu, black cutch, cutch, cachou, pegu catechu, and cashou (*A. catechu*); gambir catechu, gambir, gambier, and terra japonica (*U. gambir*).

## GENERAL DESCRIPTION

*Acacia catechu* is a spiny, deciduous medium-size tree; up to 13 m high; native to India and Burma. Part used is the heartwood, which is extracted with boiling water; the aqueous extract after filtration, evaporation, and drying yields black catechu, a shiny black mass.

*Uncaria gambir* is an evergreen woody vine, native to southeastern Asia (Malaysia, Indonesia, etc.). Parts used are the leaves and twigs, which are extracted with boiling water to yield pale catechu after filtration and evaporation of the extract to dryness; it is a pale brown to dark mass (JIANGSU, NANJING).

Both black catechu and pale catechu are incompatible with alkaloids, proteins (e.g., gelatin), and metallic salts (e.g., iron).

The terms *catechu* and *cutch* can also mean products other than black catechu and pale catechu; examples include Bombay catechu and Borneo cutch, which are derived from *Areca catechu* (betel nut) and a mangrove species, respectively.[1]

## CHEMICAL COMPOSITION

Black catechu contains 2–20% *l*- and *dl*-catechin, 20–50% catechutannic acid, *l*- and *dl*-epicatechin, quercetin, fisetin, red pigments, and others (JIANGSU).

Pale catechu contains *d*- and *dl*-catechin (30–35%) and the condensation product catechutannic acid (ca. 24%), quercetin, gallic acid, ellagic acid, catechol, pigments, and others. In addition, it contains several indole alkaloids, including gambirtannine, dihydrogambirtannine, and oxogambirtannine. Gambirine, gambirdine, and others are also found in leaves and/or stems (JIANGSU).[2–5]

## PHARMACOLOGY OR BIOLOGICAL ACTIVITIES

Because of their high tannin content, black catechu and pale catechu have astringent, antibacterial, and other pharmacological properties as well as toxicities of tannins (see ***tannic acid***).

*d*-Catechin has been reported to cause constriction of isolated rabbit ear blood vessels and suppression followed by enhancement of the amplitude of the isolated toad heart (JIANGSU).

Gambirine has been reported to have hypotensive properties (GLASBY 1).

An aqueous extract of *A. catechu* (small branches) has been shown to have hypotensive effects in anesthetized dogs and rats.[6]

Preliminary studies have shown fisetin to have liver protectant properties and to have been effective in icteric viral hepatitis.[7]

(+)-Catechin derived from the leaves and twigs of *Uncaria gambir* and the heartwood of *Acacia catechu* have been shown to protect against experimentally induced ulcers in animals.[8]

Anithepatotoxic activity has also been observed in animals with (+)-catechin or (+)-cyanidanol-3, leading to a dose-dependent increase in hepatic alkaline phosphatase levels, suggesting the compounds may affect alkaline phosphatase formation through oxidative phosphorylation.[8]

## USES

**Medicinal, Pharmaceutical, and Cosmetic.** Both black and pale catechus are used primarily as an astringent in certain antidiarrheal preparations and in mouthwashes.

**Food.** Both are used as flavor components in major categories of food products, including alcoholic and nonalcoholic beverages, frozen dairy desserts, candy, baked goods, and gelatins and puddings. Black catechu extract (type of extract not specified) has been reported as the more commonly used, with highest average maximum use levels of 0.01% and 0.016% reported in candy and alcoholic beverages, respectively.

**Traditional Medicine.** Both black and pale catechus are used in stopping nosebleeding and in treating boils, sores, ulcers, hemorrhoids, and others; black catechu is also reported used in cancers.[9]

In China, one of the major uses of black catechu is in treating indigestion in children.

**Others.** As a source of tannic acid.

## COMMERCIAL PREPARATIONS

Crude and extracts (e.g., tincture). Black catechu was formerly official in U.S.P., while pale catechu was official in N.F.

*Regulatory Status*. Both black catechu and pale catechu have been approved for food use (§172.510).

## REFERENCES


See the General References for CLAUS; FEMA; GRIEVE; KARRER; JIANGSU; MARTINDALE; MERCK; NANJING; TERRELL.

1. W. Gardner, *Chemical Synonyms and Trade Names*, 6th ed., The Technical Press, London, 1968, p. 182.
2. L. Merlini et al., *Tetrahedron*, **23**, 3129 (1967).
3. C. Cardani, *Corsi Semin. Chim.*, **11**, 131 (1968); through *Chem. Abstr.*, **72** 32098r (1970).
4. L. Merlini et al., *Phytochemistry*, **11**, 1525 (1972).
5. K. C. Chan, *Tetrahedron Lett.*, **30**, 3403 (1968).
6. J. S. K. Sham et al., *Planta Med.*, **50**, 177 (1984).
7. Z. C. Zhang et al., *Zhongyao Tongbao*, **13**(3), 34 (1988).
8. D. Pathak et al., *Fitoterapia*, **62**(5), 371 (1991).
9. J. L. Hartwell, *Lloydia*, **33**, 97 (1970).


# CATNIP

**Source:** ***Nepeta cataria*** L. (Family Labiatae or Lamiaceae).

*Synonym*. Catnep.

## GENERAL DESCRIPTION

Gray, hairy, erect, branched perennial, 40–100 cm high; leaves ovate, crenate, base cordate; 2–8 cm long; flowering in spike, white, tinged with purple; native to southern and eastern Europe; widely naturalized elsewhere in Europe and North America; commercially harvested from naturalized populations in Virginia, North Carolina; cultivated in Washington. Part used is the flowering tops and the essential oil obtained from steam distillation.

## CHEMICAL COMPOSITION

Contains 0.3–1% essential oil with nepetalic acid (10%), $\beta$-caryophyllene (14%), nepetalic anhydride (36%), plus up to 42% $\alpha$- and $\beta$-nepetalactone (GUENTHER), and its isomers epinepetalactone, 5,9-dehydronepetalactone, dihydronepetalactone, isodihydronepetalactone, neonepetalactone;[1,2] carvacrol, citronellal, nerol, geraniol, pulegone, thymol; tannins; iridoids, including epideoxyloganic acid and 7-deoxyloganic acid.

## PHARMACOLOGY OR BIOLOGICAL ACTIVITIES

Best known for its ability to elicit unique sequence in Felidae, including sniffing; licking and chewing with head shaking; chin and cheek rubbing; head-over rolling and body rubbing; known as "the catnip response."[2] The response is observed in domestic and large cats, such as lions, jaguars, tigers, etc. Nepetalactone and its isomers, with the exception of epinepetalactone (questionable) and 5,9-dehydronepetalactone (never tested) are responsible. Nepetalactone mildly sedative, antispasmodic, herbicidal, and insecticidal.

## USES

**Food.** The leaves and flowering tops have been used as a flavoring in sauces and cooked foods; dried in mixtures for soups, stews, etc.

**Health Food/Herb Teas.** Tops in teas; pleasant-tasting, mintlike, but characteristic (FOSTER).

**Traditional Medicine.** Flowering tops used in infusions for colds, colic, fever, flu, diarrhea, headache, menstrual irregularities; reputed to have diaphoretic, febrifuge, spasmolytic, mild sedative (especially to induce sleep in children) (FOSTER; FOSTER AND DUKE).

**Others.** Oil formerly used as an attractant in wild cat traps.

## COMMERCIAL PREPARATIONS

Crude herb; extracts, essential oil; formerly official in both N.F. and U.S.P.

*Regulatory Status.* Presently undetermined. Formerly included in U.S.P. (1840–1870) and N.F. (IV–VII).

## REFERENCES

See the General References for CSIR VII; DUKE 2; FOSTER; FOSTER AND DUKE; GUENTHER; HARBOURNE AND BAXTER; SIMON, STEINMETZ; TUTIN 3; TYLER; UPHOF; WREN.

1. S. D. Sastry et al. *Phytochemistry*, **11**, 453 (1972).
2. A. O. Tucker and S. S. Tucker, *Econ. Bot.*, **42**, 214 (1988).

# CEDAR LEAF OIL

**Source:** ***Thuja occidentalis*** L. (Family Cupressaceae).

*Synonyms.* White cedar, northern white cedar, eastern white cedar, American arborvitae, and thuja oils.

## GENERAL DESCRIPTION

*Thuja occidentalis* is a small- to medium-size tree belonging to the cypress family, up to about 20 m high; native to northeastern North America (Nova Scotia south to North Carolina and west to Illinois). There are many cultivated varieties. Parts used are the fresh leaves and twigs, from which cedar leaf oil is obtained by steam distillation. Major producers of the oil are Canada and the United States.

## CHEMICAL COMPOSITION

Contains mainly thujone, isothujone, *l*-fenchone, borneol, *l*-bornyl acetate, *dl*-limonene, *d*-sabinene, *d*-terpinen-4-ol, pinene, camphor, myrcene, and *l*-$\alpha$-thujene, among others, with thujone in major concentration and accounting for up to 65% (w/w) of the oil (KARRER).[1–4]

## PHARMACOLOGY OR BIOLOGICAL ACTIVITIES

Oil is believed to have expectorant, uterine stimulant, emmenagogue, anthelmintic, and counterirritant properties.

Due to its high thujone content, the oil is poisonous when ingested in large quantities, producing symptoms such as hypotension and convulsions and eventually death (see ***absinthium***) (MERCK).

Antiviral activity has been demonstrated *in vitro*.[5]

Leaf extract stimulates phagocytosis (erythrocytes) of Kupfer cells in isolated rat liver.[6]

The oil is reported to be nontoxic when applied externally.[4]

### USES

**Medicinal, Pharmaceutical, and Cosmetic.** Used primarily as a counterirritant in certain analgesic ointments and liniments. In Europe, tincture used externally for its antifungal and antiviral activity in treating warts (WREN). Used in German phytomedicine for nonspecific immunostimulant therapy.[7] Principal use in the United States is as a fragrance ingredient in soaps, detergents, creams, lotions, and perfumes, with maximum use level of 0.4% reported in perfumes.[4]

**Food.** Oil is used as a flavor ingredient in most categories of foods, including alcoholic and nonalcoholic beverages, frozen dairy desserts, candy, baked goods, gelatins and puddings, meat and meat products, condiments and relishes, and others. Reported average maximum use levels are quite low, with the highest being 0.002% in condiments and relishes.

**Traditional Medicine.** Ointment or decoction of fresh leaves is used to treat rheumatism, coughs, fever, gout, and other ailments. Oil is used internally as an expectorant, antirheumatic, diuretic, and emmenagogue and externally to treat skin diseases and as insect repellent; also used in treating condyloma and cancers.[8]

### COMMERCIAL PREPARATION

Essential oil; official in F.C.C.

*Regulatory Status.* Has been approved for food use, provided that the finished food is thujone free (§172.510).

### REFERENCES

See the General References for ARCTANDER; BAILEY 1; FEMA; GUENTHER; KROCHMAL AND KROCHMAL; LUST; MERCK; TERRELL; UPHOF; WREN; YOUNGKEN.

1. A. C. Shaw, *Can. J. Chem.*, **31**, 277 (1953).
2. R. M. Ideda et al., *J. Food Sci.*, **27**, 455 (1962).
3. D. V. Banthorpe et al., *Planta Med.*, **23**, 64 (1973).
4. D. L. J. Opdyke, *Food Cosmet. Toxicol.*, **12**(Suppl.), 843 (1974).
5. N. Beuscher and L. Kopanski, *Planta Med.*, **52**(6), 111P (1986).
6. T. Vomel, *Arzneim. Forsch.*, **35**II(9), 1437 (1985).
7. H. Wagner and A. Proksch in N. Farnsworth, N. H. Hikino, and H. Wagner, eds., *Economic and Medicinal Plant Research*, Vol. 1., Academic Press, New York, 1985, p. 113.
8. J. L. Hartwell, *Lloydia*, **33**, 288 (1970).

## CEDARWOOD OIL

**Source:** ***Cedarwood oil Virginia:*** *Juniperus virginiana* L. (Family Cupressaceae); ***cedarwood oil Texas:*** *Juniperus mexicana* Spreng. (Family Cupressaceae); ***cedarwood oil Atlas:*** *Cedrus atlantica* Manetti (Family Pinaceae).

*Synonyms.* Cedar oil and red cedarwood oil (cedarwood oil Virginia); cedarwood oil Moroccan (cedarwood oil Atlas).

### GENERAL DESCRIPTION

There are several cedarwood oils with different physical and chemical properties.

They are often referred to in the literature simply as cedarwood oil. The most common ones, cedarwood oil Virginia, cedarwood oil Texas, and cedarwood oil Atlas, are derived from *J. virginiana*, *J. mexicana*, and *C. atlantica*, respectively. Others such as cedarwood oil Himalaya, cedarwood oil East Africa, and cedarwood oil Japanese are obtained from other conifers (ARCTANDER). *Juniperus ashei* Buchh. is also used as a source of cedarwood oil; and other species (e.g., *J. erythrocarpa* Cory and *J. scopulorum* Sarg.) containing high oil content are potential sources.[1]

Cedarwood oil Virginia is obtained by steam distillation of the wood (sawdust, shavings, and other lumber wastes) of *J. virginiana*, commonly known as red cedar, eastern red cedar, and savin, which is a tree up to about 33 m high growing in North America east of the Rocky Mountains. The tree has many cultivated varieties. This oil is primarily produced in the United States and is most commonly referred to as cedarwood oil or cedar oil; it has a sweet "pencil wood" and balsamic odor.

Cedarwood oil Texas is prepared by steam distillation of the wood (shavings, etc.) of *J. mexicana*, which is a small tree up to about 6 m high growing in mountains of southwestern United States, Mexico, and Central America. The oil is produced in Texas; it has an odor similar to that of Virginia cedarwood oil.

Cedarwood oil Atlas is obtained by steam distillation of the wood of *C. atlantica*, which is a pyramidal tree closely related to the pines up to about 40 m high and growing in the Atlas Mountains of Algeria. This oil is produced primarily in Morocco; it has different odor characteristics than the Virginian and Texan oils.

## CHEMICAL COMPOSITION

Cedarwood oil Virginia contains mainly $\alpha$- and $\beta$-cedrene (ca. 80%), cedrol (3–14%), and cedrenol. Other sesquiterpenes present include thujopsene, $\beta$-elemene, caryophyllene, cuparene, $\alpha$-acoradiene ("acorene"), and others. Monoterpenes are also present (mostly sabinene and sabinyl acetate) (MASADA).[2–8]

Cedarwood oil Texas contains similar major constituents as cedarwood oil Virginia (ARCTANDER, KARRER).[3,5,9]

Cedarwood oil Atlas contains as its major odoriferous components $\alpha$- and $\gamma$-atlantone. Other constituents include acetone, $\alpha$-ionone, and $\alpha$-caryophyllene, among others.[10–12]

## PHARMACOLOGY OR BIOLOGICAL ACTIVITIES

Cedarwood oil, most likely Virginia, has been reported to have tumor-producing properties on mouse skin.[13]

Cedarwood oil (Virginia and/or Texas) has been reported to have slight local allergenic (acute and chronic) and acute local irritant properties (SAX).

Dermatological data have indicated cedarwood oils (Virginia, Texas, and Atlas) to be generally nontoxic.[6,9,12]

## USES

**Medicinal, Pharmaceutical, and Cosmetic.** All three types of cedarwood oils (Virginia, Texas, and Atlas) are primarily used as fragrance components or fixatives in cosmetic and household products, particularly soaps and detergents; others include creams, lotions, and perfumes. The maximum use level reported is 0.8% for all three in perfumes.[6,9,12]

**Traditional Medicine.** Cedarwood oil Virginia has been used as an insect repellent. Decoctions of the leaves, bark, twigs, and seeds of *J. virginiana* are used to treat various illnesses including coughs, bronchitis, rheumatism, venereal warts, and skin rash, among others.[14]

**Others.** Cedarwood oil Virginia is used in microscopy as a clearing agent and, together with resins, as an immersion oil. It can also serve as source of cedrene, a starting material for fragrance chemicals.

### COMMERCIAL PREPARATIONS

The essential oils.

### REFERENCES

See the General References for ARCTANDER; BAILEY 1; FERNALD; GUENTHER; KROCHMAL AND KROCHMAL; MARTINDALE; POUCHER; SAX.

1. R. P. Adams, *Econ. Bot.*, **41**, 48 (1987).
2. J. A. Wenninger et al., *J. Ass. Offic. Anal. Chem.*, **50**, 1304 (1967).
3. W. D. Fordham in L. W. Codd et al., eds., *Chemical Technology: An Encyclopedic Treatment*, Vol. 5, Barnes & Noble, New York, 1972, p. 1.
4. D. V. Banthorpe et al., *Planta Med.*, **23**, 64 (1973).
5. G. C. Kitchens et al., *Givaudanian*, **1**, 3 (1971); through *Chem. Abstr.*, **75**, 40237n (1971).
6. D. L. J. Opdyke, *Food Cosmet. Toxicol.*, **12**(Suppl.), 845 (1974).
7. J. A. Marshall et al. in W. Herz et al., eds., *Progress in The Chemistry of Organic Natural Products*, Vol. 31, Springer-Verlag, Vienna, 1974, p. 283.
8. G. C. Kitchens et al., *Ger. Offen.* 2,202,249 (1972); through *Chem. Abstr.*, **77**, 126896e (1972).
9. D. L. J. Opdyke, *Food Cosmet. Toxicol.*, **14**, 711 (1976).
10. A. S. Pfau and P. Plattner, *Helv. Chim. Acta*, **17**, 129 (1934).
11. D. R. Adams et al., *Tetrahedron Lett.*, **44**, 3903 (1974).
12. D. L. J. Opdyke, *Food Cosmet. Toxicol.*, **14**, 709 (1976).
13. F. J. C. Roe and W. E. H. Field, *Food Cosmet. Toxicol*,. **3**, 311 (1965).
14. J. L. Hartwell, *Lloydia*, **33**, 288 (1970).

# CELERY SEED

**Source:** ***Apium graveolens*** L. (Family Umbelliferae or Apiaceae).

*Synonym.* Celery fruit.

### GENERAL DESCRIPTION

An erect biennial herb, up to about 1 m high; native to southern Europe; extensively cultivated. There are many varieties. *Apium graveolens* var. *dulce* (Mill.) Pers. yields the celery vegetable which is its leafstalk (petiole), and *A. graveolens* var. *rapaceum* (Mill.) Gaudich., the turnip-rooted celery, yields celeriac. The seeds (dried ripe fruits) used for oil production or as spices are produced from other varieties. Major seed-producing countries are France and India. Celery seed oil is obtained by steam distillation of the whole or crushed seeds in about 2% yield. An oleoresin and extracts are also prepared by extracting the seeds with solvents.

### CHEMICAL COMPOSITION

Celery seed contains coumarins, coumarin glycosides, including bergapten, apiumoside, vellein, celereoin, nodakenin, and celereoside.[1]

Major components of the oil are reported to be *d*-limonene (ca. 60%), selinene (ca. 10%), and about 3% phthalides. Other constituents include santalol, $\alpha$- and $\beta$-eudesmol, dihydrocarvone, and fatty acids (palmitic, petroselinic, oleic acids, etc.), among others (JIANGSU, LIST AND HÖRHAMMER).[2–7]

The phthalides are the odoriferous principles that consist mostly of 3-*n*-butylphthalide, sedanenolide (3-*n*-butyl-4,5-dihydrophthalide), sedanolide, and sedanonic anhydride, with several others in minor amounts;[4–6] presence of sedanolide and sedanonic anhydride is disputed.[4,5]

Celery seed oleoresin contains more odoriferous principles and less terpenes; it also contains apiin and other flavonoids (JIANGSU).

## PHARMACOLOGY OR BIOLOGICAL ACTIVITIES

Phthalides present in celery seed oil have been reported to have sedative activities on mice.[4]

3-*n*-Butylphthalide has been shown to have anticonvulsant effects in experimental chronic epilepsy induced by coriaria lactone in rats. Its anticonvulsant effects were weaker than those of diazepam, but its ability in counteracting the learning and memory impairment caused by coriaria lactone was greater than that of diazepam, causing no damage to brain cells.[8,9] It also has low acute and chronic toxicities and no teratogenic activity in experimental animals.[10]

A celery extract has been claimed to aid the suntanning of the skin,[11] and the petroleum ether-soluble fraction of celery seed has been reported to exhibit antioxidative properties on lard.[12]

An extract of celery (*A. graveolens* L. var. *dulce* DC) is reported to have distinct hypotensive effects in rabbits and dogs when administered intravenously. Clinically, celery juice has been reported effective in lowering the blood pressure of 14 out of 16 patients with high blood pressure; the juice was mixed with equal amounts of honey or syrup, and 40 mL were taken orally three times a day for up to several days (JIANGSU).

Significant antiinflammatory activity of aqueous extracts of celery stem has been found in two animal models, suggesting a rational basis for traditional use in rheumatoid arthritis.[13]

Diuretic activity of the seed has been reported in animal experiments.[14]

Celery seed oil has been reported to be generally nonirritating, nonsensitizing, and nonphototoxic, though cases of mild to severe dermatitis resulting from contact with celery plants have been well documented.[15,16] A case of allergic reaction ending in anaphylactic shock has also been reported.[14]

## USES

**Medicinal, Pharmaceutical, and Cosmetic.** Oil is used in certain tonic, sedative, and carminative preparations and as a fragrance component in soaps, detergents, creams, lotions, and perfumes. Use levels in cosmetics range from a low of 0.0003% (3 ppm) in detergents to a maximum of 0.4% in perfumes.[15]

**Food.** Celery seed oil, celery seed, and celery seed extracts are all extensively used as flavoring ingredients in all major food products, including alcoholic and nonalcoholic beverages, frozen dairy desserts, candy, baked goods, gelatins and puddings, meat and meat products, condiments and relishes, soups, gravies, snack foods, and others. Use levels reported for the oil are usually very low, with the highest average maximum of about 0.005% (46.6 ppm) in condiments and relishes.

**Health Food/Herb Teas.** Celery seed or celery seed extracts are used as flavoring or for biological effect in herbal combination dietary supplements; also in antirheumatic formulations (WREN).

**Traditional Medicine.** Oil reportedly used as diuretic in dropsy and bladder ailments, as a nervine and antispasmodic, and in rheumatoid arthritis. In European tradition seeds reportedly used as carminative, stomachic, emmenagogue, diuretic, laxative; for glandular stimulation, gout, kidney stones, rheumatic complaints, nervous unrest, loss of appetite and exhaustion.[14] Leaves and petioles are used for skin problems in addition to above uses.

**COMMERCIAL PREPARATIONS**

Oil and oleoresin (extracts); seed was formerly official in N.F. and oil is official in F.C.C. Extracts come in various forms with strengths (see ***glossary***) expressed in weight-to-weight ratios or in flavor intensities. Celery (including stems, roots, herb, and seed) are the subject of a German therapeutic monograph. Effectiveness of traditional claims is not documented; therefore, use is not recommended.[14]

*Regulatory Status*. GRAS (§182.10 and §182.20).

**REFERENCES**

See the General References for ARCTANDER; BAILEY 2; FEMA; GUENTHER; JIANGSU; LIST AND HÖRHAMMER; LUST; MARTINDALE; ROSENGARTEN; TERRELL; UPHOF; WREN.


1. A. K. Jain et al., *Planta Med.*, 246 (1986).
2. M. M. Ahuja and S. S. Nigam, *Riechst., Aromen, Körperpflegem.*, **21**, 281 (1971).
3. S. I. Balbaa et al., *Egypt. J. Pharm. Sci.*, **16**, 383 (1976).
4. L. F. Bjeldanes and I. Kim, *J. Org. Chem.*, **42**, 2333 (1977).
5. D. H. R. Barton and J. X. De Vries, *J. Chem. Soc. (London),* 1916 (1963).
6. H. J. Gold and C. W. Wilson III, *J. Org. Chem.*, **28**, 985 (1963).
7. R. M. Ideda et al., *J. Food Sci.*, **27**, 455 (1962).
8. S. R. Yu et al., *Yaoxue Xuebao*, **23**, 656 (1988).
9. S. R. Yu et al., *Acta Pharmacol. Sin.*, **9**, 385 (1988).
10. S. R. Yu et al., *Yaoxue Tongbao*, **20**, 187 (1985).
11. J. Courtin, Fr. Demande, 2,295,735 (1976); through *Chem. Abstr.*, **86**, 95874a (1977).
12. Y. Saito et al., *Eiyo To Shokuryo*, **29**, 505 (1976); through *Chem. Abstr.*, **87**, 150314r (1977).
13. D. A. Lewis et al., *Int. J. Crude Drug Res.*, **23**, 1, 27 (1985).
14. Monograph *Celery*, *Bundesanzeiger* (July 12, 1991).
15. D. L. J. Opdyke, *Food Cosmet. Toxicol.*, **12**(Suppl.), 849 (1974).
16. R. M. Adams, *Occupational Contact Dermatitis*, J. B. Lippincott, Philadelphia, 1969, p. 190.


# CENTAURY

**Source:** ***Centaurium erythraea*** Rafn. (syn. *C. umbellatum* Gilib., *C. minus* Moench and *Erythraea centaurium* Pers.) (Family Gentianaceae).

*Synonyms*. European centaury, common centaury, minor centaury, lesser centaury, drug centaurium, and bitter herb.

**GENERAL DESCRIPTION**

Annual, mostly biennial herb with upright stem branching near the top; up to 0.5 m high; leaves opposite; flowers in corymbiform cymes, sessile, rose-purple; native to Europe, western Asia, and northern Africa and naturalized in North America in dry grassland, scrub, and mountain slopes. Variable (stem branching, leaf shape and size, flower size, etc.), separated into six subspecies in Europe, more or less restricted by geographic region (TUTIN 3). Part used is the dried flowering herb. Not to be confused with the genus *Centaurea* (Compositae).

## CHEMICAL COMPOSITION

Contains several bitter glucosides (gentiopicrin, centapicrin, swertiamarin, gentioflavoside, and sweroside), alkaloids, (gentianine, gentianidine, gentioflavine, etc.), phenolic acids (protocatechuic, *m*- and *p*-hydroxybenzoic, vanillic, syringic, *p*-coumaric, ferulic, 3,4-dihydroxyphenylacetic, sinapic, caffeic acids, etc.), triterpenes ($\alpha$- and $\beta$-amyrin, erythrodiol, crataegolic acid, oleanolic acid, oleanolic lactone, etc.), sterols (sitosterol, campesterol, etc.), fatty acids (palmitic and stearic acids, etc.), *n*-alkanes (nonacosane and heptacosane, etc.), wax, and others.[1-10]

## PHARMACOLOGY OR BIOLOGICAL ACTIVITIES

Reputedly has bitter tonic, sedative, antipyretic, and other properties.

Its antipyretic activity has been reported to be due to its content of phenolic acids.[11]

Gentiopicrin is reported to have antimalarial properties (MERCK).

## USES

**Medicinal, Pharmaceutical, and Cosmetic.** Used in some bitter tonic preparations in Europe to increase gastric secretions for dyspeptic discomfort and loss of appetite.[12] Also reportedly used in some cosmetic and toiletry preparations for its alleged soothing and astringent properties.[13]

**Food.** Used in bitters and vermouth formulations; average maximum use level reported is very low, about 0.0002% (2.29 ppm). Also reportedly used in nonalcoholic beverages at an average maximum use level of 0.0008%.

**Traditional Medicine.** Used since ancient times in Egypt to treat hypertension and to eliminate kidney stones, and in Europe as a tonic, stomachic, febrifuge, and sedative. Used in lotions to remove freckles, spots, and other skin blemishes and in treating cancers.[14]

## COMMERCIAL PREPARATIONS

Mainly the crude herb.

*Regulatory Status.* Has been approved for use in alcoholic beverages only (§172.510). Herb subject of a German therapeutic monograph (in daily dose of 1–2 g); use not recommended since effectiveness is not verified.[12]

## REFERENCES

See the General References for BIANCHINI AND CORBETTA; BLUMENTHAL: FEMA; LIST AND HÖRHAMMER; LUST; STAHL; TUTIN 3; UPHOF.

1. K. Sakina and K. Aota, *Yakugaku Zasshi*, **96**, 683 (1976); through *Chem. Abstr.*, **85**, 74968y (1976).
2. S. S. Popov et al., *Dokl. Bolg. Akad. Nauk*, **25**, 1225 (1972); through *Chem. Abstr.*, **78**, 40382t (1973).
3. D. W. Bishay et al., *Planta Med.*, **33**, 422 (1978).

4. F. Rulko and K. Witkiewicz, *Diss. Pharm. Pharmacol.*, **24**(1), 73 (1972).
5. F. Rulko, *Pr. Nauk. Akad. Med. Wroclawiu*, **8**, 3 (1976); through *Chem. Abstr.*, **86**, 103037x (1977).
6. M. Hatjimanoli and A. M. Debelmas, *Ann. Pharm. Fr.*, **35**(3–4), 107 (1977).
7. V. Bellavita et al., *Phytochemistry*, **13**, 289 (1974).
8. J. H. Zwaving, *Pharm. Weekbl.*, **101**, 605 (1966).
9. N. Marekov and S. Popov, *C. R. Acad. Bulg. Sci.*, **20**, 441 (1967); through *Chem. Abstr.*, **67**, 105950w (1967).
10. R. Lacroix et al., *Fitoterapia*, **5**, 213 (1983).
11. R. Lacroix et al., *Tunisie Med.*, **51**, 327 (1973); through *Chem. Abstr.*, **82**, 7590u (1975).
12. Monograph *Centaurii herba*, *Bundesanzeiger*, no. 122 (July 6, 1988); revised (Mar. 13, 1990).
13. H. B. Heath, *Cosmet. Toilet.*, **92**, 19 (1977).
14. J. L. Hartwell, *Lloydia*, **32**, 153 (1969).

# CHAMOMILE (GERMAN AND ROMAN)

**Source:** ***German chamomile*** *Matricaria recutita* L. (syn. *Matricaria chamomilla* L., *Chamomilla recutita* (L.) Rauschert) (Family Compositae or Asteraceae); ***Roman chamomile*** *Chamaemelum nobile* (L.) All. (syn. *Anthemis nobilis* L.) (Family Compositae or Asteraceae).

*Synonyms*. Matricaria, Hungarian chamomile, wild chamomile, and sweet false chamomile (*M. recutita*); English chamomile and garden chamomile (*C. nobile*); camomile.

## GENERAL DESCRIPTION

German chamomile is a fragrant, low annual herb, with ligulate flowerheads about 2 cm broad; up to about 0.6 m high; native to Europe and northern and western Asia; naturalized in North America; extensively cultivated, particularly in Hungary, Romania, Bulgaria, the Czech Republic, Slovakia, Germany, Greece, Argentina, and Egypt. Parts used are the dried flowerheads. An essential oil is obtained from the flowers by steam distillation; it is blue when fresh.

Frequent changes in interpretation of the scientific name of German chamomile have led to great confusion over the past two decades. The currently accepted scientific name is *Matricaria recutita*, though *Chamomilla recutita* and, to a lesser extent, *Matricaria chamomilla* are still commonly seen in the literature.[1]

Roman chamomile is a strongly fragrant, hairy, half-spreading, and much-branched perennial, with flowerheads about 2.5 cm across; up to about 0.3 m high; native to southern and western Europe; naturalized in North America; cultivated in England, Belgium, the United States, Argentina, and other countries. Parts used are the dried expanded flowerheads, from which a blue essential oil is obtained by steam distillation.

## CHEMICAL COMPOSITION

German chamomile contains variable amounts of volatile oil (0.24–1.9%); flavonoids including apigenin, apigetrin (apigenin-7-D-glucoside), apigenin-7-acetylglucoside, apiin (apigenin-7-apiosylglucoside), rutin (quercetin-3-rutinoside), luteolin, and quercimeritrin (quercetin-7-D-glucoside), among others; coumarins, including umbelliferone (7-hydroxycoumarin) and its methyl ether (herniarin);

proazulenes (matricin, matricarin, etc.); plant acids and fatty acids; a polysaccharide containing D-galacturonic acid as the major component; choline; amino acids; and others (JIANGSU, LIST AND HÖRHAMMER, STAHL).[2–5]

The volatile oil contains chamazulene, farnesene, α-bisabolol oxide A, α-bisabolol oxide B, α-bisabolone oxide A, matricin, and en-yn-dicycloether as major constituents, with their relative concentrations varying considerably, depending on the sources (JIANGSU, LIST AND HÖRHAMMER, STAHL).[6–14]

Roman chamomile contains up to 1.74% volatile oil; about 0.6% of bitter sesquiterpene lactones (germacranolides), which include nobilin, 3-epinobilin, 1,10-epoxynobilin, and 3-dehydronobilin; flavonoids (apigenin, apigetrin, apiin, quercitrin, luteolin-7-glucoside, etc.); coumarins (e.g., scopoletin-7-β-glucoside); choline; fatty acids; and others (LIST AND HÖRHAMMER).[15]

The volatile oil contains mainly (ca. 85%) esters of angelic and tiglic acids (e.g., butyl, amyl, isoamyl, and hexyl angelates or tiglates). Other constituents reported to be present include α-pinene, farnesol, nerolidol, chamazulene, *l-trans*-pinocarveol, *l-trans*-pinocarvone, and 1,8-cineole, among others. Relative concentrations of the constituents vary, depending on sources of the oil (ARCTANDER, MARTINDALE).[16–19]

## PHARMACOLOGY OR BIOLOGICAL ACTIVITIES

German chamomile has been reported to have numerous pharmacological properties, some of which are the following. The oil has bactericidal and fungicidal activities, particularly against Gram-positive bacteria (e.g., *Staphylococcus aureus*) and *Candida albicans*. It also reduced blood urea concentration in rabbits to a normal level.[20,21] Chamazulene, a major component of the oil, has pain-relieving, wound-healing, antispasmodic, antiinflammatory, and antimicrobial properties; α-bisabolol, another constituent of the oil has antiinflammatory, antimicrobial, and antipeptic activities; the cyclic ethers (e.g., en-yn-dicyloether) also have antimicrobial, antiinflammatory, antianaphylactic, and antispasmodic properties (JIANGSU, LIST AND HÖRHAMMER, MARTINDALE).[20,22–27] The tea has marked hypnotic effect (MARTINDALE). Umbelliferone has fungistatic properties.[22] Matricin has been found to have a significantly stronger antiinflammatory effect than chamazulene (ESCOP 1).

Infusions and standardized preparations considered antiphlogistic, musculotropic, spasmolytic, promotes wound healing, antibacterial and inhibitory to bacteriotoxins.[28]

Roman chamomile is emetic in large doses (MARTINDALE). Three sesquiterpenes lactones (nobilin, 1,10-epoxynobilin, and 3-dehydronobilin) isolated from Roman chamomile have been reported to exhibit antitumor activities *in vitro* against human tumor cells.[15]

Both German chamomile and Roman chamomile have been reported to cause contact dermatitis in humans.[29] Nevertheless, available data indicate both German and Roman chamomile oils to be generally nontoxic when applied externally.[30,31]

## USES

**Medicinal, Pharmaceutical, and Cosmetic.** Both German chamomile and Roman chamomile extracts are used in pharmaceutical preparations, with the former more frequently used; they are used in antiseptic ointments, creams, and gels to treat cracked nipples, sore gums, inflammations, irritation of the skin and mucosa, respiratory tract inflammation, and for wound healing. The volatile oils are used in carminative, antispasmodic, and tonic preparations, among others. An infusion of 2–3 g of the flowers or 1–4 mL of tincture (1:5), t.i.d., are used for gastrointestinal spasms, inflammatory conditions of the gastrointestinal tract, and peptic ulcers, in addition to mild sleep disorders, especially in children (ESCOP 1).

Extracts of both German and Roman chamomiles are used in cosmetics including bath preparations, hair dye formulas (for blond hair), shampoos, preparations to prevent sunburn, mouthwashes, and others. The oils are used as fragrance components or active ingredients in soaps, detergents, creams, lotions, and perfumes. Use levels reported range from a low of 0.0005% in detergents to a maximum of 0.4% in perfumes. Considered deodorant and stimulative to skin metabolism.[27,28,30,31]

**Food.** The essential oils and extracts of both German and Roman chamomiles are used as flavor components in most major food categories, including alcoholic (bitters, vermouths, Benedictine liqueurs, etc.) and nonalcoholic beverages, frozen dairy desserts, candy, baked goods, and gelatins and puddings. Average maximum use levels reported are usually less than 0.002% for the oils.

**Health Food/Herb Teas.** German chamomile and to a lesser extent Roman chamomile crude flowers or extracts are one of the most widely used herb tea ingredients, singly or in combination with other ingredients. Topical products as cosmetics for inflammation. Tinctures and extracts used as mild sleep aids, antispasmodics, digestive aids (FOSTER).

**Traditional Medicine.** German chamomile has been used since ancient times in treating colic, diarrhea, indigestion, insomnia, infantile convulsions, toothache, bleeding and swollen gums, and other ailments, usually in the form of an infusion, decoction, or tincture. Also used for sciatica, gout, lumbago, skin problems, and inflammation, in the form of compresses.

Roman chamomile is used essentially for the same purposes. Both German and Roman chamomiles have been reportedly used in cancers.[32]

## COMMERCIAL PREPARATIONS

Crude, extracts, and volatile oils. German chamomile was formerly official in N.F., and English chamomile (i.e., Roman) in U.S.P.; both oils are prone to insect infestation on storage. Strengths (see ***glossary***) of extracts are expressed in weight-to-weight ratios.

*Regulatory Status.* GRAS (§182.10 and §182.20). German chamomile flowers subject of a positive German therapeutic monograph; preparations allowed; internally for gastrointestinal spasms and inflammatory diseases of the gastrointestinal tract; externally for skin and mucous membrane inflammation, bacterial skin disease or the oral cavity and gums; inflammatory disease of the respiratory tract (as inhalations); bath and irrigation for inflammation of the genital and anal areas.[28]

## REFERENCES

See the General References for ARCTANDER; BAILEY 2; BIANCHINI AND CORBETTA; BLUMENTHAL; ESCOP 1; FEMA; FOSTER; GUENTHER; JIANGSU; LEWIS AND ELVIN-LEWIS; LUST; MASADA; STAHL; TERRELL; TYLER; UPHOF.

1. S. Foster, *Chamomiles*, *Botanical Series*, *no. 307*, American Botanical Council, Austin, Tex., 1991.
2. D. Kustrak and F. Benzinger, *Farm. Glas.*, **33**, 331 (1977); through *Chem. Abstr.*, **88**, 78956c (1978).
3. A. G. Gorin and A. I. Yakovlev, *Sb. Nauchn. Tr., Ryazan. Med. Inst.*, **50**, 9

(1975); through *Chem. Abstr.*, **84**, 79626z (1976).
4. J. Hölzl and G. Demuth, *Planta Med.*, **27**, 37 (1975).
5. C. Redaelli et al., *Planta Med.*, **42**, 288 (1981).
6. J. Hölzl et al., *Z. Naturforsch.*, **30**, 853 (1975).
7. J. Reichling and H. Becker, *Dtsch. Apoth.-Ztg.*, **117**, 275 (1977).
8. O. Motl et al., *Arch. Pharm. (Weinheim)*, **310**, 210 (1977).
9. H. Schilcher, *Planta Med.*, **23**, 132 (1973).
10. G. Verzar-Petri et al., *Herba Hung.*, **15**, 69 (1976); through *Chem. Abstr.*, **85**, 37084m (1976).
11. G. Verzar-Petri et al., *Herba Hung.*, **12**, 119 (1973); through *Chem. Abstr.*, **82**, 95345w (1975).
12. L. Z. Padula et al., *Planta Med.*, **30**, 273 (1976).
13. O. Isaac et al., *Dtsch. Apoth.-Ztg.*, **108**, 293 (1968).
14. O. Motl et al., *Arch. Pharm. (Weinheim)*, **311**, 75 (1978).
15. M. Holub and Z. Samek, *Collect. Czech. Chem. Commun.*, **24**, 1053 (1977).
16. S. I. Balbaa et al., *Egypt J. Pharm. Sci.*, **16**, 161 (1975).
17. Y. Chretien-Bessiere et al., *Riv. Ital. Essenze, Profumi, Piante Offic., Aromi, Saponi, Cosmet., Aerosol*, **52**, 211 (1970); through *Chem. Abstr.*, **74**, 15680y (1971).
18. G. M. Nano et al., *Sixth Int. Cong. Essent. Oils (Pap)*, 114 (1974).
19. G. M. Nano et al., *Essenze Deriv. Agrum.*, **46**, 171 (1976); through *Chem. Abstr.*, **86**, 177153d (1977).
20. M. E. Aggag and R. T. Yousef, *Planta Med.*, **22**, 140 (1972).
21. A. Grochulski and B. Borkowski, *Planta Med.*, **21**, 289 (1972).
22. M. Szalontai et al., *Parfüm Kosmet.*, **58**, 121 (1977).
23. O. Isaac and K. Thiemer, *Arzneim. Forsch.*, **25**, 1352 (1975).
24. H. Wirth, *Am. Perfum. Cosmet.*, **82**(10), 81 (1967).
25. N. R. Farnsworth and B. M. Morgan, *J. Am. Med. Assc.*, **221**, 410 (1972).
26. A. Tubaro et al., *Planta Med.*, **50**, 359 (1984).
27. C. Mann and E. J. Staba in L. E. Craker and J. E. Simon, eds., *Herbs, Spices, and Medicinal Plants: Recent Advances in Botany, Horticulture, and Pharmacology*, Vol. 1, Oryx Press, Phoenix, 1984, p. 235.
28. Monograph *Matricariae flos*, *Bundesanzeiger*, no. 228 (Dec. 5, 1984).
29. J. C. Mitchell in V. C. Runeckles, ed., *Recent Advances in Phytochemistry*, Vol. 9, Plenum Press, New York, 1975, p. 119.
30. D. L. J. Opdyke, *Food Cosmet. Toxicol.*, **12**(Suppl.), 851 (1974).
31. D. L. J. Opdyke, *Food Cosmet. Toxicol.*, **12**(Suppl.), 853 (1974).
32. J. L. Hartwell, *Lloydia*, **31**, 71 (1968).

# CHAPARRAL

**Source:** ***Larrea tridentata*** (DC) Cov. (Family Zygophyllaceae).

*Synonym.* Creosote bush.

### GENERAL DESCRIPTION

Erect to prostrate evergreen shrub, 1–3 m high; resinous, distinctively aromatic leaves, lanceolate to curved, to 18 mm long, 8.5 mm wide; flowers solitary in axils, yellow, five-petaled, twisted or propeller-like; to 2.5 cm wide; dominant shrub of desert scrub in much of the arid western United States (southwest Utah to California and Texas); also central Mexico. Clones known to live 10,000+ years; longer than any other plants (HICKMAN). *Larrea* is represented by five species (one North American; four South American); cytologically distinct *L. tridentata* populations in the Sonoran, Chihuahuan, and Mojave deserts often considered conspecific with South American *L. divaricata* Cav.[1] Parts used are leaves and stems.

## CHEMICAL COMPOSITION

Lignans dominate chemistry of resin, stems, and leaves, especially nordihydroguaiaretic acid (NDGA), at 1.6–6.55%;[2] plus dihydroguaiaretic acid, mesodihydroguaiaretic acid, 3′-methoxyisoguaiacin, 3′-demethoxyisoguaiacin,[3,4] 6,3′-di-*O*-demethylisoguaiacin (previously designated 3′-hydroxynorisoguaiacin), 6-*O*-demethylisoguaiacin (norisoguaiacin), didehydro-3′-demethoxy-6-*O*-demethylguaiacin, 3′-demethoxy-6-*O*-demethylguaiacin,[5] 4-*epi*-larreatricin, larreatricin, 3′,3″-dimethoxylarreatricin, 3,4-dehydrolarreatricin, larreatridenticin, and others;[6] flavonoids include 2,6-di-*C*-glucopyranosylapigenin, 6,8-di-*C*-glucopyranosylchrysoeriol, gossypetin 3,7-dimethyl ether, 5,8,4′-trihydroxy-3,7,3′-trimethoxyflavone,[7] quercetin, kaempferol, rhamnetin, rutin;[8] triterpenes, including larreagenin A, larreic acid, erythrodiol-3-$\beta$-(4-hydroxy-cinnamdyl), erythrodiol-3-$\beta$-(4-dihydroxy cinnamdyl); essential oil, containing $\alpha$-pinene, $\Delta$-3-carene, limonene, camphene, linalool, borneol, camphor, bornyl acetate, etc.[9]

## PHARMACOLOGY OR BIOLOGICAL ACTIVITIES

Early studies report uterine relaxation activity *in vitro*, leading to bioassay-directed isolation of an antiimplantation agent (3′-demethoxyisoguaiacin);[5,10] NDGA, a relatively nontoxic (acute) food antioxidant from 1940s to 1970, was removed from GRAS status after chronic feeding study in rats (0.5–1% NDGA for 74 weeks) showed lymph node and kidney lesions (TYLER 1).

NDGA antimutagenic, anticarcinogenic; an observation with one patient, who experienced a remission of a malignant melanoma prompted a 1968–1969 clinical trial on the effect of NDGA on 59 patients with tumors of various etiology. No statistically significant anticancer activity was observed. All 59 patients entered study with advanced malignancies deemed incurable.[11,12]

Other activities reported for *Larrea* extracts include cytotoxic, insecticidal (against American cockroach *Periplaneta americana*). NDGA reported to inhibit respiratory enzymes, antigalactogogue, antiinflammatory, larvicidal, antithyrotropic; inhibits ocular inflammation *in vitro*.[9]

A 1983 case (reported in 1990) described subacute liver necrosis in a 33-year-old female) diagnosed with a benign breast lump: On a friend's advice she consumed 15 tablets of chaparral leaf a day, producing anorexia, nausea, dark urine, and burning pain in the sternum after 3 days. Symptoms persisted, patient hospitalized; condition attributed to chaparral leaf.[13] Two additional cases reported by CDC on October 30, 1992. A fourth case reported in November 1992 by a Wisconsin physician; in early December a fifth case reported from Illinois. A December 10, 1992 press release from the FDA Center for Food Safety and Applied Nutrition warned of the potential link between chaparral use and liver toxicity, prompting removal of products from store shelves. Mechanism of toxicity and toxic compound unknown.[14–16]

## USES

**Medicinal, Pharmaceutical, and Cosmetic.** Rarely used biological additive to cosmetics.

**Food.** While NDGA is prohibited as an antioxidant in products under FDA jurisdiction, use continues as antioxidant in animal fat products (lard, animal shortenings at 0.01%) under USDA authority (TYLER 3).

**Health Food/Herb Teas.** Leaves and stems in capsules, tablets, teas, and other products (MOORE 1).

**Traditional Medicine.** Chaparral claimed to have antimicrobial, diuretic, expectorant, emetic, tonic, anodyne (as poultice), antiparasitic, carminative, and "blood purifier" activity in: genitourinary infections (dysuria, gonorrhea, kidney tract infections); respiratory tract infections (bronchitis, colds, coughs); inflammation of the musculoskeletal system (arthritis, rheumatism, etc.); skin diseases (sores, wounds, burns, bruises, chicken pox, impetigo, acne, warts, ulcers, insect bites, snake bites, etc.); alimentary tract conditions (bowel cramps, diarrhea, dysentery, enteritis); nervous system conditions (neuritis, sciatica); and a widely used folk cancer remedy (in the United States) for facial, stomach, liver, lung, kidney, skin cancers; melanoma; leukemia.[9] Leaves used in Mexico as a contraceptive agent.

## COMMERCIAL PREPARATIONS

Crude herb; extracts.

*Regulatory Status*. Considered unsafe; oral forms removed from market.[15]

## REFERENCES

See the General References for DER MARDEROSIAN AND LIBERTI; DUKE 2; HARBOURNE AND BAXTER; GLASBY 1; HICKMAN; MOORE 1; NIKITAKIS; STEINMETZ; TYLER 1; TYLER 3; UPHOF.


1. D. M. Porter, *Taxon*, **23**, 339 (1974).
2. J. L. Valentine et al., *Anal Lett.*, **17**, 1617 (1984).
3. F. R. Fronczek et al., *J. Nat. Prod.*, **40**, 497 (1987).
4. O. Gisvold and E. Thaker, *J. Pharm. Sci.*, **63**, 1905 (1974).
5. C. Konno et al., *J. Nat. Prod.*, **52**, 1113 (1989).
6. C. Konno et al., *J. Nat. Prod.*, **53**, 396 (1990).
7. M. Sakakibrara et al. *Phytochemistry*, **15**, 727 (1976).
8. J. T. Mabry et al., ed. *Creosote Bush, Biology and Chemistry of Larrea in New World Deserts*, Dowden, Hutchinson and Ross, Stroudsburg, Pa., 1977.
9. F. Brinker in E. K. Alstat, ed., *Eclectic Dispensatory of Natural Therapeutics*, Vol. 1, Eclectic Medical Publications, Portland, Ore., 1989.
10. C. Konnon et al., *Proceedings International Congress on Natural Products*, **2**, 328 (1987).
11. C. R. Smart et al., *Cancer Chemother. Rep. Pt 1*, **53**, 147 (1969).
12. C. R. Smart et al., *Rocky Mt. Med. J.*, **7**, (1970).
13. M. Katz Miriam and F. Saibil, *J. Clin. Gastroenterol.*, **12**, 203 (1990).
14. M. Blumenthal, *HerbalGram*, **28**, 38 (1993).
15. FDA, HHS News (Dec. 10, 1992); FDA Talk Paper (Dec. 11, 1992); *Food Drug Cosmet. Law Rep.*, 1577; §43,127 (Dec. 28, 1992).
16. Centers for Disease Control and Prevention, *Morbid. Mortal. Weekly*, **41**, 812 (1992).

# CHASTE-TREE

**Source:** ***Vitex agnus-castus*** L. (Family Verbenaceae).

*Synonym.* Monk's pepper.

## GENERAL DESCRIPTION

Deciduous shrub to 6 m; leaves palmately compound; leaflets linear lanceolate, entire, white tomentose beneath, glabrous above, to 10 cm long; flowers blue to pink in spikelike panicles; indigenous to southern Europe, naturalized in north America (Florida, Georgia, Alabama, Mississippi, Louisiana, Arkansas, Texas, southeast Oklahoma, north to Maryland). Part used is the dried fruit (a small reddish black drupe).

## CHEMICAL COMPOSITION

Essential oil of dried fruits (0.5–1.22%) contains $\alpha$-pinene and $\beta$-pinene (39.7%), cineole (25.5%), limonene (14.8%);[1] flavonoids, including casticin, penduletin, 6-hydroxykaempferol-3,6,7,4′-tetramethylether, chrysosplenol-D, isovitexin, luteolin-7-glucoside, orientin, isoorientin, homoorientin, etc.;[2–5] iridoid glycosides include aucubin, eurostoside, agnuside;[6,7] triterpenoids include 3$\beta$-acetoxyolean-12-en-27-oic acid; 2$\alpha$,3$\alpha$-dihydroxyolean-5,12-dien-28-oic acid; 2$\beta$,3$\alpha$-diacetoxyolean-5,12-dien-28-oic acid; and 2$\alpha$,3$\beta$-diacetoxy-18-hydroxyolean-5,12-dien-28-oic acid;[8] ketosteroids detected in flower extracts include progesterone, 17-$\alpha$-hydroxyprogesterone, testosterone, and epitestosterone; leaf extracts yield androstenedione.[9]

## PHARMACOLOGY OR BIOLOGICAL ACTIVITIES

Increases luteinizing hormone production, inhibiting release of follicle-stimulating hormone, leading to positive ratio shift of estrogens over gestagens, producing corpus luteum hormone effect (WEISS). Used in modern Europe for menstrual and menopausal disorders; upon termination of use of "the Pill," used to help reestablish normal menstruation and ovulation. Comparative clinical studies show extracts to increase or stimulate milk flow; reduces water retention during menstruation; allays effusions in the knee joints associated with premenstrual syndrome.[10,11]

Other activities reported included antiinflammatory, antibacterial (against *Bacillus cereus*, *B. megaterium*, *Escherichia coli*, *Staphylococcus aureus*), and antifungal (against *Candida albicans*).[5,12]

Generally considered safe; rare occurrences of itching and urticaria have been reported.[13]

## USES

**Health Food/Herb Teas.** Dried fruits in capsules, tea, etc. (FOSTER).

**Traditional Medicine.** In small doses, for impotence, nervousness, mild dementia; anaphrodisiac; tea of the fruits for the treatment of rheumatic conditions and colds; digestive carminative. Eclectic medical practitioners (19th century) used fresh fruit tincture as galactagogue, emmenagogue (FELTER AND LLOYD).

## COMMERCIAL PREPARATIONS

Crude herb, alcoholic and aqueous extracts of pulverized fruits and their formulations.

*Regulatory Status.* Undetermined in United States. In Germany, formulation indicated for menstrual disorders due to primary or secondary corpus luteum insufficiency; premenstrual syndrome, mastodynia, inadequate lactation, menopausal symptoms.[13]

## REFERENCES

See the General References for BLUMENTHAL; FELTER AND LLOYD; FOSTER; STEINMETZ; TUTIN 3; UPHOF; WEISS.

1. E. Winder and R. Hansel, *Arch. Pharm. (Weinheim)*, **293**, 556 (1960).
2. G. Hahn et al., *Notabene Med.*, **16**, 233, 297 (1986).
3. E. Wollenweber and K. Mann, *Planta Med.*, **48**, 126 (1983).
4. R. Hansel and H. Rimpler, *Arch. Pharm. (Wienhiem)*, **296**, 598 (1963).
5. C. S. Gomaa et al., *Planta Med.*, **52**, 277 (1978).
6. R. Hansel and E. Winder, *Arzneim. Forsch*, **9**, 180 (1959).
7. K. Gorler et al., *Planta Med.*, **51**, 530 (1985).
8. A. S. Chawala et al., *J. Nat. Prod.*, **55**, 163 (1992).
9. M. Saden-Krehula et al., "*Short Reports of Short Lectures and Poster Presentations*," *Bonn BACANS Symposium*, $P_1$77, July 17–22, 1990, p. 59.
10. D. Propping, *Therapiewoche*, **38**, 2992 (1988).
11. D. Propping and T. Katzorke, *Z. Allg.*, **63**, 932 (1987).
12. S. S. Mishurova et al., *Rast. Resur.*, **22**, 526 (1986).
13. Monograph *Agni casti fructus*, *Bundesanzeiger*, no. 90 (May 15, 1985); replaced (Dec. 2, 1992).

# CHENOPODIUM OIL

**Source:** ***Chenopodium ambrosioides*** L. var. ***anthelminticum*** (L.) A. Gray or ***C. ambrosioides*** L. (Family Chenopodiaceae).

*Synonyms*. Oils of American wormseed (*C. ambrosioides* var. *anthelminticum*), Mexican tea, and epazote (*C. ambrosioides*).

### GENERAL DESCRIPTION

Strongly aromatic, hairy annual or perennial herb up to about 1.5 m high; native to tropical America; naturalized and cultivated worldwide. Part used is the fresh aboveground flowering and fruiting plant, from which the volatile oil (with a disagreeable odor and bitter taste) is obtained by steam distillation. Major producing countries include India, China, Brazil, and the United States.

Due to its high ascaridole (a peroxide) content, chenopodium oil may explode when heated or treated with acids and should be handled with caution.

### CHEMICAL COMPOSITION

Contains variable amounts of ascaridole (17–90%, usually 60–80%), *l*-limonene, myrcene, *p*-cymene, $\alpha$-terpinene, saturated hydrocarbons ($C_{21}$ to $C_{31}$, with $C_{29}$ predominant), triacontyl alcohol, $\alpha$-spinasterol, and others (JIANGSU, LIST AND HÖRHAMMER).[1–4]

### PHARMACOLOGY OR BIOLOGICAL ACTIVITIES

Ascaridole, the active principle of the oil, has anthelmintic properties, particularly against roundworms (*Ascaris*); it is also effective against hookworms and dwarf tapeworms but not large tapeworms.

The oil is considered as very toxic. Toxic effects include irritation of skin and mucous membranes, vomiting, headache, vertigo, kidney and liver damage, temporary deafness, and circulatory collapse, among others. Effects may be cumulative. Cases of death have also been reported. (GOSSELIN, MARTINDALE).[4]

### USES

**Medicinal, Pharmaceutical, and Cosmetic.** Now seldom (if at all) used in pharmaceutical preparations as it is largely replaced by synthetic anthelmintics such as piperazine and other compounds. Major use is as a fragrance component in soaps, detergents, creams, lotions, and perfumes, with maximum use level of 0.4% reportedly used in perfumes.[4]

**Food.** The leaves and seeds of *C. ambrosioides* are used in Mexican cooking as a carminative flavoring with bean dishes (MOORE 1).

**Traditional Medicine.** Used as an anthelmintic for roundworms, hookworms, and dwarf tapeworms, among others. Leaf, root, and plant of *C. ambrosioides* have been used in tumors.[5] In China, the fresh root is used to treat articular rheumatism.

### COMMERCIAL PREPARATION

Essential oil. It was formerly official in N.F. and U.S.P.

*Regulatory Status.* Not permitted in foods.

### REFERENCES

See the General References for FERNALD; GUENTHER; JIANGSU; MOORE 1; NANJING; TYLER 3; USD 26th; YOUNGKEN.

1. L. Bauer et al., *Rev. Bras. Farm.*, **54**, 240 (1973); through *Chem. Abstr.*, **84**, 35177z (1976).
2. G. S. Gupta and M. Behari, *J. Indian Chem. Soc.*, **49**, 317 (1972).
3. G. S. Gupta and M. Behari, *Indian Perfum.*, **18**(Pt. 2), 40 (1975).
4. D. L. J. Opdyke, *Food Cosmet. Toxicol.*, **14**, 713 (1976).
5. J. L. Hartwell, *Lloydia*, **31**, 71 (1968).

# CHEROKEE ROSEHIP

**Source:** ***Rosa laevigata*** Michx. (syn. *R. sinica* Murr.; *R. cherokensis* Donn.; *R. ternata* Poir.; *R. nivea* DC.; *R. Camellia* Hort.) (Family Rosaceae).

*Synonyms.* Chinese rosehip, Fructus Rosae Laevigatae, and *jinyingzi*.

### GENERAL DESCRIPTION

High-climbing shrub (up to 5 m), with slender green prickly branches; flowering in May and fruiting in September to October. Native to China and Japan; naturalized in the southern United States; now widely distributed throughout China.

Part used is the dried ripe prickly fruit (hip) collected when it ripens and turns red in autumn. After partially sun dried, the fruits are placed in a barrel and their prickles removed by stirring with a wooden bat. They are then further dried to yield whole *jinyingzi*. Whole hips are normally further processed to yield *jinyingzi rou* (meat) by soaking in water until soft, slicing in half, removing the seeds and again sun drying. Commercial Cherokee rosehips come in either whole or sliced, with the latter of better quality, yielding more extractives.

## CHEMICAL COMPOSITION

Contains saponin glycosides (ca. 17%); vitamin C (ca. 1.5%); sugars; plant acids (citric, malic, etc.); starch; tannins; pigments; resin; and others (IMM-3, JIANGSU).

Although not specifically reported, it probably also contains other chemical constituents found in other rose hips (see ***rose hips***).

## PHARMACOLOGY OR BIOLOGICAL ACTIVITIES

In addition to the antiscorbutic activities of its vitamin C, Cherokee rosehip has been shown to have a wide variety of biological activities, including antibacterial [decoction (of fruit or root not clear) active vs. *Escherichia coli*, *Staphylococcus aureus*, *Pseudomonas aeruginosa*, and *Bacillus dysenteriae*; aqueous and alcoholic extracts active vs. *Clostridium tetani*], antiviral [aqueous extract strongly active vs. influenza virus (strain PR8 and others)], antiatherosclerotic and antilipemic in rabbits; and digestive aid and antidiarrheal in humans (IMM-3, JIANGSU, NATIONAL).

Its hot water extract has recently been shown to have strong antimutagenic activities using the Ames test.[1]

Toxicities of its polyhydroxy pigments: $LD_{50}$ (white mice) = 519 ± 105 mg/kg i.p.; and s.c. injection of 500 mg/kg and 1100 mg/kg in white rats (observed 1–2 weeks) retarded weight gain and caused an increase in white blood cells and decrease in red cells but no pathological changes in heart, liver, kidney, spleen, intestine, and adrenal tissues.[2]

## USES

**Health Food/Herb Teas.** Used in tonic (especially male tonic) preparations in various forms (drinks, soup packets, tablets and capsules).

**Traditional Medicine.** Considered one of the most important Chinese health tonics. First described around A.D. 500, traditional Chinese medicine considers it to have *gu jing* (strengthens male essence), *suo niao* (antidiuretic), *yi shen* (tonic; invigorates urinary and reproductive functions), and *se chang* (intestinal astringent) properties. Traditionally used to treat male sexual inadequacies, including nocturnal emission and spermatorrhea; female problems (e.g., uterine bleeding and leukorrhea); chronic diarrhea and enteritis; sweating and night sweating; polyuria and enuresis; also used in sexual neurasthenia, hypertension, and chronic cough (LU AND LI).

## COMMERCIAL PREPARATIONS

Mainly crude (both whole and sliced). Extracts available on limited basis.

*Regulatory Status.* It is an ethnic food. Although other *Rosa* spp. are GRAS, its U.S. regulatory status is not clear.

## REFERENCES

See General References for BAILEY 2; CHEUNG AND LI; CHP; FOSTER AND DUKE; IMM-3; JIANGSU; LU AND LI; NATIONAL.

1. X. B. Ni, *Zhongcaoyao*, **22**, 429 (1991).
2. L. Sun et al., *Jiangxi Yixueyuan Xuebao*, **30**(3), 5 (1990).

# CHERRY BARK, WILD

**Source:** ***Prunus serotina*** Ehrh. (*P. virginiana* Mill. not L.) (Family Rosaceae).

*Synonyms.* Wild black cherry, black cherry, and rum cherry bark.

## GENERAL DESCRIPTION

Large tree with rough dark trunk and reddish brown branches; up to about 1.5 m in diameter and 30 m in height; native to North America (Nova Scotia to Florida and west to Nebraska and Texas.) Part used is the dried stem bark, free of borke (the rough outer bark), preferably collected in the fall.

## CHEMICAL COMPOSITION

Contains prunasin (*d*-mandelonitrile glucoside); emulsin; eudesmic acid (3,4,5-trimethoxybenzoic acid); *p*-coumaric acid; scopoletin; tannins that contain nonhydrolyzable monomeric and polymeric leucocyanidins; sugars; and others (KARRER).

Prunasin is a cyanogenic glucoside which is hydrolyzed by the enzyme prunase into hydrocyanic acid (HCN, prussic acid), glucose, and benzaldehyde. The yield of HCN by the bark varies with the times of collection and the thickness and types of bark. Bark collected in the fall has the highest HCN yield (ca. 0.15%), while that collected in the spring has only about 0.05% yield of HCN. In contrast, leaves have been reported to yield the highest amounts of HCN in the spring, up to about 0.25% of potential HCN yield in fresh leaves (LIST AND HÖRHAMMER).[1,2]

Hydrocyanic acid has been reported present in extracts (MARTINDALE).

## PHARMACOLOGY OR BIOLOGICAL ACTIVITIES

Believed to have astringent, sedative, and antitussive properties. Hydrocyanic acid is a lethal poison.

## USES

**Medicinal, Pharmaceutical, and Cosmetic.** Extracts used extensively in cold and cough preparations, particularly in formulations based on White Pine Compound.

**Food.** Extracts used in most major food products, including alcoholic and nonalcoholic beverages, frozen dairy desserts, candy, baked goods, gelatins and puddings, processed fruits, and others. Highest average maximum use level of the extract (strength and type unspecified) reported is 0.06% in alcoholic beverages.

**Health Food/Herb Teas.** Used in cough syrups or bronchial formulations; tea ingredient (FOSTER AND DUKE).

**Traditional Medicine.** Reportedly used to treat diarrhea, colds, coughs, and other lung problems and to relieve pain, usually in the form of a tea or infusion. Also reportedly used in cancers.[3]

In China, the stem bark and root of a related *Prunus* species (*P. armeniaca* L., the apricot) has been used for centuries in treating apricot kernel poisoning. Recent clinical reports have substantiated this usage. Decoctions of the fresh bark were used to treat 80 cases of apricot kernel poisoning; all patients were reported to recover completely within 4 h (JIANGSU).

## COMMERCIAL PREPARATIONS

Crude and extracts; crude and syrup were formerly official in U.S.P., and fluid extract was official in N.F. The crude comes in two types: thin and thick with the former being considered superior in quality. Strengths (see ***glossary***) of extracts are expressed in weight-to-weight ratios; extracts may contain detectable amounts of prunasin or its hydrolysis product, HCN.

*Regulatory Status.* GRAS (§182.20); no HCN limits are specified.

## REFERENCES

See the General References for BAILEY 2; CLAUS; FEMA; FERNALD; FOSTER AND DUKE; KARRER; MARTINDALE; UPHOF; USD 26th; YOUNGKEN.

1. L. Buchalter, *J. Pharm, Sci.*, **58**, 1272 (1968).
2. D. M. Smeathers et al., *Agron. J.*, **65**, 775 (1973).
3. J. L. Hartwell, *Lloydia*, **34**, 103 (1971).


# CHERRY LAUREL LEAVES

**Source:** ***Prunus laurocerasus*** L. (syn. *Laurocerasus officinalis* Roem.) (Family Rosaceae).

*Synonyms*. Common cherry laurel and laurocerasus leaves.

### GENERAL DESCRIPTION

An evergreen bush to small tree with oblong leathery leaves 7.5–15 cm long; up to about 6 m high; native to western Asia; widely cultivated. Parts used are the fresh leaves. The oil is obtained by steam distillation of the warm-water macerated leaves. During maceration the enzyme prunase (or emulsin) hydrolyzes the cyanogenic glucoside present to yield benzaldehyde and hydrocyanic acid (HCN), which are volatile and distilled with steam. Most of the HCN is removed by neutralization and washing of the oil. Cherry laurel water is the water distillate adjusted to contain 0.1% HCN.[1]

### CHEMICAL COMPOSITION

The leaves contain variable amounts (usually ca. 1.5%) of prunasin (*d*-mandelonitrile glucoside), with the young and small leaves containing the highest concentrations. During isolation prunasin is partially converted to its isomer sambunigrin (*l*-mandelonitrile glucoside), resulting in a racemic mixture of the two isomers known as prulaurasin (*dl*-mandelonitrile glucoside). Other constituents present in the leaves include 1% ursolic acid, wax, tannin, emulsin, and others (KARRER).[2,3]

The oil, like bitter almond oil, is composed almost entirely of benzaldehyde and HCN, with small amounts of benzyl alcohol.

Cherry laurel oil (FFPA) for food use should not contain HCN.

### PHARMACOLOGY OR BIOLOGICAL ACTIVITIES

Hydrocyanic acid is a deadly poison.

Cherry laurel oil (FFPA) is practically equivalent to pure benzaldehyde and has the pharmacological and toxicological properties of benzaldehyde (see ***bitter almond***).

### USES

**Medicinal, Pharmaceutical, and Cosmetic.** Not used in the United States, but cherry laurel water is used in Europe as sedative, anodyne, and antispasmodic and in eye lotions.

**Food.** Cherry laurel oil (FFPA) is used as a flavor component in numerous food products, including alcoholic (liqueurs such as cordials, etc.) and nonalcoholic beverages, frozen dairy desserts, candy and baked goods. Highest average maximum use level reported is 0.014% in candy.

**Traditional Medicine.** Leaves used in treating coughs, insomnia, stomach and intestinal spasms, vomiting, and other ailments; also reportedly used in cancers.[4]

**COMMERCIAL PREPARATION**

Cherry laurel oil (FFPA).

*Regulatory Status.* Has been approved for food use provided HCN does not exceed 25 ppm (§172.510).

**REFERENCES**

See the General References for ARCTANDER; BAILEY 2; BIANCHINI AND CORBETTA; FEMA; LIST AND HÖRHAMMER; MERCK; MARTINDALE; UPHOF.

1. A. Puech et al., *Trav. Soc. Pharm. Montpellier*, **36**, 101 (1976).
2. M. Henriet et al., *J. Pharm. Belg.*, **29**, 437 (1974).
3. L. P. Miller in L. P. Miller, ed., *Phytochemistry*, Vol. 1, Van Nostrand Reinhold, New York, 1973, p. 297.
4. J. L. Hartwell, *Lloydia*, **34**, 103 (1971).

# CHERVIL

**Source: *Anthriscus cerefolium*** (L.) Hoffm. (syn. *A. longirostris* Bertol.) (Family Umbelliferae or Apiaceae).

*Synonyms.* Garden chervil and salad chervil.

**GENERAL DESCRIPTION**

Slender annual with small leaves and erect branching stem, hairy near the nodes; up to about 0.8 m high; native to Europe (the Caucasus and south Russia) and western Asia; naturalized in North America (Quebec to Pennsylvania), Australia, and New Zealand; widely cultivated. Parts used are the leaves (fresh or dried) and the dried flowering herb.

**CHEMICAL COMPOSITION**

Contains a volatile oil (ca. 0.03% in the herb and 0.9% in the fruits), apiin (apigenin-7-apiosylglucoside), bitter principles, and high concentrations of potassium (ca. 4.7%), calcium (ca.1.3%), magnesium (130 mg/100 g), phosphorus (450 mg/100 g), and others. Fruits (seeds) contain luteolin-7-glucoside and about 13% fixed oils, which are composed of petroselinic acid and linoleic acid as the main components, with minor concentrations of palmitic acid and short-chain hydrocarbons ($C_{23}$ or less, mainly branched-chain $C_{17}$). The volatile oil contains estragole (methyl chavicol) and 1-allyl-2,4-dimethoxybenzene as major constituents, with anethole also reported to be present in the oil of Indian origin (KARRER, LIST AND HÖRHAMMER, MARSH).[1–4]

**PHARMACOLOGY OR BIOLOGICAL ACTIVITIES**

Estragole, the major component of the volatile oil, has been reported to produce tumors in mice (see ***sweet basil***).

**USES**

**Food.** Used as a flavor ingredient in food products, including nonalcoholic beverages, frozen dairy desserts, candy, baked goods, meat and meat products, and condiments and relishes. Highest average maximum use level reported is 0.114% of the herb in meats and meat products.

The leaves (particularly when fresh) are used as a domestic spice in soups, salads, vinegar for salad dressings, omelets, and other dishes.

**Traditional Medicine.** Used as diuretic, expectorant, and digestive; also to lower blood pressure, in the form of an infusion. The juice from the fresh herb is used to treat eczema, gout stones, and abscesses, among others.

**COMMERCIAL PREPARATION**

Available mainly as the crude.

*Regulatory Status.* GRAS (§182.10 and §182.20).

**REFERENCES**

See the General References for BAILEY 2; FEMA; LUST; ROSENGARTEN; TERRELL; UPHOF.

1. J. H. Zwaving et al., *Pharm. Weekbl.*, **106**, 182 (1971).
2. J. Van Loon, *Z. Lebesm. Unters. Forsch.*, **153**, 289 (1973).
3. S. O. Brown et al., *Phytochemistry*, **14**, 2726 (1975).
4. J. B. Harborne and C. A. Williams, *Phytochemistry*, **11**, 1741 (1972).

# CHESTNUT LEAVES

**Source:** ***Castanea dentata*** (Marsh.) Borkh. (syn. *C. americana* (Michx.) Raf.) (Family Fagaceae).

*Synonym.* American chestnut leaves.

**GENERAL DESCRIPTION**

Deciduous tree with rough bark and glabrous mature leaves that reach about 25 cm in length and 5 cm in width; up to about 30 m high; native to North America. Parts used are the dried leaves.

The American chestnut has been extensively destroyed by a fungal disease during recent years. The leaves used in commerce are mostly derived from *C. sativa* Mill., a native of the Mediterranean region, or from other *Castanea* species.

**CHEMICAL COMPOSITION**

Contains 8–9% tannins, mucilage, resins, and others.

**PHARMACOLOGY OR BIOLOGICAL ACTIVITIES**

Considered to have tonic and astringent properties; also antitussive, antirheumatic. Offers no advantage over other antitussives (WEISS, WREN).

**USES**

**Food.** Used in alcoholic and nonalcoholic beverages in the form of an extract; reported use levels are low, less than 0.0075%.

**Traditional Medicine.** Used for coughs (particularly whooping cough) and other respiratory ailments; infusion a gargle in pharyngitis; also as sedative, tonic, astringent, and antirheumatic.

**COMMERCIAL PREPARATIONS**

Crude and extracts; crude was formerly official in N.F. and U.S.P. Strengths (see ***glossary***) of extracts are expressed either in weight-to-weight ratios or, when intended for food use, in flavor intensities.

*Regulatory Status.* Has been approved for food use (§172.510); only *C. dentata* is listed. *Castanea sativa* leaves are the subject of a German therapeutic monograph; however, use is not recommended since efficacy is not well documented.[1]

## REFERENCES

See the General References for BAILEY 2; BIANCHINI AND CORBETTA; BLUMENTHAL; FEMA; FERNALD; KROCHMAL AND KROCHMAL; STAHL; UPHOF; WEISS; WREN.

1. Monograph *Castaneae folium*, *Bundesanzeiger*, no. 76 (Apr. 23, 1987).

# CHICKWEED

**Source: *Stellaria media*** (L.) Vill. (Family Caryophyllaceae).

*Synonyms*. Starweed and star chickweed.

### GENERAL DESCRIPTION

Prostrate to decumbent annual herb to 40 cm; leaves ovate, sessile, glabrous; flower white, star-shaped, petals two lobed, to 3 mm long, sepals to 5 mm; mainly a weed of cultivated ground around human dwellings; throughout Europe, North America; cosmopolitan elsewhere; part used is the herb.

### CHEMICAL COMPOSITION

Poorly studied species; carboxylic acids, coumarins, hydroxycoumarins; glycosides; flavonoids (rutin); saponins, steroids, triterpene glycosides;[1] vitamin C (150–375 mg/100 g), thiamine (0.02 mg), riboflavin (0.14 mg), niacin (0.51 mg), carotene (35.2 mg); linolenic acids[2] (DUKE 2). Octadecatetraenic acid in leaf lipids concentrated in monogalactoxyl diglyceride fraction; $\gamma$-linolenic acid in polar fraction.[3]

*Stellaria dichotoma* var. *lanceolata* contains wogonin (flavone derivative), $\alpha$-spinasterol, stimast-7-enol, palmitate and furan 3-carboxylic acid, and *C*-glycosylflavonoids.[4] *C*-glycosylflavones (related to apigenin) have been isolated from *S. media*.[5]

### PHARMACOLOGY OR BIOLOGICAL ACTIVITIES

This ubiquitous plant has not been the subject of pharmacological and biological studies.

### USES

**Food.** Formerly a source of vitamin C; traditionally a pot herb, emergency food; seeds once a commercial bird seed source.

**Health Food/Herb Teas.** Leaves in capsules, teas; widely used emollient in salves and ointments (WREN).

**Traditional Medicine.** Antipruritic, antirheumatic, demulcent, emollient and vulnerary activity reported; externally poulticed for boils, eczema, inflammation, psoriasis, sores, swelling, ulcers; ointment also used to allay itching. WEISS reports negative results as antirheumatic.

### COMMERCIAL PREPARATIONS

Crude herb; extracts.

*Regulatory Status*. Undetermined.

## REFERENCES

See the General References for DUKE 2; FOSTER AND DUKE; STEINMETZ; TUTIN 1; TYLER 1; UPHOF; WEISS; WREN.

1. G. G. Tsotsoriua et al., *Kromatogr. Metody. Farm.*, 172 (1977); through *Chem. Abstr.*, **90**, 51421.
2. M. L. Salo and T. Makinen, *Maataloust. Aikakousk*, **37**, 127 (1965).
3. G. R. Jamieson and E. H. Reid, *Phytochemistry*, **10**, 1575 (1971).
4. A. M. Rizk, *The Phytochemistry of the Flora of Qatar*, Scientific and Applied Research Centre, University of Qatar, Doha, 1987, p. 26.
5. J. Budzianowski and G. Pakulski, *Planta Med.*, **57**, 290 (1991).

# CHICLE

**Source:** ***Manilkara zapota*** (L.) van Royen (syn. *M. zapotilla* (Jacq.) Gilly; *M. achras* (Mill.) Fosb.; *Sapota achras* Mill.; *Achras sapota* L.; *A. zapotilla* (Jacq.) Nutt.) (Family Zapotaceae).

## GENERAL DESCRIPTION

An evergreen tree with shiny leaves, up to about 33 m high; native to tropical America (Mexico, Guatemala, Venezuela, etc.); now extensively cultivated in the tropics for its edible fruit. Part used is the latex present in the bark, pith, and leaves; it is collected by making multiple incisions in the trunk and dried by careful boiling to remove excess water. Crude chicle is purified by repeatedly washing with strong alkali and neutralizing with sodium acid phosphate, followed by drying and powdering. The resulting product is a water-insoluble, amorphous powder that softens on heating.[1,2]

## CHEMICAL COMPOSITION

Crude chicle contains 15–20% hydrocarbons that are polyisoprenes (mixture of low molecular weight *cis*-1,4 and *trans*-1,4 units in an approximately 2:7 ratio); up to 55% of a yellow resin, consisting primarily of lupeol acetate with minor amounts of $\beta$-amyrin and $\alpha$-spinasterol acetates; a gum composed of a $(1\rightarrow4)$-linked xylan backbone highly substituted with oligosaccharide chains; sugar; inorganic salts; and others (KARRER).[1–7]

Refined chicle for use in chewing gums does not contain the water-soluble constituents present in crude chicle. However, data on its precise chemical composition are limited.

## PHARMACOLOGY OR BIOLOGICAL ACTIVITIES

Limited available data indicate it to be nontoxic.[8]

## USES

**Medicinal, Pharmaceutical, and Cosmetic.** Has been claimed to be a useful ingredient in hair preparations (dressings and pomades).[9,10]

**Food.** The primary use of chicle is as the "gum" base in chewing gum; its use level in chewing gum is about 20%. The rest of the chewing gum is sugar and corn syrup, with small amounts of flavorings. This "gum" is not a true gum (see ***glossary***) but is close in chemical and physical nature to natural rubber and resins; hence it is soft and plastic when chewed and is reportedly not soluble in saliva.[11]

## COMMERCIAL PREPARATIONS

Mainly crude.

*Regulatory Status.* Has been approved for foods use as a chewing gum base (§172.615).

## REFERENCES

See the General References for HORTUS 3rd; LIST AND HÖRHAMMER; TERRELL; UPHOF; YOUNGKEN.

1. B. L. Archer and B. G. Audley in L. P. Miller, ed., *Phytochemistry*, Vol. 2, Van Nostrand Reinhold, New York, 1973, p. 310.
2. P. D. Strausbaugh and E. L. Core in D. N. Lapedes et al., eds., *McGraw-Hill Encyclopedia of Science and Technology*, Vol. 3, McGraw-Hill, New York, 1977, p. 63.
3. F. W. Stavely et al., *Rubber Chem. Technol.*, **34**, 423 (1961); through *Chem. Abstr.*, **55**, 25309h (1961).
4. E. Azpeitia et al., *Can. J. Chem.*, **39**, 2321 (1961).
5. Y. Tanaka and H. Sato, *Polymer*, **17**, 113 (1976).
6. G. G. S. Dutton and S. Kabir, *Carbohydr. Res.*, **28**, 187 (1973).
7. E. Anderson and H. D. Ledbetter, *J. Am. Pharm. Ass.*, **40**, 623 (1951).
8. T. Shoji et al., *Shokuhin Eiseigaku Zasshi.*, **6**, 27 (1965); through *Chem. Abstr.*, **62**, 16871b (1965).
9. E. M. Mendez, U.S. Pat. 3,453,361 (1969).
10. M. Fujiwara, Jpn. Kokai 72 47,665 (1972); through *Chem. Abstr.*, **80**, 40936t (1974).
11. H. W. Conner in D. N. Lapedes et al., eds., *McGraw-Hill Encyclopedia of Science and Technology*, Vol. 5, McGraw-Hill, New York, 1977, p. 447.

# CHICORY ROOT

**Source: *Cichorium intybus*** L. (Family Compositae or Asteraceae).

*Synonyms*. Succory, blue sailors, wild chicory, and common chicory root.

### GENERAL DESCRIPTION

Biennial or perennial herb with spindle-shaped taproot, bright blue flowers, and cauline hairy leaves (borne on stem) resembling those of dandelion; up to 2 m high; believed to be native of Europe and Asia; naturalized and weedy in North America (FOSTER). Parts used are the dried root and the dried aboveground parts, collected in autumn.

### CHEMICAL COMPOSITION

The root contains a high concentration (up to 58% in fresh cultivated root) of inulin, which yields on hydrolysis mostly fructose, with glucose in minor amounts; bitter principles lactucin and lactucopicrin (intybin); cichoriin (esculetin-7-glucoside); $\alpha$-lactucerol (taraxasterol); tannins; sugars (fructose, mannose, etc.); pectin; fixed oils; choline; and others (JIANGSU, LIST AND HÖRHAMMER).[1–4]

The roasted root contains a steam-distillable fraction (aroma), which is composed of pyrazines, benzothiazoles, aldehydes, aromatic hydrocarbons, furans, phenols, organic acids, and others, totaling 33 identified compounds, among which acetophenone is a characteristic component of roasted chicory not previously reported as a component of aroma of any heated food products such as coffee.[5]

Small amounts of two indole alkaloids ($\beta$-carbolines), harman and norharman, have also been isolated from roasted chicory root.[6]

The herb (leaves, flowers, shoots, etc.) contains inulin, fructose, choline, resin, chicoric acid (dicaffeoyl tartaric acid), esculetin, esculin (esculetin-6-glucoside), cichoriin, and others (LIST AND HÖRHAMMER).[7–9]

## PHARMACOLOGY OR BIOLOGICAL PROPERTIES

A substance thought to be a sesquiterpene isolated from extracts of the roasted root has been reported to have bacteriostatic properties.[10]

Alcoholic extracts of chicory root have been demonstrated to exhibit marked depression of amplitude and rate on the isolated toad heart, simulating the effects of quinidine on the heart (see ***cinchona***). This has led to the suggestion that tinctures of chicory root may be of value in diseases characterized by tachycardia, arrhythmias, and fibrillations.[2]

An alcoholic extract of the whole plant (after petroleum ether treatment) has been demonstrated to have antiinflammatory activities in rats.[11]

Chicory root is reported to have sedative, mild laxative, tonic, and other properties (JIANGSU).

The herb (vegetable) has been reported to cause contact dermatitis in humans.[12,13]

## USES

**Medicinal, Pharmaceutical, and Cosmetic.** In Germany, cut herb and root used in infusion or extract for loss of appetite and dyspeptic disorders. Contraindicated in allergies to chicory or other Compositae; in gallstones, only after consultation with physician.[14]

**Food.** Extracts are used extensively as a flavor ingredient in major food products, including alcoholic (primarily bitter formulations) and nonalcoholic beverages (e.g., instant coffee substitutes), frozen dairy desserts, candy, baked goods, and gelatins and puddings, among others. Highest average maximum use level reported for the extract (type unspecified) is about 0.61% (6,116 ppm) in frozen dairy desserts, though use level in instant coffee substitutes could be much higher.

Ground roasted root is increasingly more used admixed with coffee to impart "richer" flavor and to decrease the caffeine content of the resulting coffee formulation. This use is very common in Europe.

Chicory leaf buds known as "chicons," usually obtained from the Witloof variety, are used as a vegetable and salad.[15]

**Health Food/Herb Teas.** The root and leaves reportedly used as a flavor component in herb teas; also in diuretic and digestive formulations (FOSTER).

**Traditional Medicine.** Both root and herb reportedly used as bitter tonics to increase appetite and to treat digestive problems, usually in the form of a tea or as the juice. Also used as diuretics and in treating gallstones, liver ailments (e.g., hepatitis), and cancers, among others (JIANGSU).[16]

## COMMERCIAL PREPARATIONS

Crudes (both roasted and unroasted) and their extracts; strengths (see ***glossary***) of extracts are either expressed in weight-to-weight ratios or in terms of flavor potencies.

*Regulatory Status.* GRAS (§182.20). Root and herb subject of a combined positive German therapeutic monograph for mild dyspeptic disorders and loss of appetite.[14]

## REFERENCES

See the General References for BAILEY 2; BIANCHINI AND CORBETTA; BLUMENTHAL; JIANGSU; FEMA; FERNALD; FOSTER; GRIEVE; LIST AND HÖRHAMMER; LUST.

1. M. P. J. Kierstan, *Biotechnol. Bioeng.*, **20**, 447 (1978).
2. S. I. Balbaa et al., *Planta Med.*, **24**, 133 (1973).
3. J. Promayon et al., *Cafe, Cacao, The*, **20**, 209 (1976); through *Chem. Abstr.*, **86**, 41883j (1977).
4. M. Blanc, *Lebensm. Wiss. Technol.*, **11**, 19 (1978).
5. S. Kawabata and M. Deki, *Kanzei Chuo Bunsekishoho*, **17**, 63 (1977); through *Chem. Abstr.*, **87**, 66760a (1977).
6. A. Proliac and M. Blanc, *Helv. Chim. Acta*, **59**, 2503 1976.
7. G. F. Fedorin et al., *Rastit. Resur.*, **10**, 573 (1974); through *Chem. Abstr.*, **82**, 90001w (1975).
8. V. G. Dem'yanenko et al., USSR 577, 033 (1977); through *Chem. Abstr.*, **88**, 27793q (1978).
9. H. Schmidtlein and K. Herrmann, *Z. Lebesm. Unters. Forsch.*, **159**, 255 (1975).
10. Z. Grodzinska-Zachwieja et al., *Bull. Acad. Polon. Sci., Ser. Sci. Biol.*, **10**, 513 (1962); through *Chem. Abstr.*, **59**, 5535c (1963).
11. P. S. Benoit et al., *Lloydia*, **39**, 160 (1976).
12. J. C. Mitchell in V. C. Runeckles, ed., *Recent Advances in Phytochemistry*, Vol. 9, Plenum Press, New York, 1975, p. 119.
13. R. M. Adams, *Occupational Contact Dermatitis*, J. B. Lippincott, Philadelphia, 1969, p. 188.
14. Monograph *Cichorium intybus*, *Bundesanzeiger*, no. 76 (Apr. 23, 1987); revised (Sept. 1, 1990).
15. J. Swabey in L. W. Codd et al., eds., *Chemical Technology: An Encyclopedic Treatment*, Vol. 7, Barnes & Noble, New York, 1975, p. 287.
16. J. L. Hartwell, *Lloydia*, **31**, 71 (1968).

# CHIRATA

**Source:** ***Swertia chirata*** Buch.-Ham. (Family Gentianaceae).

*Synonyms*. Chirayta, chiretta, bitter stick, and East Indian balmony.

### GENERAL DESCRIPTION

Annual herb with opposite leaves and branching, four-angled stem, with large continuous pith; about 1 m high; leaves broadly lanceolate, subsessile; flowers in large panicles, greenish yellow, purple-tinged; native to northern India, Nepal, and Pakistan in temperate altitudes of the Himalayas from 1200 to 3000 m. Part used is the whole dried herb. Other *Swertia* species appear as adulterants (especially *S. angustifolia*; also *Andrographis paniculata*, and roots of *Rubia cordifolia*); *S. chirata* is distinguished by the large dark stem pith, and intensely bitter flavor.

### CHEMICAL COMPOSITION

Contains a bitter glucoside amarogentin (chiratin); numerous tetraoxygenated xanthones, including swertinin (7,8-dihydroxy-1,3-dimethoxyxanthone), swertianin (1,7,8-trihydroxy-3-methoxyxanthone), swerchirin (1,8-dihydroxy-3,5-dimethoxyxanthone), decussatin (1 - hydroxy - 2, 6, 8 - trimethoxyxanthone), isobellidifolin (1,6,8-trihydroxy-4-methoxyxanthone), 1,3,7,8-tetrahydroxyxanthone, 1, 8 - dihydroxy - 3, 7 - dimethoxyxanthone, 1-hydroxy-3,5,8-trimethoxyxanthone, mangiferin (1, 3, 6, 7 - tetrahydroxyxanthone-$C_2$-β-D-glucoside), and several others; triterpenes (lupeol, β-amyrin, etc.); and monoterpene alkaloids, among others (KARRER).[1–5]

## PHARMACOLOGY OR BIOLOGICAL ACTIVITIES

Chirata is considered to have bitter tonic and febrifuge properties.

Some of its xanthones (swertianin, 1,3,7,8-tetrahydroxyxanthone, and 1,8-dihydroxy-3,7-dimethoxyxanthone) have been claimed to have antituberculous activities.[2]

Amarogentin is hepatoprotective *in vitro* against carbon tetrachloride toxicity.[6]

Swerchirin is antimalarial *in vivo*.[7]

Significant antiinflammatory activity (in various acute, subacute, and chronic experimental models with rodents) of a total benzene extract. Close resemblance of physiological, biochemical, and immunological (adjuvant arthritis) alterations suggest possible use, both local and systematic in human rheumatoid arthritis.[8]

## USES

**Medicinal, Pharmaceutical, and Cosmetic.** Used in certain bitter tonic preparations.

**Food.** Reportedly used in alcoholic (bitters) and nonalcoholic beverages. Average maximum use levels reported are 0.0016 and 0.0008%, respectively.

**Health Food/Herb Teas.** Sometimes used in bitter tonic formulations; tea (GRIEVE).

**Traditional Medicine.** Used in India as a bitter tonic, febrifuge, laxative, anthelmintic, and for skin diseases.[9] Also reported used in cancers.[10]

## COMMERCIAL PREPARATIONS

Limited availability as crude.

*Regulatory Status.* Has been approved for use in alcoholic beverages only (§172.510).

## REFERENCES

See the General References for CSIR 10; FEMA; GRIEVE; UPHOF; YOUNGKEN.

1. K. K. Purushothaman et al., *Leather Sci. (Madras)*, **20**, 132 (1973); through *Chem. Abstr.*, **79**, 113218b (1973).
2. M. Komatsu et al., Jpn. Kokai, 7127,558 (1971); through *Chem. Abstr.*, **75**, 143990g (1971).
3. S. Ghosal et al., *J. Pharm. Sci.*, **62**, 926 (1973).
4. L. Bennaroche et al., *C. R. Acad. Sci. Ser. D*, **280**, 2493 (1975); through *Chem. Abstr.*, **83**, 111152j (1975).
5. L. Bennaroche et al., *Planta Med. Phytother.*, **8**, 15 (1974).
6. H. Goyal et al., *J. Res. Ayur. Siddha*, **2**(3), 286 (1981).
7. H. Hikino et al., *Shoyakugky Zasshri*, **38**, 359 (1984).
8. S. Mandal et al., *Fitoterapia*, **63**(2), 122 (1992).
9. S. Presad et al., *J. Sci. Ind. Res.*, **19C**, 119 (1960); through *Chem. Abstr.*, **54**, 25583e (1960).
10. J. L. Hartwell, *Lloydia*, **32**, 153 (1969).

# CINCHONA (RED AND YELLOW)

**Source:** ***Red cinchona***: *Cinchona succirubra* Pav. ex Klotsch (syn. *C. pubescens* Vahl.) and its hybrids; ***Yellow cinchona***: *Cinchona calisaya* Wedd., *C. ledgeriana* Moens ex Trim., and their hybrids with other *Cinchona* species. (Family Rubiaceae).

*Synonyms*. Red bark, red Peruvian bark, and cinchona rubra (*C. succirubra*); yellow bark, calisaya bark, ledger bark, brown bark, and cinchona flava (*C. calisaya* and *C. ledgeriana*); Jesuit's bark, Peruvian bark, China bark, cortex chinae, and fever tree.

## GENERAL DESCRIPTION

Evergreen shrubs or trees, up to about 30 m high; *C. calisaya* being the tallest, while *C. succirubra* reaching about 24 m and *C. ledgeriana* only up to 6 m; all native to mountains of tropical America (Bolivia, Costa Rica, Ecuador, Guatemala, Peru, etc.) between altitudes of about 900 and 3400 m; extensively cultivated in Central and South America, Southeast Asia (India, Java, Sumatra, China, etc.), and Africa. Part used is the dried bark.

## CHEMICAL COMPOSITION

Contains up to about 16% (average 6–10%) total quinoline alkaloids that consist mainly of quinine, quinidine, cinchonine, and cinchonidine, with quinine usually in major concentration. Other alkaloids in minor amounts include quinamine, epiquinamine, epiquinine, hydroquinine, hydroquinidine, and many others, totaling over three dozen. Contents of the total alkaloids vary, depending on the sources, with *C. ledgeriana* generally containing a higher amount than *C. succirubra*.

Other constituents present include norsolorinic acid (an anthraquinine), $\beta$-sitosterol, tannins, quinovin (bitter glycosides), quinic acid, starch, resin, wax, and others (JIANGSU; LIST AND HÖRHAMMER, MORTON 3, NANJING, STAHL, USD 26th).[1–4]

## PHARMACOLOGY OR BIOLOGICAL ACTIVITIES

Cinchona has astringent and bitter tonic properties; also reportedly has analgesic and local anesthetic properties, among others (GOODMAN AND GILMAN, JIANGSU, NANJING).

The alkaloids of cinchona have antimalarial and antipyretic activities, with quinine being the most potent. Certain strains of malarial parasites, particularly those of Vietnamese origin, that have become resistant to synthetic antimalarials are still susceptible to quinine treatment.[5]

Quinidine and quinine have cardiac-depressant properties, with quinidine being twice as active as quinine.

Ground cinchona bark and quinine have been reported to cause urticaria, contact dermatitis, and other hypersensitive reactions in humans.

Cinchona alkaloids are toxic. Poisoning (cinchonism) is usually due to overdosage or hypersensitivity, with symptoms including blindness, deafness, severe headache, delirium, abdominal pain, diarrhea, convulsions, paralysis, and collapse. A single oral dose of 8 g quinine may be fatal to an adult (GOODMAN AND GILMAN, MARTINDALE, USD 26th).

## USES

**Medicinal, Pharmaceutical, and Cosmetic.** Besides being used as an antimalarial, quinine has been used for treating various conditions, including hemorrhoids and varicose veins (as hardening agent), and in eye lotions for its astringent, bactericidal, and anesthetic effects. Its current use, besides in malaria, is primarily as the sulfate salt in preparations for treating cold and leg cramps, mostly as prescription drugs.

In European phytomedicine fluid extract of crude bark used to stimulate saliva and gastric secretions for loss of appetite and dyspeptic discomfort.[6]

Quinidine is used in prescription preparations mainly for treating cardiac arrhythmias.

In cosmetics, extracts of cinchona are primarily used in hair tonics, reportedly for stimulating hair growth and controlling oiliness (deNAVARRE).

**Food.** Quinine and extracts of cinchona (mostly red cinchona) are extensively used as a bitter in tonic water and alcoholic bitters and liqueurs; highest average maximum use level reported for red cinchona extract (type not given) is 0.028% (278 ppm) in alcoholic beverages.

Other food products in which red cinchona extract has been reported used include frozen dairy desserts, candy, baked goods, and condiments and relishes. Use levels reported are lower than those reported in beverages.

**Traditional Medicine.** Cinchona is used in treating malaria, fevers, indigestion, and for mouth and throat problems, usually in the form of an infusion; has been used in China to treat hangovers; also reportedly used in cancers.[7]

## COMMERCIAL PREPARATIONS

Available as crude and extracts (fluid extract, solid extract, etc.). Both red and yellow cinchonas were formerly official in U.S.P. and N.F.; quinine sulfate and quinidine sulfate and gluconate are official in U.S.P.; quinine hydrochloride and sulfate are also official in F.C.C.

*Regulatory Status.* Both red and yellow cinchona barks have been approved for use in beverages only, with the limitation that the total cinchona alkaloids not to exceed 83 ppm (0.0083%) in the finished beverage (§172.510 and §172.575). Crude drug subject of a positive German therapeutic monograph.[6]

## REFERENCES

See the General References for BLUMENTHAL; FEMA; GOSSELIN; JIANGSU; LUST; MERCK; NANJING; TERRELL; UPHOF; YOUNGKEN.

1. A. Haznagy, *Acta Pharm. Hung.*, **47**, 249 (1977); through *Chem. Abstr.*, **88**, 158517a (1978).
2. P. Niaussat et al., *P. V. Seances Soc. Sci. Phys. Nat. Bordeaux*, **35** (1974); through *Chem. Abstr.*, **83**, 111166s (1975).
3. R. Adamski and J. Bitner, *Farm. Pol.*, **32**, 661 (1976); through *Chem. Abstr.*, **86**, 86126f (1977).
4. S. Dorairaj et al., *Planta Med.*, **54**, 469 (1988).
5. N. R. Farnsworth in L. P. Miller, ed., *Phytochemistry*, Vol. 3, Van Nostrand Reinhold, New York, 1973, p. 351.
6. Monograph *Cinchonae cortex*, *Bundesanzeiger*, no. 22 (Feb. 1, 1990).
7. J. L. Hartwell, *Lloydia*, **34**, 103 (1971).

# CINNAMON (AND CASSIA)

**Source:** ***Cinnamon*** *Cinnamomum zeylanicum* Blume (syn. *C. verum* J. S. Presl), *C. loureirii* Nees (syn. *C. obtusifolium* Nees var. *loureirii* Perr. et Eb.), and *C. burmanii* (Nees & T. Nees) Blume (syn. *C. pedunculata* J. Presl); ***cassia*** *Cinnamomum cassia* J. Presl (syn. *C. aromaticum* Nees) (Family Lauraceae).

*Synonyms*. Ceylon cinnamon and true cinnamon (*C. zeylanicum*); Saigon cinnamon and Saigon cassia (*C. loureirii*); Batavia cassia, Batavia cinnamon, Padang-cassia, and Panang cinnamon (*C. burmanii*); Chinese cinnamon, false cinnamon, and cassia lignea (*C. cassia*).

## GENERAL DESCRIPTION

Medium-size evergreen trees up to about 10–20 m high, with cassia (*C. cassia*) being a bigger tree than Ceylon cinnamon (*C. zeylanicum*) and Saigon cinnamon (*C. loureirii*); the tallest tree, Batavia cinnamon (*C. burmanii*), may reach 20 m high; all species are native to southeast Asia. Parts used are the dried bark, leaves, and twigs. The trees are mostly cultivated for commercial production of cinnamon and are usually cut back (coppiced) to form bushes or shrubs. The essential oils are obtained by steam distillation. Cassia oil (Chinese cinnamon oil) is obtained from leaves, bark, and twigs of *C. cassia* (JIANGSU); cinnamon bark oil (Ceylon or Seychelles) from the dried inner bark of *C. zeylanicum*; and cinnamon leaf oil (Ceylon or Seychelles) from the leaves and twigs of *C. zeylanicum*.

Cinnamon bark (particularly from *C. zeylanicum*) and its oils are generally considered superior in flavor characteristics to cassia bark and cassia oil.

## CHEMICAL COMPOSITION

Cinnamon bark contains up to 4% volatile oil (usually ca. 1%); tannins, consisting of polymeric 5,7,3′,4′-tetrahydroxyflavan-3,4-diol units;[1] catechins and proanthocyanidins;[2] resins; mucilage; gum; sugars; calcium oxalate; two insecticidal compounds (cinnzelanin and cinnzelanol);[3] coumarin (lowest concentration in Ceylon cinnamon); and others (LIST AND HÖRHAMMER).[1,3–5]

Cinnamon bark oil contains as its major component cinnamaldehyde (usually 60–75%); other constituents include eugenol, eugenol acetate, cinnamyl acetate, cinnamyl alcohol, methyl eugenol, benzaldehyde, cuminaldehyde, benzyl benzoate, linalool, monoterpene hydrocarbons (e.g., pinene, phellandrene, and cymene), carophyllene, safrole, and others (LIST AND HÖRHAMMER, MASADA).[6–10]

Cinnamon leaf oil contains high concentrations of eugenol (Ceylon type 80–88%; Seychelles type 87–96%); it also contains many of the major constituents present in cinnamon bark oil (e.g., cinnamaldehyde, cinnamyl acetate, eugenol acetate, and benzaldehyde) as well as other minor compounds, including humulene, isocaryophyllene, α-ylangene, coniferaldehyde, methyl cinnamate, and ethyl cinnamate (MASADA).[5,6,10]

A cinnamon leaf oil of Chinese origin (*C. japonicum* Sieb.) is reported to contain only about 3% eugenol but with a high content (ca. 60%) of safrole (JIANGSU).

Cassia bark contains 1–2% of volatile oil and other similar constituents as cinnamon bark, including tannins, resins, sugars, calcium oxalate, coumarin, and mucilage; its mucilage content is significantly higher than that of cinnamon (LIST AND HÖRHAMMER, NANJING).

Cassia oil and cassia bark oil contain mainly cinnamaldehyde (75–90%); other constituents present include salicylaldehyde, methylsalicylaldehyde, and methyl eugenol, with methyl eugenol in higher concentration than in Ceylon cinnamon bark oil (JIANGSU, LIST AND HÖRHAMMER, NANJING).[9]

Of the various types of cinnamon bark oils, that of *C. zeylanicum* has the largest amount of eugenol.[8] Eugenol is reportedly absent in cassia bark oil (JIANGSU).[4,9,11]

Cinnamaldehyde present in the oils is in the *trans* form.[8]

The petroleum ether-soluble fraction of cinnamon has been reported to exhibit antioxidative activities on lard.[12]

## PHARMACOLOGY OR BIOLOGICAL ACTIVITIES

Both cinnamon bark and cassia bark have carminative and astringent properties. The essential oils have carminative as well as antiseptic properties.

Cinnamaldehyde, a major component of cassia oil and cinnamon bark oil, has been reported to have sedative activities in animals as well as hypothermic and antipyretic properties (FARNSWORTH 1, JIANGSU).[13] It can cause dermatitis in humans[14] and is also reported to have mutagenic activities.[15]

Cinnamon oil (presumably bark oil) has exhibited antifungal, antiviral, bactericidal, and larvicidal activities (FARNSWORTH).[16–18]

A liquid carbon dioxide extract of cinnamon bark at a 0.1% concentration has been demonstrated to suppress completely the growth of numerous microorganisms, including *Escherichia coli*, *Staphylococcus aureus*, and *Candida albicans*.[19]

Cinnamon also reportedly has strong lipolytic properties.[20,21]

An aqueous extract of Chinese cinnamon exhibited antiulcerogenic activities in rats, being as effective as cimetidine.[22] Two of the active agents have been isolated and identified as 3-(2-hydroxyphenyl)-propanoic acid and its *O*-glucoside.[23]

Eugenol, a major constituent of cinnamon leaf oil and other oils (see ***allspice***, ***basil***, and ***clove***), has antiseptic, irritant, and local anesthetic properties as well as weak tumor-promoting activity on mouse skin and weak cytotoxic activity against HeLa cells (GOSSELIN, MARTINDALE).[24,25]

Eugenol as well as eugenol acetate and methyl eugenol, which are also constituents of cloves and clove oil, have been reported to enhance trypsin activity *in vitro* (see ***allspice***).

## USES

**Medicinal, Pharmaceutical, and Cosmetic.** Cassia, cinnamon, and their bark oils are frequently used either as flavors or as carminative, stomachic, tonic, or counter-irritants in pharmaceutical and cosmetic preparations, including liniments, suntan lotions, nasal sprays, mouthwashes or gargles, and toothpaste, among others.

In European phytomedicine, cassia and cinnamon bark (2.0–4.0 g daily) or the essential oils (0.05–0.2 g daily) used in teas and other galenicals as antibacterial, carminative, fungistatic; also for loss of appetite and dyspeptic disturbances.

Ceylon cinnamon leaf oil is used as a fragrance component in soaps, detergents, creams, lotions, and perfumes, with highest reported maximum use level of 0.8% in perfumes.[24]

**Food.** Cassia, cinnamon (Ceylon, Saigon, Batavia), or their oils and extracts are extensively used as flavor ingredients in most major categories of food products, including alcoholic (bitters, vermouths, etc.) and nonalcoholic beverages (in cola drinks, etc.), frozen dairy desserts, candy, baked goods (cinnamon buns, etc.), gelatins and puddings, meat and meat products, condiments and relishes, soups, gravies, and others. Highest average maximum use levels for crudes are about 1.3–1.4% in baked goods; and for the oils they are about 0.057% (cinnamon bark oil) and 0.047% (cassia bark oil) in alcoholic beverages and baked goods, respectively.

Cinnamon (usually cassia or Saigon cinnamon) is extensively used as a domestic spice in home cooking.

**Health Food/Herb Teas.** Ground bark widely used as flavor ingredient in numerous herbal tea formulations and herbal tonics; also in digestive and stimulant capsulated, tableted products, tinctures, etc. (DUKE 2).

**Traditional Medicine.** Both cassia and cinnamon bark have been used for several thousand years in Eastern and Western cultures in treating chronic diarrhea, rheumatism, colds, abdominal and heart pains, kidney troubles, hypertension, female disorders (amenorrhea, cramps, etc.), and cancer, among others (BIANCHINI AND CORBETTA, FARNSWORTH 1, FOGARTY, JIANGSU, NANJING).[28] Flowers reportedly used in European tradition as "blood purifier."[29]

**Others.** A major use of cinnamon leaf oil is for the isolation of eugenol.

**COMMERCIAL PREPARATIONS**

Bark, extracts, and oils; oils are of various types and qualities. Cassia oil (Chinese cinnamon oil) is official in N.F., where it is simply monographed as cinnamon oil; it is also official in F.C.C. Ceylon cinnamon bark oil and cinnamon leaf oil (both Ceylon and Seychelles types) are official in F.C.C. Both cassia and cinnamon bark are subjects of positive German therapeutic monographs indicated for treatment of loss of appetite and dyspeptic discomfort such as mild gastrointestinal spasms.[26,27] Cinnamon flowers are the subject of a neutral monograph, since efficacy is not established, and risks (allergic skin reactions and mucosal irritation) outweigh benefits.[29]

*Regulatory Status.* GRAS (§182.10 and §182.20).

**REFERENCES**

See the General References for ARCTANDER; BLUMENTHAL; DUKE 2; FURIA; GOSSELIN; GUENTHER; KARRER; JIANGSU; LIST AND HÖRHAMMER; MARTINDALE; ROSENGARTEN; TERRELL; UPHOF.


1. L. Buchalter, *J. Pharm. Sci.*, **60**, 144 (1971).
2. J. M. Schulz and K. Herrmann, *Z. Lebensm. Unters. Forsch.*, **171**, 278 (1980).
3. A. Isogai et al., *Agr. Biol. Chem.*, **41**, 1779 (1977).
4. Y. S. Lewis et al., *Curr. Sci.*, **46**, 832 (1977).
5. J. E. Angmor et al., *Planta Med.*, **21**, 416 (1972).
6. R. O. B. Wijeskera et al., *J. Sci. Food Agr.*, **25**, 1211 (1974).
7. M. S. F. Ross, *J. Chromatogr.*, **118**, 273 (1976).
8. J. S. Chou, *T'ai-wan Ko Hsueh*, **31**(2), 8 (1977); through *Chem. Abstr.*, **88**, 20757x (1978).
9. A. Herisset et al., *Plant. Med. Phytother.*, **6**, 11 (1972).
10. A. Bhramaramba and G. S. Sidhu, *Perfum. Essent. Oil Rec.*, **54**, 732 (1963).
11. F. Karig, *Deut. Apoth. Ztg.*, **115**, 1781 (1975).
12. Y. Sato et al., *Eiyo To Shokuryo*, **29**, 505 (1976); through *Chem. Abstr.*, **87**, 150314r (1977).
13. M. Harada and Y. Ozaki, *Yakugaku Zasshi*, **92**, 135 (1972); through *Chem. Abstr.*, **77**, 729j (1972).
14. G. Angelini et al., *J. Appl. Cosmetol.*, **3**, 223 (1985); through *Chem. Abstr.*, **104**, 192897h (1986).
15. M. Ishidate Jr. et al., *Food Chem. Toxicol.*, **22**, 623 (1984).
16. K. Oishi et al., *Nippon Suisan Gak-*

*kaishi*, **40**, 1241 (1974); through *Chem. Abstr.*, **82**, 84722r (1975).
17. N. Mukherjee, *Z. Pflanzenkr. Pflanzenschutz*, **81**, 468 (1974); through *Chem. Abstr.*, **82**, 150330q (1975).
18. A. C. Pizsolitto et al., *Rev. Fac. Farm. Odontol. Araraquara*, **9**, 55 (1975); through *Chem. Abstr.*, **86**, 12226s (1977).
19. M. L. Khanin et al., *Khim. Farm. Zh.*, **2**, 40 (1968); through *Chem. Abstr.*, **69**, 944k (1968).
20. G. Paulet et al., *Rev. Fr. Corps Gras*, **21**, 415 (1974); through *Chem. Abstr.*, **82**, 2691t (1975).
21. E. Halbert and D. G. Weeden, *Nature (London)*, **212**, 1603 (1966).
22. T. Akira et al., *Planta Med.*, **52**, 440 (1986).
23. S. Tanaka et al., *Planta Med.*, **55**, 245 (1989).
24. D. L. J. Opdyke, *Food Cosmet. Toxicol.*, **13**, 545 (1975).
25. D. L. J. Opdyke, *Food Cosmet. Toxicol.*, **13**(Suppl.), 749 (1975).
26. Monograph *Cinnamomi cassiae cortex*, *Bundesanzeiger*, no. 22 (Feb. 1, 1990).
27. Monograph *Cinnamomi ceylanici cortex*, *Bundesanzeiger*, no. 22 (Feb. 1, 1990).
28. J. L. Hartwell, *Lloydia*, **32**, 247 (1969).
29. Monograph *Cinnamomi flos*, *Bundesanzeiger* (Mar. 11, 1992).

# CITRONELLA OIL (CEYLON AND JAVA)

**Source:** ***Cymbopogon nardus*** (L.) Rendle (syn. *Andropogon nardus* L.); ***C. winterianus*** Jowitt, and their varieties. (Family Gramineae).

*Synonyms*. Ceylon or Lenabatu citronella oil (*C. nardus*); Java or Maha Pengiri citronella oil (*C. winterianus*).

## GENERAL DESCRIPTION

*Cymbopogon nardus* (Ceylon citronella) and *C. winterianus* (Java citronella) are both perennial grasses. The former is extensively cultivated in southern Sri Lanka, while the latter is widely cultivated in many parts of the tropical world (e.g., Java, Taiwan, Hainan Island, Indonesia, India, Nepal, Africa, Vietnam, Guatemala, Brazil, Paraguay and Argentina).

The essential oils are obtained by steam distillation of the fresh, partly dried, or dried grass. The Java-type oil is generally considered to be of superior quality to the Ceylon oil.[1]

## CHEMICAL COMPOSITION

Both Ceylon and Java citronella oils contain citronellal, geraniol, and citronellol as the major components, with the Java type having a higher concentration of these constituents than the Ceylon type; the relative proportions of these components vary greatly, depending on the sources (ARCTANDER, LIST AND HÖRHAMMER, MARTINDALE, MASADA, YOUNGKEN).[1–9]

Other constituents include esters (acetates, propionates, etc.) of geraniol, citronellol, and linalool; monoterpene hydrocarbons (limonene, pinene, camphene, etc.); sesquiterpene hydrocarbons and alcohols (bourbonene, caryophyllene, elemol, farnesol, etc.); phenols (eugenol, methyl eugenol, etc.); and free acids, among others. Java citronella oil contains higher amounts of sesquiterpenes, while the Ceylon type contains much larger amounts of monoterpene hydrocarbons (ARCTANDER, LIST AND HÖRHAMMER).[1,2,4–6,8]

## PHARMACOLOGY OR BIOLOGICAL ACTIVITIES

Citronella oil has been reported to have antibacterial and antifungal activities *in vitro*, the Ceylon oil being as active as penicillin against certain Gram-positive bacteria.[10,11]

Citronella oil has been reported to cause contact dermatitis in humans (LEWIS AND ELVIN-LEWIS).[12]

## USES

**Medicinal, Pharmaceutical, and Cosmetic.** Both oils are used as a component in certain insect repellent formulations. Major current use is as a fragrance component in soaps, brilliantines, disinfectants, and perfumes, among others. Maximum use levels reported for the Ceylon oil were 0.6% in soaps and 0.8% in perfumes.[12]

**Food.** The Ceylon oil is reported used as flavor ingredient in numerous food products, including alcoholic and nonalcoholic beverages, frozen dairy desserts, candy, baked goods, gelatins and puddings, and breakfast cereals. Highest average maximum use level reported is about 0.005% in candy and baked goods (45.9 and 47.6 ppm, respectively).

**Health Food/Herb Teas.** Essential oil of citronella widely available in health food stores, primarily used as an insect repellent for humans and pets (ROSE).

**Traditional Medicine.** Leaves of Ceylon citronella are used in medicinal and aromatic tea, as vermifuge, febrifuge, stomachic, diaphoretic, diuretic, emmenagogue, antispasmodic, and stimulant in various cultures (LIST AND HÖRHAMMER, ROSE).

**Others.** Java citronella oil is used as a major source for the isolation of citronellal and geraniol.

## COMMERCIAL PREPARATIONS

Both oils are available.

*Regulatory Status.* GRAS, with only *C. nardus* listed (§182.20). Subject of a German therapeutic monograph as a mild astringent and stomachic; efficacy not documented.[13]

## REFERENCES


See the General References for ARCTANDER; FEMA; GUENTHER; MASADA; ROSE; TERRELL; UPHOF.

1. W. D. Fordham in L. W. Codd et al., eds., *Chemical Technology: An Encyclopedic Treatment*, Vol. 5, Barnes & Noble, New York, 1972, p. 1.
2. R. O. B. Wijesekera et al., *Phytochemistry*, **12**, 2697 (1973).
3. E. Guenther, *Am. Perfum. Cosmet.*, **83**, 57 (1968).
4. B. C. Gulati and Sadgopal, *Indian Oil Soap J.*, **37**, 305 (1972); through *Chem. Abstr.*, **80**, 40923m (1974).
5. T. K. Razdan and G. L. Koul, *Indian Chem. J.*, **8**, 27 (1973).
6. H. S. Singh et al., *Indian Perfum.*, **20**(1-B), 77 (1976).
7. B. L. Kaul et al., *Ind. J. Pharm.*, **39**, 42 (1977).
8. B. M. Lawrence, *Perfum. Flavor.*, **2**(2), 3 (1977).
9. R. M. Ideda et al., *J. Food Sci.*, **27**, 455 (1962).

10. C. K. Kokate and K. C. Varma, *Sci. Cult.*, **37**, 196 (1971).
11. B. G. V. N. Rao and P. L. Joseph, *Riechst., Aromen, Körperpflegem.*, **21**, 405 (1971).
12. D. L. J. Opdyke, *Food Cosmet. Toxicol.*, **11**, 1067 (1973).
13. Monograph *Cymbopogon species*, *Bundesanzeiger*, no. 22 (Feb. 1, 1990).

# CIVET

**Source:** ***Viverra civetta*** Schreber (syn. *Civettictis civetta* Schreber), ***V. zibetha*** L., and other related species (Family Viverridae).

*Synonyms.* African civet (*V. civetta*), large Indian civet (*V. zibetha*); and zibeth.

## GENERAL DESCRIPTION

Civets, also known as civet cats, are not related to cats and have shorter legs and longer muzzles than do cats. Both the African civet and the large Indian civet have gray coats with black markings, erectile manes, and short tails; their overall length is about 1.2 m.[1-3] Part used is the secretion from their anal glands, which is called civet and is collected by curetting (scraping) the glands with a wood or horn spatula at regular intervals (about once a week for the African civet and two to three times a week for the Indian civet). The civet cats are raised in captivity for this purpose (JIANGSU).[4] A concrète is prepared from crude civet by extracting with hydrocarbons; from this, the absolute is obtained by alcohol extraction. Major suppliers of crude civet are African countries (primarily Ethiopia but also Belgian Congo, Kenya, etc.); minor suppliers include India, Indonesia, Malaya, and China (ARCTANDER, JIANGSU).[4]

## CHEMICAL COMPOSITION

Contains civetone (9-*cis*-cycloheptadecenone) as its major aromatic principle; others include butyric acid, skatole, cycloheptadecanone, cyclononadecanone, and various saturated and unsaturated cyclic ketones and alcohols.[1,5,6]

## PHARMACOLOGY OR BIOLOGICAL ACTIVITIES

Limited available data indicate civet (absolute) to be nontoxic.[7]

## USES

**Medicinal, Pharmaceutical, and Cosmetic.** Civet absolute and tincture are extensively used as fixatives and fragrance components in perfumes (especially Oriental and rose types), with maximum use level of 0.4% reported for the absolute. Other cosmetic products in which the absolute is reported to be used include soaps, detergents, creams, and lotions.[7]

**Food.** Civet absolute has been reported used as a flavor component in most major food products, which include alcoholic and nonalcoholic beverages, frozen dairy desserts, candy, baked goods, and gelatins and puddings. Average maximum use levels reported are very low, less than 0.0014% (14.2 ppm).

**Traditional Medicine.** Used in Chinese medicine for centuries to relieve pain and as cardiac and neural sedatives, among others (JIANGSU).

## COMMERCIAL PREPARATIONS

Crude and extracts; crude has been reported to be frequently adulterated (ARCTANDER).[4]

*Regulatory Status.* GRAS (§182.50).

**REFERENCES**

See the General References for ARCTANDER; FEMA; GUENTHER; JIANGSU; YOUNGKEN.

1. E. Shiftan in A. Standen, ed., *Kirk-Othmer Encyclopedia of Chemical Technology*, Vol. 14, 2nd ed., Interscience, New York, 1976, p. 717.
2. *The Larousse Encyclopedia of Animal Life*, McGraw-Hill, New York, 1967, p. 564.
3. R. F. Ewer, *The Carnivores*, Cornell University Press, Ithaca, N.Y., 1973, p. 400.
4. C. L. Fischbeck, *Am. Perfum. Cosmet.*, **82**(12), 45 (1967).
5. Y. Ohno and S. Tanaka, *Bunseki Kagaku*, **26**, 232 (1977); through *Chem. Abstr.*, **87**, 44124u (1977).
6. D. A. Van Drop et al., *Recl. Trav. Chim. Pays-Bas*, **92**, 915 (1973); through *Chem. Abstr.*, **79**, 112605p (1973).
7. D. L. J. Opdyke, *Food Cosmet. Toxicol.*, **12**(Suppl.), 863 (1974).

# CLARY SAGE

**Source:** ***Salvia sclarea*** L. (Family Labiatae or Lamiaceae).

*Synonyms*. Clary, clary wort, muscatel sage, clear eye, see bright, and eyebright.

## GENERAL DESCRIPTION

Erect biennial or perennial herb with large hairy leaves and stout hairy stem; up to about 1 m high; native to southern Europe; cultivated worldwide (e.g., Mediterranean region, central Europe, Russia, the UK, and the United States). Parts used are the flowering tops and leaves, from which an essential oil is obtained by steam distillation and an absolute obtained by solvent extraction (ARCTANDER, POUCHER).[1]

## CHEMICAL COMPOSITION

Contains 0.1–0.15% volatile oil, which is composed primarily of linalyl acetate (up to ca. 75%), linalool, $\beta$-pinene, $\beta$-myrcene, and phellandrene, with minor constituents such as $\alpha$-terpineol, geraniol, *l*-terpinen-4-ol, benzaldehyde, cumene aldehyde, limonene, cineole, camphor, nerolidol, acetic, propionic, butyric, and valeric acid esters, and *trans*- and *cis*-alloocimene, among others, also present. The yield of the volatile oil and the relative concentrations of its components vary with the sources (MASADA, POUCHER).[2–5]

Other constituents present in clary include waxes, acids, and two diterpene alcohols (manool and sclareol).[1,6]

## PHARMACOLOGY OR BIOLOGICAL ACTIVITIES

Clary sage oil has been reported to show anticonvulsive activity in animals; it also potentiated the narcotic effects of Evipam and chloral hydrate.[7]

Except being moderately irritating to rabbit skin, available data indicate clary sage oil to be generally nontoxic.[8]

## USES

**Medicinal, Pharmaceutical, and Cosmetic.** Both oil and absolute are used as fragrance components in soaps, detergents, creams, lotions, and perfumes (e.g., eau de cologne). Maximum use level reported for the oil is 0.8% in perfumes.[8]

**Food.** While clary sage is reportedly used

only in beverages (e.g., wines and liqueurs with muscatel flavor), clary sage oil is used rather extensively in major food products such as alcoholic (vermouths, etc.) and nonalcoholic beverages, frozen dairy desserts, candy, baked goods, gelatins and puddings, and condiments and relishes. Highest average maximum use level reported is about 0.016% (155 ppm) for the oil in alcoholic beverages.

**Traditional Medicine.** Herb is used as a stomachic, in digestive disorders, and in kidney diseases. Mucilage of seeds is used in tumors and in removing dust particles from the eyes, among others.[9]

**Others.** After the essential oil is removed by distillation, the crude material is used as a source of sclareol, which can be solvent extracted from the plant and converted to sclareolide; both are used in flavoring tobaccos. Sclareolide is also used in the production of an ambergris substitute.[1]

### COMMERCIAL PREPARATIONS

Crude and oil; oil is official in F.C.C.

*Regulatory Status.* GRAS (§182.10 and §182.20).

### REFERENCES

See the General References for ARCTANDER; BAILEY 1; BIANCHINI AND CORBETTA; FEMA; GRIEVE; GUENTHER; POUCHER; TERRELL.

1. S. E. Allured, *Cosmet. Perfum.*, **90**(4), 69 (1975).
2. H. B. Heath, *Cosmet. Toilet.*, **92**(1), 19 (1977).
3. S. Chorbadzhiev et al., paper given at the 4th Mezhdunar. Kongr. Efirnym Maslam, (Mater.), 1968; through *Chem. Abstr.*, **78**, 128344t (1973).
4. G. Petri Verzar and M. Then, *Herba Hung.*, **13**, 51 (1974); through *Chem. Abstr.*, **83**, 128640q (1975).
5. A. I. Karetnikova et al., *Maslo Zhir. Prom.* **7**, 29 (1974); through *Chem. Abstr.*, **81**, 111394y (1974).
6. D. P. Popa and L. A. Salei, *Khim. Prir. Soedin.*, **3**, 405 (1974); through *Chem. Abstr.*, **81**, 166356t (1974).
7. S. Atanasova-Shopova and K. S. Rusinov, *Izv. Acad. Nauk. Inst. Fiziol., Bulg.*, **13**, 89 (1970); through *Chem. Abstr.*, **74**, 123533m (1971).
8. D. L. J. Opdyke, *Food Cosmet. Toxicol.*, **12**(Suppl.), 865 (1974).
9. J. L. Hartwell, *Lloydia*, **32**, 247 (1969).

# CLOVES

**Source:** ***Syzygium aromaticum*** (L.) Merr. et Perry (syn. *Eugenia aromatica* (L.) Baill., *E. caryophyllata* Thunb., and *E. caryophyllus* (Spreng.) Bull. et Harr.) (Family Myrtaceae).

### GENERAL DESCRIPTION

The clove is an evergreen tree with narrowly elliptic, pinkish (young) to dark green (mature) leaves; up to about 12 m high; believed to be native of Southeast Asia (eastern Indonesia); now cultivated worldwide (tropical Asia, Africa, tropical America, etc.). Parts used are the buds (cloves), stems, and leaves from which their respective essential oils are produced; bud and leaf oils by water distillation and stem oil by steam distillation (ARCTANDER). The major clove-producing country is Tanzania; other producers include Malagasy Republic, Indonesia, Malaysia, and Sri Lanka. Clove bud oil is considered more valuable than stem and leaf oils in flavor applications.

## CHEMICAL COMPOSITION

Clove buds yield 15–18% volatile oil; clove stems yield 4–6%; and clove leaves yield 2–3%.

Other constituents present in clove buds include glucosides of sterols (sitosterol, stigmasterol, and campesterol), crataegolic acid methyl ester, oleanolic acid, quercetin, eugeniin, kaempferol, rhamnetin, about 6% protein, 20% lipids, 61% carbohydrates, vitamins, and others (JIANGSU, MARSH).[1–3]

Clove bud oil contains 60–90% eugenol, 2–27% eugenol acetate, and 5–12% $\beta$-caryophyllene, with minor constituents such as methyl salicylate, methyl eugenol, benzaldehyde, methyl amyl ketone, $\alpha$-ylangene, and chavicol also present (ARCTANDER, MASADA).[4,5]

Clove stem oil usually contains 90–95% and clove leaf oil, 82–88% eugenol (ARCTANDER, MASADA); they contain little or no eugenyl acetate. Naphthalene (not present in the bud oil) is reportedly present in both oils in trace amounts; also many of the minor constituents in the bud oil are absent or present in much smaller concentrations in the leaf and stem oils.[5]

## PHARMACOLOGY OR BIOLOGICAL ACTIVITIES

A tincture of cloves (15% in 70% alcohol) has been reported to be effective in treating ringworms such as athlete's foot (JIANGSU).

Clove oil has antihistaminic and spasmolytic (musculotropic) properties, the latter probably due to its content of eugenyl acetate (see ***balm***).[6–8]

Clove oil (due to its eugenol) has anodyne and mildly antiseptic properties, exhibiting broad antimicrobial activities (against Gram-positive, Gram-negative, and acid-fast bacteria, and fungi)[5,7–10] as well as anthelmintic and larvicidal properties.[11] It is also reported to cause skin irritation and sensitization in humans.[7,8] Despite its possible toxicity in high dosage levels, eugenol (and presumably cloves and clove derivatives) is considered nontoxic at normal use levels.[5] No data are available that correlate the pharmacological properties of noneugenol clove constituents such as eugenyl acetate, methyl eugenol (see ***sweet bay***), and caryophyllene, which are often present in relatively large amounts in cloves and clove derivatives.

Aqueous extracts of cloves, clove oil, eugenol, eugenyl acetate, and methyl eugenol all have trypsin-potentiating activity (see ***cinnamon***).

Eugeniin exhibited strong antiviral activity against herpes simplex virus.[3]

## USES

**Medicinal, Pharmaceutical, and Cosmetic.** Clove bud oil (or eugenol) is used for the symptomatic relief of toothache; the oil is applied directly without pressure on the carious tooth with a small piece of cotton. It is also extensively used as a major component in preparations for the treatment of postextraction alveolitis (dry socket) and in dental cements and fillings, among others.

Clove bud and stem oils are used extensively as fragrance components in dentifrices, soaps, detergents, creams, lotions, and perfumes. Maximum use levels reported for the bud and stem oils are, respectively, 0.15 and 0.25% in soaps, and 0.7 and 1.0%, respectively, in perfumes.[7,8] Clove leaf oil is primarily used in soaps and low-cost perfumes, and to a much lesser extent than the other oils (ARCTANDER).

**Food.** Cloves, clove bud oil, clove stem oil, clove leaf oil, and eugenol are widely used in flavoring many food products, with cloves and clove bud oil by far the most used. Clove bud extract and oleoresin are also used, though to a lesser scale. Major food products in which cloves and their derivatives are used include alcoholic (bitters, vermouths, etc.) and nonalcoholic beverages, frozen dairy desserts, candy, baked goods, gelatins and puddings, meat and meat products, condiments and relishes, and gravies, among others. Highest average maximum use level reported for cloves is 0.236% in condiments and relishes; that for the oils is 0.06% of clove stem oil in alcoholic beverages, and that for clove bud oleoresin is about 0.078% (775 ppm) in alcoholic beverages.

**Health Food/Herb Tea.** Powdered cloves are used as a flavoring ingredient in Oriental-type herb teas (DUKE 2).

**Traditional Medicine.** Cloves are used as a carminative, antiemetic, and counterirritant. Clove tea is used to relieve nausea. Clove oil is used also as an antiemetic as well as in relieving toothache.

In Chinese medicine clove oil is used in diarrhea, hernia, and bad breath, in addition to the above and other uses (JIANGSU, NANJING).

**Others.** Clove leaf oil is used as a source for the isolation of eugenol.

A large portion of the world's clove production goes to Indonesia for use in Kretak cigarettes, which consist of a mixture of two parts tobacco and one part ground cloves and when smoked produce a crackling noise (ROSENGARTEN).

Clove extracts and oil have been demonstrated to have strong antioxidative properties.[12–14] Clove oil (also eugenol) and clove aqueous extract also markedly increase trypsin activity.[15] These properties could be useful in food and drug applications.

## COMMERCIAL PREPARATIONS

Cloves, extracts (e.g., oleoresin), and oils (bud, stem, and leaf). Clove bud oil is official in N.F. and all three oils are official in F.C.C.

*Regulatory Status.* Cloves and their derivatives (oils, extracts, etc.) have been affirmed as GRAS (§184.1257).[5] Subject of a German therapeutic monograph indicated for inflamed oral and pharyngeal mucosa; topical anesthesia in dentistry.[16]

## REFERENCES

See the General References for ADA; ARCTANDER; DUKE 2; GUENTHER; JIANGSU; FEMA; FURIA AND BELLANCA; LUST; MARTINDALE; MASADA; ROSENGARTEN; USD 26th.


1. C. H. Brieskorn et al., *Phytochemistry*, **14**, 2308 (1975).
2. B. Voesgen and K. Herrmann, *Z. Lebensm. Unters. Forsch.*, **170**, 204 (1980).
3. M. Takechi and Y. Tanaka, *Planta Med.*, **42**, 69 (1981).
4. E. Cerma and B. Stancher, paper given at the 4th Atti Conv. Reg. Aliment., 1st Conv. Naz. Qual., Trieste, 1965; through *Chem. Abstr.*, **68**, 48441a (1968).
5. Anon., *Fed. Regist.*, **42**(146), 38613 (1977).
6. A. M. Debelmas and J. Rochat, *Plant. Med. Phytother.*, **1**, 23 (1967).
7. D. L. J. Opdyke, *Food Cosmet. Toxicol.*, **13**, 761 (1975).
8. D. L. J. Opdyke, *Food Cosmet. Toxicol.*, **13**, 765 (1975).
9. N. G. Martinez Nadal et al., *Cosmet. Perfum.*, **88**(10), 37 (1973).
10. F. M. Ramadan et al., *Chem. Mikrobiol. Technol. Lebensm.*, **1**, 96 (1972).

11. K. Oishi et al., *Nippon Suisan Gakkaishi*, **40**, 1241 (1974); through *Chem. Abstr.*, **82**, 84722r (1975).
12. Y. Saito et al., *Eiyo To Shokuryo*, **29**, 505 (1976); through *Chem. Abstr.*, **87**, 150314r (1977).
13. H. Fujio et al., *Nippon Shokuhin Kogyo Gakkaishi*, **16**, 241 (1969); through *Chem. Abstr.*, **74**, 2846g (1971).
14. F. Hirahara et al., *Eiyogaku Zasshi*, **32**, 1 (1974); through *Chem. Abstr.*, **82**, 2764u (1975).
15. Y. Kato, *Koryo*, **113**, 17, 24 (1975); through *Chem. Abstr.*, **84**, 178343m, 149393x (1976).
16. Monograph *Caryophylli flos*, *Bundesanzeiger*, no. 223 (Nov. 30, 1985).

# CLOVER TOPS, RED

**Source:** ***Trifolium pratense*** L. (Family Leguminosae).

*Synonyms.* Trifolium, meadow clover, purple clover, and cow clover.

## GENERAL DESCRIPTION

A biennial or perennial herb with rose-purple flowers and leaves consisting of three, often hairy, white-blotched leaflets; up to 0.8 m high; native to Europe and naturalized in North America. Parts used are the flowering tops (inflorescence).

## CHEMICAL COMPOSITION

Contains isoflavones such as biochanin A, formononetin, genistein, daidzein, pratensein, and trifoside (5-hydroxy-7-methoxyisoflavone-4′-*O*-*β*-D-glucopyranoside); flavones (e.g., pectolinarin); coumarins (coumarin, medicagol, and coumestrol, etc.),[1–4] with presence of coumestrol disputed (JIANGSU).[5]

Other constituents include *trans*- and *cis*-clovamide (L-dopa conjugated with *trans*- and *cis*-caffeic acids), trifoliin (isoquercitrin), phaselic acid, a galactoglucomannan composed of a backbone of $\beta$-(1→4) linked D-glucose and D-mannose units with $\alpha$-(1→6)-linked D-galactose side chains, sugars, protein, a volatile oil containing furfural, resins, fat, minerals (particularly rich in magnesium, copper, and calcium), phosphorus, vitamins, and others (DUKE 1; KARRER, JIANGSU).[6–9]

Red clover produces phytoalexins (pterocarpan types) in response to viral or fungal infections.[4,10–12]

Commercial solid extracts of red clover have been reported to contain traces of cannabinol, caffeine, scopolamine, isocoumarin, phenylpentadienal, phenylhexadiene, and nepetalactone (see ***alfalfa***).[13]

## PHARMACOLOGY OR BIOLOGICAL ACTIVITIES

Red clover has estrogenic properties in animals, which are due to its isoflavones (see ***alfalfa***).[1,4,14–18]

It is also reported to have antispasmodic and expectorant properties (MERCK).

## USES

**Food.** The solid extract is reportedly used as a flavor ingredient in many food products, including nonalcoholic beverages, frozen dairy desserts, candy, baked goods, gravies, and jams and jellies. Average maximum use levels reported are usually below 0.002%, except in jams and jellies, where it is about 0.053% (525 ppm).

**Health Food/Herb Teas.** Whole or ground flowering tops used as an herb tea ingredient; also in capsules, tablets, tinctures, etc., primarily as an "alterative" (blood purifier) used for skin ailments, such as psoriasis, eczema; ingredient in unconventional anti-cancer formulas including the Hoxsey Formula (FOSTER AND DUKE). Genistein, a weakly estrogenic compound in the flower may block off estrogen receptors, thwarting tumor development.[18]

**Traditional Medicine.** Dried inflorescence and whole herb are used in both Eastern and Western cultures as diuretic, sedative, and antitussive and in treating whooping cough, asthma, bronchitis, skin sores and ulcers, sore eyes, and burns, among others, usually in the form of a tea, infusion, or salve.

The whole plant or its various parts (leaves, flowers, roots, etc.) have also been extensively used in cancers.[19]

## COMMERCIAL PREPARATIONS

Crude and extracts (solid, fluid, etc.); crude and fluid extract were formerly official in N.F. Strengths (see ***glossary***) of extracts are expressed in weight-to-weight ratios.

## REFERENCES

See the General References for DUKE 1; FEMA; FOSTER AND DUKE; GRIEVE; HORTUS 3rd; JIANGSU; KROCHMAL AND KROCHMAL; LUST; ROSE; UPHOF; WREN; YOUNGKEN.

1. Z. Rolinski, *Ann. Univ. Mariae Curie Sklodowska, Sect. D D*, **24**, 165 (1970); through *Chem. Abstr.*, **75**, 72481d (1971).
2. G. Schultz, *Deut. Tieraerztl. Wochenschr.*, **74**, 118 (1967); through *Chem. Abstr.*, **67**, 21074a (1967).
3. N. S. Kattaev et al., *Khim. Prir. Soedin.*, **6**, 806 (1972); through *Chem. Abstr.*, **78**, 94807d (1973).
4. P. M. Dewick, *Phytochemistry*, **16**, 93 (1977).
5. W. Dedio, *Diss. Abstr. Int., B*, **34**, 5281 (1974).
6. T. Yoshihara et al., *Agr. Biol. Chem.*, **38**, 1107 (1974).
7. A. J. Buchala and H. Meier, *Carbohydr. Res.*, **31**, 87 (1973).
8. D. Smith et al., *Agron. J.*, **66**, 817 (1974).
9. B. D. E. Guillard and R. W. Bailey, *Phytochemistry*, **7**, 2037 (1968).
10. P. M. Dewick, *Phytochemistry*, **14**, 979 (1975).
11. J. L. Ingham, *Phytochemistry*, **15**, 1489 (1976).
12. J. N. Bitton et al., *Phytochemistry*, **15**, 1411 (1976).
13. S. R. Srinivas, *Dev. Food Sci.*, **18**, 343 (1988).
14. A. K. Tuskaev, *Rast. Resur.*, **7**, 295 (1971); through *Chem. Abstr.*, **75**, 85222v (1971).
15. E. Krause, *Acta Vet. (Brno)*, **39**, 279 (1970); through *Chem. Abstr.*, **74**, 85885t (1971).
16. Z. Rolinski, *Ann. Univ. Mariae Curie Sklodowska, Sect. D D*, **24**, 187 (1970); through *Chem. Abstr.*, **75**, 150167x (1971).
17. F. Garcia and P. H. Reinshagen, *Nutr. Bromatol. Toxicol.*, **5**, 67 (1966); through *Chem. Abstr.*, **67**, 71485f (1967).
18. J. M. Bergeron and M. Goulet, *Can. J. Zool.*, **58**, 1575 (1980).
19. J. L. Hartwell, *Lloydia*, **33**, 97 (1970).
20. J. A. Duke, *Bus. Herbs*, 8 (Sept./Oct. 1990).

# COCA

**Source:** ***Erythroxylum coca*** Lam. (syn. *Erythroxylon coca* Lam.); ***E. novogranatense*** (Morris) Hieron. (Family Erythroxylaceae).

*Synonyms*. Bolivian and Huanuco coca (*E. coca*); Peruvian, Truxillo, and Java coca (*E. novogranatense* and varieties); cocaine plant and spadic.

## GENERAL DESCRIPTION

Leafy evergreen shrubs to small trees, with slender branches; up to about 5 m high at lower altitudes and 2 m at higher altitudes; native to the South American Andes; now primarily cultivated at altitudes between 450 and 1800 m. *Erythroxylum coca* is cultivated throughout the wet, tropical valleys of the eastern Andes (Ecuador, Peru and Bolivia) and the Amazon basin, while *E. novogranatense* is grown mainly in drier regions of Venezuela, Colombia, and Peru.[1] They are also grown in Asia (China, India, Indonesia, Japan, etc.) and Australia. Part used is the leaf. Major producers include Bolivia, Colombia, and Peru (LIST AND HÖRHAMMER, MORTON 3, NANJING).

## CHEMICAL COMPOSITION

Coca leaves contain as active constituents alkaloids, including cocaine, cinnamoylcocaine, benzoylecgonine, ecgonine, tropacocaine, hygrine, cuscohygrine, nicotine, and others. The concentrations of these alkaloids and their relative proportions vary widely, depending on the sources of the leaves, their age when harvested, and storage age, among other factors. Thus the contents of alkaloids can range from 0.002 to 2.5%, with the usual range being 0.5–1.5%, of which cocaine accounts for 70–80% of the total alkaloids in Bolivian coca and only about 50% in Peruvian leaves (LIST AND HÖRHAMMER, MARTINDALE).[1-3]

Other constituents present include 0.02–0.13% of a volatile oil composed mainly of methyl salicylate, two dihydrobenzaldehydes (tentatively identified), *cis*-3-hexen-1-ol, *trans*-2-hexenal, 1-hexanol, and *n*-methylpyrrole;[6] $\alpha$- and $\beta$-truxillic acids; rutin and isoquercitrin (LIST AND HÖRHAMMER); about 19% protein and 44% carbohydrates; and high contents of calcium, iron, vitamin A, riboflavin, and phosphorus.[4]

## PHARMACOLOGY OR BIOLOGICAL ACTIVITIES

The pharmacological activity and toxicity of coca are generally attributed to cocaine. In addition to its local anesthetic, central nervous system stimulant, and addictive (similar to amphetamines) properties, it has many other activities (GOODMAN AND GILMAN, LIST AND HÖRHAMMER, MARTINDALE, MORTON 3).

The fatal dose of cocaine in humans is reported to be about 1.2 g, but a dose as low as 20 mg (0.02 g) has been reported to cause severe toxic effects (GOODMAN AND GILMAN).

## USES

**Medicinal, Pharmaceutical, and Cosmetic.** Coca leaf and its extracts are not used in pharmaceutical preparations in the United States.

Cocaine (free base or salt form) is used as a local anesthetic, mainly for eye (cornea), nose, and throat mucosae during surgery.

**Food.** Coca extract, from which cocaine is removed, is used together with extracts of kola, cinnamon, ginger, lime, orange peel, and others as a flavor component in cola drinks; average maximum use level is reported to be 0.02%. Other food products in which the decocainized extract is used include alcoholic beverages, frozen dairy desserts, and candy. The highest average maximum use level is 0.055% in frozen dairy desserts.

**Traditional Medicine.** Used in South America by natives to relieve hunger and fatigue; the leaves are usually chewed with lime. During this process cocaine is reportedly degraded to ecgonine.[5]

Extracts (e.g., infusion) are used as a stomachic and sedative, and in treating asthma, colds, and other ailments.

### COMMERCIAL PREPARATIONS

Coca and cocaine are controlled as narcotic agents in the United States.

Coca was formerly official in U.S.P. Cocaine and cocaine hydrochloride are official in N.F. and U.S.P., respectively.

*Regulatory Status.* Coca extract (decocainized) is GRAS (§182.20).

### REFERENCES

See the General References for FEMA; GOODMAN AND GILMAN; HORTUS 3rd; MORTON 3; NANJING; TERRELL; UPHOF; USD 26th.

1. B. Holmstedt et al., *Phytochemistry*, **16**, 1753 (1977).
2. G. H. Anilian et al., *J. Pharm. Sci.*, **63**, 1938 (1974).
3. G. Espinel Ovalle and I. Guzman Parra, *Rev. Colomb. Cienc. Quim. Farm.*, **1**, 95 (1971); through *Chem. Abstr.*, **76**, 89994s (1972).
4. J. A. Duke et al., *Bot. Mus. Leafl. Harv. Univ.*, **24**, 113 (1975).
5. O. Neischulz and P. Schmersahl, *Planta Med.*, **17**, 178 (1969).
6. M. Novak and C. A. Salemink, *Planta Med.*, **53**, 113 (1987).

# COCILLANA BARK

**Source:** ***Guarea rusbyi*** (Britt.) Rusby (syn. *Sycocarpus rusbyi* Britt.) and closely related species (Family Meliaceae).

*Synonyms.* Guapi, Upas, Trompillo, and grape bark.

### GENERAL DESCRIPTION

Trees native to the South American Andes, especially in Bolivia. Part used is the dried bark. The original cocillana bark is believed to be derived from *G. rusbyi*. However, there is evidence that the current drug is obtained from other closely related *Guarea* species and is not the same as the cocillana first introduced into modern medicine (USD 23rd).[1]

### CHEMICAL COMPOSITION

Chemical studies on cocillana are limited. Cocillana bark is reported to contain small amounts (0.003–0.023%) of alkaloid(s) (rusbyine) of which the chemical structure(s) has not been determined. It also contains $\beta$-sitosterol, a volatile oil, tannin, anthraquinones, flavonols, and others (LIST AND HÖRHAMMER).[1]

### PHARMACOLOGY OR BIOLOGICAL ACTIVITIES

Cocillana is reported to have expectorant and, in higher doses, emetic properties similar to those of ipecac, but is more stimulating (LIST AND HÖRHAMMER).[1] These properties are based on findings reported at the end of the last century. No recent pharmacological or toxicological data on cocillana are available.

## USES

**Medicinal, Pharmaceutical, and Cosmetic.** Used rather extensively in cough syrups and similar preparations, being particularly popular in the British Commonwealth countries.

**Health Food/Herb Teas.** Seldom seen on the American market; common cough syrup ingredient in the UK (WREN).

**Traditional Medicine.** Used by natives in South America as an expectorant for alleviating coughs.[1]

Root bark of *G. spiciflora* A. Juss. and leaf of *G. trichiloides* L. have been reportedly used in treating skin indurations and tumors, respectively, in South America.[2]

## COMMERCIAL PREPARATIONS

Crude and extracts; the bark and fluid extract were formerly official in N.F. Strengths (see ***glossary***) of extracts are expressed in weight-to-weight ratios.

## REFERENCES

See the General References for GOSSELIN; MARTINDALE; YOUNGKEN.

1. E. B. Ritchie and J. W. Steel, *Planta Med.*, **14**, 247 (1966).
2. J. L. Hartwell, *Lloydia*, **33**, 97 (1970).

# COCOA (CACAO)

**Source: *Theobroma cacao*** L. subsp. ***cacao*** (Family Sterculiaceae or Byttneriaceae).

*Synonym*. Theobroma.

## GENERAL DESCRIPTION

Evergreen tree with leathery oblong leaves, about 8 m high; fruits are berries borne directly on trunk and branches, with seeds within a mucilaginous pulp. Parts used are the seeds, which are commonly called cacao or cocoa beans. Cacao is generally used to describe the crude materials (e.g., cacao tree and cacao beans), while cocoa is used to describe the processed products. However it is increasingly common to use the term cocoa for both crude and processed products, thus cocoa tree, cocoa beans, cocoa powder, cocoa butter, and so on.[1,2]

There are three varieties of cacao: forastero, criollo, and trinitario. Forastero accounts for more than 90% of the world's usage and is produced primarily in West African countries (e.g., Ghana, Nigeria, Cameroon, and the Ivory Coast), while the criollo variety is produced in Venezuela and Central America as well as Papua New Guinea, Java, and Samoa. Trinitario is believed to be a hybrid of the other two varieties and is produced in Venezuela, Trinidad, Sri Lanka, and other countries. Both criollo and trinitario cacao are considered to have better flavor qualities than forastero cacao. Thus criollo is blended with forastero to improve the flavor of forastero in the manufacture of cocoa and, like trinitario, is also used in certain high-quality eating chocolates.[1]

Three main types of ingredients are produced from cacao seeds: cocoa powder cocoa butter, and cocoa extracts. For the manufacture of these products the cacao beans are first cured by fermentation and drying, during which time the pulp surrounding the seeds is decomposed and removed and flavor precursors develop in the seeds. The dried beans, now called raw cocoa containing about 6–8% moisture, are roasted to produce the required flavor, aroma, and color and to facilitate removal of the seed coat (shell); temperatures vary from 100 to 150°C, depending on the types of beans and the products to be made. Beans for manufacturing cocoa butter or chocolate are roasted at lower temperatures, while those for cocoa powder production are roasted at higher temperatures. After roasting, the shell and hypocotyl are separated from the cotyledons (called nibs).

The nib, containing about 55% cocoa butter, is ground while hot to a liquid mass called cocoa or chocolate liquor, from which variable amounts of the cocoa butter is removed by hydraulic pressing. The cocoa cake left on the filter is cooled and then ground to a fine powder under controlled cool temperatures to yield cocoa powder that has cocoa fat contents of up to 22% or more.

Currently most cocoa powders are produced by the so-called Dutch or alkalized process, in which the nib is treated with a warm aqueous solution of up to 3 parts of anhydrous potassium carbonate to 100 parts of nib (or equivalent amounts of other alkalis such as potassium bicarbonate and hydroxide; carbonates, bicarbonates, and hydroxides of sodium, magnesium, and ammonium; or their combinations). After the alkali is completely absorbed, the nib is processed as in the above method to yield alkalized cocoa powder. Alkalized cocoa is considered to have improved dispersibility, color, and flavor over unalkalized cocoa.

Cacao nibs, cocoa powder, and certain other cocoa products (e.g., chocolates) are governed by standards of identity set forth in the Code of Federal Regulations (21 CFR §§163.110–163.155). For example, cacao nibs used for cocoa manufacture are required to have no more than 1.75% cacao shell. Breakfast cocoa (or high-fat cocoa) must have at least 22% cacao fat (cocoa butter), cocoa (or medium-fat cocoa) must contain less than 22% but not less than 10%, and low-fat cocoa must contain less than 10% cacao fat. Alkalized cocoa must be labeled "processed with alkali," and so on.

Cocoa butter (also called cacao butter and theobroma oil) is produced commonly by three methods: hydraulic pressing, extrusion or expeller pressing, and solvent extraction. Cocoa butter produced by the first two methods has a faint chocolate flavor and aroma that can be removed by steam distillation under vacuum; it is brittle at temperatures below 25°C and melts at 34–35°C.[1,2]

Cocoa extracts are generally prepared by extraction with hydroalcoholic solvents of the roasted seeds (nibs); an essential oil is also produced by steam distillation.

## CHEMICAL COMPOSITION

Cocoa contains more than 300 volatile compounds, including hydrocarbons, monocarbonyls, pyrroles, pyrazines, esters, lactones, and others.

The important flavor components are reported to be aliphatic esters, polyphenols, unsaturated aromatic carbonyls, pyrazines, diketopiperazines, and theobromine.[3–5] Cocoa also contains about 18% proteins (ca. 8% digestible);[2,6,7] fats (cocoa butter); amines and alkaloids, including theobromine (0.5–2.7%), caffeine (ca. 0.25% in cocoa; 0.07–1.70% in fat-free beans, with forasteros containing less than 0.1% and criollos containing 1.43–1.70%),[8–10] tyramine, dopamine, salsolinol,[11,12] trigonelline, nicotinic acid, and free amino acids;[9] tannins; phospholipids;[13] starch and sugars;[2] minerals (particularly high in sodium or potassium in alkalized cocoa); and others (MARTINDALE, MORTON 3, WATT AND MERRILL).[14]

The characteristic bitter taste of cocoa is reported to be due to the diketopiperazines (especially those containing phenylalanine) reacting with the theobromine present during roasting.[5]

Cocoa butter contains mainly triglycerides of fatty acids that consist primarily of oleic ($C_{18:1}$), stearic ($C_{18:0}$), and palmitic ($C_{16:0}$) in decreasing concentrations, with small amounts of linoleic ($C_{18:2}$) and arachidic ($C_{20:2}$) acids. Over 73% of the glycerides are present as monounsaturated forms (oleopalmitostearin and oleodistearin), the remaining being mostly diunsaturated glycerides (palmitodiolein and stearodiolein) with lesser amounts of fully saturated and triunsaturated (triolein) glycerides. Linoleic acid levels have been reported to be up to 4.1% (MERCK, MORTON 3).[2,15,16] Also present in cocoa butter are small amounts of sterols and methylsterols; sterols consist mainly of $\beta$-sitosterol, stigmasterol, and campesterol, with a small quantity of cholesterol (0–0.28%) present.[17–19]

In addition to alkaloids (mainly theobromine), tannins, and other constituents, cocoa husk contains a pigment that is a polyflavone glucoside with a molecular weight of over 1500. This pigment is claimed to be heat and light resistant, highly stable at pH 3–11, and useful as a food colorant; it was isolated at a 7.9% yield.[20]

Depending on the alkali used, cocoa powder produced by the Dutch process may contain relatively high concentrations of sodium; this may cause problems in persons who are on a low-sodium diet.

## PHARMACOLOGY OR BIOLOGICAL ACTIVITIES

Theobromine, the major alkaloid in cocoa, has similar pharmacological activities as caffeine. However its stimulant activities on the central nervous system, respiration, and skeletal muscles are much weaker than those of caffeine; but its cardiac stimulant, coronary dilating, smooth muscle relaxant, and diuretic properties are stronger (GOODMAN AND GILMAN).

Cocoa butter has been reported to have skin allergenic and comedogenic (forming blackheads) properties in animals.[21,22]

## USES

**Medicinal, Pharmaceutical, and Cosmetic.** Cocoa powder (or cocoa syrup) is used in flavoring pharmaceutical preparations.

Cocoa butter is used extensively as a suppository and ointment base; also used as emollient, skin softener, and skin protectant in creams (e.g., massage), lotions, lipsticks, and soaps, among others.

**Food.** Beverages made from cacao flavored with vanilla and other spices have been used by native Mexicans (Aztecs) for centuries.

Cocoa powder is used extensively as a flavor or nutrient component in nonalcoholic beverages, ice cream, cakes, biscuits, and others.

Cocoa butter is extensively used in chocolate manufacture, where it is mixed with cocoa liquor (ground cacao nibs), sugar, milk, and other ingredients such as flavors. Dark chocolate does not contain milk.

Cocoa extract is used in both alcoholic (liqueurs such as creme de cacao) and nonalcoholic beverages, frozen dairy desserts, candies, baked goods, and others.

**Health Food/Herb Teas.** Cocoa butter is used in creams, massage oils, and other cosmetic preparations sold in health food stores (ROSE).

**Traditional Medicine.** Cocoa butter is used to treat wrinkles on neck (turkey neck), around the eyes, and at the corners of the mouth (ROSE). Reportedly used in European tradition in combination with other ingredients for infectious intestinal disease, diarrhea; bronchial expectorant in asthma, bronchitis, irritating cough and lung congestion; to regulate function of endocrine glands, especially the thyroid.[23]

**Others.** Cocoa and cocoa butter have been reported to contain fat-soluble antioxidants and could be a source of such substances.[24]

## COMMERCIAL PREPARATIONS

Cocoa powders, cocoa butter, cocoa syrup, and cocoa extracts. Cocoa (10–22% fat), cocoa butter, and cocoa syrup are official in N.F.

*Regulatory Status.* Standards of identity for cocoa products apply (§§163.110–163.155). Cocoa extracts are GRAS (§182.20). Cocoa seed is the subject of a German therapeutic monograph; not recommended as claimed efficacy is unsubstantiated; allowed as flavoring agent.[23]

## REFERENCES

See the General References for BAILEY 2; BLUMENTHAL; GOSSELIN; HORTUS 3rd; KARRER; MARTINDALE; TERRELL; USDA.

1. R. J. Clarke and J. W. Drummond in L. W. Codd et al., eds., *Chemical Technology: An Encyclopedic Treatment,* Vol. 7, Barnes & Noble, New York, 1975, p. 645.
2. B. D. Powell and T. L. Harris in A. Standen, ed., *Kirk-Othmer Encyclopedia of Chemical Technology*, Vol. 5, Wiley Interscience, New York, 1964, p. 363.
3. D. Reymond, *Chemtech*, **7**, 664 (1977).
4. P. G. Keeney, *J. Am. Oil Chem. Soc.*, **49**, 567 (1972).
5. W. Pickenhagen et al., *Helv. Chim. Acta*, **58**, 1078 (1975).
6. D. J. Timbie and P. G. Keeney, *J. Agr. Food Chem.*, **25**, 424 (1977).
7. D. L. Zak and P. G. Keeney, *J. Agr. Food Chem.*, **24**, 483 (1976).
8. Y. Asamoa and J. Wurziger, *Gordian*, **76**, 138 (1976); through *Chem. Abstr.*, **85**, 76524z (1976).
9. G. Barbiroli, *Atti Cong. Qual.*, **6**(1967), 149 (1968); through *Chem. Abstr.*, **72**, 77599 (1970).
10. M. Mironescu, *Rev. Fiz. Chim. Ser. A*, **11**, 218 (1974); through *Chem. Abstr.*, **82**, 84684e (1975).
11. T. M. Kenyhercz and P. T. Kissinger, *Phytochemistry*, **16**, 1602 (1977).
12. R. M. Riggin and P. T. Kissinger, *J. Agr. Food Chem.*, **24**, 900 (1976).
13. L. Biino and E. Clabot, *Atti Soc. Peloritana Sci. Fis. Mat. Natur.*, **16**, 257 (1970); through *Chem. Abstr.*, **78**, 1996b (1973).
14. T. M. Kenyhercz and P. T. Kissinger, *Lloydia*, **41**, 130 (1978).
15. R. F. Looney in L. W. Codd et al., eds., *Chemical Technology: An Encyclopedic Treatment,* Vol. 8, Barnes & Noble, New York, 1975, p. 1.
16. Y. Asamoa and J. Wurziger, *Gordian*, **74**, 280 (1974); through *Chem. Abstr.*, **82**, 96606u (1975).
17. H. Chaveron, *Choc. Confiserie Fr.*, **273**, 12 (1971); through *Chem. Abstr.*, **76**, 84643s (1972).
18. T. Itoh et al., *J. Am. Oil Chem. Soc.*, **50**, 300 (1973).
19. T. Itoh et al., *Oleagineux*, **29**, 253 (1974).
20. K. Kimura et al., Jpn. Kokai 73 17, 825 (1973); through *Chem. Abstr.*, **79**, 124920a (1973).
21. V. V. Ivanov, *Vestn. Dermatol. Venerol.*, **3**, 57 (1976); through *Chem. Abstr.*, **86**, 111084a (1977).
22. O. H. Mills et al., *Br. J. Dermatol.*, **98**, 145 (1978).

23. Monograph *Cacao semen*, *Bundesanzeiger*, no. 40 (Feb. 27, 1991).
24. P. A. Dewdney and M. L. Meara, *Sci. Tech. Surv. Br. Food Manuf. Ind. Res. Assoc.*, 96 (1977).

# CODONOPSIS

**Source:** ***Codonopsis pilosula*** (Franch.) Nannf., ***C. pilosula*** Nannf. var. ***modesta*** (Nannf.) L. T. Shen, ***C. tangshen*** Oliv., ***C. tubulosa*** Kom., and many other ***Codonopsis*** species (Family Campanulaceae).[1]

*Synonyms*. Radix codonopsis, bonnet bellflower, bastard ginseng, and *dangshen*.

## GENERAL DESCRIPTION

Mostly small herbaceous perennials, strongly scented, with thick fleshy cylindrical to slightly spindle-shaped roots; native to Asia; distributed throughout China, including the provinces of Shanxi, Shaanxi, Sichuan, Yunnan, Xinjiang, Gansu, Jilin, and Liaoning; now extensively cultivated, also as ornamental in the United States. Part used is the root, collected in autumn from wild or cultivated plants at least 3 years old, washed clean. It is sorted and strung out to sun dry to about half dry, then massaging or rubbing by hand or between two wood boards to bring internal tissues together, followed by further drying and rubbing until completely dry; this process eliminates or minimizes the presence of air space, cracks, or holes in the dried herb, which would not keep well (CMH). There are five major types each with different grades: *xidang* (western codonopsis) from Gansu; *dongdang* (eastern codonopsis) from Jilin, Liaoning, and Heilongjiang; *ludang* (Shanxi codonopsis) from Shanxi; *chuandang* or *tiaodang* (Sichuan codonopsis) from Sichuan, Hubei, and Shaanxi; and *baidang* or *guanhua dangshen* (white or tubular-flowered codonopsis) from Guizhou and Yunnan (CMH, ZHU). *Ludang* is the most often encountered in the United States.

Considered as the poor man's ginseng, it is frequently used as a substitute for ginseng.

## CHEMICAL COMPOSITION

Most of the chemical studies on codonopsis were performed during the last decade. Compounds identified include polysaccharides and sugars (e.g., inulin, starch, glucose, sucrose, fructose);[2,3] 1.24–10.38% saponins (tangshenosides I, II, III, and IV);[3–6] amino acids (1.47–5.33%), with ca. 0.1% as free amino acids;[4,5,7] a $\beta$-carboline alkaloid (perlolyrine) and other nitrogen compounds (choline, nicotinic acid, *n*-butylallophanate, etc.);[8] triterpenes (e.g., taraxeryl acetate, taraxerol, friedelin) and sterols (stigmasterol and spinasterol) and their glucosides;[9–11] atractylenolides II and III (see ***baizhu***);[3] scutellarein glucoside; oroxylin A;[12] a volatile oil (ca. 0.12%) composed of 50% acidic compounds, with palmitic acid and methyl palmitate predominant;[13] and trace minerals, among others.[4,5]

Four polysaccharides (CP-1, CP-2, CP-3, and CP-4) have been isolated with molecular weights of 10,500, 12,000, 14,000 and 79,000, respectively; the first two have $\beta$-glycosidic linkages, while the latter two have $\alpha$-linkages, involving an unusually large number of sugars (glucose, fructose, galactose, arabinaose, mannose, xylose, and rhamnose).[2]

No ginseng saponins have been found in codonopsis.

## PHARMACOLOGY OR BIOLOGICAL ACTIVITIES

As a frequent substitute for ginseng, *dangshen* has many of the properties of ginseng. Its biological activities include central stimulant in mice, weight gain in rabbits, increased swimming time in mice, prevention of leukocytosis induced by turpentine oil in experimental animals, increased tolerance to anoxia and elevated temperatures in mice and prolonged survival, radioprotective, enhanced phagocytosis of macrophages, immunoregulating, improved blood picture in mice (red and white cells and hemoglobin all increased), increased serum corticosterone in mice, hypotensive and peripheral vasodilatory as well as adrenolytic; and stimulation as well as relaxation of isolated guinea pig ileum (IMM-2, WANG).[5,14,15]

*Dangshen* polysaccharides have exhibited immunomodulating effects in guinea pigs, inhibited experimental ulcers (stress, indomethacin, acetic acid, and pyloric ligation models) in rats, and had antistress effects (prolonging swimming time and increasing tolerance to anoxia and to elevated temperatures) in mice, etc.[16–18]

Oroxylin A was shown to have antihistaminic effects on isolated guinea pig ileum.[12]

## USES

**Health Food/Herb Teas.** Powdered herb and extracts are used in tonic formulas (in tablet, capsule or liquid form), often as Oriental ginseng substitute, for boosting one's immune system and replenishing *qi* (vital energy); cut or teabag cut herb is used in tea or soup mixes (FOSTER AND YUE).

**Traditional Medicine.** A relatively recent addition to Chinese materia medica, *dangshen* was first described in the *Ben Jing Feng Yuan* (A.D. 1695) as a lung-clearing (*qing fei*) drug with a sweet taste and neutral nature. Its tonic properties were later described in the *Ben Cao Cong Xin* (1757) and has since become a highly valued *qi* tonic of equal status as some ancient ones such as astragalus, ginseng, and common jujube (see those entries). It is used to treat many of the same conditions as these tonics, including general weakness, lack of appetite, chronic diarrhea, shortness of breath, palpitations, asthma, cough, thirst and diabetes, conditions due to Spleen and blood deficiencies, and damaged *qi*.

## COMMERCIAL PREPARATIONS

Crude (whole or powdered) and extracts, with no uniform standards.

*Regulatory Status.* U.S. regulatory status not determined; sold as nutritional supplement.

## REFERENCES

See the General References for CHP; CMH; FOSTER AND YUE; IMM-2; JIANGSU; LU AND LI; NATIONAL; WANG; ZHU.

1. Z. T. Wang and G. J. Xu, *Zhongcaoyao*, **23**, 144 (1992).
2. S. J. Zhang and S. Y. Zhang, *Zhongcaoyao*, **18**(3), 2 (1987).
3. D. G. Cai, *Zhongguo Zhongyao Zazhi*, **16**, 376 (1991).
4. S. M. Wang et al., *Shanxi Zhongyi*, **6**(3), 35 (1990).
5. S. M. Wang and Y. Yang, *Shanxi Zhongyi*, **5**(1), 37 (1989).
6. G. R. Han et al., *Zhongguo Zhongyao Zazhi*, **15**(2), 41 (1990).

7. Y. X. Gong, *Zhongcaoyao*, **18**(11), 37 (1987).
8. T. Liu et al., *Planta Med.*, **54**, 472 (1988).
9. M. P. Wong et al., *Planta Med.*, **49**, 60 (1983).
10. G. R. Han et al., *Zhongcaoyao*, **22**, 422 (1991).
11. Y. Z. Wang et al., *Zhongcaoyao*, **17**(5), 41 (1986).
12. D. X. Zhou et al., *Zhongguo Zhongyao Zazhi*, **16**, 564 (1991).
13. J. Liao and Y. Q. Lu, *Zhongcaoyao*, **18**(9), 2 (1987).
14. X. L. Mao et al., *Chin. J. Integr. Trad. Western Med.*, **5**, 739 (1985).
15. C. Q. Ling, *Henan Zhongyi*, **13**, 94 (1993).
16. M. X. Zhuang et al., *Zhongguo Yaoxue Zazhi*, **27**, 653 (1992).
17. J. C. Cui et al., *Zhongcaoyao*, **19**(8), 21 (1988).
18. W. Li et al., *Jilin Zhongyiyao*, (6), 33 (1990).

# COFFEE

**Source:** ***Coffea arabica*** L., ***C. canephora*** Pierre ex Froehner (syn. *C. robusta* Linden ex De Wild.), and other ***Coffea*** species, varieties or hybrids. (Family Rubiaceae).

*Synonyms*. Arabica, Arabian, Colombian, or Santos coffee (*C. arabica*); and robusta coffee (*C. robusta*).

### GENERAL DESCRIPTION

Evergreen shrubs to small trees with two-seeded, deeply crimson fruits (berries) that are commonly called "cherries;" *C. arabica* up to 6 m and *C. canephora* to 8 m high; believed to be native to Ethiopia; now extensively cultivated in tropical and subtropical countries. Parts used are the roasted seeds, commonly called "beans."

During the production of coffee beans, the freshly picked ripe cherries are either sun dried (requiring 2–3 weeks) followed by mechanical removal of the dried husk (pulp, skin, etc.) and seed coat, or they are placed in water and subjected to pulping machines to remove most of the pulp, followed by fermentation (requiring up to several days), drying, and mechanical removal of the silver skin. The former method is called the dry process, and the latter is called the wet process, producing respectively the so-called natural and washed coffees. The dried beans at this stage are known as green coffee and are exported.[1,2]

Arabica coffee is produced mostly in South and Central America, particularly Brazil, Colombia, Mexico, and Guatemala, while robusta coffee is produced mainly by African countries (Ivory Coast, Uganda, Angola, etc.).[1,2]

In the United States, Colombian and Central American coffees are preferred over Brazilian and African coffees.

To develop the characteristic coffee aroma and taste, the green coffee is roasted to the required time at temperatures up to about 220°C, depending on the types of coffee beans to be produced. During roasting, the beans acquire their typical flavor, at the same time turning dark and slightly increasing in size (swelling) as well as losing weight (due to loss of moisture, carbon dioxide, and other volatile compounds from pyrolysis). Coffee beans are often blended before or after roasting to produce various commercial grades or brands.[1,2]

Decaffeinated coffee is produced by removing most of its caffeine content while at the green coffee stage, generally by extraction of the whole beans with organic (e.g., chlorinated) solvents. The beans are then rid of solvent and roasted.[1,2]

Instant coffee is produced by extracting ground roasted coffee with hot water, often under pressure. The extract is concentrated and freeze-dried or spray-dried to produce a granular or powdered product. For instant coffee manufacture, robusta coffees are more commonly used due probably to their higher contents of soluble materials and thus giving higher yields.[1,2]

Coffee extracts for flavoring purposes are prepared by extracting roasted coffee with water or water–alcohol mixtures.[1,2]

## CHEMICAL COMPOSITION

Green coffee contains 0.6–3.2% (usually 1.5–2.5%) caffeine; 0.3–1.3% trigonelline; 5–10% chlorogenic acid (robusta more than arabica);[3] 7.4–17% oil called coffee oil (arabica more than robusta);[1,2,4] up to 60% carbohydrates (mostly a galactomannan);[5-7] protein (ca. 12%); about 2% free amino acids consisting mainly of glutamic and aspartic acids and asparagine;[8] polyamines (putrescine, spermine, and spermidine);[9] tannins (ca. 9%); B vitamins and trace of niacin; and others (LIST AND HÖRHAMMER, MERCK, MORTON 3, WATT AND MERRILL).[1,2,10]

Coffee oil contains mainly glycerides of fatty acids (e.g., linoleic, palmitic, oleic, and stearic acids, with the first two in predominant concentrations) and 5–8% of unsaponifiable matter, which consists of squalene, *n*-nonacosane, lanosterol, cafestol, cahweol, sitosterol, stigmasterol, methylsterols, tocopherols ($\alpha$-, $\beta$-, and $\gamma$-, with the last two being predominant), and others (LIST AND HÖRHAMMER).[4,11-13]

Roasted coffee contains slightly less caffeine than green coffee, but contains much lower concentrations of trigonelline, chlorogenic acid, tannins, polyamines, proteins, and sugars, which are degraded and involved in flavor formation during roasting. The most important flavor precursors are reported to be trigonelline, sugars, free amino acids, and peptides.[1,9,10,14]

More than 100 aroma compounds have been identified in roasted coffee, including such important flavor contributors as furan derivatives, pyrazines, pyrroles, oxazoles, and acids (LIST AND HÖRHAMMER).[1,10]

Roasted coffee contains a relatively high content of niacin, and coffee has been suggested as a source of niacin, for treating pellagra or niacin deficiency.[10,15] A cup of coffee contains about 100 mg caffeine, which is within the therapeutic dose range.

## PHARMACOLOGY OR BIOLOGICAL ACTIVITIES

The physiological activities of coffee are generally attributed to its caffeine. Caffeine is a powerful stimulant of the central nervous system, respiration, and skeletal muscles; other activities include cardiac stimulation, coronary dilation, smooth muscle relaxation, and diuresis.

The fatal dose of caffeine in humans is reported to be 10 g. A dose of 1 g or more would produce toxic effects, including headache, nausea, insomnia, restlessness, excitement, mild delirium, muscle tremor, tachycardia, and extrasystoles.

In addition to the above well-known activities, caffeine has been reported to have many other activities, including mutagenic, teratogenic, and carcinogenic activities; it is also reported to cause temporary increase in intraocular pressure, to have calming effects on hyperkinetic children (effect similar to methyl phenidate or dextroamphetamine), and to cause chronic recurrent headache, among others.[16,17] More than 50% of the total mutagenic activity of coffee can be attributed to the activity of methylglyoxal.[18] Coffee drinking has also been linked to myocardial infarction (a kind of blood clot in blood vessels that supply blood to heart muscles), cancer of the lower urinary tract (e.g., bladder), ovaries, prostate, and others.[19,20] However, most of these findings are disputed by later reports (GOODMAN AND GILMAN, MARTINDALE, USD 23rd).[15,21,22]

Mutagenic activity of coffee is inactivated by sodium sulfite (completely suppressing the mutagenicities of the 1,2 dicarbonyls, diacetyl and glyoxal) as well as by sodium bisulfite and metabisulfite. Sodium sulfite also inactivates the phage-inducing activity of coffee. It has been suggested that sulfites be added to coffee to reduce mutagenicity.[23]

Coffee (even decaffeinated) is reported to stimulate gastric secretion and should be taken only with proper precautions (e.g., with cream or during meals) by individuals with peptic ulcer (GOODMAN AND GILMAN, MARTINDALE).

Apart from the biological activities of caffeine, those of chlorogenic acid (which is present in substantial quantities in coffee) should not be ignored, as chlorogenic acid is reported to have stimulant, diuretic, and choleretic properties (see ***artichoke*** and ***honeysuckle***); it also has allergenic properties (MORTON 3).

## USES

**Medicinal, Pharmaceutical, and Cosmetic.** In addition to its use as a central and respiratory stimulant, usually as caffeine, U.S.P., and caffeine and sodium benzoate injection, U.S.P., caffeine is extensively used as an ingredient in many types of pharmaceutical preparations, particularly internal analgesics, cold and allergy products, weight-control formulations (appetite depressants), and others.

**Food.** Coffee has been used for centuries by various cultures as a beverage to stay alert and to improve work efficiency.

Coffee extract (type not specified) is widely used as a flavor ingredient in many food products, including alcoholic (e.g., liqueurs) and nonalcoholic beverages, frozen dairy desserts, candy, baked goods, gelatins and puddings, sweet sauces, and milk products. Highest average maximum use level reported is about 2.8% (28,216 ppm) in baked goods.

Caffeine is extensively used in nonalcoholic beverages (particularly colas), with reported average maximum use level of about 0.014% (141 ppm). It is also used in frozen dairy desserts, candy, gelatins and puddings, and baked goods. Average maximum use level reported is 0.04% in all except the last category which is about 0.007% (68 ppm).

## COMMERCIAL PREPARATIONS

Extracts (e.g., fluid, solid, and tincture) and natural caffeine. Caffeine is official in U.S.P. and F.C.C.

*Regulatory Status*. Coffee extracts are GRAS (§182.20), but the GRAS status of caffeine (§182.1180) is being reassessed.[17] Coffee charcoal, consisting of the milled, roasted to blackened, carbonized outer parts of green dried fruits is the subject of a German therapeutic monograph for treatment of nonspecific acute diarrhea.[24]

## REFERENCES

See the General References for APhA; BAILEY 1; BLUMENTHAL; FEMA; GOSSELIN; LIST AND HÖRHAMMER; MARTINDALE; TERRELL; USDA; YOUNGKEN.

1. R. G. Moores and A. Stefanucci in A. Standen, ed., *Kirk-Othmer Encyclopedia of Chemical Technology*, Vol. 5, Wiley-Interscience, New York, 1964, p. 748.
2. R. J. Clarke and J. W. Drummond in L. W. Codd et al., eds., *Chemical Technology: An Encyclopedic Treatment*, Vol. 7, Barnes & Noble, New York, 1975, p. 645.
3. H. Vilar and L. A. B. Ferreira, *Coll. Int. Chim. Cafes, (C.R.)* **6**, 135 (1974);

through *Chem. Abstr.*, **83**, 204919e (1975).
4. P. Folstar et al., *J. Agr. Food Chem.*, **25**, 283 (1977).
5. V. Ara and H. Thaler, *Z. Lebensm. Unters. Forsch.*, **161**, 143 (1976).
6. M. Asante and H. Thaler, *Chem. Mikrobiol., Technol. Lebensm.*, **4**, 110 (1975).
7. V. Ara and H. Thaler, *Z. Lebensm. Unters. Forsch.*, **164**, 8 (1977).
8. W. Walter et al., *Naturwissenschaften*, **57**, 246 (1970).
9. H. V. Amorim et al., *J. Agr. Food Chem.*, **25**, 957 (1977).
10. D. Reymond, *Chemtech*, **7**, 664 (1977).
11. E. Cerma and P. Baradel, *Atti Cong. Qual.*, **6**, 321 (1967); through *Chem. Abstr.*, **72**, 77600q (1970).
12. J. Wurziger, *Fette, Seifen, Anstrichmit.*, **79**, 334 (1977).
13. B. A. Nagasampagi et al., *Phytochemistry*, **10**, 1101 (1971).
14. R. Viani and I. Horman, *J. Food Sci.*, **39**, 1216 (1974).
15. G. Czok, *Z. Ernaehrungswiss.*, **16**, 248 (1977); through *Chem. Abstr.*, **88**, 103571h (1978).
16. N. Loprieno et al., *Mutat. Res.*, **21**, 275 (1973).
17. M. T. O'Brien, *Food Prod. Dev.*, **12**(9), 86 (1978).
18. H. Kasai et al., *Gann*, **73**, 381 (1982).
19. P. Stocks, *Br. J. Cancer*, **24**, 215 (1970).
20. D. H. Shennon, *Br. J. Cancer*, **28**, 473 (1973).
21. D. Simon et al., *J. Natl. Cancer Inst.*, **54**, 587 (1975).
22. H.-P. Wuerzner et al., *Food Cosmet. Toxicol.*, **15**, 289 (1977).
23. Y. Suwa et al., *Mutat. Res*, **102**, 383 (1982).
24. Monograph *Coffea carbo*, *Bundesanzeiger*, no. 85 (May 5, 1988).

# COMFREY

**Source: *Symphytum officinale*** L.; ***Symphytum* × *uplandicum*** Nym.; ***S. asperum*** Lepechin. (Family Boraginaceae).

*Synonyms*. Common comfrey, Russian comfrey (*S.* × *uplandicum*), prickly comfrey (*S. asperum*) and blackwort.

## GENERAL DESCRIPTION

Perennial herbs with branching stems and thick root. Parts used are the dried rhizome and root; also the leaves.

Common comfrey (*S. officinale*) is an erect, stout, often branched perennial, to 1 m; leaves broadly lance shaped; middle and upper ones sessile, but at point of insertion extend downward on stalk; stalk distinctly winged; flowers variable from white or cream, yellowish to rose, pink, or light to dark violet; anther about as wide as filament; petal lobes recurved; calyx segments distinctly lanceolate; nutlets smooth; occurs in moist grasslands and riverbanks most of Europe, is rare in extreme south, naturalized alien in northern Europe; eastern North America.

Russian comfrey (*S.* × *uplandicum*), a hybrid of *S. officinale* and *S. asperum*, robust perennial, 1–2 m. Intermediate between parents. Leaves narrow winged on the main stalk, ending between internodes; flowers dark violet to blue, or pinkish to pink-blue. Native to northern Europe, mostly persisted after cultivation.

Prickly comfrey, (*S. asperum*), stems not winged, upper leaves on short stalks; flowers rose to bluish; anthers significantly shorter than filaments; from Southwest

Asia; naturalized in Europe, and eastern United States (rare).

## CHEMICAL COMPOSITION

*S. officinale* root contains 0.75–2.55% allantoin;[1–3] about 0.3% alkaloids, including the pyrrolizidine alkaloids symphytine, echimidine, heliosupine, viridiflorine, echinatine, 7-acetyllycopsamine, 7-angelylretronecine viridiflorate, lasiocarpine, and acetylechimidine;[3–5] the presence of lasiocarpine is questioned;[6] lithospermic acid;[7] 29% mucopolysaccharide that is composed of glucose and fructose;[8] a gum consisting of L(−)-xylose, L-rhamnose, L-arabinose, D-mannose, and D-glucuronic acid;[9] pyrocatechol tannins (2.4%); 0.63% carotene; glycosides, sugars; isobauerenol, β-sitosterol, and stigmasterol; steroidal saponins; triterpenoids; rosmarinic acid, and others.[1,4,10,11]

*S. asperum* contains the pyrrolizidine alkaloids asperumine, echinatine, heliosupine, 7-acetyllycopsamine, and acetylechimidine.[5] *S.* × *uplandicum* contains symphytine, symlandine, echimidine; 7-acetyllycopsamine, 7-angelylintermidine, uplandicine, lycopsamine, and intermedine.[5]

Leaves also contain substantial quantities of allantoin,[2,3] alkaloids (ca. 0.15%),[12] and possibly other similar constituents as the root.

## PHARMACOLOGY OR BIOLOGICAL ACTIVITIES

Allantoin is reported to have healing properties (MARTINDALE).

An aqueous extract of comfrey containing lithospermic acid and other common plant acids has been reported to exhibit antigonadotropic activity in mice,[13] though lithospermic acid itself has no such activity unless oxidized by a plant phenol oxidase preparation.[7]

Comfrey is reported to have antiinflammatory properties.[14] It is also considered to have many beneficial properties, including astringent, demulcent, emollient, hemostatic, and expectorant properties, among others.

Antiinflammatory efficacy has been confirmed, correlating clinical and analytic data of topical comfrey products by measuring redness and pain sensitivity *in vivo*, to the allantoin and rosmarinic acid fractions of preparations.[10]

The root and leaves of *S. officinale* have been found to be carcinogenic in rats.[15] Pyrrolizidine alkaloids from *S.* × *uplandicum* have also been found to cause chronic hepatotoxicity.[15] Venoocclusive disease from ingestion of various comfrey species, including leaves and roots, has been clearly documented in humans in recent years.[6,16–18]

## USES

**Medicinal, Pharmaceutical, and Cosmetic.** Comfrey root and leaves and their extracts are used as ingredients in various types of cosmetic preparations such as lotions, creams, ointments, eyedrops, hair products, and others.

**Food.** Young shoot and leaves have been used as vegetables, though that use is currently discouraged because of pyrrolizidine alkaloid toxicity.

**Health Food/Herb Teas.** The root and leaves of various *Symphytum* species, have been sold and labeled as "*Symphytum officinale*."[19] Formerly widely available in teas, capsules, tablets, tinctures, extracts, etc. Many product manufacturers have withdrawn comfrey products from sale because of toxicity. Topical products, including salves, ointments, and balms, are still widely available (FOSTER).

**Traditional Medicine.** A root decoction is reportedly used as a gargle or mouthwash for throat inflammations, hoarseness, and bleeding gums. The root, in one form or another, is also used to treat a wide variety of ailments such as gastrointestinal problems (e.g., ulcers), excessive menstrual flow, diarrhea, dysentery, bloody urine, persistent cough, bronchitis, cancers, and others.[20]

Externally, the powdered root is used as a hemostatic and in poulticing wounds, bruises, sores, and insect bites. The mucilage is believed to help soften the skin when used in baths.

**COMMERCIAL PREPARATIONS**

Mainly as crude; no uniform standards in extracts (see "Strength" in ***glossary***).

*Regulatory Status.* Unclear in the United States. Leaf and root subjects of German therapeutic monograph for external use on bruises and sprains. Daily applied dose not to exceed 100 $\mu$g of pyrrolizidine alkaloids.[21,22]

**REFERENCES**


See the General References for BAILEY 1; BLUMENTHAL; FOSTER; GOSSELIN; LUST; MARTINDALE; ROSE; UPHOF.

1. G. V. Makarova et al., *Farm. Zh. (Kiev),* **21**(5), 41 (1966); through *Chem. Abstr.*, **66**, 49229h (1967).
2. D. Fijalkowski and M. Seroczynska, *Herba Pol.*, **23**, 47 (1977); through *Chem. Abstr.*, **88**, 119979h (1978).
3. T. Furuya and K. Araki, *Chem. Pharm. Bull.*, **16**, 2512 (1968).
4. T. Furuya and M. Hikichi, *Phytochemistry*, **10**, 2217 (1971).
5. L. W. Smith and C. C. J. Culvenor, *J. Nat. Prod.*, **44**, 2, 29 (1981).
6. D. V. C. Awang, *Can. Pharm. J.*, 101 (Feb. 1987).
7. H. Wagner et al., *Arzneim. Forsch.*, **20**, 705 (1970).
8. G. Franz, *Planta Med.*, **17**, 217 (1969).
9. Z. Michalska and T. Jakimowicz, *Farm. Pol.*, **25**, 185 (1969); through *Chem. Abstr.*, **71**, 88407n (1969).
10. R. Andres et al., *Planta Med.*, **55**, 643 (1989).
11. V. U. Ahmad et al., *J. Nat. Prod.*, **56**, 3, 329 (1993).
12. I. V. I. Man'ko et al., *Rast. Resur.*, **5**, 508 (1969); through *Chem. Abstr.*, **72**, 87175u (1970).
13. I. S. Kozhina et al., *Rast. Resur.*, **6**, 345 (1970); through *Chem. Abstr.*, **74**, 72812s (1971).
14. G. Furnadzhiev et al., *Stomatologiya (Sofia),* **58**, 37 (1976); through *Chem. Abstr.*, **85**, 13897c (1976).
15. I. Hirono et al., *J. Natl. Cancer Inst.*, **61**(3), 865 (1978).
16. P. M. Ridker et al., *Gastroenterology*, **88**, 1050 (1985).
17. R. Huxtable et al., *N. Engl. J. Med.*, **315**, 1095 (1986).
18. P. M. Ridker, *Lancet*, **1**, 657 (1989).
19. R. Huxtable, *Am. J. Med*, **89**, 548 (1990).
20. J. L. Hartwell, *Lloydia*, **31**, 71 (1968).
21. Monograph *Symphyti herba-folium*, *Bundesanzeiger*, no. 138 (July 27, 1990).
22. Monograph *Symphyti radix*, *Bundesanzeiger*, no. 138 (July 27, 1990).

# CORIANDER

**Source:** ***Coriandrum sativum*** L. (Family Umbelliferae or Apiaceae).

*Synonyms*. Cilantro and Chinese parsley.

## GENERAL DESCRIPTION

Strong-smelling annual herb with erect hollow stem, up to about 1 m high; native to Europe and Western Asia; naturalized in North America; widely cultivated. Parts used are the dried ripe fruits (commonly called coriander seeds) and leaves (both fresh and dried). An essential oil is obtained by steam distillation of the crushed fruits; it is mainly produced in Europe.

## CHEMICAL COMPOSITION

Fruits contain 0.2–2.6% (usually 0.4–1.0%) volatile oil. The major component of the oil is *d*-linalool (coriandrol), which is present in 55–74%, depending on the ripeness of the fruits, geographical locations, and other factors.[1–4] Other compounds present in the oil include decyl aldehyde, *trans*-tridecene-(2)-al-(1), borneol, geraniol, geranyl acetate, camphor, carvone, anethole, caryophyllene oxide, elemol, and monoterpene hydrocarbons (mainly γ-terpinene, and α- and β-pinene, *d*-limonene, *p*-cymene, β-phellandrene, and camphene, with relative proportions varying considerably with sources).[2,3,5–11]

Other constituents present in fruits include up to 26% fats made up of glycerides (primarily of oleic, petroselinic and linolenic acids), a small amount of unsaponifiable matter (containing β-sitosterol, δ-sitosterol, triacontane, triacontanol, tricosanol, etc.), and $\Delta^{5,6}$-octadecenoic acid; proteins (11–17%); about 1.0% starch and 20% sugars; coumarins (psoralen, angelicin, scopoletin, umbelliferone, etc.); flavonoid glycosides, including quercetin-3-glucuronide, isoquercitrin, coriandrinol (β-sitosterol-D-glucoside), and rutin; tannins; chlorogenic and caffeic acids; and others (JIANGSU, LIST AND HÖRHAMMER, WATT AND MERRILL).[4,8,12–16]

Leaves contain less volatile oil than fruits; about 5% fats; about 22% proteins; sugars; coumarins and flavonoid glycosides similar to those in fruits; chlorogenic and caffeic acids; vitamin C; and others. The volatile oil contains mainly decyl and nonyl aldehydes, and linalool, among others (JIANGSU, WATT AND MERRILL).[15,16]

Coriander has been reported to have strong lipolytic activity.[17] Its petroleum ether-soluble fraction is reported to have antioxidative activity when mixed with lard.[18]

## PHARMACOLOGY OR BIOLOGICAL ACTIVITIES

Coriander possesses hypoglycemic activities in experimental animals.[19]

Coriander oil is reported to have larvicidal properties (see ***allspice*** and ***clove***) as well as bactericidal and weakly cytotoxic activities.[20,21]

A liquid carbon dioxide extract of coriander seeds has been reported to exhibit antibacterial and antifungal activities.[22]

Coriander oil when tested at a concentration of 6% in petrolatum on human subjects (25 per test) did not produce skin irritation or sensitization reactions.[23]

An aqueous extract of fresh coriander seeds produced a dose-dependent significant antiimplantation effect in rats (related to a significant decrease in serum progesterone levels after 5 days).[24]

## USES

**Medicinal, Pharmaceutical, and Cosmetic.** Oil is used mainly as a flavoring agent in pharmaceutical preparations (e.g., Aromatic Cascara Sagrada Fluid extract); fruits are used as aromatic and carminative and in laxative preparations to prevent griping (MARTINDALE).

In cosmetics, oil is used as a fragrance component in soaps, creams, lotions, and perfumes, with maximum use level of 0.6% in perfumes.[23]

Oil is also used in flavoring tobacco.

**Food.** The young leaves are widely used as a garnish in cooking (e.g., Chinese, Armenian, Spanish, etc.); they are known as Chinese parsley in Chinese cuisine and cilantro in Spanish cooking.

The seeds (fruits) and oil are extensively used as flavor ingredients in all types of food products, including alcoholic (vermouths, bitters, gin, etc.) and nonalcoholic beverages, frozen dairy desserts, candy, baked goods, gelatins and puddings, meat and meat products, condiments and relishes, and others. Highest average maximum use levels reported for seeds and oil were 0.52 and 0.012%, respectively, in meat and meat products and in alcoholic beverages.

**Health Food/Herb Teas.** Fruits sometimes used in carminative and digestive products (FOSTER).

**Traditional Medicine.** Fruits are used as an aromatic carminative, stomachic, and antispasmodic, usually in the form of an infusion.

In Chinese medicine, in addition to being used as a stomachic, they are used in measles, dysentery, hemorrhoids, and other ailments; a decoction is also used as a gargle to relieve toothache. The whole herb is also used in stomachache, nausea, measles, and painful hernia (JIANGSU).

## COMMERCIAL PREPARATIONS

Crude and oil; crude was formerly official in N.F., and oil is official in N.F. and F.C.C.

*Regulatory Status.* GRAS (§182.10 and §182.20). Fruits subject of a German therapeutic monograph indicated for dyspeptic complaints and loss of appetite.[25]

## REFERENCES


See the General References for ARCTANDER; BLUMENTHAL; FEMA; FOGARTY; FOSTER; JIANGSU; LUST; MASADA; NANJING; ROSENGARTEN.

1. E. Gliozheni, *Bul. Shkencave Nat., Univ. Shteteror Tiranes*, **28**(3), 41 (1974); through *Chem. Abstr.*, **83**, 15477y (1975).
2. G. Jukneviciene et al., *Liet. TSR Mokslu Akad. Darb., Ser. C*, **3**, 9 (1977); through *Chem. Abstr.*, **88**, 3090g (1978).
3. N. N. Glushchenko et al., *Maslo-Zhir. Prom.*, **6**, 28 (1977); through *Chem. Abstr.*, **87**, 73220c (1977).
4. H. Karow, *Riechst., Aromen, Körperpflegem.*, **19**(2), 60 (1969).
5. S. K. Chogovadze and D. M. Bakhtadze, *Lebensm. Ind.*, **24**, 513 (1977).
6. E. Schratz and S. M. J. S. Qadry, *Planta Med.*, **14**, 310 (1966).
7. S. Rasmussen et al., *Medd Nor. Farm. Selsk.*, **34**(3–4), 33 (1972); through *Chem. Abstr.*, **80**, 124564h (1974).
8. G. K. Gupta et al., *Indian Perfum.*, **21**, 86 (1977).
9. R. M. Ikeda et al., *J. Food Sci.*, **27**, 455 (1962).
10. J. S. T. Chou, *Koryo*, **106**, 55 (1974); through *Chem. Abstr.*, **81**, 82235q (1974).
11. J. Taskinen and L. Nykanen, *Acta Chem. Scand., Ser. B*, **29**, 425 (1975).

12. G. A. Stepanenko et al., *Khim. Prir. Soedin.*, **10**, 37 (1974); through *Chem. Abstr.*, **81**, 54277m (1974).
13. L. T. Lee and M. W. Wah, *Hua Hsueh*, **2**, 52 (1973); through *Chem. Abstr.*, **80**, 45638g (1974).
14. A. R. S. Kartha and Y. Selvaraj, *Chem. Ind. (London)*, **25**, 831 (1970).
15. N. V. Sergeeva, *Khim. Prir. Soedin.*, **10**, 94 (1974); through *Chem. Abstr.*, **81**, 60801w (1974).
16. J. Kunzemann and K. Herrmann, *Z. Lebensm. Unters. Forsch.*, **164**, 194 (1977).
17. G. Paulet et al., *Rev. Fr. Corps Gras.*, **21**, 415 (1974); through *Chem. Abstr.*, **82**, 2691t (1975).
18. Y. Saito et al., *Eiyo To Shokuryo*, **29**, 505 (1976); through *Chem. Abstr.*, **87**, 150314r (1977).
19. N. R. Farnsworth and A. B. Segelman, *Tile Till*, **57**, 52 (1971).
20. K. K. Abdullin, *Uch. Zap. Kazansk. Vet. Inst.*, **84**, 75 (1962); through *Chem. Abstr.*, **60**, 11843b (1964).
21. K. Silyanovska et al., *Parfüm. Kosmet.*, **50**, 293 (1969).
22. M. L. Khanin et al., *Khim. Farm. Zh.*, **2**, 40 (1968); through *Chem. Abstr.*, **69**, 944k (1968).
23. D. L. J. Opdyke, *Food Cosmet. Toxicol.*, **11**, 1077 (1973).
24. M. S. Al-Said et al, *J. Ethnopharmacol*, **21**, 165 (1987).
25. Monograph *Coriandri fructus*, *Bundesanzeiger*, no. 173 (Sept. 18, 1986).

# CORN SILK

**Source:** ***Zea mays*** L. subsp. ***mays*** (Family Poaceae or Gramineae).

*Synonyms*. Stigmata maydis and zea.

## GENERAL DESCRIPTION

Coarse erect annual with prop (adventitious) roots near the ground and long sword-shaped leaves, one at each node; up to 4 m high; generally thought to be a native of tropical America. Parts used are the long styles and stigmata of the pistils called corn silk; the dried product is normally used.

## CHEMICAL COMPOSITION

Corn silk is reported to contain 2.5% fats, 0.12% volatile oil, 3.8% gums, 2.7% resin, 1.15% bitter glucosidic substances, 3.18% saponins, 0.05% allaloids, cryptoxanthin, vitamins C and K, sitosterol, stigmasterol, plant acids (malic, tartaric, etc.), anthocyanins, and others (JIANGSU).[1,2]

## PHARMACOLOGY OR BIOLOGICAL ACTIVITIES

Corn silk has diuretic, hypoglycemic, and hypotensive activities in experimental animals. A dialyzed methanol-insoluble fraction of its aqueous extract has been demonstrated to be strongly diuretic in humans and rabbits; the toxicity of this fraction was low when compared with its effective dose. Lethal intravenous dose in rabbits was 250 mg/kg, while effective dose was 1.5 mg/kg (JIANGSU).

A crystalline constituent from an aqueous extract of corn silk has also been reported to be hypotensive and to stimulate uterine contraction in rabbits.[3]

Corn silk and its aqueous extracts are also reported to be effective in kidney and other diseases during clinical trials (JIANGSU).

USES

**Medicinal, Pharmaceutical, and Cosmetic.** Both crude and extracts are used as an ingredient in certain diuretic preparations; crude is also used in face powders, among others.

**Food.** Extracts are used as flavor components in major food products such as alcoholic and nonalcoholic beverages, frozen dairy desserts, candy, baked goods, and others. Use levels are generally lower than 0.002%.

**Traditional Medicine.** Used as a diuretic in urinary problems (cystitis, pyelitis, etc.); also as a demulcent.

In Chinese medicine, in addition to being used as a diuretic in dropsy, corn silk is used to treat sugar diabetes (diabetes mellitus) in the form of a decoction and to treat hypertension when decocted with watermelon peel and banana, as well as other ailments.

COMMERCIAL PREPARATIONS

Crude and extracts; extracts come in varying strengths (see ***glossary***), with those for food use expressed in flavor intensities and those for pharmaceutical applications expressed in weight-to-weight ratios. Crude and fluid extract were formerly official in N.F.

*Regulatory Status.* GRAS (§182.20).

REFERENCES

See the General References for APhA; FEMA; GOSSELIN; JIANGSU; LUST; ROSE; TERRELL; UPHOF.

1. N. E. Bobryshev, *Kukuruza*, **9**, 59 (1962); through *Chem. Abstr.*, **61**, 1117h (1964).
2. E. D. Styles and O. Ceska, *Phytochemistry*, **14**, 413 (1975).
3. S. J. Hahn, *K'at'ollik Taehak Uihakpu Nonmunjip*, **25**, 127 (1973); through *Chem. Abstr.*, **80**, 116288f (1974).

# COSTUS OIL

**Source:** ***Saussurea lappa*** Clarke (syn. *Aucklandia costus* Falc.) (Family Compositae or Asteraceae).

GENERAL DESCRIPTION

Large erect perennial herb with a thick taproot; up to about 2 m high; native to the mountains of northern India (the Himalayas); cultivated in India and southwestern China. Part used is the dried root, from which a volatile oil is obtained by steam distillation followed by solvent extraction of the distilled water. India is the major producer of the oil.

CHEMICAL COMPOSITION

Root contains 0.3–3% volatile oil; saussurine (an alkaloid); betulin; stigamsterol; about 18% inulin; and resins (JIANGSU, NANJING, WILLAMAN AND SCHUBERT).

The major components in the oil are sesquiterpene lactones, including the crystalline dehydrocostus lactone and costunolide, which together make up about 50% of the oil; $\alpha$- and $\beta$-cyclocostunolide; alantolactone; isoalantolactone; dihydrodehydrocostus lactone; cynaropicrin, and others.[1–6] Also present are other sesquiterpenes such as β-costol, elema-1,3,11(13)-trien-12-ol, α-costol, γ-costol, β-selinene, β-elemene, elemol, caryophyllene, caryophyllene oxide, *ar*-curcumene, α-selinene, α-costal, β-costal, γ-costal, and others, with their concentrations in decreasing order and the first four accounting for about 18% of commercial root oil;[7,8] (*Z*,*Z*,*Z*)-1,8,11,14-heptadecatetraene;[9,10] an unusual terpenoid $C_{14}$-ketone ((*E*)-9-isopropyl-6-methyl-5,9-decadien-2-one);[11] α- and β-ionones; dihydro-α-ionone; (*E*)-geranylacetone; aplotaxene and dihydroaplotaxene; 3,9,11-guaiatriene-12-carboxylic acid;[12] costic acid; palmitic, linoleic, and oleic acids; friedelin; β-sitosterol; and others (JIANGSU, MASADA).[2–4,7]

The sesquiterpene lactones (especially alantolactone, dehydrocostus lactone, and costunolide) have plant growth–regulating activities.[1,13]

## PHARMACOLOGY OR BIOLOGICAL ACTIVITIES

Various fractions of costus oil have been reported to have hypotensive activities in anesthetized dogs, with 12-methoxydihydrocostunolide and the delactonized oil being the most potent, acting through direct peripheral vasodilation and cardiac depression. Most fractions were also effective in relieving bronchial spasm induced by histamine and acetylcholine in guinea pigs, but none had antitussive activity.[5]

Costus and its derivatives (e.g., absolute used in perfumes) are known to cause allergic reactions (e.g., contact dermatitis) in humans.[13–15]

A decoction of costus root is reported to exhibit weak inhibitory activities on paratyphoid A bacterium and certain other pathogenic bacteria (JIANGSU).

## USES

**Medicinal, Pharmaceutical, and Cosmetic.** Costus and its derivatives (essential oil, absolute, and concrète) are used as fixatives and fragrance components in creams, lotions, and perfumes (e.g., Oriental types); reported maximum use level is 0.4% (no specific product form given) in perfumes.[14]

**Food.** Oil is used as a flavor component in most major food products, including alcoholic and nonalcoholic beverages, frozen dairy desserts, candy, baked goods, gelatins and puddings, and confectioner's frosting. Use levels are low, with highest average maximum of about 0.0004% (4.04 and 4.16 ppm) reported for alcoholic beverages and baked goods.

**Traditional Medicine.** Root has been used for millennia in China and India as a tonic, stomachic, carminative, and stimulant and in treating asthma, cough, dysentery, and cholera, among others; also used in incense.

## COMMERCIAL PREPARATIONS

Mainly oil; it is official in F.C.C.

*Regulatory Status.* Has been approved for food use (§172.510.)

## REFERENCES

See the General References for ARCTANDER; FEMA; FOGARTY; GUENTHER; JIANGSU; NANJING.

1. P. S. Kalsi et al., *Phytochemistry*, **16**, 784 (1977).
2. S. B. Mathur, *Phytochemistry*, **11**, 449 (1972).
3. S. V. Govindan and S.C. Bhattacharyya, *Indian J. Chem., Sect. B*, **15**, 956 (1977).
4. S. B. Mathur and S. C. Bhattacharyya,

*Int. Cong. Essent. Oils (Pap.)*, **6**, 126 (1974).
5. O. P. Gupta and B. J. R. Ghatak, *Indian J. Med. Res.*, **55**, 1078 (1967).
6. F. Bohlmann et al., *Planta Med.*, **51**, 74 (1985).
7. B. Maurer and A. Grieder, *Helv. Chim. Acta*, **60**, 2177 (1977).
8. A. S. Bawdekar et al., *Tetrahedron*, **23**, 1993 (1967).
9. R. G. Binder et al., *Phytochemistry*, **14**, 2085 (1975).
10. M. Romanuk et al., *Collect. Czech. Chem. Commun.*, **24**, 2018 (1959).
11. B. Maurer and G. Ohloff, *Helv. Chim. Acta*, **60**, 2191 (1977).
12. E. Klein and F. Thoemel, *Tetrahedron*, **32**, 163 (1976).
13. E. Rodriguez et al., *Phytochemistry*, **15**, 1573 (1976).
14. D. L. J. Opdyke, *Food Cosmet. Toxicol.*, **12**(Suppl.), 867 (1974).
15. J. C. Mitchell, *Arch. Dermatol.*, **109**, 572 (1974).

# CRANBERRY

**Source:** ***Vaccinium macrocarpon*** Aiton (Family Ericaceae).

*Synonym.* Low-bush cranberry.

### GENERAL DESCRIPTION

Trailing, evergreen, slender-stemmed shrub; leaves leathery, flower white to pink, 1 cm wide; fruit glossy red, 1–1.5 cm wide, tart flavored; occurs in bogs from Newfoundland to Manitoba south to Virginia, Ohio and northern Illinois; also grown in northern and central Europe; locally naturalized. Part used is the fruit from which a juice (cranberry juice) is produced and extensively marketed in the United States; Massachusetts is the major producer. *Vaccinium oxycoccus* L., found in peat bogs in northern and central Europe used as a cranberry source in Europe. *V. macrocarpon* should not be confused with *Viburnum opulus* L., sometimes known as "high-bush cranberry."

### CHEMICAL COMPOSITION

Fruits contain anthocyanins; flavonol glycosides (leptosine); catechin; triterpenoids; citric, malic and quinic acids; smaller levels of benzoic and glucuronic acids; trace of alkaloids; carbohydrates (10%), protein, and vitamin C.[1] Quinic acid 0.5–0.9%, conjugated with glycine to produce hippuric acid, to which biological activity was formerly attributed.

### PHARMACOLOGY OR BIOLOGICAL ACTIVITIES

Fruit widely regarded as possessing bacteriostatic activity for urinary tract infections. Antibacterial activity has been variously, though not conclusively attributed to anthocyanins; flavonol glycosides; catechin; volatile components; and benzoic, quinic, malic, and citric acids (CRELLIN AND PHILPOTT). It has been suggested and disputed that the urinary antiseptic effect is due to the action of hippuric acid.[2,3] Recent studies suggest cranberry juice possesses antiadhesion activity to mucous membrane surfaces.[4–6] *Escherichia coli*, the most common bacteria producing urinary tract infections adheres to cells of the urinary and alimentary tracts, enhancing their capacity to withstand nutrient deprivation and cleansing mechanisms, enhancing both toxicity and colonization. Blocking the adhesion of *E. coli* to urinary bladder mucosal cells has been shown to prevent development of urinary tract infections in mice.

Cranberry juice fructose and an uncharacterized high molecular weight polymeric compound inhibit cellular adhesion of uropathogenic strains of *E. coli*. The antiadhesive agent of cranberry juice may prevent *E. coli* colonization in the gut, in the bladder or both.[7] Dosage of juice ranges from 5–20 oz daily (6 oz are equivalent to 90 g of fresh fruit). A recent study found that drinking 4–6 oz of cranberry juice daily for 7 weeks appeared to prevent urinary tract infections in 19 of 28 nursing home patients; a preventative rather than curative effect was suggested.[8]

## USES

**Food.** Fruit juice, jelly, sauce; commonly eaten with poultry. Leaves a folk tea substitute. Cranberry juice cocktail is a 33% dilution of pure juice with added fructose. Approximately 1500 g of fresh fruit produce 1 L of juice. Anthocyanin pigment from fruit pulp used as commercial food coloring.

**Health Food/Herb Teas.** Fruit juice concentrates or dried fruit in capsules and tablets, intended for relief of urinary tract infections (LUST).

**Traditional Medicine.** Fruit juice traditionally considered diuretic, antiseptic, febrifuge, refrigerant; a home remedy for treatment of urinary tract infections; folk cancer remedy in eastern Europe (STEINMETZ).

## COMMERCIAL PREPARATIONS

Fruit juice, juice concentrates, dried fruit, juice concentrate in capsules.

## REFERENCES

See the General References for CRELLIN AND PHILPOTT; GLEASON AND CRONQUIST; LUST; STEINMETZ; TUTIN 3; UPHOF.

1. L. Liberti, *Lawrence Rev.* (Aug. 1987).
2. P. T. Bodel et al., *J. Lab. Clin. Med.*, **54**, 881 (1959).
3. P. Sternlieb, *N. Engl. J. Med.*, **268**, 57 (1963).
4. A. E. Sobata, *J. Urol.*, **131**, 1013 (1984).
5. D. R. Schmidt and A. E. Sobata, *Microbios*, **55**, 173 (1988).
6. D. Zafriri et al., *Antimicrob. Agents Chemother.*, **33**, 92 (1989).
7. I. Ofek et al., *N. Eng. J. Med.*, **324**(22), 1599 (1991).
8. L. Gibson et al., *J. Naturopathic Med.*, **2**(1), 45 (1991).

# CUBEBS

**Source:** ***Piper cubeba*** L. f. (syn. *Cubeba officinalis* Miq.) (Family Piperaceae).

*Synonyms*. Cubeba, cubeb berries, and tailed pepper.

## GENERAL DESCRIPTION

Evergreen shrub or climbing vine, up to about 6 m high, often grown with other economic plants such as coffee; native to Indonesia and cultivated throughout Southeast Asia. Part used is the dried, fully grown but unripe fruit, commonly called cubeb berry, with its attached peduncle or stem (hence also called tailed pepper); cubeb oil is produced by steam distillation of the crushed fruits, usually in Europe and in the United States.

## CHEMICAL COMPOSITION

Fruits contain 10–20% volatile oil; about 2.5% cubebin;[1] 1–1.7% amorphous cubebic acid of undetermined structure; an acidic resin and 3–3.7% neutral resins, 8% gum; fats; and others (JIANGSU, LIST AND HÖRHAMMER).

The essential oil contains mainly sesquiterpenes and monoterpenes and their alcohols, with their relative concentrations varying considerably according to different reports. Sesquiterpene hydrocarbons present include caryophyllene, cadinene, α- and β-cubebene,[2] copaene, and 1-isopropyl-4-methylene-7-methyl-1,2,3,6,7,8,9-heptahydronaphthalein; monoterpene hydrocarbons including sabinene, α-thujene, β-phellandrene, α-pinene, myrcene, β-pinene, α-phellandrene, γ- and α-terpinene, limomene, and ocimene, among others;[3] oxygenated terpenes include 1,4-cineole, α-terpineol, cadinol, and cubebol; and others (JIANGSU, LIST AND HÖRHAMMER, STAHL).[1,4]

## PHARMACOLOGY OR BIOLOGICAL ACTIVITIES

Fruit is reported to have local stimulant effect on mucous membranes (urinary and respiratory tract). Cubebic acid (not cubebin) is mainly responsible for its physiological properties (JIANGSU).

Ground cubeb was found 90% clinically effective in treating amebic dysentery (JIANGSU).

Cubeb oil has been reported to exhibit antiviral activities in rats as well as weak to strong antibacterial activities *in vitro*; it also has certain urinary antiseptic properties (JIANGSU, MERCK).[5]

## USES

**Medicinal, Pharmaceutical, and Cosmetic.** Berry and oil have been used in diuretic and urinary antiseptic preparations. Oil is used as a fragrance component in soaps, detergents, creams, lotions, and perfumes, with highest maximum use level of 0.8% reported in perfumes.[5] Also used in flavoring tobacco.

**Food.** Oil is used as a flavor ingredient in most major categories of food products, including alcoholic (liqueurs) and nonalcoholic beverages, frozen dairy desserts, candy, baked goods, gelatins and puddings, meat and meat products, condiments and relishes, and others. Highest average maximum use level reported is about 0.004% (38.2 ppm) in condiments and relishes. Fruits are reportedly used only in nonalcoholic beverages at an average maximum use level of 0.085%.

**Traditional Medicine.** Used as a diuretic, urinary antiseptic, carminative, and stimulating expectorant, among others. Also used in treating gonorrhea and cancer.[6]

## COMMERCIAL PREPARATIONS

Crude and oil; crude was formerly official in N.F., and oil is official in F.C.C.

*Regulatory Status.* Has been approved for food use (§172.510).

**REFERENCES**

See the General References for ARCTANDER; FEMA; GRIEVE; GUENTHER; JIANGSU; LUST; MASADA; TERRELL.

1. C. K. Atal et al., *Lloydia*, **38**, 256 (1975).
2. Y. Ohta et al., *Tetrahedron Lett.*, **51**, 6365 (1966).
3. R. M. Ikeda, *J. Food Sci.*, **27**, 455 (1962).
4. S. J. Terhune et al., *Int. Congr. Essent. Oils (Pap.)*, **6**, 153 (1974).
5. D. L. J. Opdyke, *Food Cosmet. Toxicol.*, **14**, 729 (1976).
6. J. L. Hartwell, *Lloydia*, **33**, 288 (1970).

# CUMIN

**Source:** ***Cuminum cyminum*** L. (syn. *C. odorum* Salisb.) (Family Umbelliferae or Apiaceae).

*Synonyms.* Cummin and cumin seed.

## GENERAL DESCRIPTION

Small annual with a slender stem, much branched above; up to about 0.6 m high; native to the Mediterranean region, now extensively cultivated there (Morocco, Turkey, Greece, Egypt, etc.) and in Iran, India, and other countries. Part used is the dried ripe fruit, commonly called "seed." An essential oil (cumin oil) is obtained by steam distillation of the crushed fruit. Major cumin seed producers include Egypt, Iran, India, Morocco, Turkey, and the former U.S.S.R.; major oil producers include India and the United States.

## CHEMICAL COMPOSITION

Contains 2–5% volatile oil;[1,2] up to about 22% fats with a small amount of $\Delta^{5,6}$-octadecenoic acid;[3] 14 free amino acids, including five essential ones; about 18% protein;[4] flavonoid glycosides, including apigenin-7-glucoside (apigetrin), apigenin-7-glucuronosyl glucoside, luteolin-7-glucoside, and luteolin-7-glucuronosyl glucoside;[5] tannin; resin; gum; and others (LIST AND HÖRHAMMER, MARSH).

The volatile oil contains aldehydes (up to 60%) as its major components, which consist mainly of cuminaldehyde, 1,3-*p*-menthadien-7-al, 1,4-*p*-menthadien-7-al, and 3-*p*-menthen-7-al; commercial volatile oil and the volatile oil from previously ground commercial cumin contain more cuminaldehyde than the other aldehydes, with the absence of 1,4-*p*-menthadien-7-al, while the essential oil from freshly ground cumin contains primarily 1,4-*p*-menthadien-7-al, with cuminaldehyde in a much smaller amount and the other two aldehydes only in traces. Other major components of the oil are monoterpene hydrocarbons (up to 52%) composed mainly of $\beta$-pinene, $\gamma$-terpinene, and *p*-cymene, with $\alpha$- and $\beta$-phellandrene, myrcene, $\alpha$-terpinene, and limonene also present. Minor constituents include sesquiterpene hydrocarbons ($\beta$-farnesene, $\beta$-caryophyllene, $\beta$-bisabolene, etc.); cuminyl alcohol (believed to be an artifact, as it is present only in trace quantities in the volatile oil from freshly ground cumin); perillaldehyde; phellandral; *cis*- and *trans*-sabinene hydrate; cryptone; and others (LIST AND HÖRHAMMER, MASADA).[1,6,7]

Fine milling of cumin is reportedly responsible for up to 50% loss of its essential oil content, with the greatest loss occurring during the first hour of storage after milling.[8]

Cuminaldehyde, 1,4-*p*-menthadien-7-al, and 1,3-*p*-menthadien-7-al have indistinguishable odors; they appear to be mostly responsible for the characteristic aroma of unheated whole cumin seeds. The chief odor characteristics of heated cumin are due to 3-*p*-menthen-7-al in combination with the other three aldehydes.[2]

The petroleum ether-soluble fraction of cumin reportedly has antioxidative activity when mixed in lard.[9]

## PHARMACOLOGY OR BIOLOGICAL ACTIVITIES

Cumin oil (especially cuminaldehyde) has been reported to exhibit strong larvicidal activities (see ***cinnamon***, ***clove***, and ***coriander***);[10] it also has antibacterial properties.[11] It was rapidly absorbed through the shaved intact abdominal skin of the mouse.[12]

Undiluted cumin oil has been demonstrated to have distinct phototoxic effects which were not due to cuminaldehyde, its principal component.[13]

## USES

**Medicinal, Pharmaceutical, and Cosmetic.** Oil is used as a fragrance component in creams, lotions, and perfumes, with a maximum use level of 0.4% reported in perfumes.[13]

**Food.** Cumin is a major flavor component of curry and chili powders. It is also used in other food products, including baked goods, meat and meat products, condiments and relishes, processed vegetables, soups, gravies, snack foods, with the highest maximum use level of about 0.4% (4308 ppm) reported in soups.

The oil is used as a flavor component in alcoholic and nonalcoholic beverages, frozen dairy desserts, candy, baked goods, gelatins and desserts, meat and meat products, condiments and relishes, gravies, snack foods, and others. Highest average maximum use level is reported to be about 0.025% (247 ppm) in condiments and relishes.

**Health Food/Herb Teas.** Commonly used in specialty curry products, also as carminative and tea ingredient.

**Traditional Medicine.** Used as a stimulant, antispasmodic, carminative, diuretic, aphrodisiac, and emmenagogue, among others.

## COMMERCIAL PREPARATIONS

Crude and oil; oil is official in F.C.C.

*Regulatory Status.* GRAS (§182.10 and §182.20).

## REFERENCES

See the General References for ARCTANDER; BAILEY 1; FEMA; GRIEVE; GUENTHER; LIST AND HÖRHAMMER; MARSH; ROSENGARTEN; TERRELL.

1. H. Karow, *Riechst., Aromen, Körperpflegem.*, **19**(2), 60 (1969).
2. C. G. Tassan and G. F. Russell, *J. Food Sci.*, **40**, 1185 (1975).
3. A. R. S. Kartha and Y. Selvaraj, *Chem. Ind. (London)*, **25**, 831 (1970).
4. F. Toghrol and H. Daneshpejouh, *J. Trop. Pediatr. Environ. Child Health*, **20**, 109 (1974).

5. J. B. Harborne and C. A. Williams, *Phytochemistry*, **11**, 1741 (1972).
6. P. T. Varo and D. E. Heinz, *J. Agr. Food Chem.*, **18**, 234 (1970).
7. P. T. Varo and D. E. Heinz, *J. Agr. Food Chem.*, **18**, 239 (1970).
8. E. Georgiev and Van Hong Tam, *Nauch. Tr., Vissh Inst. Khranit. Vkusova Prom. Plovdiv*, **20**, 99 (1973); through *Chem. Abstr.*, **83**, 168312r (1975).
9. Y. Saito et al., *Eiyo To Shokuryo*, **29**, 505 (1976); through *Chem. Abstr.*, **87**, 150314r (1977).
10. K. Oishi et al., *Nippon Suisan Gakkaishi*, **40**, 1241 (1974); through *Chem. Abstr.*, **82**, 84722r (1975).
11. F. M. Ramadan et al., *Chem. Mikrobiol. Technol. Lebensm.*, **2**, 51 (1972).
12. F. Meyer and E. Meyer, *Arzneim. Forsch.*, **9**, 516 (1959).
13. D. L. J. Opdyke, *Food Cosmet. Toxicol.*, **12**(Suppl.), 869 (1974).

# DAMIANA

**Source:** ***Turnera diffusa*** Willd. (syn. *T. aphrodisiaca* L.F. Ward and *T. microphylla* Desv.) (Family Turneraceae).

## GENERAL DESCRIPTION

A shrub with small aromatic leaves (mostly 1–2 cm long), up to 2 m high; native to tropical America (Mexico, Texas, Central America, South America, and the West Indies).[1] Part used is the dried leaf.

## CHEMICAL COMPOSITION

Contains 0.5–1% volatile oil;[2,3] triacontane; hexacosanol-1; β-sitosterol; gonzalitosin I (5-hydroxy-7,3′,4′-trimethoxyflavone);[3] a cyanogenic glycoside;[4] arbutin;[5] resin; tannin; a bitter substance (called damianin) of undetermined structure; and others.[2]

The volatile oil is composed of two main fractions. A low-boiling fraction contains mainly 1,8-cineole, α- and β-pinenes, and *p*-cymene, while the higher-boiling fraction consists primarily of thymol and sesquiterpenes (α-copaene, δ-cadinene, and calamenene).[5,6] The presence of *p*-cymene and 1,8-cineole has been disputed.[3]

## PHARMACOLOGY OR BIOLOGICAL ACTIVITIES

Damiana is widely believed to have aphrodisiac properties and has also been reported to have diuretic, laxative, and stimulant properties, among others (LEWIS AND ELVIN-LEWIS, UPHOF, TYLER 1).[2]

## USES

**Food.** Used as a flavor ingredient in major food products, including alcoholic and nonalcoholic beverages, frozen dairy desserts, candy, baked goods, and gelatins and puddings. Highest maximum use level reported to be 0.125% for the crude in baked goods.

**Health Food/Herb Teas.** Various products, including capsules, tablets, and drinks; tea ingredient; primarily for alleged aphrodisiac reputation (TYLER 1).[7]

**Traditional Medicine.** Reportedly used as a laxative, nervous stimulant, tonic, and aphrodisiac; also used in coughing, nephritis, menstrual disorders, and other ailments.[2]

## COMMERCIAL PREPARATIONS

Crude and extracts; crude was formerly official in N.F. Extracts do not have uniform standards.

*Regulatory Status.* Has been approved for food use (§172.510). Subject of a German therapeutic monograph; not recommended as claimed efficacy is unproven.[8]

## REFERENCES

See the General References for BLUMENTHAL; CLAUS; LUST; TERRELL; TYLER 1; UPHOF.

1. P. C. Standley, Vol. 23, Part 3 of *Contributions from the United States National Herbarium, Trees and Shrubs of Mexico*, Smithsonian Press, Washington, D.C., 1923, p. 848.
2. E. F. Steinmetz, *Acta Phytother.*, **7**, 1 (1960).
3. X. A. Dominguez and M. Hinojosa, *Planta Med.*, **30**, 68 (1976).

4. B. Tantisewie et al., *Pharm. Weekbl.*, **104**, 1341 (1969).
5. H. Auterhoff and H. P. Häufel, *Arch. Pharm. (Weinheim)*, **301**, 537 (1968).
6. H. Auterhoff and H. Momberger, *Arch. Pharm. (Weinheim)*, **305**, 455 (1972).
7. V. E. Tyler, *Pharmacy in History*, **25**, 55, 1983.
8. Monograph *Turnerae diffusae folium et herba*, *Bundesanzeiger*, no. 43 (Mar. 2, 1989).

# DANDELION ROOT

**Source: *Taraxacum officinale*** Wiggers (syn. *T. vulgare* (Lam.) Schrank) and other ***Taraxacum*** species (Family Compositae or Asteraceae).

*Synonyms*. Common dandelion, lion's tooth, and taraxacum.

## GENERAL DESCRIPTION

*Taraxacum officinale* is a perennial herb with deeply cut leaves forming a basal rosette in the spring and flower heads borne on long stalks, up to about 45 cm high; native to Europe and naturalized in North America, occurring as a common weed on lawns. Related species are found worldwide. Parts used are the dried rhizome and root; leaves and flowers are sometimes used for direct domestic consumption.

## CHEMICAL COMPOSITION

Root contains several triterpenes, including taraxol, taraxerol, taraxasterol, $\psi$-taraxasterol, and $\beta$-amyrin; sterols (stigmasterol, $\beta$-sitosterol); inulin (ca. 25%); sugars (fructose, glucose, sucrose, etc.); pectin; glucosides; choline; phenolic acids (e.g., caffeic and *p*-hydroxyphenylacetic acids; gum; resins; vitamins; and others (CLAUS, KARRER, JIANGSU).[1]

Flowers contain carotenoids (e.g., lutein, lutein epoxide, cryptoxanthin, cryptoxanthin epoxide, flavoxanthin, chrysanthemaxanthin, and violaxanthin) and their monoesters and diesters with fatty acids (mainly myristic; also lauric, palmitic, and stearic acids);[2–4] arnidiol; and others (JIANGSU).

Leaves contain lutein, violaxanthin, and other carotenoids; bitter substances; vitamins A, B, C, and D; and others. Vitamin A content (14,000 IU/100 g) is higher than that in carrots (11,000 IU/100 g) (JIANGSU, WATT AND MERRILL).

Coumesterol (see ***alfalfa*** and ***red clover***) has been reported present in dandelion, with plant part not specified (JIANGSU).[5]

## PHARMACOLOGY OR BIOLOGICAL ACTIVITIES

Dandelion (plant part not specified) has been reported to exhibit hypoglycemic effects in experimental animals and to cause contact dermatitis in humans.[6,7]

Dandelion root is generally considered to have diuretic, choleretic, tonic and laxative properties, among others (ESCOP 3, MARTINDALE)[8]

Rodent experiments have confirmed diuretic activity of a fluid extract of the herb (8 g dried/kg body weight); greater activity than that of the root; comparable with furosemid (80 mg/kg body weight); high potassium content (4% in dried leaves) replaces that eliminated in urine (ESCOP 3).

Bitter sesquiterpene lactones in the root increase bile secretion in rats by more than 40%; and increases gastric secretions (ESCOP 3).

Traditional use for rheumatic conditions could be related to observed antiinflammatory activity.[9]

Numerous clinical studies using Chinese

*Taraxacum* species have been reported. Root, leaves, juice, and extracts were effective in treating infections of various kinds (e.g., upper respiratory infections, pneumonia, chronic bronchitis, hepatitis, etc.) with few side effects (JIANGSU).

## USES

**Medicinal, Pharmaceutical, and Cosmetic.** Extracts are quite extensively used in tonics (especially those for female ailments). Also used in diuretic, laxative, and antismoking preparations as well as in cosmetic and toiletry formulations, presumably for their tonic properties.[10] In Germany, the root is used for disturbances in bile flow, as diuretic, and as an appetite stimulant. The herb is used for appetite and dyspeptic disorders, such as abdominal fullness and flatulence. Products are contraindicated in obstruction of bile ducts, gallbladder empyema, and in gallstones, only under advice of physician.[11,12]

**Food.** Extracts are used as flavor components in various food products, including alcoholic (e.g., bitters) and nonalcoholic beverages, frozen dairy desserts, candy, baked goods, gelatins and puddings, and cheese. Highest average maximum use levels reported are about 0.014% (143 ppm) for the fluid extract in cheese and 0.003% (33.3 ppm) for the solid extract in baked goods.

The roasted root and its extract are used as coffee substitutes or in instant coffee substitute preparations.

Young leaves, particularly those of cultivated forms, are used as salad or vegetables.

Flowers are used in home wine making.

**Health Food/Herb Teas.** Root used as a flavoring ingredient in tea formulations; capsules, tincture, tablet formulations for choleretic activity; leaf in tea as flavoring, bulk filler; capsules, tablets, tinctures, etc., primarily as diuretics (FOSTER).

**Traditional Medicine.** The root is reportedly used as a laxative, tonic, and diuretic and to treat various liver and spleen ailments. Root and leaves are also used for heartburn and bruises and in treating chronic rheumatism, gout, and stiff joints as well as eczema, and other skin problems, and cancers.[13]

In China, *Taraxacum mongolicum* Hand.-Mazz. and other *Taraxacum* species have been used for more than 1100 years in treating breast cancer and other breast problems (inflammation of the mammary glands, lack of milk flow, etc.), liver diseases (e.g., hepatitis), stomach problems, and others (JIANGSU; NANJING).

**Others.** Due to its high content of inulin, juice of root can serve as source of a special high fructose syrup; a very light-colored syrup containing 71% total sugars of which 77% was fructose has been produced from dandelion root by hydrolysis and other treatment.[14]

## COMMERCIAL PREPARATIONS

Crude and extracts (solid, fluid, tincture, etc.); crude and fluid extract were formerly official in N.F. Strengths (see ***glossary***) of extracts are expressed in weight-to-weight ratios or flavor intensities. In Europe the drug consists of the root, with herb, gathered while blooming. Dandelion herb consists of the fresh or dried aboveground portions of the plant.[11,12]

*Regulatory Status*. GRAS (§182.20). Dandelion root (with herb) and dandelion leaves are subjects of positive German therapeutic monographs, allowed as a diuretic, for loss of appetite, dyspeptic problems, and disturbances in bile flow.[11,12]

## REFERENCES

See the General References for BAILEY 1; BLUMENTHAL; ESCOP 3; FEMA; FOSTER; FOSTER AND DUKE; GOSSELIN; JIANGSU; KROCHMAL AND KROCHMAL; LUST; MARTINDALE; NANJING; ROSE; TERRELL.

1. B. Proda and E. Andrzejewska, *Farm. Polska*, **22**, 181 (1966); through *Chem. Abstr.*, **65**, 9341c (1966).
2. H. Kleinig and H. Nietsche, *Phytochemistry*, **7**, 1171 (1968).
3. V. H. Booth, *Phytochemistry*, **3**, 229 (1964).
4. R. Buchecker et al., *Helv. Chim. Acta*, **59**, 1360 (1976).
5. J. Chury and F. Prosek, *Vet. Med. (Prague)*, **13**, 305 (1968); through *Chem. Abstr.*, **70**, 34902f (1969).
6. N. R. Farnsworth and A. B. Segelman, *Tile Till*, **57**, 52 (1971).
7. J. C. Mitchell in V. C. Runeckles, ed., *Recent Advances in Phytochemistry*, Vol. 9, Plenum Press, New York, 1975, p. 119.
8. K. Faber, *Pharmazie*, **13**, 423 (1958).
9. N. Mascolo, *Phytother. Res.*, **1**, 28 (1987).
10. H. B. Heath, *Cosmet. Toilet.*, **92**(1), 19 (1977).
11. Monograph *Taraxaci radix cum herba*, *Bundesanzeiger*, no. 228 (Dec. 5, 1984); corrected (Sept. 1, 1990).
12. Monograph *Taraxaci radix*, *Bundesanzeiger*, no. 228, (Aug. 29, 1992).
13. J. L. Hartwell, *Lloydia*, **31**, 71 (1968).
14. V. F. Belyaev and P. V. Golovin, *Uch. Zap., Beloruss. Gos. Univ. V. I. Lenina, Ser. Khim.*, **20**, 220 (1954); through *Chem. Abstr.*, **51**, 11495c (1957).

# DEERTONGUE

**Source:** ***Trilisa odoratissima*** (J. F. Gmel.) Cass. (syn. *Carphephorus odoratissimus* (J.F. Gmel.) Hebert; *Liatris odoratissima* Michx.) (Family Compositae or Asteraceae)

*Synonyms*. Deer's tongue, Carolina vanilla, wild vanilla, liatris, vanilla trilisa, vanilla plant, vanilla leaf, and hound's tongue.

## GENERAL DESCRIPTION

A perennial herb with large, thick leaves and a stem that is branched near the top; up to about 1.2 m high; native to eastern United States. Part used is the dried leaf which has a characteristic coumarin-like (new-mown hay) fragrance.

## CHEMICAL COMPOSITION

Contains about 1.6% coumarin as the major aromatic constituent.[1] Other volatile components include dihydrocoumarin, 2,3-benzofuran, terpenes, and straight-chain aldehydes and ketones, among others, totaling more than 90 identified compounds.[2]

Nonvolatile compounds reported present include triterpenes (e.g., $\beta$-amyrin, lupeol, lupenone, 11-oxo-$\beta$-amyrin, 11-oxo-$\alpha$-amyrin, and others); sesquiterpenes (e.g., ($\pm$)-eudesmin and ($\pm$)-epieudesmin); and others.[1,3]

Fresh leaves contain high concentrations of *cis*- and *trans*-*O*-hydroxycinnamic acids, present mostly as glucosides. It is postulated that during drying (curing) *cis*-*O*-hydroxycinnamic acid glucoside is converted to coumarin by hydrolysis followed by cyclization.[4]

### PHARMACOLOGY OR BIOLOGICAL ACTIVITIES

Coumarin has toxic properties, including liver injury and hemorrhages (GOSSELIN, LEWIS AND ELVIN-LEWIS).[5]

Coumarin and related compounds have been reported to be effective in reducing high-protein edemas, especially lymphedema.[6]

Some studies have shown coumarin to be nonteratogenic in animals.[7]

The charcoal-decolorized extract (deertongue incolore) has been reported to be nonirritating, nonsensitizing, nonphototoxic, and nonphotoallergenic when applied to the skin of animals and/or humans.[8]

Reported to have diaphoretic, demulcent, and febrifuge activity (WREN).

### USES

**Medicinal, Pharmaceutical, and Cosmetic.** Extracts are used as fixatives or fragrance components in perfumes and other cosmetic products (e.g., soap, detergent, creams, and lotions), with maximum use level of 0.5% of the incolore being reported in perfumes.[8]

**Traditional Medicine.** Used as a tonic in treating malaria.

**Others.** Extracts are used extensively in flavoring various types of tobaccos.

### COMMERCIAL PREPARATIONS

Leaves and extracts (oleoresin, solid extract, etc.).

*Regulatory Status*. Not permitted in foods.

### REFERENCES

See the General References for BAILEY 2; KROCHMAL AND KROCHMAL; MARTINDALE; WREN.

1. R. A. Appleton and C. R. Enzell, *Phytochemistry*, **10**, 447 (1971).
2. K. Karlsson et al., *Acta Chem. Scand.*, **26**, 2837 (1972).
3. I. Wahlberg et al., *Acta Chem. Scand.*, **26**, 1383 (1972).
4. F. A. Haskins et al., *Econ. Bot.*, **26**, 44 (1972).
5. D. L. J. Opdyke, *Food Cosmet. Toxicol.*, **12**, 385 (1974).
6. T. Bolton and J. R Casley-Smith, *Experientia*, **31**, 271 (1975).
7. W. Grote and I. Weinmann, *Arzneim. Forsch.*, **23**, 1319 (1973).
8. D. L. J. Opdyke, *Food Cosmet. Toxicol.*, **14**(Suppl.), 743 (1976).

# DEVIL'S CLAW

**Source:** ***Harpagophytum procumbens*** DC (Family Pedaliaceae)

*Synonyms*. Grapple plant, and wood spider.

### GENERAL DESCRIPTION

Herbaceous trailing perennial; flowers red in axils; fruit with pointed and barbed woody grapples to 2.5 cm long; occurring in steppes on red sand in south tropical Africa, especially in the Kalahari desert and in the Namibian steppes plus Madagascar.

The secondary tuber, about 6 cm in diameter, 20 cm long, is the part used.

## CHEMICAL COMPOSITION

Iridoid glycosides, including harpagoside (0.1–3%), found at twice the concentration in secondary tubers than primary root, in trace amounts in leaves; also harpagide and procumbide; phytosterols ($\beta$-sitosterol and stigmasterol);[1–3] triterpenes; flavonoids, including kaempferol and luteolin glycosides; cinnamic, chlorogenic, oleanolic, and ursolic acids; harpagoquinone; and others (WEISS; WREN).

## PHARMACOLOGY OR BIOLOGICAL ACTIVITIES

Antiinflammatory, analgesic, antirheumatic, and reputed sedative activity.

Antiinflammatory activity evaluated in the carrageenan edema assay in rats showed no significant activity (6% inhibition) compared with indomethacin (63% inhibition) in oral administration of an aqueous extract of the roots; positive results were reported, however, in acute inflammatory reactions with intravenous administration.[4] A lack of significant experimental antiinflammatory activity in orally administered extracts in the carrageenan-induced edema test has been reported by several research groups.[5–8] Oral administration of dried aqueous extract in 13 arthritic patients showed no significant improvement after 6 weeks.[8] Experimental antiinflammatory activity, however, was recently demonstrated in dose-dependent intraperitoneal administration of a dried aqueous extract, at a dose corresponding to 100 mg of the dried secondary root/kg, comparable to the efficacy of 2.5 mg/kg of indomethacin. The same tuber extract at a dose corresponding to 400 mg of the dried secondary root/kg proved more efficient than indomethacin at 10 mg/kg. Harpagoside administered intraperitoneally at 10 mg/kg (5 mg harpagoside equals the content in 400 mg of dried secondary tubers) did not have significant antiinflammatory effects. It has been suggested that simultaneous action of various principles other than iridoid glycosides are responsible for positive experimental antiinflammatory activity.[9]

Peripheral analgesic activity of an aqueous extract of the tuber against a chemical stimulus has been observed. Other compounds besides harpagoside are responsible for analgesic activity. Conflicting results in various studies have been attributed to non-efficacy of oral versus intraperitoneal administration.[9] Analgesic activity, with a reduction of high uric acid and cholesterol levels, has also been observed.[10]

Extracts have exhibited an arterial blood pressure reduction in rats, a decrease in heart rate in rabbits, and a protective effect against arrythmias.[11]

## USES

**Health Food/Herb Teas.** Oral dosage forms, including tablets, capsules, and extracts (e.g. tinctures) are sold in natural food or health food stores; primarily for relief of arthritic symptoms (WEISS, WREN).

**Traditional Medicine.** Used in Africa and since the early 20th century in Europe for indigestion (bitter tonic), blood diseases, headache, allergies, rheumatism, arthritis, lumbago, neuralgia; also as febrifuge, purgative; externally for sores, ulcer, boils, and skin lesions; folk cancer remedy.

## COMMERCIAL PREPARATIONS

Crude and extracts; extracts do not have uniform standards. In Germany, ampoules are available for i.v. and i.m. administration.

*Regulatory Status.* Undetermined in the United States. Root the subject of an official monograph in Germany, indicated for loss of appetite, dyspeptic discomfort, and others.[12]

## REFERENCES

See the General References for BLUMENTHAL; STEINMETZ; TYLER 1; UPHOF; WEISS; WREN.

1. H. Litichi and A. Wartburg, *Tetrahedron Lett*, **15**, 835 (1964).
2. O. Sticher, *Deut. Apoth. Ztg.*, **32**, 1279 (1977).
3. F. C. Cyzgan and A. Kreuger, *Planta Med.*, **31**, 305 (1977).
4. A. Erdos et al., *Planta Med.*, **34**, 97 (1978).
5. O. Eichler and C. Koch. *Arzneim. Forsch.*, **20**, 107 (1970).
6. D. McLeod et al., *Br. J. Pharmacol.*, **6**, 140 (1979).
7. I. Whitehouse et al., *Can. Med. Assoc.*, **129**, 249 (1983).
8. R. Graham, and B. V. Robinson, *Ann. Rheum. Dis.*, **40**, 632 (1981).
9. M. C. Lanhers et al., *Planta Med.*, **58**, 117 (1992).
10. R. Kampf, *Schweiz Apothek Zeitung*, **114**, 337 (1976).
11. C. Circost et al. *J. Ethnopharmacol.*, **11**, 259 (1984).
12. Monograph *Harpagophyti radix*, *Bundesanzeiger*, no. 43 (Mar. 2, 1989).

# DILL AND INDIAN DILL

**Source:** ***Dill*** *Anethum graveolens* L.; ***Indian dill*** *Anethum sowa* Roxb. (Family Umbelliferae or Apiaceae)

*Synonyms*. European dill and American dill (*A. graveolens*); East Indian dill (*A. sowa*).

### GENERAL DESCRIPTION

Dill (*A. graveolens*) is an annual or biennial herb with a smooth and erect stem; up to 1 m high; native to Mediterranean region and Asia (southern Russia); now cultivated worldwide (Germany, the Netherlands, the UK, Italy, the United States, India, China, etc.). Parts used are the dried ripe fruit (commonly called "seed") and the whole aboveground herb (dill herb) harvested immediately before the fruits mature or, for best quality, before flowering. Dill seed oil is obtained by steam distillation of the crushed dried fruits, and dillweed oil (dill oil or dill herb oil) is obtained by steam distillation of the freshly harvested herb. Dill seed oil is produced mainly in Europe, while dillweed oil is produced primarily in the United States.

Indian dill (*A. sowa*) is a smooth perennial herb, up to about 1 m high; native to tropical Asia and widely cultivated in India and Japan. Part used is the dried ripe fruit (commonly called "seed"); Indian dill seed oil (Indian dill oil or East Indian dill seed oil) is obtained by steam distillation of the crushed fruit.

### CHEMICAL COMPOSITION

Dill seeds contain 1.2–7.7% (usually 2.5–4%) volatile oil, with concentrations varying according to geographical origin and seasons;[1–3] dillanoside (a xanthone glucoside);[4] coumarins (scopoletin, esculetin, bergapten, umbelliferone, umbelliprenine, etc.); kaempferol and its 3-glucuronide;[5,6] vicenin (6,8-di-*C*-glucosyl-5,7,3′-trihydroxyflavone) and other flavonoids;[7] petroselinic acid triglyceride and β-sitosterol glucoside;[8] phenolic acids (caffeic, ferulic, and chlorogenic); protein (ca. 16%); fats (ca. 15%); and others (JIANGSU, LIST AND HÖRHAMMER, MARSH).

Dill seed oil contains mainly carvone (35–60%), *d*-limonene, and α-phellandrene, which together can account for 90% of the oil.[1,2,9] Other components present include dihydrocarvone, eugenol, β-phellandrene, α-pinene, anethole, dillapiole, myristicin, carveol, β-caryophyllene and others.[1,2,9,10]

Dillweed oil contains α-phellandrene, limonene, and carvone as its major components, usually with carvone in lesser amount than that in dill seed oil.[1,3,11,12] Other constituents include terpinene, α-pinene, dillapiole, myristicin, and two coumarans, among others (LIST AND HÖRHAMMER).[1,12–14]

Indian dill seeds contain 2–6 % volatile oil;[6] kaempferol, isorhamnetin, quercetin, and their 3-glucuronides;[5,6] petroselinic acid triglyceride and β-sitosterol;[8] fats; proteins; and others (LIST AND HÖRHAMMER).[15]

Indian dill seed oil consists mainly of carvone, dihydrocarvone, and limonene in varying amounts;[6] dillapiole (often in significant quantities) and apiole;[6,16–19] myristicin;[18] and others (LIST AND HÖRHAMMER).

Egyptian dill seed was found to contain limonene (30.3%), dillapiole (26.8%), carvone, (22%), piperitone (8.2%); reported in the oil for the first time were $D^8$-dehydro-*p*-cymene, camphor, and linalylacetate.

## PHARMACOLOGY OR BIOLOGICAL ACTIVITIES

Dill seed oil has been reported to have spasmolytic effects on smooth muscles. A 5% emulsion in physiological saline administered intravenously to cats (5–10 mg/kg) increased respiratory volume and lowered blood pressure, while guinea pigs receiving a higher dose (35 mg/kg) via intraperitoneal injection went into anaphylactic shock.[20]

Dill seed oil has antibacterial properties, and Indian dill seed oil has antifungal activities.[21,22]

An infusion of the young dill herb when administered intravenously to animals is reported to lower the blood pressure, dilate blood vessels, stimulate respiration, slow the heart rate, and other activities (JIANGSU).

Ethanolic extracts of dill seeds as well as the volatile oil have been shown to produce diuresis in dogs, while significantly increasing $Na^+$ and $Cl^-$ excretion.[23]

At ordinary use levels, dill and Indian dill oils are considered nontoxic.[17]

## USES

**Medicinal, Pharmaceutical, and Cosmetic.** Dill seed and dill seed oil are occasionally used in digestive preparations. Dill weed oil is used as a fragrance component in cosmetics, including soaps, detergents, creams, lotions, and perfumes. Maximum use level reported is 0.4% in perfumes.[24]

**Food.** Dill, Indian dill, and their oils are extensively used in many food products. Dill and Indian dill are reported used in baked goods, meat and meat products, condiments and relishes, fats and oils, and others; highest average maximum use level is about 2.9% (28,976 ppm) for Indian dill in condiments and relishes. Dill oil is used in all above foods as well as in alcoholic and nonalcoholic beverages, frozen dairy desserts, cheese, snack foods, gravies, and others, with highest average maximum use level of 0.075% reported in snack foods.

**Health Food/Herb Teas.** Bulk dillweed and dill seeds are available at natural food stores valued as a carminative (FOSTER).

**Traditional Medicine.** Both in domestic Western and Chinese medicine, dill seed and dill seed oil are used as aromatic carminative and stumulant in the treatment of flatulence, especially in children. Reported to have carminative, antispasmodic, sedative, lactagogue, and diuretic properties. Used in India, Africa, and elsewhere for hemorrhoids, bronchial asthma, neuralgias, renal colic, dysuria, genital ulcers, dysmenorrhea, and others.[25] In European tradition dill herb is reportedly used as an antispasmodic for conditions of the gastrointestinal tract, kidney and urinary tract; also for sleep disorders.[26]

**COMMERCIAL PREPARATIONS**

Crude and oils. Dill seed oil, dillweed oil, and Indian dill seed oil are official in F.C.C., with ketones (as carvone) contents highest in dill seed oil and lowest in Indian dill seed oil.

*Regulatory Status.* GRAS status affirmed (§184.1282). Dill seed is the subject of a positive German therapeutic monograph, indicated as a spasmolytic and bacteriostatic for dyspeptic disorders, with a mean daily dosage of 3 g (or 0.1–0.3 g of the essential oil).[27] Dillweed is not allowed to carry therapeutic claims, since efficacy for reported uses has not been documented.[26]

**REFERENCES**


See the General References for ARCTANDER; BLUMENTHAL; FOSTER; GUENTHER; JIANGSU; LIST AND HÖRHAMMER; MARTINDALE; ROSENGARTEN; UPHOF.

1. H. Karow, Riechst., *Aromen, Körperpflegem.*, **19**, 60 (1969).
2. M. B. Embong et al., *Can. Inst. Food Sci. Technol. J.*, **10**, 208 (1977).
3. R. Gupta, *Cultiv. Util. Med. Aromat. Plants*, 337 (1977); through *Chem. Abstr.*, **88**, 47583z (1978).
4. M. Kozawa et al., *Chem. Pharm. Bull.*, **24**, 220 (1976).
5. J. B. Harborne and C.A. Williams, *Phytochemistry*, **11**, 1741 (1972).
6. C. S. Shah, *Cultiv. Util. Med. Aromat. Plants*, 335 (1977); through *Chem. Abstr.*, **88**, 47519h (1978).
7. L. I. Dranik, *Khim. Prir. Soedin.*, **6**, 268 (1970); through *Chem. Abstr.*, **73**, 84627c (1970).
8. M. Bandopadhyay et al., *Curr. Sci.*, **41**, 50 (1972).
9. R. K. Baslas et al., *Flavour Ind.*, **2**, 241 (1971).
10. M. Miyazawa and H. Kameoka, *Yukagaku*, **23**, 746 (1974); through *Chem. Abstr.*, 83, 136696c (1975).
11. S. Zlatev, *Riv. Ital. Essenze, Profumi, Piante Offic., Aromi, Saponi, Cosmet., Aerosol*, **58**, 553 (1976); through *Chem. Abstr.*, **86**, 145787x (1977).
12. K. Belafi-Rethy et al., *Acta Chim. Acad. Sci. Hung.*, **83**, 1 (1974); through *Chem. Abstr.*, **82**, 28600n (1975).
13. K. Belafi-Rethy and E. Kerenyi, *Acta Chim. Acad. Sci. Hung.*, **94**, 1 (1977); through *Chem. Abstr.*, **88**, 110365x (1978).
14. E. P. Lichtenstein et al., *J. Agr. Food Chem.*, **22**, 658 (1974).
15. A. R. S. Kartha and Y. Selvaraj, *Chem. Ind.* (*London*), **25**, 831 (1970).
16. B. C. Gulati et al., *Perfum. Essent. Oil Record*, **60**, 277 (1969).
17. Anon., *Fed. Regist.*, **39**, 34211 (1974).
18. J. B. Harborne et al., *Phytochemistry*, **8**, 1729 (1969).
19. R. K. Sahdev et al., *J. Inst. Chem., Calcutta*, **47**, 234 (1975); through *Chem. Abstr.* **84**, 150546f (1976).
20. T. Shipochliev, *Vet. Med. Nauki*, **5**, 63 (1968); through *Chem. Abstr.*, **70**, 86144e (1969).
21. F. M. Ramadan et al., *Chem. Mikrobiol. Technol. Lebensm.*, **1**, 96 (1972).
22. B. Dayal and R. M. Purohit, *Flavour Ind.*, **2**, 484 (1971).
23. G. H. Mahran et al., *Phytother. Res.*, **5**, 169 (1992).
24. D. L. J. Opdyke, *Food Cosmet. Toxicol.*, **14**(Suppl.), 747 (1976).

25. G. H. Mahran et al., *Int. J. Pharmacognosy*, **30**, 2, 139 (1992).
26. Monograph *Anethi herba*, *Bundesanzeiger*, no. 193 (Oct. 15, 1987).
27. Monograph *Anethi fructus*, *Bundesanzeiger*, no. 193 (Oct. 15, 1987); revised (Mar. 13, 1990).

# DOGGRASS

**Source:** ***Elytrigia repens*** (L.) Desv. ex B. D. Jackson (syn. *Agropyron repens* (L.) Beauv.; *Triticum repens* L.) (Family Gramineae or Poaceae)

*Synonyms*. Quack grass, couch grass, witchgrass, quick grass, quitch grass, triticum, and agropyron.

## GENERAL DESCRIPTION

A perennial grass, up to 1.5 m high; native to Eurasia, naturalized in North America; widely distributed as a weed. Parts used are the dried rhizomes and roots collected in the fall or early spring.

## CHEMICAL COMPOSITION

Contains up to about 8% triticin (a fructosan); 2–3% inositol and mannitol; 1.5% fixed oil; vitamins A and B; glycosides (e.g., vanillin glucoside); mucilaginous substances; ash with high silicon and iron contents; and small amount of volatile oil (up to ca. 0.05%); among others (LIST AND HÖRHAMMER).[1,2]

The volatile oil is composed of up to 95% agropyrene (1-phenyl-2,4-hexadiyne) and monoterpenes, including carvacrol (10.81%), *trans*-anethole (6.80%), carvone (5.5%), thymol (4.30%), menthol (3.5%), menthone, (1.40%), and *p*-cymene (1.10%), among others (LIST AND HÖRHAMMER).[3]

## PHARMACOLOGY OR BIOLOGICAL ACTIVITIES

Agropyrene and its oxidation product, 1-phenylhexa-2,4-diyne-1-one, have been reported to have broad antibiotic activities (LIST AND HÖRHAMMER).[3,4]

A doggrass infusion has been demonstrated to have pronounced sedative effects in mice.[5]

Both the aqueous and hydroalcoholic extracts of doggrass have diuretic activities in rats.[6]

## USES

**Medicinal, Pharmaceutical, and Cosmetic.** Extracts used in certain diuretic preparations. Used in European phytomedicine for irrigation therapy in the treatment of inflammatory diseases of the urinary tract, and the prevention of kidney stones.

**Food.** Extracts are used as flavor components in nonalcoholic beverages, frozen dairy desserts, candy, baked goods, and gelatins and puddings. Use levels reported are low, with the highest average maximum of about 0.003% (32.4 ppm) being reported (type of extract not specified) in baked goods.

**Traditional Medicine.** Used as a diuretic and expectorant, to reduce blood cholesterol, and in treating nephrolithiasis (kidney stones), diabetes, chronic skin diseases, and liver ailments, among others.[1]

## COMMERCIAL PREPARATIONS

Crude and extracts (fluid, solid, powdered, etc.); crude and fluid extract were formerly official in N.F. and U.S.P. Strengths (see ***glossary***) of extracts are expressed in weight-to-weight ratios.

*Regulatory Status.* GRAS (§182.20). Root subject of a positive German therapeutic monograph for inflammatory diseases of the kidney tract.[7]

## REFERENCES

See the General References for APhA; BLUMENTHAL; FEMA; GOSSELIN; HORTUS 3rd; LIST AND HÖRHAMMER; WREN; YOUNGKEN.

1. S. Paslawska and R. Piekos, *Planta Med.*, **30**, 216 (1976).
2. D. Smith and R. D. Grotelueschen, *Crop Sci.*, **6**, 263 (1966).
3. R. Soiesel and H. Schilcher, *Planta Med*, **55**, 399 (1989).
4. M. Hejtmanek and V. Dadak, *Cesk. Mykol.*, **13**, 183 (1959); through *Chem. Abstr.*, **56**, 5208e (1962).
5. R. Kiesewetter and M. Müller, *Pharmazie*, **13**, 777 (1958).
6. E. Racz-Kotilla and E. Mozes, *Rev. Med. (Tirgu-Mures)*, **17**, 82 (1971); through *Chem. Abstr.*, **75**, 128341c (1971).
7. Monograph *Graminis rhizoma*, *Bundesanzeiger*, no. 22 (Feb. 1, 1990).

# DOGWOOD, JAMAICAN

**Source:** ***Piscidia piscipula*** (L.) Sarg. (syn. *P. communis* Harms, *P. erythrina* L., and *Ichthyomethia piscipula* (L.) A. S. Hitchc. ex. Sarg.) (Family Leguminosae or Fabaceae)

*Synonyms.* West Indian dogwood, fishfuddle, and fish-poison tree.

## GENERAL DESCRIPTION

Evergreen tree with scaly gray bark, up to about 15 m high; native to tropical America (West Indies, south Florida, Mexico, etc.). Part used is the dried root bark.

Jamaican dogwood bark is different from the bark of flowering dogwood (also known as common dogwood, boxwood, and simply, dogwood). The latter is *Cornus florida* L. (Family Cornaceae), which is widely distributed in eastern United States; its dried bark is usually used in domestic medicine as a febrifuge, tonic, and astringent (KROCHMAL AND KROCHMAL, LUST).

## CHEMICAL COMPOSITION

Contains ichthynone and jamaicin (isoflavones);[1–3] rotenone and related compounds, inclucling milletone, isomilletone, and dehydromilletone;[1,2,4] piscidic acid (*p*-hydroxy-benzyltartaric acid) and its monoethyl and diethyl esters, fukiic acid, and 3′-*O*-methyl-fukiic acid;[5,6] β-sitosterol;[1,4] simple plant acids (e.g., malic, succinic, and tartaric acids); a saponin glycoside; tannin; and others (LIST AND HÖRHAMMER).

## PHARMACOLOGY OR BIOLOGICAL ACTIVITIES

An extract of Jamaican dogwood has been reported to exhibit a sedative effect in cats and guinea pigs as well as marked antitussive and antipyretic activities; it also has antiinflammatory properties and antispasmodic action on smooth muscles. Its toxicity was reported to be very low in most of the animal species tested.[7]

## USES

**Medicinal, Pharmaceutical, and Cosmetic.** Extracts are widely used in certain female tonic preparations.

**Traditional Medicine.** It is used as analgesic, narcotic, and antispasmodic; in promoting sleep; and in treating whooping cough, toothache, asthma, and other ailments.

## COMMERCIAL PREPARATIONS

Crude and extracts (solid, powdered, etc.); there are no uniform standards.

## REFERENCES

See the General References for HORTUS 3rd; LIST AND HÖRHAMMER; MORTON 2; SARGENT; YOUNGKEN.

1. J. Buechi et al., *Arch. Pharm. Chemi*, **68**, 183 (1961); through *Chem. Abstr.*, **55**, 12774h (1961).
2. J. S. P. Schwarz et al., *Tetrahedron*, **20**, 1317 (1964).
3. O. A. Stamm et al., *Helv. Chim. Acta*, **41**, 2006 (1958).
4. A. L. Kapoor et. al., *Helv. Chim. Acta*, **40**, 1574 (1957).
5. W. Heller and C. Tamm, *Helv. Chim. Acta*, **58**, 974 (1975).
6. A. Nordal et al., *Acta Chem. Scand.* **20**, 1431 (1966).
7. M. Aurousseau et al., *Ann. Pharm. Fr.*, **23**, 251 (1965); through *Chem. Abstr.* **63**, 8909a (1965).

# ECHINACEA

**Source:** ***Echinacea angustifolia*** DC, ***E. pallida*** (Nutt.) Nutt., and ***E. purpurea*** (L.) Moench. (Family Compositae or Asteraceae).

*Synonyms*. Purple coneflower, common purple coneflower (*E. purpurea*), and pale purple coneflower (*E. pallida*); mixed lots of *E. angustifolia* and *E. pallida* ambiguously traded as "Kansas snakeroot."

## GENERAL DESCRIPTION

The genus *Echinacea* has nine indigenous North American herbaceous perennial species. *E. angustifolia*, to 60 cm high; leaves lanceolate; flowers violet, ray florets as long as or less than width of receptacle; found in barrens, dry prairies; Minnesota to Texas, west to eastern Colorado and Montana; taproot wild dug; little cultivation. *E. pallida*, to 120 cm, leaves lanceolate; purple ray flowers to 9 cm; in glades, prairies; Wisconsin to Arkansas, eastern Texas to Iowa; taproot wild dug; cultivation in United States and Europe. *E. purpurea*, to 90 cm, leaves ovate, coarsely toothed, basal ones often cordate; root fibrous; widely distributed in Midwestern United States; entire market supply cultivated in Europe, North America, and Australia.[1]

Commercial supplies involve the roots of the above three species, dried tops or fresh flowering herbage of *E. purpurea*, and to a lesser extent, dried tops of *E. angustifolia* and *E. pallida*. Endemic or rare species, including *E. atrorubens*, *E. paradoxa*, and *E. simulata*, documented in commercial supplies, vicariously harvested; misidentified or intentionally substituted. Root supply of *E. purpurea*, also *E. angustifolia*, historically and persistently adulterated with *Parthenium integrifolium*, traded as the ambiguous "Missouri snakeroot."[1]

## CHEMICAL COMPOSITION

Misidentification of source plants involved in chemical analysis before 1986, except for authenticated cultivated *E. purpurea*, render earlier chemical studies unreliable. Components attributed to *E. angustifolia*, which may instead have involved *E. pallida*, include flavonoid components of the leaves; essential oil constituents, including echinolone, humulene, caryophyllene epoxide, and various polyacetylene components.[2] Four sesquiterpene esters (cinnamates of echinadiol, epoxyechinadiol, echinaxanthol and dihydroxynardol), attributed to *E. purpurea* in fact do not occur in *Echinacea* but are constituents of *Parthenium integrifolium*.[2] Chemical work by R. Bauer and co-workers at Munich and Düsseldorf now makes distinction of source species in commercial supplies possible.

Echinacoside, once believed a marker compound for *E. angustifolia* (0.3–1.3% roots dry weight, flowers 0.1–1%), has also been found in *E. atrorubens*, *E. pallida* (0.4–1.7% roots), *E. paradoxa*, *E. simulata*, and tissue cultures of *E. purpurea*.[2]

Essential oil components common to the aerial parts of *E. pallida*, *E. purpurea*, and *E. angustifolia* include borneol, bornylacetate, pentadeca-8-en-2-one, germacrene D, caryophyllene, caryophyllene epoxide and palmitic acid.[2]

*E. angustifolia* root contains cynarin (1,5-di-*O*-caffeoylquinic acid); cichoric acid (trace amounts); essential oil (less than 0.1%), containing dodeca-2,4-dien-l-yl-isovalerate, penta-(1,8*Z*)-diene, l-pentadecene, palmitic and linolenic acid; alkylamides, including dodeca-(2*E*,6*Z*,8*E*,10*E*)-tetraenoic acid isobutylamide (echinacein, at 0.01% dried roots);[3] and 14 additional isobutylamides from the dried roots of *E. angustifolia* at 0.009–0.151%.[4] *E. angustifolia* polysaccharides include inulin (5.9%) and fructans. Other constituents include a resin (yielding oleic, linoleic, cerotic, and palmitic acids on hydrolysis), myristic and linolenic acids, *n*-triacontanol, $\beta$-sitosterol, stigmasterol, sitosterol-3-$\beta$-*O*-glucoside (leaves and stems), behenic acid ethyl ester (roots), and three glycoproteins. Chlorogenic acid and isochlorogenic acids are found in the leaves and stems of *E. angustifolia* and *E. pallida*.[2]

Rutoside is the major leaf flavonoid of *E. angustifolia*, *E. pallida*, and *E. purpurea*. Flavonoids from *E. angustifolia* (possibly involving *E. pallida*) leaves include luteolin, kaempferol, quercetin, quercetagetin-7-glucoside, luteolin-7-glucoside, kaempferol-3-glucoside, apigenin, and isorhamnetin. *E. purpurea* leaves contain quercetin, quercetin-7-glucoside, kaempferol-3-rutinoside, rutin, and others.

*E. pallida* root essential oil (0.2–2.0%) contains a series of ketoalkynes and ketoalkenes, including tetradeca-8*Z*-en-11,13-diyn-2-one; pentadeca-8*Z*-en,11,13- diyn-2-one; pentadeca-8*Z*,13*Z*-dien-11-yn-2-one; pentadeca-8*Z*,11*Z*,13*E*-trien-2-one; pentadeca-8*Z*,11*E*,13*Z*-trien-2-one; pentadeca-8*Z*,11*Z*-dien-2-one; pentadeca-8*Z*-en-2-one; and heptadeca-8*Z*,11*Z*-dien-2-one. In stored roots (commercial dried root), these components are oxidized by atmospheric oxygen to the hydroxylated derivatives 8-hydroxy-tetradeca-9*E*-en-11,13-diyn-2-one; 8-hydroxy-pentadeca-9*E*-13*Z*-dien-11-yn-2-one; and 8-hydroxy-pentadeca-9*E*-en-11,13-diyn-2-one. These compounds differentiate *E. pallida* extracts from *E. angustifolia* (absent).[5]

*E. purpurea*, the best studied species, contains 1.2–3.1% (flowers) and 0.6–2.1% (roots) cichoric acid, and other caffeic acid derivatives; essential oil (in addition to above reported essential oil components) contains vanillin, *p*-hydroxycinnamic acid methyl ester, and germacrene alcohol (characteristic of fresh plant extracts). Alkylamides include isomeric dodeca-(2*E*,4*E*,8*Z*,10*E*/*Z*)-tetraenoic acid isobutylamides, along with 10 additional alkylamides possessing a 2,4-diene structure (compared with one double bond in conjugation with the carbonyl group in *E. angustifolia* alkylamides). The aerial parts of *E. pallida*, *E. angustifolia*, and *E. purpurea* have a comparable alkylamide spectrum, mainly of the 2,4-diene type.[2]

Various polysaccharides from *E. purpurea* dried root include a heteroxylan (mean mol. wt. 35,000), rhamnoarabinogalactan (mean mol. wt. 450,000), and a xyloglucan.[6] An acidic arabinogalactan and two neutral fucogalactoxyloglucans are found in *E. purpurea* cell tissue cultures.[2]

Alkaloids include glycine betaine in *E. purpurea*.[2] The pyrrolizidine alkaloids tussilagine (0.006%) and isotussilagine have been identified from *E. angustifolia* and *E. purpurea* dried roots, neither of which possesses a 1,2-unsaturated necine ring structure associated with hepatotoxicity; therefore, they are not considered to be problematic.[7]

## PHARMACOLOGY OR BIOLOGICAL ACTIVITIES

Numerous studies report on the wound-healing mechanism of a preparation of the expressed juice of fresh flowering *E. purpurea* on local tissues:[1,2,8,9] inhibits hyaluronidase (both direct and indirect via formation of a complex with hyaluronic acid, causing depolymerization of hyaluronidase), promotes the formation of mesenchymal mucopolysaccharides, stimulates histogenic and haematogenic phagocytes, promotes differentiation of fibrocytes from fibroblasts, stimulates the anterior pituitary–adrenal cortex, and has antiinflammatory activity.

Bacteriostatic and fungistatic activity of isolated compounds and/or plant extracts are reported against *Staphylococcus aureus* (weak *in vitro*), *Escherichia coli*, *Pseudomonas aeruginosa*, *Trichomonas vaginalis* (weak *in vitro*), and *Epidermophyton interdigitale*.

*E. purpurea* extracts have shown indirect antiviral activity against encephalomyocarditis, vesicular stomatitis, influenza, herpes, and poliovirus; described as interferon-like.

Weak oncolytic activity has been reported from pentane extracts against Walker carcinoma 256 and P-388 lymphocytic leukemia, but not lymphoid leukemia.[8] The acidic arabinogalactan isolated from *E. purpurea* tissue culture activated macrophages to cytotoxicity against tumor cells and *Leishmania enriettii*, in addition to stimulating macrophages to produce tumor-necrosis factor, interleukin-1, and interferon $\beta_2$, and a slight increase in T-cell proliferation[10]

Major activity for *Echinacea* species and chemical fractions is nonspecific stimulation of the immune system. Immunostimulant activity involves an overall increase in phagocytosis by macrophages and granulocytes. Oral dosage is as effective as parenteral dosage forms, though acts more slowly.[1,2,8,9]

Immunostimulatory principles have been demonstrated both in lipophilic and polar fractions of extracts of various species and plant parts. Active components of lipophilic (chloroform) extracts may include polyacetylenes, alkylamides, and essential oils. Enriched alkylamide fractions of *E. purpurea*, *E. angustifolia*, and *E. pallida* root ethanol extracts eliminate carbon particles (carbon clearance test) by a factor of 1.5–1.7. Cichoric acid from the polar water-soluble fraction of *E. purpurea* increased carbon elimination by a factor of 2.1. In the granulocyte smear test all ethanolic root extracts of the three species increased *in vitro* phagocytosis by 20–30% (*E. purpurea* most active). Lipophilic (chloroform) fractions of *E. angustifolia* and *E. pallida* were more active than hydrophilic fractions, while hydrophilic fractions of *E. purpurea* stimulate phagocytosis by 40%.[8,11]

Two high-molecular-weight polysaccharides from an aqueous extract of *E. purpurea* stimulated T-lymphocyte activity 20–30% more than a potent T-cell stimulator, with a concurrent enhancement of phagocytosis in the carbon clearance test.[12] These compounds are found only in low concentrations in the expressed juice of the herb and precipitate out of ethanol extracts.[2]

Bauer and Wagner conclude, based on present knowledge, that the immunostimulatory activity of alcoholic and aqueous extracts depend on the combined action of several constituents. In lipophilic fractions, alkylamides and the polar caffeic acid derivative cichoric acid contribute to activity of alcoholic extracts. Polysaccharides are implicated in the expressed juice of *E. purpurea* and aqueous extracts as well as orally administered powdered whole drug.[2,13]

The caffeic acid glycoside echinacoside, once regarded as a significant active principle (based on one report of weak antibacterial activity) and inactive in the carbon clearance test and granulocyte smear test, does not appear to possess immunostimulatory activity. Currently, considered a component of minor importance.[2,9]

Further studies have suggested that echinacea could play a therapeutic role in a wide range of clinical disciplines, including dermatology, pediatrics, gynecology, surgery, urology, and the treatment of allergies.[14] A recent double-blind, placebo-controlled study indicates a dose of 450 mg/day of *E. purpurea* root extract (1:5 in 55% ethanol) significantly relieved the severity and duration of flu symptoms.[15] A double-blind, monocentric, placebo-controlled clinical trial examined the immunostimulating influence of an expressed fresh juice *E. purpurea* preparation on the course and severity of colds and flulike symptoms with patients deemed to have greater susceptibility to infections. At a dose of 2–4 mL/day, patients with diminished immune response (expressed by a low T4:T8 cell ratio), were found to benefit significantly from preventative treatment with the echinacea preparation.[16]

## USES

**Medicinal, Pharmaceutical, and Cosmetic.** Echinacea preparations—especially oral (liquid extract), topical (ointment), and parenteral products of the fresh aboveground preparations of *E. purpurea* and to a lesser extent, the roots of *E. angustifolia* and *E. pallida*—are used in Germany for the external treatment of hard-to-heal wounds, eczema, burns, psoriasis, herpes simplex, etc. As immunostimulants; internally a prophylactic at the onset of cold and flu symptoms and for treatment of *Candida albicans* infections, chronic respiratory infections, prostatitis, polyarthritis (rheumatoid arthritis), etc.[13]

Echinacea extracts, increasingly seen in cosmetics, are used in lip balms, shampoos, toothpaste, and other product categories.

**Health Food/Herb Teas.** Numerous oral dosage products, vicariously positioned as cold and flu preventatives.[1]

**Traditional Medicine.** "This plant [*E. angustifolia*] was universally used as an antidote for snake bite and other venomous bites and stings and poisonous conditions. Echinacea seems to have been used as a remedy for more ailments than any other plant."[17] Diseases and conditions for which echinacea was employed by physicians (1887–1939) included old sores, wounds, snakebite, gangrene, and as a local antiseptic; internally for diphtheria, typhoid conditions, cholera infantum, syphilis, and blood poisoning.[1]

## COMMERCIAL PREPARATIONS

Ointment, oral liquid, intravenous and intramuscular ampoules from expressed juice of fresh-flowering *E. purpurea*; tinctures, extracts, capsules, tablets, etc., of *E. angustifolia*, *E. pallida* and *E. purpurea*. More than 280 echinacea pharmaceutical products are available in Europe.[2] Echinacea extracts "standardized" to echinacoside persist in the marketplace, despite the fact that the compound has insignificant biological activity. *E. angustifolia* and *E. pallida* crude were formerly official in N.F.

*Regulatory Status*. Undetermined in the United States; formerly *E. angustifolia*/*E. pallida* official in N.F. (1916–1950). *E. purpurea* fresh aboveground parts,[18] and *E. pallida* root[19] subjects of positive therapeutic monographs for human use in Germany.

Based on historical confusion of plant identity, persistence of adulterated commercial supplies of *E. purpurea* root,[20] and lack of current clinical studies on *E. angustifolia* root and *E. angustifolia*/*E. pallida* aerial parts, negative (not recommended) therapeutic monographs have recently been published in Germany.[21]

## REFERENCES

See the General References for FOSTER; FOSTER AND DUKE; STEINMETZ; TYLER 1; UPHOF; WEISS; WREN.

1. S. Foster, *Echinacea—Nature's Immune Enhancer*, Healing Arts Press, Rochester, Vt., 1991.
2. R. Bauer and H. Wagner in H. Wagner and N. R. Farnsworth, eds., *Economic and Medicinal Plant Research*, Vol. 5, Academic Press, New York, 1991, p. 253.
3. M. Jacobson, *J. Org. Chem.* **32**, 1646 (1967).
4. R. Bauer et al., *Phytochemistry*, **28**, 505 (1989).

5. R. Bauer et al., *Planta Med.* **54**, 426 (1988).
6. H. Wagner and A. Proksch in N. Farnsworth, N. H. Hikino and H. Wagner, eds., *Economic and Medicinal Plant Research*, Vol. 1., Academic Press, New York, 1985, p. 113.
7. E. Röder et al., *Deut. Apoth. Ztg.*, **124**, 2316 (1984).
8. C. Hobbs, *The Echinacea Handbook*, Eclectic Medical Publications, Portland, Oreg., 1989.
9. D. V. C. Awang and D. G. Kindack, *Can. Pharm. J.*, **124**(11), 512 (1991).
10. B. Luetting et al., *J. Natl. Cancer Inst.*, **81**, 669 (1989).
11. R. Bauer et al., *Z. Phytother.*, **10**, 43 (1989).
12. H. Wagner and A. Proksch, *Angew. Phytother.*, **2**(5), 166 (1981).
13. S. Foster. *Echinacea—The Purple Coneflowers, Botanical Series, no.* 301, Austin, Tex., American Botanical Council, 1991.
14. G. Hahn and A. Mayer, *Öster. Apoth. Ztg.*, **38**, 1040 (1984).
15. B. Bräunig et al., *Z. Phytother.*, **13**, 7 (1992).
16. D. Schöneberger, *For. Immunol.*, **8**, 2 (1992).
17. M. R. Gilmore in *Thirty-Third Annual Report of the Bureau of American Ethnology*, U.S. Government Printing Office, Washington, D.C., 1919, p. 145.
18. Monograph *Echinaceae purpureae herba*, *Bundesanzeiger*, no. 43 (Mar. 2, 1989).
19. Monograph *Echinaceae pallidae radix*, *Bundesanzeiger*, no. 44 (Aug. 29, 1992).
20. Monograph *Echinaceae purpureae radix*, *Bundesanzeiger*, no. 44 (Aug. 29, 1992).
21. Monograph *Echinaceae angustifoliae/pallidae herba; Echinacea angustifoliae radix*, *Bundesanzeiger*, no. 44 (Aug. 29, 1992).

# ELDER FLOWERS (AMERICAN AND EUROPEAN)

**Source:** ***American elder*** *Sambucus canadensis* L.; ***European elder*** *Sambucus nigra* L. (Family Caprifoliaceae).

*Synonyms.* Sweet elder, common elder, and American elderberry (*S. canadensis*); sambucus.

## GENERAL DESCRIPTION

American elder (*S. canadensis*) is a tall shrub with white pith, spreading by suckers; up to about 4 m high; native to eastern North America. Part used is the flower.

European elder (*S. nigra*) is a tall shrub or small tree; up to about 10 m high; native to Europe and naturalized in the United States. Parts used are the dried flower and leaf.

## CHEMICAL COMPOSITION

Most of the chemical work has been performed on the European elder.

European elder flowers contain a small amount of an essential oil (ca. 0.3%), which is composed primarily of free fatty acids (66%) and alkanes (ca. 7%), with palmitic and linolenic acids being the major acids and $C_{19}$, $C_{21}$, $C_{23}$, and $C_{25}$ alkanes the major alkanes;[1] triterpenes, including $\alpha$-amyrin and $\beta$-amyrin (mainly as fatty acid esters), ursolic acid, 30$\beta$-hydroxyursolic acid, and oleanolic acid;[2,3] sterols (as free sterols, esters, and glycosides);[3]

flavonoids, flavone glycosides, and phenolic acids, including quercetin, kaempferol, isoquercitrin, rutin (up to 1.9%), and chlorogenic acid;[4–6] pectin; sugar; and others.[6]

Elder leaf contains sambunigrin (a cyanogenic glucoside) at 0.042% concentration, according to one source;[7] choline; rutin and quercetin;[7,8] sterols (sitosterol, stigmasterol, and campesterol); triterpenes (α- and β-amyrin palmitates, oleanolic acid, and ursolic acid); alkanes (mainly *n*-nonacosane and *n*-hentriacontane); fatty acids (stearic, oleic, linoleic, etc.);[8,9] tannins; resins; fats; sugars; vitamin C; and others.[7,10]

## PHARMACOLOGY OR BIOLOGICAL ACTIVITIES

Elder flowers are generally considered to have diuretic, diaphoretic, and laxative properties.

A recent study found that sambuculin A and α- and β-amyrin palmitate from *S. formosana* to have antihepatotoxic activity against carbon tetrachloride induced liver damage.[11]

Compounds from *Sambucus* species contain a number of plant lectins with hemagglutinin characteristics, which could prove useful in blood typing and hematological tests.[12,13]

## USES

**Medicinal, Pharmaceutical, and Cosmetic.** European elder flower water (water phase from steam distillate) has been used as a vehicle for eye and skin lotions.

Extracts (e.g., absolute) of the flowers are used in perfumes.

**Food.** Flowers are reported used as flavor components in numerous food products, including alcoholic (bitters and vermouths) and nonalcoholic beverages, frozen dairy desserts, candy, baked goods, and gelatins and puddings. Highest average maximum use level reported is 0.049% in nonalcoholic beverages.

**Health Food/Herb Teas.** The flowers of *S. canadensis* are used as a tea ingredient, also in capsules, tablets, etc; primarily as diaphoretic for colds and flu (FOSTER AND DUKE).

**Traditional Medicine.** Flowers are used as diuretic, laxative, and diaphoretic as well as a gentle astringent for the skin and in treating rheumatism, usually in the form of a tea, infusion, poultice, or water distillate.

Bark of American elder and bark, leaf, flower, root, and fruit of European elder have been reported used in cancers.[14]

In Chinese medicine a related species, *Sambucus williamsii* Hance, is widely used. Flowers are used as diaphoretic and diuretic; twigs, leaves, and roots in treating rheumatoid arthritis, among other conditions, usually in the form of a decoction taken internally (JIANGSU).

## COMMERCIAL PREPARATIONS

Crude and extracts; flowers were formerly official in N.F. Strengths (see ***glossary***) of extracts are expressed in weight-to-weight ratios or in flavor intensities.

*Regulatory Status.* Flowers are GRAS (American, §182.10 and §182.20; European, §182.20); leaves have been approved for use in alcoholic beverages only, with the provision that hydrocyanic acid (HCN) not exceed 25 ppm (0.0025%) in the flavor (§172.510). *S. nigra* flowers are subject of a German therapeutic monograph; allowed as a diaphoretic and to increase bronchial secretion in the treatment of colds.[15]

## REFERENCES

See the General References for ARCTANDER; BAILEY 1; BLUMENTHAL; FEMA; FERNALD; FOSTER AND DUKE; FURIA AND BELLANCA; KROCHMAL AND KROCHMAL; MARTINDALE; ROSE; TERRELL; UPHOF.

1. W. Richter and G. Willuhn, *Deut. Apoth. Ztg.*, **114**, 947 (1974).
2. R. Hänsel and M. Kussmaul, *Arch. Pharm. (Weinheim)*, **308**, 790 (1975).
3. W. Richter and G. Willuhn, *Pharm. Ztg.*, **122**, 1567 (1977).
4. A. Radu et al., *Farmacia (Bucharest)*, **24**, 9 (1976); through *Chem. Abstr.*, **85**, 74954r (1976).
5. I. Leifertova et al., *Acta Fac. Pharm., Univ. Comeniana*, **20**, 57 (1971); through *Chem. Abstr.*, **77** 16587x (1972).
6. I. I. Hajkova and V. Brazdova, *Farm. Obzor.*, **32**, 343 (1963); through *Chem. Abstr.*, **61**, 7358g (1964).
7. Z. N. Guseinova, *Azerb. Med. Zh.*, **42**(6), 29 (1965); through *Chem. Abstr.*, **63**, 18644h (1965).
8. T. Inoue and K. Sato, *Phytochemistry*, **14**, 1871 (1975).
9. A. M. Al-Moghazy Shoaib, *Egypt. J. Pharm. Sci.*, **13**, 255 (1972).
10. S. R. Jensen and B. J. Nielsen, *Acta Chem. Scand.*, **27**, 2661 (1973).
11. C. N. Lin and W. P. Tome, *Planta Med.*, **54**, 3, 223 (1988).
12. L. Maklc et al., *Bichem. J.*, **278**, 667 (1991).
13. H. Kaku et al., *Arch Biochem. Biophys.*, **277**(2), 255 (1990).
14. J. L. Hartwell, *Lloydia*, **31**, 71 (1968).
15. Monograph *Sambuci flos*, *Bundesanzeiger*, no. 50 (Mar. 13, 1986); revised (Mar. 13, 1990).

# ELECAMPANE

**Source:** ***Inula helenium*** L. (syn. *Helenium grandiflorum* Gilib.; *Aster officinalis* All.; *A. helenium* (L.) Scop.) (Family Compositae or Asteraceae).

*Synonyms*. Scabwort, inula, alant, horseheal, and yellow starwort.

## GENERAL DESCRIPTION

A perennial herb, covered with soft short hairs; up to about 1.8 m high; native to Europe and Asia; naturalized in North America; cultivated in Europe (Belgium, France, Germany, etc.) and Asia (e.g., China). Parts used are the dried roots and rhizomes collected in late fall or early winter. An essential oil is obtained by steam distillation.

## CHEMICAL COMPOSITION

Contains 1–4% volatile oil, which is composed primarily of sesquiterpene lactones, including alantolactone (also called helenin, elecampane camphor, and alant camphor), isoalantolactone, dihydroisoalantolactone, and dihydroalantolactone; alantic acid; and azulene (JIANGSU, LIST AND HÖRHAMMER).[1–5]

Other constituents reported present include up to about 44% inulin; sterols (stigmasterol, $\beta$- and $\gamma$-sitosterols, damaradienol, etc.);[6] friedelin; resin; pectic substances, and others (LIST AND HÖRHAMMER).

Alantolactone was reported to have strong inhibitory effects on seed germination and seedling growth.[7]

## PHARMACOLOGY OR BIOLOGICAL ACTIVITIES

An infusion of elecampane has been reported to have a pronounced sedative effect on mice.[8]

Much of the pharmacological properties of elecampane is due to alantolactone, which has been reported to have anthelmintic activities in humans (similar to santonin but better and less toxic) and hypotensive, hyperglycemic (in large doses), and hypoglycemic (smaller doses) effects in experimental animals. It also has antibacterial and antifungal properties, among others (JIANGSU).

Among 105 plant lactones studied, alantolactone and isoalantolactone are among the few that have been reported to exhibit the highest bactericidal and fungicidal properties *in vitro*.[9]

Alantolactone reported as an immunostimulant.[10]

Some individuals are extremely sensitive to the oil when applied to the skin.[11]

## USES

**Medicinal, Pharmaceutical, and Cosmetic.** Alantolactone is used as an anthelmintic, primarily in the UK and Europe.

The oil is used as a fragrance component in cosmetic products, including soaps, detergents, creams, lotions, and perfumes (e.g., Oriental types). Highest maximum use level reported is 0.4% in perfumes.[11]

**Food.** Reportedly used as a flavor ingredient in major food products, including alcoholic (aromatic bitters, vermouths, etc.) and nonalcoholic beverages, frozen dairy desserts, candy, baked goods, and gelatins and puddings. Highest average maximum use level reported is 0.08% for the crude in baked goods.

**Health Food/Herb Teas.** Powdered root used in tea formulations; sometimes in other product forms for lung conditions (FOSTER).

**Traditional Medicine.** Reportedly used in treating asthma, bronchitis, whooping cough, nausea, diarrhea, and other ailments. Also used as a diuretic, stomachic, and anthelmintic both in Western domestic medicine and in Chinese medicine, usually in the form of a decoction or tea. It has reportedly been used in treating cancers.[12]

## COMMERCIAL PREPARATIONS

Crude, oil, and extracts; crude was formerly official in N.F.

*Regulatory Status.* Has been approved for use in alcoholic beverages only (§172.510). Root subject of a German therapeutic monograph; not recommended due to lack of evidence of efficacy and risk for allergic reactions.[13]

## REFERENCES

See the General References for ARCTANDER; BAILEY 1; BIANCHINI AND CORBETTA; BLUMENTHAL; FEMA; FOSTER; FURIA AND BELLANCA; GUENTHER; JIANGSU; LUST; MARTINDALE; NANJING; TERRELL; UPHOF, WREN.

1. A. Boeva et al., *Farmatsiya (Sofia)*, **21**, 34 (1971); through *Chem. Abstr.*, **76**, 131416g (1972).
2. S. S. Kerimov and O. S. Chizhov, *Khim. Prir. Soedin.*, **10**, 254 (1974); through *Chem. Abstr.*, **81**, 60894b (1974).
3. P. P. Khvorost and N. F. Komissarenko, *Khim. Prir. Soedin.*, **6**, 820 (1976); through *Chem. Abstr.*, **86**, 136307p (1977).
4. V. G. Sinitsina and Z. I. Boshko, *Nek. Probl. Farm. Nauki Prakt., Mater. S'ezda Farm. Kaz.*, **1**, 87 (1975); through *Chem. Abstr.*, **87**, 2355q (1977).

5. Y. Kashman et al., *Isr. J. Chem.*, **5**, 23 (1967).
6. W. Olechnowicz-Stepien et al., *Rocz. Chem.*, **49**, 849 (1975); through *Chem. Abstr.*, **83**, 75401d (1975).
7. E. Rodríguez et al., *Phytochemistry*, **15**, 1573 (1976).
8. R. Kiesewetter and M. Müller, *Pharmazie*, **13**, 777 (1958).
9. S. A. Vichkanova et al., *Rastit. Resur.*, **13**, 428 (1977); through *Chem. Abstr.*, **87**, 162117s (1977).
10. H. Wagner and A. Proksch in N. Farnsworth, N. H. Hikino and H. Wagner, eds., *Economic and Medicinal Plant Research*, Vol. 1, Academic Press, New York, 1985, p. 113.
11. D. L. J. Opdyke, *Food Cosmet. Toxicol.*, **14**, 307 (1976).
12. J. L. Hartwell, *Lloydia*, **31**, 71 (1968).
13. Monograph *Helenii radix*, *Bundesanzeiger*, no. 85 (May 5, 1988).

# ELEMI GUM

**Source:** ***Canarium commune*** L. and ***C. luzonicum*** Miq. (Family Burseraceae).

*Synonyms*. Manila elemi, elemi oleoresin, and elemi resin.

## GENERAL DESCRIPTION

Trees up to about 30 m high; native to the Philippines and Moluccas. Part used is their resinous pathological exudation from which an essential oil (elemi oil) is obtained by steam distillation.

## CHEMICAL COMPOSITION

Elemi gum (oleoresin) contains 65–75% triterpenoid resinous compounds, which include amyrin, brein, maniladiol, elemadienolic acid, and elemadienonic acid; and 10–25% volatile oil, consisting mainly of phellandrene, dipentene, elemol, elemicin, terpineol, carvone, and terpinolene (KARRER, POUCHER).[1–4]

## PHARMACOLOGY OR BIOLOGICAL ACTIVITIES

Available data indicate elemi and elemi oil to be relatively nontoxic.[5]

## USES

**Medicinal, Pharmaceutical, and Cosmetic.** Elemi resinoid and elemi oil are used as fixatives and fragrance components in soaps, detergents, creams, lotions, and perfumes. Maximum use level of the oil reported is 0.6% in perfumes.

**Food.** Elemi oil is reported used as a flavor component in major categories of food products, including alcoholic and nonalcoholic beverages, frozen dairy desserts, candy, baked goods, gelatins and puddings, meat and meat products, and condiments and relishes. Average maximum use levels are low, with highest reported being about 0.002% (17.2 ppm) in baked goods.

Elemi gum is also used but much less frequently.

**Traditional Medicine.** Resin is reported used as a stomachic and as an expectorant; also used externally as a local stimulant (MARTINDALE).[2]

## COMMERCIAL PREPARATIONS

Gum and oil.

*Regulatory Status*. Has been approved for food use (§172.510).

**REFERENCES**

See the General References for ARCTANDER; FEMA; GUENTHER; POUCHER; UPHOF.

1. A. F. Summa, *Diss. Abstr.*, **21**, 1772 (1961).
2. R. Pernet, *Lloydia*, **35**, 280 (1972).
3. G. D. Manalo and A. P. West, *Philippine J. Sci.*, **78**, 111 (1949); through *Chem. Abstr.*, **44**, 7564f (1950).
4. M. Mladenović and D. Fodor-Mandušić, *Acta Pharm. Jugoslav.*, **8**, 59 (1958); through *Chem. Abstr.*, **52**, 17614i (1958).
5. D. L. J. Opdyke, *Food Cosmet. Toxicol.*, **14**(Suppl.), 755 (1976).

# ELEUTHERO

**Source:** ***Eleutherococcus senticosus*** (Rupr. and Maxim.) Maxim. (syn. *Acanthopanax senticosus* Harms) (Family Araliaceae).

*Synonyms.* Siberian ginseng, eleutherococ, eleuthero ginseng, and Ussurian thorny pepperbush.

## GENERAL DESCRIPTION

Deciduous shrub, 1–3 m high, branches beset with numerous small sharp spines; leaves palmate; northeast Asia, including much of far southeastern Russia (middle Amur region in the north to Sakhalin), northeast China (Heilongjiang, Jilin, Liaoning, Nei Monggol, Hebei, Shanxi); abundant in the Xiaoxinganling Mountains of Heilongjiang, adjacent Korea, and Japan (Hokkaido). In traditional Chinese medicine, the bark of the root historically used; currently root, rhizome, stems, and leaves enter commerce.[1]

## CHEMICAL COMPOSITION

Russian workers initially isolated seven compounds from a methanol extract of roots, deemed eleutherosides A–G ( in a ratio of 8:30:10:12:4:2:1), ranging from 0.6 to 0.9% (roots); 0.6 to 1.5% (stems); seven additional eleutherosides have been identified.[2] Eleutheroside A is the sterol daucosterol; eleutheroside B is the phenylpropanoid syringin; eleutheroside $B_1$ is isofraxidin-7-*O*-α-L-glucoside (β-calycanthoside); eleutheroside $B_2$ and $B_3$ (a coumarin) are not characterized. Eleutheroside $B_4$ ((−)-sesamin), eleutheroside D ((−)-syringaresinol-di-*O*-β-D-glucoside), and eleutheroside E (acanthoside D) are lignans. Triterpenes include eleutherosides I–M, (eleutheroside I is mussenin B; eleutheroside M is hederasaponin), senticosides A–F (incompletely characterized oleanolic acid glycosides), and oleanolic acid. Additional phenylpropanoids include caffeic acid, caffeic acid ethyl ester, coniferyl aldehyde, and sinapyl alcohol; other components include β-sitosterol, galactose, α- and β-glucose, α- and β-maltose, sucrose, vitamin E, β-carotene,[2] polysaccharides (eleutherans A–G),[3] and two glucose-, galactose-, and arabinose-containing polysaccharides.[4] Saponin glycosides, well known in various *Panax* species, are absent in *E. senticosus*.

## PHARMACOLOGY OR BIOLOGICAL ACTIVITIES

*In vivo* animal studies of root ethanol extracts have been evaluated for adaptogenic activity in hyperthermia, electroshock-induced convulsions, toxic cardioglycoside dose resistance, cortisone-induced lymphatic stress, gastric ulcers, x-ray irradiation, increased metabolic efficiency in swimming-induced stress, increased conditioned response to stimuli, and inhibition of conditioned avoidance response.[2]

Antioxidant (free-radical scavenging) activity has been demonstrated in animals.[2]

A hypoglycemic effect has been demonstrated in animals, though the mechanism of action is unclear.[2]

Antiedema, antiinflammatory, diuretic, gonadotropic activity, estrogenic activity, and antihypertensive activity also reported *in vivo*.[2]

Polysaccharides are responsible for immunostimulatory activity (carbon clearance and granulocyte tests); lessened thioacetamide, phytohemagglutin, and x-ray toxicity; antitumor activity; and hypoglycemic activity.[3,4]

A carcinostatic effect, slowing spread of metastases, is suggested by Russian research (DUKE 2). A comprehensive review of Russian studies is available.[2]

## USES

**Medicinal, Pharmaceutical, and Cosmetic.** Used to increase general resistance as an adaptogenic and immunostimulant.[2] Hydroalcoholic and glycolic extracts increasingly used in skin care products, including creams and lotions.

**Health Food/Herb Teas.** Numerous oral dosage products, including liquid extract, tinctures, capsules, tablets, crude herb, etc., as tonic tea (FOSTER AND YUE).

**Traditional Medicine.** In China, the whole root and rhizome is known as *ciwujia*; reportedly used as stimulant, tonic, adaptogenic, diuretic; for sleeplessness, lower back or kidney pain, lack of appetite, rheumatoid arthritis; to enhance overall resistance to disease or adverse physical influences or stress (FOSTER AND YUE).

Bark of *E. senticosus* (*jiapi* or *ciwujiapi*) is a substitute to the bark of *E. gracilistylus* (*Acanthopanax gracilistylus*), source of *wujiapi* (also *jiapi*). Bark of *Periploca sepium* (Asclepidaceae), also known as "*wujia*" (*xiangjiapi*, *gangliupi*, and *beiwujiapi*) has prompted market confusion, and consequently entered trade as an adulterant to *E. senticosus*.[1] There is a report of a purported case of neonatal androgenization associated with maternal "ginseng" use; attributed to "pure Siberian ginseng."[5] Follow-up revealed that the product in question did not contain eleuthero but *Periploca sepium* instead, prompting a pharmacological study in which no androgenicity was observed.[6–8]

## COMMERCIAL PREPARATIONS

Crude (bark of stem, whole root and rhizome, rarely bark of root) and extracts (liquid, solid, powder, etc.).

*Regulatory Status.* Industry self-GRAS determination rejected by FDA on basis that it was not materially marketed in the United States, hence not subject to grandfathering under GRAS regulations. However, a landmark court decision, *Fmali Herb v. Heckler*, 715 F. 2d 1385 (9th Cir. 1983), required FDA to consider use outside the United States. Root subject of a German regulatory monograph indicated as a tonic for invigoration during fatigue, debility, declining work capacity and concentration, and during convalescence.[9]

## REFERENCES

See the General References for BLUMENTHAL; DUKE 2; FOSTER AND YUE; TYLER 1; WEISS.

1. S. Foster, *Siberian ginseng—Eleutherococcus senticosus*, *Botanical Series*, *no. 302*, Austin, Tex., American Botanical Council, 1991.

2. N. R. Farnsworth et al. in H. Wagner, H. Hikino and N. R. Farnsworth, eds., *Economic and Medicinal Plant Research*, Vol. 1, Academic Press, New York, 1985, p. 155.
3. H. Hikino et al., *J. Nat. Prod.*, **49**(2), 293 (1986).
4. H. Wagner and A. Proksch in N. Farnsworth, N. H. Hikino and H. Wagner, eds., *Economic and Medicinal Plant Research*, Vol. 1., Academic Press, New York, 1985, p. 113.
5. G. Koren et al., *J. Am. Med. Assoc.*, **264**(22), 2866 (Dec. 12, 1990).
6. D. V. C. Awang, *J. Am. Med. Assoc.*, **265**(14), 1828 (Apr. 10, 1991).
7. D. V. C. Awang, *J. Am. Med. Assoc.*, **266**(3), 363 (July 17, 1991).
8. D. P. Waller et al., *J. Am. Med. Assoc.*, **267**(17), 2329 (May 6, 1992).
9. Monograph *Eleutherococci radix*, *Bundesanzeiger*, no. 11 (Jan. 17, 1991).

# EPHEDRA

**Source:** ***Chinese ephedra*** *Ephedra sinica* Stapf.; ***Intermediate ephedra*** *Ephedra intermedia* Shrenk et C. A. Mey.; ***Mongolian ephedra*** *Ephedra equisetina* Bge.; ***Other*** *Ephedra* spp. (Family Ephedraceae).

*Synonyms*. *Mahuang*, *Herba Ephedrae*, *cao mahuang* (Chinese ephedra), *zhong mahuang* (intermediate ephedra), and *muzei mahuang* (Mongolian ephedra); *mahuanggen* (root).

## GENERAL DESCRIPTION

Low shrubs with scalelike leaves, 1.5–3.3 m high; stems herbaceous above and woody below; herbaceous stems greenish, with those of *E. intermedia* and *E. equisetina* often covered with a white powder. *E. equisetina* is the largest among the three while *E. sinica* is the smallest, with herbaceous features; all flowering in spring and fruiting in late summer. Native to central Asia, now distributed throughout northern China from Xinjiang to Inner Mongolia and Jilin; also cultivated.

Parts used are the herbaceous green stems (*mahuang*) and the root (*mahuanggen*, ephedra root). Stems are collected in autumn either by cutting the green parts aboveground or the whole plant is pulled out and rid of dirt, and then the stems and roots separated and sun dried.

## CHEMICAL COMPOSITION

Stem (*mahuang*) contains 1–2% alkaloids composed mainly of *l*-ephedrine and *d*-pseudoephedrine, with ephedrine ranging from 30 to 90%, depending on the source. Thus *E. sinica* contains ca. 1.3% alkaloids with more than 60% ephedrine; *E. intermedia* contains ca. 1.1% alkaloids with 30–40% ephedrine; and *E. equisetina* contains ca. 1.7% alkaloids with 85–90% ephedrine. Other alkaloids include *l*-*N*-methylephedrine, *d*-*N*-methylpseudoephedrine, *l*-norephedrine, *d*-norpseudoephedrine (cathine), ephedine, ephedroxane, and pseudoephedroxane (IMM-4).[1,2] The alkaloids are concentrated in the internodes, with lesser amount (ca. 50%) in the nodes, and none in the root (ZHOU).

Other compounds present include glycans (ephedrans A, B, C, D, and E with mol. wt. of $1.2 \times 10^6$, $1.5 \times 10^6$, $1.9 \times 10^4$, $6.6 \times 10^3$ and $3.4 \times 10^4$, respectively);[3] a volatile oil;[4,5] catechin, gallic acid, and condensed tannin; flavonoid glycosides; inulin, dextrin, starch, and pectin; and other common plant constituents, including plant acids (citric, malic, oxalic, etc.), sugars and trace minerals.

Active components in the volatile oil

include limonene, caryophyllene, phellandrene, linalool, *l*-$\alpha$-terpineol, and 2,3,5,6-tetramethylpyrazine. Concentration and composition vary considerably, depending on botanical sources (e.g., 0.250% volatile oil in Chinese ephedra vs. 0.124% in Mongolian ephedra)[4] and type of processing. For example, honey-cured and stir-fried *mahuang* contain higher concentration of antiasthmatic (*l*-$\alpha$-terpineol, caryophyllene and tetramethylpyrazine) and antitussive, expectorant, antibacterial, and antiviral components (limonene and linalool) but at the same time is devoid of other compounds (nerolidol, farnesol, selinene, nonadecane, eicosane, octadecane, dodecanoic acid, tetradecanoic acid, 1,2-benzenedicarboxylic acid dibutyl ester, etc.) that are present in raw *mahuang*.[5]

Root (*mahuanggen*) contains macrocyclic spermine alkaloids (ephedradines A, B, C, and D); an imidazole alkaloid (feruloylhistamine); *l*-tyrosine betaine (maokonine); and diflavonols (mahuannins A and B) (IMM-4, WANG).[6–10]

## PHARMACOLOGY OR BIOLOGICAL ACTIVITIES

The pharmacological effects of *mahuang* are generally attributed to ephedrine, which include central nervous system (CNS) stimulation, peripheral vasoconstriction, elevation of blood pressure, bronchodilatation, cardiac stimulation, decrease of intestinal tone and motility, mydriasis, and tachycardia, among others (ZHOU). The central stimulant action of ephedrine appears to be mediated by 1-adrenoceptors and not by dopamine receptors.[11]

*d*-Pseudoephedrine has similar activities as ephedrine except that its pressor (hypertensive) and CNS effects are weaker; it also has strong diuretic action in animals (dog and rabbit) (MARTINDALE, ZHOU).

*d*-Norpseudoephedrine (major active chemical present also in the African stimulant khat) is also a CNS stimulant (MARTINDALE, TYLER 3).

Ephedrine, pseudoephedrine, ephedroxane, and pseudoephedroxane have been shown to have antiinflammatory effects on experimental edema in animals.[2]

The glycans (ephedrans A, B, C, D, and E) exhibited marked hypoglycemic effects in normal and alloxan-induced hyperglycemic mice.[3]

*Mahuang* decoction and volatile oil have diaphoretic action in humans, and its decoction and alcoholic extract have antiallergic effects *in vitro*. When administered subcutaneously, the volatile oil was effective in treating mice infected with the Asian influenza virus strain $AR_8$ (ZHOU).

*Mahuanggen* extract when injected intravenously into cats and rabbits caused vasodilation and lowered blood pressure as well as stimulated respiration and inhibited perspiration; it also contracted isolated guinea pig and rabbit smooth muscle preparations (e.g., uterus and intestine) (HU). The ephedradines, tyrosine betaine, and feruloylhistamine are the hypotensive principles (IMM-4).[7,8]

Reported side effects of abuse include insomnia, motor disturbances, high blood pressure, glaucoma, impaired cerebral circulation, urinary disturbances, and others.[12]

## USES

**Health Food/Herb Teas.** Used in diet formulas for its appetite suppressive effect and in "energy" formulas for its central nervous system stimulant action; also used in cold and flu remedies, usually in tablet, capsule and tea forms.

**Traditional Medicine.** *Mahuang* and *mahuanggen* (root) are traditionally used for different purposes: the former as diaphoretic and the latter as antisudorific and antiperspirant.

*Mahuang* is traditionally considered to have diaphoretic, diuretic, antiasthmatic, cold-dispersing, lung-soothing (*xuan fei*), and antiswelling (*xiao zhong*) properties

(CHP); used for more than 2000 years to treat bronchial asthma, cold and flu, fever, chills, lack of perspiration, headache, nasal congestion, aching joints and bones, cough and wheezing, and edema, among others.

*Mahuanggen* is traditionally used to treat spontaneous and night sweating due to body deficiency (*ti xu*); now also used externally to treat excessive perspiration (e.g., foot), where the powder is topically applied (NATIONAL).

**Others.** *Mahuang* serves as raw material for the extraction and production of natural ephedrine and pseudoephedrine. *Mahuanggen*, due to its antiperspirant properties, is a potential ingredient in antiperspirant preparations.

**COMMERCIAL PREPARATIONS**

Crude (mainly raw, cut, and powdered) and extracts; extracts normally come in 5–9% total alkaloid content. Powdered crude *mahuang* claimed to contain 6–8% ephedrine has been offered, which should be considered adulterated, as commercial crude *mahuang* normally contains only about 1% ephedrine alkaloids. Ephedrine, ephedrine hydrochloride, and ephedrine sulfate formerly official in U.S.P.

*Regulatory Status*. Although ephedrine alkaloids are controlled drugs in the United States, the status of the herb *mahuang* is not clear. Herb subject of German therapeutic monograph for treatment of diseases of the respiratory tract with mild bronchospasms in adults.[12]

**REFERENCES**

See the General References for BLUMENTHAL; CHP; HU; IMM-4; JIANGSU; NATIONAL; TYLER 2; TYLER 3; WANG; ZHOU AND WANG.


1. C. Konno et al., *Phytochemistry*, **18**, 697 (1979).
2. Y. Kasahara et al., *Planta Med.*, 325 (1985).
3. C. Konno et al., *Planta Med.*, 162 (1985).
4. Y. Y. Jia et al., *Zhongguo Yaoxue Zazhi*, **24**, 402 (1989).
5. Q. Zeng et al., *Zhongguo Zhongyao Zazhi*, **17**, 83 (1992).
6. H. Hikino et al., *Planta Med.*, 478 (1984).
7. H. Hikino et al., *Planta Med.*, **48**, 108 (1983).
8. H. Hikino et al., *Planta Med.*, **48**, 290 (1983).
9. H. Hikino et al., *Heterocycles*, **17**, 155 (1982).
10. H. Hikino et al., *Heterocycles*, **19**, 1381 (1982).
11. Q. Li and B. H. Li, *Zhongguo Yaoli Xuebao*, **12**, 468 (1991).
12. Monograph *Ephedrae herba*, *Bundesanzeiger*, no. 11 (Jan. 17, 1991).

# EPIMEDIUM

**Source:** ***Epimedium brevicornum*** Maxim., ***E. pubescens*** Maxim., ***E. koreanum*** Nakai, ***E. wushanense*** T. S. Ying, ***E. acuminatum*** Franch., ***E. sagittatum*** (Sieb. et Zucc.) Maxim., and nine other ***Epimedium*** species (Family Berberidaceae).[1]

*Synonyms*. Herba epimedii, *yinyanghuo* and *xian ling pi*.

## GENERAL DESCRIPTION

Perennial herbs, most under 50 cm high, with compound leaves, leaflets thin leathery; native to China and Korea; now widely distributed in China; can be differentiated microscopically by the characteristic features of their hairs, especially nonglandular hairs.[2] The species of most commercial importance are *E. brevicornum*, *E. pubescens*, *E. koreanum*, *E. wushanense* and *E. acuminatum*, in decreasing order.[1] Although *E. sagittatum* is often described as a major source, it is in fact a minor source. Also, although *E. grandiflorum* Morr. (syn. *E. macranthum* Komarov.) has been listed as a major source, it is a Japanese species not commercially available in China.[1]

Part used is the aerial portion (mostly leaves) collected from wild plants in summer or autumn when leaves are bright green; then rid of thick petioles and impurities and sun dried or dried in the shade. Normally exported in neatly tied rectangular bundles; sometimes also in loose form, with stems and petioles.

## CHEMICAL COMPOSITION

Leaves and stems from different species (including *E. brevicornum*, *E. pubescens*, *E. davidii*, *E. wushanense*, *E. acuminatum*, and *E. sagittatum*) contain flavonoid glycosides, mostly rhamnosides (ca. 4.5% in *E. brevicornum*):[3] icariin and epimedosides, wushanicariin (3′,5′,7-trihydroxy-4′ - methoxy - 6 - (3, 3 -dimethylallyl) - flavone - 7 - β - D-glucopyranoside), hyperin (hyperoside, quercetin galactoside), quercitrin (quercetin rhamnoside), kaempferol-3-*O*-α-L-rhamnopyranoside, acuminatin (6″,6″-dimethylpyrano-(2″,3″,7,8)-4′-methyl kaempferol-3-*O*-α-L-rhamnopyranoside), kaempferol-3-dirhamnoside, baohuosides, sagittatins A and B, and others.[3–9]

Traditional curing of epimedium herb by stir-frying with lamb fat (20% w/w) followed by drying only slightly reduces total flavonoids content but significantly improves their water extractability (MA).[10]

Other chemical constituents present include flavonols (icaritin, baohuosu, tricin, desmethylicaritin, etc.);[5,7] polysaccharides;[11] volatile oil, phytosterols (daucosterol), tannin, fatty acids, and others (JIANGSU).[9]

## PHARMACOLOGY OR BIOLOGICAL ACTIVITIES

Epimedium is a highly valued tonic herb in traditional Chinese medical practice and has been shown to have numerous pharmacologic effects in humans and experimental animals, including stimulation or improvement of male sexual function in experimental animals, with cured (fried with 20% w/w lamb fat) but not raw epimedium (*E. brevicornum*) being the active herb;[12] broad cardiovascular effects (hypotensive, peripheral vasodilatory, increasing peripheral and coronary blood flow volumes, stimulating ADP-induced platelet aggregation, etc.);[11] catecholamine inhibition;[13] promotion of growth of chick embryonic femur and its protein and polysaccharide synthesis *in vitro*;[14] immunomodulating; regulating nucleic acid metabolism; antiviral and antibacterial; antiinflammatory; antitussive and expectorant; and others (JIANGSU, WANG).[15]

Thus far, the total flavonoids, icariin, and polysaccharides have been shown to be the active constituents of epimedium.

Acute toxicity of epimedium is low: the $LD_{50}$ of its total flavonoids (i.p.) in white mice was $2.99 \pm 0.14$ g/kg.[15]

## USES

**Medicinal, Pharmaceutical, and Cosmetic.** Extracts used in personal care products (e.g., disinfectant sprays) for its antimicrobial effects.

**Health Food/Herb Teas.** Powdered herb and extracts used in tonic formulas and teas for its traditional male tonic (aphrodisiac) properties (JIANGSU).

**Traditional Medicine.** First recorded use dates back to the *Shen Nong Ben Cao Jing* (ca. 200 B.C.–A.D. 100). Traditionally regarded as acrid and sweet tasting; warming; invigorating kidney *yang* (*bu shen yang*); benefiting *jing* (semen, life essence); strengthening bones, tendons, and muscles (*qiang jin gu*); and antirheumatic and antiarthritic (*qu feng shi*). Used in impotence, spermatorrhea, weak back and knees, rheumatism and arthritic pain, mental fatigue and poor memory, and postmenopausal hypertension, among others.

In recent years extensively used in China in the treatment of coronary heart disease, hypertension (including postmenopausal), bronchitis, and neurasthenia;[16,17] also used in chronic hepatitis, poliomyelitis, chronic leukopenia, and others (JIANGSU, WANG).[16,18]

## COMMERCIAL PREPARATIONS

Crude and extracts. Crude (raw herb) comes in whole bundled form composed of mostly leaflets with little or no petiolules or in loose form containing leaves with stems and petioles, also in powdered form. Cured form not available in bulk quantities. Extracts occasionally come with assay for flavonoids.

*Regulatory Status.* U.S. regulatory status not determined.

## REFERENCES

See the General References for CHP; IMM-4; JIANGSU; LU AND LI; MA; NATIONAL; WANG; ZHU.

1. H. R. Liang et al., *Zhongyao Tongbao*, **13**(12), 7 (1988).
2. H. R. Liang et al., *Beijing Zhongyi Xueyuan Xuebao*, **13**(2), 42 (1990).
3. L. X. Xu and X. Q. Zhang, *Yaoxue Xuebao*, **24**, 606 (1989).
4. H. R. Liang et al., *Yaoxue Xuebao*, **23**, 34 (1988).
5. F. Li and Y. L. Liu, *Yaoxue Xuebao*, **23**, 739 (1988).
6. Y. S. Li and Y. L. Liu, *Zhongcaoyao*, **23**, 8 (1992).
7. F. Li and Y. L. Liu, *Yaoxue Xuebao*, **23**, 672 (1988).
8. Y. Oshima et al., *Planta Med.*, **55**, 309 (1989).
9. B. H. Hu et al., *Yaoxue Xuebao*, **27**, 397 (1992).
10. R. Niu et al., *Zhongchengyao*, **13**(1), 18 (1991).
11. R. S. Li et al., *Zhongyao Tongbao*, **12**(8), 40 (1987).
12. R. Niu, *Zhongguo Zhongyao Zazhi*, **14**(9), 18 (1989).
13. H. C. Li and G. H. Huang, *Zhongcaoyao*, **15**(2), 26 (1984).
14. Z. F. Gao et al., *Chin. J. Integr. Trad. Western Med.*, **5**, 172 (1985).
15. J. H. Liu and L. S. Shen, *Beijing Zhongyi Xueyuan Xuebao*, **16**(1), 29 (1993).
16. F. C. Liu, *Zhongcaoyao*, **16**(10), 44 (1985).

17. L. Yu et al., *Zhongyi Zazhi* (3), 36 (1990).

18. P. G. Xiao and K. J. Chen, *Phytother. Res.*, **2**(2), 55 (1988).

# EUCALYPTUS

**Source: *Eucalyptus globulus*** Labill. (Family Myrtaceae).

*Synonyms*. Blue gum, Tasmanian blue gum, fever tree, and gum tree.

## GENERAL DESCRIPTION

Evergreen tree with bluish green leaves often covered with a white powder; up to about 90 m high; native to Australia; extensively cultivated worldwide (e.g., Europe, United States, China, Africa, and South America). Part used is the fresh or partially dried leaf from which the essential oil is produced by steam distillation. Major oil-producing countries include Spain, Portugal, and Brazil.[1]

Essential oils from other *Eucalyptus* species are also used, some of which may have quite different chemical compositions.

## CHEMICAL COMPOSITION

Eucalyptus leaves contain 0.5–3.5% volatile oil, tannins, polyphenolic acids (gallic, caffeic, ferulic, gentisic, protocatechuic acids, etc.), flavonoids (quercetin, quercitrin, rutin, hyperoside, eucalyptin, etc.), wax, and others (JIANGSU, LIST AND HÖRHAMMER).[2–6]

Eucalyptus oil contains usually 70–85% of eucalyptol (1,8-cineole);[1,2,7] other constituents present are mostly monoterpene hydrocarbons ($\alpha$-pinene, *d*-limonene, *p*-cymene, $\beta$-pinene, $\alpha$-phellandrene, camphene, $\gamma$-terpinene, etc., with the first three in major amounts), with lesser amounts of sesquiterpenes (e.g., aromadendrene, allo-aromadendrene, globulol, epiglobulol, ledol, and viridiflorol), aldehydes (e.g., myrtenal), ketones (e.g., carvone and pinocarvone), and others (JIANGSU).[1,8,9] A recent study reported eucalyptol content of 61.2% in Brazilian eucalyptus;[10,11] in Chinese eucalyptus, 83.9% reported.[11]

The rectified oil contains little or no unpleasant smelling lower aliphatic aldehydes.

## PHARMACOLOGY OR BIOLOGICAL ACTIVITIES

Eucalyptus oil and eucalyptol reportedly have antiseptic (antibacterial) and expectorant properties;[12–14] strongly antibacterial against several strains of *Streptococcus*.[15]

A crude extract of *E. globulus* leaves rich in phenolic glycoside(s) has been reported to have antihyperglycemic activity in rabbits; a loss of this activity resulted upon purification of this material.[5]

The flavonoids quercitrin and hyperoside have been reported to have eliminated influenza type A viral infections in mouse tissue and in chick embryos.[16]

Eucalyptus oil has been reported to be rapidly absorbed through the intact, shaved abdominal skin of the mouse and to promote the formation of tumors (papillomas) by 9,10-dimethyl-1,2-benzanthracene.[17,18] More recent studies have demonstrated eucalyptus oil and eucalyptol to be generally nonirritating, nonsensitizing, and nonphototoxic to the skin.[19,20]

When taken internally, eucalyptus oil is toxic, and ingestion of as little as 3.5 mL has been reported as fatal (JIANGSU).

Rare instances of nausea, vomiting, and diarrhea have been reported after ingestion of non-fatal doses of leaf preparations or essential oil.[21,22]

## USES

**Medicinal, Pharmaceutical, and Cosmetic.** Both eucalyptus oil and eucalyptol are extensively used as expectorants and/or flavoring agents in cold and cough medicines (e.g., cough drops and syrups), vaporizer fluids, antiseptic liniments, ointments, toothpastes, and mouthwashes. Also widely used as fragrance components in soaps, detergents, creams, lotions, and perfumes, with maximum use levels of 1.0 and 1.6% in perfumes reported for eucalyptus oil and eucalyptol, respectively.[19,20] Eucalyptus oil and eucalyptol are used in dentistry as components of certain root canal sealers; also used as solvents for root canal fillings.

**Food.** Both eucalyptus oil and eucalyptol are used as flavor ingredients in most food products, including alcoholic and nonalcoholic beverages, frozen dairy desserts, candy; baked goods, gelatins and puddings, meat and meat products, and others. Average maximum use levels reported are generally low, with the highest being about 0.002% (19.5 ppm) for eucalyptol in candy.

**Health Food/Herb Teas.** Leaves used in tea; oil a fragrance ingredient in topical balms and massage oils (ROSE).

**Traditional Medicine.** Leaves and oil are reportedly used as antiseptic and febrifuge, and as expectorant and stimulant in respiratory ailments; also used for wounds, burns, ulcers, and cancers.[23] In Chinese medicine leaves and oil are used for similar purposes. In addition, aqueous extracts and decoctions of the leaves are used to treat aching joints, bacterial dysentery, ringworms, pulmonary tuberculosis, and others; successful clinical studies on some of these uses have been reported (JIANGSU).

## COMMERCIAL PREPARATIONS

Leaves, oil, and eucalyptol. The leaves were formerly official in U.S.P., while eucalyptol was official in N.F. Eucalyptus oil and eucalyptol are currently official in F.C.C., and eucalyptus oil is official in N.F.

*Regulatory Status.* Has been approved for food use (§172.510); eucalyptol is listed as a synthetic flavoring agent (§172.515). Leaves and essential oil subjects of German therapeutic monographs, indicated for catarrhs of the upper respiratory tract;[22] oil topically for rheumatic complaints.[21]

## REFERENCES

See the General References for ADA; ARCTANDER; BAILEY 1; BIANCHINI AND CORBETTA; BLUMENTHAL; FEMA; GUENTHER; JIANGSU; LIST AND HÖRHAMMER; LUST; ROSE; UPHOF.

1. W. D. Fordham in L. W. Codd et al., eds., *Chemical Technology: An Encyclopedic Treatment*, Vol. 5, Barnes & Noble, New York, 1972, p. 1.
2. R. K. Baslas, *Indian Oil Soap J.*, **35**, 136 (1969); through *Chem. Abstr.*, **73**, 59210d (1970).
3. K. Boukef et al., *Plant. Med. Phytother.*, **10**, 24 (1976).
4. K. Boukef et al., *Plant. Med. Phytother.*, **10**, 30 (1976).
5. K. Boukef et al., *Plant. Med. Phytother.*, **10**, 119 (1976).
6. M. A. Elkeiy et al., *Bull. Fac. Pharm.*, **31**, 83 (1964); through *Chem. Abstr.*, **64**, 16277f (1966).
7. C. H. Brieskorn and W. Schlicht, *Pharm. Acta Helv.*, **51**, 133 (1976).
8. R. M. Ikeda et al., *J. Food Sci.*, **27**, 455 (1962).
9. J. De Pascual Teresa et al., *An. Quim.*, **73**, 751 (1977); through *Chem. Abstr.*, **87**, 156994j (1977).
10. T. Nakashima et al., *Trib. Farm.*, **53**(1), 29 (1985).
11. B. M. Lawrence, *Perfum. Flavor*, **15**(6), 45 (1990).

12. A. C. Pizsolitto et al., *Rev. Fac. Farm. Odontol. Araraquara*, **9**, 55 (1975); through *Chem. Abstr.*, **86**, 12226s (1977).
13. S. Prakash et al., *Indian Oil Soap J.*, **37**, 230 (1972); through *Chem. Abstr.*, **79**; 727y (1973).
14. J. C. Maruzella and P. A. Henry, *J. Am. Pharm. Assoc.*, **47**, 294 (1958).
15. A. Benouda et al., *Fitoterapia*, **59**, 115 (1988).
16. S. A. Vichkanova and L. V. Goryunova, *Tr. Vses. Nauch. Issled. Inst. Lek. Rast.*, **14**, 212 (1971); through *Chem. Abstr.*, **79**, 728z (1973).
17. F. Meyer and E. Meyer, *Arzeim. Forsch.*, **9**, 516 (1959).
18. F. J. C. Roe and W. E. H. Field, *Food Cosmet. Toxicol.*, **3**, 311 (1965).
19. D. L. J. Opdyke, *Food Cosmet. Toxicol.*, **13**, 107 (1975).
20. D. L. J. Opdyke, *Food Cosmet. Toxicol.*, **13**, 105 (1975).
21. Monograph *Eucalypti aetheroleum*, *Bundesanzeiger*, no. 177a (Sept. 24, 1986); revised (Mar. 6, 1990).
22. Monograph *Eucalypti folium*, *Bundesanzeiger*, no. 177a (Sept. 24, 1986); revised (Mar. 6, 1990).
23. J. Hartwell, *Lloydia*, **33**, 288 (1970).

# EUPHORBIA

**Source:** ***Euphorbia pilulifera*** L. (syn. *E. hirta* L.; *E. capitata* Lam.) (Family Euphorbiaceae).

*Synonyms*. Snake weed and pill-bearing spurge.

## GENERAL DESCRIPTION

An upright hairy annual; up to 0.5 m high; native to India. Part used is the whole flowering or fruiting plant.

## CHEMICAL COMPOSITION

Contains choline and shikimic acid as active constituents.[1] Other compounds present include triterpenes (e.g., free taraxerol and $\beta$-amyrin; esters of taraxerone, $\alpha$-amyrin, and $\beta$-amyrin; friedelin), sterols (campesterol, sitosterol, stigmasterol, etc.);[2,3] flavonoids (quercitrin, quercetin, leucocyanidin, xanthorhamnin, etc.);[4] *n*-alkanes (e.g., hentriacontane);[3] phenolic acids (e.g., gallic and ellagic), *l*-inositol, sugars (glucose, fructose, and sucrose), resins; and others (LIST AND HÖRHAMMER).[1]

## PHARMACOLOGY OR BIOLOGICAL ACTIVITIES

Euphorbia has been reported to have antispasmodic and histamine-potentiating properties[5] as well as antitumor activities in laboratory animals.[6]

Choline produces contraction of isolated guinea pig ileum, while shikimic acid produces a relaxation of guinea pig ileum.[1]

Shikimic acid, a ubiquitous constituent of higher plants, has been reported to have carcinogenic properties in mice,[7,8] though no mutagenic activities have been observed using the Ames assay.[9]

## USES

**Health Food/Herb Teas.** Used primarily in certain cough preparations.

**Traditional Medicine.** Reportedly used mainly in treating respiratory ailments (e.g., asthma, bronchitis, coughs, and hay fever); also in tumors.[10]

In India it is used in treating worms in children and for dysentery, gonorrhea, digestive problems, and others.

In China numerous *Euphorbia* species (e.g., *E. humifusa* Willd., *E. pekinensis* Rupr., *E. lunulata* Bunge, *E. lathyris* L., and *E. sieboldiana* Morr. et Decne.) are traditionally used in treating conditions that include dysentery, enteritis, ascites, bleeding, dropsical nephritis, and chronic bronchitis. Some of these uses have been clinically substantiated (JIANGSU).

**COMMERCIAL PREPARATIONS**

Crude and extracts (fluid, solid, etc.); crude and fluid extract were formerly official in N.F. Strengths (see ***glossary***) of extracts are expressed in weight-to-weight ratios.

**REFERENCES**

See the General References for FARNSWORTH 3; FOGARTY; FOSTER AND DUKE; GOSSELIN; JIANGSU; LIST AND HÖRHAMMER; YOUNGKEN.

1. L. El-Naggar et al., *Lloydia*, **41**, 73 (1978).
2. A. Atallah and H. Nicholas, *Phytochemistry*, **11**, 1860 (1972).
3. D. Gupta and S. Garg, *Bull. Chem. Soc. Jpn.*, **39**, 2532 (1966); through *Chem. Abstr.*, **66**, 26559b (1967).
4. P. Blanc and G. Sannes, *Plant. Med. Phytother.*, **6**, 106 (1972).
5. R. C. Hellerman and L. Hazelton, *J. Am. Pharm. Assoc.*, **39**, 142 (1950).
6. M. Belkin and D. B. Fitzgerald, *J. Natl. Cancer Inst.*, **13**, 139 (1952).
7. I. A. Evans and M. A. Osman, *Nature (London)*, **250**, 348 (1974).
8. B. Stavric and D. R. Stoltz, *Food Cosmet. Toxicol.*, **14**, 141 (1976).
9. L. B. Jacobsen, et al., *Lloydia*, **41**, 450 (1978).
10. J. L. Hartwell, *Lloydia*, **32**, 153 (1969).

# EVENING PRIMROSE

**Source:** ***Oenothera biennis*** L. (Family Onagraceae).

**GENERAL DESCRIPTION**

Annual or biennial, 1–3 m. Leaves in basal rosette before anthesis, lanceolate, 10–22 cm long, 1 cm wide, margins undulate or minutely toothed; flowers four-merous, yellow; fruit a dry pod to 4 cm, with numerous minute seeds; throughout North America, pastures, old fields, roadsides; cultivated in Europe, North America, and elsewhere for seed oil.

**CHEMICAL COMPOSITION**

Seed contains about 14% fixed oil (evening primrose oil, EPO), with about 50–70% *cis*-linolenic acid, and 7–10% *cis*-γ-linolenic acid (GLA); *cis*-6,9,12-octadecatrienoic acid, plus small amounts of oleic, palmitic, and stearic acid; steroids campesterol and β-sitosterol.[1]

**PHARMACOLOGY OR BIOLOGICAL ACTIVITIES**

GLA inhibits platelet aggregation, reduces blood pressure, restores motility of red blood cells in multiple sclerosis. Exaggerated claims to efficacy in obesity unsubstantiated (WREN). EPO of interest for GLA content as a prostaglandin precursor, especially for $PGE_1$; prostaglandins help regulate metabolic functions. Normal synthesis of GLA from linoleic acid via δ-6-desaturase may be blocked or diminished in mammalian systems as the result of aging, diabetes, excessive carbohydrate intake, or fasting. Therapeutic use for atopic eczema produced modest, but significant, improvement (20–25% over controls),[2] premenstrual syndrome (PMS), diabetes, alcoholism, inflammation and multiple sclerosis; preventive in heart disease and stroke.[3] A prospective randomized, double-blind, placebo-controlled, cross-over trial evaluated efficacy in relief of PMS symptoms showed improvement, though statistically insignificant.[4] However, other double-blind, placebo-controlled studies showed that evening primrose oil significantly reduced irritability, breast pain and tenderness, and mood changes associated with PMS.[5] GLA supplement is valid for increased demand for GLA in alcoholism.[6]

## USES

**Medicinal, Pharmaceutical, and Cosmetic.** Evening primrose oil used clinically in the UK for the treatment of atopic eczema, mastalgia, premenstrual syndrome; increasingly seen in cosmetic products, including hand lotions, soaps, shampoos, etc.

**Health Food/Herb Tea.** Capsulated seed oil products widely available; dietary supplement for addition of essential fatty acids to diet (FOSTER).

**Traditional Medicine.** Whole plant infusion as astringent, sedative, antispasmodic in asthmatic coughs, gastrointestinal disorders, whooping cough; poulticed to enhance wound healing, anodyne; root rubbed on muscles to give athletes strength. Leaves, shoots, root, and seeds also used as food by American Indians.[6]

## COMMERCIAL PREPARATIONS

Seed oil; seed oil capsules with GLA and vitamin E. Products also combined with linseed oil, safflower oil.

*Regulatory Status.* In the UK, approved therapeutic agent for treatment of atopic eczema. In Canada, a dietary supplement for increased essential fatty acid intake. Undetermined in the United States; the FDA has treated EPO both as a "misbranded drug" and "unsafe food additive," seizing product by treating EPO as a "food additive," rather than a food. However, a recent decision of the U. S. Court of Appeals for the 7th Circuit (decision rendered Jan. 27, 1993), involving seizure of black currant oil (also a GLA source), found in favor of the defendant (Traco Labs, Inc.). Writing for the court, Judge Cudahy stated that "the [Food, Drug, and Cosmetic] Act distinguishes between food additives and food in the generic sense, and this distinction is critical in allocating the burden of proof. The FDA's food additive definition is so broad, however, that it would blur this distinction. It would classify every component of food—even single, active ingredients—as food additives. Thus, it would seem, even the addition of water to food would make the food a food additive. The only justification for this Alice-in-Wonderland approach is to allow the FDA to make an end-run around that statutory scheme and shift to the processors the burden of proving the safety of a substance in all circumstances."[7]

The court's decision positively affects the dietary supplement status of EPO.

## REFERENCES

See the General References for DUKE 2; FOSTER; FOSTER AND DUKE; GLASBY; MARTINDALE; TYLER; WEISS; WREN.

1. Fedeli et al., *Riv. Ital. Sostanze Grasse*, **53**, 23 (1976).
2. C. R. Lovell et al., *Lancet*, **1**(8214), 278 (1981).

3. J. Janick et al. in L. E. Craker and J. E. Simon, eds., *Herbs, Spices, and Medicinal Plants: Recent Advances in Botany, Horticulture, and Pharmacology*, Vol. 4, Oryx Press, Phoenix, Ariz., 1989, p. 145.
4. S. K. Khoo et al., *Med. J. Aust.*, **152**, 189 (1990).
5. J. K. Pyke et al., *Lancet*, **2**, 373 (1985).
6. C. J. Briggs, *Can. Pharm. J.*, 250 (May 1986).
7. M. Blumenthal, *HerbalGram*, **29**, 38 (1993).

# EYEBRIGHT

**Source:** ***Euphrasia rostkoviana*** F. Hayne and other ***Euphrasia*** spp. (***E. officinalis*** L.) (Family Scrophulariaceae).

## GENERAL DESCRIPTION

*Euphrasia officinalis* has been used by modern authors to refer collectively to the genus; hence is a "collective species" and, as such, a *nomen ambiguum*. Close to 450 species described; many into hardly distinguishable microspecies; *Euphrasia officinalis* probably the most useful designation for commercial supplies.

Small, hemiparasitic, herbaceous, mostly annual herbs to 4 dm, simple or freely branched; leaves opposite sessile, ovate to rotund, palmately veined, coarsely toothed above; bracteal leaves tend to alternate; flowers small, four lobed, deeply cleft above; corolla bilabiate, upper lip concave, two lobed or notched. Cold temperate regions, Northern and Southern Hemisphere; subarctic, alpine areas of tropical mountains (GLEASON AND CRONQUIST, TUCKER 2, TYLER 1). Part used is the whole herb. Most commercial supply from Europe, limited wild harvest in North America.

## CHEMICAL COMPOSITION

*E. rostkoviana* iridoid glycosides including aucubin, catapol, and erostoside; eukovoside; geniposide and luproside; gallotannins; caffeic and ferulic acids; trace amounts of an essential oil at 0.017 %; choline; $\beta$-sitosterol; oleic, linoleic, linolenic, palmitic, and stearic acids.[1,2]

## PHARMACOLOGY OR BIOLOGICAL ACTIVITIES

Astringent ophthalmic; infusion internally and externally (as eye wash) recommended historically and in modern literature for eye irritations, particularly conjunctivitis, without scientific substantiation (TYLER 1).

A report of clinical success with compresses of an eyebright decoction to provide surprisingly rapid relief of redness, swelling, and visual disturbances in acute and subacute eye inflammations, particularly conjunctivitis, blepharitis, and recent eye injuries with risk of serpiginous corneal ulcers developing; used internally at the same time (WEISS).

## USES

**Medicinal, Pharmaceutical, and Cosmetic.** Used mainly in Europe as rinse, compress, or eye bath for eye-related inflammatory and vascular conditions, including eye lid inflammation, conjunctivitis, secreting and inflamed eyes, catarrh of eyes, prevention of mucous secretion from eyes.[3]

**Health Food/Herb Teas.** Tea, capsules, tablets, tincture, etc., presumably for traditional applications.

**Traditional Medicine.** Use as ophthalmic for eye inflammations with mucous discharge; a folk remedy for allergy, cancer, coughs, earache, headache with congestion, hoarseness, inflammation, jaundice, rhinitis, and sore throat (DUKE 2, FOSTER AND DUKE). In European tradition also reported as stomachic and for skin diseases.[3]

*Regulatory Status*. Undetermined in the United States. Subjects of a German therapeutic monograph; use not recommended for eye conditions because of hygienic concerns and efficacy not documented.[3]

## COMMERCIAL PREPARATIONS

Crude herb, extracts, etc.

## REFERENCES

See the General References for BLUMENTHAL; DUKE 3; FOSTER AND DUKE; GLASBY; GLEASON AND CRONQUIST; LUST; MARTINDALE; TUCKER 2; STEINMETZ; TYLER 1; WEISS; WREN.

1. O. Salama and O. Sticher, *Planta Med.*, **47**, 90 (1983).
2. O Sticher et al., *Helv. Chim. Acta*, **65**, 1538 (1982).
3. Monograph *Euphrasia*, *Bundesanzeiger* (Aug. 29, 1992).

# FANGFENG

**Source:** ***Saposhnikovia divaricata*** (Turcz.) Schischk. (syn. *Ledebouriella divaricata* (Turcz.) Hiroe; *L. seseloides* (Hoffm.) Wolff; *Siler divaricatum* (Turcz.) Benth. et Hook. f.) (Family Umbelliferae or Apiaceae).

*Synonyms. Guan fangfeng, saposhnikovia,* siler, ledebouriella, radix saposhnikoviae, radix ledebouriellae, and radix sileris.

## GENERAL DESCRIPTION

Herbaceous glabrous perennial, 30–80 cm high, with thick root; base of stem covered with brown fibrous remains of petioles. Plant native to China and is distributed throughout northern and northeastern provinces, with some fields still in their virgin primitive state existent in Heilongjiang and Inner Mongolia (HU). Part used is the root collected in spring or fall (when it reaches about 30 cm long and 1.2 cm thick) from cultivated (2–3 years old) or wild plants, rid of rootlets and dirt, and sun dried. Heilongjiang is the largest producer.

Substitutes produced in other regions of China are not saposhnikovia but are from other species; they include *chuan fangfeng* from Sichuan (*Peucedanum dielsianum* Fedde ex Wolff) and *yun fangfeng* from Yunnan (*Seseli mairei* Wolff, *Pimpinella candolleana* Wight et Arn. or *Seseli yunnanense* Franch.) (CMH, ZHU).[1–3]

## CHEMICAL COMPOSITION

Contains coumarins (anomalin, bergapten, imperatorin, phellopterin, xanthotoxin, psoralen, scopoletin, etc.); chromones (5-*O*-methylvisamminol, 4-*O*-$\beta$-glucopyranosyl-5-*O*-methylvisamminol, hamaudol, *sec*-*O*-glucosylhamaudol, 3′-*O*-acetylhamaudol, 3′-*O*-angeloylhamaudol, cimifugin, *prim*-*O*-glucosylcimifugin, mannitol and sucrose); lignoceric acid; a volatile oil; mannitol; and sucrose; $\beta$-sitosterol and its glucoside; and others.[2,4–6]

## PHARMACOLOGY OR BIOLOGICAL ACTIVITIES

Both aqueous and alcoholic extracts have marked antipyretic, analgesic, and antiinflammatory effects in mice and rats (WANG).[7] Alcohol extractives had mild antihistaminic activity on isolated guinea pig trachea;[8] also had tyrosinase inhibitory effects.[9–10] Aqueous extract also exhibited immunopotentiating effects in mice, markedly increasing phagocytosis by macrophages.[11]

Certain coumarins (e.g., bergapten and xanthotoxin) are phototoxic (see ***angelica*** and ***bergamot oil***).

## USES

**Medicinal, Pharmaceutical, and Cosmetic.** Listed in Li Shi-Zhen's *Ben Cao Gang Mu* (ca. 1590) as one of the herbs for removing facial dark spots as well as having antipruritic properties, *fangfeng* is used in skin care products (creams, lotions, bath preparations, and antiallergic ointments) and in hair tonics for these properties and for its antiinflammatory and whitening (tyrosinase inhibitory) effects.

**Health Food/Herb Teas.** Powdered herb and extracts used extensively as ingredients in formulas for treating the common cold, influenza, arthritis, rheumatism, pruritus, and urticaria, usually in capsule or tablet form (JIANGSU).

**Traditional Medicine.** First described in the *Shen Nong Ben Cao Jing* (ca. 200 B.C.–A.D. 100), *fangfeng* is one of the major wind- and dampness-dispelling drugs (*qu feng chu shi*; antiinflammatory) and has since been used in countless prepared formulas for relieving pain due to arthritis, rheumatism, common cold, and influenza. Traditionally considered pungent and sweet tasting, warming, and also to have diaphoretic (*jie biao*) properties, it is commonly used in treating cold and flu and their associated headaches, migraine, rheumatism, and arthritis, urticaria; pruritus, hard-to-heal carbuncles and tetanus. It is one of the three ingredients (with astragalus and *baizhu*) of the famous 15th-century formula, *yu ping fang san* ("jade screen powder") for fortifying body defense against outside pathogens, whose efficacy in enhancing immune functions and preventing colds and flus has been well documented (DENG).

### COMMERCIAL PREPARATIONS

Crude comes in sticks (20–30 cm long), slices or as a powder. Extracts (water or hydroalcoholic) do not have uniform strengths nor assays of chemical components.

*Regulatory Status*. U.S. regulatory status not determined.

### REFERENCES

See the General References for CHP; CMH; HU; JIANGSU; NATIONAL; WANG; ZHU.

1. J. H. Wang and Z. C. Lou, *Zhongyao Tongbao*, **13**(1), 9 (1988).
2. J. H. Wang and Z. C. Lou, *Zhongguo Yaoxue Zazhi*, **27**(6), 323 (1992).
3. J. H. Wang and Z. C. Lou, *Zhongyao Tongbao*, **13**(2), 5 (1988).
4. A. R. Ding et al., *Zhongcaoyao*, **18**(6), 7 (1987).
5. G. Z. Jin et al., *Zhongguo Zhongyao Zazhi*, **17**(1), 38 (1992).
6. H. Sasaki et al., *Chem. Pharm. Bull.*, **30**, 3555 (1982).
7. J. H. Wang et al., *Acta Med. Sinica*, **4**(1), 20 (1989).
8. R. D. Xiang et al., *Zhongcaoyao*, **15**(2), 22 (1985).
9. X. T. Liu, *Zhongchengyao*, **13**(3), 9 (1991).
10. Y. Masamoto et al., *Planta Med.*, **40**, 361 (1980).
11. S. Y. Zhang et al., *Zhongcaoyao*, **18**(9), 9 (1987).

# FENNEL

**Source:** ***Foeniculum vulgare*** Mill. (syn. *F. officinale* All.; *F. capillaceum* Gilib.; *Anethum foeniculum* L.) (Family Umbelliferae or Apiaceae).

*Synonyms*. Florence fennel and finocchio.

### GENERAL DESCRIPTION

Perennial herb with erect stem; up to about 1.5 m high; generally considered to be native of the Mediterranean region; cultivated as an annual or a perennial worldwide (Argentina, Hungary, Bulgaria, Germany, France, Italy, Greece, China, India, etc.). Part used is the dried ripe fruit (commonly called seed) from which an essential oil is obtained by steam distillation.

There are two commonly used varieties of fennel: common fennel (or bitter fennel) and sweet fennel, with the latter occurring only in the cultivated form.[1,2] Common fennel appears to be the more commonly used fennel whenever the spice is called for. However, although fennel oils are official in the N.F. and F.C.C. (without specific distinctions between them), sweet fennel oil is reported to be the one generally used; bitter fennel oil is used only to a limited extent, mainly in cosmetics (ARCTANDER, FEMA, FURIA AND BELLANCA, ROSENGARTEN).[3,4]

## CHEMICAL COMPOSITION

Fruits contain 1.5–8.6% (usually 2–6%) volatile oil;[1,5–8] 9–28% (usually 17–20%) fixed oil composed primarily of petroselinic acid (60–75%), oleic acid, and linoleic acid with a relatively high concentration of tocopherols (mostly γ-tocotrienol);[7–9] flavonoids (mainly quercetin-3-glucuronide, rutin, isoquercitrin, and quercetin-3-arabinoside, with minor amounts of kaempferol-3-arabinoside and kaempferol-3-glucuronide;[10,11] umbelliferone (7-hydroxycoumarin); stigmasterol; protein (16–20%); sugars; vitamins; minerals (relatively high in calcium and potassium); and others (JIANGSU, LIST AND HÖRHAMMER, MARSH).[12]

The volatile oil contains mostly *trans*-anethole, with lesser amounts of fenchone, estragole (methyl chavicol), limonene, camphene, and α-pinene. Other compounds present include more monoterpene hydrocarbons (β-pinene; α-thujene, α-fenchene, 3-carene, sabinene, α-phellandrene, myrcene, α- and β-terpinene, *cis*- and *trans*-ocimenes, terpinolene, and *p*-cymene), fenchyl alcohol, anisaldehyde, *p*-anisic acid, *trans*-1,8-terpin, myristicin, and apiole, the last two reportedly only present in the cultivated sweet variety.[1,2,6,7,13–16] The concentrations of *trans*-anethole in the oil vary widely, with reported values ranging from 50 to 90%, depending on the varieties, sources, ripeness of fruits, and other factors.[1,2,6,7,14] There are also considerable variations in the amounts of fenchone (0–22%) and estragole.[2,6,7,14,17] Common fennel is reported to contain usually lower amounts of anethole but higher amounts of fenchone than sweet fennel (JIANGSU, LIST AND HÖRHAMMER).

## PHARMACOLOGY OR BIOLOGICAL ACTIVITIES

Fennel and its volatile oil have carminative and stimulant properties.

Fennel oil has been reported to have spasmolytic effects on smooth muscles of experimental animals.[18] It also exhibited antibacterial activities *in vitro*.[19]

A terpene fraction of fennel oil has shown strong cytotoxic properties.[20]

Aqueous extracts of fennel experimentally increase ciliary action of ciliary epithelium in frogs.[21]

Anethole is reported to have allergenic, weakly insecticidal, and toxic properties (see ***anise***). It also stimulates secretions of the upper respiratory tract, stimulating ciliary action and ciliary epithelium in frogs.[21] Recent research suggests polymers of anethole, such as dianethole and photoanethole, are active estrogenic compounds.[22] Anethole and fenchone experimentally reduce secretions of upper respiratory tract.[21]

Estragole has been reported to cause tumors in animals (see ***sweet basil***).

## USES

**Medicinal, Pharmaceutical, and Cosmetic.** Fennel and sweet fennel oil are used as a carminative or flavoring agent in certain laxative preparations.

In Germany, the fruits used in phytomedicines for dyspeptic disorders; mild gastrointestinal antispasmodic, also upper respiratory tract conditions (expectorant); in syrup for children's coughs.[21,23]

Bitter (common) fennel and sweet fennel oils are used as fragrance components in cosmetics, including soaps, detergents, creams, lotions, and perfumes, with highest maximum use levels of 0.4% reported for both oils in perfumes.

**Food.** Common fennel is used as a flavor component in alcoholic beverages (especially liqueurs), baked goods, meat and meat products, fats and oils, snack foods, and gravies, with highest average maximum use level of about 0.119% (1186 ppm) reported in meat and meat products.

Sweet fennel is reported used in nonalcoholic beverages, candy, baked goods, meat and meat products, condiments and relishes, gravies, and processed vegetables. Highest average maximum use level reported is about 0.305% (3049 ppm) in meat and meat products.

Sweet fennel oil is widely used in most major food products, including alcoholic (e.g., liqueurs) and nonalcoholic beverages, frozen dairy desserts, candy, baked goods, gelatins and puddings, meat and meat products, and condiments and relishes, among others. Highest average maximum use level reported is about 0.023% (234 ppm) in alcoholic beverages.

**Health Food/Herb Teas.** Crushed or ground fruit in teas, tincture, or in honey syrup (FOSTER).[23]

**Traditional Medicine.** Fennel fruits and oil are reportedly used as a stomachic and as a carminative in treating flatulence and other stomach troubles as well as for catarrhs of the upper respiratory tract.[21,23]

Traditionally fennel fruits have been used for many similar purposes as dill fruits, both in Western traditional medicine and in Chinese medicine.

In Chinese medicine fennel has also been used for centuries in treating hard-to-heal snakebites, for which the powdered drug is used as a poultice; and for cholera, backache, and bedwetting, usually decocted with other drugs (JIANGSU).

## COMMERCIAL PREPARATIONS

Crude and oil. Crude was formerly official in N.F. and U.S.P. Oil official in N.F. and F.C.C.

*Regulatory Status.* GRAS: common fennel and sweet fennel (§182.10); sweet fennel (§182.20). Seeds and essential oil subjects of German therapeutic monographs.[21,23] The oil (0.1–0.6 mL daily dose) and fruits (5–7 g daily dose) allowed for stimulation of gastrointesinal motility (or spasmolytic effect at high end of dosage) for dyspeptic discomfort, gastrointestinal spasms and congestion of upper respiratory tract.[21,23]

## REFERENCES

See the General References for BAILEY 1; BIANCHINI AND CORBETTA; BLUMENTHAL; FEMA; FOSTER; GUENTHER; JIANGSU; LIST AND HÖRHAMMER; MARTINDALE; ROSENGARTEN; TERRELL; UPHOF.

1. M. B. Embong et al., *Can. J. Plant Sci.*, **57**, 829 (1977).
2. J. Karlsen et al., *Planta Med.*, **17**, 281 (1969).
3. D. L. J. Opdyke, *Food Cosmet. Toxicol.*, **12**(Suppl.), 879 (1974).
4. D. L. J. Opdyke, *Food, Cosmet. Toxicol.*, **14**, 309 (1976).
5. M. A. Wahid and M. Ikram, *Pak. J. Sci. Ind. Res.*, **4**, 40 (1961); through *Chem. Abstr.*, **60**, 15678d (1964).
6. M. Ashraf and M. K. Bhatty, *Pak. J. Sci. Ind. Res.*, **18**, 236 (1975); through *Chem. Abstr.*, **87**, 28848c (1977).
7. M. R. I. Saleh et al., *J. Pharm. Sci. U. Arab. Rep.*, **5**, 55 (1964); through *Chem. Abstr.*, **65**, 14096a (1966).
8. A. Seher and S. Ivanov, *Fette, Seifen, Anstrichmit.*, **78**, 224 (1976).
9. J. P. Moreau et al., *J. Am. Oil Chem. Soc.*, **43**, 352 (1966).
10. J. B. Harborne and C. A. Williams, *Phytochemistry*, **11**, 1741 (1972).
11. J. Kunzemann and K. Herrmann, *Z.*

*Lebensm. Unters. Forsch.*, **164**, 194 (1977).
12. A. Z. Abyshev et al., *Farmatsiya (Moscow)*, **26**, 42 (1977); through *Chem. Abstr.*, **87**, 2351k (1977).
13. H. Rothbächer and A. Kraus, *Pharmazie*, **25**, 566 (1970).
14. G. A. de A. Brasil e Silva and L. Bauer, *Rev. Bras. Farm.*, **54**, 143 (1973); through *Chem. Abstr.*, **84**, 21978z (1976).
15. J. B. Harborne et al., *Phytochemistry* **8**, 1729 (1969).
16. L. Peyron et al., *Bull. Soc. Chim. Fr.*, **1**, 339 (1969); through *Chem. Abstr.*, **70**, 112337s (1969).
17. C. S. Shah et al., *Planta Med.*, **18**, 285 (1970).
18. T. Shipochliev, *Vet. Med. Nauki*, **5**, 63 (1968); through *Chem. Abstr.*, **70**, 86144e (1969).
19. F. M. Ramadan et al., *Chem. Mikrobiol. Technol. Lebensm.*, **2**, 51 (1972).
20. K. Silyanovska et al., *Parfüm. Kosmet.*, **50**, 293 (1969).
21. Monograph *Foeniculi fructus*, *Bundesanzeiger*, no. 74 (Apr. 19, 1991).
22. M. Albert-Puleo, *J. Ethnopharmacol.*, **2**, 337 (1980).
23. Monograph *Foeniculi aetheroleum*, *Bundesanzeiger*, no. 74 (Apr. 19, 1991).

# FENUGREEK

**Source: *Trigonella foenum-graecum*** L. (Family Leguminosae or Fabaceae).

*Synonyms*. Foenugreek and Greek hay.

## GENERAL DESCRIPTION

An annual herb with alternate leaves each consisting of three ovate leaflets; up to 0.6 m high; native to western Asia and southeastern Europe; cultivated worldwide (e.g., Mediterranean region, northern Africa, South America, China, and India). Part used is the dried ripe seed; it is hard, smooth, and oblong, somewhat flattened, resembling a triangle.

## CHEMICAL COMPOSITION

Contains simple alkaloids consisting mainly of trigonelline (up to 0.13%), choline (0.05%), gentianine, and carpaine; much of the trigonelline is degraded during roasting to nicotinic acid and other pyridines and pyrroles, which probably account for much of the flavor of roasted fenugreek (JIANGSU, MARTINDALE).[1]

Other constituents include (*1*) saponins which yield on hydrolysis 0.6–1.7% steroid sapogenins consisting mainly of diosgenin and its isomer yamogenin usually in a 3:2 ratio, with tigogenin and neotigogenin also present;[2–7] treatment of the seeds with enzymes before acid hydrolysis has increased the yield of diosgenin and yamogenin by 10–90%;[5,8,9] yamogenin tetrosides B and C have been reported to be two of the glycosides (saponins) present.[10] (*2*) Flavonoids, including vitexin, vitexin-7-glucoside, orientin arabinoside, homoorientin, saponaretin (isovitexin), vicenin-1, vicenin-2, quercetin, luteolin, and vitexin cinnamate.[11–13] (*3*) Fixed oils (5–8%), which on extraction with fat solvents yield an extract with a strong odor; varying from fishy to nutty, depending on age of the extract (MARSH, ROSENGARTEN).[3,14] (*4*) Considerable amount of a mucilage, which appears to be mostly a galactomannan and is probably responsible for swelling of the seed in water. (*5*) Protein (23–25%) which is low in *S*-amino acids but high in lysine and trytophan; it has been suggested as a supplement of cereal proteins.[15] (*6*) Free amino acids, including (2*S*,3*R*,4*R*)-4-hydroxyisoleucine, histidine, lysine, and arginine, with the first one isolated at 0.09% yield as the major component.[16] (*7*) Vitamins, especially A, $B_1$, and C.[17] (*8*) Minerals (especially calcium and iron). (*9*) volatile components (more than 50) of which 39 have been identified, which include *n*-alkanes, sesquiterpenes, and oxygenated compounds (undecane to hexadecane, elemenes, muurolenes, γ-nonalactone, 5-methyl-δ-caprolactone, etc.); and others (JIANGSU, NANJING).[18–21]

## PHARMACOLOGY OR BIOLOGICAL ACTIVITIES

Both water and alcoholic extracts have been reported to have a stimulating effect on the isolated guinea pig uterus, especially during the last period of pregnancy, indicating that these extracts may have a highly oxytocic activity; they were suggested as possible replacements for oxytocin. The water extract has also been reported to have accelerating effects on the heartbeats of the isolated mammalian heart.[22]

Trigonelline and fenugreek infusion have been shown to have hypoglycemic effects in animals; however the effects of trigonelline in diabetics have been inconclusive (MARTINDALE).[23,24]

When fed both before and after experimental diabetes induction, fenugreek has antidiabetic activities in rats.[25]

A report has indicated that fenugreek absolute is nonirritating, nonsensitizing, and nonphototoxic to human skin.[24]

## USES

**Medicinal, Pharmaceutical, and Cosmetic.** Fenugreek extracts are used in certain perfume bases as well as in soaps, detergents, creams, and lotions, with maximum use level of 0.2% reported in perfumes.[24]

**Food.** Used as an ingredient of curry powder and many spice blends. Its major use in the United States is in imitation maple syrups for which solid extracts are mostly employed; flavor of the extracts varies with the extent of roasting and the solvents used. Other food products in which it is used include alcoholic and nonalcoholic beverages, frozen dairy desserts, candy, baked goods, gelatins and puddings, meat and meat products, and others. Use levels for extracts are usually below 0.05%.

**Traditional Medicine.** Has been used for millennia as a drug and a food or spice in Egypt, India, and the Middle East. Its medicinal uses include fever reducing and treating mouth ulcers, bronchitis, chronic coughs, and chapped lips, for milk promotion, as digestive aid, for cancers, and others;[26,27] also reported used in Java in hair tonics and to cure baldness (ROSENGARTEN).

Fenugreek was first introduced into Chinese medicine in the Sung Dynasty (ca. 1057) and has since been used as a nutrient and in treating kidney ailments, beriberi, hernia, impotence, other male problems, and others. Both unroasted and roasted (fried and sprayed with salt water) seeds are used.

**Others.** Extracts used in flavoring tobacco. Used extensively in foreign countries as a feed for livestock.

Due to its content of sapogenins, particularly diosgenin, fenugreek seed is a potential source of sapogenins for the manufacture of steroid hormones and related drugs. Because it is an annual herb, the time required for its planting to seed harvesting is much shorter than that for *Dioscorea* species and may prove to have a distinct advantage.

## COMMERCIAL PREPARATIONS

Crude and extracts in liquid and spray-dried forms. Strengths (see ***glossary***) of extracts are expressed in flavor intensities.

*Regulatory Status*. GRAS (§182.10 and §182.20). Seeds subject of a German therapeutic monograph; allowed internally for loss of appetite; externally as poultice for local inflammation.[28]

## REFERENCES


See the General References for ARCTANDER; BLUMENTHAL; FEMA; HORTUS 3rd; JIANGSU; LUST; NANJING; POUCHER; ROSENGARTEN; UPHOF; YOUNGKEN.

1. D. Reymond, *Chemtech*, **7**, 664 (1977).
2. H. S. Puri et al., *Planta Med.*, **30**, 118 (1976).
3. T. M. Jefferies and R. Hardman, *Analyst*, **101**, 122 (1976).
4. M. B. Bohannon et al., *Phytochemistry*, **13**, 1513 (1974).
5. R. Hardman and F. R. Y. Fazli, *Planta Med.*, **21**, 322 (1972).
6. R. Hardman and K. R. Brain, *Planta Med.*, **21**, 426 (1972).
7. J. C. Knight, *J. Chromatogr.*, **133**, 222 (1977).
8. D. A. Voloshina et al., *Prikl. Biokhim. Mikrobiol.*, **11**, 896 (1975); through *Chem. Abstr.*, **84**, 56692d (1976).
9. A. A. Elujoba and R. Hardman, *Planta Med.*, **51**, 113 (1985).
10. N. G. Bogacheva et al., *Khim. Farm. Zh.*, **11**, 65 (1977); through *Chem. Abstr.*, **87**, 180685e (1977).
11. H. Wagner et al., *Phytochemistry*, **12**, 2548 (1973).
12. M. Adamksa and J. Lutosmski, *Planta Med.*, **20**, 224 (1971).
13. A. R. Sood et al., *Phytochemistry*, **15**, 351 (1976).
14. F. R. Y. Fazli and R. Hardman, *Phytochemistry*, **10**, 2497 (1971).
15. I. Elmadfa, *Nahrung*, **19**, 683 (1975).
16. L. Fowden et al., *Phytochemistry*, **12**, 1707 (1973).
17. N. Saleh et al., *Z. Ernährungswiss.*, **16**, 158 (1977); through *Chem. Abstr.*, **88**, 4998w (1978).
18. S. Ghosal et al., *Phytochemistry*, **13**, 2247 (1974).
19. R. C. Badami and G. S. Kalburgi, *J. Karnatak Univ.*, **14**, 16 (1969); through *Chem. Abstr.*, **74**, 115791d (1971).
20. D. K. Bhardwaj et al., *Indian J. Chem. Sect.*, *B*, **15**, 94 (1977).
21. P. Girardon et al., *Planta Med.*, **51**, 533 (1985).
22. M. S. Abdo and A. A. Al-Kafawi, *Planta Med.*, **17**, 14 (1969).
23. N. R. Farnsworth and A. B. Segelman, *Tile Till*, **57**, 52 (1971).
24. D. L. J. Opdyke, *Food Cosmet. Toxicol.*, **16**(Suppl.1), 755 (1978).
25. M. A. Riyad et al., *Planta Med.*, **54**, 286 (1988).
26. J. L. Hartwell, *Lloydia*, **33**, 97 (1970).
27. S. B. Vohora et al., *Planta Med.*, **23**, 381 (1973).
28. Monograph *Foenugraeci semen*, *Bundesanzeiger*, no. 22 (Feb. 1, 1990).

# FEVERFEW

**Source:** ***Tanacetum parthenium*** (L.) Schulz. Bip. (syn. *Chrysanthemum parthenium* (L.) Bernh., *Leucanthemum parthenium* (L.) Gren. et Godron, *Pyrethrum parthenium* L. Sm.) (Family Compositae or Asteraceae).

## GENERAL DESCRIPTION

Strongly aromatic perennial, stems ridged, to 1 m; leaves yellowish green, pinnatisect into 3–7 oblong to ovate segments, divided into crenate-toothed to entire lobes; Flower heads 5–30, in dense corymb, ligules white, disk yellow to white; single, double (both ligulate and disk) forms common in horticulture; indigenous to rocky mountain scrub of the Balkan peninsula; cultivated for many centuries, naturalized throughout Europe, occasionally escaped in eastern North America, Central and South America, and elsewhere (TUTIN 4, GLEASON AND CRONQUIST). Parts used are the leaves and/or stems.

## CHEMICAL COMPOSITION

Leaves contain sesquiterpenoids: artecanin, canin, chrysanthemolide, chrysanthemonin, 10-epi-canin, 1β-hydroxyarbusculin, 8β-hydroxyreynosin, 3β-hydroxyparthenolide, magnoliolide, parthenolide (up to 85% of sesquiterpene content), reynosin, santamarin, *seco*-tanapartholide A, tanaparthin, tanaparthin-1α,4α-epoxide, tanaparthin-1β,4β-epoxide (GLASBY 2).[1]

Parthenolide, deemed the active sesquiterpene lactone, is highly variable in quantity or absent, depending on chemotype and geographic location of originating material; sesquiterpene fraction dominated by parthenolide contains smaller levels of other germacranolides and guaianolides. A Mexican chemotype is dominated by the eudesmanolides reynosin and santamarin as well as the guaianolides canin and artecanin; parthenolide absent. Other chemotypes (Balkan) also dominated by eudesmanolides and guaianolides; parthenolide absent.[2–6] A cultivar *T. parthenium* f. *flosculosum* (DC) Beck. was found to contain as much as 1.27% parthenolide in leaves.[7]

## PHARMACOLOGY OR BIOLOGICAL ACTIVITIES

Crude extracts inhibit both human blood platelet aggregation and secretory activity in platelets and neutrophils (polymorphonuclear leucocytes). *In vitro*, extracts protect endothelial cell wall of rabbit aortas from perfusion-induced injury and inhibit deposition of human platelets on collagen.[3]

Ability of feverfew extracts to inhibit release of serotonin (5-hydroxytryptamine) from blood platelets has been suggested as mechanism of action in treatment of migraines. Inhibition of serotonin release *in vitro* correlates well with parthenolide levels.[6] A minimum level of 0.2% parthenolide has been proposed by the Health Protection Branch of Health and Welfare Canada (roughly half the parthenolide content of feverfew samples used in clinical trials).[7]

A 1985 double-blind placebo-controlled trial on the use of feverfew as a prophylactic treatment for migraine assessed a 25-mg dose of freeze-dried leaf capsules on 17 patients, concluding that when taken prophylactically, feverfew reduces frequency and severity of migraine symptoms.[8] A 1988 randomized, double-blind, placebo-controlled trial involving 72 volunteers, clearly associated feverfew treatment with a reduction in the frequency of and vomiting associated with migraine attacks as well as a reduction in their severity. However, duration of individual attacks was unaltered.[9]

Based on previous reports of antiinflammatory activity, a 1989 double-blind, placebo-controlled, randomized study evaluated the effect of dried leaves (70–86 mg) in the treatment of rheumatoid arthritis. Over the 6-week trial, 41 female patients with symptomatic rheumatoid arthritis received feverfew or placebo. More than 13 laboratory and/or clinical parameters were assessed. The authors concluded that there were no important differences between the control group and those receiving feverfew. Participating patients, however, had not previously responded to conventional therapies. The results do not preclude possible benefits for the use of feverfew in osteoarthritis and soft tissue lesions.[10]

## USES

**Medicinal, Pharmaceutical, and Cosmetic.** Feverfew extracts, standardized to contain 0.1% parthenolide (leaf and stem; France) or 0.2% parthenolide (leaf only; Canada) used for the prophylactic treatment of migraines.

**Health Food/Herb Teas.** Tea; capsules, tablets, tincture, etc.[3]

**Traditional Medicine.** In the UK lay use of fresh or dried leaf primarily for allaying rheumatic and arthritic joint inflammation is more widespread than use in migraine prophylaxis.[10]

Antiinflammatory, anodyne, antipyretic, antispasmodic, carminative, emmenagogic, febrifuge, stimulant, tonic, and vermifuge activity; used in migraine headache, asthma, rheumatism, gynecological problems, etc. In Latin America used to promote functional activity of digestion, for colic, earache, stomachache, morning sickness, kidney pains, etc.[11]

## COMMERCIAL PREPARATIONS

Crude herb, extracts, freeze-dried leaf capsules, dried leaf capsules, tinctures, etc.

*Regulatory Status.* Not yet determined in the United States. Health and Welfare Canada issued a Drug Identification Number (D.I.N. 01958712) to a feverfew leaf capsules product, standardized to 0.2% parthenolide. Product allowed to carry the claim "used as a prophylactic against migraines."[11]

## REFERENCES

See the General References for DUKE 2; FOSTER AND DUKE; GLASBY; GLEASON AND CRONQUIST; MARTINDALE; TUCKER 2; STEINMETZ; TUTIN 4; TYLER 1; WEISS; WREN.

1. C. Hobbs, *HerbalGram*, **20**, 26 (1989).
2. D. V. C. Awang, *Pharm. J.*, **239**, 487 (1987).
3. D. V. C. Awang, *Can. Pharm. J.*, **122**(5), 266 (1989).
4. D. V. C. Awang, paper presented to the 57 Congrès de l'Association Canadienne Francaise pour l'Advancment des Sciences, Montréal, Québec, May 15–19, 1989.
5. D. V. C. Awang, *HerbalGram* **22**(2), 34, 42 (1990).
6. S. Heptinstall et al., *J. Pharm. Pharmacol.*, **44**, 391 (1992).
7. D. V. C. Awang et al., *J. Nat. Prod.*, **54**(6), 1516 (1991).
8. E. S. Johnson et al., *Br. Med. J.*, **291**, 589 (1985).
9. J. J. Murphy et al., *Lancet*, **22**, 189 (July 23, 1988).
10. M. Pattrick et al., *Ann. Rheumatic Dis.*, **48**, 547 (1989).
11. S. Foster. *Feverfew—Tanacetum parthenium, Botanical Series, no. 310,* Austin Tex., American Botanical Council (1991).

# FICIN

**Source:** ***Ficus insipida*** Willd. (syn. *F. glabrata* H.B.K., *F. anthelmintica* Mart., and *F. laurifolia* Hort. ex. Lam.) (Family Moraceae).

*Synonyms*. Leche de oje and leche de higueron.

## GENERAL DESCRIPTION

A tree with tall, erect cylindrical trunk of over 1 m in diameter; up to about 45 m high; native to tropical South America; growing in Peru, Colombia, Venezuela, and Central America. Part used is the latex from the tree. It is generally collected by felling the tree, making incisions all over the surface, and allowing the latex to drip into wooden or noniron containers placed beneath the incised areas. To prevent coagulation, acetic acid is generally added; sodium benzoate (1%) is also added as a preservative. This is the usual form of crude ficin imported into the United States. The major producing countries are Peru and Colombia; the Peruvian material has a higher solids content.[1]

Crude ficin (latex) is cream to pinkish in color and has an acidic pH (usually 3–4). It is usually purified by filtration, followed by spray-drying. Commercial purified ficin is not pure ficin but is a mixture of several proteases and small amounts of other enzymes (e.g., peroxidases) in addition to diluents (e.g., lactose, dextrose, or starch) and other constituents.[1–3]

Commercial purified ficin is a beige to light-brown powder, odorless or with a putrid odor, depending on the quality. It is relatively soluble in water but insoluble in most organic solvents. It is active over a pH range of 4–9. It can withstand an acidic pH of 2 and its solutions are reported to be stable over a pH range of 3.5–9.5, with maximum stability at pH 5–8.[1,4,5] It is inactivated at 80.6°C.[6]

The optimal pH for ficin activity varies with the substrate, ranging from 5 (for gelatin and elastin) to 9.5 (for casein). With gelatin there is another optimum at pH 7.5, and casein has another optimum at pH 6.7.[4,6]

The optimal temperature range for ficin activity is generally considered to be between 50 and 65°C, depending on the substrates and the pH as well as the purity of the commercial ficin preparations. Thus with gelatin at pH 7.5, it is reported to be 62.6°C.[1,6]

## CHEMICAL COMPOSITION AND PROPERTIES

Ficin is a sulfhydryl proteinase containing a carbohydrate moeity as bromelain.[7] It has been reported to consist of three major components and 248 amino acid residues, with a molecular weight of 23,800–26,000 and an isoelectric point at pH 9 or 10.[1,3,7,8] It is similar to papain in its chemical properties, being affected similarly by the usual papain activators (e.g., cysteine, sulfide, bisulfite, and cyanide) and inhibitors (e.g., methyl bromide, iodoacetate, and hydrogen peroxide); its activity has been reported to be reduced by ethanol and propanol.[9] It is also easily inactivated by metals (e.g., iron, copper, aluminum, and mercury) and sorbic and maleic acids.[1] Like bromelain and papain, the lost activity due to inactivation by certain metals can be restored by EDTA and a reducing agent such as cysteine, mercaptoethanol, or 1,2-dimercaptopropanol.

Like papain, ficin has broad specificity in hydrolyzing proteins, amides, esters, and small peptides.

## PHARMACOLOGY OR BIOLOGICAL ACTIVITIES

Ficin (also bromelain but not papain) when administered orally to rats has been demonstrated to have antiinflammatory activity against paw edema induced by serotonin, egg white, dextran, brewer's yeast, and carrageenan.[10]

Ficin is well known for its ability to digest intestinal worms *in vitro*.

Crude ficin (latex) is corrosive to the skin and may cause bleeding on prolonged contact.

Like other proteases, ficin may cause contact allergies in certain individuals, and large oral doses are reported to cause catharsis (MERCK).

## USES

**Medicinal, Pharmaceutical, and Cosmetic.** Used in antiinflammatory preparations, primarily in Europe; also used as digestive aid.

**Food.** Used mainly in meat tenderizers (usually in combination with papain and/or bromelain) and in the preparation of protein hydrolysates, edible collagen films, and sausage casings. Also used in cheese making (curdling milk) and in chillproofing beer.

**Traditional Medicine.** The latex is used in South America by natives as a vermifuge.

**Others.** Due to its increasingly high cost and limited availability, ficin is used in special applications where the other proteases are less suitable. Such uses include cleaning and preparation of intestinal submucosa in the manufacture of sutures; cleaning and preparation of animal arteries for implantation in humans; and in serology (e.g., determination of the Rh factor).[1]

## COMMERCIAL PREPARATIONS

Crude (latex) and purified; purified ficin is official in F.C.C. Available grades and activities vary, depending on suppliers (see ***bromelain*** and ***papain***).

## REFERENCES

See the General References for MERCK; TYLER 3.

1. E. R. L. Gaughran, *Q. J. Crude Drug Res.*, **14**, 1 (1976).
2. J. P. G. Malthouse and K. Brocklehurst, *Biochem. J.*, **159**, 221 (1976).
3. I. K. Jones and A. N. Glazer, *J. Biol. Chem.*, **245**, 2765 (1970).
4. A. Yamamoto in G. Reed, ed., *Enzymes in Food Processing*, 2nd ed., Academic Press, New York, 1975, p. 123.
5. I. E. Liener and B. Friedenson in G. E. Perlmann and L. Lorand, eds., *Methods in Enzymology*, Academic Press, New York, 1970, p. 261.
6. J. R. Whitaker, *Food Res.*, **22**, 468 (1957).
7. I. E. Liener, *Adv. Chem. Ser.*, **136**, 202 (1974).
8. T. Murachi and N. Takahashi in P. Desnuelle et al., eds., *Structure-Function Relationships of Proteolytic Enzymes*, Academic Press, New York, 1970, p. 298.
9. E. Van den Eeckhout and R. Ruyssen, *Farm. Tijdschr. Belg.*, **47**, 73 (1970); through *Chem. Abstr.*, **74**, 576v (1971).
10. C. Netti et al., *Farmaco, Ed. Prat.*, **27**, 453 (1972); through *Chem. Abstr.*, **77**, 109523p (1972).

# FO-TI (RAW AND CURED)

**Source:** ***Polygonum multiflorum*** Thunb. (Family Polygonaceae).

*Synonyms*. *Heshouwu*, *shouwu*, polygonum, Radix Polygoni Multiflori; *zhiheshouwu*, *zhishouwu*, Radix Polygoni Multiflora Preparata (cured fo-ti); also *ho-shou-wu*, *ho-show-wu*, *ho-shau-wu* and *ho-shao-wu*.

## GENERAL DESCRIPTION

A perennial vine, reaching 3–4 m long, with hollow stems, somewhat woody at the base and mostly branched at the top; the branches herbaceous; roots slender, bearing reddish brown to dark brown thick tubers near the tip; native to China; now distributed mainly along coastal provinces, from Guangxi to Hebei, and extending inland to Sichuan and Yunnan. Part used is the tuberous root collected from autumn through spring (preferably autumn and winter) and processed into different types of fo-ti. Since the early 1970s, fo ti has been increasingly produced from cultivated plants, for which root tubers of 3- to 4-year old plants are used (IMM-1, IMM-CAMS, JIANGSU).

Although easy to pronounce, the term fo-ti does not mean anything and was originally coined in America for marketing purposes only. Depending on methods of processing, there are at least four types of fo-ti: raw fo-ti, wine fo-ti, steamed fo-ti and prepared or cured fo-ti (ICMM). Raw and cured fo-ti are the most commonly used and are the ones imported into the United States.

To prepare raw fo-ti, freshly collected tubers are washed with water. After having both ends removed, the larger tubers are normally cut in half or sliced and sun or oven dried (CHP, IMM-1, JIANGSU).

To prepare cured fo-ti, raw fo-ti slices are stirred into black soybean broth in a noniron container, sealed and cooked in a water or steam bath until all liquid is absorbed and the fo-ti slices turn dark brown to reddish brown. This heating may take many hours; the optimal heating time for yielding fo-ti with the best immunomodulating effects (tonic properties) being 32 h.[1] For every 100 kg of raw fo-ti slices, broth from 10 kg of black soybean is used. The slices are then dried (CHP, ICMM).

Much confusion exists in the Western literature regarding fo-ti, with some major or popular works making no mention of the existence of different types (BENSKY AND GAMBLE, CHEUNG AND LI, LIST AND HÖRHAMMER, TYLER 1).[2–5] For instance, in one bilingual reference, even though raw fo-ti is mentioned in the Chinese version as having lubricating effects on the intestines as well as antitoxic and nodule-dispersing (anti-swelling) properties, these properties are simply stated in the English version, without giving any hints as to the existence of the two types of fo-ti with distinct differences in properties and uses (CHEUNG AND LI). This ambiguity has caused problems in fo-ti products as manufacturers not familiar with Chinese herbs have been using the cheaper raw fo-ti in place of cured fo-ti in their tonic formulas. Adding to this confusion is the lack of specificity in reporting findings on fo-ti, particularly in the already meager English literature, where authors frequently do not specify type of fo-ti used in their research or report, thus making the reported information of dubious value (LIST AND HÖRHAMMER).[6]

Thus the two types of fo-ti have distinctly different chemical compositions, traditional properties, and modern pharmacologic effects and uses; they should be sourced carefully and used specifically.

## CHEMICAL COMPOSITION

Many chemical studies have been performed on the genus *Polygonum* but few on *Polygonum multiflorum* itself (LIST AND HÖRHAMMER). Even with the few that have been reported on fo-ti, the results are confusing as most did not specify the types of fo-ti used, especially in the non-Chinese literature. Thus one report (consisting of only three lines) mentions the presence of anthraquinones in the rhizome, rootstock, and stem (LIST AND HÖRHAMMER), which has been the only chemical study quoted in a well-known herbal, attributing cathartic activity as solely responsible for the action of fo-ti (TYLER 1). Two other reports simply mention the dried roots or *heshouwu* as being used,[7,8] with no indication as to what type of fo-ti; the only clue to their using raw fo-ti was the use description "for the treatment of suppurative dermatitis, gonorrhea, favus athlete's foot, inflammation," for which raw fo-ti is traditionally used. Nevertheless, from these few reports and from the Chinese literature that does describe raw and cured fo-ti, the following chemical profile of fo-ti has emerged.

Both raw and cured fo-ti contain anthraquinones, including chrysophanol, chrysophanic acid anthrone, emodin, rhein, physcion, and digitolutein, which exist both free and as glucosides (mainly glucosides) (LIST AND HÖRHAMMER);[9–11] the concentration can reach 1.1% in raw fo-ti.[11] Curing by successively steaming for 12 h, standing overnight and sun drying for 8 h (repeated nine times), or according to the *Chinese Pharmacopeia* for 32 h, reduced the concentrations of both the free and conjugated anthraquinones by 42–96%.[9,10] In addition, the proportion of free to conjugated anthraquinones is also greatly increased in cured fo-ti, thus further reducing the laxative effects.[11,12]

Fo-ti (type not specified) has been reported to contain up to 3.7% lecithins (JIANGSU).[11] Curing was found to increase the phosphorus (presumably lecithins) content by 36.9%;[13] it also increases the sugar content.[12] However, a more recent study found curing to reduce the amount of phospholids in fo-ti.[14]

Other chemical constituents reported to be present in fo-ti (mainly raw fo-ti) include rhapontin, β-sitosterol, catechins, cyanidins, stilbene glycoside gallates, 2,3,4′,5-tetrahydroxystilbene-2-*O*-β-D-glucoside (>1.2%)[7,15,16] and relatively high concentrations of calcium (2225 ppm), iron (350 ppm), zinc (24.5 ppm), manganese (18.5 ppm), and other trace minerals[17] as well as copious amounts of starch and others (WANG).[18]

## PHARMACOLOGY OR BIOLOGICAL ACTIVITIES

Fo-ti has exhibited numerous pharmacologic effects, including the following:

*1. Antiaging effects.* Cured fo-ti was found to influence favorably various biological processes related to aging. Thus its decoction (p.o.) significantly increased the levels of superoxide dismutase (SOD), biogenic amines (5-hydroxytryptamine, norepinephrine, and dopamine) and proteins but decreased the levels of monoamine oxidase-B (MAO-B), lipid peroxide, and malonyl dialdehyde (MDA; a product of destructive lipid peroxidation) in key organs of aging mice; extracts of fo-ti (type not specified) showed strong stimulating effects on SOD activity in red blood cells and strong inhibition of thiobarbituric acid reactive material formation (indicator of lipid peroxidation) in the liver.[19] Decoction of cured fo-ti (p.o) significantly inhibited the increase of serum ceruloplasmin level in aging mice; it also significantly inhibited atrophy of the thymus gland of mice due to natural aging or induced by hydrocortisone as well as inhibited atrophy of their adrenal glands.[11,20]

*2. Immunologic effects.* Fo-ti, cured per method of *Chinese Pharmacopeia* for 32 h, was shown to enhance nonspecific immunity and cellular immunity in mice; along with fo-ti cured by pressure-cooking for 6 h at 120°C, it also had antagonistic activity against the immunosuppressive effects of prednisolone or hydrocortisone. In contrast, raw fo-ti had little or no such activities.[1,21]

*3. Hypolipemic and antiatherosclerotic effects.* In a study using Japanese quails, an alcoholic extract of cured fo-ti raised blood plasma high-density lipoprotein cholesterol: total cholesterol ratio; markedly reduced plasma total cholesterol, free cholesterol, and triglyceride levels; inhibited hyperlipemia; and retarded the development of atherosclerosis in the animals.[21] Earlier studies in other animal species (e.g., rabbits, pigeons and rats) also showed that fo-ti (type not reported) or its formulations exhibited similar effects (WANG).[11,23]

*4. Liver protectorant effects.* Stilbene glucosides (e.g., 2,3,5,4′-tetrahydroxystilbene-2-*O*-D-glucoside) isolated from fo-ti were found to partially inhibit the deposition of lipid peroxides in the liver of rats fed peroxidized corn oil. They inhibited the elevation of GOT and GPT levels in the serum of the rats. They also inhibited lipid peroxidation induced by ADP and NADPH in rat liver microsomes.[8]

*5. Resistance to cold.* Daily intragastric administration of fo-ti extract (type not specified) at a dose of 0.5 mL (equivalent to 0.2 g crude drug) for 14 days markedly decreased the mortality rate in mice induced by refrigeration at −5°C (WANG).

*6. Antimicrobial activities.* Among 80 herbs commonly used in treating hepatitis tested against hepatitis B virus *in vitro*, water extract of cured fo-ti was found to be one of the eight most active in inhibiting HBV DNA replication.[24]

Raw fo-ti and four different types of cured fo-ti all exhibited varying degrees of inhibition against nine species of bacteria, but there was no predictable pattern in their activities.[25]

*7. Others.* Fo-ti (type not specified) is reported to be nonmutagenic per the Ames *Salmonella*/microsome assay;[6] the hydroxyanthraquinones in its aqueous extract (decoction) had antimutagenic effects.[26]

Raw fo-ti has cathartic activities.

*8. Toxicities.* Compared to cured fo-ti, raw fo-ti is relatively toxic. Thus the $LD_{50}$ of an alcoholic percolate of raw fo-ti per i.p. administration was 2.7 g/kg while the $LD_{50}$ of an alcoholic percolate of cured fo-ti was 169.4 g/kg (animal species not reported).[11]

One case of allergic reaction (cystitis, patient recovered) due to cured fo-ti has been reported.[27] Gastrointestinal disturbances (diarrhea, abdominal pain, nausea, and vomiting) reported for fo-ti were most likely due to raw or improperly cured fo-ti (WANG). Other reported adverse effects due to fo-ti (or could be due to other ingredients present in the formulas) include numbness of the extremities and skin rashes (a few cases) (WANG).

## USES

**Medicinal, Pharmaceutical and Cosmetic.** Extract of cured fo-ti is a popular ingredient in hair-care products (e.g., shampoos and tonics) for its alleged hair-darkening and growth-promoting properties, especially in China and Hong Kong (ZHOU); also used in skin-care products (e.g., creams and lotions) for its traditional detoxicant (antiallergic) and nourishing properties.

**Health Food/Herb Teas.** Powder and occasionally extracts are used in tonic formulas in America, though often with no distinction between the raw and cured forms; also used in sliced form in soup mix packets.

**Traditional Medicine.** Fo-ti was first described in *He Shou Wu Lu* (before 10th century) and later in *Kai Bao Ben Cao* (10th century, in the Song Dynasty). Raw fo-ti is traditionally considered a detoxicant and laxative and is used to treat scrofula (lymph node tuberculosis), sores, carbuncles, skin eruptions (*feng zhen*), pruritus, and constipation, among other conditions. On the other hand, cured fo-ti is traditionally considered to have slightly warming properties and a liver and kidney tonic; it is believed to tone up the vital essence and blood and fortify the muscles, tendons, and bones. Cured fo-ti is traditionally used to treat dizziness with tinnitus, insomnia, premature graying, soreness and weakness of lower back and knees, and numbness of limbs and others. In recent years, both raw and cured fo-ti are also used in treating hyperlipemia (CHP, JIANGSU).

## COMMERCIAL PREPARATIONS

Crude, powder and extracts. Raw fo-ti comes in whole, halved or thick slices, light brown to brown in color; cured fo-ti in very dark brown to dark reddish brown slices. Due to the much lower price of raw fo-ti, this is normally the powder available in America. Also, there is no simple practical assay method to determine whether a given extract from a supplier is genuine fo-ti and which type it is.

*Regulatory Status.* U.S. regulatory status not yet determined.

## REFERENCES

See the General References for CHEUNG AND LI; CHP; FOSTER AND YUE; ICMM; IMM-1; IMM-CAMS, JIANGSU; LIST AND HÖRHAMMER; WANG.


1. D. J. Ye et al., *Zhongyao Tongbao*, **12**(3), 21 (1987).
2. *A Barefoot Doctor's Manual*, Running Press, Philadelphia, 1977, p. 743.
3. J. D. Keys, *Chinese Herbs*, Charles E. Tuttle Co., Rutland, Vt., 1976, p. 151.
4. *Herbal Pharmacology in the People's Republic of China*, National Academy of Sciences, Washington, D.C., 1975, p. 186.
5. M. Tierra, *The Way of Herbs*, Pocket Books, New York, 1989, p. 183.
6. J. K. Kam, *Am. J. Chin. Med.*, **9**, 213 (1981).
7. K. Hata et al., *Yakugaku Zasshi*, **95**, 211 (1975).
8. Y. Kimura et al., *Planta Med.*, **49**, 51 (1983).
9. D. J. Ye et al., *Zhongyao Tongbao*, **11**(12), 23 (1986).
10. G. G. Yao et al., *Zhongcaoyao*, **14**(6), 15 (1983).
11. W. L. Deng and S. R. Gong, *Zhongcaoyao*, **18**(3), 42 (1987).
12. L. Q. Ling et al., *Shanghai Zhongyiyao Zazhi* (2), 78 (1966); through *Zhongyao Yanjiu Wenxian Zhaiyao*, **1962–1974**, 407.
13. M. X. Chang et al., *Zhongcaoyao*, **19**(7), 17 (1988).
14. C. H. Ma and J. S. Wang, *Zhongguo Zhongyao Zazhi*, **16**, 662 (1991).
15. X. Z. Yan, *Shanghai Ti I I Hsueh Pao*, **8**(2), 123 (1981); through *Chem. Abstr.*, **95**, 121033h.
16. M. Yoshizaki et al., *Planta Med.*, **53**, 273 (1987).
17. Z. X. Li et al., *Zhongcaoyao*, **16**(12), 15 (1985).
18. J. B. Li and M. Lin, *Zhongcaoyao*, **24**, 115 (1993).
19. Y. R. Dai et al., *Planta Med.*, **53**, 309 (1987).
20. M. C. Yao et al., *Yaoxue Tongbao*, **19**(11), 28 (1984).
21. J. H. Ying et al., *Zhongguo Zhongyao Zazhi*, **17**, 722 (1992).
22. W. Wang and H. D. Jing, *Chin. J. Integr. Trad. Western Med.*, **4**, 748 (1984).
23. M. Z. Mei, *Yao Hsueh Hsueh Pao*, **14**(1), 8 (1979); through *Chem. Abstr.*, **92**, 33601z.
24. J. Y. Yang et al., *Chin. J. Integr. Trad. Western Med.*, **9**, 494 (1989).
25. H. S. Zhen et al., *Zhongyao Tongbao*, **11**(3), 53 (1986).
26. Z. H. Xie et al., *J. Zhejiang Coll. Trad. Chin. Med.*, **14**(1), 22 (1990).
27. K. G. Ren, *Zhongcaoyao*, **16**(10), 40 (1985).

# GALBANUM

**Source:** ***Ferula gummosa*** Boiss. (syn. *F. galbaniflua* Boiss. et Buhse) and other ***Ferula*** species (Family Umbelliferae or Apiaceae).

*Synonyms*. Galbanum resin, galbanum gum, galbanum oleoresin, galbanum gum resin, and galbanum oleogum resin.

## GENERAL DESCRIPTION

Resinous perennial herbs with resin ducts throughout the plants; native to the Middle East and western Asia (e.g., Iran, Turkey, and Afghanistan). Part used is the dried resinous exudation obtained by incising the stems near the ground. There are two types of galbanum: soft galbanum (or Levant galbanum) is a viscous liquid, while hard galbanum (or Persian galbanum) is a solid. The former contains more volatile oil and is used for the production of galbanum oil. A resinoid is prepared by solvent extraction; it often contains a high-boiling odorless solvent such as diethyl phthalate, diethyl sebacate, or propylene glycol used as diluent (ARCTANDER, GUENTHER).

## CHEMICAL COMPOSITION

Galbanum contains from 5% (Persian) to more than 26% (Levant) volatile oil; about 60% resin consisting mainly of resinic acids; 30–40% gummy substances containing galactose, arabinose, galacturonic acid, and 4-methylglucuronic acid residues; and umbelliferone and its esters, among others (ARCTANDER, CLAUS, GUENTHER, LIST AND HÖRHAMMER, KARRER, MARTINDALE).[1–3]

The volatile oil contains 63–75% monoterpene hydrocarbons (mostly $\beta$-pinene, 3-carene, and $\alpha$-pinene, with small amounts of *d*-limonene, terpinolene, etc.);[4,5] monoterpene alcohols (linalool, terpineol, borneol, fenchol, etc.) and their acetates;[1] sesquiterpenes (e.g., cadinene, guaiol, bulnesol, galbanol, and 10-epijunenol);[1,6,7] azulenes (e.g., guaiazulene and isoguaiazulene); thiol esters (e.g., *S*-isopropyl-3-methylbutanethioate and *S*-*sec*-butyl-3-methylbutanethioate);[8] polysulfanes;[9] pyrazines (e.g., tetramethylpyrazine, 2,6-diethyl-3-methylpyrazine, and 2-methoxy-3-*sec*-butylpyrazine);[10,11] and (*E*,*Z*)-1,3,5-undecatriene and (*E*,*E*)-1,3,5-undecatriene (LIST AND HÖRHAMMER).[12–14]

The (*E*,*Z*)-isomer of *n*-1,3,5-undecatriene is reported to be the major odor principle of galbanum.[12]

## PHARMACOLOGY OR BIOLOGICAL ACTIVITIES

Extracts (aqueous, hydroalcoholic, and chloroform) of galbanum gum have been reported to have antimicrobial properties (especially against *Staphylococcus aureus*) *in vitro*; they were also effective in preserving emulsions for up to 6 months without change in physical or organoleptic characteristics.[15]

A report indicates galbanum oil to be nonirritating and nonsensitizing to human skin.[16]

## USES

**Medicinal, Pharmaceutical, and Cosmetic.** Now rarely used in pharmaceuticals. Galbanum oil and resinoid are used as fragrance components and fixatives in cosmetics, including soaps, detergents, creams, lotions, and perfumes, with a maximum use level of 0.7% reported for the oil in perfumes.[16]

**Food.** Galbanum resin and galbanum oil are used as flavor components in most food products, including nonalcoholic beverages, frozen dairy desserts, candy, baked goods, gelatins and puddings; and condiments and relishes. Galbanum oil is also used in alcoholic beverages, meat and meat products, snack foods, and gravies. Highest average maximum use level reported is about 0.003% (33 ppm) for the resin in candy and gelatins and puddings.

**Traditional Medicine.** Used as a carminative, stimulant, expectorant, and antispasmodic for purposes similar to those of ***asafetida***; also used in treating wounds.

**COMMERCIAL PREPARATIONS**

Crude, oil and resinoid.

*Regulatory Status.* Has been approved for food use (§172.510).

**REFERENCES**

See the General References for ARCTANDER; CLAUS; FEMA; GRIEVE; GUENTHER; LEWIS AND ELVIN-LEWIS; LIST AND HÖRHAMMER; MARTINDALE; TERRELL; UPHOF.

1. P. Teisseire, *Recherches (Paris)*, **14**, 81 (1964).
2. M. G. Jessenne et al., *Plant. Med. Phytother.*, **8**, 241 (1974).
3. E. Graf and M. Alexa, *Planta Med.*, **51**, 428 (1985).
4. R. M. Ikeda et al., *J. Food Sci.*, **27**, 455 (1962).
5. Y. R. Naves, *Parfum. Cosmet. Savons*, **12**, 586 (1969); through *Chem. Abstr.*, **72**, 82898x (1970).
6. M. Wichtl, *Planta Med.*, **11**, 53 (1963).
7. A. G. Thomas et al., *Tetrahedron*, **32**, 2261 (1976).
8. J. W. K. Burrell et al., *Tetrahedron Lett.*, **30**, 2837 (1971).
9. Z. D. Min et al., *Planta Med.*, **53**, 300 (1987).
10. J. W. K. Burrell et al., *Chem. Ind.*, **44**, 1409 (1970).
11. A. F. Bramwell et al., *Tetrahedron Lett.*, **37**, 3215 (1969).
12. F. Naef et al., *Helv. Chim. Acta*, **58**, 1016 (1975).
13. Y. Chretien-Bessiere et al., *Bull. Soc. Chim. Fr.*, **1**, 97 (1967); through *Chem. Abstr.*, **67**, 4332p (1967).
14. Y. R. Naves, *Bull. Soc. Chim. Fr.*, **9**, 3152 (1967); through *Chem. Abstr.*, **68**; 6097n (1968).
15. A. Vaziri, *Planta Med.*, **28**, 370 (1975).
16. D. L. J. Opdyke, *Food Cosmet. Toxicol.*, **16**(Suppl.1), 765 (1978).

# GANODERMA

**Source:** ***Ganoderma lucidum*** (Leyss. ex Fr.) Karst. (syn. *Polyporus japonicus* Fr.) and ***G. japonicum*** (Fr.) Lloyd. (syn *G. sinense* Zhao, Xu et Zhang) (Family Polyporaceae).

*Synonyms.* Reishi, lingzhi, *ling zhi cao*, *ling chi*, mannentake, holy mushroom, *chizhi* (red *lingzhi* or *G. lucidum*), *zizhi* (purple *lingzhi* or *G. japonicum*), etc.

**GENERAL DESCRIPTION**

Fungi of the polypore family. Part used is the fruiting body.

The pileus (cap) of *G. lucidum* is corky, kidney shaped to semicircular, with a hard upper surface, yellow at first but gradually changing to reddish brown, reddish purple, or dull purple; shiny, with annular grooves or ridges and radial wrinkles; edges thin, often curved downward. Cap sizes vary considerably, ranging from $4 \times 3$ cm to

20 × 10 cm in area and 0.5–2 cm in thickness. Woody stalk (stipe) is mostly lateral, 0.5–2.5 cm thick and up to 19 cm long, purplish brown to black and shellacked.[1,2]

The fruiting body of *G. japonicum* (purple *lingzhi*) resembles closely that of *G. lucidum* (red *lingzhi*), with cap sizes ranging from 2 × 1.4 cm to 20 × 20 cm; stalk up to 15 cm long and 0.9 cm thick. The only major difference is the dark purple to black colored cap and stalk of *G. japonicum*. However, some old specimens of red *lingzhi* also have dark purple caps and stalks and thus cannot be readily distinguished from purple *lingzhi* (JIANGSU).[1–3]

Both ganodermas are widely distributed in China, especially along coastal provinces, growing at stumps and decaying logs of oak and other broad-leaf trees as well as on decaying conifers, especially *Tsuga chinensis* (Franch.) Pritz., which is parasitized by *G. lucidum*. The latter can also be found on hardwoods in North America as well as in Japan and Korea.[4,5]

The mushrooms are collected in autumn, washed to rid of dirt and dried under the sun.[1] They are not processed further.

Although now commercially cultivated in China, much of the ganoderma is still gathered wild. The type imported into the United States is mainly red *lingzhi* (*G. lucidum*); this species is now mostly cultivated in America as well as in China, Taiwan, Japan, and Korea (WANG).

In addition to above two *Ganoderma* species, other species of polypores are occasionally used as substitutes of *lingzhi*, including *Ganoderma applanatum* (Pers. ex Gray), *G. lobatum* (Schw.) Atk., *G. capense* (Lloyd) Teng, *Fomes pinicola* (Swartz ex Fr.) Cke., *Trametes dickinsii* Berk., *Polyporus montanus* (Quel.) Freey., *P. grammocephalus* Berk., and *Polysticus vernicipes* (Berk.) Cke.[2,6]

*Ganoderma lucidum*, *G. japonicum*, *G. capense*, and *G. applanatum* are used as fungal sources for the fermentative production of *lingzhi* biomass (WANG).

## CHEMICAL COMPOSITION

Most of the recent chemical studies have been performed on red *lingzhi* (*G. lucidum*), including its spores and cultivated biomass (mycelium). Hence unless otherwise stated, the data reported here are from this fungus and its strains, either wild crafted or cultivated.

Chemical constituents present include ergosterol (0.3–0.4%), β-sitosterol, 24-methylcholesta-7,22-dien-3-β-ol, and other sterols; fungal lysozyme, acid protease, and other enzymes (laccase, endopolygalacturonase, cellulase, amylase, etc.); water-soluble protein, polypeptides; amino acids; trehalose and other sugars; mannitol; betaine; adenosine; alkanes (tetracosane, hentriacontane); and fatty acids (tetracosanoic, stearic, palmitic, nonadecanoic, and behenic acids) (JIANGSU, WANG).[7–16]

Triterpenes (mainly lanostane type) present include ganoderic acids A, B, C, D, E, F, G, H, I, J, K, L, M, N, O, P, Q, R, S, T, U, V, W, X, Y, and Z;[17–33] lucidenic acids A, B, C, D, E, F, G, and others;[19,20,24,28,29,34] lucidones A, B, and C;[28,34] ganolucidic acids A, B, C, D, and E;[23,27,28,35] ganoderal A;[18] ganoderiols A, B, C, D, E, F, G, H and I;[35,36] ganoderols A (ganodermanonol) and B (ganodermadiol);[18,37] ganodermanontriol and ganodermatriol, among others.[36–38]

Polysaccharides present include a water-soluble branched arabinoxyloglucan (polysaccharide GL-1) of mol. wt. 40,000;[39] an alkali-extracted, water-soluble heteroglycan of mol. wt. 38,000;[40] several water-insoluble heteroglucans of mol. wt.10,000–100,000;[41] ganoderans A and B of mol. wt. 7,400–23,000;[42] and others.[5,43–48]

Major inorganic elements present include Ca, Mg, Na, Mn, Fe, Zn, Cu, and Ge.[49]

Purple *lingzhi* (*G. japonicum*) contains ergosterol in lesser amount (0.03%), ergosta-7,22-dien-3-β-ol, trehalose, alkaloids (betaine, γ-butyrobetaine, etc.), 14 free amino acids, glucosamine, resin, and polysaccharides, including a water-insoluble alkali-soluble glucan (WANG).[50,51]

Mycelium of cultivated *G. capense* contains adenine, adenosine, uridine, and uracil; D-mannitol; ergosterol, β-sitosterol, soyasapogenol B and other steroids; stearic, docosanoic, tricosanoic and tetracosanoic acids; nicotinic acid, and furans, among others (WANG).[52,53]

## PHARMACOLOGY OR BIOLOGICAL ACTIVITIES

Most of the pharmacological studies have been performed on various extracts of red *lingzhi* (*G. lucidum*). Hence unless otherwise stated, the data reported here are for this species.

Bioactivities exhibited by ganoderma in humans and/or experimental animals include (*1*) central effects (sedative, analgesic and anticonvulsive), which are due to adenosine.[54,55] (*2*) Cardiovascular effects (both hypertensive and hypotensive as well as increase of coronary blood flow); hypotensive effects due to triterpenoids, including ganoderols A and B, and ganoderic acids B, D, F, H, K, S, and Y, with ganoderic acid F being the most active (WANG).[18] (*3*) Antiallergic effects (inhibition of passive cutaneous anaphylaxis reaction, inhibition of histamine release and prevention of experimental asthma and contact dermatitis); antihistaminic effects due to ganoderic acids C and D.[22,56,57] (*4*) Protective activities against carbon tetrachloride-induced liver damage, with fat-soluble extractives of both red and purple *lingzhi* as well as *G. capense* being active.[58,59] (*5*) Hypoglycemic activities, with polysaccharides (especially ganoderans A and B) being the active principles.[42,60,61] (*6*) Antitumor or cytotoxic activities against experimental tumors (solid sarcoma 180, hepatoma cells, and Ehrlich ascites carcinoma), with some of the polysaccharides and triterpenoids being the active principles (e.g., GL-1, ganoderans A and B, ganoderic acids T and Z, etc.).[5,39,41,43,44,46,62,63] (*7*) Platelet aggregation inhibitory effects, with adenosine being the active agent.[10,64,65] (*8*) Hypolipemic as well as hypercholesterolemic effects;[59] anticholinergic,[58] inhibition of cholesterol synthesis (active compounds: ganoderic acid derivatives);[66] free-radical scavenging (antioxidant) effects;[67] immunomodulating effects,[47–49] smooth muscle relaxant, antitussive, vasodilative, diuretic, anabolic, antiinflammatory, antifatigue, and others (JIANGSU, WANG).[54,68–70]

Toxicities of ganoderma are very low, with $LD_{50}$ (i.p.) in mice varying considerably among different ganoderma preparations, ranging from $3.42 \pm 0.11$ to $38.3 \pm 1.04$ g/kg. Oral doses of up to an equivalent of 1.88 g/kg of crude material could be tolerated by mice for 20 days with no toxic effects. The sensitivity to ganoderma toxicity varies with the animal species, with rabbits being the least sensitive and mice, the most sensitive (JIANGSU, WANG, ZHOU AND WANG).

Human toxicities are rare and not serious, consisting of dizziness, dry mouth and nasal passage, dry throat, nosebleeds, pruritus, stomach upset and bloody stools, which have been observed after long-term (3–6 month) continuous oral use. Clinical examinations revealed no toxic effects on vital organs such as heart, liver, and kidney (ZHOU AND WANG). Recent documented adverse side effects include a case of skin rash after drinking 200 mL of a *lingzhi* wine and a case of allergic shock due to use of a *lingzhi* injection (i.m.).[71,72]

## USES

**Medicinal, Pharmaceutical, and Cosmetic.** Extracts are used in skin-care products, especially creams and lotions for their traditional moisturizing, nourishing and whitening properties (ETIC). Known for at least 2000 years for its complexion-benefiting properties, *lingzhi* has been demonstrated to contain a wide spectrum of bioactive chemical constituents.

**Health Food/Herb Teas.** Powdered crude and/or extracts are used singly or in combination with other herbs in capsule, tablet, or liquid (syrup or drink) form as a general (*qi*) tonic to improve energy, stamina, and resistance to stress and diseases; hydroalcoholic extracts also used to flavor instant soup mixes and herbal drinks because of their mushroom aroma (similar to shiitake).

**Traditional Medicine.** *Lingzhi* was first described in the *Shen Nong Ben Cao Jing* (ca. 200 B.C.–A.D. 100) under the superior category of drugs as good for deafness, beneficial to the joints, calming, benefiting vital energy (*jing qi*), strengthening tendons and bones, and good for one's complexion. It is generally considered sweet and slightly bitter tasting and neutral or warm, and nontoxic. One of the major Chinese tonics, it was once considered the "elixir of life" and was for centuries reserved for emperors and glorified in Chinese literary classics. Although six types of *lingzhi* have been recorded and used since ancient times, only red *lingzhi* and purple *lingzhi* are currently used, and interchangeably.

Major traditional uses include treatment of general weakness (*xu lao*), cough, asthma, insomnia, and indigestion. Modern uses also include the treatment of excessive dreams (nightmares), lack of appetite, neurasthenia, chronic hepatitis, pyelonephritis, mushroom poisoning (large single dose of 120 g used),[1] coronary heart disease, arrhythmia, hyperlipemia, hypertension, chronic bronchitis, rhinitis (topical treatment), acute altitude sickness, Keshan disease, leukocytopenia, and others (JIANGSU, NATIONAL, WANG).

## COMMERCIAL PREPARATIONS

Crude (mainly *G. lucidum*, both cultivated and wildcrafted) and extracts. Extracts come in various types with different physicochemical and biological properties, depending on extraction menstruums used.

*Regulatory Status.* Ganoderma is an ethnic food/herb; its U.S. regulatory status not clear.

## REFERENCES

See the General References for CHEUNG AND LI; JIANGSU; LU AND LI; NATIONAL; WANG; ZHOU AND WANG.

1. B. Liu, *Zhongguo Yao Yong Zhen Jun (Chinese Medicinal Fungi),* Shanxi People's Press, Taiyuan, 1984, p. 70.
2. R. A. Ren, ed., *Zhong Yao Jian Ding Xue (Identification of Chinese Drugs),* Shanghai Scientific and Technical Publications, Shanghai, 1986, p. 517.
3. W. S. Mao et al., eds., *Zhong Yao Zhen Wei Jian Bie (Differentiation of Genuine and Adulterated Chinese Drugs),* Shaanxi Scientific and Technical Publications, Xian, 1987, p. 540.
4. O. K. Miller, *Mushrooms of North America*, E. P. Dutton & Co., New York, 1973, p. 178.
5. Y. Sone et al., *Agr. Biol. Chem.*, **49**, 2641 (1985).
6. Institute for Assay of Drugs and Biologicals, Ministry of Health, and Institute of Botany, Academia Sinica, eds., *Zhong Yao Jian Bie Shou Ce (Manual of Chinese Drug Identification),* Science Publishers, Beijing, 1981, p. 174.
7. D. Kac et al., *Phytochemistry*, **23**, 2686 (1984).

8. C. Y. Hou et al., *Zhiwu Xuebao*, **30**, 66 (1988).
9. S. S. Subramanian and M. N. Swamy, *J. Sci. Ind. Res. (India)*, **20B**, 39 (1961); through *Chem Abstr.*, **55**, 16677h.
10. A. Shimizu et al., *Chem. Pharm. Bull.*, **33**, 3012 (1985).
11. L. Lukacs and J. Zellner, *Monatsh. Chem.*, **62**, 214 (1933); through *Chem. Abstr.*, **27**, 45578.
12. H. L. Kumari and M. Sirsi, *Arch. Mikrobiol.*, **84**, 350 (1972); through *Chem. Abstr.*, **77**, 71781f.
13. H. L. Kumari and M. Sirsi, *J. Gen. Microbiol.*, **65**(Pt. 3), 285 (1971); through *Chem Abstr.*, **75**, 71669j.
14. T. Terashita et al., *Agr. Biol. Chem.*, **48**, 1029 (1984).
15. J. H. Do and S. D. Kim, *Han'guk Kyunhakhoechi*, **14**(1), 79 (1986); through *Chem. Abstr.*, **105**; 167658u.
16. J. H. Do and S. D. Kim, *Sanop Misaengmul Hakhoechi*, **13**, 173 (1985); through *Chem Abstr.*, **104**, 2589p.
17. T. Kubota et al., *Helv. Chim. Acta*, **65**, 611 (1982).
18. A. Morigiwa et al., *Chem. Pharm. Bull.*, **34**, 3025 (1986).
19. T. Nishitoba et al., *Agr. Biol. Chem.*, **48**, 2905 (1984).
20. T. Nishitoba et al., *Agr. Biol. Chem.*, **49**, 1793 (1985).
21. M. Hirotani et al., *Phytochemistry*, **24**, 2055 (1985).
22. H. Kohda et al., *Chem. Pharm. Bull.*, **33**, 1367 (1985).
23. T. Nishitoba et al., *Agr. Biol. Chem.*, **49**, 3637 (1985).
24. T. Kikuchi et al., *Chem. Pharm. Bull.*, **33**, 2624 (1985).
25. Y. Komoda et al., *Chem. Pharm. Bull.*, **33**, 4829 (1985).
26. M. Hirotani and T. Furuya, *Phytochemistry*, **25**, 1189 (1986).
27. T. Kikuchi et al., *Chem. Pharm. Bull.*, **33**, 2628 (1985).
28. T. Nishitoba et al., *Agr. Biol. Chem.*, **50**, 809 (1986).
29. T. Nishitoba et al., *Phytochemistry*, **26**, 1777 (1987).
30. M. Hirotani et al., *Phytochemistry*, **26**, 2797 (1987).
31. M. Hirotani et al., *Chem. Pharm. Bull.*, **34**, 2282 (1986).
32. J. Toth et al., *J. Chem. Res., Synop.*, **12**, 299 (1983); through *Chem. Abstr.*, **100**, 117512t.
33. J. Toth et al., *Tetrahedron Lett.*, **24**, 1081 (1983).
34. T. Nishitoba et al., *Agr. Biol. Chem.*, **49**, 1547 (1985).
35. T. Nishitoba et al., *Agr. Biol. Chem.*, **52**, 367 (1988).
36. H. Sato et al., *Agr. Biol. Chem.*, **50**, 2887 (1986).
37. M. Arisawa et al., *J. Nat. Prod.*, **49**, 621 (1986).
38. L. J. Lin et al., *Phytochemistry*, **27**, 2269 (1988).
39. T. Miyazaki and M. Nishijima, *Chem. Pharm. Bull.*, **29**, 3611 (1981).
40. T. Miyazaki and M. Nishijima, *Carbohydr. Res.*, **109**, 290 (1982); through *Chem. Abstr.*, **97**, 212191r.
41. T Mizuno et al., *Nippon Nogei Kagaku Kaishi*, **59**, 1143 (1985); through *Chem. Abstr.*, **104**, 135927u.
42. H. Hikino et al., *Planta Med.* (4), 339 (1985).
43. T. Miyazaki, *Shinkin to Shinkinsho*, **24**(2), 95 (1983); through *Chem. Abstr.*, **100**, 79551v.
44. T. Mozuno et al., *Nippon Nogei Kagaku Kaishi*, **58**, 871 (1984); through *Chem. Abstr.*, **101**, 226886j.
45. C. H. Wang et al., *Linchan Huaxue Yu Gongye*, **4**(3), 42 (1984); through *Chem. Abstr.*, **102**, 151189u.
46. T. Mizuno and S. Sakamura, *Kagaku*

*to Seibutsu*, **23**, 797 (1985); through *Chem. Abstr.*, **104**, 230280m.

47. L. S. Lei and Z. B. Lin, *Yaoxue Xuebao*, **27**, 331 (1992).
48. Y. Q. He et al., *Zhongguo Zhongyao Zazhi*, **17**, 226 (1992).
49. H. W. Shin et al., *Saengyak Hakhoechi*, **16**, 181 (1986); through *Chem. Abstr.*, **105**, 29874k.
50. S. Ukai et al., *Chem Pharm. Bull.*, **31**, 741 (1983).
51. Z. P. Fang et al., *Zhongchengyao*, **11**(7), 36 (1989).
52. G. D. Zhang et al., *Yaoxue Xuebao*, **21**, 35 (1986).
53. J. G. Yu et al., *Zhongcaoyao*, **14**(10), 6 (1983).
54. Y. Kasahara and H. Hikino, *Phytother. Res.*, **1**(1), 17 (1987).
55. Y. Kasahara and H. Hikino, *Phytother. Res.*, **1**(4), 173 (1987).
56. M. Nogami et al., *Yakugaku Zasshi*, **106**, 594 (1986).
57. M. Nogami et al., *Yakugaku Zasshi*, **106**, 600 (1986).
58. G. T. Liu et al., *Chin. Med. J.*, **92**, 496 (1979).
59. M. J. Lee and M. H. Chung, *Korean J. Pharmacog.*, **18**, 254 (1987); through *Napralert*, T15474.
60. H. Hikino and T. Mizuno, *Planta Med.*, **55**, 385 (1989).
61. Y. Kimura et al., *Planta Med.*, **54**, 290 (1988).
62. H. Ito et al., *Mie Med. J.*, **26**(2–3), 147 (1977); through *Chem. Abstr.*, **87**, 177873j.
63. T. Kosuge et al., *Yakugaku Zasshi*, **105**, 791 (1985).
64. M. Kubo et al., *Yakugaku Zasshi*, **103**, 871 (1983).
65. J. Tao and K. Y. Feng, *Chin. J. Integr. Trad. Western Med.*, **9**, 733 (1989).
66. Y. Komoda et al., *Chem. Pharm. Bull.*, **37**, 531 (1989).
67. J. F. Wang et al., *J. Trad. Chin. Med.*, **5**(1), 55 (1985).
68. F. C. Wan and D. Z. Huang, *Zhongguo Zhongyao Zazhi*, **17**, 619 (1992).
69. C. Lin et al., *Zhongchengyao*, **14**(7), 31 (1992).
70. M. H. Jiang et al., *Zhongchengyao*, **13**(1), 24 (1991).
71. Z. M. Zeng, *Chin. J. Integr. Trad. Western Med.*, **6**, 494 (1986).
72. Z. T. Yan, *Zhongguo Yaoxue Zazhi*, **24**, 166 (1989).

# GARLIC

**Source:** ***Allium sativum*** L. (Family Amaryllidaceae or Liliaceae).

*Synonyms*. Common garlic and allium.

### GENERAL DESCRIPTION

A strong scented perennial herb with long, flat, and firm leaves, 0.5–1.5 cm wide; flowering stem up to 1.2 m high; bulbs with several bulblets (cloves), all enclosed in membranous skins; origin unclear; a variable cultigen (found only in cultivation), garlic's wild progenitor, *Allium longicuspis*, is thought to have originated in the high plains of west-central Asia perhaps in the Kirgiz Desert; spread east and west with nomadic tribes; known to be cultivated in the Middle East more than 5000 years ago; naturalized in North America; cultivated worldwide. Part used is the fresh or dehydrated bulb. Garlic oil is obtained by steam distillation of the crushed fresh bulbs; powdered garlic is derived from the dried bulbs.

## CHEMICAL COMPOSITION

Contains 0.1–0.36% (usually ca. 0.2%) volatile oil, alliin (*S*-allyl-L-cysteine sulfoxide), *S*-methyl-L-cysteine sulfoxide, enzymes (e.g., alliinase, peroxidase, and myrosinase), ajoenes (*E*,*Z*-ajoene, *E*,*Z*-methylajoene, and dimethylajoene), protein (16.8%, dry weight basis), minerals, vitamins (thiamine, riboflavin, niacin, etc.), lipids, amino acids, and others (JIANGSU, KARRER, LIST AND HÖRHAMMER, MARSH, MARTINDALE).[1,2]

The volatile oil contains allicin (diallyldisulfide-*S*-oxide; diallyl thiosulfinate), allylpropyl disulfide, diallyl disulfide, and diallyl trisulfide as the major components, with lesser amounts of dimethyl sulfide, dimethyl disulfide, dimethyl trisulfide, allylmethyl sulfide, 2,3,4-trithiapentane, and other related sulfur compounds.[1] Most published data since 1892 have indicated diallyl disulfide to be the main compound in garlic oil (60%). However, a recent study indicated that diallyl trisulfide dominated in freshly distilled oils.[3] Other volatile compounds present include citral, geraniol, linalool, and $\alpha$- and $\beta$-phellandrene (JIANGSU, KARRER, LIST AND HÖRHAMMER).[4]

Prostaglandins $A_2$ and $F_{1a}$ were recently isolated from a homogenized garlic extract.[2]

Allicin is the major odor principle that is produced by the enzymatic action of alliinase on alliin; it is decomposed by heat and alkali but is unaffected by dilute acids in solution (JIANGSU, MERCK).

Composition of garlic products depends on product form. Thiosulfinates (e.g., allicin) were found to be released only from garlic cloves and garlic powder products; vinyl dithiins and ajoenes were found only in products containing garlic macerated in vegetable oil; diallyl, methyl allyl, and dimethyl sulfide series components were exclusive to products containing the oil of steam distilled garlic.[5]

## PHARMACOLOGY OR BIOLOGICAL ACTIVITIES

Garlic and garlic oil have been reported to exhibit numerous pharmacological properties, including hypoglycemic activities in rabbits;[6–8] lowering of serum cholesterol (or lipids) in rabbits and humans (including lowering triglycerides and total low-density lipoprotein cholesterol), while raising levels of high-density lipoprotein cholesterol;[7,9–16] hypotensive properties in humans and animals (MARTINDALE);[15] antibacterial and antifungal properties, with the hydroalcoholic extract reportedly being much more potent than the essential oil;[1,15,17–20] larvicidal and insecticidal activities;[20,21] antitumor activities;[22] amebicidal activities; antihepatotoxic activity in rats;[23] antimycotic and antiviral (*in vitro* and *in vivo*); lowers blood viscosity; improves microcirculation; and expectorant, diaphoretic, and diuretic properties, among others (ESCOP 2, JIANGSU, LIST AND HÖRHAMMER, MARTINDALE).[24]

Antithrombotic activity has been attributed to various fractions of garlic. Ajoene has antithrombotic activity, triggered by inhibiting exposure of fibrinogen receptors on platelet membranes. Diallyl sulfide and methyl allyl sulfide, claimed responsible for antithrombotic activity, were found to be inactive in inhibiting platelet aggregation.[25] However, a recent study attributes higher thrombocyte aggregation–inhibiting potential to chloroform fractions of thiosulfinates than to the ajoenes.[2] The platelet aggregation–inhibiting activity may be responsible for the potential utility of garlic as a useful protective agent in atherosclerosis, coronary thrombosis, and stroke (TYLER 1).[26]

The volatile sulfur compounds (especially allicin, diallyl disulfide, and diallyl trisulfide) are generally considered to be responsible for much of garlic's pharmacological activities (e.g., hypoglycemic, hypocholesterolemic, antimicrobial, insecticidal, and larvicidal).[6,10,17,20,21] Allicin, believed to be the most important biologically active compound in garlic is primarily responsible for antibiotic and antimutagenic effects of garlic.[27,28]

Allergic contact dermatitis due to garlic has been reported (MARTINDALE).

A recent study found that thiosulfinates (e.g., allicin) were not formed below pH 3.6, thus alliinase is completely and irreversibly inhibited by stomach acid. A second (unidentified) enzyme, in addition to alliinase is involved in thiosulfinate formation. A stomach acid–resistant coating on garlic powder tablets is necessary for thiosulfinate release, which if prepared carefully can release amounts of total thiosulfinates similar to whole garlic cloves.[29]

Garlic oil acts as a gastrointestinal smooth muscle relaxant, suggested for further research in patients with hypermotile intestinal disorders.[30]

A pilot study in 10 AIDS patients found that a garlic extract produced an improved helper: suppresser ratio in natural killer cell activity, with a concurrent improvement in AIDS-related conditions, including diarrhea, genital herpes, candidiasis, and pansinusitis with recurrent fever.[26,31]

## USES

**Medicinal, Pharmaceutical, and Cosmetic.** Rarely used in the United States. However, strong popular interest has made garlic preparations the best-selling over-the-counter drugs in Germany, with sales topping $250 million per year in Europe.[16]

**Food.** Fresh garlic and powdered garlic are widely used as domestic spices. Garlic oil is extensively used as a flavor ingredient in most food products, including nonalcoholic beverages, frozen dairy desserts, candy, baked goods, gelatins and puddings, condiments and relishes, meat and meat products, fats and oils, snack foods, and gravies, with highest average maximum use levels generally much below 0.003% (34.4 ppm).

**Health Food/Herb Teas.** Use of various dosage forms and formulations, including fresh, dried, powdered, or freeze-dried garlic, essential oil (garlic oil), and various proprietary preparations of powdered, aged, or extracted fresh or dried garlic (FOSTER).

**Traditional Medicine.** Has been used for thousands of years in treating coughs, colds, chronic bronchitis, toothache, earache, dandruff, high blood pressure, arteriosclerosis, hysteria; also used extensively in cancers; generally as the juice, cold infusion, or tincture.[32] In addition, the fresh cloves, garlic tea, syrup, tincture, and other preparations have been used as an aphrodisiac; to treat fever, flu symptoms, shortness of breath, sinus congestion, headache, stomachache, hypertension, gout, rheumatism, pinworms, old ulcers, and snakebites; and for numerous other ailments, conditions, and applications (FOSTER AND DUKE).[26]

In Chinese medicine, in addition to above uses, garlic is used for diarrhea, dysentery (amebic and bacterial), pulmonary tuberculosis, bloody urine, diphtheria, whooping cough, typhoid, hepatitis, trachoma, scalp ringworm, hypersensitive dentin, vaginal trichomoniasis, and others, some of which have been reported successful in clinical trials (JIANGSU).

## COMMERCIAL PREPARATIONS

Crude, powder, and oil; crude was formerly official in N.F. and U.S.P. and oil is official in F.C.C. Composition of organosulfur compounds varies greatly with the method of processing, and changes occur in their chemistry when garlic is crushed, cooked, ingested, metabolized or commercially processed. Analysis of the compounds has only recently been achieved, necessitating standardization of these products, especially in commercial products and for biological studies.[5,27,33] However, before meaningful standardization can be established, a consensus must be reached as to what constitute the active components of garlic.

*Regulatory Status.* Has been affirmed as GRAS (§184.1317), but only as oil, extract, or oleoresin.[24]

A German therapeutic monograph allows use of the minced bulb and preparations calculated to an average daily dose of 4 g (fresh garlic) or 8 mg (essential oil) for supportive dietary measures to reduce blood lipids and as a preventative for age-dependent vascular changes.[34]

A proposed European monograph indicates use for prophylaxis of atherosclerosis; treatment of elevated blood lipid levels influenced by diet; improvement of arterial vascular disease blood flow; and use for relief of coughs, colds, catarrh, and rhinitis (ESCOP 2).

## REFERENCES

See the General References for ARCTANDER; BAILEY 1; BLUMENTHAL; ESCOP 2; FEMA; FERNALD; FOSTER; FOSTER AND DUKE; GUENTHER; JIANGSU; LEWIS AND ELVIN-LEWIS; LIST AND HÖRHAMMER; LUST; MABBERLY; NANJING; ROSE; ROSENGARTEN; TERRELL; TYLER 1; UPHOF; YOUNGKEN.


1. J. R. Whitaker, *Adv. Food Res.*, **22**, 73 (1976).
2. S. A. Al-Nagdy et al., *Phytother. Res.*, **2**(4), 196 (1988).
3. H. Miething, *Phytother. Res.*, **2**(3), 149 (1988).
4. O. E. Schultz and H. L. Mohrmann, *Pharmazie*, **20**, 441 (1965).
5. L. D. Lawson et al., *Planta Med.*, **57**, 363 (1991).
6. P. T. Mathew and K. T. Augusti, *Indian J. Biochem. Biophys.*, **10**, 209 (1973); through *Chem. Abstr.*, **81**, 45370j (1974).
7. R. C. Jain, *Lancet*, **1**, 1240 (1975).
8. R. C. Jain and C. R. Vyas, *Am. J. Clin. Nutr.*, **28**, 684 (1975).
9. A. Bordia et al., *Atherosclerosis*, **21**, 15 (1975).
10. K. T. Augusti, *Indian J. Exp. Biol.*, **15**, 489 (1977).
11. R. C. Jain and C. R Vyas, *Medikon*, **6**(5), 12 (1977); through *Chem. Abstr.*, **87**, 116795p (1977).
12. A. Bordia et al., *Atherosclerosis*, **26**, 379 (1977).
13. K. K. Sharma et al., *Indian J. Nutr. Diet.*, **13**, 7 (1976); through *Chem. Abstr.*, **85**, 41021f (1976).
14. D. Kritchevsky, *Artery*, **1**, 319 (1975); through *Chem. Abstr.*, **84**, 41631j (1976).
15. V. Petkov, *Deut. Apoth. Ztg.*, **106**, 1861 (1966).
16. P. Mansell and J. P. D. Reckless, *Br. Med. J.*, **303**, 379 (1991).
17. F. E. Barone and M. R. Tansey, *Mycologia*, **69**, 793 (1977).
18. A. V. Kolodin, *Sovrem. Metody Issled.*, **1**, 101 (1968); through *Chem. Abstr.*, **73**, 13029y (1970).
19. Y. Kominato et al., *Oyo Yakuri*, **11**, 941 (1976); through *Chem. Abstr.*, **88**, 99210v (1978).
20. N. B. K. Murthy and S. V. Amonkar, *Indian J. Exp. Biol.*, **12**, 208 (1974).
21. S. V. Amonkar and A. Banerji, *Science*, **174**, 1343 (1971).
22. Y. Kimura and K. Yamamoto, *Gann*, **55**, 325 (1964); through *Chem. Abstr.* **63**, 1089d (1965).
23. H. Hikino et al., *Planta Med.*, **52**(3), 163 (1986).
24. Anon., *Fed. Regist.*, **39**(185), 34213 (1974).

25. E. Block in R. P. Steiner, ed., *Folk Medicine the Art and the Science*, American Chemical Society, New York, 1986, p.125.
26. S. Foster, *Garlic—Allium sativum, Botanical Series, no. 311*, American Botanical Council, Austin, Tex. 1991.
27. L. D. Lawson et al., *Planta Med.*, **57**, 263 (1991).
28. H. Koch and G. Hahn. *Garlic—Fundamentals of the Therapeutic Application of Allium sativum*, Munich, Urban & Schwarzenberg, 1988.
29. L. D. Lawson and B. G. Hughes, *Planta Med.*, **58**, 347 (1992).
30. D. J. Joshi et al., *Phtyother Res.* **1**(3), 140 (1987).
31. T. H. Abdullah et al., *Deut. Z. Onkologie* **21**, 52 (1989).
32. J. L. Hartwell, *Lloydia*, **33**, 97 (1970).
33. L. D. Lawson, in A. D. Kinghorn and M. F. Balandrin, Eds., *Human and Medicinal Agents from Plants*, American Chemical Society, Washington, D.C., in press.
34. Monograph *Allii sativi bulbus*, *Bundesanzeiger*, no.122 (July, 6 1988).

# GELSEMIUM

**Source:** ***Gelsemium sempervirens*** (L.) Ait. f. (syn. G. *nitidum* Michx. and *Bignonia sempervirens* L.) (Family Loganiaceae or Spigeliaceae)

*Synonyms*. Yellow jasmine, wild jessamine, woodbine, Carolina yellow jessamine, and evening trumpet flower.

## GENERAL DESCRIPTION

Evergreen woody vine with very fragrant, bright yellow flowers; sometimes climbing to about 6 m; native to southwestern United States; also grows in Mexico and Guatemala; widely cultivated as an ornamental plant. Parts used are the dried rhizome and roots.

## CHEMICAL COMPOSITION

Contains as active constituents about 0.5% alkaloids, consisting mainly of gelsemine, with lesser amounts of gelsemicine, gelsedine, gelsevirine, sempervirine, gelsemidine, 1-methoxygelsemine, 21-oxo-gelsemine, and 14-hydroxygelsemicine (GLASBY 2, LIST AND HÖRHAMMER).[1–4]

Other constituents present include the coumarin scopoletin (also called gelsemic acid and $\beta$-methylesculetin), a volatile oil (0.5%), fatty acids (palmitic, stearic, oleic, and linoleic acids), *n*-pentatriacontane, and tannins (KARRER, LIST AND HÖRHAMMER).

## PHARMACOLOGY OR BIOLOGICAL ACTIVITIES

Gelsemium and gelsemine, its major alkaloid, have been reported to have central stimulant and analgesic properties; they also potentiated the analgesic effects of aspirin and phenacetin (JIANGSU, LIST AND HÖRHAMMER).

Gelsemium alkaloids are very toxic. Ingestion of as little as 4 mL of a fluid extract (1:1) or a tea made from as few as three leaves has been reported as fatal. Children have been severely poisoned by chewing leaves or sucking flower nectar. Toxic symptoms include giddiness, weakness, ptosis, double vision, dilated pupils, and respiratory depression (GOSSELIN, HARDIN AND ARENA, MARTINDALE). Gelsemicine is reportedly more toxic than glesemine.[5]

## USES

**Medicinal, Phamaceutical and Cosmetic.** Used as an ingredient in some analgesic, sedative, and antispasmodic preparations.

**Traditional Medicine.** Due to its high toxicity, gelsemium is rarely used as a domestic medicine in the United States. However when it is used, it is usually for treating nervous heart conditions, migraine, neuralgia, and sciatica; also used in cancer.[6]

Widely used in the 19th century, especially by eclectic practitioners; root tincture used for fevers, inflammations of the spinal column, and neuralgia; believed to diminish blood to the cerebrospinal centers, reducing spasmodic action.[7]

A related species (*G. elegans* Benth.) is used in China to treat conditions, including neuralgia, rheumatic pain, various kinds of sores, scrofula, and wounds. It is used only externally; all parts of the plant are employed (JIANGSU).

A rare species *G. rankinii* Small occurs from Alabama and Florida to North Carolina and may have been involved in traditional usage in the southern United States.[7]

## COMMERCIAL PREPARATIONS

Crude and extracts (fluid, solid, etc.); strengths (see ***glossary***) of extracts are expressed in weight-to-weight ratios. Crude and fluid extract were formerly official in N.F.

## REFERENCES

See the General References for BAILEY 1; GOSSELIN; HARDIN AND ARENA, LIST AND HÖRHAMMER; LUST; MARTINDALE; RAFFAUF; UPHOF; WILLAMAN AND SCHUBERT.


1. A. Nikiforov et al., *Monatsh. Chem.*, **105**, 1292 (1974).
2. E. Wenkert et al., *Experimentia*, **28**, 377 (1971).
3. M. Wichtl et al., *Monatsh. Chem.*, **104**, 87 (1973).
4. M. Wichtl et al., *Monatsh. Chem.*, **104**, 99 (1973).
5. B. R. Olin, ed., *Lawrence Rev. Nat. Prod.* (Feb. 1993).
6. J. L. Hartwell, *Lloydia*, **33**, 97 (1970).
7. S. Foster. *East-West Botanicals: Comparisons of Medicinal Plants Disjunct between Eastern Asia and Eastern North America*, Ozark Beneficial Plant Project, Brixey, Mo., 1986.


# GENET

**Source:** ***Spartium junceum*** L. (syn. *Genista juncea* Lam. (Family Leguminosae or Fabaceae)

*Synonyms*. Spanish broom and weavers broom.

## GENERAL DESCRIPTION

Upright shrub with green branches, up to about 3 m high; few leaves and showy, fragrant yellow flowers; native to the Mediterranean region, southwest Europe, and the Canary Islands; extensively cultivated, especially as an ornamental plant in western United States. Part used is the dried flower from which genet absolute and other extracts are produced by solvent extraction.

## CHEMICAL COMPOSITION

Contains alkaloids (mainly anagyrine, cytisine, thermopsine, *N*-methylcytisine, and sparteine), *n*-nonacosane, 1,18-octadecanediol, 1,26-hexacosanediol, lupeol, sitosterol, phenols, acids, terpenes, and trace of volatile oil (KARRER, WILLAMAN AND SCHUBERT).[1–3]

Major constituents of the absolute include ethyl palmitate (14.56%), linalool (10.91%), methyl linoleate (7.13%), ethyl stearate (3.67%), ethyl oleate (4.86%), methyl linolenate (1.53%), phenethyl alcohol (1.30%), and at least 50 additional, mostly ubiquitous, volatile components; plus caproic, caprylic, capric, lauric, myristic, palmitic, stearic, oleic, linoleic, and linolenic acids. Occurrence of pentanal diethyl acetal in the absolute has been attributed to formation during ethanol back extraction of the concrète or added as part of a coupage. $\beta$-terpineol and diethyl phthalate occurrence in the absolute may result from addition of synthetic additives.[4]

Due to the extraction methods used, genet absolute (see ***glossary***), the more commonly used form, should not contain any alkaloids. However, genet extracts obtained by other methods may contain alkaloids.

## PHARMACOLOGY OR BIOLOGICAL ACTIVITIES

Sparteine has oxytocic properties (see ***broom tops***).

There are no available data on the pharmacology of genet.

Available dermatological data indicate genet absolute to be nonphototoxic to mice and swine and nonirritating and nonsensitizing to humans.[5]

## USES

**Medicinal, Pharmaceutical, and Cosmetic.** Genet absolute is used as a fragrance ingredient in soaps, detergents, creams, lotions, and perfumes, with maximum use level Of 0.2% reported in perfumes.[5]

**Food.** Genet absolute and genet extract (type not specified) are used as flavor components in major categories of food products, including alcoholic and nonalcoholic beverages, frozen dairy desserts, candy, baked goods, and gelatins and puddings. Reported average maximum use levels are usually below 0.001%.

**Traditional Medicine.** Reportedly laxative, slightly narcotic, diuretic. Stems have appeared as an adulterant to *Cytisus scoparius* (STEINMETZ) (see ***broom tops***).

**Others.** Fiber formerly used for ropes, mattress filler, pillows, paper making, rubberized belts for mine conveyors; maintains strength under high humidity; high-grade cellulose extract from dry stems (CSIR X).

## COMMERCIAL PREPARATIONS

Crude and extracts.

*Regulatory Status.* Has been approved for food use (§172.510).

## REFERENCES

See the General References for BAILEY 2; CSIR X: FEMA; GUENTHER; HAY; STEINMETZ; TUTIN 2; UPHOF.

1. A. D. Dauksha and E. K. Denisova, *Nauch. Dokl. Vysshei Shkoly, Biol. Nauki*, **3**, 178 (1966); through *Chem. Abstr.*, **65**, 17362e (1966).

2. N. L. Gurvich and Z. I. Abasova, *Maslo. Zhir. Promy.*, **33**, 27 (1967); through *Chem. Abstr.*, **66**, 79481b (1967).
3. G. Faugergas et al., *Plant. Med. Phytother.*, **7**, 68 (1973).
4. B. M. Lawrence, *Perfum. Flavor.*, **6**(2), 59 (1981).
5. D. L. J. Opdyke, *Food Cosmet. Toxicol.*, **14**(Suppl.), 779 (1976).

# GENTIAN

**Source**: ***Gentian*** *Gentiana lutea* L. (Family Gentianaceae); ***Stemless gentian*** *Gentiana acaulis* L.)

*Synonyms*. Yellow gentian and bitter root (*G. lutea*).

## GENERAL DESCRIPTION

Gentian (*G. lutea*) is a perennial herb with simple erect stems and large oval leaves; up to about 1.8 m high; flowers yellow; native to mountains of central and southern Europe and western Asia. Parts used are the dried rhizome and roots.

Stemless gentian (*G. acaulis*) is a small variable taxonomic complex of perennial herbs with basal rosettes; leaves lance shaped; up to about 10 cm high; native to Europe (Alps and Pyrenees). Part used is the whole herb.

## CHEMICAL COMPOSITION

Gentian contains bitter principles (mainly amarogentin, also gentiopicroside or gentiopicrin and swertiamarin);[1–3]; alkaloids (mainly gentianine and gentialutine);[4,5] xanthones (gentisein, gentisin, isogentisin, 1,3,7-trimethoxyxanthone, 1-hydroxy-3,7-dimethoxyxanthone, 2,3′,4,6-tetrahydroxybenzophenone, etc.;[6,7] triterpenes (lupeol and $\beta$-amyrin);[8] gentianose (glucosyl gluconyl fructose), gentiobiose, and common sugars;[3,9] and traces of volatile oil (KARRER, LIST AND HÖRHAMMER).[3]

It has been reported that content of gentiopicroside does not vary significantly on drying or at various stages of vegetative growth.[10] However, another study found that amarogentin and gentiopicroside content were highest in spring, with concurrent decreased levels of sugar accumulation in the roots.[11]

It has been observed that cultivated plants are much richer in amarogentin than wild ones.[10]

*G. lutea* ecotypes have been found to retain morphological and chemical characteristics, with amarogentin content ranging from 0.05 to 0.33% in 3-year-old ecotypes, with varying differences in total sugar and essential oil composition.[3] Wild plants harvested at lower altitudes in Italy were found to have higher levels of bitter principles.[12]

Stemless gentian contains gentiacauloside (a xanthone glycoside) and gentisin. It probably also contains similar constituents as gentian (LIST AND HÖRHAMMER).[13]

Amarogentin is one of the most bitter glycosides known (STAHL). Amarogentin and gentiopicrin are the main compounds responsible for the bitter taste (TYLER 1).

## PHARMACOLOGY OR BIOLOGICAL ACTIVITIES

Gentian extracts have been reported to exhibit choleretic activities in animals. Gentianine also exhibited strong antiinflammatory properties in laboratory animals (FARNSWORTH 1).[5,14,15]

Gentiopicrin has been reported to be lethal to mosquito larvae (JIANGSU).

The root may not be well tolerated by patients with extreme high blood pressure or by pregnant women (TYLER 1).

Highly toxic *Veratrum album* L., growing wild in proximity to gentian has resulted in several cases of vicarious accidental poisoning in Europe among persons making gentian preparations for home use.[16]

## USES

**Medicinal, Pharmaceutical, and Cosmetic.** Gentian extracts are used in tonic preparations; also used in certain antismoking formulas. They are also used in cosmetics.[17]

**Food.** Gentian and stemless gentian are widely used in alcoholic bitters and vermouth formulations; average maximum use levels reported are about 0.02% (199 ppm) for gentian extract (type not specified) and 0.001% (12.9 ppm) for stemless gentian. Other food products in which both are used include nonalcoholic beverages, frozen dairy desserts, candy, baked goods, and gelatins and puddings, with highest average maximum use level of about 0.015% (153 ppm) reported for gentian extract (type not specified) in baked goods.

Gentian is also a major constituent of "angostura bitters" (see ***angostura***).

**Health Food/Herb Teas.** Dried root and extract used in various digestive formulations; tea flavoring; also in nervine formulations. In England the most popular of gastric stimulants (WREN).

**Traditional Medicine.** Yellow gentian as well as related gentians reportedly used to stimulate the appetite, improve digestion, and to treat numerous gastrointestinal problems (stomachache, heartburn, gastritis, diarrhea, vomiting, etc.) as well as externally for wounds; also used in cancers.[18]

In Chinese medicine several related *Gentiana* species (e.g., G. *scabra* Bunge, *G. triflora* Pall., and *G. rigescens* Franch.) are used for similar purposes. In addition, they are used for treating jaundice, sore throat, headache, sores, inflammations, and rheumatoid arthritis, among others, both internally and externally (JIANGSU, NANJING).

## COMMERCIAL PREPARATIONS

Crude and extracts. Gentian and its fluid extract, compound tincture, and glycerinated elixir were formerly official in N.F. Strengths (see ***glossary***) of extracts are expressed in weight-to-weight ratios.

*Regulatory Status.* Gentian and stemless gentian both have been approved for food use, the latter in alcoholic beverages only (§172.510). The root is the subject of a German therapeutic monograph, approved for digestive disorders, such as appetite loss and flatulence; use contraindcated in gastric and duodenal peptic ulcers.[19]

## REFERENCES

See the General References for BAILEY 1; BIANCHINI AND CORBETTA; BLUMENTHAL; FARNSWORTH 3; FEMA; GOSSELIN; JIANGSU; LUST; NANJING; POLUNIN AND SMYTHIES; RAFFAUF; TERRELL; TUTIN 3; TYLER 1; WREN.

1. J. Bricout, *Phytochemistry*, **13**, 2819 (1974).
2. Y. Hatakeyama et al., *Shoyakugaku Zasshi.*, **26**, 75 (1972); through *Chem. Abstr.*, **79**, 45712f (1973).
3. C. Franz and D. Fritz, *Planta Med.*, **28**, 289 (1975).
4. F. Rulkco, *Pr. Nauk. Akad. Med. Wroclawiu*, **8**, 3 (1976); through *Chem. Abstr.*, **86**, 103037x (1977).

5. F. Sadritdinov, *Farmakol. Alkaloidov Serdechnykh Glikozidov*, **146** (1971); through *Chem. Abstr.*, **78**, 79634b (1973).
6. J. R. Lewis and P. Gupta, *J. Chem. Soc. C* (4), 629 (1971).
7. J. E. Atkinson at al., *Tetrahedron*, **25**, 1507 (1969).
8. L. Benarroche et al., *Plant. Med. Phytother.*, **8**, 15 (1974).
9. P. Rivaille and D. Raulais, *C. R. Acad. Sci., Ser. D*, **269**, 1121 (1969); through *Chem. Abstr.*, **72**, 39744g (1970).
10. V. Rossetti et al., *Q. J. Crude Drug Res.*, **19**(1), 37 (1981).
11. Ch. Franz, et al., *Sci. Pharm.*, **53**, 31 (1985).
12. V. Rossetti et al., *Plant. Med. Phytother.*, **18**(1), 15 (1984).
13. V. Plouvier et al., *C. R. Acad. Sci., Ser. D*, **264**, 1219 (1967); through *Chem. Abstr.*, **67**, 64182b (1967).
14. H. C. Chi et al., *Sheng Li Hsueh Pao*, **23**, 151 (1959); through *Chem. Abstr.*, **57**, 11821g (1962).
15. B. R. Olin, ed., *Lawrence Rev. Nat. Prod.* (Feb. 1993).
16. R. Garnier et al., *Ann. Med. Interne (Paris)*, **136**, 125 (1985).
17. H. B. Heath, *Cosmet. Toilet.*, **92**(1), 19 (1977).
18. J. L. Hartwell, *Lloydia*, **32**, 153 (1969).
19. Monograph *Gentianae radix*, *Bundesanzeiger*, no. 223 (Nov. 30, 1985); revised (Mar. 13, 1990).

# GERANIUM OIL, ROSE

**Source**: ***Pelargonium graveolens*** (L.) L'Her. ex Ait. (Family Geraniaceae)

*Synonyms*. Algerian geranium oil, Bourbon geranium oil, and Moroccan geranium oil.

## GENERAL DESCRIPTION

A perennial shrubby, hairy, and glandular plant with fragrant leaves; up to about 1 m high; native to South Africa; widely cultivated in Africa (Algeria, Morocco, etc.) and Europe (Spain, Italy, France, etc.). Parts used are the fresh leaves and stems, from which geranium oil is obtained by steam distillation, generally in a 0.08–0.4% yield.[1,2]

There are several types of geranium oils produced from cultivated forms, varieties, and hybrids of *P. graveolens*, and other *Pelargonium* species such as *P. odoratissimum* Ait., *P. capitatum* Ait., *P. crispum* (L.) L'Her., and *P. radula* (Cav.) L'Her. ex Ait. (syn. *P. roseum* Willd.). The more commonly used ones are Algerian or African geranium oil, Reunion or Bourbon geranium oil, and Moroccan geranium oil. Despite mention of several commercial source species in the literature, *P. graveolens* appears to be the only one commercially cultivated (TUCKER AND LAWRENCE).

Geranium oils should not be confused with East Indian or Turkish geranium oil; the latter is palmarosa oil and is derived from a completely different plant source.

## CHEMICAL COMPOSITION

All three major types of geranium oil (Algerian, Bourbon, and Moroccan) contain large amounts of alcohols (primarily *l*-citronellol and geraniol, with linalool and phenethyl alcohol in lesser amounts);[1–7] esters (e.g., geranyl tiglate, geranyl acetate, citronellyl formate, and citronellyl acetate);[1,6] and aldehydes and ketones (*l*-isomenthone, citronellal, citral, decyl aldehyde, etc.).[4–7] The total alcohols content is usually 60–70%, and the esters content is normally 20–30% but they vary considerably, depending on the type of oil.[1,4–6] The proportions of alcohols (especially *l*-citronellol, geraniol, and phenethyl alcohol) also vary with the type of oil.[2,3,5,6]

Other compounds reported present in geranium oils include sesquiterpene hydrocarbons ($\alpha$- and $\beta$-bourbonene, $\alpha$-santalene, $\beta$-elemene, caryophyllene, $\gamma$-muurolene, humulene, $\delta$-cadinene, *ar*-curcumene, selina-3,7(11)-diene, selina-4(14),7(11)-diene, germacratriene, 6,9-guaiadiene, etc.);[8–10] sesquiterpene alcohols (10-epi-$\gamma$-eudesmol, junenol, *l*-selin-11-en-4-ol, etc.);[10,11] acids (formic, acetic, propionic, caprylic, tiglic acids, etc.);[6,7] dimethyl sulfide, *cis*- and *trans*-dehydrocitronellol, menthol, citronellyl-diethylamine, and others.[6,12,13]

## PHARMACOLOGY OR BIOLOGICAL ACTIVITIES

A geranium oil (from leaves of *P. roseum*) with its alcohol component consisting almost entirely of *l*-citronellol has been reported to exhibit *in vitro* inhibitory activities against several fungi that are pathogenic to humans.[2] Other rose geranium oils have also been reported to have antibacterial and antifungal activities *in vitro*.[14–16]

Cases of dermatitis in hypersensitive individuals caused by geranium oil present in cosmetics have been well documented, though some data indicate geranium oil to be nonsensitizing, nonirritating, and nonphototoxic to human skin.[14,17,18]

## USES

**Medicinal, Pharmaceutical, and Cosmetic.** Rose geranium oil (especially Bourbon) is widely used as a fragrance component in all kinds of cosmetic products, including soaps, detergents, creams, lotions, and perfumes, with maximum use level of 1.0% reported in perfumes.[14,17,18]

**Food.** Rose geranium oil is extensively used in most major food products, including alcoholic and nonalcoholic beverages, frozen dairy desserts, candy, baked goods, and gelatins and puddings. Use levels are generally below 0.001%.

**Health Food/Herb Teas.** Rose geranium oil is commonly available; primarily used as an inexpensive substitute for rose oil in massage oil formulations, etc. (ROSE).

**Traditional Medicine.** In Africa the roots of various *Pelargonium* spp. are reported used as astringent; in diarrhea, dysentery, etc. (WATT AND BREYER-BRANDWIJK).

## COMMERCIAL PREPARATIONS

Most common types (e.g., Algerian, Bourbon, Moroccan, and Spanish). Algerian geranium oil is official in F.C.C.

*Regulatory Status.* GRAS: rose geranium (§182.20); other types (§182.10 and §182.20).

## REFERENCES

See the General References for ARCTANDER; BAILEY 2; FEMA; FURIA AND BELLANCA; GUENTHER; LIST AND HÖRHAMMER; MARTINDALE; MERCK; ROSE; TERRELL; TUCKER AND LAWRENCE; UPHOF; WATT AND BREYER-BRANDWIJK.

1. S. Angjeli, *Bull. Univ. Shteteror Tiranes, Ser. Shkencat Natyrore*, **1**, 64 (1964); through *Chem. Abstr.*, **61**, 15925h (1964).
2. H. Wollman et al., *Pharmazie*, **28**, 56 (1973).
3. E. Gliozheni and H. Ahmataj, *Bull. Univ. Shteteror Tiranes, Ser. Shkencat Natyrore*, **21**, 129 (1967); through *Chem. Abstr.*, **70**, 17534v (1969).
4. L. Peyron, *Compt. Rend.*, **255**, 2981 (1962).
5. J. Y. Conan et al., *Riv. Ital. Essenze, Profumi, Piante Off, Aromi, Saponi, Cosmet., Aerosol*, **58**, 556 (1976); through *Chem. Abstr.*, **86**, 95858y (1977).
6. V. T. Gogiya and L. I. Ivanova, *Mezhdunar. Kongr. Efirnym. Maslam, (Mater.)*, **1**, 71 (1968); through *Chem. Abstr.*, **79**, 83391h (1973).
7. C. de la Torre et al., *An. Acad. Bras. Cienc.*, **44**(Suppl.), 168 (1972); through *Chem. Abstr.*, **83**, 136700z (1975).
8. J. Krepinsky et al., *Tetrahedron Lett.* (3), 359 (1966).
9. E. Tsankova and I. Ognyanov, *Dokl. Bolg. Akad. Nauk*, **25**, 1229 (1972); through *Chem. Abstr.*, **78**, 47635m (1973).
10. P. Pesnelle et al., *Riv. Ital. Essenze, Profumi, Piante Off, Aromi, Saponi, Cosmet., Aerosol*, **54**, 92 (1972); through *Chem. Abstr.*, **77**, 66096g (1972).
11. P. Pesnelle et al., *Recherches*, **18**, 45 (1971); through *Chem. Abstr.*, **76**, 131362m (1971).
12. B. H. Kingston, *Manuf. Chem.* (10), 463 (1961).
13. W. Rojahn and E. Klein, *Dragoco Rep. (Ger. Ed.)*, **24**, 150 (1977); through *Chem. Abstr.*, **87**, 156990e (1977).
14. D. L. J. Opdyke, *Food Cosmet. Toxicol.*, **14**(Suppl.), 781 (1976).
15. J. C. Maruzzella and P. A. Henry, *J. Am. Pharm. Assoc.*, **47**, 294 (1958).
16. J. C. Maruzzella and L. Liquori, *J. Am. Pharm. Assoc.*, **47**, 250 (1958).
17. D. L. J. Opdyke, *Food, Cosmet. Toxicol.*, **12**(Suppl.), 883 (1974).
18. D. L. J. Opdyke, *Food, Cosmet. Toxicol.*, **13**, 451 (1975).

# GINGER

**Source**: ***Zingiber officinale*** Roscoe (Family Zingiberaceae)

*Synonym*. Common ginger.

### GENERAL DESCRIPTION

An erect perennial herb with thick tuberous rhizomes (underground stems) from which the aerial stem grows up to about 1 m high; native to southern Asia; extensively cultivated in the tropics (e.g., India, China, Jamaica, Haiti, and Nigeria). Part used is the pungent rhizome commonly called "root," both in fresh and dried forms. Ginger oil is usually produced from freshly ground, unpeeled dried ginger by steam distillation. Extracts and oleoresin are produced from dried unpeeled ginger, as peeled ginger loses much of its essential oil content.

### CHEMICAL COMPOSITION

Ginger has been reported to contain 0.25–3.3% (usually 1–3%) volatile oil; pungent principles (gingerols and shogaols);[1–5] about 6–8% lipids composed of triglycerides, phosphatidic acid, lecithins, free fatty acids (lauric, palmitic, stearic, oleic, linoleic, etc.), and others;[6] protein (ca. 9%); starch (up to 50%); vitamins (especially niacin and A); minerals; amino acids; resins; and others (JIANGSU, MARSH, ROSENGARTEN, STAHL).[7]

Ginger oil contains as its major components the sesquiterpene hydrocarbons zingiberene and bisabolene.[1,3] Other sesquiterpene hydrocarbons and alcohols present include *ar*-curcumene, $\beta$-sesquiphellandrene, sesquithujene, zingiberol, zingiberenol, *cis*-sesquisabinene hydrate, and *cis*- and *trans*-$\beta$-sesquiphellandrol.[1,3,8–10] It also contains monoterpene hydrocarbons, alcohols, and aldehydes (e.g., phellandrene, camphene, geranial, neral, linalool, and *d*-borneol),[3,7,8] methylheptenone, nonyl aldehyde; gingediacetate and gingerol;[11] and others (GUENTHER, JIANGSU).

Ginger oleoresin contains mainly the pungent principles gingerols and shogaols as well as zingerone.[1,2,5] Shogaols and zingerone are dehydration and degradation products, respectively, of gingerols; the latter have been reported to constitute about 33% of a freshly prepared oleoresin.[3] Shogaols have recently been found to be twice as pungent as gingerols.[4] The pungency of ginger oleoresin is lost on prolonged contact with alkalis (ARCTANDER).

A protease has been isolated from fresh ginger; yield of the crude enzyme was 2.26% from the fresh rhizome.[12]

## PHARMACOLOGY OR BIOLOGICAL ACTIVITIES

Ginger has carminative properties.

Ginger extracts are reported to exhibit numerous pharmacological properties, including stimulating the vasomotor and respiratory centers of anesthetized cats as well as direct heart stimulation;[7] lowering of serum and hepatic cholesterol in rats previously fed cholesterol;[13] and killing vaginal trichomonads *in vitro* (JIANGSU).

Dried ethanol and acetone extracts of fresh rhizome given orally significantly inhibited gastric secretion of induced acute stress in rats, producing an antiulcer effect.[14]

Powdered ginger root (250 mg q.i.d.) over 4 days was better than placebo in diminishing or eliminating symptoms of hyperemesis gravidarum.[15]

Ginger oil is reported to be nonirritating and nonsensitizing in humans, and its low phototoxicity is not considered significant.[16]

Fresh ginger (juice, aqueous extract, in poultice or sliced form) has been reported highly effective in China in the clinical treatment of rheumatism, acute bacterial dysentery, malaria, and orchitis (inflammation of the testicles) (JIANGSU).

Ginger (dried) has been the subject of several clinical trials to assess efficacy in motion sickness with positive results.[17,18]. Its antiemetic effect is attributed to an effect on gastric activity rather than a CNS mechanism, characteristic of conventional antimotion sickness drugs.[19]

## USES

**Medicinal, Pharmaceutical, and Cosmetic.** Ginger (mainly as the oleoresin) is used as an ingredient in certain digestive, laxative, antitussive, carminative, and antacid preparations. Ginger oil is used as a fragrance component in cosmetic products, including soaps, detergents, creams, lotions, and perfumes (especially Oriental and men's fragrances). Maximum use level is 0.4% reported in perfumes.[16]

**Food.** Ginger is widely used as a domestic spice, especially in Oriental cooking. It is used commercially in many foods, including nonalcoholic beverages, baked goods (e.g., cookies), gelatins and puddings, meat and meat products, and condiments and relishes, with the highest average maximum use level of about 0.525% (5248 ppm) reported in baked goods.

Ginger oil, oleoresin, and extract are widely used in soft drinks (e.g., gingerales and colas). In addition to the above uses, they are used in alcoholic beverages (liqueurs, bitters, etc.), frozen dairy desserts, and candy. Highest average maximum use levels reported are about 0.004% for the oil in baked goods (36.9 ppm) and gelatins and puddings (37.9 ppm); about 0.01% for the oleoresin in alcoholic beverages (99.8 ppm), baked goods (104 ppm), and condiments and relishes (108 ppm); and about 0.023% (233 ppm) for the extract (type not given) in baked goods.

**Health Food/Herb Tea.** Dried powdered root and extracts used in teas, capsules, and drinks, tablets, singly or in combination as a digestive aid, or antinauseant; also for colds, flu, and as general stimulant (FOSTER AND YUE).

**Traditional Medicine.** Ginger is usually used as a carminative and diaphoretic and to stimulate the appetite. Dried ginger has been used for thousands of years in China to treat numerous ailments, including stomachache, diarrhea, nausea, cholera, and bleeding. In addition to the above uses, fresh ginger is used to treat rheumatism, poisonous snakebite, baldness, toothache, and other conditions (JIANGSU).

Recently, fresh ginger juice has been reported to be used in treating thermal burns with considerable success.[20]

**Others.** Ginger and its extracts have strong antioxidative activities on various foods (e.g., lard, cookies, potato chips, oils, and fats) and could serve as potential source of food antioxidants.[21–24]

Due to its reportedly high content of a protease, fresh ginger could serve as a source of this protease, which could have applications similar to those of papain or other plant proteases (see ***bromelain***, ***ficin***, and ***papain***).

## COMMERCIAL PREPARATIONS

Dried crude in peeled (e.g., Jamaican), rough-peeled (e.g., Cochin), and unpeeled (e.g., Nigerian) forms; extracts; oils; and oleoresin. Strengths (see ***glossary***) of extracts are expressed either in weight-to-weight ratios or in flavor intensities. Crude, fluid extract, and oleoresin were formerly official in N.F.; oil is official in F.C.C.

*Regulatory Status.* GRAS (§182.10 and §182.20). Subject of a German therapeutic monograph; allowed for dyspeptic complaints and prevention of motion sickness.[25]

## REFERENCES

See the General References for ARCTANDER; BAILEY 1; BLUMENTHAL; FEMA; FOSTER AND YUE; GOSSELIN; GUENTHER; JIANGSU; LUST; MARTINDALE; NANJING; ROSENGARTEN; TERRELL.

1. C. R. Mitra, *Riechst., Aromen, Körperpflegem.*, **25**, 170 (1975).
2. S. M. Annathakrishna and V. S. Govindarajan, *Lebensm. Wiss. Technol.*, **7**, 220 (1974); through *Chem. Abstr.*, **81**, 150390p (1974).
3. D. W. Connell, *Aust. Chem. Process. Eng.*, **24**, 27 (1971); through *Chem. Abstr.*, **76**, 139120b (1972).
4. S. Narasimhan and V. S. Govindarajan, *J. Food Technol.*, **13**, 31 (1978).
5. D. W. Connell and R. McLachlan, *J. Chromatogr.*, **67**, 29 (1972).
6. I. P. Singh et al., *Indian J. Agr. Sci.*, **45**, 545 (1975).
7. M. M. Ally, *Proc. Pan Indian Ocean Sci. Congr.*, **4**, 11 (1960); through *Chem. Abstr.*, **61**, 6047e (1964).

8. D. W. Connell and R. A. Jordan, *J. Sci. Food Agr.*, **22**, 93 (1971).
9. S. J. Terhune et al., *Can. J. Chem.*, **53**, 3285 (1975).
10. A. A. Bednarczyk et al., *J. Agr. Food Chem.*, **23**, 499 (1975).
11. Y. Masada et al., *Int. Congr. Essent. Oils (Pap.)* **6**, 97 (1974).
12. E. H. Thompson et al., *J. Food Sci.*, **38**, 652 (1973).
13. S. Gujral et al., *Nutr. Rep. Int.*, **17**, 183 (1978).
14. J. A. A. Sertie et al., *Fitoterapia*, **63**(1), 55 (1992).
15. W. Fischer-Rasmussen et al., *J. Obstet. Gynecol. Reprod. Biol.*, **38**, 19 (1990).
16. D. L. J. Opdyke, *Food Cosmet. Toxicol.*, **12**(Suppl.), 901 (1974).
17. D. B. Mowrey and D. E. Clayson, *Lancet* (20), 655 (1982).
18. A. Grøntved et al., *Acta Otolaryngol. (Stockh.)*, **105**, 45 (1988).
19. S. Holtmann et al., *Acta Otolaryngol. (Stockh.)*, **108**, 168 (1989).
20. N. X. Cui, *Zhejiang Zhongyi Zazhi*, (10), 451 (1990).
21. Y. Saito et al., *Eiyo To Shokuryo*, **29**, 505 (1976); through *Chem. Abstr.*, **87**, 150314r (1977).
22. Y. Kihara and T. Inoue, *Nippon Shokuhin Kogyo Gakkaishi*, **9**, 290 (1962); through *Chem. Abstr.*, **59**, 13276b (1963).
23. F. Hirahara et al., *Eiyogaku Zasshi*, **32**, 1 (1974); through *Chem. Abstr.*, **82**, 2764u (1975).
24. H. Fujio, *Nippon Shokuhin Kogyo Gakkaishi*, **16**, 241 (1969); through *Chem. Abstr.*, **74**, 2846g (1971).
25. Monograph *Zingiberis rhizoma*, *Bundesanzeiger*, no. 85 (May 5, 1988).

# GINKGO

**Source**: ***Ginkgo biloba*** L. (Family Ginkgoaceae)

*Synonyms*. Maidenhair Tree.

## GENERAL DESCRIPTION

Monotypic deciduous tree to 40 m; leaves alternate or borne on spurs in clusters of 3–5; parallel veined, broad fan shaped, up to 12 cm, with notch at apex, forming two distinct lobes, hence the species name '*biloba*.' Flowers dioecious; male flowers on pendulous catkins with numerous, loosely arranged anthers in stalked pairs on a slender axis; female flowers are in pairs on long foot-stalks. The drupelike fruits have an acrid, foul-smelling pulp (likened to dog droppings) surrounding a single smooth oval, thin-shelled, semiedible nut (seed).

Ginkgo is known only from cultivation; widely planted ornamental tree worldwide; limited occurrence in undisturbed forests in Zhejiang province disputed as natural spontaneous specimens or progeny of planted specimens.

The leaves are used in Western pharmaceuticals; seeds and leaves traditionally used in China. Leaves are grown on a commercial scale in China, South Carolina and Maryland in the United States, and in the Bordeaux region of France (FOSTER AND YUE).

## CHEMICAL COMPOSITION

Leaves and root bark contain terpenoids,

including the monomethyl-mononorditerpenes: ginkgolide A (3-$\alpha$OH,10-$\beta$OH), ginkgolide B (1,3-$\alpha$OH,10-$\beta$OH), ginkgolide C (1,3-$\alpha$OH, 7,10-$\beta$OH) in the root bark and leaves; ginkgolide M (1-$\alpha$OH, 7,10-$\beta$OH) in the root bark; ginkgolide J(3-$\alpha$OH, 7,10-$\beta$OH) in the leaves; and the sesquiterpene bilobalide (10-$\alpha$OH, 8-$\beta$OH) in the leaves. The ginkgolides differ in the number and position of hydroxyl groups present on C1, C3 and/or C7 of the spirononane framework[1,2]

Flavonoids from the leaves include the flavones luteolin and tricetin (dalphidenon). Biflavones including amentoflavone, bilobetin, ginkgetin; isoginkgetin, sciadopitysin, 5′-methoxybilobetin; flavonols including kaempferol, kaempferol-3-rutinoside, kaempferol-3-*O*-$\alpha$(6‴-*p*-coumaroyl-glucosyl-$\beta$-1,4-rhamnoside), quercetin, quercetin-3-rutinoside (rutin), quercetin-3-glucoside (isoquercitrin), quercetin 3-*O*-$\alpha$ (6‴-*p*-coumaroyl-glucosyl-$\beta$-1,4-rhamnoside), isorharmnetin, and 3-*O*-methylmyricetin-3-rutinoside. Catechins include (+)-catechin, (−)-epicatechin, (+)-gallocatechin, and (−)-epigallocatechin. Proanthocyanidins include gallocatechin-4,8″-catechin (procyanidin), and gallocatechin-4,8″-gallocatechin (prodelphinidin).[1,2]

Other leaf components include the lignin *Z*,*Z*,-4,4′-(1,4-pentadien-1,5-diyl)diphenol; steroids including sitosterol and its glucoside (ipuranol); starch in sieve tube plastids; oil, balsam, and benzoic acid in cavities; calcium oxalate raphides; wax (0.7–1% in dry leaves), consisting of alkanes and alcohols (75%), esters (15%), and free acids (10%); pinitol, sequoyitol, D-glucaric acid, plus shikimic and succinic acids, and others.[1,2]

The nut (seed) contains 67.9% starch, 13.1% proteins (globulins, glutelin, and albumins), 2.9% lipids (composed of linoleic, oleic, palmitic, stearic, linolenic, and $\alpha$-hydroxypalmitic acids), 1.6% pentosans; plus sucrose, glucose, fructose, citric and quinic acids, fiber, and others.[1]

## PHARMACOLOGY OR BIOLOGICAL ACTIVITIES

Ginkgo leaf extract has vascular tone–regulating properties, an antihypoxic effect, experimental inhibitory activity in cerebral edema and neurotoxicity; modulates cerebral energy metabolism;[3] free-radical scavenging properties as effective as uric acid; inhibits lipid peroxidation of membranes, helping to maintain integrity and permeability of cell walls;[4] antiasthmatic, bronchodilator, etc. (WREN).

Numerous pharmacological and clinical studies of ginkgo leaf extract have demonstrated a positive effect in increasing vasodilation and peripheral blood flow rate in capillary vessels and end arteries in various circulatory disorders, Reynaud's disease, varicose conditions, postthrombotic syndrome (WEISS),[5] chronic cerebral vascular insufficiency,[6] short-term memory improvement,[7] cognitive disorders secondary to depression, dementia,[8] tinnitus,[9] vertigo, obliterative arterial disease of the lower limbs,[10] and other conditions.[5]

A recent retrospective critical review of the quality and methodology of 40 trials (published since 1975) on the use of ginkgo extracts in cerebral insufficiency has found 8 such trials to be of high quality.[11]

Ginkgolide B is a selective antagonist of platelet aggregation induced by platelet-activating factor (PAF). PAF, an inflammatory autacoid, is involved in various inflammatory, cardiovascular, and respiratory disorders.[12–14] Total synthesis of ginkgolide B has been achieved.[15]

## USES

**Medicinal, Pharmaceutical, and Cosmetic.** A standardized extract of the dried leaves is among, the best-selling phytomedicines in Europe; used clinically for heart disease, eye ailments, tinnitus, cerebral and peripheral vascular insufficiency, injuries involving brain trauma, dementias, short-term memory improvement, cognitive disorders secondary to depression, vertigo, and various conditions associated with senility.[5]

Leaf extracts used in cosmetics including shampoos, creams, and lotions.

**Food.** Seeds considered a delicacy in Japan and China; edible after acrid, foul-smelling pulp is removed; seeds are then boiled or roasted and eaten sparingly (no more than 8–10 per day). Fresh seeds are toxic and have reportedly caused death in children; pulp may cause contact dermatitis similar to poison ivy rash; handled with rubber gloves. Ingestion of fresh seeds may cause stomachache, nausea, diarrhea, convulsions, weak pulse, restlessness, difficult breathing, and shock. Dyed-red, the nuts were traditionally eaten at weddings (FOSTER AND YUE).

**Health Food/Herb Teas.** In the United States, various ginkgo leaf preparations or crude leaf are sold as dietary supplements in the form of tablets, capsules, tincture, standardized extracts, tea, etc.

**Traditional Medicine.** In China the dried, processed seed (*baiguo*), is used in prescriptions for asthma, coughs with phlegm, enuresis, mucous vaginal discharges, bronchitis with asthma, chronic bronchitis, tuberculosis, frequent urination, seminal emissions, turbid urine, etc. Externally, seed poulticed for scabies and sores.

The leaves (*bai guo ye*) used in prescriptions for arteriosclerosis, angina pectoris, high serum cholesterol levels, dysentery, and filariasis. An infusion of the boiled leaves used as wash for chilblains (FOSTER AND YUE).

**Other.** The root and inner bark are Chinese folk medicines; bark used in prescriptions for mucous vaginal discharges, seminal emission, or weak, convalescing patients; dried bark, burned to ash, mixed with vegetable oil as poultice for neurodermatitis.

## COMMERCIAL PREPARATIONS

Crude herb, extracts, dried leaf in capsules, tinctures, etc. Purified ginkgolide B is known commercially as BN 52021. A complex standardized extract of the dried leaves (EGb 761), produced by a German/French consortium, standardized to 24% flavone glycosides and ginkgolide B is widely sold in Europe (especially in Germany and France).

*Regulatory Status.* Undetermined in United States.

## REFERENCES

See the General References for FOSTER AND YUE; GLASBY 2; TYLER 1; WEISS; WREN.


1. N. Boralle et al. in P. Braquet, ed., *Ginkgolides—Chemistry, Biology, Pharmacology, and Clinical Perspectives*, Vol. 1, J. R. Prous Science Publishers, Barcelona, 1988, p. 9.
2. H. Huh and E. J. Staba. *J. Herbs, Spices Med. Plants*, **1**(1/2), 91 (1992).
3. S. S. Chatterjee in A. Agnoli et al., eds., *Effects of Ginkgo Biloba Extract on Organic Cerebral Impairment*, John Libbey Eurotext, Ltd., London, 1985.
4. J. Pincemail and C. Deby in E.W. Fünfgeld, ed., *Rökan (Ginkgo biloba), Recent Results in Pharmacology, and Clinic*, Springer-Verlag, Berlin, 1988, p. 5
5. S. Foster. *Ginkgo biloba*, *Botanical Series*, *no. 304*, American Botanical Council, Austin, Tex. 1991.
6. G. Vorberg, *Clin. Trials J.*, **22**, 149 (1985).
7. I. Hindmarch in E. W. Fünfgeld, ed.,

*Rökan (Ginkgo biloba), Recent Results in Pharmacology, and Clinic*, Springer-Verlag, Berlin, 1988, p. 321.
8. D. M. Warburton in E. W. Fünfgeld, ed., *Rökan (Ginkgo biloba), Recent Results in Pharmacology, and Clinic*, Springer-Verlag, Berlin, 1988, p. 327.
9. B. Meyer in E. W. Fünfgeld, ed., *Rökan (Ginkgo biloba), Recent Results in Pharmacology, and Clinic*, Springer-Verlag, Berlin, 1988, p. 245.
10. U. Bauer, *Arzneim. Forsch*, **34**, 716 (1984).
11. J. Kleijnen and P. Knipschild, *Br. J. Clin. Pharmacol.* **34**, 352, (1992).
12. P. Braquet, ed., *Ginkgolides—Chemistry, Biology, Pharmacology and Clinical Perspectives*, Vol. 1, J. R. Prous Science Publishers, Barcelona, 1988.
13. P. Braquet, ed., *Ginkgolides—Chemistry, Biology, Pharmacology and Clinical Perspectives*, Vol. 2, J. R. Prous Science Publishers, Barcelona, 1989.
14. F. V. DeFeudis, ed., *Ginkgo biloba Extract (EGb 761): Pharmacological Activities and Clinical Applications*, Elsevier, New York, 1991.
15. E. Cory et al., *J. Am. Chem. Soc.*, **110**, 649 (1988).

# GINSENG (ASIAN AND AMERICAN)

**Source**: ***Asian ginseng*** *Panax ginseng* C. A. Mey. (syn. *P. schinseng* Nees); ***American ginseng*** *Panax quinquefolius* L. (Family Araliaceae)

*Synonyms*. Chinese ginseng, Korean ginseng, and Japanese ginseng (*P. ginseng*); Western ginseng (*P. quinquefolius*); seng and sang.

## GENERAL DESCRIPTION

Both are perennial herbs with simple single stems bearing at flowering a whorl of three to six long-petioled compound leaves at the top.

Asian ginseng bears only a single leaf with three leaflets the 1st year. In the 2nd year it bears a single leaf with five leaflets, and in its 3rd year two leaves with five leaflets. It usually starts flowering at its 4th year when bearing three leaves (JIANGSU).[1–5] Asian ginseng is native to northeastern China and extensively cultivated there and in nearby Russia as well as Korea and Japan.

American ginseng is native to eastern North America, from Quebec to Manitoba, south to northern Florida, Alabama, and Oklahoma. Once considered abundant in eastern North America, it is now considered a threatened, rare, or endangered species in many areas due to overzealous harvest of the root for commercial purposes. It is cultivated in Canada (Quebec, Ontario, British Columbia) and in the United States (New England, Wisconsin, Illinois, North Carolina, Tennessee, Georgia, Missouri, etc.) as well as in China; Chinese material enters commerce in Hong Kong under the ambiguous name "China White."[6,7]

Parts used are the dried, often specially treated (cured) roots; normally roots of plants about 6 years old are used.

There are many types and grades of Asian ginseng, depending on the sources, ages, and parts of the roots as well as methods of preparation. Old, wild well-farmed roots are most valued, while rootlets of cultivated plants are considered the lowest grade. Powdered ginseng currently imported from Korea for use in cosmetics and health foods is probably from the latter.

American ginseng does not undergo special curing as Asian ginseng and there are considerably fewer grades, mostly separated on the basis of "wild" or "cultivated." The international trade of American ginseng is regulated under the provisions of the Convention on International Trade in Endangered Species (CITES), which regulates trade through permit requirements for imports, exports, and reexports of listed species.[6,7]

Siberian ginseng is the dried root of *Eleutherococcus senticosus* (Rupr. and Maxim.) Maxim. It does not have a long history of usage as Asian ginseng (or even American ginseng), but it is reported to have similar properties as Asian ginseng and is consequently gaining popularity in the United States and Canada (see ***eleuthero***).[8,9]

## CHEMICAL COMPOSTION

Asian ginseng contains numerous saponins, which are one of the major groups of active constituents;[8–16] a trace of volatile oil consisting mainly of panacene ($C_{15}H_{24}$); sterols (e.g., $\beta$-sitosterol and its $\beta$-glucoside); 8–32% starch;[17] 7–9% ginseng polysaccharides (panaxans A-U) and pectin[18–21] free sugars (e.g., glucose, fructose, sucrose, maltose, trisaccharides, etc.); pectin; vitamins (e.g., vitamins $B_1$, $B_2$, and $B_{12}$, nicotinic acid, pantothenic acid, and biotin); 0.1–0.2% choline; fats; minerals (Zn, Cu, Mn, Ca, Fe, etc.); polyacetylenes (e.g., panaxynol); a polypeptide;[22] and others (JIANGSU, LIST AND HÖRHAMMER).[8,23]

The saponins are called ginsenosides by Japanese and panaxosides by Russian workers. According to most recent reports, there are at least 18 saponins found in Asian ginseng, including ginsenosides $R_0$, $R_{b\text{-}1}$, $R_{b\text{-}2}$, $R_{b\text{-}3}$, $R_c$ $R_d$, $R_e$, $R_f$, $R_{f\text{-}2}$, $R_{20\text{-gluco-}f}$, $R_{g\text{-}1}$, and $R_{g\text{-}2}$, They are all triterpenoids. Ginsenoside $R_0$ is an oleanane type; the rest are all damarane type. The sapogenin of ginsenoside $R_0$ is oleanolic acid, that of ginsenosides $R_{b\text{-}1}$ to $R_d$ is 20-*S*-protopanaxadiol, and that of ginsenosides $R_e$ to $R_{g\text{-}2}$ is 20-*S*-protopanaxatriol. Ginsenosides $R_{b\text{-}1}$, $R_{b\text{-}2}$ $R_c$, $R_e$, and $R_{g\text{-}1}$ are present in major concentrations in Asian ginseng. American ginseng contains primarily ginsenosides $R_{b\text{-}1}$ and $R_e$; it does not contain ginsenosides $R_{b\text{-}2}$, $R_f$, and $R_{g2}$, and in some instances $R_{g\text{-}1}$.[6,7,11,12]

Prized American ginseng contains a high ratio of ginsenosides $R_{b1}$ to $R_{g1}$; it should not contain $R_f$ and $R_{g2}$.

Six panaxosides (A, B, C, D, E, and F) have been reported, with panaxosides A, B, and C having panaxatriol as their sapogenin and the sapogenin of panaxosides D, E, and F being panaxadiol.[8,9] Since the genuine aglycones reported for ginsenosides are 20-*S*-protopanaxadiol and 20-*S*-protopanaxatriol, they are most likely the sapogenins of panaxosides also.[15,16]

Panaxoside A is reported to be the same as ginsenoside $R_{g\text{-}1}$ (LIST AND HÖRHAMMER).

Ginseng herb oil (*P. ginseng*) has been reported to contain sesquiterpenes, including bicyclogermacrene, $\alpha$- and $\beta$-panasinsenes, caryophyllene, $\alpha$- and $\beta$-humulenes, $\alpha$- and $\beta$-neoclovenes, $\beta$-farnesene, and $\alpha$-, $\beta$-, and $\gamma$-selinenes.[24]

A recent study showed that highest yields of ginsenosides were obtained at the end of summer of the 5th year; the root doubles in weight between the 4th and 5th years.[25]

At least 56 closely related saponins, called gynosaponins, have recently been isolated from *Gymnostemma pentaphyllum* (Thunb.) Makino of the gourd family, four of which (gynosaponins 3, 4, 8, and 12) are identical to ginsenosides $R_{b1}$, $R_{b3}$, $R_d$, and $R_{f2}$.[26–28]

## PHARMACOLOGY OR BIOLOGICAL ACTIVITIES

The pharmacological properties of Asian ginseng are multiple, which are due to its various components, notably saponins and polysaccharides.

Asian ginseng has been reported to have numerous pharmacological activities in humans and in laboratory animals, including general stimulatory effect, raising mental and physical capacity for work;[29] remedial effect on radiation sickness, alloxan diabetes, experimental neurosis, and cancer; altering carbohydrate and albumin metabolism;[8] decreasing sensitivity to stress;[30] lowering blood sugar level and liver glycogen content;[31] promoting the biosynthesis of cholesterol, lipid, RNA, DNA, and protein;[17,32,33] antagonizing the effects of depressants such as alcohol, chloral hydrate, and barbiturates;[34,35] and antihepatotoxic effects *in vitro*.[36] Ginsenosides stimulated the sexual response in both male and female animals, but they themselves exhibited no sex hormonal effects;[37] ginsenosides $R_{b-1}$ and $R_d$ potentiated nerve growth factor, among others (JIANGSU).

Some members of the ginseng saponins produce effects directly opposed to those produced by others, and under certain conditions, ginseng acts in opposite directions (JIANGSU).[12,29,35] Ginsenoside $R_{b-1}$ reportedly is CNS-tranquilizing, hypotensive, antipyretic, antipsychotic, and ulcer protective; inhibits conditioned avoidance response; is weakly anti-inflammatory; antihemolytic; increases gastrointestinal motility; accelerates glycolysis; and accelerates serum and liver cholesterol, nuclear RNA, and serum protein synthesis.[6,38,39] Ginsenoside $R_{g-1}$ reportedly has weak CNS-stimulant, hypertensive, and antifatigue activity; aggravates stress ulcer; and increases motor activity. In behavioral tests it accelerates discrimination behavior in pole-climbing tests and Y-maze tests, a reversal learning response in the Y-maze test, and one-trial passive avoidance learning using the step down method.[6,38,39]

Conflicting results of various studies are attributed to type of preparation, route of administration, dosage, and presence or absence of biologically active compounds, among other factors.[6,7,38–42]

American ginseng was found to be nonmutagenic in a recent study.[43]

Ginseng polysaccharides (especially panaxan A) and the polypeptide lowered blood sugar in experimental animals[21,22] the polysaccharides also normalized the immune function of guinea pigs treated i.p. with cobra venom factor.[44]

Oleanolic acid has antiallergenic activities in experimental animals (see also ***ligustrum***).[45]

An aqueous extract of Asian ginseng had antioxidant activities in human erythrocytes *in vitro*.[46]

## USES

**Medicinal, Pharmaceutical, and Cosmetic.** Used in all kinds of cosmetic products such as lotions, creams, soaps, bath preparations, and perfumes. Ginseng oil and extracts (probably both American and Asian) are used.

**Food.** Used in soft drinks; those manufactured in America are mostly made from American ginseng while those manufactured overseas are from Asian ginseng, although type of ginseng is not always labeled.

**Health Food/Herb Teas.** Both American and Asian ginseng are available in a wide variety of product forms, including powdered root (or leaf) as single or combination teas, capsules, tablets, liquid extracts, chewing gum, extract, and instant tea (up to 5% extract on fructose carrier). Some products are standardized to 4–7% ginsenoside content (FOSTER).

**Traditional Medicine.** In Chinese medicine, Asian ginseng is considered to have warming properties while American ginseng is said to have cooling properties. They are generally used for different purposes. Thus American ginseng is normally used for its cooling and thirst-quenching effects in summer and as a febrifuge. This is also true with Asian ginseng leaf, which is considered to have similar properties as American ginseng (cooling and thirst quenching) and is similarly used; both are also used to treat hangovers (JIANGSU, LEUNG).

Asian ginseng is used generally as a tonic, for its revitalizing properties, especially after a long illness. Either alone or in combination with other drugs, it is used to treat a wide variety of conditions, including amnesia, dizziness, headache, tiredness, convulsions, impotence, vomiting, rheumatism, dysentery, lack of appetite, difficulties in pregnancy and childbirth, internal hemorrhage, nosebleed, and cancers, among others (JIANGSU, LEUNG, NANJING).[2]

Primarily consumed by Asians, American ginseng has always been regarded as an export commodity. The root was official in the *United States Pharmacopoeia* from 1842 to 1882; primarily used as a stimulant and a stomachic.[7]

## COMMERCIAL PREPARATIONS

Crude, extracts, and oils. Currently there are no standards for ginseng. Powdered ginseng and ginseng extracts should be tested for ginsenosides and ginseng polysaccharides as well as diluents such as dextrose, lactose, corn syrup, and caramel. Chromatographic methods are available and can be used.[11–13]

*Regulatory Status.* Not yet determined. Asian ginseng is the subject of a German therapeutic monograph. The root is used as a tonic for invigoration for fatigue and reduced work capacity and concentration and during convalescence. Daily dosage is 1–2 g of root in appropriate formulations.[47]

## REFERENCES

See the General References for BAILEY 1; BLUMENTHAL; FOSTER; FOSTER AND YUE; JIANGSU; LEUNG; LIST AND HÖRHAMMER; NANJING; TYLER 1.

1. S. Y. Hu, *Econ. Bot.*, **30**, 11 (1976).
2. S. Y. Hu, *Am. J. Chin. Med.*, **5**, 1 (1977).
3. B. Goldstein, *Am. J. Chin. Med.*, **3**, 223 (1975).
4. T. E. Hemmerly, *Econ. Bot.*, **31**, 160 (1977).
5. L. Veninga, *The Ginseng Book*, Big Trees Press, Santa Cruz, Calif., 1973.
6. S. Foster, *Asian Ginseng—Panax ginseng*, *Botanical Series*, *no. 303*, American Botanical Council, Austin, Tex., 1991.
7. S. Foster, *American Ginseng—Panax quinquefolius*. *Botanical Series*, no. 308, American Botanical Council, Austin, Tex., 1991.
8. I. I. Brekham and I. V. Dardymov, *Lloydia*, **32**, 46 (1969).
9. F. Sandberg, *Planta Med.*, **24**, 392 (1973).
10. G. I. Shaposhnikova et al., *Carbohydr. Res.*, **15**, 319 (1970).
11. E. Bombardelli et al., *Fitoterapia*, **3**, 99 (1976).
12. H. Otsuka et al., *Planta Med.*, **32**, 9 (1977).
13. H. Wagner and A. Wurmböck, *Deut. Apoth. Ztg.*, **117**, 743 (1977).
14. T. Komori et al., *Org. Mass Spectrom.*, **9**, 744 (1974); through *Chem. Abstr.*, **82**, 156519e (1975).
15. S. Hiai et al., *Planta Med.*, **28**, 131 (1975).
16. S. Hiai et al., *Planta Med.*, **28**, 363 (1975).
17. K. Y. Yim, *Hakhoe Chi*, **10**, 26 (1977); through *Chem. Abstr.*, **87**, 90643k (1977).

18. C. Konno et al., *Planta Med.*, **50**, 434 (1984).
19. Y. Oshima et al., *J. Ethnopharmacol.*, **14**, 255 (1985).
20. C. Konno et al., *Int. J. Crude Drug Res.*, **25**, 53 (1987)
21. S. J. Wu and D. Y. Li, *Zhongcaoyao*, **23**, 549 (1987).
22. B. X. Wang et al., *Yaoxue Xuebao*, **25**, 401 (1990).
23. H. O. Cho et al., *Hanguk Sikp'um Kwahakhoe Chi*, **8**, 95 (1976); through *Chem. Abstr.*, **86**, 21736k (1977).
24. K. Yoshihara and Y. Hirose, *Bull. Chem. Soc. Jpn.*, **48**, 2078 (1975).
25. F. Soldati and O. Tanaka. *Planta Med.* **50**, 351 (1984).
26. G. Q. Liu et al., *Zhongcaoyao*, **18**(10), 47 (1987).
27. S. Z. Guo et al., *Zhongcaoyao*, **18**(7), 37 (1987).
28. S. L. Ding and Z. L. Zhu, *Zhongcaoyao*, **23**, 627 (1992).
29. K. Takagi, *Yakhak Hoeji*, **17**, 1 (1973); through *Chem. Abstr.*, **82**, 261d (1975).
30. S. L. Friedman and A. N. Khlebnikov, *Biol. Akt. Veshchestva (Mikroelem., Vitam. Drugie) Rastenievod., Zhivotnovod. Med.*, **113** (1975); through *Chem. Abstr.*, **87**, 307b (1977).
31. T. Yokozawa et al., *Chem. Pharm. Bull.*, **23**, 3095 (1975).
32. H. Oura et al., *J. Biochem.*, **77**, 1057 (1975).
33. M. Yamamoto et al., *Arzneim. Forsch.*, **27**, 1404, (1977).
34. H. S. Kang et al., *Hanguk Saenghwa Hakhoe Chi*, **8**, 225 (1975); through *Chem. Abstr.*, **85**, 73225s (1976).
35. S. Fulder, *New Scientist* (1), 138 (1977).
36. H. Hikino et al., *Planta Med.*, **51**, 62 (1985).
37. B. X. Wong, *Yaoxue Tongbao*, **19**(5), 41 (1984).
38. S. O. Shibata et al. in H. Wagner, H. Hikino and N. R. Farnsworth, eds., *Economic and Medicinal Plant Research*, Vol. 1, Academic Press, Orlando, Fl., 1985, p. 218.
39. T. B. Ng and H. W. Yeung, in R. P. Steiner, ed., *Folk Medicine, The Art and the Science*, American Chemical Society, Washington, D.C., 1986, p.139.
40. J. A. Duke, *Ginseng—A Concise Handbook*, Reference Publications, Algonac, Mich. 1989.
41. H. Saito in *Proceedings of the 3rd International Ginseng Symposium*, Korean Ginseng Research Institute, Seoul, Korea, 1980, p. 181.
42. W. H. Lewis, in N. L. Etkin, ed., *Plants in Indigenous Medicine and Diet: Biobehavioral Approaches*, Redgrave Publishing Co., Bedford Hills, N.Y., 1986, p. 290.
43. Y. S. Chang et al., *Planta Med.*, **52**, 338 (1986).
44. M. X. Zhuang et al., *Zhongguo Yaoxue Zazhi*, **27**, 653 (1992).
45. Y. Dai et al., *Zhongguo Yaoli Xuebao*, **9**, 562 (1988).
46. L. M. Feng et al., *Yaoxue Xuebao*, **25**, 401 (1990).
47. Monograph *Ginseng radix*, *Bundesanzeiger*, no. 11, (Jan. 17, 1991).

# GOLDENSEAL

**Source**: ***Hydrastis canadensis*** L. (Family Ranunculaceae)

*Synonyms*. Orange root, yellow root, jaundice root, Indian turmeric, eye root, and eye balm.

## GENERAL DESCRIPTION

A perennial herb with a knotty yellow rhizome (rootstock) from which arise a single leaf (radical leaf) and an erect hairy stem in early spring bearing two five- to nine-lobed rounded leaves near the top, terminated by a single greenish white flower; up to about 30 cm high; native to rich, moist, deciduous forests, Vermont to Georgia, west to Alabama and Arkansas, north to eastern Iowa and Minnesota; formerly cultivated in Oregon and Washington. Various botanical writers note rarity where it once flourished due to overcollection of the root.[1] Parts used are the dried rhizome and roots.

## CHEMICAL COMPOSITION

Contains as active principles isoquinoline alkaloids consisting mainly of hydrastine (1.5–4%) and berberine (0.5–6%), with lesser amounts of canadine (tetrahydroberberine), canadaline, 1-$\alpha$-hydrastine, 5-hydroxytetrahydroberberine, and other related alkaloids also present.[2–6] Other constituents include meconin, chlorogenic acid, lipids with 75% unsaturated and 25% saturated fatty acids, resin, starch, sugar, and a small amount of volatile oil (LIST AND HÖRHAMMER).

## PHARMACOLOGY OR BIOLOGICAL ACTIVITIES

The alkaloids present in goldenseal have been reported to have anticonvulsive activity on mouse intestine and uterus.[7]

Berberine is reported to have pharmacological properties resembling those of hydrastine.[8] Its many activities include stimulating secretion of the bile in humans, sedative effect on cats and mice upon intraperitoneal administration, lowering the blood pressure of laboratory animals, and strong antibacterial effects, among many others (see also ***barberry***).[5,8]

Numerous modern secondary reports of contact ulceration and inflammation in topical use can be traced to an ointment of goldenseal containing zinc chloride and *Datura stramonium*.[1]

## USES

**Medicinal, Pharmaceutical, and Cosmetic.** Hydrastis extracts and hydrastine hydrochloride have been used for stopping uterine hemorrhage and in relieving menstrual pain. Current use is mainly as components in eyewashes and in certain bitter tonic preparations.

**Health Food/Herb Teas.** One of the most popular indigenous North American botanicals in health and natural food markets in the United States. Root in capsules, tablets, tinctures, extracts, teas, and various other product forms, single or in combination (often with *Echinacea*). Presence of green mass in powdered root may indicate presence of goldenseal leaf as filler (FOSTER).

Uses are numerous including, but not limited to, that as antiseptic, hemostatic, diuretic, laxative, and tonic; antiinflammatory for inflammations of the mucous membranes.[1]

Goldenseal consumption has increased due to word-of-mouth circulation of the belief that root products may be used to mask urine tests for illicit drugs. This grows out of the fictional plot of *Stringtown on the Pike* (1900), a novel by pharmacist John Uri Lloyd (1849–1936). This use has persisted throughout this century; especially in attempts to mask morphine detection in race horses. There is no scientific evidence to support this use, in fact, it may instead promote false-positive readings (TYLER 2).[9]

Despite goldenseal's continued popularity, it has been poorly researched.[1]

**Traditional Medicine.** Reportedly used as antiperiodic, antiseptic, hemostatic, diuretic, laxative, and tonic. Conditions for which it is used include inflammation of mucous membranes (vaginal and uteral), hemorrhoids, nasal congestion, sore gums, sore eyes, wounds, sores, acne, dandruff, and ringworm, among others. It has also been used in cancers.[10]

American Indian groups used the roots as a wash for local inflammations; decoction for general debility, dyspepsia, whooping cough, diarrhea, jaundice, fever, sour stomach, flatulence, pneumonia, and with whisky, for heart disease (FOSTER AND DUKE; MOERMAN).[1]

## COMMERCIAL PREPARATIONS

Crude, extracts, and hydrastine salts. Crude, fluid extract, and tincture were formerly official in N.F. Strengths (see ***glossary***) of extracts are expressed in weight-to-weight ratios or in total alkaloids content.

*Regulatory Status*. Undetermined in the United States. Official in the first revision (1830) of New York edition of U.S.P., but absent in the Philadelphia 1830 U.S.P. Dropped in 1840, then official in U.S.P. from 1860 to 1926. Included in N.F. 1888; also 1936–1955.[11]

In 1982 the alkaloid, hydrastine, was still official in the pharmacopoeias of nine countries.[11]

## REFERENCES

See the General References for BAILEY 1; FERNALD; FOSTER; FOSTER AND DUKE; GOSSELIN; KROCHMAL AND KROCHMAL; LUST; MARTINDALE; MERCK; MOERMAN; ROSE; TERRELL; TYLER 1; UPHOF.

1. S. Foster, *Goldenseal—Hydrastis canadensis, Botanical Series, no. 309*, American Botanical Council, Austin, Tex., 1991
2. J. Gleye and E. Stanislas, *Plant. Med. Phytother.*, **6**, 306 (1972).
3. G. Caille et al., *Can. J. Pharm. Sci.*, **5**, 55 (1970).
4. J. Gleye et al., *Phytochemistry*, **13**, 675 (1974).
5. K. Genest and D. W. Hughes, *Can. J. Pharm. Sci.*, **4**, 41 (1969).
6. S. El-Masry et al., *J. Pharm. Sci.*, **69**, 5, 597 (1980).
7. J. Haginiwa and M. Harada, *Yakugaku Zasshi*, **82**, 726 (1962); through *Chem. Abstr.*, **57**, 9145b (1962).
8. V. Preininger in R. H. F. Manske, ed., *The Alkaloids*, Vol. 15, Academic Press, New York, 1975, p. 207.
9. S. Foster, *HerbalGram* **21**, 7 (1989).
10. J. L. Hartwell, *Lloydia*, **34**, 103 (1971).
11. C. Hobbs, *Pharm. Hist.*, **32**(2), 79 (1990).

# GOTU KOLA

**Source**: ***Centella asiatica*** (L.) Urban (syn: *Hydrocotyle asiatica* L. and *Centella coriacea* Nannfd.) (Family Umbelliferae or Apiaceae)

*Synonyms*. Indian Pennywort.

## GENERAL DESCRIPTION

Weakly aromatic prostrate perennial; leaves orbicular, reniform to 4 cm in diameter, growing near water or marshy places, moist rocky outcrops to 700 m, in India, China, Indonesia, Sri Lanka, western South Sea Islands, Australia, Madagascar, southern Africa, Hawaii, etc.[1] Parts used are the fresh or dried leaves., aboveground herb, or whole herb with root.

## CHEMICAL COMPOSITION

Leaves contain triterpenoid saponins, highly variable (1.1–8.0%), including asiaticoside, oxyasiaticoside, and madecassoside (Madagascar chemotype), centelloside (Sri Lanka chemotype), brahmoside, brahminoside, thankunoside, isothankunoside (India chemotype); sapogenins from various chemotypes including asiatic, madecassic, centellic, indocentoic, brahmic, thankunic, and iosthankunic acids.[1]

Volatile oil with *trans*-$\beta$-farnesene, germacrene D, $\beta$-caryophyllene, camphor, cineole, *n*-dodecane, *p*-cymol, $\alpha$-pinene, methanol, allyl mustard, and an unidentified terpene acetate (36% of oil content).[1]

Other components include rhamnose, arabinose, glucose, fructose, sucrose, and raffinose; an oligosaccharide, centellose; a fatty oil containing glycerides of oleic, linoleic, lignoceric, palmitic, stearic, linolenic, and elaidic acids; steroids, including $\beta$-sitosterol, stigmasterol, campesterol, and sitosterol; amino acids, including glutamic acid, serine, and alanine; flavonols, including kaempferol, quercetin, 3-glycosylkaempferol, and 3-glycosylquercetin; polyphenols, tannins, carotenoids, vellarin and ascorbic acid (13.8 mg/100 g dry weight).[1,2]

## PHARMACOLOGY OR BIOLOGICAL ACTIVITIES

Antiinflammatory, CNS depressant, anticonvulsant, antidepressant, analgesic,[3] etc.

Topically, asiaticoside or leaf extracts standardized to asiaticoside have been shown to accelerate the wound-healing process and significantly improved tensile strength of tissues, promoting keratization and stimulating rapid and healthy growth of the reticuloendothelium. It is suggested that inhibition of the biosynthesis of collagen and acidic mucopolysaccharides is involved in the mechanism of action.[1]

A leaf extract (standardized to asiaticoside) was evaluated in clinical patients with soiled wounds and chronic atony, resistant to treatment; results showed complete healing in 64% and improvement in 16% of 20 patients.[4]

Based on traditional use in India claiming the herb improves intelligence, various studies (both animal and human) have suggested the herb to be beneficial in improving memory, adaptogenic in fatigue and stress, tranquilizing effect in rats (alcoholic extract), increase general mental ability, behavioral patterns, and increase I.Q. in mentally retarded children. A two-compartment passive avoidance task test (with rats) showed an improvement in 24-h retention. Assessment of turnover of biogenic amines (norepinephrine, dopamine, and serotonin) showed significant reductions of these amines and their metabolites in the brain following oral administration of a fresh juice (1 mL = 0.38 g fresh leaves), at a dose of 0.18 g/kg for 15 days. The decrease of amine levels was correlated to improved learning and memory in rats.[5]

A water-soluble fraction was shown to have an antianxiety effect in animals comparable to diazepam and inhibit hepatic enzymes responsible for barbiturate metabolism, hence may prolong pentobarbitone-induced sleep.[6]

## USES

**Medicinal, Pharmaceutical, and Cosmetic.** Oral extracts, injectable extracts, and ointments standardized to asiaticoside have been used successfully in India for the treatment of mal perforant lesions in leprosy patients; in Europe used clinically for leg ulcers in postphlebitic patients (injectable extract); accelerating healing of superficial postsurgical wounds; inhibiting hypertrophic formation of scar tissue in the treatment of second- and third-degree burns; and general use in wounds, ulcer, and scleroderma.[1]

Leaf extracts used in cosmetics, including hand creams and lotion, hair conditioners, and shampoos.

**Food.** In Bangladesh, Thailand, and Sri Lanka, the leaves are sold as a leafy vegetable, rich in digestible protein, carotene, and vitamin C.[7]

**Health Food/Herb Teas.** In the United States various gotu kola leaf preparations or crude leaf are sold as dietary supplements in the form of tablets, capsules, tincture, standardized extracts, tea, etc., used for memory improvement; topically for the treatment of wounds (WEISS; WREN).

**Traditional Medicine.** In Chinese folk medicine, a decoction of the whole aboveground herb is used for treatment of colds, sunstroke, tonsillitis, pleurisy, urinary tract infections, infectious hepatitis, jaundice, and dysentery; as an antidote for arsenic poisoning, poisoning by *Gelsemium elegans*, and toxic mushrooms; external poultice for snakebites, scabies, traumatic injuries, and herpes zoster (JIANGSU).

In India, used as a folk remedy for leprosy, lupus, syphilis, tuberculosis, improving mental function, and others (CSIR II).

In Uttar Pradesh, fresh leaf juice used externally for elephantiasis, inflammations, swelling; whole plant decoction used for skin disease (itch, fungal infections), chronic rheumatism, amenorrhea, and as "blood purifier."[8]

In East Africa, leaf used for fevers, bowel complaints, and syphilitic and scrofulous conditions; in Central Africa, widely used as a folk medicine, particularly for leprosy (WATT AND BREYER-BRANDWIJK).

**Other.** Insecticidal properties have been reported from leaf extracts.

## COMMERCIAL PREPARATIONS

Crude (leaf, whole herb, etc.), tincture, extracts (calculated to contain 70% total triterpenes), etc.

*Regulatory Status.* Undetermined in the United States; official in the French pharmacopoeia as early as 1884; official in Indian pharmacopoeia; also once official in the Dutch, Mexican, Spanish, and Venezuelan pharmacopoeias.

## REFERENCES

See the General References for CHP; CSIR II; GLASBY 2; TYLER 1; WATT AND BREYER-BRANDWIJK; WEISS; WREN.

1. T. Kartnig in L. E. Craker and J. E. Simons, eds., *Herbs, Spices, and Medicinal Plants—Recent Advances in Botany, Horticulture and Pharmacology*, Vol. 3. Oryx Press, Phoenix, 1988, p. 145.
2. P. S. Rao and T. R. Seshardi, *Curr. Sci.*, **38**, 77 (1969).
3. M. R. Sakin and P. C. Dandiya, *Fitoterapia*, **61**, 291 (1990).

4. R. Morisset et al., *Phytother. Res.*, **1**(3), 117 (1987).
5. K. Nalin et al., *Fitoterapia*, **63**(3), 232 (1992).
6. P. V. Diwa et al., *Fitoterapia*, **62**(3), 253 (1991).
7. H. S. Puri and D. C. Bagchi, *Am. Herb Assn. Nltr.*, **8**(12), 4 (1991).
8. M. Badruzzaman Siddiqui and W. Husain, *Fitoterapia*, **63**(3), 245 (1992).

# GRAPEFRUIT OIL

**Source**: ***Citrus* × *paradisi*** Macf. (syn. *C. racemosa* (Risso et Poit.) Marcov. ex Tanaka; C. *decumana* var. *racemosa* (Risso et Poit.) Roem.; *C. maxima* (L.) Osbeck var. *racemosa* (Roem.) Stone) (Family Rutaceae)

*Synonyms*. Expressed grapefruit oil, cold-pressed grapefruit oil, and shaddock oil.

## GENERAL DESCRIPTION

A cultivated tree with large fruits, often over 10 m high; is considered to be a relatively recent hybrid of *C. maxima* and *C. sinensis*. Numerous cultivars are grown commercially; 'Duncan' is the standard grown in Florida (TUCKER AND LAWRENCE); cultivated in the United States (esp. California, Florida, and Texas), the West Indies (e.g., Jamaica and the Dominican Republic), Nigeria, Brazil, and Europe (e.g., Israel and Portugal). Part used is the fresh peel of the fruit from which grapefruit oil is produced by cold expression. Naringin extract is a bitter flavoring material prepared by extraction of the expressed peel; it is not pure naringin.

## CHEMICAL COMPOSITION

Grapefruit oil contains mostly the monoterpene hydrocarbon, limonene (ca. 90%). Other volatile constituents present include sesquiterpenes (e.g., cadinene and paradisiol or intermedeol);[1,2] aldehydes ($C_7$ to $C_{12}$ aldehydes, neral, geranial, perillaldehyde, citronellal, $\alpha$-sinensal, and $\beta$-sinensal); esters (e.g., geranyl acetate, neryl acetate, perillyl acetate, octyl acetate, decyl acetate, citronellyl acetate, *trans*-carvyl acetate, 1,8-*p*-menthadien-2-yl acetate, and 1,8-*p*-menthadien-9-yl acetate); and nootkatone (a bicylcic sesquiterpene ketone); among others (GUENTHER, LIST AND HÖRHAMMER).[1–6]

It also contains sizable amounts (ca. 1.4%) of coumarins and furocoumarins (bergaptens) composed mainly of 7-geranoxycoumarin, with marmin, osthol, limettin, 7-methoxy-8-(2-formyl-2-methylpropyl) coumarin, bergapten, bergamottin, bergaptol, byakangelicin, and 5-[(3,6-dimethyl-6-formyl-2-heptenyl)-oxy]psoralen also present.[7,8]

The characteristic grapefruit aroma and flavor of grapefruit oil is reported to be due primarily to nootkatone and other carbonyls present (e.g., geranyl acetate, neryl acetate, octyl acetate, 1,8-*p*-menthadien-2-yl acetate, and others).[1,9–11]

The characteristic bitter taste of grapefruit is due to naringin present mainly in the peel.[12,13]

## PHARMACOLOGY OR BIOLOGICAL PROPERTIES

Grapefruit oil has been reported to promote tumor formation on mouse skin by the primary carcinogen, 9,10-dimethyl-1,2-benzanthracene.[14] It also exhibited antibacterial activities.[15]

Certain bergaptens are known to be phototoxic and allergenic to humans (see ***bergamot oil***).

Dermatological studies have indicated grapefruit oil to be nonirritating, nonsensitizing, and nonphototoxic to humans.[15]

## USES

**Medicinal, Pharmaceutical, and Cosmetic.** Grapefruit oil is used as a fragrance component in soaps, detergents, creams, lotions, and perfumes, with maximum use level of 1.0% reported in perfumes.[15]

**Food.** Grapefruit oil is extensively used as a flavor ingredient in alcoholic and nonalcoholic beverages (especially soft drinks), frozen dairy desserts, candy, baked goods, gelatins and puddings, and milk products, with highest maximum average use level of about 0.108% (1084 ppm) reported in candy.

Naringin extract is used mainly in soft drinks. Other foods in which it is also used include alcoholic beverages, frozen dairy desserts, candy, baked goods, and gelatins and puddings. Highest average maximum use level is about 0.018 (175 ppm) reported in alcoholic beverages; average maximum use level reported in soft drinks (nonalcoholic beverages) is about 0.004% (38.4 ppm).

## COMMERCIAL PREPARATION

Grapefruit oil (also partially deterpenized) and naringin extract. Oil is official in F.C.C.

*Regulatory Status.* GRAS with both grapefruit and naringin listed (§182.20).

## REFERENCES

See the General References for ARCTANDER; BAILEY 1; FEMA; GUENTHER; KARRER; JIANGSU; POLUNIN AND SMYTHIES; TERRELL; TUCKER AND LAWRENCE; UPHOF.

1. H. Sulser et al., *J. Org. Chem.*, **36**, 2422 (1971).
2. J. W. Huffman and L. H. Zalkow, *Tetrahedron Lett.* (10), 751 (1973).
3. M. Koketsu et al., *Bol. Pesgui EMBRAPA Cent. Technol. Agr. Ailment*, **7**, 21 (1983).
4. J. A. Remar in J. Erghese, ed., *On Essential Oils*, Synthie Industrial Chemical Pte, Ltd., Lolenchery, India, 1986, p. 123.
5. G. Dugo et al., *Flav. Frag. J.*, **5**, 205 (1990).
6. W. A. König et al., *J. High Res. Chromatogr.*, **13**, 328 (1990).
7. W. L. Stanley and L. Jurd, *J. Agr. Food Chem.*, **19**, 1106 (1971).
8. J. F. Fisher and H. E. Nordby, *J. Food Sci.*, **30**, 869 (1965).
9. M. G. Moshonas, *J. Agric. Food Chem.*, **19**, 769 (1971).
10. J. L. K. Hunter and W. B. Brogden, *J. Food Sci.*, **30**, 383 (1965).
11. J. L. K. Hunter and M. G. Moshonas, *J. Food Sci.*, **31**, 167 (1966).
12. G. M. Fishman and M. N. Gumanitskaya, *U.S.S.R.* 261,166 (1970); through *Chem. Abstr.*, **73**, 2838k (1970).
13. V. A. Bandyukova and G. M. Fishman, *Subtrop. Kul't.*, **5–6**, 137 (1976); through *Chem. Abstr.*, **87**, 180664x (1977).
14. F. J. C. Rose and W. E. H. Field, *Food Cosmet. Toxicol.*, **3**, 311 (1965).
15. D. L. J. Opdyke, *Food Cosmet. Toxicol.*, **12**, 723 (1974).

# GRAPE SKIN EXTRACT (ENOCIANINA)

**Source**: Varieites of ***Vitis vinifera*** L. (Family Vitaceae)

## GENERAL DESCRIPTION

Grape skin extract or enocianina is the coloring matter derived from the skin of certain varieties of the wine grape.

If is commonly obtained by acidic aqueous extraction of fermented grape skin (or marc) after the juice has been expressed from it.

The major producer of enocianina is Italy; United States is a potential producer.

## CHEMICAL COMPOSITION

The pigments present in enocianina are anthocyanins. They are glycosides of polyhydroxy derivatives of 2-phenylbenzopyrylium salts. Their aglycones are generally called anthocyanidins.[1,2] The most common anthocyanidins in grape skin extract are peonidin, malvidin, delphinidin, and petunidin. The grape anthocyanins are usually either monoglycosides or diglycosides[3,4]

In addition to anthocyanins, grape skin extract contains plant acids (mainly tartaric acid), tannins, sugars, amino acids, minerals, and other constituents present in grapes. It may also contain diluents used in processing.

Grape skin anthocyanins are soluble in water and water–alcohol mixtures to yield red to magenta solutions with an acidic pH (ca. 3). The color varies with the pH, from red to purple, and finally to almost blue as the pH is changed from 1 to 8. The pigments are most stable below pH 3 and are not stable at higher pH values: They are also sensitive to heat, oxygen, and light.[5,6]

## PHARMACOLOGY OR BIOLOGICAL ACTIVITIES

Anthocyanins from grapes have been reported to have antifungal activities (e.g., against *Penicillium notatum* and *Aspergillus* species)[7] and strong "vitamin P" (bioflavonoids) activity, strengthening blood capillaries of guinea pigs.[8] A grape anthocyanin (delphinidin-3-monoglucoside) has also been reported to inhibit the growth of *Lactobacillus acidophilus*.[9]

A tannic substance (leucocyanidin) present in some grape juices has been reported to be toxic to laboratory animals; symptoms included cardiac failure and hepatic lesions.[10]

## USES

**Food.**, Used in coloring alcoholic and nonalcoholic beverages. The powder is also used in drink mixes.

## COMMERCIAL PREPARATIONS

Liquid and powder (vacuum dried or spray dried); the powder is more stable than the liquid. Strengths are expressed in color intensities (absorbance values).

***Regulatory Status.*** Has been approved for food use in beverages only, with specific restrictions (§73.170); exempt from certification.

## REFERENCES

1. H. B. Hass, *Chemtech* (9), 525 (1977).
2. J. B. Harborne, in L. Zechmeister, ed., *Fortschritte der Chemie Organischer Naturstoffe*, Vol. 20, Springer-Verlag, Vienna, Austria, 1962, p. 165.
3. V. M. Malikov, *Vinogradarstvo*, 158

(1973); through *Chem. Abstr.*, **83**, 75374x (1975).
4. L. S. Diaz et al., *Rev. Agroquim.*, *Tecnol. Aliment.*, **16**, 509 (1976); through *Chem. Abstr.*, **88**, 150891x (1978).
5. C. Skalski and W. A. Sistrunk, *J. Food Sci.*, **38**, 1060 (1974).
6. L. Szechenyi, *Ind. Aliment Agr. (Paris)*, **80**, 521 (1963); through *Chem. Absr.*, **61**, 11249f (1964).
7. K. Rizvanov and B. Karadimcheva, *Lozar. Vinar.*, **21**, 26 (1973); through *Chem. Abstr.*, **79**, 112264b (1973).
8. A. S. Sturua et al., *Prikl. Biokhim. Mikrobiol.*, **7**, 606 (1971); through *Chem. Abstr.*, **76**, 21575a (1972).
9. D. E. Pratt et al., *Food Res.*, **25**, 26 (1960).
10. R. Patay et al., *C. R. Congr. Natl. Soc. Savantes, Paris Dept. Sect. Sci.*, **86,** 69 (1961); through *Chem. Abstr.*, **58,** 7284c (1963).

# GUAR GUM

**Source**: ***Cyamopsis tetragonoloba*** (L.) Taub. (syn. *C. psoralioides* DC.) (Family Leguminosae or Fabaceae)

*Synonyms*. Guar flour and jaguar gum.

## GENERAL DESCRIPTION

Guar gum is derived from the seed of the guar plant. The guar plant is a small nitrogen-fixing annual that bears fruits known as legumes (pods) containing five to nine seeds per pod; up to about 1.8 m high. It is believed to be native to tropical Asia and has been grown in India and Pakistan for centuries as food for both humans and animals. It was introduced into the United States in the early 1900s and is now grown in Texas and Oklahoma. Major guar producers are India, Pakistan, and the United States.

Part used is the endosperm of the seed. The endosperm constitutes 35–42% of the seed; it is separated from the outer components of the seed (seed coat or hull and embryo or germ) during processing. Processing involves hull removal by water or acid soaking and grinding, followed by preferential grinding to remove the embryo. The endosperm left is then ground to a fine powder, which is commercial guar gum.

There are different grades of guar gum with varying amounts of the hull and germ present as the main impurities. Food- and pharmaceutical-grade guar gum is a white to yellowish white, nearly odorless powder. It is easily dispersed in cold or hot water (see ***locust bean gum***) to form solutions (sols) with a slightly acidic to almost neutral pH (5.4–6.4). The rate of hydration (dispersion) is dependent on its particle size, the water temperature, and the rate of agitation. Its optimal rate of hydration occurs between pH 7.5 and 9. Even at low concentrations (1–2%), guar sols have high viscosity and form gels with borate ions at alkaline pH values as locust bean gum and aloe vera gel; these gels can be liquified by lowering the pH below 7, by heating, or by adding simple polyols (glycerol, mannitol, etc.) that can react with the borate ions. The borate gels are not edible. Guar gum is not soluble or dispersible in organic solvents.

The viscosity of guar gum sols is unaffected by pH changes between 4 and 10.5.

Guar gum is reported to be compatible with gelatin, starch, and most water-soluble gums (e.g., acacia, agar, algin, karaya gum, locust bean gum, pectin, and tragacanth).

## CHEMICAL COMPOSITION

Commercial food-grade guar gum is reported to contain usually about 80% guaran (a galactomannan), 5–6% crude protein, 8–15% moisture, 2.5% crude fiber, 0.5–0.8% ash, and small amounts of lipids composed mainly of free and esterified fatty acids.[1]

Guaran (the pure galactomannan from guar gum) is a polysaccharide with a primary structure consisting of regular repeating units of the trisaccharide 4-*O*-(6-*O*-α-D-galactopyranosyl-β-D-mannopyranosyl)-β-D-mannopyranose, that is, a straight chain of D-mannose units with a D-galactose unit attached to every other mannose unit. Its molecular weight has been reported to be around 220,000 (WHISTLER AND BEMILLER).[2]

## PHARMACOLOGY OR BIOLOGICAL ACTIVITIES

Guar gum has been reported to lower the serum and liver cholesterol levels in chickens and rats as well as the serum cholesterol and postprandial (after meal) blood glucose in humans (MARTINDALE)[3–5] When included at different levels in the diets of chickens, guar gum has been demonstrated to cause growth depression, though with inconsistent results. It also reduced the metabolizable energy of the diets in which it was included.[3,6,7]

Guar gum does not seem to be digested by animals.[3]

In women, one study found ingestion of guar gum led to permanent weight loss, but did not influence serum lipids in hypercholesterolemia.[8] Similar results were noted in male patients and elderly patients.[9–11] However, positive results are reported in use of guar gum as a long-term dietary supplement in control of hypercholesterolemia in diabetics.[10,12–14] Long-term administration (21 g/day) produced a sustained improvement in control of Type 2 diabetes, with significantly lower serum total and LDL cholesterol concentrations.[15] An average reduction of 14% total cholesterol levels was observed in doses of 10 g b.i.d. immediately before meals as well as a reduction in postprandial glucose levels.[16] Other studies have produced similar positive results.[10,12]

A blood pressure lowering effect (8% systolic; 7% diastolic) has been observed in overweight men with mild hypertension.[17]

Guar gum did not produce teratogenic effects in rats.[18] A human study found guar gum consumption in diabetic mellitus patients did not adversely affect mineral balance.[19] Flatulence has frequently been reported as a side effect to guar gum dietary supplementation.[20] Occupational asthma has been reported in subjects working with industrial production of guar gum.[21]

## USES

**Medicinal, Pharmaceutical, and Cosmetic.** It is used as a binding and disintegrating agent in tablets and as a thickener in lotions and creams;[22] also used as an appetite depressant and in certain antihypercholesterolemic preparations.

**Food.** Used extensively as a thickener, stabilizer, suspending agent, and binder of free water in many food products, including nonalcoholic beverages (e.g., fruit drinks), frozen dairy desserts (especially ice cream and sherbets where it binds free water to prevent ice crystals formation), baked goods, gelatins and puddings, meat and meat products, condiments and relishes, breakfast cereals, cheeses (especially soft cheeses and spreads), milk products, soups, sweet sauces, gravies, snack foods, and processed vegetables, among others.[23,24] Highest average maximum use level reported is about 1% in breakfast cereals (11,260 ppm), sweet sauces (9,000 ppm), and processed vegetables (10,747 ppm).

**Health Food/Herb Teas.** Capsules, tablets, powder, and other product forms have, until recently, been widely used in weight loss formulations. Guar gum was blamed in causing esophageal obstruction. A death has been attributed to the use of one guar gum tablet product, which apparently swelled in the esophagus, indirectly resulting in complications that caused the fatality.[25] Major adverse reactions appear to result from product formulations (tablets) that dissolve in the mouth or esophagus before they reach the stomach. The FDA issued regulatory letters to manufacturers of guar gum capsules or tablets. Products that included claims implying use in weight loss, appetite suppression, or cholesterol or blood glucose lowering effects were deemed misbranded drugs.[25]

**Others.** Technical guar gum is extensively used in other industries (especially paper, oil drilling, and textile).

**COMMERCIAL PREPARATIONS**

Various grades with different particle sizes and viscosities. It is official in N.F. and F.C.C.

*Regulatory Status.* Has been affirmed as GRAS (§184.1339), as a thickener, stabilizer, suspending agent or binder in food products.

Based on "numerous adverse reaction reports of esophageal, gastric, or intestinal obstruction associated with the use of guar gum in a weight control drug," the FDA now considers guar gum, when labeled as a drug, to be a hazardous ingredient.[25]

**REFERENCES**

See the General References for FEMA; FURIA; GLICKSMAN; LAWRENCE; MARTINDALE; MERCK; TERRELL; UPHOF; WHISTLER AND BEMILLER.

1. J. Gynther et al., *Planta Med.*, **46**, 60 (1982).
2. C. W. Baker and R. L. Whistler, *Carbohydr. Res.*, **45**, 237 (1975).
3. S. E. Davis and B. A. Lewis, *ACS Symp. Ser.*, **15**, 296 (1975).
4. D. J. A. Jenkins et al., *Lancet*, 1351 (1976).
5. D. J. A. Jenkins et al., *Ann. Intern. Med.*, **86**, 20 (1977).
6. F. H. Kratzer et al., *Poultry Sci.*, **46**, 1489 (1967).
7. R. S. Thakur and K. Pradham, *Indian J. Anim. Sci.*, **45**, 880 (1975); through *Chem. Abstr.*, **87**, 150635w (1977).
8. J. Tuomilehto et al., *Acta Med. Scand.*, **208**, 45 (1980).
9. A. Aro et al., *Am. J. Clin. Nut.* **39**, 911 (1984).
10. A. Lakdhar et al., *Br. Med. J.*, **296**, 1471 (1988).
11. S. A. Rajala et al., *Compr. Gerantol [A].*, **2**, 83 (1988).
12. J. Tuomilehto et al., *Atherosclerosis*, **76**, 71 (1989).
13. M. Uusitupa et al., *Int. J. Clin. Pharm. Ther. Toxicol.*, **28**, 4, 153 (1990)
14. P. R. Turner et al., *Atherosclerosis*, **81**, 145 (1990).
15. A. Aro et al., *Diabetologia*, **21**, 29, (1981).
16. U. Smith and G. Holm, *Atherosclerosis*, **45**, 1, (1982).
17. M. Krotkiewski, *Acta Med. Scand.*, **222**, 43 (1987).
18. T. F. X. Collins, et al., *Fd. Chem. Toxicol.*, **25**, 11, 807 (1987).
19. K. M. Behall, *Diabetes Care*, **12**(5), 357 (1989).
20. P. A. Todd et al., *Drugs*, **39**(6), 917 (1990).
21. F. Lagier et al., *J. Allergy Clin. Immunol.*, **85**(4), 785 (1990).
22. B. N. Patel, *Drug Cosmet. Ind.*, **95**(3), 337 (1964).

23. G. Meer et al., *Food Technol.*, **29**(11), 22 (1975).
24. P. Kovacs and R. S. Igoe, *Food Prod. Dev.* **10**(8), 32 (1976).
25. Anon., *Aust. Adv. Drug React. Bull.* (Aug. 1989).

# GUAIAC WOOD OIL

**Source**: ***Bulnesia sarmienti*** Lorentz ex Griseb. (Family Zygophyllaceae)

*Synonym.* Champaca wood oil.

## GENERAL DESCRIPTION

A tree native to South America, growing in Brazil, Paraguay, and Argentina. Guaiac wood oil is obtained by steam distillation of the comminuted wood and sawdust; it is a thick semisolid mass with an odor resembling tea roses and may sometimes have an undesirable "smoked ham" note.

Guaiac wood oil is different from ***guaiac resin*** or ***guaiac gum*** products. The latter are obtained from the wood of other trees (*Guaiacum officinale* L. and G. *santum* L. (Family Zygophyllaceae)). This guaiac contains a small amount of $\alpha$-guaiaconic acid mixed with large amounts of other phenolic lignans (formerly collectively called guaiaconic acid) and other substances. It was formerly used in treating rheumatism and gout but currently is mainly used as a diagnostic reagent (e.g., in testing for occult blood).[1] Nevertheless, it has been approved for food use as an equivalent of guaiac wood oil or products derived from *B. sarmienti*; it is used in foods mainly as an antioxidant and is official in F.C.C.

## CHEMICAL COMPOSITION

Guaiac wood oil contains 42–72% guaiol, bulnesol, $\delta$-bulnesene ($\delta$-guaiene), $\beta$-bulnesene, $\alpha$-guaiene, $\beta$-patchoulene, and guaioxide (LIST AND HÖRHAMMER).

## PHARMACOLOGY OR BIOLOGICAL ACTIVITIES

One available report indicates guaiac wood oil to be nonirritating, nonsensitizing, and nonphototoxic to human skin.[2] another report indicates guaiac wood oil to be nontoxic to rats on short-term feeding.[3]

Guaiazulene, obtained from guaiac wood by dehydration of guaiol has been demonstrated to have antiinflammatory activity in animal studies. It has been used in Germany as an ingredient in combination bath products, claimed for efficacy in a wide variety of skin conditions and inflammation, but without clear demonstration of pharmacological or clinical efficacy.[4]

## USES

**Medicinal, Pharmaceutical, and Cosmetic.** Guaiac wood oil is used as a fixative, modifier, or fragrance component in soaps, detergents, creams, lotions, and perfumes, with maximum use level of 0.8% reported in perfumes.

**Food.** Guaiac wood oil is used as a flavor component in most categories of food products, including alcoholic and nonalcoholic beverages, frozen dairy desserts, candy, baked goods, gelatins and puddings, and meat and meat products, with highest average maximum use level of about 0.002% (22 ppm) reported in meat and meat products.

**Health Food/Herb Teas.** Guiac wood preparations are occasionally used in formulation for antiinflammatory activity, mostly in Europe, including gout and rheumatism formulations (WREN).

## COMMERCIAL PREPARATIONS

Oil.

*Regulatory Status*. Has been approved for food use (§172.510); gum guaiac is GRAS when used in edible oils or fats in accordance with good manufacturing practice (§182.3336).

Guaic wood is the subject of a German therapeutic monograph, used for supportive therapy for rheumatic complaints.[4]

## REFERENCES

See the General References for ARCTANDER; BLUMENTHAL; FEMA; GUENTHER; MERCK; TERRELL; UPHOF; WREN.

1. J. F. Kratochvil et al., *Phytochemistry*, **10**, 2529 (1971).
2. D. L. J., Opdyke, *Food Cosmet. Toxicol.*, **12**(Suppl.), 905 (1974).
3. B. L. Oser et al., *Food Cosmet. Toxicol.*, **3**, 563 (1965).
4. Monograph *Guajaci lignum Bundesanzeiger*, no. 76 (April 23, 1987).

# GUARANA

**Source**: ***Paullinia cupana*** Kunth ex H.B.K. (syn. *P. sorbilis* (L.) Mart.) (Family Sapindaceae)

*Synonyms*. Guarana paste and guarana gum.

## GENERAL DESCRIPTION

A climbing evergreen liana native to South America in the Amazon region (e.g., Brazil and Venezuela); under cultivation it becomes a shrub to 2 m high. Guarana paste (also called gum) is prepared from the pulverized and roasted seeds by mixing with water to form a paste, which is then molded into bars and dried. Commercial production is in the middle Amazon in northern Brazil, with the city and county of Maués accounting for 80% of the world's supply. It is cultivated there for its seeds.[1]

## CHEMICAL COMPOSITION

Contains usually 2.6–7% caffeine as its active constituent, together with traces of related alkaloids (theophylline, theobromine, xanthine, adenine, guanine, hypoxanthine, etc.). Other constituents present include tannins (ca. 12%), *d*-catechin, starch (5–6%), fats (ca. 3%), resin (ca. 7%), saponins, mucilage, red pigment, and choline (KARRER, LIST AND HÖRHAMMER, MERCK).

## PHARMACOLOGY OR BIOLOGICAL ACTIVITIES

Guarana has stimulant and astringent properties due to its caffeine and tannins contents (see ***coffee*** and ***cocoa***).

## USES

**Food.** Guarana extract (especially fluid extract) is widely used as a flavor ingredient of cola drinks. It is also reported used in alcoholic beverages (e.g., liqueurs and cordials) and in candy. Use levels are generally below 0.002%, reported for guarana gum (paste). In Brazil, a carbonated soft drink made from the seeds is considered a national beverage, offered commercially since 1909.

**Health Food/Herb Teas.** Powdered seeds are used in tablets, capsules, combinations, chewing gum, and tea, primarily as a stimulant; also in weight loss formulations (DUKE 3, TYLER 1).

**Traditional Medicine.** Guarana paste is used by South American natives mainly as a stimulant, astringent, in treating chronic diarrhea. The seeds, grated into water are also used for fevers, heart problems, headache (associated with menstrual or rheumatic conditions), rheumatism, lumbago, migraine, and reduction of heat stress; diuretic (DUKE 2; DUKE 3).[1]

## COMMERCIAL PREPARATION

Guarana seeds, guarana paste and extracts.

*Regulatory Status.* Has been approved for food use (§172.510).

## REFERENCES

See the General References for ARCTANDER; DUKE 2; DUKE 3; FEMA; FURIA AND BELLANCA; STAHL; TERRELL; TYLER 1; UPHOF.

1. H. T. Erickson et al., *Econ. Bot.* **38**(3), 273 (1984).

# HAWTHORN

**Source:** ***Crataegus laevigata*** (Poir) DC (syn. *C. oxyacantha* L.), ***C. monogyna*** Jacq., ***C. pinnatifida*** Bge., ***C. pinnatifida*** var. ***major*** N.E. Br., ***C. cuneata*** Sieb. et Zucc. and other ***Crataegus spp.*** (Family Rosaceae).

*Synonyms. Bei shanzha* or northern Chinese hawthorn (*C. pinnatifida*), *nan shanzha* or southern Chinese hawthorn (*C. cuneata*).

## GENERAL DESCRIPTION

The genus *Crataegus* includes approximately 280 species primarily from northern temperate zones in East Asia (20), Europe (20) and Eastern North America (200+).

*C. laevigata* is a spiny shrub; leaves mostly smooth obovate, three to five lobed, serrulate; flowers white to pink with red anthers; fruit globose or ellipsoid, deep red; found in woods from northwest and central Europe, from England to Latvia, west to the Pyrenees and northern Italy; naturalized in eastern North America and India. Many smooth-leaved or slightly pubescent species in Europe have been treated as *C. oxyacantha*, a synonym for the official European source, resulting in much confusion in the botanical and pharmacy literature.

*C. pinnatifida*, *C. pinnatifida* var. *major*, and *C. cuneata* are source plants for the Chinese drug shanzha (dried hawthorn fruit). *Crataegus pinnatifida* is distributed in northeast China, Shanxi, and Jiangsu provinces. *Crataegus pinnatifida* var. *major* ("northern *shanzha*") is distributed in north and northeast China; supply mostly cultivated, in Shangdong, Henan, and Hebei. *Crataegus cuneata* "Southern *shanzha*," is grown the Changjian Valley in Guangdong and Guangxi (TUCKER 3, JIANGSU).[1]

The part used is the dried fruits (China); or flowering tops, leaves, or fruits (Europe).

## CHEMICAL COMPOSITION

*C. laevigata* and *C. monogyna* contain flavonoids, including hyperoside (hyperin), quercetin, vitexin, vitexin-4′-L-rhamno-D-glucoside, vitexin-4′-L-rhamnoside, vitexin-4′-7-di-D-glucoside, rutin, quercetin-3-rhamno-galactoside, and others;[1–4] pentacyclic triterpenes (0.5–1.4% in fruits), including oleanolic acid, ursolic acid, acantolic acid, neotegolic acid, 2-$\alpha$-hydroxyoleanolic acid (crataegolic acid);[15] xanthine derivatives such as adenosine, adenine, guanine, and uric acid;[1] amines, including ethylamine, dimethylamine, trimethylamine, isobutylamine, isoamylamine, ethanolamine, $\beta$-phenylethylamine, choline, acetylcholine, *O*-methoxyphenethylamine, tyramine, and others;[1,6] proanthocyanidins; plus $\beta$-sitosterol, chlorogenic acid, caffeic acid, (+)-catechin, (−)-epicatechin, vitamins $B_1$, $B_2$, and C; calcium, iron, phosphorus, fructose, traces of an essential oil; and others (LIST AND HÖRHAMMER).[1]

## PHARMACOLOGY OR BIOLOGICAL ACTIVITIES

Described activities of hawthorn and its preparations include cardiotonic, hypotensive, antiarrhythmic, antilipemic, antibacterial, stomachic, and analgesic. Pharmacological studies report enhanced coronary blood flow and myocardial perfusion;[7,8] improvement of cardiac muscle contractility;[8] increased left ventrical output velocity; lowering of blood pressure;[9,10] an antiarrhythmic effect; increased myocardium tolerance to oxygen deprivation under hypoxic conditions; stimulation of revascularization after myocardial ischemia (ESCOP 2).[11,12]

Various clinical studies reveal increased cardiac performance and output; decrease in peripheral vascular resistance; decrease in pulmonary arterial and capillary pressures; reductions in blood pressure at rest and during exercise; and improved metabolic parameters (ESCOP 2).

Compounds associated with cardiotonic activity include hyperoside, vitexin, vitexin-2′-rhamnoside, oligomeric procyanidins, and (−)-epicatechin. The flavonoids and oligomeric procyanidins have a tonic effect on the cardiac muscles, are negatively chronotropic and dromotropic, and also show the bradycardiac effect commonly noted for *Crataegus* (LIST AND HÖRHAMMER).[1]

Similar activity has been reported for *C. pinnatifida*; also, oral administration of the fruit extract facilitates the clearance of serum cholesterol but does not prevent absorption (WANG).

Hawthorn extract or decoction is antibacterial against *Shigella flexneri*, *S. sonneni*, *Proteus vulgaris*, and *Escherichia coli* (WANG).

Free-radical scavenging activity of *C. pinnatifida* water extracts have been reported.[13]

No toxic effects, contraindications, or drug interactions are known (ESCOP 2). A single case of immediate-type hypersensitivity to *C. monogyna* has been reported.[14]

## USES

**Medicinal, Pharmaceutical, and Cosmetic.** Various drug preparations (oral or parenteral as i.m. or i.v. injections) are used in Europe for declining cardiac performance, corresponding to stages I and II of the New York Heart Association (NYHA) classification, senile heart conditions not requiring digitalis, and mild stable forms of angina pectoris, and mild forms of dysrythmia. Flowering tops are used in sleep-inducing preparations (ESCOP 2).

**Food.** Fruits of various hawthorn species have served as food in Europe, Asia, and at least a dozen species were used by American Indian groups. Candied fruits slices, jam, jelly, and wine available in major American Chinatowns.

**Health Food/Herb Teas.** Fruits (or flowers) used in tea, tablets, capsules, tinctures, etc.[1]

**Traditional Medicine.** In traditional Chinese medicine hawthorn fruits (dried, stir-fried, or charred) are used to stimulate digestion, promote function of the stomach, and stimulate blood circulation in epigastric distension, diarrhea, abdominal pain, amenorrhea, abdominal colic, indigestion, enteritis, acute bacillus dysentery, hypertension, hyperlipemia, and coronary heart disease. Charred fruits are used to promote digestion in stagnation of undigested meat, diarrhea, and with inadequate discharge from the bowels (TU). Up to 500 g of the fruits are eaten to treat tapeworm infections; externally as a wash for lacquer sores, itching and frost bite.

In European tradition the fruits, flowers, leaves or a combination thereof reportedly used as astringent, antispasmodic, cardiotonic, diuretic, hypotensive, and antisclerotic (STEINMETZ).

American Indian groups reportedly used a poultice of the leaves for boils, sores, ulcers; root decoction a gastrointestinal aid, diuretic, and to increase circulation (MOERMAN).

## COMMERCIAL PREPARATIONS

Crude, and extracts (e.g., powdered, solid, and liquid); tablets and parenteral dosage forms with at least 5 mg of flavones (calculated as hyperoside), or 10 mg total flavonoid fraction (total phenols, calculated as hyperoside), or 5 mg oligomeric procyanidins (calculated as epicatechol).

*Regulatory Status.* Undetermined in the United States. Subject of a positive German therapeutic monograph,[15] and a proposed European Union monograph (ESCOP 2).

## REFERENCES

See the General References for BLUMENTHAL; CSIR II; ESCOP 2; HSU; LIST AND HÖRHAMMER; JIANGSU; MABBERLY; MARTINDALE; MOERMAN; STEINMETZ; TU; TUCKER 3; TYLER 1; WANG; WEISS; WREN.

1. C. Hobbs and S. Foster, *HerbalGram*, **22**, 19 (1990).
2. J. Fisel, *Arzneim. Forsch.*, **15**, 1417 (1965).
3. P. Ficarra et al., *Farmaco Ed. Prat.*, **39**, 148 (1984).
4. M. Simova and T. Pangarova, *Pharmazie*, **38**, 791 (1983).
5. W. Braasch and W. Bienroth, *Arzneim. Forsch.*, **10**, 127 (1960).
6. T. Hockerts and G. Mülke, *Arnzeim. Forsch.*, **5**, 755 (1955).
7. C. Roddewig and H. Hensel, *Arnzeim. Forsch.*, **27**, 1407 (1977).
8. F. Occhiuto et al., *Plant. Med. Phytother.*, **20**, 52 (1986).
9. W. Stepka and A. D. Winters, *Lloydia*, **36**, 436 (1973).
10. H. P. T. Ammon and M. Händel, *Planta Med.*, **43**, 105, 209, 313 (1981).
11. B. Gabhard and E. Schuler in N. Rietbrock et al., eds., *Wandlungen in der Therapie der Herzinsuffizienz*, Friedr. Vieweg & Soyn, Wiesbaden, 1983, p. 43.
12. J. Guendjev, *Arnzeim. Forsch.*, **27**, 1576 (1977).
13. Y. R. Dai et al., *Planta Med.*, **53**, 309 (1987).
14. H. K. Steinman et al., *Contact Dermatitis*, **5**, 321 (1984).
15. Monograph *Crataegus*, *Bundesanzeiger*, no. 85 (Dec. 22, 1983); revised (May 5, 1988).

# HENNA

**Source:** ***Lawsonia inermis*** L. (syn. *L. alba* Lam.) (Family Lythraceae).

*Synonym.* Egyptian privet.

### GENERAL DESCRIPTION

A shrub with opposite leaves and very fragrant flowers; up to about 6 m high; generally considered a native of Africa and Asia; widely cultivated in tropical regions of the world (e.g., Egypt, Sudan, China, India, Florida, and the West Indies). Part used is the dried leaf. Major producing countries include Sudan, Egypt, and India.

### CHEMICAL COMPOSITION

Contains 0.55–1.0% lawsone (2-hydroxy-1,4-naphthoquinone);[1,2] 1,4-naphthoquinone;[3] 5–10% gallic acid and tannin; about 11% sugars; resin; and others (LIST AND HÖRHAMMER).[2,3]

Two xanthones (laxanthone-I and laxanthone-II) and a substituted coumarin named lacoumarin (5-allyloxy-7-hydroxycoumarin) have also been isolated from the whole plant and are probably present in the leaves.[4,5]

Lawsone is the major active principle (coloring and pharmacological) in henna. It is not present in the bark, stem, or root of the henna plant. Its concentrations in the leaves vary with climatic conditions; hot localities yield henna with higher lawsone content than temperate areas.[1] There has been a report indicating lawsone to be a degradation and autoxidation product of primary glycosides called hennosides A, B, and C.[6]

## PHARMACOLOGY OR BIOLOGICAL ACTIVITIES

Henna leaves (but not the seeds) have been reported to exhibit antifertility activity in female rats.[7]

Lawsone has various biological activities, including antifungal (fungicidal and fungistatic) activities toward *Alternaria*, *Aspergillus*, *Absidia*, *Penicillium*, and other species, being effective at 0.1% (1000 ppm) concentrations;[8,9] antibacterial activities toward *Brucella*, *Neisseria*, *Staphylococcus*, *Salmonella*, *Streptococcus*, and others, with a concentration of 0.005–0.02% (50–200 $\mu$g/mL) being effective against the first two groups;[1,3,10] antitumor activities (e.g., against sarcoma 180 in mice and Walker 256 carcinosarcoma in rats);[10] and antispasmodic properties as well as weak vitamin K activity.[1,2]

In addition to the antibacterial properties of lawsone, other fractions from a henna extract containing gallic acid and 1,4-naphthoquinone have also exhibited antibacterial activities.[3]

An ethanol extract containing luteolin, $\beta$-sitosterol, and lawsone has claimed antiinflammatory, anti-hyaluronidase, and analgesic activity.[11,12]

## USES

**Medicinal, Pharmaceutical, and Cosmetic.** Henna is used in numerous hair care products (e.g., dyes, conditioners, rinses, etc.). Prolonged use of henna on the hair would turn the hair orange red, unless henna is mixed with other dyes such as indigo and logwood to obtain different shades. To obtain a long-lasting color, the henna preparation must be rendered slightly acid (ca., pH 5.5) by adding a weak acid (e.g., citric, boric, or adipic).

**Traditional Medicine.** Leaves have been extensively used for centuries in the Middle East, the Far East, and northern Africa as a dye for nails, hands, hair, clothing, and others; they are also used in treating skin problems, headache, jaundice, amebiasis, enlargement of the spleen, and cancers, among others.[1,13]

**Others.** Lawsone can be used as an acid-base indicator for the titration of strong acids with weak bases.[14]

## COMMERCIAL PREPARATIONS

Mainly the crude.

*Regulatory Status.* Has been approved for use as a color additive exempt from certification, to be used in cosmetics (hair) only (§73.2190).

## REFERENCES

See the General References for BALSAM AND SAGARIN; JIANGSU; LUST; MARTINDALE; MERCK; MORTON 2; ROSE; TERRELL; UPHOF; WREN.

1. M. S. Karawya et al., *Lloydia*, **32**, 76 (1969).
2. A. Latif, *Indian J. Agr. Sci.*, **29**, 147 (1959).
3. Y. Abd-el-Malek et al., *Zentralbl. Bakteriol., Parasitenk., Infektionskr. Hyg., Abt. 2*, **128**, 61 (1973); through *Chem. Abstr.*, **79**, 62033c (1973).
4. D. K. Bhardwaj et al., *Phytochemistry*, **16**, 1616 (1977).
5. D. K. Bhardwaj et al., *Phytochemistry*, **15**, 1789 (1976).
6. G. J. Kapadia et al., *Lloydia*, **32**, 523 (1969).
7. S. R. Munshi et al., *Planta Med.*, **31**, 73 (1977).

8. R. D. Tripathi et al., *Experientia*, **34**, 51 (1978).
9. N. R. Farnsworth and G. A. Cordell, *Lloydia*, **39**, 420 (1976).
10. O. Goncalves de Lima et al., *Rev. Inst. Antibiot., Univ. Fed. Pernambuco, Recife*, **11**, 21 (1971); through *Chem. Abstr.*, **77**, 29629n (1972).
11. A. Gupta et al., *Indian J. Pharmacol.*, **18**, 113 (1986).
12. S. B. Vohora and P. C. Dandiya, *Fitoterapia*, **63**(3), 195 (1992).
13. J. L. Hartwell, *Lloydia*, **33**, 97 (1970).
14. K. C. Joshi et al., *Z. Naturforsch. B*, **32B**, 890 (1977); through *Chem. Abstr.*, **88**, 31484g (1978).

# HONEY

**Source:** Sugar secretions collected in honeycomb by honeybees (*Apis* spp.).

*Synonyms*. Purified honey, mel depuratum, clarified honey, strained honey, etc.

## GENERAL DESCRIPTION

Sugar secretions collected and stored in honeycomb by *Apis mellifera* L., and other *Apis* species (Family Apidae). A thick, syrupy, transparent liquid, honey is extracted from bee hives, then strained through a sieve and allowed to sit in settling tanks for 24 h to allow air bubbles to rise to the surface. Depending on pasturage source, color varies from amber to reddish brown to black.

## CHEMICAL COMPOSITION

Honey consists chiefly of dextrose and levulose (70–80%) with smaller amount of water, sucrose (2–10%), dextrin, wax, proteins, volatile oil, minerals, acids, and coloring and flavoring components, based on derivative plant source; contains vitamin $B_1$, vitamin $B_2$, vitamin C, nicotinic acid, and formic acid (CSIR I; MARTINDALE).

## PHARMACOLOGY OR BIOLOGICAL ACTIVITIES

Antibacterial, generally attributed to production of locally high osmolality due to water activity of honey; antimicrobial activity may be produced enzymatically by liberating hydrogen peroxide. Individual source plants may also contribute to antimicrobial activity. A recent assessment of 26 honeys in New Zealand found that of *Leptospermum scoparium* to have high antibacterial activity against *Escherichia coli* and *Staphylococcus aureus*; weaker against *Streptococcus pyogenes*, and *Salmonella typhimurium*.[1]

Spores of *Clostridium botulinum*, responsible for infant botulism, are often contained in honey, which may germinate in adults without adverse effects, but may cause serious illness in infants. In 1976, of 43 cases of infant botulism in California, 13 involved honey (*Clostridium botulinum* found in 13% of 60 tested samples). It has been recommended that honey not be given to infants under 1 year old.[2]

## USES

**Medicinal, Pharmaceutical, and Cosmetic.** Demulcent and sweetener in cough mixtures; in China as pill binder.

Honey is used as a fragrance ingredient, and humectant (in skin conditioners), also a biological additive in shampoos; face, body, and hand creams and lotions; bath products, hair conditioners; cleansing products; moisturizing creams and lotions; and paste masks (mud packs) (NIKITAKIS).

**Food.** Raw honey or honey in honeycomb sections (in beeswax) used as a sweetener.

**Health Food/Herb Teas.** A Vermont physician, D. C. Jarvis, catapulted honey into the health food realm by claiming that it improved digestion; facilitated wound healing; and had sedative, antiarthritic, and antibacterial effects (TYLER 1). The primary use, however, is as a sweetener for herb teas.

**Traditional Medicine.** In Indian folk medicine honey is considered demulcent; used for the treatment of eye ailments; sore throat, coughs, colds and constipation. Heated or mixed with hot liquids it is used in conjunction with emetic and enema formulations.

In Chinese medicine used as a nutritive, demulcent, emollient laxative, in cough due to lung dryness, constipation, stomachache, sinusitis, oral ulcerations, scalds, and as a detoxicant for aconitine (HSU).

Traditionally considered useful topically for indolent skin ulcers where antibiotics fail to achieve results. Has also been used to soothe dermal lesions and necrotic malignant breast ulcers.

### COMMERCIAL PREPARATIONS

Raw honey, honey in beeswax comb.

*Regulatory Status*. A common food; GRAS.

### REFERENCES

See the General References for CSIR I; HSU; MARTINDALE; NIKITAKIS; TYLER 1.

1. D. Greenwood. *Lancet*, **341**, 191 (1993).
2. R. A. Mangione, *Am. Pharm.*, **NS23**, 5 (1983).

# HOPS

**Source:** ***Humulus lupulus*** L. (Family Moraceae or Cannabaceae).

*Synonyms*. European hops and common hops.

### GENERAL DESCRIPTION

A twinning perennial herb with male and female flowers on separate plants (dioecious); up to about 8 m high; native to Eurasia and North America; extensively cultivated worldwide (e.g., United States, Germany, and the Czech Republic). Part used is the female membranous conelike inflorescence (strobile) with its glandular hairs, collected in the fall and carefully dried, often bleached with sulfur dioxide from burning sulfur; an essential oil is obtained by steam distillation of the freshly dried cones. The glandular hairs, separated from the strobiles, compose lupulin, which contains more resins and volatile oil than hops and is also used like hops.

Major producers of hops include the United States, Germany, and the Czech Republic.

### CHEMICAL COMPOSITION

Contains 0.3–1% volatile oil; 3–12% resinous bitter principles composed of $\alpha$-bitter acids (humulone, cohumulone, adhumulone, prehumulone, posthumulone, etc., with first three in predominance) and $\beta$-bitter acids (lupulone, colupulone, adlupulone, etc., in decreasing concentration); other resins, some of which are oxidation products of the $\alpha$- and $\beta$-acids; xanthohumol (a chalcone); flavonoid glycosides (astragalin, quercitrin, isoquercitrin, rutin, kaempferol-3-rutinoside, etc.); phenolic acids; tannins; lipids; amino acids; estrogenic substances; and many other compounds (KARRER, JIANGSU, LIST AND HÖRHAMMER).[1-3]

The volatile oil is made up mostly of humulene ($\alpha$-caryophyllene), myrcene, $\beta$-caryophyllene, and farnesene, which together may account for more than 90% of the oil.[1] Other compounds present number more than 100, including germacratriene, $\alpha$- and $\beta$-selinenes, selina-3,7(11)-diene, selina-4(14),7(11)-diene, $\alpha$-copaene, $\alpha$- and $\beta$-pinenes, limonene, *p*-cymene, linalool, nerol, geraniol, nerolidol, citral, methylnonyl ketone, other oxygenated compounds, 2,3,4-trithiapentane (present only in oil of unsulfured hops in ca. 0.01%), *S*-methylthio-2-methylbutanoate, *S*-methylthio-4-methylpentanoate, and 4,5-epithiocaryophyllene.[1,4-9]

## PHARMACOLOGY OR BIOLOGICAL ACTIVITIES

Hops extracts have been reported to have various biological activities, including antimicrobial activities, which are due to the bitter acids (especially lupulone and humulone), the more hydrophobic ones being the more active;[10-13] strong spasmolytic effects on isolated smooth muscle preparations;[14] hypnotic and sedative effects, which were disputed by one report;[15] estrogenic properties, which were not observed in a later study;[16] and allergenic activity on humans, causing contact dermatitis due to the pollen (JIANGSU, LIST AND HÖRHAMMER, MERCK).

Alcoholic extracts of hops in various dosage forms have been used clinically in treating numerous forms of leprosy, pulmonary tuberculosis, and acute bacterial dysentery, with varying degrees of success (JIANGSU).

2-methyl-3-butene-2-ol (present in hops up to 0.15%) has sedative effects in rats.[17,18]

A stimulant effect on gastric secretion has been demonstrated for hops in laboratory animals.[19]

## USES

**Medicinal, Pharmaceutical, and Cosmetic.** Extracts are used in certain skin creams and lotions, especially in Europe, for their alleged skin-softening properties.

**Food.** Major use is in beer, with the bitter taste derived primarily from oxidation products of humulone. Extracts and oil are also used as flavor components in nonalcoholic beverages, frozen dairy desserts, candy, baked goods, and gelatins and puddings, with the highest average maximum use level of 0.072% reported for an extract (type not indicated) in baked goods.

**Health Food/Herb Teas.** Used in sleeping preparations. Cut strobiles, powdered, or dried extract powder for tea, tincture, capsules, tablets, etc. Also used in "dream pillows," to promote sleep; and bath preparations (FOSTER). Use in medicinal bath preparations in Germany has been disallowed.[20]

**Traditional Medicine.** Used as a diuretic and anodyne and in treating nervous diarrhea, insomnia, restlessness, and other nervous conditions as well as intestinal cramps and lack of appetite, among others, usually in the form of a tea; also used in Chinese medicine for pulmonary tuberculosis and cystitis (JIANGSU).

It has reportedly been used in cancers.[21]

## COMMERCIAL PREPARATIONS

Crude, extracts, and oil; crude was formerly official in N.F., and oil is official in F.C.C. Strengths (see ***glossary***) are either expressed in flavor intensities or in weight-to-weight ratios.

*Regulatory Status.* GRAS with both hops and lupulin listed (§182.20). The strobiles are the subject of a German therapeutic monograph, with use approved for mood disturbance (unrest, anxiety) and sleep disturbances. The proposed ESCOP monograph indicated usage for nervous tension, excitability, restlessness, sleep disturbances, and lack of appetite (ESCOP 2).

## REFERENCES

See the General References for ARCTANDER; BAILEY 2; BIANCHINI AND CORBETTA; BLUMENTHAL; ESCOP 2; FEMA; FOSTER; FOSTER AND DUKE; GUENTHER; JIANGSU; LEWIS AND ELVIN-LEWIS; LIST AND HÖRHAMMER; LUST; TERRELL; UPHOF; WREN; YOUNGKEN.


1. A. Mijavec and P. Spevok, *Bilt. Hmelj. Sirak.*, **7**(24), 23 (1975); through *Chem. Abstr.*, **85**, 92161j (1976).
2. M. Anguelakova et al., *Riv. Ital. Essenze, Profumi, Piante Offic., Aromi, Saponi, Cosmet., Aerosol*, **53**, 275 (1971); through *Chem. Abstr.*, **75**, 143906j (1971).
3. A. Strenkovskaya, *Mezhdunar. Kongr. Efirnym Maslam (Mater.)*, **4**(1), 325 (1971); through *Chem. Abstr.*, **78**, 140348e (1973).
4. R. G. Buttery and L. C. Ling, *Brewers Dig.*, **8**, 71 (1966).
5. R. G. Buttery and L. C. Ling, *J. Agr. Food Chem.*, **15**, 531 (1967).
6. R. D. Hartley and C. H. Fawcett, *Phytochemistry*, **8**, 1793 (1969).
7. R. G. Buttery et al., *Chem. Ind. (London)* (28), 1225 (1966).
8. A. S. Morieson, *Inst. Brewing (Austral. Sect.), Proc. Conv.*, **6**, 102 (1960); through *Chem. Abstr.*, **61**, 12330c (1964).
9. T. L. Peppard and F. R. Sharpe, *Phytochemistry*, **16**, 2020 (1977).
10. M. Teuber and A. F. Schmalreck, *Arch. Mikrobiol.*, **94**, 159 (1973).
11. Y. Kuroiwa et al., Jpn. Kokai 73 58, 114 (1973); and 73 58,115 (1973); through *Chem. Abstr.*, **80**, 6908a and 6907z (1974).
12. A. F. Schmalreck et al., *Can. J. Microbiol.*, **21**, 205 (1975).
13. J. Boatwright, *J. Inst. Brew.*, **82**, 334 (1976); through *Chem. Abstr.*, **86**, 70045t (1977).
14. F. Caujolle et al., *Agressologie*, **10**, 405 (1969); through *Chem. Abstr.*, **72**, 41267x (1970).
15. R. Hänsel and H. H. Wagener, *Arzneim. Forsch.*, **17**, 79 (1967).
16. C. Fenselau and P. Tatalay, *Food Cosmet. Toxicol.*, **11**, 597 (1973).
17. R. Hänsel et al., *Planta Med.*, **45**, 224 (1982).
18. R. Wohlfart et al., *Planta Med.*, **48**, 120 (1983).
19. St. Tamasdan, *Farmacia (Bucharest)*, **29**, 71 (1985).
20. Monograph *Lupuli strobulus*, *Bundesanzeiger*, no. 228 (Dec. 5, 1984); revised (Mar. 13, 1990).
21. J. L. Hartwell, *Lloydia*, **33**, 97 (1970).

# HOREHOUND

**Source:** ***Marrubium vulgare*** L. (Family Labiatae or Lamiaceae)

*Synonyms*. Marrubium, hoarhound, common hoarhound, and white horehound.

## GENERAL DESCRIPTION

A perennial aromatic herb with hairy stems and leaves; up to about 1 m high; native to Europe and Asia; naturalized in North America. Parts used are the dried leaves and flowering tops.

*Ballota nigra* L and its six subspecies, traded (primarily in Europe) as black horehound, has similar indications to horehound, though use is limited because of extreme bitter taste (WREN).

## CHEMICAL COMPOSITION

Contains 0.3–1% of a bitter principle called marrubiin (a diterpene lactone);[1,2] several diterpene alcohols (e.g., marrubiol, marrubenol, peregrinol, and vulgarol);[1,3,4] small amounts of alkaloids (ca. 0.3% betonicine, its stereoisomer turicine, and ca. 0.2% choline);[5] trace of a volatile oil containing monoterpenes ($\alpha$-pinene, camphene, limonene, sabinene, *p*-cymene, etc.) and a sesquiterpene;[6] $C_{27}$ to $C_{34}$ alkanes (normal and branched);[7] free phytol;[4] and tannin, pectic substances, saponin, resin, $\beta$-sitosterol, and others (LIST AND HÖRHAMMER).[8]

## PHARMACOLOGY OR BIOLOGICAL ACTIVITIES

Marrubiin is reported to have expectorant properties, and when its lactone ring is opened, the resulting acid (marrubic acid) has strong choleretic activity (TYLER 1). It also has a normalizing effect on extrasystolic arrythmias but in large doses disturbs the heart rhythm, which can be counteracted by atropine (LIST AND HÖRHAMMER).

The volatile oil of horehound has also been reported to have vasodilative and expectorant properties.[6]

An aqueous extract of horehound has been reported to be antagonistic toward serotonin *in vitro*, and is antiinflammatory in the rat paw edema test (WREN).[9]

Weak antioxidant activity is linked to total hydroxycinnamic derivatives (0.6%).[10]

## USES

**Medicinal, Pharmaceutical, and Cosmetic.** Extracts used as an expectorant in cold and cough medicines, especially in Europe.

**Food.** Extracts are used in flavoring numerous food products, especially candy and alcoholic beverages (e.g., bitters and liqueurs). Other foods in which horehound extracts are used include nonalcoholic beverages, frozen dairy desserts, baked goods, and gelatins and puddings. Highest average maximum use level is 0.073% reported in candy.

**Health Food/Herb Teas.** Herb used as a minor (bitter) flavoring component in some tea formulations; also in confectionaries intended to have a soothing effect on coughs and colds (FOSTER).

**Traditional Medicine.** It is used in treating sore throat, colds, coughs, and other respiratory ailments; also used as a diuretic, bitter tonic, and diaphoretic as well as in treating cancers.[11]

## COMMERCIAL PREPARATIONS

Crude and extracts; crude was formerly official in U.S.P. Strengths (see ***glossary***) of extracts are expressed in flavor intensities or weight-to-weight ratios.

*Regulatory Status*. GRAS (§182.10 and §182.20). The leaves are the subject of a

German therapeutic monograph, with approval of use of herb and extracts for coughs, colds, and as digestive aid and appetite stimulant.[12] Declared ineffective as an OTC cough suppressant and expectorant by the FDA (TYLER 1).

## REFERENCES

See the General References for BAILEY 1; BIANCHINI AND CORBETTA; BLUMENTHAL; FEMA; FOSTER; GOSSELIN; KROCHMAL AND KROCHMAL; LEWIS AND ELVIN-LEWIS; LUST; TERRELL; TYLER 1; UPHOF; WREN; YOUNGKEN.

1. D. P. Popa et al., *Khim. Prir. Soedin.*, **4**, 345 (1968); through *Chem. Abstr.*, **70**, 93276c (1969).
2. H. J. Nicholas, *J. Pharm. Sci.*, **53**, 895 (1964).
3. D. P. Popa and G. S. Pasechnik, *Khim. Prir. Soedin.*, **11**, 722 (1975); through *Chem. Abstr.*, **84**, 150776f (1976).
4. D. P. Popa and L. A. Salei, *Rast. Resur.*, **9**, 384 (1973); through *Chem. Abstr.*, **80**, 12454r (1974).
5. W. W. Paudler and S. Wagner, *Chem. Ind. (London)* (42),1693 (1963).
6. M. O. Karryev et al., *Izv. Akad. Nauk Turkm. SSR, Ser. Biol.*, **3**, 86 (1976); through *Chem. Abstr.*, **86**, 2355u (1977).
7. C. H. Brieskorn and K. Feilner, *Phytochemistry*, **7**, 485 (1968).
8. I. M. Bartarelli, *Boll. Chim. Farm.*, **105**, 787 (1966); through *Chem. Abstr.*, **66**, 88626e (1967).
9. R. Cahen, *CR Soc. Biol.*, **164**, 1467 (1970); through *Chem. Abstr.*, **75**, 2494m (1971).
10. J. L. Lamaison et al., *Fitoterapia*, **62**(2), 166 (1991).
11. J. L. Hartwell, *Lloydia*, **32**, 247 (1969).
12. Monograph *Marrubii herba*, *Bundesanzeiger*, no. 22 (Feb. 1, 1990).

# HORSECHESTNUT

**Source:** ***Aesculus hippocastanum*** L. (Family Hippocastanaceae).

## GENERAL DESCRIPTION

Deciduous tree to 25 m; leaves opposite, digitate with five to seven obovate, irregularly crenate-serrate leaflets; glabrous above, tomentose beneath. Flowers white, with yellow to pink spot at base, in large cylindrical panicle. Spiny globose fruits 2 to 6 cm in diameter with large brown smooth seed, 2–4 cm in diameter; found in mountain woods, indigenous to central Balkan peninsula, widely planted and established throughout the northern hemisphere as a shade and ornamental tree (TUTIN 2). The parts used are the seed, branch bark, and leaves.

## CHEMICAL COMPOSITION

The seeds and bark contain a mixture of triterpene saponins known as aescin (escin), composed of acylated glycosides of protoeasigenin and barringtogenol-C, hippoaesculin and others;[1] quinones, including plastoquinone 8; flavones, including 3,5-dihydroxy-3′,4′,7-trimethoxyflavone, myricetin 3′,4′,7-trimethyl ether; sterols, including stigmasterol, α-spinasterol, and β-sitosterol; linolenic, palmitic, and stearic acids; and others. The glycoside aesculin (esculin) (7-hydroxycoumarin 6-β-glucoside) is considered the most toxic component of the seed (GLASBY 2; WREN).[2]

## PHARMACOLOGY OR BIOLOGICAL ACTIVITIES

Horsechestnut extracts, notably aescin, has antiinflammatory, antiedema, antiexudative, and venotonic activity.

Aescin was found to be responsible for antiexudative and edema-protective activity.[3] Aescin acts on the capillary membrane, normalizing vascular permeability, enhancing capillary resistance, and reducing the outflow of fluid into the extracapillary space.[4] Aescin has a "sealing" effect on the capillaries and reduces the number and diameter of the small pores of the capillary wall by which exchange of water occurs. Antiinflammatory activity at the initial exudation phase of inflammation has been confirmed in various *in vitro* and *in vivo* models.[5,6]

*In vitro* aescin has been found to stimulate an increase in venous tone, with a decrease in the volume of venous district of the saphenous vein and its collaterals, facilitating return blood flow to the heart.[2,7]

Efficacy of a 2%-aescin-containing gel in reducing tenderness of experimentally induced hematoma has recently been confirmed.[8]

*In vitro* antitumor activity has been observed from hippoaesculin and barringtogenol-C-21-angelate.[2,9]

Horsechestnut seeds are considered inedible and poisonous. The bitter flavor prevents consumption of large amounts. The leaves, flowers, young sprouts and seeds are toxic. Symptoms of poisoning include nervous muscle twitching, weakness, dilated pupils, vomiting, diarrhea, depression, paralysis and stupor (HARDIN AND ARENA). Incidents of anaphylactic shock after i.v. injections of horsechestnut have been reported, along with renal toxicity or failure (FROHNE AND PFÄNDER).[2]

## USES

**Medicinal, Pharmaceutical, and Cosmetic.** Horsechestnut extract or aescin (0.25–0.5%) has reportedly been used in shampoos, shower foams, foam baths, skin care products, body and hand creams, lotions, and toothpastes. Cosmetic use in Europe has been based on its clearing and redness-reducing properties, and its effectiveness in preventing cellulitis.[10]

Numerous clinical studies and published case reports confirm the efficacy of aescin-containing topical products, especially in the treatment of sports injuries, including blunt trauma of the lower limbs,[11] joint sprains, tendonitis, hematomas, muscle strain, traumatic edema,[12] Achilles' tendonitis; surgical outpatient trauma, including fractures, sprains, crush injuries, and contusion;[13] postoperative or postpartum edema in gynecology and obstetrics;[4] and others.

Intravenous (never extravenous) administration of aescin in ampoules is used clinically by physicians in Germany and other European countries for treatment of posttraumatic, intraoperative or postoperative conditions of cerebral edema, and other surgical specialties.[14]

**Traditional Medicine.** Fruits, bark, or seed has reportedly been used externally for ulcers; a folk cancer remedy. Seeds used for gastritis, enteritis, and hemorrhoids (DUKE 2). Bark tea astringent, used in malaria, dysentery; externally for lupus and skin ulcers (FOSTER AND DUKE).

Leaf preparations used in European traditions for eczema, varicose veins, supportive treatment of varicose ulcers, phlebitis, thrombophlebitis, hemorrhoids, menstrual spastic pain, soft tissue swelling from bone fracture and sprains, and other uses. Effectiveness of leaf preparation claims is unsubstantiated.[15]

## COMMERCIAL PREPARATIONS

In Germany and other countries topical gels contain 1% aescin; ampoules containing 5.1 mg sodium aescinate (equivalent to 5 mg aescin); sugar-coated tablets; and liquid oral preparations are available. Topical products are available in Canada.[2] No horsechestnut or aescin-containing drug formulations are available in the United States, though aescin is available in bulk.

*Regulatory Status.* Undetermined in the United States. In Germany, horsechestnut seeds are the subject of a positive therapeutic monograph, indicated for chronic venous insufficiency, including edema, cramps in the calves, itching, pain and sensations of heaviness in the legs, varicose veins, postthrombotic syndrome, plus posttraumatic and postoperative swelling of soft tissue, in average daily doses equivalent to 30–150 mg of aescin in liquid or solid preparations for oral administration.[16] Horsechestnut leaf preparations claims are not substantiated, therefore, therapeutic use is not recommended.[15]

## REFERENCES

See the General References for BLUMENTHAL; DUKE 2; FOSTER AND DUKE; FROHNE AND PFÄNDER; GLASBY 2; HARDIN AND ARENA; LIST AND HÖRHAMMER; MABBERLY; MARTINDALE; STEINMETZ; TUTIN 2; WEISS; WREN.

1. T. Konoshima and K. H. Lee, *J. Nat. Prod.*, **49**, 4, 650.
2. R. F. Chandler, *Can. Pharm. J.*, **297** (July/Aug. 1993).
3. D. M. Lorenz and M. L. Marek, *Arzneim. Forsch.*, **10**, 263 (1960).
4. D. Tenhaeff, *Arztliche Praxis*, **24**(12), 559 (1972).
5. G. Vogel et al. *Arzneim. Forsch.*, **20**(5), 699 (1970).
6. M. I. Rothkopf and G. Vogel, *Arzneim. Forsch.*, **26**(2), 225 (1976).
7. F. Annoni et al., *Arzneim. Forsch.*, **29**(4), 672 (1979).
8. C. C. Calabrese and P. Preston, *Planta Med.*, **59**(5), 394 (1993).
9. J. De Meirsman and N. Rosselle, *Ars. Med.*, **9**(3), 247 (1980).
10. G. Properpio et al., *Fitoterapia*, **2**, 113 (1980).
11. J. Rothhaar and W. Theil, *Med. Welt.*, **33**(27), 1006 (1982).
12. C. Zuinen, *Rev. Med. Liege*, **31**(5), 169 (1976).
13. F. Jakob and B. Fassbender, *Fortschr. Med.*, **87**(20/21), 893 (1969).
14. F. Heppner et al., *Wein. Med. Wschr.*, **117**, 706 (1967).
15. Monograph *Hippocastani folium*, *Bundesanzeiger* (July 14, 1993).
16. Monograph *Hippocastani semen*, *Bundesanzeiger*, no. 228 (Dec. 5, 1984).

# HORSETAIL

**Source:** ***Equisetum arvense*** L.; ***E. hymale*** L. (Family Equisetaceae).

*Synonyms.* Common horsetail, field horsetail, running clubmoss, and *shenjincao* (*E. arvense*); rough horsetail, common scouring rush, and *muzei* (*E. hymale*).

## GENERAL DESCRIPTION

*E. arvense* stems are branched, jointed, with 6–16 grooves and ridges; leaves minute, in whorls united into a sheath at the base. Spore-bearing cones on separate stalks; *E. hymale* stems are unbranched, with 10–30 grooves, 30–100 cm high; both species occur in moist soils; common in much of temperate Northern Hemisphere

(Asia, Europe, North America).[1] The part used is the dried stems.

## CHEMICAL COMPOSITION

*E. arvense* contains 0.3–1% flavonoids, including quercetin-3-glucoside, luteolin-5-glucoside, protogenkwanin-4″-*O*-glucoside, kaempferol, apigenin, isoquercitrin, and others (depending on chemotype).[2–4] Two distinct chemotypes are recognized, one from Europe, characterized by quercetin-3-*O*-sophoroside, protogenkwanin-4′-*O*-β-D-glucopyranoside, and genkwanin-4′-*O*-β-D-glucopyranoside. North American and Asian materials are characterized by the presence of flavone-5-glucosides and their 6″-malonyl esters, especially luteolin. Both chemotypes contain quercetin-3-*O*-β-D-glucopyranoside and its malonyl ester as the major flavonoids.[2,3] *Equisetum* hybrids are extremely variable in morphological features. It has been suggested that hybrid parent species can be verified with analysis of flavonoid patterns.[5]

Phenolic acids of *E. arvense* include di-*E*-caffeoyl-*meso*-tartaric acid, and methyl esters of protocatechuic and caffeic acids; aconitic, oxalic, malic, tannic, arabinoic, and threonic acids;[6,7] minerals, including silicic acid and silicates (5–8%) water soluble up to 80%,[8] potassium, aluminum, and manganese; sterols, including campesterol, isocuosterol, and brassinosteroids;[7] and others. Trace amounts of nicotine have been found in *E. arvense*.[9]

## PHARMACOLOGY OR BIOLOGICAL ACTIVITIES

Horsetail is considered mildly diuretic, hemostyptic, and vulnerary.

A study of four Mexican *Equisetum* species revealed that chloroform extracts of *E. hymale* var. *affine* had the greatest diuretic activity, more effective than spironolactone, furosemide, and hydrochlorothiazide. An increase in excretion of sodium, chloride, and potassium, with a rise in urine pH was also observed.[7,10]

Other activities include a hemostyptic effect observed in animals, and a strengthening and regenerating effect on connective tissue (BRADLY, WEISS).

Silica, necessary for the formation of articular cartilage and connective tissue, is taken up by the plant in the form of a bioavailable monosilicic acid,[7] perhaps accounting for traditional uses of the plant.

*E. arvense* and other species are known to cause toxicity in livestock, including horses, sheep, and rarely cattle. Toxicity, similar to nicotine poisoning has been reported in children who have chewed the stems. Horsetail may also caused seborrheic dermatitis.[7] An antithiamine action has been recognized in *Equisetum* ingestion in horses, destroying thiamine (vitamin $B_1$) in the stomach of monogastric animals, including humans.[7,11]

*E. palustre* L., which contains the alkaloid palustrine, is toxic to livestock. The German Pharmacopoeia requires examination of *E. arvense* for adulteration with other *Equisetum* species, especially *E. palustre* (FROHNE AND PFÄNDER). A critical review of the German Pharmacopeial methods, along with new analytical procedures, investigations, and improvements, have recently been proposed.[12]

## USES

**Medicinal, Pharmaceutical, and Cosmetic.** Horsetail extract is used as a biological additive in shampoos, skin care products, etc.

Preparations of the herb are used in German phytotherapy for posttraumatic and static edema and in irrigation therapy for bacterial and inflammatory conditions of the lower urinary and renal tract (contraindicated in cases of impaired heart or kidney function). Externally, compresses or poultices are used for the supportive treatment of poorly healing wounds.[13]

**Health Food/Herb Teas.** Horsetail is primarily used as a dietary supplement for mineral content, also in diuretic formulations, including teas, tinctures, capsules, tablets, etc. (FOSTER AND DUKE).

**Traditional Medicine.** In Europe, the herb has reportedly been used to promote renal function; digestive elimination; an adjuvant in weight loss products (BRADLEY). American Indian groups used plant tea for kidney and bladder ailments and constipation. In India, horsetail is used as a diuretic, hemostatic; root as an analgesic for teething babies. A folk remedy for bloody urine, gout, gonorrhea, stomach disorders; poulticed for wounds (FOSTER AND DUKE).

A French patent exists for the use of isolated silica compounds from *E. arvense* for the treatment of bone fractures, osteoporosis, connective tissue, and tooth and nail injuries.[7]

In traditional Chinese medicine, *E. hymale*, and *E. debile* Roxb. (substitute in Yunnan) have been used to treat bloody stools, dysentery with blood, anal prolapse, malaria, sore throat; externally poulticed for sores; an injectable drug has been used in China to treat neurodermatitis (JIANGSU).

## COMMERCIAL PREPARATIONS

Crude dried herb, fresh stems and extracts.

*Regulatory Status.* Undetermined in the United States. *E. arvense* is the subject of a positive German monograph, as a mild diuretic; externally for supportive treatment of poorly healing wounds.[13] Health and Welfare Canada requires manufacturers to prove *E. arvense* products are free of thiaminase-like activity (though the compound responsible for antithiamine action has not been identified). The action is based on the concern that irreversible brain damage may occur in thiamine-deficient individuals.[7]

## REFERENCES

See the General References for BLUMENTHAL; BRADLY, DUKE 2; FOSTER AND DUKE; FROHNE AND PFÄNDER; GLASBY 2; HARDIN AND ARENA; JIANGSU; LIST AND HÖRHAMMER; MABBERLY; MARTINDALE; STEINMETZ; TUTIN 1; WEISS; WREN.

1. R. L. Hauke, *New Bot.*, **1**, 89 (1974).
2. M. Veit et al., *Planta Med.*, **55**, 214 (1989).
3. M. Veit et al., *Phytochemistry*, **29**, 2555 (1990).
4. N. A. M. Saleh et al., *Phytochemistry*, **11**, 1095 (1972).
5. M. Veit et al., paper presented at the Bonn BACANS Symposium, July 17–22, 1990.
6. I. L. F. Bakke et al., *Acta Pharm. Suec.*, **15**, 141 (1978).
7. N. W. Hamon and D. V. C. Awang, *Can. Pharm. J.*, 399 (Sept. 1992).
8. R. Piekos et al., *Planta Med.*, **27**, 145, (1975).
9. J. D. Phillipson and C. Mellville, *J. Pharm. Pharmacol.*, **12**, 506 (1960).
10. R. M. Perez-Gutierrez, *J. Ethnopharmacol.*, **14**, 269 (1985).
11. J. A. Henderson et al., *J. Am. Vet. Med. Assoc.*, 375 (June 1952).
12. M. Viet et al., *Deut. Apoth. Ztg.*, **129**, 1591 (1989).
13. Monograph *Equisete herba*, *Bundesanzeiger*, no. 173, (Sept. 18, 1986).

# HYDRANGEA

**Source:** ***Hydrangea arborescens*** L. (Family Saxifragaceae or Hydrangeaceae).

*Synonyms*. Smooth hydrangea, mountain hydrangea, wild hydrangea, and seven barks.

## GENERAL DESCRIPTION

An erect shrub with large ovate leaves (to 20 cm long); up to about 3 m high; native to the eastern United States, growing from New York south to Florida and west to Iowa and Oklahoma. Part used is the dried root, collected in the fall.

## CHEMICAL COMPOSITION

Although much work has been done on other *Hydrangea* species, especially those of Asian origin such as *H. macrophylla* (Thunb.) Ser. and *H. paniculata* Sieb., chemical data relating to *H. arborescens* have been very limited. Constituents reported present in *H. arborescens* have included hydrangin, saponin, resin, rutin, starch, a fixed oil, and a volatile oil. The identity of hydrangin is unclear; it has been reported in the literature as a glucoside, an alkaloid, as synonymous with umbelliferone, and as a glycoside with formula of $C_{34}H_{25}O_{11}$ (GRIEVE, LIST AND HÖRHAMMER, MERCK, UPHOF).

Compounds found in other *Hydrangea* species (*H. macrophylla*, *H. paniculata*, etc.) include hydrangenol, hydrangeic acid, phyllodulcin, lunularic acid, their glycosides, lunularin, and 3,4′-dihydroxystilbene;[1] paniculatan (a mucous polysaccharide);[2,3] and hydrangetin (7-hydroxy-8-methoxycoumarin), rutin, umbelliferone, febrifugine (an alkaloid), gum, resin, and others (JIANGSU, LIST AND HÖRHAMMER).

Of the leaves of numerous *Hydrangea* and *Viburnum* species tested for tannic substances and flavonoids (quercetin, cyanidin, kaempferol, and luteolin), only those of *H. arborescens* and a subspecies of *H. macrophylla* were found to lack tannins and cyanidin but to contain comparable quantities of quercetin and kaempferol.[4] This perhaps can give some indication as to the possible similarity in other constituents.

## PHARMACOLOGY OR BIOLOGICAL ACTIVITIES

Is believed to have diuretic properties. Hydrangenol and its derivatives have antiallergic activities (antihistaminic and hyaluronidase inhibition).[5]

## USES

**Medicinal, Pharmaceutical, and Cosmetic.** Used in certain diuretic preparations, often in combination with corn silk, doggrass, and others.

**Traditional Medicine.** Used as a tonic, diaphoretic, and diuretic as well as in removing and preventing kidney and bladder stones. Used in American Indian tradition as diuretic, cathartic, emetic; externally bark for wounds, burns, sore muscles, sprains; folk cancer remedy (FOSTER AND DUKE).

In Chinese medicine, the roots, leaves, and flowers of certain *Hydrangea* species (e.g., *H. macrophylla*, *H. strigosa* Rehd., *H. umbellata* Rehd., and *H. paniculata*) are used to treat malaria and as diuretics and antitussives (JIANGSU).

**Health Food/Herb Teas.** Rarely used, tincture available from small manufacturers (LUST).

**Others.** The leaves of *H. macrophylla* var. *thunbergii* Makino, *H. strigosa*, and *H. umbellata* are sweet; the sweet principle from the first species has been isolated and found to be phyllodulcin (an isocoumarin). They are a potential source of natural sweeteners (JIANGSU).[6]

**COMMERCIAL PREPARATIONS**

Crude and extracts; crude was formerly official in N.F. Strengths (see ***glossary***) of extracts are expressed in weight-to-weight ratios.

**REFERENCES**

See the General References for APhA; BAILEY 1; FOSTER AND DUKE; GOSSELIN; GRIEVE; JIANGSU; KROCHMAL AND KROCHMAL; LIST AND HÖRHAMMER; LUST; UPHOF.

1. J. Gorham, *Phytochemistry*, **16**, 249 (1977).
2. M. Tomoda and N. Satoh, *Chem. Pharm. Bull.*, **24**, 230 (1976).
3. M. Tomoda and N. Satoh, *Chem. Pharm. Bull.*, **25**, 2910 (1977).
4. E. C. Bate-Smith, *Phytochemistry*, **17**, 267 (1978).
5. H. Kakegawa et al., *Planta Med.*, **54**, 385 (1988).
6. G. E. Inglett, personal communication.

# HYPERICUM

**Source: *Hypericum perforatum*** L. (Family Hypericaceae or Guttiferae).

*Synonyms*. St. Johns Wort, Klamath weed, and goatweed.

**GENERAL DESCRIPTION**

Perennial herb to 3–9 dm high; stems ridged or two sided, smooth, erect, branching toward top; dark glands along ridges, with distinct dark rings at the lower nodes; leaves simple, oblong, entire, opposite, sessile 2.5 cm long; flowers five-merous, yellow, 2.5 cm wide, petal margins glandular, in terminal corymbs; June through August; indigenous to Europe; naturalized in waste places and along roadsides in Asia, Africa, North America, and Australia; serious weed in range lands of western North America and Australia. Two introduced leaf-feeding beetles *Chrysolina quadrigemina* and *C. hyperici* have been used successfully in Canada as a biological control.[1]

The part used is the fresh or dried flowering tops or herb.

**CHEMICAL COMPOSITION**

The herb contains naphthodianthrones, chiefly hypericin (0.5–0.7%) and pseudohypericin; flavonols, including 0.5–2% of the glucoside hyperin (hyperoside), quercetin, isoquercetin, quercitrin, isoquercitrin, rutin, kaempferol; flavanols, including (+)-catechin, leucocyanidin, (−)-epicatechin; acylphloroglucinols, hyperforin and adhyperforin;[2] volatile oil (0.05%–0.9%) with monoterpenes, including $\alpha$-pinene, $\beta$-pinene, myrcene, and limonene, and the sesquiterpenes caryophyllene and humulene; xanthones including kielcorin, magniferin, 1,3,6,7-tetrahydroxy-xanthone; I3-II8-biapigenin; tannins (3.8–10%, up to 16% in flowers); nicotinic acid; vitamins C and A; $\beta$-sitosterol; choline; pectin; phlobaphene; rhodan; and others.[3–5]

**PHARMACOLOGY OR BIOLOGICAL ACTIVITIES**

Described activities for hypericum include anxiolytic, antiinflammatory, astringent, antiviral, and antiseptic.

Hypericin has been found to inhibit *in vitro* almost irreversibly both type A and B monoamine oxidase (MAO) in rat brain mitochondria. Type A MAO (serotonin) inhibition was greatest.[6]

Often described as a "sedative" in the literature, hypericum is now classified as an antidepressant; this activity has been demonstrated in a number of animal models. It also enhances mice exploratory activity in a foreign environment, extends narcotic sleeping time (dose dependent), is a reserpine antagonist, and decreases aggression in socially isolated male mice.[7,8] Limited clinical trials have demonstrated an improvement in symptoms of anxiety, dysphoric mood, hypersomnia, anorexia, depression, insomnia, psychomotor retardation, and other subjective indicators.[7,9]

Hyperforin is antibacterial against staphylococci and streptococci.[2]

Hepatoprotective activity of hypericin has been demonstrated in mice and rats.[10]

Potent antiretroviral activity has been demonstrated for hypericin and pseudohypericin *in vivo* and *in vitro*. It is postulated that the compounds interfere with viral infections and/or spread by direct inactivation of the virus or prevent virus shedding, budding, or assembly at the cell membrane.[11] The National Institutes of Health, which has been sponsoring human clinical trials with hypericin in HIV-positive subjects, has initiated phase I/II human clinical trials with oral doses of hypericin. The compound is also being developed as an antiretroviral agent for the transfusion blood supply. Hypericin achieves complete inactivation of more than five logs (more than 100,000 HIV particles/mL of blood) of infectious HIV. The hypericin dose used for inactivation has been shown to be safe to normal blood cells and does not interfere with standard blood tests.[12]

Hypericin is absorbed in the intestine and concentrates near the skin. Photosensitization has been described for cattle, sheep, and other livestock. Moderate doses, up to 4 g crude, 30 mL of 1:5 tincture (40% EtOH), or 240 mg of 1:5 powdered extract (calculated at 0.125% hypericin), have been reported not to cause photodermatitis in humans, however light-skinned persons should be cautioned to avoid direct sunlight after ingestion.[7]

## USES

**Medicinal, Pharmaceutical, and Cosmetic.** Hypericum extract is reportedly used in various types of skin care products, including face, body, and hand creams and lotions, night creams and lotions, skin fresheners, and skin cleansers.

In European phytomedicine, a dose of 2–4 g of herb (0.2–1.0 mg hypericin) is used for mild antidepressant action (MAO inhibitor) or nervous disturbances; externally oil is used for the treatment of wounds, abrasions, myalgias, and first-degree burns; antiphlogistic activity for topical preparations.[13]

Synthetic hypericin is currently under development as an antiretroviral agent for the transfusion blood supply.

**Food.** Hypericin-free extract used in alcoholic beverages.

**Health Food/Herb Teas.** Hypericum is used in herbal teas, tinctures, capsules, tablets, and other product forms; fresh flowers soaked in mineral or olive oil used topically (FOSTER).

**Traditional Medicine.** Fresh flowers in tea, tincture, or olive oil, reportedly a popular domestic medicine for treatment of external ulcers, wounds (especially with severed nerve tissue), sores, cuts, bruises, etc. Tea, a folk remedy for bladder ailments, depression, dysentery, diarrhea, worms; folk cancer remedy (FOSTER AND DUKE).

Modern applications of hypericum (except antiviral use) date back 2000 years.[14]

## COMMERCIAL PREPARATIONS

Crude dried or fresh herb, fresh or dried

flowers in olive oil, extracts, tinctures, synthetic hypericin, etc.

Hypericum oil, prepared by maceration of fresh flowers in sunlight for several weeks (usually in olive oil) acquires a red hue usually ascribed to hypericin or lipophilic substituted compounds with a hypericin-like color and fluorescence.[15] Hyperforin is found in freshly macerated oil but is unstable, breaking down within 30–90 days; stability can be increased to 6 months in storage by excluding air and using alternate preparation methods.[15]

*Regulatory Status*. Hypericin-free extract has been affirmed GRAS for use in alcoholic beverages (§121.1163). Herb is subject of a positive German therapeutic monograph allowed as a mild antidepressant for depression, anxiety, or nervous unrest. Oil allowed internally for dyspeptic complaints; externally for treatment of injuries, myalgia, and first degree burns.[13]

## REFERENCES

See the General References for BLUMENTHAL; DUKE 2; FOSTER; FOSTER AND DUKE; LIST AND HÖRHAMMER; MARTINDALE; NIKITAKIS; STEINMETZ; WEISS; WREN.

1. C. W. Crompton et al., *Can. J. Plant Sci.*, **68**, 149 (1988).
2. P. Maisenbacher and K.-A. Kovar, *Planta Med.*, **58**, 291 (1992).
3. G. Hahn, *J. Naturopathic Med.*, **3**(1), 94 (1992).
4. D. V. C. Awang, *Can. Pharm. J.*, **124**, 33 (1991).
5. E. V. Rao and Y. R. Prasad, *Fitoterapia*, **63**(5), 473 (1992).
6. O. Suzuki et al., *Planta Med.*, **50**, 272 (1984).
7. C. Hobbs, *HerbalGram*, **18/19**, 24 (1989).
8. S. N. Okpanyi, *Arneim. Forsch.*, **43**, 10 (1987).
9. H. Muldner and M. Zoller, *Arzneim. Forsch.*, **34**, 918 (1984).
10. Y. Özturk et al., *Phyother. Res.*, **6**, 44 (1992).
11. D. Meruelo et al., *Proc. Natl. Acad. Sci.*, **85**, 5320 (1988).
12. America On-Line Business News Wire, VIMRx Pharmaceuticals, Inc., Stamford, Conn. (Nov. 23, 1993).
13. Monograph *Hyperici herba*, *Bundesanzeiger*, no. 228 (Dec. 5, 1984); revised (Mar. 2, 1989).
14. C. Hobbs. *Pharm. Hist.*, **32**(4), 166 (1990).
15. P. Maisenbacher and K.-A. Kovar, *Planta Med.*, **58**, 351 (1992).

# HYSSOP

**Source:** ***Hyssopus officinalis*** L. (Family Labiatae or Lamiaceae).

### GENERAL DESCRIPTION

A perennial aromatic subshrub with slender herbaceous stems arising from a woody base; up to about 0.5 m high; native to southern Europe and temperate Asia; naturalized in the United States. Parts used are the leaves and flowering tops from which hyssop oil is obtained by steam distillation. Major producing countries include France, Hungary, and Holland.

### CHEMICAL COMPOSITION

Contains 0.3–2% volatile oil,[1] hyssopin (a glucoside), 5–8% tannin, flavonoid glycosides (5–6% hesperidin and 3–6% diosmin), ursolic acid, oleanolic acid, $\beta$-sitosterol, marrubiin (a bitter substance, see ***horehound***), resin, gum, and others (KARRER; LIST AND HÖRHAMMER).

The essential oil contains mainly pinocamphone, isopinocamphone, $\alpha$- and $\beta$-pinenes, camphene, and $\alpha$-terpinene, which together constitute about 70% of the oil.[2] Other constituents present include pinocampheol, cineole, linalool, terpineol, terpinyl acetate, bornyl acetate, *cis*-pinic acid, *cis*-pinonic acid, myrtenic acid, myrtenol methyl ether, *d*-2-hydroxy-isopinocamphone, methyl myrtenate, cadinene, and other unidentified compounds totalling more than 50 (KARRER, LIST AND HÖRHAMMER).[1–4]

Crude also contains 0.5% rosmarinic acid and total hydroxycinnamic derivatives at 2.2%.[5]

## PHARMACOLOGY OR BIOLOGICAL ACTIVITIES

Hyssop extracts have been reported to exhibit antiviral activities (against herpes simplex virus), which may be due to certain tannins present (see ***balm***).[6]

A report indicates hyssop oil to be nonirritating and nonsensitizing to human skin as well as nonphototoxic to mice and swine skin.[7]

Hyssop has mild antioxidant activity, due probably to its rosmarinic acid.[5]

## USES

**Medicinal, Pharmaceutical, and Cosmetic.** Hyssop oil is used as a fragrance component in soaps, creams, lotions, and perfumes, with a maximum use level of 0.4% in perfumes.[7]

**Food.** Hyssop, its extracts, and its oil are used in the formulation of bitters and liqueurs as well as in pickles and meat sauces. The extract is also used as a flavor component in candy. The oil is also used in nonalcoholic beverages, candy, baked goods, and gelatins and puddings. Highest average maximum use level reported for hyssop herb used in alcoholic beverages is 0.06%, that reported for hyssop extract (type not indicated) is 0.03% both in alcoholic beverages and in candy, and that reported for the oil is about 0.004% (36.9 ppm) in alcoholic beverages.

**Health Food/Herb Teas.** Rarely used in tea formulations (FOSTER).

**Traditional Medicine.** Reportedly used in treating sore throat, coughs, colds, breast and lung problems, digestive disorders, intestinal ailments, menstrual complaints, and others, usually in the form of a tea or gargle; externally as a diaphoretic (in baths) and in treating skin irritations, burns, bruises, frostbite.[8] It has also been used in tumors.[9]

## COMMERCIAL PREPARATIONS

Mainly crude and oil.

*Regulatory Status.* GRAS (§182.10 and §182.20). Subject of a German therapeutic monograph; not recommended since claimed efficacy is not documented; below 5% in tea mixtures allowed as a flavor corrigent.[8]

## REFERENCES

See the General References for ARCTANDER; BAILEY 2; BIANCHINI AND CORBETTA; BLUMENTHAL; FEMA; FOSTER; GUENTHER; LIST AND HÖRHAMMER; LUST; ROSE; TERRELL; UPHOF.

1. K. K. Khodzhimatov and N. Ramazanova, *Rastit. Resur.*, **11**, 238 (1975); through *Chem. Abstr.*, **83**, 84697g (1975).
2. D. Joulain, *Riv. Ital. Essenze Profumi, Piante Off,, Aromi, Saponi, Cosmet., Aerosol*, **58**, 479 (1976); through *Chem. Abstr.*, **86**, 95852s (1977).

3. D. Joulain and M. Ragault, *Riv. Ital. Essenze, Profumi, Piante Off., Aromi, Saponi, Cosmet., Aerosol*, **58**, 129 (1976); through *Chem. Abstr.*, **85**, 51615s (1976).
4. L. N. Misra et al., *Planta Med.*, **54**, 165 (1988).
5. J. L. Lamaison et al., *Fitoterapia*, **62**(2), 166, (1991).
6. E. C. Herrmann Jr. and L. S. Kucera, *Proc. Soc. Exp. Biol. Med.*, **124**, 874 (1967).
7. D. L. J. Opdyke, *Food Cosmet Toxicol.*, **16**(Suppl. 1), 783 (1978).
8. Monograph *Hyssopi herba/Hyssopi aetheroleum*, *Bundesanzeiger* (Aug. 29, 1992).
9. J. L. Hartwell, *Lloydia*, **32**, 247 (1969).

# IMMORTELLE

**Source:** ***Helichrysum angustifolium*** DC. (syn. *H. italicum* G. Don; *H. italicum* (Roth) Guss.) (Family Compositae or Asteraceae).

*Synonyms*. Helichrysum and everlasting.

## GENERAL DESCRIPTION

A strongly aromatic subshrub with much branched stems that are woody at the base; up to about 0.6 m high; native to the Mediterranean region, especially the eastern part, and growing in dry sandy soil. Parts used are the fresh flowers or flowering tops. Extracts (e.g., concrètes and subsequently absolute) are prepared from the flowers by solvent extraction, and the essential oil is prepared from the flowering tops by steam distillation.

Other *Helichrysum* species (e.g., *H. orientale* Gaertn., *H. arenarium* (L.) Moench, and *H. stoechas* (L.) DC.) are also used: *H. orientale* in the production of the essential oil and *H. stoechas* the absolute.

Major producing countries include Italy, Spain, and France.

Should not be confused with southwestern United States native *Asclepias asperula* (Dcne.) Woods, also known as immortal.

## CHEMICAL COMPOSITION

Immortelle (*H. angustifolium*) contains a small amount (0.075–0.2%) of a volatile oil, two phthalides (5-methoxy-7-hydroxyphthalide and 5,7-dimethoxyphthalide),[1] helipyrone,[2] triterpenes ($\alpha$-amyrin, uvaol, and ursolic acid) and a triterpene lactone (ursolic acid lactone),[3] wax consisting mostly of $C_{31}$ and $C_{29}$ $n$-alkanes (ca. 2:1 ratio),[4] flavonoids, $\beta$-sitosterol, caffeic acid, and others (LIST AND HÖRHAMMER).[5]

The volatile oil contains 30–50% of nerol and neryl acetate (the major components); $\alpha$- and $\beta$-pinenes, geraniol, isovaleric aldehyde, myrcene, limonene, 1,8 cineole, borneol, linalool (not present in a stem oil according to one report),[5] 4,7-dimethyl-6-octen-3-one, several $\beta$-diketones (with 3,5-dimethyloctane-4,6-dione and 2,4-dimethylheptane-3,5-dione reported to be the odor principles), sesquiterpenes, furfurol, and eugenol (LIST AND HÖRHAMMER).[5–10]

Numerous flavonoids including naringenin, apigenin, luteolin, kaempferol, helichrysin, isohelichrysin, naringenin-5-glucoside, luteolin-7-glucoside, and quercitrin have been isolated from various *Helichrysum* species.[10–13]

## PHARMACOLOGY OR BIOLOGICAL ACTIVITIES

The volatile oil of *H. italicum* flowers has been reported to exhibit antimicrobial properties *in vitro* against *Staphylococcus aureus*, *Escherichia coli*, a *Mycobacterium* species, and *Candida albicans*.[5,6] High activities were observed in oil samples containing higher concentrations of nerol, geraniol, eugenol, $\beta$-pinene, and furfural.[6]

Certain flavonoids isolated from *Helichrysum* flowers (especially quercitrin, kaempferol, naringenin, and isohelichrysin) have been reported to increase bile secretion in experimental animals. Quercitrin also increased the detoxifying function of the liver and exhibited an antiinflammatory activity.[12]

In *H. picardii* Boiss et Reuter phenolic fractions are responsible for antimicrobial activity.[13]

## USES

**Medicinal, Pharmaceutical, and Cosmetic.** The absolute is used as a fixative and fragrance component in perfumes. The essential oil is also used as a fragrance ingredient in perfumes.

Extracts have also been reported useful in before- and after-sun products making use of the UV absorption properties of the flavonoids present.[11]

**Food.** Extracts are used as flavor components (fruit types) in major food products, including alcoholic and nonalcoholic beverages, frozen dairy desserts, candy, baked goods, and gelatins and puddings. Use levels are generally below 0.003% reported for immortelle extract (type not given).

**Traditional Medicine.** Immortelle is used as an expectorant, antitussive, choleretic, diuretic, antiinflammatory, and antiallergic agent in Europe. Conditions for which it is used include chronic bronchitis, asthma, whooping cough, psoriasis, burns, rheumatism, headache, migraine, allergies, and liver ailments, among others; usually in the form of a decoction or infusion.[5]

**Others.** The absolute is used in flavoring certain tobaccos.

### COMMERCIAL PREPARATIONS

Extracts and oil.

*Regulatory Status.* GRAS (§182.20).

### REFERENCES

See the General References for ARCTANDER; BIANCHINI AND CORBETTA; FEMA; GUENTHER; KARRER; LIST AND HÖRHAMMER; POLUNIN AND SMYTHIES.

1. L. Opitz and R. Hänsel, *Arch. Pharm. (Weinheim)*, **304**, 228 (1971).
2. L. Opitz and R. Hänsel, *Tetrahedron Lett.* (38), 3369 (1970).
3. T. Mezzitti et al., *Planta Med.*, **18**, 326 (1970).
4. C. Bicchi et al., *Planta Med.*, **28**, 389 (1975).
5. P. Manitto et al., *Phytochemistry*, **11**, 2112 (1972).
6. N. N. Chirkina and E. A. Osipova, *Biol. Nauki*, **17**, 86 (1974); through *Chem. Abstr.*, **81**, 45837s (1974).
7. S. Tira et al., *Tetrahedron Lett.*, **2**, 143 (1967).
8. L. Trabaud, *Fr. Ses Parfums*, **12**, 215 (1969); through *Chem. Abstr.*, **71**, 128581y (1969).
9. L. Peyron and M. Roubaud, *Soap, Perfum. Cosmet.*, **43**, 726 (1970).
10. L. Peyron et al., *Perfum. Flavor.*, **3**(5), 25 (1978).
11. G. Prosperio, *Cosmet. Toilet.*, **91**(3), 34 (1976).
12. O. P. Prokopenko et al., *Farm. Zh. (Kiev)*, **27**, 3 (1972); through *Chem. Abstr.*, **78**, 52637t (1973).
13. F. Tomas-Lorente et al., *Fitoterapia*, **62**(6), 521 (1991).

# IPECAC

**Source:** ***Cephaelis ipecacuanha*** (Brot.) A. Rich. (syn. *Uragoga ipecacuanha* Baill.; *Psychotria ipecacuanha* Stokes) and ***C. acuminata*** Karsten (syn. *Uragoga granatensis* Baill.) (Family Rubiaceae).

*Synonyms.* Rio ipecac, Brazilian ipecac, and Brazil root (*C. ipecacuanha*); Cartagena, Nicaragua, and Panama ipecac (*C. acuminata*).

### GENERAL DESCRIPTION

Small evergreen shrubs (or herbs according to certain authors), up to about 0.5 m high. *Cephaelis ipecacuanha* is native to the tropical forests of South America (Brazil and Bolivia), and *C. acuminata* is native to tropical Central America (Nicaragua, Panama, etc.) and Colombia. *Cephaelis ipecacuanha* is also cultivated in southern Asia (e.g., India, Burma, and Malaysia). Parts used are the dried roots and rhizome collected during the dry season; roots are generally annulated. Cartagena ipecac roots are thicker than those of Rio ipecac.

## CHEMICAL COMPOSITION

Contains 1.8–4% (usually 2–3%) alkaloids that are mainly emetine and cephaeline with minor amounts of *O*-methylpsychotrine, psychotrine, emetamine, protoemetine, and others; most of the alkaloids are present in the cortex below the cork. Emetine constitutes 60–75% of the total alkaloids present in *C. ipecacuanha* and only 30–50% of the total alkaloids in *C. acuminata*, the remaining being mostly cephaeline. *Cephaelis acuminata* generally contains higher amounts of total alkaloids than *C. ipecacuanha*.

Other constituents include starch (30–40%), tarnnins (e.g., ipecacuanhin), choline, glycosides (e.g., ipecoside and saponins), resins, an allergen composed of a mixture of glycoproteins with average molecular weight of about 35,000–40,000,[1,2] and others (LIST AND HÖRHAMMER, MORTON 3, STAHL).

## PHARMACOLOGY OR BIOLOGICAL ACTIVITIES

Ipecac has emetic properties, acting both centrally and locally to cause vomiting.[3] Emetic action usually takes 15–30 min to occur. In small doses it has diaphoretic, expectorant, and stimulant properties (USD 26th).

It also has amebicidal activities that are due to its contained alkaloids. Emetine is much more active than cephaeline and less toxic, causing less nausea and vomiting than cephaeline.

Toxic effects of ipecac and its alkaloids include gastrointestinal irritation (nausea, vomiting, diarrhea, abdominal pain, etc.), dizziness, hypotension, dyspnea, and tachycardia, among others (GOODMAN AND GILMAN, USD 26th).[3]

An ipecac extract has been reported to exhibit antiinflammatory effects in Wistar rats.[4]

Severe allergenic reactions (e.g., asthmatic attacks) have been reported in people who have inhaled ipecac powder (dust); the allergen has been identified as a mixture of glycoproteins.[1]

## USES

**Medicinal, Pharmaceutical, and Cosmetic.** Ipecac Syrup U.S.P. is widely used as a domestic emetic for children, but it should not be used if the swallowed poisons are alkalis, strong acids, strychnine, petroleum distillates, and cleaning fluids.

Crude and extracts are used as components in cold and cough medicines as well as in diaphoretic, analgesic, and antipyretic preparations.

Emetine hydrochloride is used to treat amebic infections; however, the clinical use of emetine and related compounds is limited because of severe side effects.[5]

**Traditional Medicine.** Used in treating amebic dysentery and, in small doses, to stimulate the appetite. Also has been reported used in cancers (LEWIS AND ELVIN-LEWIS).[6]

## COMMERCIAL PREPARATIONS

Crude and extracts. Crude and syrup as well as emetine hydrochloride are official in U.S.P.; fluid extract was formerly official in U.S.P., and tincture was formerly official in N.F. The tincture is reported to be unstable.[7,8]

## REFERENCES

See the General References for CLAUS; GOODMAN AND GILMAN; GOSSELIN; LIST AND HÖRHAMMER; MARTINDALE; MORTON 3; NANJING; UPHOF; YOUNGKEN.

1. L. Berrens and E. Young, *Int. Arch. Allergy Appl. Immunol.*, **21**, 335 (1962).
2. L. Berrens and E. Young, *Int. Arch. Allergy Appl. Immunol.*, **22**, 51 (1963).
3. B. R. Mano and J. E. Manno, *Clin. Toxicol.*, **10**, 221 (1977).
4. M. Kroutil and J. Kroutilova, *Acta Univ. Palacki. Olomuc., Fac. Med.*, **48**, 55 (1968); through *Chem. Abstr.*, **71**, 59307z (1969).
5. M. H. Ansari and S. Ahmad, *Fitoterapia*, **62**(2), 171 (1991).
6. J. L. Hartwell, *Lloydia*, **34**, 103 (1971).
7. P. Rózsa, *Acta Pharm. Hung.*, **31**, 97 (1961); through *Chem. Abstr.*, **55**, 19138e (1961).
8. I. P. Ionescu-Stoian et al., *Farmacia (Bucharest)*, **15**, 333 (1967); through *Chem. Abstr.*, **67**, 84807n (1967).

# JASMINE

**Source: *Jasminum officinale*** L., ***J. grandiflorum*** L. (syn. *J. officinale* L. var. *grandiflorum* Bailey), and other ***Jasminum*** species (Family Oleaceae).

*Synonyms*. Royal jasmine, Italian jasmine, and Catalonian jasmine (*J. grandiflorum*); poet's jessamine and common jasmine (*J. officinale*).

## GENERAL DESCRIPTION

Evergreen or deciduous shrubs or shrubby vines with very fragrant flowers; most currently used species are native to China, India, or western Asia. *Jasminum officinale* and *J. grandiflorum* are cultivated extensively in Mediterranean countries. *Jasminum officinale* L. var. *grandiflorum* (L.) Kobuski, *J. sambac* (L.) Ait., and other *Jasminum* species are cultivated in China and India. Part used is the flower.

The concrète (see ***glossary***) is obtained from the flowers in about 3% yield, and the absolute (see ***glossary***) is obtained from the concrète in approximately 50–60% yield. The essential oil is obtained from the absolute by steam distillation.

Concrète-producing countries include Italy, France, Morocco, Egypt, China, and Japan. The major absolute producer is France.

## CHEMICAL COMPOSITION

The aroma chemicals of jasmine essence (concrète, absolute, and oil) have been reported to number more than 100.[1,2] The compound present in the highest concentration is benzyl acetate.[3–5] Other compounds include phenylacetic acid, linalool, benzyl alcohol, methyl anthranilate, methyl heptenone, farnesol, *cis*-3-hexenyl acetate, *cis*-linalool oxide, *trans*-linalool oxide, nerolidol, *cis*-jasmone, *trans*-methyl jasmonate, *cis*- and *trans*-ethyl jasmonates, jasmolactone, δ-jasmonic acid lactone, and methyl dehydrojasmonate, among others.[1,2,4,6–9]

The absolute contains about 2% acids including benzoic, myristic, palmitic, stearic, oleic, linolenic, arachidic, pelargonic, lauric, tetradecanoic, ursolic acids, and others.[10]

Among the carbonyl compounds, jasmone, jasmolactone, and methyl jasmonate predominate and have been reported to account for up to 94% of the total carbonyls present in jasmine absolute.[8]

## PHARMACOLOGY OR BIOLOGICAL ACTIVITIES

Available data indicate jasmine absolute to be nonirritating, nonsensitizing, and nonphototoxic.[5,11] However, allergenic components of the absolute have been characterized as coniferyl acetate and coniferyl benzoate.[12]

## USES

**Medicinal, Pharmaceutical, and Cosmetic.** Due to its strong power in rounding off undesirable rough notes, jasmine has been widely used in fragrance formulations. Jasmine absolute is reported used in creams, lotions, and perfumes, with maximum use level of 0.3% in perfumes.[5]

**Food.** Jasmine absolute and oil are widely used as flavor ingredients in most major food products, including alcoholic (e.g., liqueurs) and nonalcoholic beverages, frozen dairy desserts, candy, baked goods, and gelatins and puddings. The concrète is also used in all the above foods, but less extensively. Average maximum use levels are generally much lower than 0.001% (usually 1–3 ppm).

Dried jasmine flowers (*J. sambac*) are used as an ingredient of Chinese jasmine tea.

**Traditional Medicine.** The flowers and volatile oil of several *Jasminum* species are used in Western culture mainly as a calmative as well as an aphrodisiac.

In China, numerous *Jasminum* species are also used in medicine. Different parts of the plants (flowers, leaves, roots, etc.) are used and often for different purposes. Thus, flowers of *J. officinale* var. *grandiflorum* (L.) Kobuski are used mainly to treat hepatitis, pain due to liver cirrhosis, and abdominal pain due to dysentery, while flowers of *J. sambac* are used to treat conjunctivitis, skin ulcers and tumors, as well as abdominal pain due to dysentery. The root of *J. sambac* is used to treat headaches, insomnia, and pain due to dislocated joints and broken bones; it is reported to have anesthetic properties. The root and stem of *J. lanceolarium* Roxb. are also used to alleviate pain around the waist and near the joints due to rheumatism and to treat headaches and the other conditions as *J. sambac*.

Several *Jasminum* species have been used in cancers.[13]

## COMMERCIAL PREPARATIONS

Concrète, absolute, and oil.

*Regulatory Status*. GRAS (§182.20).

## REFERENCES

See the General References for ARCTANDER; BAILEY 1; BAUER; FEMA; GUENTHER; JIANGSU; LUST; MORTON 2; NANJING; ROSE; TERRELL; UPHOF.

1. R. Kaiser and D. Lamparsky, *Tetrahedron Lett.* (38), 3413 (1974).
2. E. H. Polak, *Cosmet. Perfum.*, **88**(6), 46 (1973).
3. L. Peyron and J. Acchiardi, *Riv. Ital. Essenze, Profumi, Piante Offic., Aromi, Saponi, Cosmet., Aerosol*, **58**, 2 (1976); through *Chem. Abstr.*, **84**, 169548e (1976).
4. M. S. Karawya et al., *Bull. Fac. Pharm., Cairo Univ.*, **13**, 183 (1974); through *Chem. Abstr.*, **86**, 60396u (1977).
5. D. L. J. Opdyke, *Food Cosmet. Toxicol.*, **14**, 337 (1976).
6. S. Lemberg, *Int. Congr. Essent. Oils (Pap.)*, **6**, 91 (1974).
7. M. Calvarano, *Essenze Deriv. Agrum.*, **36**, 237 (1966); through *Chem. Abstr.*, **67**, 93900q (1967).
8. B. D. Mookherjee et al., *Int. Congr. Essent. Oils (Pap.)* **6**, 150 (1974).
9. E. P. Demole in E. T. Theimer, ed., *Fragrance Chemistry—The Science of the Sense of Smell*, Academic Press, New York, 1982, p. 349.
10. J. Garnero et al., *Riv. Ital.*, **62**, 8 (1980).
11. J. P. Guillot et al., *Perfums. Cosmet., Aromes*, **18**, 61 (1977); through *Chem. Abstr.*, **88**, 145869s (1978).
12. S. Kato, *Perfum. Flavor.*, **9**(2), 137 (1984).
13. J. L. Hartwell, *Lloydia*, **33**, 288 (1970).

# JOB'S TEARS

**Source:** ***Coix lachryma-jobi*** L. and ***C. lachryma-jobi*** L. var. ***ma-yuen*** (Roman.) Stapf (syn. *C. ma-yuen* Roman.; *C. lachryma-jobi* L. var. *frumentacea* Makino) (Family Gramineae).

*Synonyms*. Coix seeds, Chinese pearl barley, pearl barley, Semen Coicis, *yokuinin*, *yiyiren*, and *yimi*.

## GENERAL DESCRIPTION

Annual or perennial herbs, 1–1.5 m high, flowering from July to September and fruiting from September to October; native to East Asia, but now distributed worldwide, especially in tropical regions; cultivated in China, India, and other Asian countries (BAILEY 1).[1]

Part used is the seed, collected when it ripens in late summer to autumn (depending on latitude), and sun dried. After removing husks, seed coats and impurities, the resulting raw kernels constitute Job's tears. Two types are commercially available in the United States: raw and roasted (fried), usually labeled as pearl barley. However, occasionally it is substituted with barley (*Hordeum* spp.), especially in certain packaged tonic tea formulas.

## CHEMICAL COMPOSITION

Seed contains 50–79% starch, 16–19% proteins, 2–7% fixed oil, lipids (5.67% glycolipids, 1.83% phospholipids, sterols, etc.), thiamine, amino acids, adenosine and trace minerals, among others.

Coixenolide (ca. 0.25%) and coixol have been isolated from the fixed oil; coixans A, B, and C (glycans with 2.4%, 33.5%, and 3.1% peptides, respectively) from the aqueous extract; and feruloyl stigmasterol and feruloyl campesterol are two of the sterols isolated.[2,3]

## PHARMACOLOGY OR BIOLOGICAL ACTIVITIES

Both alcoholic and acetone extractives of Coix seed inhibited growth of Ehrlich ascites carcinoma (ECA) in mice; acetone extractives also had marked inhibitory effects against murine uterine cervix carcinoma-14 (U-14) and hepatoma (HCA) (activity due to coixenolide) as well as antimutagenic effects against aflatoxin $B_1$ mutagenesis (WANG).[3,4]

Coixol has antiinflammatory, antihistaminic,[5] and numerous other activities, including inhibiting frog muscle contraction, tranquilizing in mice and rabbits, and lowering normal body temperature and experimental fever in rats. Its toxicity is very low: daily oral administration of 500 mg/kg to mice for 1 month did not produce any toxic reaction; a single i.p. dose of 500 mg/kg in mice only elicited temporary tranquilizing effects but no deaths and a single i.v. dose of 100 mg/kg also did not result in any fatalities or obvious abnormalities (IMM-3, JIANGSU, WANG).

Coixans A, B, and C exhibited marked hypoglycemic effects in normal and hyperglycemic mice treated with alloxan; coixan A showed the strongest activity.[3]

Alcohol extractives retarded the oxidation of linoleic acid *in vitro*.[6]

## USES

**Medicinal, Pharmaceutical, and Cosmetic.** Mainly used in Asia: fixed oil primarily in hair care products; aqueous and alcoholic extracts in facial creams (especially for acne and freckles), body lotions and bath preparations for their traditional/alleged skin softening, soothing, whitening, moisturizing and sunscreening properties (ETIC, ZHOU).

**Health Food/Herb Teas.** Powdered roasted seed and extracts of both raw and roasted seed are used in tonic, antioxidant, and pain-relief formulas in liquid (e.g., soy milk and herbal drinks), dried tea, tablet, or capsule form.

**Traditional Medicine.** Traditional Chinese medicine considers Job's tears to have spleen-invigorating, diuretic, heat-dissipating, and pus-expelling (*pai nong*) properties. Has been on Chinese medical records for more than 2000 years for treating stiff and painful joints, rheumatism, and edema; also used in chronic enteritis, diarrhea (*pi xu xie xie*), lung abscess, and acute appendicitis (*chang yong*) as well as flat wart and eczema (CHEUNG AND LI, CHP). It is often cooked and eaten with or without other herbs or meat for its nutritional qualities as well as to help ease movement of stiff joints.

**Others.** Concentrated boiling water extract has metal-cleansing (e.g., stainless steel) properties that could be utilized for producing a biodegradable natural cleansing product.

**COMMERCIAL PREPARATIONS**

Crude: both raw and roasted. Care should be taken to ensure that what is offered is not regular barley (*Hordeum* spp.).

*Regulatory Status.* It is an ethnic food; U.S. regulatory status not clear.

**REFERENCES**


See the General References for BAILEY 1; CHEUNG AND LI; CHP; IMM-3; JIANGSU; WANG; ZHOU.

1. R. K. Arora, *Econ. Bot.*, **31**, 358 (1977).
2. G. Y. Gu, *Zhongchengyao*, **12**(12), 38 (1990).
3. M. Takahashi et al., *Planta Med.*, **50**, 64 (1986).
4. C. C. Ruan et al., *Chin. J. Cancer*, **8**(1), 29 (1989).
5. H. Otsuka et al., *J. Nat. Prod.*, **51**, 74 (1988).
6. Y. L. Zhou and R. X. Xu, *Zhongguo Zhongyao Zazhi*, **17**, 368 (1992).


# JOJOBA

**Source:** ***Simmondsia chinensis*** (Link) C. Schneid. (syn. *Buxus chinensis* Link; *Simmondsia californica* Nutt.) (Family Simmondsiaceae).

*Synonyms.* Goatnut, pignut, and deernut.

**GENERAL DESCRIPTION**

Evergreen, much branched shrub, 1–2 m high; leaves opposite, entire, oblong-ovate, 2–4 cm long, dull green, subsessile, erect; flowers unisexual, small in peduncles; fruit nutlike, ovoid, leathery, three-angled, 2.5 cm long; seed contains waxy liquid; occurs in arid habitats, dry slopes, along washes, below 1500 m, often abundant; southern Arizona, southern California, Sonora, and Baja California. Now grown in different parts of the world; the largest grower in Catamarca, Argentina.

The part used is seed and its liquid wax, traded as jojoba oil, obtained by expression or solvent extraction.

**CHEMICAL COMPOSITION**

Upon expression the seeds yield about 50% (by weight) of a liquid wax composed of high molecular weight $C_{20}$ and $C_{22}$ esters of straight long-chain, monounsaturated fatty acids and alcohols (up to 85% of the oil), including monoethylenic acids composed mainly of eicosenoic acid (34%), docosenoic acid (14%); alcohols, including eicosenol (22%), docosenol (21%); trace amounts of oleic and palmitoleic acids; the seed contains a glucoside (simmondsin).[1–5]

Physical properties of jojoba oil resemble those of sperm oil; hydrogenated to produce a solid wax very similar to spermaceti. Unlike other vegetable oils, it is not a triglyceride fat.[5]

The oil is highly stable, does not oxidize, volatilize or become rancid after standing for long periods of time. Repeated heating to temperatures above 285°C for four days and exposure to high pressure does not alter its properties. Oil, isomerized at the double bonds, produces a cream (jojoba butter). The hardness of the hydrogenated oil is second only to carnauba wax.[5]

## PHARMACOLOGY OR BIOLOGICAL ACTIVITIES

Oil is considered emollient, skin, and hair conditioning.

Orally administered wax in mice has been reported to be mostly excreted; only a small amount absorbed by internal organs and epididymal fat, which diminished with time.[3]

The seed oil is considered nontoxic, though toxicity and appetite-depressant activity has been reported for simmondsin in the seed and seed meal (DUKE 3).

## USES

**Medicinal, Pharmaceutical, and Cosmetic.** Jojoba oil is used as a skin- and hair-conditioning agent and occlusive in shampoos, lipsticks, makeup products, cleansing products, face, body and hand creams and lotions, moisturizing creams and lotions (NIKITAKIS). Jojoba wax beads are used as an exfoliating agent in facial scrubs, body polishing preparations, soaps, and shower gels. Hydrogenated jojoba wax is used in lipsticks, lotions, creams, ointments, hair styling gels, and other products; provides lubricity and emolliency in creams and lotions; forms matrix with other waxes for holding pigments and oils in lipsticks; increases viscosity; as hair- and skin-conditioning agent; replaces petroleum waxes.[4]

The oil has been suggested as an antifoaming agent in penicillin and cephalosporin fermentations.[1]

**Foods.** Indian groups and early settlers ate seeds as a survival food; seeds roasted as a coffee substitute.

**Health Food/Herb Teas.** Used primarily in cosmetic products available through health and natural food outlets; crude oil topically as a skin moisturizer and emollient (ROSE).

**Traditional Medicine.** Indian groups in Mexico use the oil as a hair restorer, also a folk cancer remedy; externally applied to head sores (DUKE 3).

**Others.** The oil is sulfurized to produce a lubricating oil and a factice for use in print ink and linoleum. Sulfurized oil is stable, remains liquid, and darkens only slightly. Incompletely hydrogenated oil is used for polishes, coatings, impregnation, and carbon paper.[1,5]

## COMMERCIAL PREPARATIONS

Crude wax (as "jojoba oil"), hydrogenated wax, jojoba beads, jojoba butter (isomerized oil).

## REFERENCES

See the General References for DUKE 2; DUKE 3; HICKMAN; MARTINDALE; NIKITAKIS; TYLER 1.

1. U.S. National Research Council, *Jojoba: New Crop for Arid Lands, New Material for Industry,* National Academy Press, Washington, D.C., 1985.
2. H. S. Gentry, *Econ. Bot.*, **12**, 261 (1958).
3. A. Yaron et al., *Lipids*, **17**(3), 169 (1982).
4. R. Wilson, *Drug Cosmetic Ind.*, 43 (Nov. 1992).
5. A. Benzioni and M. Forti in G. Röbbelen et al., eds., *Oil Crops of the World*, McGraw-Hill, New York, p. 448.

# JUJUBE, COMMON

**Source:** ***Ziziphus jujuba*** Mill. (syn. *Zizyphus jujuba*; *Z. sativa* Gaertn.) and ***Z. jujuba*** Mill. var ***inermis*** (Bge.) Rehd. (syn. *Z. vulgaris* Lam. var. *inermis* Bge.) (Family Rhamnaceae).

*Synonyms*. Chinese jujube, red date, black date, jujube plum, *da zao*, *hong zao*, *hei zao* and *zao*.

## GENERAL DESCRIPTION

Deciduous shrub or small tree, up to 10 m high, with short spines, straight or curved; stems glabrous; fruit (drupe) oval, 1.5–5 cm long, deep red when ripe, sweet and edible; native to southern Europe and Asia; escaped in Gulf States; now mostly cultivated. Part used is the ripe fruit collected in autumn, rid of impurities and sun dried or first briefly boiled in water and then sun dried (CMH, JIANGSU). Produced throughout China, with major production in the northern and central provinces. Several types of jujube are produced, depending on the process used; the red type (red date) is the most commonly available in the United States.

## CHEMICAL COMPOSITION

Jujube fruit is rich in nutrients, including common ones such as vitamins (A, $B_2$, C, etc.), free amino acids, sugars, proteins, trace elements, fats, and $\beta$-carotene. Other constituents include plant acids (malic, tartaric), sterols, coumarins, flavonoids (kaempferol, myricetin), triterpenes and triterpenic glycosides (oleanolic acid; oleanonic acid; ursolic acid; maslinic acid; betulin; betulinic acid; betulonic acid; zizyphus saponins I, II, and III; jujuboside B, etc.), isoquinoline alkaloids (stepharine, asimilobine, *N*-nor-nuciferine, etc.), and cyclic AMP (adenosine-3′,5′-monophosphate) (in the largest amount ever reported in plants or animals), among others.[1–7]

## PHARMACOLOGY OR BIOLOGICAL ACTIVITIES

Decoction (per os) has been shown to increase body weight and prolong swimming time in mice as well as protect liver damage from carbon tetrachloride in rabbits (JIANGSU).

Ethanolic extract has exhibited antiinflammatory (vs. carrageenan-induced paw edema and cotton-pellet granuloma in rats) and analgesic effects; it also inhibited growth of *Bacillus subtilis*.[4]

Methanolic extract markedly inhibited water-insoluble glucan synthesis (an initial step in tooth decay) by the cariogenic bacterium *Streptococcus mutans*; the active principles were found to be oleanolic acid (see also ***ligustrum***) and ursolic acid.[6]

Common jujube has no significant toxic effects on experimental animals.[4]

## USES

**Medicinal, Pharmaceutical, and Cosmetic.** Water and hydroalcoholic extracts of common jujube are used in skin-care products for its antiinflammatory and traditional moisturizing, soothing, antiwrinkle and sunburn-relieving properties (ETIC).

**Health Food/Herb Teas.** Extracts used as an ingredient in herbal tonic formulas in capsule, tablet or liquid form; crude used in soup mixes.

**Traditional Medicine.** *Da zao* is one of the major Chinese *qi* (*chi*, energy) tonics whose recorded use dates back to the *Shan Hai Jing* (ca. 800 B.C.). It is a common food, normally eaten in candied form or in soups, often in winter, to normalize dry skin and to relieve itching. Traditionally regarded as sweet tasting and warming, invigorating vital energy (*bu qi*), promoting the secretion of body fluids (*sheng jin*), regulating body nutritional balance and defense (*tiao he ying wei*), tonifying blood and tranquilizing, and neutralizing drug toxicities. Traditionally used in treating lack of appetite, fatigue, and diarrhea due to spleen deficiency; hysteria; also more recently in treating anemia, hypertension, and purpura.

In Arab system of medicine, common jujube is used in treating fever, wounds and ulcers, inflammatory conditions, asthma, and eye diseases and as a blood purifier.[4]

### COMMERCIAL PREPARATIONS

Mainly crude, whole. Due to its high content of sugar and other nutrients, common jujube is prone to mold and insect attack if not treated and stored properly. One effective method is to store it in a wood container in an airy place mixed with black ashes of rice husks.[7]

*Regulatory Status.* It is an ethnic food; U.S. regulatory status not clear.

### REFERENCES

See the General References for BAILEY 1; CHP; CMH; FOSTER AND YUE; IMM-3; JIANGSU; LU AND LI; PETRIDES.

1. S. Z. Li and B. Zhang, *Zhongcaoyao*, **14**(10), 39 (1983).
2. N. Okamura et al., *Chem. Pharm. Bull.*, **29**, 676 (1984).
3. R. Q. Zhang and Q. Z. Yang, *Zhongcaoyao*, **23**, 609 (1992).
4. A. H. Shah et al., *Phytother. Res.*, **3**, 232 (1989).
5. K. Hanabusa et al., *Planta Med.*, **42**, 380 (1981).
6. H. Kohda et al., *Planta Med.*, **52**, 119 (1986).
7. H. P. Huang, *Zhongguo Yaoxue Zazhi*, **25**, 142 (1990).

# JUNIPER BERRIES

**Source:** ***Juniperus communis*** L. and its varieties (Family Cupressaceae).

*Synonym.* Common juniper berries.

### GENERAL DESCRIPTION

Evergreen shrub or tree; some varieties up to 6 m or more high; native to the temperate regions of the northern hemisphere. Part used is the dried, mature female cone, which is generally called "berry" because of its berrylike appearance. An essential oil is obtained by steam distillation of the crushed, dried, partially dried, or fermented berries; the essential oil produced from unfermented berries is considered to be superior in flavor qualities.

Major producing countries include Italy, Hungary, France, Austria, Czech Republic, Slovakia, Germany, Poland, Russia, and Spain. Berries collected in northern Italy, Hungary, France, Austria, and the Czech Republic are considered superior in quality than those collected in other regions.

### CHEMICAL COMPOSITION

Berries contain 0.2–3.42% (usually 1–2%) volatile oil, depending on the geographic location, altitude, degree of ripeness, and other factors;[1–4] sugars (glucose and fructose); glucuronic acid; L-ascorbic acid;[5] resin (ca.10%); catechins, proanthocyanidins;[6] fatty acids (lauric, palmitic, oleic, linoleic, etc.), sterols ($\beta$-sitosterol, campesterol, cholesterol, etc.);[7] gallotannins;[8] geijerone (a $C_{12}$ terpenoid);[9] 1,4-dimethyl-3-cyclohexen-1-yl methyl ketone;[10] diterpene acids (myrceocommunic, *cis*- and *trans*-communic, sandaracopimaric, isopimaric, torulosic acids, etc.);[11] $\beta$-elemen-7$\alpha$-ol;[12] flavonoid glycosides (rutin, isoquercitrin etc.);[13] and others (LIST AND HÖRHAMMER, MARTINDALE).

The volatile oil is composed mainly of monoterpenes (ca. 58%), which include $\alpha$-pinene, myrcene, and sabinene as the major components, with limonene, *p*-cymene, $\gamma$-terpinene, $\beta$-pinene, $\alpha$-thujene, camphene, and others also present in minor amounts;[14] small amounts of sesquiterpenes (caryophyllene, cadinene, elemene); 1,4-cineole; terpinen-4-ol; esters; and others (LIST AND HÖRHAMMER).[14–17]

## PHARMACOLOGY OR BIOLOGICAL ACTIVITIES

Juniper berry oil is generally considered to have diuretic properties; it also has gastrointestinal antiseptic and irritant properties.

The oil has an antispasmolytic effect on smooth muscles.[17]

Juniper berry oil has been reported to be generally nonsensitizing, nonphototoxic, and only slightly irritating when applied externally to human and animal skins.[18]

Phytomedicine preparations in Europe are contraindicated in pregnancy and inflammatory kidney diseases. Prolonged use or overdose may cause kidney damage.[17]

## USES

**Medicinal, Pharmaceutical, and Cosmetic.** Berries and extracts are used as components in certain diuretic and laxative preparations. Liquid and solid product forms for oral use.[17] Oil is used as a fragrance component in soaps, detergents, creams, lotions, and perfumes, with maximum use level of 0.8% reported in perfumes.

**Food.** Berries are widely used as a flavor component in gin and also in alcoholic bitters.

Extracts and oils are used in most major food categories, including alcoholic and nonalcoholic beverages, frozen dairy desserts, candy, baked goods, gelatins and puddings, and meat and meat products. Highest average maximum use level reported for the oils is 0.006% in alcoholic beverages and 0.01% for the extract in alcoholic and nonalcoholic beverages.

**Health Food/Herb Teas.** Dried fruits (whole, crushed, powdered) used as tea flavoring; also in capsules, tablets, tincture, etc., presumed for diuretic and digestive activities, and in combination with other botanicals for bladder and kidney preparations (FOSTER AND DUKE).

**Traditional Medicine.** Used as a carminative and diuretic; to treat flatulence, colic, snakebite, intestinal worms, and gastrointestinal infections; vapor (with steam) used in bronchitis. Also reportedly used in cancers.[19]

## COMMERCIAL PREPARATIONS

Crude, extracts (e.g., fluid and solid), and oil; crude and fluid extract were formerly official in N.F., and oil is official in F.C.C. Strengths (see ***glossary***) of extracts are expressed in flavor intensities or in weight-to-weight ratios.

*Regulatory Status.* GRAS (§182.20). Subject of a German therapeutic monograph; dried berries at a daily dosage of 2–10 g, calculated to 20–100 mg of essential oil, allowed for dyspeptic complaints.[17]

## REFERENCES

See the General References for APhA; ARCTANDER; BIANCHINI AND CORBETTA; BLUMENTHAL; FEMA; FOSTER AND DUKE; GOSSELIN; GUENTHER; HORTUS 3rd; KROCHMAL AND KROCHMAL; LIST AND HÖRHAMMER; LUST; MERCK; TERRELL; UPHOF.

1. M. Mihajlov and J. Tucakov, *Bull. Acad. Serbe Sci. Arts, Cl. Sci. Med.*, **44**, 19 (1969); through *Chem. Abstr.*, **72**, 103617x (1970).
2. V. M. Staicov et al., *Riv. Ital. Essenze, Profumi, Piante Offic., Oli vegetali, Saponi*, **39**, 559 (1957); through *Chem. Abstr.*, **52**, 4934a (1958).
3. M. Mihajlov, *Lek. Sirovine*, **6**, 75 (1968); through *Chem. Abstr.*, **74**, 45566h (1971).
4. H. Hoerster, *Planta Med.*, **26**, 45 (1974).
5. I. M. Panaitov, *Bulgar. Akad. Nauk, Izvest. Khim. Inst.*, **6**, 113 (1958); through *Chem. Abstr.*, **54**, 18693d (1960).
6. J. M. Schulz and K. Herrmann, *Z. Lebensm. Unters. Forsch.*, **171**, 278 (1980).
7. G. J. Guerra Hernandez et al., *Cienc. Ind. Farm.*, **7**, 8 (1988); through *Chem. Abstr.*, **109**, 3884w (1988).
8. A. Baytop and N. Tanker, *Bull. Fac. Med. Istanbul*, **23**, 113 (1960); through *Chem. Abstr.*, **54**, 25592b (1960).
9. A. F. Thomas, *Helv. Chim. Acta*, **55**, 2429 (1972).
10. A. F. Thomas, *Helv. Chim. Acta*, **56**, 1800 (1973).
11. J. de Pascual Teresa et al., *An. Quim.*, **69**(9–10), 1065 (1973); through *Chem. Abstr.*, **81**, 74847a (1974).
12. J. de Pascual Teresa et al., *An. Quim.*, **73**(3), 463 (1977); through *Chem. Abstr.*, **87**, 114688g (1977).
13. E. Lamer-Zarawska, *Pol. J. Chem.*, **54**, 213 (1980).
14. E. Klein and H. Farnow, *Dragoco Rep.*, **11**, 223 (1964).
15. J. Taskinen and L. Nykanen, *Int. Flavours Food Addit.*, **7**, 228 (1976).
16. H. Hoerster et al., *Rev. Med. (Tirgu-Mures, Rom.)*, **20**, 215 (1974); through *Chem. Abstr.*, **82**, 167565t (1975).
17. Monograph *Juniperi fructus*, *Bundesanzeiger*, no. 228 (Dec. 5, 1984).
18. D. L. J. Opdyke, *Food Cosmet. Toxicol.*, **14**, 307 (1976).
19. J. L. Hartwell, *Lloydia*, **33**, 288 (1970).

# KARAYA GUM

**Source:** ***Sterculia urens*** Roxb. (Family Sterculiaceae).

*Synonyms*. Sterculia gum, Indian tragacanth, kadaya, and mucara.

## GENERAL DESCRIPTION

Gum karaya is the dried exudation from the trunk of *Sterculia urens*, which is a soft-wooded tree with an erect trunk and broad top, up to 9 m high. It is native to India and is cultivated there for gum karaya production.

Karaya gum is collected by blazing or charring the tree trunk and removing a piece of bark or by drilling a hole into the trunk. The gum exudes from the wound and solidifies to form large tears or wormlike strips. After being collected, the tears and strips are broken up and the fragments are graded based on color and amount of adhering bark. India is currently the only producer of karaya gum.

Food-grade karaya gum is produced from the crude gum fragments by a series of physical processes whereby most of the impurities (especially bark, wood, and soil) are removed and the gum is ground, sized, and blended to yield uniform grades of gum containing no more than 3% of water-insoluble impurities. Technical grades contain larger amounts of these impurities.

Food-grade karaya gum is usually a white to pinkish gray powder with a slightly vinegary odor and taste. The better grades are white and contain less insoluble impurities than the lower grades.

Gum karaya is the least soluble of the commercial plant exudates, but it absorbs water rapidly and swells to form viscous colloidal solutions (sols) or dispersions at low concentrations (e.g., 1%). Higher concentrations (up to 4%) when hydrated in cold water will produce a viscous gel-like paste. A 1% karaya gum dispersion has an acidic pH (usually 4.5–4.7); its viscosity is comparable to those af tragacanth, guar, and locust bean gums at the same concentration. On aging, sols of karaya gum lose viscosity and tend to develop a vinegary taste. The powdered gum behaves similarly on storage, especially under hot and humid conditions. Heat (e.g., boiling), excess acid, or electrolytes can lower the viscosity of its sols; alkalis can turn the sols stringy.

Unlike other plant gums, karaya gum swells in 60% alcohol, but it is insoluble in organic solvents as the other gums.

Gum karaya is generally compatible with proteins, carbohydrates, and other plant gums.

## CHEMICAL COMPOSITION

Commercial karaya gum contains moisture (12–14%) and impurities in addition to the polysaccharide itself.

The polysaccharide of karaya gum has been reported to have a high molecular weight (9,500,000). Its structure is complicated and has not been determined. Available data have indicated that the complex polysaccharide contains at least three different types of chains. It has been postulated that one chain (constituting 50% of the total polysaccharide) contains repeating units of four galacturonic acid residues containing β-D-galactose branches and an L-rhamnose residue at the reducing end of the unit. A second chain (17% of the polysaccharide) contains an oligorhamnan having D-galacturonic acid branch residues, and interrupted occasionally by a D-galactose residue; galacturonic acid was present in 50%, rhamnose 40%, and galactose 10% by weight. A third chain (33% of the polysaccharide) contains D-glucuronic acid residues.[1]

Karaya gum polysaccharide has also been reported to be partially acetylated and has the tendency of splitting off free acetic acid on storage (FURIA, WHISTLER AND BEMILLER).[2–4]

## PHARMACOLOGY OR BIOLOGICAL ACTIVITIES

Karaya gum is not digested or absorbed by humans. It has laxative activities, due to the ability of its granules to absorb water and swell up to 100 times their original volume, forming a discontinuous type of mucilage.

Use of the gum as an aphrodisiac in Arabic tradition lacks experimental confirmation.[5]

## USES

Most of the uses of karaya gum are based on its ability to swell in cold water, its water-absorbing and water-binding properties, and its thickening and suspending powers.

**Medicinal, Pharmaceutical, and Cosmetic.** In its larger particle size (8–30 mesh), karaya gum is used extensively as a bulk laxative in laxative preparations. Its fine powder is used in dentistry as a dental adhesive and in related preparations. Also used as a thickener and suspending agent in lotions, creams, and hair-setting preparations.

**Food.** It is used extensively as a water binder to prevent water separation or formation of ice crystals in sherbets, ice pops, and cheese spreads; as a stabilizer in French dressing, meringues, whipped cream, and toppings; and as a binder in meat products (e.g., bologna). Also used in nonalcoholic beverages and candy. Highest average maximum use level is about 0.805% (8045 ppm) reported in candy.

**Health Food/Herb Teas.** Used as a filler, and ingredient in some weight loss formulations.

**Traditional Medicine.** Gum traditionally used in aphrodisiac formulations.

**Others.** Used extensively in other industries (e.g., paper and textile), as a thickener in printing inks.

## COMMERCIAL PREPARATIONS

Various grades with different particle sizes (granules to fine powders). It was formerly official in N.F. and is currently official in F.C.C.

*Regulatory Status.* Has been affirmed as GRAS (§184.1349).[6]

## REFERENCES

See the General References for ADA; APhA; CSIR X; FEMA; FURIA; GLICKSMAN; GOSSELIN; LAWRENCE; MARTINDALE; MORTON 3; TERRELL; WHISTLER AND BeMILLER.


1. W. R. Raymond and C. W. Nagel, *Carbohydr. Res.*, **30**, 293 (1973).
2. C. R. Ordonez et al., *Rev. Farm. (Buenos Aires)*, **110**, 112 (1968); through *Chem. Abstr.*, **69**, 97850n (1968).
3. G. O. Aspinall and G. R. Sanderson, *J. Chem. Soc., C* (16), 2259 (1970).
4. G. O. Aspinall and G. R. Sanderson, *J. Chem. Soc., C* (16), 2256 (1970).
5. M. Tariq et al., "Short Reports of Short Lectures and Poster Presentations," *Bonn BACANS Symposium*, $P_3$72, July 17–22, 1990, p. 199.
6. Anon., *Fed. Regist.*, **39**(185), 34209 (1974).

# KAVA

**Source:** ***Piper methysticum*** Forst. f. (Family Piperaceae).

*Synonyms*. Kava kava.

## GENERAL DESCRIPTION

Highly variable perennial deciduous shrub to 3 m high; leaves sparse, thin, single, alternate, cordate, petiolate, 8–25 cm wide (wider than long), petioles to 6 cm long; flowers in irregular spadices; rootstock is knotty, thick, sometimes tuberous, with lateral roots up to 3 m long; under cultivation rootstocks become voluminous; average weight at 10 months is 1 kg; occurs throughout the South Pacific from Hawaii to New Guinea; exact origin unclear, present distribution result of cultivation by Polynesians. The highest degree of diversification is the Vanuatu archipelago. *P. methysticum* is a cultigen of relatively recent development (2500–3000 years); the binomial refers to sterile cultivars of *P. wichmannii* C. D.C.[1]

The part used is the rootstock.

## CHEMICAL COMPOSITION

Kava rootstock contains 43.0% starch, 20.0% fiber, 12.0% water, 3.2% sugars, 3.6% proteins, 3.2% minerals, and 3–20% kavalactones (depending on plant age and cultivar).

Kavalactones contained in the resin are $\alpha$-pyrones bearing a methoxyl group at carbon 4 and an aromatic styryl moiety at carbon 6, including kavain, 7,8-dihydrokavain, 5,6-dehydrokavain, yangonin, 5,6,7,8-tetrahydroyangonin, methysticin, dihydromethysticin, 5,6-dehydromethysticin, 5,6-dihydroyangonin, 7,8-dihydroyangonin, 10-methoxy-yangonin, 11-methoxy-yangonin, 11-hydroxy-yangonin, hydroxykavain, and 11-methoxy-12-hydroxy-dehydrokavain. The rootstock also contains flavokavins A and B; the alkaloid pipermethystin; cepharadione A;[2] ketones including cinnamalaketone, and methylene dioxy-3,4-cinnamalaketone; an alcohol, dihydrokavain-5-ol; minerals including potassium, calcium, magnesium, sodium, aluminum, iron and silica; sugars include saccharose, maltose, fructose, and glucose; over 15 amino acids; and others.[1]

## PHARMACOLOGY OR BIOLOGICAL ACTIVITIES

Anodyne, anesthetic, analgesic, antimycotic, antiseptic, antispasmodic, diuretic, expectorant, sedative, stimulant, and tonic activities have been attributed to kava. Found to be sedative, anticonvulsive, and spasmolytic in animal experiments.[3]

The analgesic effect of dihydrokavain and dihydromethysticin (120 mg/kg) is comparable to 200 mg/kg of aspirin. Local anesthesia is produced in the mouth in mastication of fresh kava, especially by kavain. However, subcutaneous injections of an alcoholic extract of kavain produces anesthesia for several hours (or days) but can cause paralysis of peripheral nerves; therefore, it is an unsuitable local anesthetic drug.[1]

Dihydrokavain and dihydromethysticin have been described as muscle relaxants superior to propanediol, benzazoles, and benzodiazepines. The muscular relaxant activity is the result of a direct effect on muscular contractility rather than inhibition of neuromuscular transmission.[1]

Dihydrokavain inhibits the growth of *Aspergillus niger*.

Among the kavalactones, dihydromethysticin has the greatest potentiating effect on barbituric narcosis.[1]

Kava has been classified as a narcotic and hypnotic; however, it is neither hallucinogenic nor stupefying, is nonaddictive, and does not cause dependency.[4]

Excessive consumption may result in photophobia and diplopia or rarely temporary oculomotor paralysis. Heavy kava consumption may also result in skin lesions and drying up of the epidermis, producing advance exanthema characterized by urticarial patches with pronounced itching.[4]

## USES

**Medicinal, Pharmaceutical, and Cosmetic.** In German phytomedicine the dried rhizome and its preparations are used for conditions of nervous anxiety, stress, and unrest. Combined with pumpkin seed oil, kava extracts have been used in the treatment of irritable bladder syndrome. Use is contraindicated during pregnancy, lactation, and depression. The German monograph on the plant also notes that continuous use can cause temporary yellow discoloration of the skin, hair, and nails, or rare allergic skin reactions. May interact with alcohol, barbiturates, and other psychopharmaceuticals or interfere with operation of machinery or vehicles (WEISS).[3]

**Food.** Melanesians, Micronesians, and Polynesians grind the fresh or dry roots to prepare a traditional beverage, often imbibed in social or ceremonial settings. Its cultural role in the Pacific societies is compared with the role of wine in southern Europe.[4]

**Health Food/Herb Teas.** Dried root used in teas, capsules, tablets, tinctures, primarily as a mild relaxant or weak euphoric (DUKE 2).

**Traditional Medicine.** In Pacific islands, a decoction of the rootstock has been used for the treatment of gonorrhea, chronic cystitis, urogenital infections, menstrual problems, migraine headache, vaginal prolapse, sleeping problems, respiratory tract infections, tuberculosis; externally juice applied to skin diseases including leprosy; fresh leaves or juice poulticed for intestinal problems, otitis, and abscesses.[4]

Kava has also been reportedly valued as an anesthetic, galactagogue, diaphoretic, diuretic, and expectorant for backache, bronchitis, chills, colds, coughs, gonorrhea, myalgia, gout, rheumatism, and others (DUKE 2, STEINMETZ).

## COMMERCIAL PREPARATIONS

Crude and extracts (e.g, ethanolic).

*Regulatory Status.* Undetermined in the United States. Kava is the subject of a positive German therapeutic monograph; allowed for conditions of nervous anxiety, stress, and unrest.[3]

## REFERENCES

See the General References for BLUMENTHAL; DUKE 2; MARTINDALE; STEINMETZ; WEISS.

1. V. Lebot et al., *Kava, the Pacific Drug*, Yale University Press, New Haven, Conn. 1992.
2. H. Jaggy and H. Achenbach, *Planta Med.*, **58**, 111 (1992).
3. Monograph *Piperis methystici rhizoma*, *Bundesanzeiger*, no. 101, (June 1, 1990).
4. V. Lebot in P. A. Cox and S. A. Banack, eds., *Islands, Plants and Polynesians*, Dioscorides Press, Portland, Oreg., 1991, p. 169.

# KOLA NUT

**Source:** ***Cola acuminata*** (Beauv.) Schott et Endl. (syn. *Sterculia acuminata* Beauv.), ***C. nitida*** (Vent.) Schott et Endl., and other ***Cola*** species (Family Sterculiaceae).

*Synonyms*. Guru nut, cola nut, and cola seed.

## GENERAL DESCRIPTION

Evergreen trees with long leathery leaves; up to about 20 m high; native to western Africa. Fruit consists of four to five leathery or woody follicles (pods), each containing one to several seeds. Part used is the dried seed from which the seed coat has been removed; it is commonly called "nut" because of its hard consistency when dried, resembling a nut.

*Cola nitida* is cultivated extensively in the tropics (e.g., Jamaica, Brazil, Nigeria, Sri Lanka, and Indonesia) and is the major source of commercial kola nuts; its nuts are larger than those from *C. acuminata*.

Kola nuts are prone to infestation by insects.

## CHEMICAL COMPOSITION

Contains 1–2.5% (usually 1.5–2%) caffeine and small amounts of theobromine (up to 0.1%) as the active principles.[1]

Other constituents include *d*-catechin, *dl*-catechin, *l*-epicatechin, betaine, a red pigment, a glucoside, an enzyme, tannin, protein (ca. 6.7%), starch (ca. 34%), fats, sugar, and cellulose, and others (KARRER, LIST AND HÖRHAMMER).[2,3]

Part of the caffeine is linked to catechin and/or tannins.[4,5]

## PHARMACOLOGY OR BIOLOGICAL ACTIVITIES

The major active principle of kola is caffeine, which has central stimulant and other properties (see ***coffee***).

## USES

**Medicinal, Pharmaceutical, and Cosmetic.** Was formerly used in central stimulant preparations and in treating migraine, neuralgia, diarrhea, and others. Now rarely used.

**Food.** Extracts are widely used as a flavor ingredient in cola drinks. Other food products in which kola extracts are used include alcoholic beverages, frozen dairy desserts, candy, baked goods, and gelatins and puddings. Highest average maximum use level is about 0.045% (446 ppm) reported in frozen dairy desserts.

**Health Food/Herb Teas.** Used in diet and "energy" formulas.

**Traditional Medicine.** Used by natives as a stimulant; the fresh nuts are chewed. Also used as a tonic and astringent.

## COMMERCIAL PREPARATIONS

Crude and extracts (alcoholic and aqueous); crude and fluid extract were formerly official in N.F. Strengths (see ***glossary***) of extracts are expressed in flavor intensities.

*Regulatory Status.* GRAS; both cola nut and kola nut are listed (§182.20). Subject of a German therapeutic monograph as an analeptic for use in mental and physical fatigue.[6]

## REFERENCES

See the General References for ARCTANDER; BAILEY 1; CLAUS; LIST AND HÖRHAMMER; MARTINDALE; UPHOF.

1. J. G. Woolley and V. A. Woolley in L. W. Codd et al., eds., *Chemical Technology: An Encyclopedic Treatment*, Vol. 5, Barnes & Noble, New York, 1972, p. 707.
2. K. Freudenberg and K. Weinges, *Bull. Natl. Inst. Sci., India*, **31**, 24 (1965); through *Chem. Abstr.*, **66**, 55333d (1967).
3. R. Paris and H. Moyse-Mignon, *Ann. Pharm. Franc.*, **14**, 464 (1956); through *Chem. Abstr.*, **51**, 3090c (1957).
4. C. Maillard et al., *Planta Med.*, **51**, 515 (1985).
5. M. A. O. Oladokon, *Econ. Bot.*, **43**(1), 17 (1989).
6. Monograph *Colae semen*, *Bundesanzeiger*, no. 127 (July 12, 1991).

# KUDZU ROOT

**Source:** ***Pueraria lobata*** (Willd.) Ohwi. (syn. *P. montana* (Lour.) Merr. var. *lobata* (Willd.) Maesen et S. Almeida, *P. thunbergiana* (Sieb. et Zucc.) Benth., *P. pseudohirsuta* Tang et Wang, and *Dolichos lobatus* Willd.) and ***P. thomsonii*** Benth. (Family Leguminosae or Fabaceae).

*Synonyms*. *Yege* (*P. lobata*); *gange* (*P. thomsonii*); *fenge* (*P. thomsonii*; also occasionally, *P. lobata*); pueraria root and radix puerariae; gegen.

## GENERAL DESCRIPTION

*Pueraria lobata* is a fast-growing, high-climbing and twining hairy perennial vine, reaching about 10 m long; mature stems woody; roots tuberous, up to 60 cm long and 45 cm in diameter;[1] native to eastern Asia (e.g., China and Japan) and widely distributed throughout China (except the provinces of Xinjiang, Xizang and Qinghai); now growing worldwide; introduced into the United States a century ago and has since run wild in the Southeast (FOSTER AND YUE).

*Pueraria thomsonii* is also a vine similar to *P. lobata*; native to China; distributed mainly in southern China, including the provinces of Guangdong, Guangxi, Hainan, Yunnan, and Sichuan.[2]

Part used is the tuberous root, collected from fall to early spring. Root of *P. lobata* is mostly wildcrafted while that of *P. thomsonii* is mainly collected from cultivated plants; the former produced throughout China (especially Hunan, Henan, Guangdong, Zhejiang, and Sichuan) while the latter mainly in the south (especially Guangxi and Guangdong) (IMM-1). After washing, the outer bark is removed (normally for *P. thomsonii*), and the root is sliced or cut into cylindrical pieces before being oven or sun dried (CHP, IMM-1).

## CHEMICAL COMPOSITION

Most chemical studies on kudzu root have been performed on *P. lobata*. Hence, unless otherwise stated, the chemical information reported here is for this species.

Contains isoflavones (daidzein or 4′,7-dihydroxyisoflavone and formononetin) and isoflavone glycosides (daidzin, daidzein-4′,7-diglucoside, puerarin, puerarin-7-xyloside, pueraria glycosides (PG)-1 to 6, etc.), flavonoids, coumarins (6,7-dimethoxycoumarin and puerarol or 6-geranyl-7,4′-dihydroxycoumestan), allantoin, sterols ($\beta$-sitosterol and $\beta$-sitosterol-$\beta$-D-glucoside), 5-methylhydantoin, lupenone, glycerol-1-monotetracosanoate, arachidic acid, large amounts of starch (up to 27% in fresh root), and others (JIANGSU, JILIN).[3–8] Puerarin is its major active principle whose water solubility is greatly enhanced by certain basic amino acids (e.g., arginine, lysine and histidine).[9]

Depending on geographic location, the total isoflavones contents vary from 1.77 to 12.0% (average: 7.64%), with puerarin in highest concentration followed by daidzin, daidzein-4′,7-diglucoside, and daidzein; the amount of total isoflavones is much less in root of *P. thomsonii*, ranging from traces to 2.22% (IMM-1).[2,5]

## PHARMACOLOGY OR BIOLOGICAL ACTIVITIES

The cardiovascular effects of kudzu root extracts have been well documented in experimental animals, with the isoflavonoids being the active principles. These effects include (*1*) Dilating coronary and cerebral vessels, increasing coronary and cerebral blood flow, decreasing vascular resistance, decreasing the oxygen consumption of the myocardium, increasing the blood oxygen supply, and depressing the production of lactic acid in oxygen-deficient heart muscles.[10,11] (*2*) Puerarin exhibited protective effects on adrenaline-induced microcirculatory disturbance in mice; these effects were stronger than those of papaverine and with much lower toxicity (i.v. $LD_{50}$ 1000 mg/kg vs. 33 mg/kg of papaverine).[12] (*3*) Extracts and total flavonoids have hypotensive effects (WANG). Specifically, puerarin decreased blood pressure (by 15%), heart rate (by 19%), and plasma renin activity (by 67%) in spontaneously hypertensive rats; these effects may be due to its adrenergic-receptor blocking action.[13,14] (*4*) Puerarin antagonized the cardiac arrhythmia induced by chloroform-epinephrine in rabbits but not aconitine-induced arrhythmia in rats.[15] Total flavonoids, daidzein, and alcoholic extracts also have antiarrhythmic activities.[11] (*5*) Puerarin inhibited ADP-induced platelet aggregation and the release of 5-hydroxytryptamine (serotonin) from blood platelets induced by thrombin.[10] (*6*) Puerarin exhibited hypoglycemic effects (potentiated by aspirin) in alloxan diabetic mice; it also decreased serum cholesterol levels and reduced capillary hyperpermeability.[16]

Other bioactivities include smooth-muscle relaxation and contraction, the former due to alcohol-soluble extractives and flavonoids, the latter due to certain water-soluble extractives; antipyretic activity (hydroalcoholic extract much stronger than decoction); antioxidant activity (pueraria glycoside (PG)-1); and others (JILIN, WANG).[10,11,17,18]

In a recent study, both daidzin and daidzein in doses of 150 mg/kg/day were found to suppress the free choice of ethanol in Syrian Golden hamsters. Daidzin and daidzein may offer therapeutic choices in the treatment of alcohol abuse. This study provides a scientific basis for traditional Chinese use of the root (as well as the flowers) for treatment of patients under influence of alcohol.[19]

Toxicities of kudzu root are very low: daily oral doses of 50–100 g in humans have not produced any adverse effects;[11] $LD_{50}$ in mice is 1.6–2.1 g/kg (i.v.) for total kudzu flavonoids and 1 g/kg (i.v.) for puerarin; daidzein 0.8 g/kg administered i.p. or 1 and 5 g/kg p.o. to mice produced no toxic symptoms.[10] Puerarin has been shown to be nonmutagenic and nonteratogenic.[20]

## USES

**Food.** Kudzu root (*fenge*) is often used by Asian-Americans in soups, for which it is cut into slices and slowly cooked for hours, sometimes together with tangerine peel, meat, and other ingredients. In Hong Kong and southern China, fresh root tubers are eaten in the form of stews. The starch (kudzu flour or *gefen*) is used in pastries and puddings, and as a thickener for sauces.[1,8]

**Health Food/Herb Teas.** Extracts used in formulas for cold, flu, and allergies and their associated symptoms; in formulas for drunkenness (tuber) and hangovers (flower); also in formulas for cardiovascular problems (especially headache).[21]

**Traditional Medicine.** Plant was first mentioned in the *Shi Jing* (ca. 5th century B.C.) and medicinal use of its root (*gegen*) first described in the *Shen Nong Ben Cao Jing* (ca. 200 B.C.–A.D. 100). *Gegen* is traditionally regarded as cool natured; tastes sweet and pungent; diaphoretic and reduces fever; promotes production of body fluids; speeds up eruption of measles (*tou zhen*); stops diarrhea; and quenches thirst (CHP). Used in treating various conditions, including cold and flu and associated fever and headache, stiffness and soreness in the neck, thirst, inadequate eruption of measles, diarrhea and dysentery, drunkenness, and others (CHP, JIANGSU). More recent clinical uses include the treatment of hypertension, angina pectoris, migraine, sudden deafness, diabetes, traumatic injuries, nasal sinusitis, urticaria, psoriasis, and pruritus, among others (JIANGSU).[11,22,23]

Root and starch as well as flower are also used to treat alcohol poisoning (e.g., unconsciousness and hangover) (JIANGSU).[24]

## COMMERCIAL PREPARATIONS

Crude (in small blocks and slices), extracts, total flavonoids, and kudzu starch (*gefen* or kudzu flour).

*Regulatory Status.* Kudzu root is an ethnic food; its U.S. regulatory status is not clear.

## REFERENCES

See the General References for CHEUNG AND LI; CHP; CMH; FOSTER AND YUE; IMM-1; JIANGSU; JILIN; NATIONAL; WANG; XIAO.

1. R. D. Tanner et al., *Econ. Bot.*, **33**, 400 (1979).
2. R. Z. Feng et al., *Zhongguo Yaoxue Zazhi*, **28**, 273 (1993).
3. M. H. Chen and S. J. Zhang, *Zhongyao Tongbao*, **10**(6), 34 (1985).
4. C. X. Zuo et al., *Zhongcaoyao*, **18**(11), 10 (1987).
5. S. P. Zhao and Y. Z. Zhang, *Yaoxue Xuebao*, **20**, 203 (1985).
6. Y. Ohshima et al., *Planta Med.*, **54**, 250 (1988).
7. L. X. Xu et al., *Yaoxue Xuebao*, **22**, 208 (1987).
8. S. Y. Zee and L. H. Hui, *Hong Kong Food Plants*, The Urban Council, Hong Kong, 1981, p. 64.
9. C. Wang et al., *Zhongguo Yaoxue Zazhi*, **28**, 294 (1993).
10. Y. P. Zhou, *Chin. J. Integr. Trad. Western Med.*, **11**, 699 (1984).
11. X. L. Lai and B. Tang, *Zhongguo Zhongyao Zazhi*, **14**(5), 52 (1989).
12. X. L. Jiang and L. N. Xu, *Yaoxue Xuebao*, **24**, 251 (1989).
13. X. P. Song et al., *Zhongguo Yaoli Xuebao*, **9**, 55 (1988).
14. X. R. Lu et al., *Zhongguo Yaoli Xuebao*, **7**, 537 (1986).
15. X. S. Chai et al., *Zhongguo Yaoli Xuebao*, **6**, 166 (1985).
16. Z. F. Shen and M. Z. Xie, *Yaoxue Xuebao*, **20**, 863 (1985).
17. A. Sato et al., *Chem. Pharm. Bull.*, **40**, 721 (1992).
18. P. G. Xiao and K. J. Chen, *Phytother. Res.*, **1**(2), 53 (1987).
19. W. M. Keung and B. L. Vallee, *Proc. Natl. Acad. Sci., U.S.A.*, **90**, 10008, (1993).
20. J. L. Shi et al., *Chin. J. Pharmacol. Toxicol.*, **6**, 223 (1992).
21. S. Dharmananda, *Bestways*, 52 (Aug. 1988).
22. C. L. Zhang et al., *Yaoyong Shucai*

*(Medicinal Vegetables)*, Guangxi People's Publications, Nanning, 1985, p. 5.

23. Y. Z. Dai and Y. M. Wang, *Fujian Zhongyiyao*, **23**(3), 12 (1992).

24. S. Meng and D. Zhang, *Shi Liao Ben Cao (Diet Therapy Herbal)*, 7th Century A.D. (reprinted), People's Health Publications, Beijing, 1984, p. 6.

# LABDANUM

**Source:** ***Cistus ladanifer*** L. (syn. *C. ladaniferus* L.) and other ***Cistus*** species, including ***C. incanus*** L. and its subspecies (syn. *C. villosus* auct. vix L., incl. *C. polymorphus* Willk). (Family Cistaceae).

*Synonyms*. Ambreine, rockrose, gum cistus, ciste, and cyste.

## GENERAL DESCRIPTION

Shrubs native to the Mediterranean region. *Cistus ladanifer* is very sticky and fragrant, with white flowers and linear, lance-shaped leaves that are viscid above and densely white woolly beneath; up to 3 m high. Parts used are the leaves and twigs.

Labdanum gum is the oleoresin obtained by boiling the plant material in water; the top and bottom layers containing the oleoresin are then separated from the water later and combined. Labdanum oil is obtained from labdanum gum by distillation. The concrète is produced by hydrocarbon solvent extraction of the dried plant materials (which usually contain flowering tops). Labdanum absolute (also called cyste absolute) is produced from the concrète by the usual method (see ***glossary***). Spain is the major producer of labdanum gum, while France is the major producer of the absolute.

Material previously identified as myrrh in the Bible (Genesis 37:25) is now realized to be correctly translated as labdanum.[1]

## CHEMICAL COMPOSITION

Labdanum gum and concrète contain volatile oil, paraffins, and resins. The resins consist of acidic constituents (e.g., labdanolic acid, 6-oxo-cativic acid, dihydrocinnamic acid, anisic acid, arachic acid, and behenic acid) and neutral labdane compounds (e.g., labdane-8α, 15-diol, labdane-8α,15,19α-triol, 15-nor-8-labdanol, labd-8(17)-en-15-ol, and 15-hydroxy-7-labden-6-one) as well as viridiflorol (KARRER, LIST AND HÖRHAMMER).[2–4]

The volatile oil has been reported to contain at least 170 compounds, including α- and β-pinenes, camphene, sabinene, myrcene, α-phellandrene, α- and β-terpinenes, limonene, *p*-cymene, 1,8-cineole, borneol, nerol, linalool, geraniol, *cis*-3-hexen-1-ol, *trans*-2-hexen-1-ol, terpinen-4-ol, eugenol, 2,2,6-trimethylcyclohexanone, fenchone, α-thujone, isomenthone, acetophenone, ledol, diacetyl, benzaldehyde, *cis*- and *trans*-citral, bornyl acetate, geranyl acetate, and fatty acids (KARRER, LIST AND HÖRHAMMER).[5–8]

α-Pinene has generally been reported to be the major component in the volatile oil (LIST AND HÖRHAMMER),[6] though a study of the oil found that no single component predominated.[7] It has also been reported that the yield of oil and its composition vary considerably with cultivation conditions, seasons, growth stages, and other factors. Field-grown plants yielded twice the amount of essential oil as greenhouse plants, with alcohol content lower in the fall than in the spring (56% versus 70%), while hydrocarbons and carbonyls content (e.g., pinene, camphene, and bornyl acetate) was highest in the fall.[5–7] Comprehensive reviews of the chemistry have been published.[8,9]

## PHARMACOLOGY OR BIOLOGICAL ACTIVITIES

The essential oil and resin extracted from *C. ladanifer* have been reported to have antimicrobial activities against *Staphylococcus aureus*, *Escherichia coli*, *Candida albicans*, and other microbes, with β-pinene, eugenol, eucalyptol (cineole), and benzaldehyde being the most active components.[10]

One study has indicated the presence in the plant (*C. ladanifer*) of an alcohol-extractable but not water-extractable nonalkaloidal substance that had toxic effects (hepatic changes, etc.) on experimental animals.[11]

Both labdanum oil and cyste absolute have been reported to be nonirritating, nonsensitizing, and nonphototoxic to human skin.[12,13]

## USES

**Medicinal, Pharmaceutical, and Cosmetic.** Labdanum absolute and oil are used as a fixative and/or fragrance component in soaps, detergents, creams, lotions, and perfumes, with maximum use levels of 0.4 and 0.8% reported for the absolute and oil, respectively, in perfumes.[12,13]

**Food.** The absolute, oil, and oleoresin are all used as flavor ingredients in major food products, including alcoholic and nonalcoholic beverages, frozen dairy desserts, candy, baked goods, and gelatins and puddings. The oil is also used in meat and meat products, condiments and relishes, and sweet sauces. Average maximum use levels reported are mostly below 0.001%, with the exception of the absolute at about 0.002% (20.3 ppm) used in candy.

**Traditional Medicine.** Used as an expectorant and in catarrh of the respiratory tract, and diarrhea; also as astringent, nervine, stimulant, hemostatic; gum used as a fumigant in Turkey (STEINMETZ).

## COMMERCIAL PREPARATIONS

Oleoresin, absolute, and oil. Labdanum oil is official in F.C.C.

*Regulatory Status.* Has been approved for food use (§172.510).

## REFERENCES

See the General References for ARCTANDER; BAILEY 1; FEMA; GUENTHER; LIST AND HÖRHAMMER; POLUNIN AND SMYTHIES; STEINMETZ; TUCKER AND LAWRENCE; TUTIN 2.


1. A. O. Tucker, *Econ. Bot.*, **40**(4), 425 (1986).
2. J. De Pascual Teresa et al., *An. Quim.*, **73**, 1024 (1977); through *Chem. Abstr.*, **88**, 191134w (1978).
3. C. Tabacik and M. Bard, *Phytochemistry*, **10**, 3093 (1971).
4. J. De Pascuel Teresa et al., *Phytochemistry*, **21**, 899 (1982).
5. P. G. Gülz, *Parfüm, Kosmet.*, **56**, 344 (1975).
6. P. G. Gülz, *Int. Congr. Essent. Oils (Pap.,)* **6**, 125 (1974).
7. R. Königs and P. G. Gülz, *Z. Pflanzenphysiol.*, **72**, 237 (1974).
8. B. M. Lawrence, *Perfum. Flav.*, **6**(6), 43 (1981–1982).
9. B. M. Lawrence, *Perfum. Flav.*, **9**(1), 49 (1984).
10. N. N. Chirkina and A. V. Patudin, *Biol. Nauki*, **14**, 100 (1971); through *Chem. Abstr.*, **76**, 95138g (1972).
11. E. Ballesteros Moreno, *An. Inst. Invest. Vet., (Madrid),* **14–15**, 77 (1964–1965); through *Chem. Abstr.*, **63**, 4851c (1965).
12. D. L. J. Opdyke, *Food Cosmet. Toxicol.*, **14**, 335 (1976).
13. D. L. J. Opdyke, *Food Cosmet. Toxicol.*, **12**, 403 (1974).

# LAVENDER

**Source:** ***Lavender*** *Lavandula angustifolia* Mill. (syn. *L. spica* L.; *L. officinalis* Chaix.; *L. vera* DC.); ***Spike lavender*** *Lavandula latifolia* Medic. or Vill. (syn. *L. spica* Cav. or DC.); ***Lavandin*** Hybrid of *L. angustifolia* and *L. latifolia* (Family Labiatae or Lamiaceae).

*Synonyms*. True lavender and garden lavender (*L. angustifolia*); broad-leaved lavender and aspic (*L. latifolia*).

## GENERAL DESCRIPTION

Aromatic evergreen subshrubs with linear or lance-shaped leaves; leaves of spike lavender broader than those of true lavender; up to about 0.9 m high; native to the Mediterranean region; cultivated elsewhere. Lavandin, being a hybrid of true lavender and spike lavender, has several forms with varying degrees of resemblance to its parents. Parts used are the fresh flowering tops from which the essential oils are obtained by steam distillation and extracts (concrète and absolute) by solvent extraction (see ***glossary***). For spike lavender oil production, the sun-dried flowers are used. Taxonomy of lavender cultivars in the United States, the UK, and the Netherlands has recently been reviewed.[1,2]

France is the major producer of all three types of lavender products. Tasmania is a major producer of true lavender oil.

The flowers of *L. stoechas* L. ("Spanish lavender") also enter commerce.[3]

## CHEMICAL COMPOSITION

Lavender contains 0.5–1.5% volatile oil,[4] tannin, coumarins (coumarin, umbelliferone, and herniarin),[5] flavonoids (e.g., luteolin), triterpenoids (e.g., ursolic acid), and others (KARRER, LIST AND HÖRHAMMER).

Spike lavender contains 0.5–1% volatile oil and triterpenoids (e.g., ursolic, and oleanolic acids) (LIST AND HÖRHAMMER).

Lavandin is reported to have a higher volatile oil content than lavender and spike lavender. Due to its hybrid nature, the composition of its essential oil is much more variable than either of its parents.

Lavender oil has been reported to contain more than 100 components, including linalool, linalyl acetate, lavandulyl acetate, terpinen-1-ol-4, 1,8-cineole, camphor, $\beta$-phellandrene, terpinolene, $\alpha$-thujene, *n*-hexanal, *n*-heptanal, methyl amyl ketone, ethyl amyl ketone, perillaldehyde, perillyl alcohol, *d*-borneol, $\alpha$-terpineol, $\alpha$-pinene, limonene, lactones (4-butanolide, dihydrocoumarin, 4-methyl-4-vinyl-4-butanolide, 5-pentyl-5-pentanolide, 4,4-dimethyl-2-buten-4-olide, etc.), sesquiterpenes (caryophyllene, cadinene, etc.), fatty acids (propionic acid, isobutyric acid, caproic acid, *p*-coumaric acid, etc.), and others (LIST AND HÖRHAMMER).[4,6–15]

Spike lavender oil and lavandin oil contain many of the constituents present in lavender oil. The important components include linalool, 1,8-cineole, camphor, and linalyl acetate.

The major distinction among the three oils is in their relative contents of linalyl acetate, linalool, 1,8-cineole, and camphor. Lavender oil contains high concentrations of linalyl acetate (ca. 40%) but only traces of 1,8-cineole and camphor (ca. 1%), while spike lavender oil contains large amounts of 1,8-cineole and camphor (40–60%) with only small amounts of linalyl acetate (ca. 1%).[7,12,16] The linalool concentration is usually higher in spike lavender oil than in lavender oil.[7,10,16] The amounts of linalool, linalyl acetate, cineole, and camphor in lavandin oil are between those of the other two oils.[7,12] Presence of lower priced lower quality lavandin oils has diluted the demand and use of higher quality oils.[17] Wide variation of constituents in various cultivars has recently been determined.[18]

## PHARMACOLOGY OR BIOLOGICAL ACTIVITIES

Lavender oil has been demonstrated to exhibit CNS-depressive activities on experimental animals (e.g., mice). Such activities include anticonvulsive effects, inhibition of the spontaneous motor activity, and potentiation of the narcotic effects of chloral hydrate.[19,20] It has also been reported to have antimicrobial activities.[21]

When tested by subcutaneous administration in mice, lavender oil exhibited low toxicity.[22]

Spike lavender oil has been reported to have spasmolytic effects on smooth muscles of laboratory animals.[23]

Available data from one source indicate spike lavender oil, lavandin oil, and lavender absolute to be nonirritating and nonsensitizing to human skin, though lavender absolute has been reported elsewhere as a sensitizer. No human phototoxicity data were reported.[24–26]

Large doses of lavender oil are considered to be a narcotic poison; one source reports the oil can cause dermatitis and that more toxicological studies are needed.[3]

## USES

**Medicinal, Pharmaceutical, and Cosmetic.** Among the several lavender products, lavender oil is usually the one used in pharmaceuticals; it is used as a fragrance component in products such as antiseptic ointments, creams, lotions, and jellies, among others.

All types of lavender products (especially essential oils) are used as fragrance ingredients (some extensively) in cosmetic products, including soaps, detergents, creams, lotions, and perfumes (e.g., lavender waters and other colognes), spike lavender oil being more extensively used in soaps and detergents. Maximum use levels are 1.2, 1.0, and 0.8% reported for lavandin oil, lavender absolute, and spike lavender oil, respectively, in perfumes.[24–26]

In Europe used in phytomedicine preparations for digestive and mild nervous disorders as an antispasmodic, carminative, and mild tranquilizer.[27]

**Food.** Lavender oil and, to a lesser extent, other lavender products (e.g., lavender absolute and concrète, lavandin oil, and spike lavender oil) are used as flavor components (e.g., fruit types) in food products, including alcoholic and nonalcoholic beverages, frozen dairy desserts, candy, baked goods, gelatins, puddings, and aromatic vinegars. Average maximum use levels reported are generally below 0.002%, except for spike lavender oil whose highest maximum use level is about 0.004% in frozen dairy desserts (35.6 ppm), candy (35.2 ppm), baked goods (43.5 ppm), and gelatins and puddings (35.0 ppm).

**Health Food/Herb Teas.** Flowers and oil used as flavoring in tea formulations; oil also used in aromatherapy (FOSTER).

**Traditional Medicine.** Lavender is reportedly used as an antispasmodic, carminative, stimulant, diuretic, sedative, tonic, and stomachic. Conditions for which it is used include flatulence, spasms, colic, giddiness, nervous headache, migraine, toothache, sprains, neuralgia, rheumatism, acne, pimples, sores, nausea, vomiting, and others, usually in the form of an infusion, a decoction, or the oil, both internally and externally.

Spike lavender has been used in Europe to promote menstruation and to treat cancers.[28]

**Other.** Lavandin oil is sometimes used as a source of linalool and linalyl acetate. Lavender oil is sometimes used externally as an insect repellent (MARTINDALE).

## COMMERCIAL PREPARATIONS

Crudes (lavender and lavandin) and oils are readily available; concrète and absolute of lavender also available. All three oils are official in F.C.C.; lavender oil is also official in N.F.

*Regulatory Status.* GRAS (lavender, §182.10 and §182.20; lavandin, §182.20; spike lavender, §182.20). *L. angustifolia* flowers subject of a German therapeutic monograph, indicated for sleep disorders, restlessness, and functional abdominal pains (e.g. nervous irritation of the stomach, Roehmheld syndrome, and nervous intestinal syndromes).[27]

## REFERENCES


See the General References for ARCTANDER; BAILEY 1; BAILEY 2; BIANCHINI AND CORBETTA; BLUMENTHAL; FEMA; FOSTER; FURIA AND BELLANCA; LEWIS AND ELVIN-LEWIS; LUST; ROSE; STAHL; TERRELL; UPHOF.

1. A. O. Tucker, *Baileya*, **21**(3), 131 (1981).
2. A. O. Tucker and K. J. W. Hensen, *Baileya*, **22**(4), 168 (1985).
3. M. De Vincenzi and M. R. Dessi, *Fitoterapia*, **62**(1), 39, (1991).
4. C. de la Torre and P. Carmen, *Int. Congr. Essent. Oils (Pap.)*, **6**, 81 (1974).
5. L. G. Ianova et al., *Khim. Prir. Soedin.*, **1**, 111 (1977); through *Chem. Abstr.*, **87**, 50192f (1977).
6. I. Ognyanov and L. Panaiotova, *Riv. Ital. Essenze, Profumi, Piante Offic., Aromi, Saponi, Cosmet., Aerosol*, **55**, 560 (1973); through *Chem. Abstr.*, **80**, 124569p (1974).
7. R. Ter Heide et al., *J. Chromatogr.*, **50**, 127 (1970).
8. R. Timmer et al., *J. Agr. Food Chem.*, **23**, 53 (1975).
9. L. Peyron, *C. R. Seances Acad. Agr. Fr.*, **57**, 1368 (1971); through *Chem. Abstr.*, **77**, 24762r (1972).
10. A. I. Karetnikova et al., *Maslo Zhir. Promy.*, **35**, 23 (1969); through *Chem. Abstr.*, **71**, 33330b (1969).
11. R. Vlakhov et al., *Riechst., Aromen, Körperpflegem.*, **19**, 293 (1969).
12. A. Herisset et al., *Plant. Med. Phytother.*, **5**, 305 (1971).
13. R. Kaiser and D. Lamparsky, *Tetrahedron Lett.* (7), 665 (1977).
14. Y. R. Naves et al., *Helv. Chim. Acta*, **44**, 316 (1961).
15. B. D. Mookherjee and R. W. Trenkle, *J. Agr. Food Chem.*, **21**, 298 (1973).
16. G. Franchi, *Riv. Ital. Essenze, Profumi, Piante Offic., Aromi, Saponi, Cosmet., Aerosol*, **53**, 245 (1971); through *Chem. Abstr.*, **75**, 121280e (1971).
17. E. F. K. Denny et al., *Perfum. Flav.* **6**(5), 23 (1981).
18. A. O. Tucker et al., *Perfum. Flav.* **9**(4), 49 (1984).
19. S. Atanasova-Shopova et al., *Izv. Inst. Fiziol. Bulg. Akad. Nauk*, **15**, 149 (1973); through *Chem. Abstr.*, **81**, 58356j (1974).
20. S. Atanasova-Shopova and K. S. Rusinov, *Isv. Inst. Fiziol. Bulg. Akad. Nauk*, **13**, 69 (1970); through *Chem. Abstr.*, **74**, 123498d (1971).
21. B. N. Uzdennikov, *Nauch. Tr. Tyumen. Sel. Khoz. Inst.*, **7**, 116 (1970); through *Chem. Abstr.*, **77**, 84292x (1972).
22. G. M. Zavarzin and I. M. Chudnova, *Vop. Med. Teor., Klin. Prakt. Kurortnogo Lech.*, **4**, 300 (1971); through *Chem. Abstr.*, **79**, 101300j (1973).
23. T. Shipochliev, *Vet. Med. Nauki*, **5**, 63

(1968); through *Chem. Abstr.*, **70**, 86144e (1969).

24. D. L. J. Opdyke, *Food Cosmet. Toxicol.*, **14**, 453 (1976).
25. D. L. J. Opdyke, *Food Cosmet. Toxicol.*, **14**, 449 (1976).
26. D. L. J. Opdyke, *Food Cosmet. Toxicol.*, **14**, 447 (1976).
27. Monograph *Lavandulae flos*, *Bundesanzeiger*, no. 228 (Dec. 5, 1984); corrected (Mar. 13, 1990).
28. J. L. Hartwell, *Lloydia*, **32**, 247 (1969).

# LEMON OIL (AND LEMON PETITGRAIN OIL)

**Source:** ***Citrus limon*** (L.) Burm. f. (syn. *C. limonum* Risso) (Family Rutaceae).

*Synonyms*. Expressed lemon oil; cedro oil (terpeneless).

## GENERAL DESCRIPTION

A small evergreen tree with very fragrant flowers and stiff thorns; up to about 6 m high; native to Asia; now cultivated worldwide, especially in the United States (e.g., California and Florida), Italy, Cyprus, and Guinea. Parts used are the peel as well as the leaves and twigs together with undeveloped fruits. Lemon oil is obtained from the peel by cold expression, while lemon petitgrain oil is produced from the leaves and twigs, sometimes including undeveloped small fruits by steam distillation. Major lemon oil producers include the United States, Italy, Guinea, and Cyprus. Major producers of lemon petitgrain oil include Guinea and Italy.

## CHEMICAL COMPOSITION

Lemon oil contains about 90% monoterpene hydrocarbons, composed mainly of limonene (ca. 70%), with lesser amounts of $\gamma$-terpinene, $\beta$-pinene, sabinene, $\alpha$-pinene, and myrcene;[1] 2–6% aldehydes (mainly citral); alcohols (linalool, octanol, nonanol, decanol, terpinen-4-ol, $\alpha$-terpineol, geraniol, etc.);[2] small amounts of sesquiterpenes (bisabolene, $\alpha$-bergamotene, and caryophyllene);[3] waxes; and 0.41–0.87% coumarins, consisting primarily of bergamottin, 7-methoxy-5-geranoxy coumarin, and limettin (citropten), with imperatorin, isoimperatorin, phellopterin, 8-geranoxypsoralen, and others also present.[4–7]

A study to examine storing comminuted lemon before oil production showed that $\alpha$-terpineol rose from 0.21 to 10% and that the chemical and organoleptic qualities of the oil changed completely.[8]

Some components in the waxes have been reported to have antioxidant properties.[6]

Lemon petitgrain oil contains large amounts of citral (up to 50%); other components present include limonene, $\alpha$-pinene, linalool, and nerol.[9]

## PHARMACOLOGY OR BIOLOGICAL ACTIVITIES

Lemon oil has been reported to promote tumor formation on the skin of mice by the primary carcinogen 9,10-dimethyl-1,2-benzanthracene.[10] It has also exhibited antimicrobial activities.[11–13]

Certain coumarin derivatives are known to be phototoxic and allergenic to humans as well as effective in treating psoriasis (see ***bergamot oil***). Lemon oil has been reported to have phototoxic effects, most likely due to its coumarins.[13]

Available data from one source indicate lemon oil to be nonirritating and nonsensitizing to human skin, though some samples have been demonstrated to be irritating to the backs of hairless mice.[13]

Lemon petitgrain oil has been reported to be nonirritating, nonsensitizing, and nonphototoxic to human skin.[9]

The flavonoid diosmin extracted from lemon is used clinically for treatment of venous insufficiency. Increased vascular tone has been observed after *in vivo* oral administration.[14] Antiinflammatory, anti histamine, and diuretic activity are also reported (WREN).

## USES

**Medicinal, Pharmaceutical, and Cosmetic.** Lemon oil is used in pharmaceuticals mainly as a flavoring agent.

Lemon oil is used as a fragrance ingredient in soaps, detergents, creams, lotions, and perfumes (e.g., colognes). Lemon petitgrain oil is used in creams, lotions, and perfumes. Maximum use levels reported are 1.0 and 0.3% for lemon oil and lemon petitgrain oil, respectively, in perfumes.[9,13]

**Food.** Lemon oil and terpeneless lemon oil are extensively used as flavor ingredients in most food products, including alcoholic (bitters, vermouths, sweet liqueurs, etc.) and nonalcoholic beverages (soft drinks, drink mixes, etc.), frozen dairy desserts, candy, baked goods, gelatins and puddings, meat and meat products, breakfast cereals, and fats and oils, among others. Lemon extract and lemon petitgrain oil are also used in many of above food categories but to a much lesser extent. Highest average maximum used levels are 0.968 and 1.208% reported for the extract (type not specified) in candy and baked goods, respectively. Average maximum use levels for the other ingredients are much lower, with the highest reported being about 0.046% (457 ppm) for lemon oil in candy.

**Health Food/Herb Teas.** Essential oil used as a flavoring ingredient in herb tea formulations.

## COMMERCIAL PREPARATION

Lemon oil (California, Italian, etc.), lemon petitgrain oil, and lemon peel extracts. Lemon oil is official in N.F. and F.C.C.

*Regulatory Status.* GRAS; listed under lemon, lemon peel, and petitgrain lemon (§182.20).

## REFERENCES

See the General References for ARCTANDER; BAILEY 1; CLAUS; FEMA; FURIA AND BELLANCA; GUENTHER; JIANGSU; TERRELL; WREN.

1. L. Pennisi and A. Di Giacomo, *Riv. Ital. Essenze Profumi, Piante Offic., Aromi, Saponi, Cosmet. Aerosol*, **47**, 370 (1965); through *Chem. Abstr.*, **67**, 36342k (1967).
2. G. L. K. Hunter and M. G. Moshonas, *J. Food Sci.*, **31**, 167 (1966).
3. G. Rispoli and A. Di Giocomo, *Riv. Ital. Essenze Profumi Piante Offic., Aromi, Saponi Cosmet.*, **47**, 650 (1965); through *Chem. Abstr.*, **64**, 17350f (1966).
4. G. Calabro and P. Curro, *Essenze Deriv. Agrum.*, **46**, 215 (1976); through *Chem. Abstr.*, **86**, 138043y (1977).
5. W. L. Stanley and L. Jurd, *J. Agr. Food Chem.*, **19**, 1106 (1971).
6. A. Di Giacomo et al., *Essenze Deriv. Agrum.*, **40**, 143 (1970); through *Chem. Abstr.*, **74**, 34543x (1971).
7. B. M. Lawrence, *Perfum. Flav.*, **14**(4), 41 (1989).
8. I. Calvarana and G. Di Giacomo, *Essenze Deriv. Agrum.*, **54**, 200 (1984).
9. D. L. J. Opdyke, *Food Cosmet. Toxicol.*, **16**(Suppl. 1), 807 (1978).

10. F. J. C. Rose and W. E. H. Field, *Food Cosmet. Toxicol.*, **3**, 311 (1965).
11. M. S. Subba et al., *J. Food Sci.*, **32**, 225 (1967).
12. A. Poretta and A. Casolari, *Ind. Conserve (Parma)*, **41**, 287 (1966); through *Chem. Abstr.*, **66**, 84879s (1967).
13. D. L. J. Opdyke, *Food Cosmet. Toxicol.*, **12**, 725 (1974).
14. A. Codignola et al., *Planta Med.*, **58**(S1), A628, (1992).

# LEMONGRASS

**Source:** ***Cymbopogon citratus*** (DC.) Stapf (syn. *Andropogon citratus* DC.) and ***C. flexuosus*** (Nees ex Steud.) W. Wats. (syn. *A. flexuosus* Nees) (Family Gramineae or Poaceae).

*Synonyms.* West Indian lemongrass, Madagascar lemongrass, and Guatemala lemongrass (*C. citratus*); East Indian lemongrass, Cochin lemongrass, native lemongrass, and British Indian lemongrass (*C. flexuosus*).

## GENERAL DESCRIPTION

Perennial grasses both native to tropical Asia. East Indian lemongrass (*C. flexuosus*) is cultivated mainly in western India and nearby countries, while West Indian lemongrass (*C. citratus*) is cultivated in the tropics worldwide (West Indies, Central and South America, Africa, and tropical Asia). Parts used are the freshly cut and partially dried leaves of cultivated plants from which the essential oils are obtained by steam distillation. India is the major producer of East Indian lemongrass oil, while major producers of the West Indian oil include Guatemala, Madagascar, the Comoro Islands, Brazil, Malayasia, and Vietnam.

## CHEMICAL COMPOSITION

West Indian lemongrass (*C. citratus*) contains a volatile oil (usually 0.2–0.4% yield from fresh grass); an unknown alkaloid; a saponin; β-sitosterol; hexacosanol and triacontanol;[1] cymbopogonol (a triterpenoid);[2] and others (KARRER, LIST AND HÖRHAMMER).

West Indian lemongrass oil contains citral (65–85%) as its major component.[3] Other compounds present include myrcene (12–20%), dipentene, methylheptenone, β-dihydropseudoionone, alcohols (linalool, methylheptenol, α-terpineol, geraniol, nerol, farnesol, citronellol, etc.), volatile acids (isovaleric, geranic, caprylic, citronellic, etc.), and others (GUENTHER, JIANGSU, LIST AND HÖRHAMMER).

East Indian lemongrass contains a volatile oil in about 0.5% yield from fresh grass. It probably also contains many of the other constituents present in West Indian lemongrass.

East Indian lemongrass oil contains normally citral as its major component in a 70–85% concentration.[4] Other components include geraniol and methyleugenol as well as many of the compounds present in the West Indian oil (e.g., dipentene, myrcene, methylheptenol, farnesol, *n*-decanal, guanic acid, and others).

Geraniol-rich strains of East Indian lemongrass have been reported to yield oils that contain citral only as a minor component (10–20%), with their major components being gerianol (35–50%) and methyl eugenol (ca. 20%).[5,6] Another type is reported to contain no citral at all but has borneol (ca. 30%) as one of its major components (LIST AND HÖRHAMMER).

East Indian lemongrass oil usually contains a slightly higher content of citral than West Indian lemongrass oil; it is also more soluble in 70% alcohol than the West Indian oil (GUENTHER).[7]

## PHARMACOLOGY OR BIOLOGICAL ACTIVITIES

West Indian lemongrass oil has antimicrobial properties, especially against Gram-positive bacteria and fungi.[3,8,9] It has been reported to have CNS-depressant effects as well as analgesic, antipyretic, and antioxidant properties.[8,10] Available data from one source indicate it to be mildly to moderately irritating to the skin of experimental animals but nonirritating and nonsensitizing to human skin. Its phototoxicity on human skin has not been determined.[11]

East Indian lemongrass oil has been reported to exhibit antifungal properties.[12,13] Available data from one source indicate it to be mildly to moderately irritating to the skin of laboratory animals but nonirritating and nonsensitizing to human skin. Its phototoxicity on human skin is not known.[14]

Citral has been reported to produce sensitization reactions in humans when applied alone but to produce no such reactions when applied as a mixture with other compounds.[15]

A recent study found that lemongrass oil has no adverse effects on the blood, liver function, kidney function, protein, carbohydrate, and lipid metabolism of rats;[16] and other studies have failed to detect mutagenic, or toxicological reactions in humans.[17–19]

## USES

**Medicinal, Pharmaceutical, and Cosmetic.** Lemongrass oil (especially the West Indian type) is used extensively as a fragrance component in soaps and detergents. Also used in creams, lotions, and perfumes, with maximum use level of 0.7% reported for both types of oil in perfumes.[11,14]

**Food.** Lemongrass oil is used in most major categories of foods, including alcoholic and nonalcoholic beverages, frozen dairy desserts, candy, baked goods, gelatins and puddings, meat and meat products, and fats and oils. Highest average maximum use levels reported are about 0.003 and 0.004%, respectively in candy (33.3 ppm) and baked goods (36.3 ppm).

**Health Food/Herb Teas.** Dried leaves widely used as a "lemon" flavor ingredient in herb teas, and other formulations.

**Traditional Medicine.** West Indian lemongrass is used in Chinese medicine to treat colds, headache, stomachache, abdominal pain, rheumatic pain, and others (JIANGSU).

**Others.** Lemongrass oils (both types) are used as starting materials for the synthesis of ionones and vitamin A as well as the production (isolation) of natural citral. The oil possesses biological activity against storage pests and has been used as a postharvest pesticide for some food commodities.[16]

## COMMERCIAL PREPARATIONS

Both East and West Indian lemongrass oils; they are official in F.C.C. Also crude.

*Regulatory Status.* GRAS (§182.20). Subject of a German therapeutic monograph as a mild astringent and stomachic; efficacy not documented.[20]

## REFERENCES

See the General References for ARCTANDER; BAILEY 1; FEMA; GUENTHER; LIST AND HÖRHAMMER; MARTINDALE.

1. A. A.. Olaniyi et al., *Planta Med.*, **28**, 186 (1975).
2. S. W. Hanson et al., *Phytochemistry*, **15**, 1074 (1976).
3. C. K. Kokate and K. C. Varma, *Sci. Cult.*, **37**, 196 (1971).
4. K. S. Ayyar et al., *Perfum. Essent. Oil Records*, **59**, 669 (1968).
5. R. K. Thappa et al., *Cultiv. Util. Med. Aromat. Plants*, 227 (1977); through *Chem. Abstr.*, **88**, 78952y (1978).
6. C. K. Atal and B. L. Bradu, *Indian J. Pharm.*, **38**, 63 (1976).
7. W. D. Fordham in L. W. Codd et al., Eds., *Chemical Technology: An Encyclopedic Treatment*, Vol. 5, Barnes & Noble, New York, 1972, p. 1.
8. D. O. Gyane, *Drug Cosmet. Ind.*, **118**(5), 36 (1976).
9. F. M. Ramadan et al., *Chem. Mikrobiol. Technol. Lebensm.*, **1**, 96 (1972).
10. G. Seth et al., *Indian J. Exp. Biol.*, **14**, 370 (1976).
11. D. L. J. Opdyke, *Food Cosmet. Toxicol.*, **14**, 457 (1976).
12. B. G. V. N. Rao and P. L. Joseph, *Riechst., Aromen, Körperpflegem.*, **21**, 405 (1971).
13. A. Dikshit and A. Husain, *Fitoterapia*, **55**, 171 (1984).
14. D. L. J. Opdyke, *Food Cosmet. Toxicol.*, **14**, 455 (1976).
15. D. L. J. Opdyke, *Food Cosmet. Toxicol.*, **14**, 197 (1976).
16. A. K. Mishra et al., *Phytother. Res.*, **6**, 279 (1992).
17. E. A. Carlinin et al., *J. Ethnopharmacol.*, **17**, 37 (1986).
18. J. R. Leite et al., *J. Ethnopharmacol.*, **17**, 75 (1986).
19. M. L. Souza et al., *J. Ethnopharmacol.*, **17**, 65 (1986).
20. Monograph *Cymbopogon species*, *Bundesanzeiger*, no. 22 (Feb. 1, 1990).

# LICORICE ROOT

**Source:** ***Spanish licorice*** *Glycyrrhiza glabra* L. var. *typica* Reg. et Herd.; ***Persian licorice*** *Glycyrrhiza glabra* L. var. *violacea* Boiss.; ***Russian licorice*** *Glycyrrhiza glabra* L. var. *glandulifera* Waldst. et Kit.; ***Chinese licorice*** *Glycyrrhiza uralensis* Fisch. (Family Leguminosae or Fabaceae).

*Synonyms*. Italian licorice (*G. glabra* var. *typica*), Turkish licorice (*G. glabra* var. *violacea*), glycyrrhiza, and sweet wood.

## GENERAL DESCRIPTION

Perennial herbs or subshrubs generally with horizontal underground stems (stolons or runners); up to 1 to 2 m high, with Chinese licorice being the smallest plant among the four; native to Eurasia and cultivated in Europe (Spain, Italy, France, etc.), Middle East (Syria, Iran, Turkey, Iraq, etc.), and Asia (e.g., China). Parts used are the dried runners and roots collected in the fall; some types of licorice are peeled. Much of the licorice used in the United States is imported in an extract form, usually in sticks or solid blocks. These extracts are usually prepared by hot water extraction followed by evaporating off the water and drying the extracts to form sticks or blocks.

A commonly used form of licorice extract is a concentrated extract containing large amounts of glycyrrhizin in its ammonium salt form; it is not pure ammonium glycyrrhizin, though the pure form is also used. Glycyrrhizin is 50 times sweeter than sucrose and has synergistic effects with sucrose and other sugars; its sweet taste is lost in an acidic medium.

Major producers of licorice include Spain, Iraq, Iran, Turkey, Russia, and China.

## CHEMICAL COMPOSITION

Licorice is one of the most extensively investigated economic plant products. It contains as its major active principle the triterpene glycoside glycyrrhizin (also known as glycyrrhizic or glycyrrhizinic acid) in concentrations ranging from 1 to 24%, depending on sources and methods of assay; a 10-fold difference in glycyrrhizin values due solely to different assay methods has been reported.[1] Glycyrrhizin on hydrolysis yields glycyrrhetinic (or glycyrrhetic) acid and two molecules of glucuronic acid.

Other constituents of licorice include flavonoids and isoflavonoids (licoflavonol, kumatakenin, licoricone, glabrol, glabrone, glyzarin, licoisoflavones A and B, licoisoflavanone, glycyrol, formononetin, liquiritigenin, liquiritin, neoliquiritin, rhamnoliquiritin, glyzaglabrin, 7-hydroxy-2-methylisoflavone, 4′,7-dihydroxyflavone, glabranine, etc.),[2–13] chalcones (isoliquiritigenin, isoliquiritin, neoisoliquiritin, licuraside, rhamnoisoliquiritin, echinatin, licochalcones A and B, 4-hydroxychalcone, etc.),[3,6,12] coumarins (umbelliferone, herniarin, liqcoumarin, glycyrin, etc.),[2,14] triterpenoids (liquiritic acid, glycyrrhetol, glabrolide, isoglabrolide, licoric acid, β-amyrin, 18-β-glycyrrhetinic acid, etc.),[15] sterols (β-sitosterol, stigmasterol, 22,23-dihydrostigmasterol etc.),[15] 2–20% starch, 3–14% sugars (glucose and sucrose), lignin, amino acids (proline, serine, aspartic acid, etc.),[16] amines (asparagine, betaine, choline), gums, wax, a volatile oil consisting of many aroma chemicals (including acetol, 2-acetylfuran, propionic acid, 2-acetylpyrrole, furfuryl alcohol, benzaldehyde, pentanol, hexanol, *trans*-hex-3-en-1-ol, oct-1-en-3-ol, linalool, linalyl oxide, α-terpineol, butyrolactone, thujone, and fenchone, among others, none of which alone can account for the licorice flavor),[17–19] and others (JIANGSU, LIST AND HÖRHAMMER).

Glycyrrhizin has been found to concentrate in the woody parts of the thickened root and stolon of *G. glabra*. Betulinic acid and soyasaponins have been previously reported from *G. glabra* cell cultures. Recently, soyasaponins (oleanane-type triterpene glycosides) have been reported in all parts of the intact plant, especially the seeds and young roots, particularly rootlets.[20]

Two new prenylflavones, licoflavones B and C, and a new dibenzoylmathane, glycyridione, have recently been isolated from *G. inflata*, along with the known flavones licoflavone A and 4′,7-dihydroxyflavone.[21]

The number of constituents and their relative concentration vary with the sources and types of licorice (JIANGSU, LIST AND HÖRHAMMER, MORTON 3, NANJING).

## PHARMACOLOGY OR BIOLOGICAL ACTIVITIES

Licorice is known to exhibit many pharmacological activities, including estrogenic in laboratory animals,[15,22–27] antiulcer,[28–31] mineralocorticoid with sodium retention and potassium loss leading to hypertension (due to glycyrrhizin),[22,32,33] inhibition of tumor growth (sarcoma 45 and Ehrlich ascites cells, due to a glycyrrhetinic acid salt and a derivative)[34] antitrichomonas,[35] antiinflammatory, antiallergic,[36] antitoxic, antitussive (comparable to codeine, due to a derivative of 18-β-glycyrrhetinic acid), anticonvulsive,[37] and antibacterial, among others (FARNSWORTH, JIANGSU, LIST AND HÖRHAMMER).

Liquiritigenin and isoliquiritigenin have MAO-inhibitory activities.[38]

Recent work has found that glycyrrhetinic acid inhibits 11-β-hydroxysteroid dehydrogenase in rats, and potentiates the action of hydrocortisone in humans.[39]

In screening for novel anxiolytics using a benzodiazepine binding assay, a dichloromethane extract of *G. uralensis* was found to stimulate the binding of benzodiazepine to rat synaptosomes.[40]

Licochalcones A and B from *G. inflata* strongly inhibited lysosomal enzyme release and degranulation from human polymorphonuclear neutrophils.[41]

Licorice extracts have been used in China in the clinical treatment of numerous illnesses (gastric and duodenal ulcers, bronchial asthma, infectious hepatitis, malaria, diabetes insipidus, contact dermatitis, etc.) with considerable success (JIANGSU).

The flavonoids have recently been shown to have strong antioxidant and antihepatotoxic activities.[42,43]

## USES

**Medicinal, Pharmaceutical, and Cosmetic.** Licorice extracts are used extensively as ingredients in cough drops and syrups, tonics, laxatives, antismoking lozenges (see ***lobelia***), and other preparations. They are also used as flavoring agents to mask bitter, nauseous, or other undesirable tastes in certain medicines (e.g., cascara, ammonium chloride, and quinine preparations).

Average daily doses of 5–15 g root (calculated to 200–600 mg of glycyrrhizin) or root juice (0.5–1 g for respiratory tract catarrhs, or 1.5–3 g for gastric duodenal ulcers) are used in European phytomedicine. Duration is limited to 4–6 weeks, because of potential adverse side effects (described above). Use is contraindicated for cholestatic liver disorders, cirrhosis, hypertonia, pregnancy, and others.[37] Known drug interactions include potassium loss due to thiazine diuretics, as well as increased sensitivity to digitalis glycosides.[37]

**Food.** Licorice is widely used in flavoring foods.[44,45]

The most well-known use of licorice and its extracts as well as ammoniated glycyrrhizin is in licorice candy where they are mixed with anise oil (see ***anise***), with average maximum use levels of about 3.279% (32,792 ppm) and 0.151% (1,512 ppm) reported for the powdered extract and ammoniated glycyrrhizin respectively. Licorice, its extracts, and ammoniated glycyrrhizin are also used in many other food products, including alcoholic (certain kinds of beer) and nonalcoholic (e.g., root beer) beverages, frozen dairy desserts, baked goods, gelatins and puddings, and meat and meat products. Average maximum use levels reported are below 0.25% for licorice and licorice extracts; the use levels reported for ammoniated glycyrrhizin are usually below 0.01%.

**Health Food/Herb Teas.** Root, powdered or cut and sifted, is widely used as tea ingredient; and in capsules, tablets, tinctures, and other dietary supplement formulations for flavoring and traditional indications; extracts also used in capsules, tablets, and drinks.

**Traditional Medicine.** The medicinal use of licorice in both Western and Eastern cultures dates back several thousand years.[46] Many of its uses are the same in both cultures. It is commonly used as a demulcent, expectorant, antitussive, and mild laxative.

Licorice is one of the most popular drugs in Chinese medicine and is often an ingredient in Chinese prescriptions. Its traditional uses include treating ulcers (gastric and duodenal), sore throat, malaria, abdominal pain, insomnia, tuberculosis (consumption), sores, abscesses, and food poisoning, among others. For food poisoning and poisoning due to swallowing of poison of undetermined nature, equal amounts of licorice and black beans or mung beans are decocted together and the liquid taken orally.

Licorice has been reported used in many countries to treat cancer.[47]

**Others.** Licorice extracts are used in flavoring tobaccos.

## COMMERCIAL PREPARATIONS

Crude, extracts, and ammoniated glycyrrhizin. Licorice syrup was formerly official in U.S.P. Licorice, licorice extract (solid), and licorice fluid extract are official in N.F. The strength (see ***glossary***) of the fluid extract is 1:1, but that of the solid extract is not precisely specified in N.F.

*Regulatory Status*. GRAS (§182.10 and §182.20) with both licorice and glycyrrhiza listed; ammoniated glycyrrhizin is also GRAS (§182.20). Licorice root is the subject of a German therapeutic monograph, indicated for catarrhs of the upper respiratory tract and gastric/duodenal ulcers.[37]

## REFERENCES

See the General References for ARCTANDER; BAILEY 1; BLUMENTHAL; FEMA; FOSTER AND YUE; JIANGSU; MARTINDALE; MORTON 3; NANJING.


1. J. Killacky et al., *Planta Med.*, **30**, 310 (1976).
2. D. K. Bhardwaj et al., *Phytochemistry*, **16**, 402 (1977).
3. M. Hoton-Dorge, *J. Pharm. Belg.*, **29**, 560 (1974).
4. T. Saitoh et al., *Chem. Pharm. Bull.*, **24**, 752 (1976).
5. T. Kinoshita et al., *Chem. Pharm. Bull.*, **26**, 141 (1978).
6. T. Saitoh et al., *Chem. Pharm. Bull.*, **26**, 144 (1978).
7. D. K. Bhardwaj and R. Singh, *Curr. Sci.*, **46**, 753 (1977).
8. D. K. Bhardwaj et al., *Phytochemistry*, **15**, 352 (1976).
9. N. S. Kattaev and G. K. Nikonov, *Khim. Prir. Soedin.*, **6**, 805 (1972); through *Chem. Abstr.*, **78**, 94806c (1973).
10. T. Saitoh et al., *Chem. Pharm. Bull.*, **24**, 1242 (1976).
11. T. Kinoshita et al., *Chem. Pharm. Bull.*, **24**, 991 (1976).
12. T. Saitoh and S. Shibata, *Tetrahedron Lett.* (50), 4461 (1975).
13. C. Van Hulle, *Pharm. Tijdschr. Belg.*, **45**, 137 (1968); through *Chem. Abstr.*, **70**, 106822e (1969).
14. T. Kinoshita et al., *Chem. Pharm. Bull.*, **26**, 135 (1978).
15. C. Van Hulle, *Pharmazie*, **25**, 620 (1970).
16. H. Nishi and I. Morishita, *Nippon Nogei Kagaku Kaishi*, **45**, 507 (1971); through *Chem. Abstr.*, **76**, 151016e (1972).
17. B. Toulemonde et al., *Ind. Aliment. Agr.*, **94**, 1179 (1977); through *Chem. Abstr.*, **88**, 150841f (1978).
18. C. Frattini et al., *Chim. Ind. (Milan)*, **59**, 522 (1977); through *Chem. Abstr.*, **88**, 4974k (1978).
19. C. Frattini et al., *J. Agr. Food Chem.*, **25**, 1238 (1977).
20. H. Hayashi et al., *Planta Med.*, **59**, 351 (1993).
21. K. Kajiyama et al., *J. Nat. Prod.*, **55**(9), 1197 (1992).
22. K. Y. Yen, *Pei I Hsueh Pao*, **5**, 23 (1973); through *Chem. Abstr.*, **81**, 85799n (1974).
23. C. H. Costello and E. V. Lynn, *J. Am. Pharm. Assoc.*, **39**, 177 (1950).
24. I. A. Murav'ev and N. F. Kononikhina, *Rast. Resur.*, **8**, 490 (1972); through *Chem. Abstr.*, **78**, 75811j (1973).
25. A. Sharaf and N. Goma, *J. Endocrinol.*, **31**, 289 (1965).
26. M. I. Elghamry et al., *Zentr. Veterinaermed.*, **11**, 70 (1964); through *Chem. Abstr.*, **61**, 1118f (1964).

27. I. M. Shihata and M. I. Elghamry, *Zentr. Veterinaermed. Ser. A.*, **10**, 155 (1963); through *Chem. Abstr.*, **59**, 7815h (1963).
28. Y. T. Wang et al., *Tai-Wan I Hsueh Hui Tsa Chih*, **71**, 256 (1972); through *Chem. Abstr.*, **77**, 96972g (1972).
29. K. Takagi et al., *Jpn. J. Pharmacol.*, **21**, 832 (1971).
30. S. Y. Chow et al., *Chung-Hua I Hsueh Tsa Chih (Taipei)*, **23**, 217 (1976); through *Chem. Abstr.*, **86**, 177203v (1977).
31. Y. Ishii and N. Sugawara, *Oyo Yakuri*, **7**, 871 (1973); through *Chem. Abstr.*, **80**, 66677j (1974).
32. R. M. Salassa et al., *J. Clin. Endocrinol. Metab.*, **22**, 1156 (1962).
33. M. T. Epstein et al., *Br. Med. J.*, **19**, 488 (1977).
34. I. F. Shvarev et al., *Vop. Izuch. Ispol'z. Solodki SSSR, Akad. Nauk SSSR*, 167 (1966); through *Chem. Abstr.*, **68**, 113187p (1968).
35. S. A. Vichkanova and M. A. Rubinchik, *Vop. Izuch. Ispol'z. Solodki SSSR, Akad. Nauk SSSR*, 176 (1966); through *Chem. Abstr.*, **68**, 113291t (1968).
36. H. Q. Zhang et al., *Zhongguo Yaoli Xuebao*, **7**, 175 (1986).
37. Monograph *Liquiritiae radix*, *Bundesanzeiger*, no. 90 (May 15, 1985); corrected (Mar. 13, 1990; Apr. 4, 1991).
38. S. Tanaka et al., *Planta Med.*, **53**, 5 (1987).
39. S. Teelucksingh et al., *Lancet*, **335**, 1060 (1990).
40. Y. K. Tony Lam et al., *Planta Med.*, **58**, 221 (1991).
41. Y. Kimura et al., *Phytother. Res.*, **7**(5), 335 (1993).
42. H. S. Ju et al., *Yaoxue Xuebao*, **24**(11), 807 (1989).
43. G. S. Wang and Z. W. Han, *Yaoxue Xuebao*, **28**(8), 572 (1993).
44. M. K. Cook, *Flav. Ind.*, **2**, 155 (1971).
45. M. K. Cook, *Food Eng.*, (5), 145 (1973).
46. M. R. Gibson, *Lloydia*, **41**, 348 (1978).
47. J. L. Hartwell, *Lloydia*, **33**, 97 (1970).

# LIGUSTRUM

**Source:** ***Ligustrum lucidum*** Ait. (Family Oleaceae).

*Synonyms*. Glossy privet, *nuzhenzi*, and *dongqingzi*.

## GENERAL DESCRIPTION

Erect large evergreen shrub or small tree, reaching over 10 m high; fruit a kidney-shaped berry with one to two seeds, bluish black when ripe, about 1 cm long and 3–4 mm in diameter; native to eastern Asia; distributed throughout China; now cultivated as an ornamental plant in the United States. Part used is the ripe fruit, mostly wildcrafted from October to December, rid of impurities and sun dried, or first briefly steamed and then sun dried; produced largely in southeastern provinces (ZHU).

Its Chinese synonym, *dongqingzi*, can also be the fruit of *Ilex chinensis* Sims, another fairly commonly used drug, which is oval and slightly smaller, with four to five seeds. The two cannot be positively differentiated and identified in the literature unless Latin binomials or other collaborative information are also given (CMH).

## CHEMICAL COMPOSITION

Contains triterpenoids, including oleanolic acid (ligustrin, 0.7–4.3%),[1–4] acetyloleanolic acid, and ursolic acid; glycosides (ligustroside, 10-hydroxyligustroside, oleuropein,

10-hydroxyoleuropein, 4-hydroxy-β-phenylethyl-β-D-glucoside, etc.); mannitol; fatty oil (10–15%), composed mainly of linoleic, linolenic, oleic, and palmitic acids, with presence of stearic acid not confirmed (HU, IMM-3); and a unique volatile oil consisting primarily of esters and alcohols, with lesser amounts of thioketones and hydrocarbons and traces of amines and aldehydes, but no terpene hydrocarbons.[5,6] Major components of the volatile oil include ethyl acetate (18.95%), thioketone (8.56%), α-butyl-benzenemethanol (5.6%), 4-acetyloxy-2-butanone (5.46%), 1-phenyl-1,2-butanediol (4.12%), 1,2-diphenyl-1,2-ethanediol (3.92%), hydrazine-methyl-oxalate (3.52%), α,α,4-trimethyl-3-cyclohexen-methanol (3.24%), 1-methyl-1-propyl-hydrazine (2.60%) and (*Z*)-1-(1-ethoxyethoxy)-3-hexene (1.89%).[6]

## PHARMACOLOGY OR BIOLOGICAL ACTIVITIES

Ligustrum is one of the highly valued tonics in traditional Chinese medicine and has recently been shown to exhibit numerous biological effects in humans and experimental animals, which include: ligustrum preparations (i.g. or i.m.) inhibited or prevented the leukopenia caused by chemotherapy (cyclophosphamide) and radiotherapy in mice and in cancer patients and are now clinically used for treating leukopenia in China; immunomodulating both in humans (phase II clinical trial) and in experimental animals;[7–9] decoction of wine-cured ligustrum (steam-heated with wine, 20% w/w, until absorbed and dried) exhibited marked antiinflammatory effects on different experimental inflammation models in animals;[10] water-soluble extractives increased blood flow volume in isolated rabbit coronary vessels; crude powder when fed to hyperlipemic rabbits (20 g/day, i.g., 30–60 days) markedly lowered blood lipids and prevented formation of atherosclerotic lesions;[11] decoction (i.g.) significantly decreased blood glucose levels in normal and alloxan diabetic mice;[12] acetone-soluble extractives had strong inhibitory effects on aflatoxin B1-induced mutation in *Salmonella typhimurium* strains $TA_{98}$ and $TA_{100}$; and others (HU, WANG).[11]

Recent studies have shown that oleanolic acid (ligustrin) is responsible for most of the biological effects of ligustrum (e.g., immunomodulating, antiinflammatory, and hypolipemic and antiatherosclerotic). In addition, it also has liver protectant effects (lowering SGPT, protecting from $CCl_4$ damage) as well as antiallergic, mild cardiotonic, diuretic, sedative, and antitumor (vs. S-180) effects.[1,7,14–16]

Oral toxicity is low: a single dose of 75 g of ripe fruit fed to rabbits did not elicit any toxic symptoms (WANG).

## USES

**Medicinal, Pharmaceutical, and Cosmetic.** Powder and extracts of ligustrum are used in hair-growth products and in formulas for removing facial dark spots primarily intended for internal use for its traditionally alleged ability to prevent premature graying and its "beautifying" properties (per Li Shi-Zhen's *Ben Cao Gang Mu*, 1590).

**Health Food/Herb Teas.** Powder and extracts are used as a tonic (usually combined with other herbs) in tablet, capsule, tea or liquid (drink) form to improve body resistance to illnesses; also used in soup mixes.

**Traditional Medicine.** First written record dates back to the *Shan Hai Jing* (ca. 800 B.C.); traditionally considered to be bitter and sweet tasting and neutral. It is one of the major *yin* tonics with vision-brightening and hair-darkening properties (*ming mu wu fa*). It invigorates the liver and kidney and is traditionally used for treating premature graying of hair, dizziness and tinnitus, sore back and knees, and blurred vision; now also used to treat habitual constipation in the elderly as well as chronic benzene poisoning (NATIONAL).

**Others.** As it contains the highest concentration of oleanolic acid (4.33%) among 216 Chinese herbal drugs from 18 genera tested, ligustrum is a potential source of this compound.[4]

**COMMERCIAL PREPARATIONS**

Mainly crude (whole or powdered) and extracts.

*Regulatory Status.* It is an ethnic food; U.S. regulatory status not clear.

**REFERENCES**


See the General References for BAILEY 1; CHP; CMH; FOSTER AND YUE; HU; IMM-3; JIANGSU; LU AND LI; NATIONAL; WANG; ZHU.

1. B. Wang and C. H. Jiang, *Zhongguo Yaoxue Zazhi*, **27**, 393 (1992).
2. Y. Rong and X. C. Ye, *Zhongguo Zhongyao Zazhi*, **14**(11), 41 (1989).
3. H. Y. Liu et al., *Zhongcaoyao*, **24**, 219 (1993).
4. N. J. Wu et al., *Zhongcaoyao*, **23**, 467 (1992).
5. N. J. Wu et al., *Zhongcaoyao*, **24**, 4 (1993).
6. K. H. Li and C. S. Li, *Zhongchengyao*, **12**(12), 32 (1990).
7. Y. Sun et al., *Chin. J. Clin. Pharmacol.*, **6**(2), 72 (1990).
8. Y. Sun et al., *J. Biol. Response Mod.*, **2**, 227 (1983).
9. B. H. S. Lau et al., *Phytother. Res.*, **3**(4), 148 (1989).
10. Y. Dai et al., *Zhongguo Zhongyao Zazhi*, **14**(7), 47 (1989).
11. B. Zhou et al., *Shiyong Zhongxiyi Jiehe Zazhi (PJCM)*, **6**, 168 (1993).
12. Z. Q. Hao et al., *Zhongguo Zhongyao Zazhi*, **17**, 429 (1992).
13. C. C. Ruan et al., *Chin. J. Cancer*, **8**(1), 29 (1989).
14. Y. Dai et al., *Zhongguo Yaoli Xuebao*, **10**, 381 (1989).
15. Y. Dai et al., *Zhongguo Yaoli Xuebao*, **9**, 562 (1988).
16. Y. Dai et al., *Chin. J. Pharmacol. Toxicol.*, **3**, 96 (1989).


# LIME OIL

**Source:** ***Citrus aurantifolia*** (Christm.) Swingle (syn. *C. medica* L. var. *acida* Brandis) (Family Rutaceae).

**GENERAL DESCRIPTION**

Evergreen tree with stiff sharp spines; up to about 4.5 m high; native to southern Asia; cultivated in south Florida, the West Indies (e.g., Cuba)[1] and Central America (e.g., Mexico). Part used is the fresh peel of the green unripe fruit from which lime oil is obtained by cold expression (expressed lime oil). Distilled lime oil is obtained by steam distillation of the whole crushed fruit or juice of the crushed fruit. Another essential oil, centrifuged lime oil, is obtained by centrifuging the pulp and oil mixture of the fruit in high-speed centrifuges, thus separating the oil from the pulp.

Distilled lime oil is of much more economic importance than the other two lime oils. Major producers of the distilled oil include Mexico and the West Indies. Terpeneless lime oil is produced from the distilled oil.

## CHEMICAL COMPOSITION

Distilled lime oil contains a large amount (ca. 75%) of terpene hydrocarbons of which *d*-limonene is a chief component, with α- and β-pinenes, camphene, sabinene, terpinolene, α-phellandrene, γ-terpenene, *p*-cymene, and others also reported present. Other types of components include oxygenated compounds (citral, α-terpineol, 1,8-cineole, 1,4-cineole, linalool, fenchol, etc.) and sesquiterpenes (e.g., α-bergamotene, β-caryophyllene, and β-bisabolene).[2–4] Germacrene B (0.35%) has been found to be an important fragrance component (with a sweet, woody-spicy, geranium-like note) for distinguishing lime oil from lemon oil.[5,6]

Expressed (cold-pressed) lime oil contains similar constituents as the distilled oil, but with lesser amounts of reaction or degradation products such as *p*-cymene. It also contains anthranilates as well as large amounts (ca. 7%) of substituted coumarins (limettin, bergapten, isoimperatorin, bergamottin, imperatorin, isopimpinellin, phellopterin, 8-geranoxypsoralen, oxypeucedanin hydrate, 5-geranoxy-7-methoxycoumarin, and 6,7-dimethoxycoumarin, with limettin being the major coumarin present.)[4,7–11]

2,4,6-Trichloroanisole, a compound with a strong musty naphthalenic odor, has been isolated in trace amounts from distilled Mexican lime oil as well as from French geranium, Italian lemon, and American peppermint oils. It is believed to be of microbial rather than pesticide origin.[12]

## PHARMACOLOGY OR BIOLOGICAL ACTIVITIES

The phototoxicity of certain substituted coumarins is well documented (see ***bergamot***). Expressed lime oil contains more of such compounds than the other citrus oils (e.g., ***bergamot*** and ***grapefruit***) and has been reported to be phototoxic to humans.[13]

Both expressed lime oil and distilled lime oil have been demonstrated to promote tumor formation on the skin and in the forestomach epithelium of mice treated with 9,10-dimethyl-1,2-benzanthracene and benzo[*a*]-pyrene, respectively.[13–15]

Distilled lime oil has been reported to be nonirritating, nonsensitizing, and nonphototoxic to human skin, though it was slightly irritating to rabbit skin.[16]

## USES

**Medicinal, Pharmaceutical, and Cosmetic.** Expressed lime oil and distilled lime oil are used as fragrance components and fixatives (expressed oil, due to its coumarins) in cosmetics, including soaps, detergents, creams, lotions, and perfumes, with maximum use level of 1.5% reported for both oils in perfumes.[13,16]

**Food.** Distilled lime oil and terpeneless lime oil are extensively used in carbonated beverages (especially the terpeneless type) and in hard candy (especially the distilled type) for the popular lemon-lime flavor. Other food products in which they are also used include alcoholic beverages, frozen dairy desserts, baked goods, gelatins and puddings, meat and meat products, and others. Highest average maximum use level reported is about 0.078% (783 ppm) for lime oil (distilled) in candy.

## COMMERCIAL PREPARATIONS

Lime oil (distilled, expressed, and terpeneless). Distilled lime oil is official in F.C.C.

*Regulatory Status.* GRAS (§182.20).

## REFERENCES

See the General References for ARCTANDER; BAILEY 2; FEMA; GUENTHER; MORTON 2.

1. J. A. Pino and R. Tapanes, *J. Food Technol.*, **18**, 523 (1983).
2. J. Perez Zayas and R. Tapanes, *Rev. CENIC, Cienc. Fis.*, **5**, 1 (1974); through *Chem. Abstr.*, **83**, 103115j (1975).
3. C. A. Slater, *J. Sci. Food Agr.*, **12**, 732 (1961).
4. C. A. Slater, *Chem. Ind. (London)*, 833 (1961).
5. B. C. Clark et al., *J. Agr. Food Chem.*, **35**, 514 (1987).
6. B. M. Lawrence, *Perfum. Flav.*, **16**(2), 59 (1991).
7. W. L. Stanley and L. Jurd, *J. Agr. Food Chem.*, **19**, 1106 (1971).
8. J. H. Tatum and R. E. Berry, *Phytochemistry*, **16**, 1091 (1977).
9. W. L. Stanley and S. H. Vannier, *Phytochemistry*, **6**, 585 (1967).
10. L. Haro-Guzman and R. Huet, *Fruits*, **25**, 887 (1970); through *Chem. Abstr.*, **75**, 25205g (1971).
11. G. L. K. Hunter and M. G. Moshonas, *J. Food Sci.*, **31**, 167 (1966).
12. J. Stoffelsma and K. B. de Roos, *J. Agr. Food Chem.*, **21**, 738 (1973).
13. D. L. J. Opdyke, *Food Cosmet. Toxicol.*, **12**, 731 (1974).
14. F. J. C. Roe and W. E. H. Field, *Food Cosmet. Toxicol.*, **3**, 311 (1965).
15. W. E. H. Field and F. J. C. Roe, *J. Natl. Cancer Inst.*, **35**, 771 (1966).
16. D. L. J. Opdyke, *Food Cosmet. Toxicol.*, **12**, 729 (1974).

# LOBELIA

**Source:** ***Lobelia inflata*** L. (Family Campanulaceae).

*Synonyms.* Indian tobacco, wild tobacco, asthma weed, gagroot, emetic herb, and vomit wort.

## GENERAL DESCRIPTION

A hairy annual or biennial herb with light blue flowers; up to about 1 m high; native to North America from Labrador to Georgia and west to Arkansas. Part used is the aboveground herb.

## CHEMICAL COMPOSITION

Contains about 0.48% pyridine (piperidine) alkaloids composed mainly of lobeline, with lesser amounts of lobelanine and lobelanidine. Other alkaloids present include norlobelanine (isolobelanine), lelobanidine, norlelobanidine, norlobelanidine, lobinine, isolobinine, lobinanidine, isolobinanidine, 8-methyl-10-ethyllobelidiol, and 8-methyl-10-phenyllobelidiol, among others.[1,2]

It also contains resin, gum, fats, chelidonic acid, and others (KARRER, LIST AND HÖRHAMMER, MARTINDALE).[3,4]

The formation of lobeline and related alkaloids by *L. inflata* in tissue culture has been reported.[5]

## PHARMACOLOGY OR BIOLOGICAL ACTIVITIES

Lobeline is reported to have many of the pharmacological properties of nicotine, although it is less potent. Like nicotine, its action is first CNS stimulation followed by severe depression (GOODMAN AND GILMAN).

Lobeline also has expectorant activities and in large doses is emetic, purgative, and diuretic. Overdosage may lead to convulsions and collapse, with (disputed) fatal results (MARTINDALE).[6]

## USES

**Medicinal, Pharmaceutical, and Cosmetic.** Lobeline salts are used as the active ingredient in several brands of antismoking (smoking deterrent) preparations. Lobelia and its extracts are used as ingredients in cough preparations and in counterirritant preparations.

**Health Food/Herb Teas.** Dried herb decreasingly available in capsules, tinctures, etc.; rare in tea formulations. Concern over potential toxicity, most unsubstantiated, has resulted in most manufacturers dropping it from product lines. Self-administration in any form is not advised (TYLER 1).

**Traditional Medicine.** Used as an antispasmodic, antiasthmatic, diaphoretic, expectorant, emetic, and sedative. Conditions for which it is used include asthma, whooping cough, bruises, sprains, ringworm, insect bites, poison ivy symptoms, and others.

In Chinese medicine related *Lobelia* species (e.g., *L. chinensis* Lour. and *L. sessilifolia* Lamb.) are also used, the former primarily for jaundice and sores, wounds, ringworm, snakebite and insect bites, and malaria; while the latter is used as an expectorant and antitussive and in treating bronchitis, ulcerous sores, and snakebite, and insect and dog bites. Both herbs also contain lobeline and related alkaloids.

Several *Lobelia* species including *L. inflata* and *L. chinensis* (syn. *L. radicans* Thunb.) have been used in cancers.[7]

**COMMERCIAL PREPARATIONS**

Crude and extracts; crude, fluid extract, and tincture were formerly official in N.F. Strengths (see ***glossary***) of extracts are expressed in weight-to-weight ratios.

*Regulatory Status.* Undefined in the United States as food ingredient; as of 1993 lobeline is no longer allowed as an ingredient in antismoking preparations.[8]

**REFERENCES**

See the General References for FOSTER; FOSTER AND DUKE; GOSSELIN; JIANGSU; LEWIS AND ELVIN-LEWIS; MARTINDALE; TYLER 1.

1. W. A. Ayer and T. E. Habgood in R. H. F. Manske, ed., *The Alkaloids*, Vol. 11, Academic Press, New York, 1968, p. 459.
2. D. Gross in W. Herz et al., eds., *Fortschritte der Chemie Organischer Naturstoffe*, Vol. 29, Springer-Verlag, Vienna, 1971, p. 1.
3. M. S. Karawya et al., *J. Assoc. Offic. Anal. Chem.*, **54**, 1423 (1971).
4. T. E. Wallis, *J. Pharm. Pharmacol.*, **9**, 663 (1957).
5. H. Wysokinska, *Farm. Pol.*, **33**, 725 (1977); through *Chem. Abstr.*, **88**, 86086j (1978).
6. E. G. C. Clarke in R. H. F. Manske, ed., *The Alkaloids*, Vol. 12, Academic Press, New York, 1970, p. 513.
7. J. L. Hartwell, *Lloydia*, **31**, 71 (1968).
8. L. Israelsen, personal communication, 1993.

# LOCUST BEAN GUM

**Source:** ***Ceratonia siliqua*** L. (Family Leguminosae or Fabaceae).

*Synonyms.* Carob bean gum, carob gum, locust gum, and carob seed gum.

Locust bean gum is derived from the seed of the carob, *Ceratonia siliqua* L. (see ***carob***). The seed consists of 33–46% endosperm, 30–33% seed coat (husk), and 23–30% embryo (germ).

The endosperm is the source of locust bean gum; it is separated from the dark seed coat and yellow embryo by a series of milling operations. The endosperm is then ground and graded according to particle size, color, and other parameters. Major producing countries include Spain, Italy, and Greece.

Food-grade locust bean gum is a white to yellowish white, nearly odorless powder. Many of its physical properties are similar to those of guar gum, but unlike guar gum it is not completely dispersed in water at room temperature. Consequently, it is customary to hydrate locust bean gum in hot water and then let its solution (sol) cool to achieve maximum viscosity.

Solutions of locust bean gum have a pH between 5 and 7 and are highly viscous like those of guar gum. They behave similarly toward pH changes (stable over pH 3–11), borate ions, and organic solvents. Locust bean gum is compatible with gelatin, starch, and other plant gums (see ***guar gum***).

Locust bean gum can improve the character and strength of carrageenan and agar gels as well as retard syneresis in these gels (see ***agar***). This property is not shared by guar gum.

**CHEMICAL COMPOSITION**

Like guar gum, commercial locust bean gum is not a pure galactomannan. It contains substances other than galactomannan (protein, moisture, ash, etc.) comparable in quantities to those of guar gum. Its molecular weight has been reported to be 310,000. It has a structure similar to that of guar gum, except that its D-galactose side chain is not attached to every other D-mannose unit as in guar but is attached to every consecutive mannose unit in blocks of 25 units separated by unsubstituted blocks of 85 mannose units each. Thus locust bean gum contains fewer galactose units than guar gum.[1]

**PHARMACOLOGY OR BIOLOGICAL ACTIVITIES**

Like guar gum, locust bean gum does not seem to be digested by animals. Its growth-depressing effects in laboratory animals have been reported, with inconclusive results as in guar gum (which see).[2]

**USES**

Locust bean gum has similar uses as guar gum, many of which are interchangeable (see ***guar gum***).

It is also used in place of tragacanth in some pharmaceutical and cosmetic applications (see ***tragacanth***).

**Food.** Gum is used in ice cream, cheese, meat and fish sauces, pie filling, soups, bakery goods, etc.[3]

**Others.** Gum used in manufacture of papers as a sizing, bonding fibers together more efficiently than starch.[3]

**COMMERCIAL PREPARATIONS**

Various grades with different viscosities and particle sizes. It is official in F.C.C.

*Regulatory Status.* Has been affirmed as GRAS (§184.1343).

**REFERENCES**

See the General References for FURIA; GLICKSMAN; LAWRENCE; MARTINDALE; MERCK; WHISTLER AND BeMILLER.

1. C. W. Baker and R. L. Whistler, *Carbohydr. Res.*, **45**, 237 (1975).
2. S. E. Davis and B. A. Lewis, *ACS Symp. Ser.*, **15**, 296 (1975).
3. R. L. Whistler, *Econ. Bot.*, **36**(2), 195 (1982).

# LOVAGE ROOT

**Source:** ***Levisticum officinale*** W. D. J. Koch (syn. *Angelica levisticum* Baill.) (Family Umbelliferae or Apiaceae).

*Synonyms*. Smellage, smallage, and maggi herb.

## GENERAL DESCRIPTION

A large, nonhairy perennial herb with a stout hollow stem; up to about 2 m high; native to the mountains of southern Europe; naturalized in North America; cultivated in central and southern Europe (e.g., France, Belgium, Czech Republic, Hungary, Slovakia, and Germany). Parts used are the dried rhizome and roots of 2- to 3-year old plants, collected in the spring. Lovage oil is obtained by steam distillation of the fresh root.

## CHEMICAL COMPOSITION

Contains up to 1.8% (usually 0.5–1.0%) volatile oil composed of 70% phthalides (butylidene-, dihydrobutylidene-, and butylphthalides; sedanonic anhydride; ligustilide; senkyunolide; validene-4,5-dihydrophthalide; etc.), with lesser amounts of terpenoids ($\alpha$- and $\beta$-pinenes, $\alpha$- and $\beta$-phellandrenes, $\gamma$-terpinene, carvacrol, and *l*-$\alpha$-terpineol), volatile acids (butyric acid, isovaleric acid, maleic acid, angelic acid, etc.),[1–4] 0.1–4.3% coumarins (coumarin, umbelliferone, bergapten, psoralen, etc.);[5–8] $\beta$-sitosterol and its glucoside,[9] gum, and resin, among others (KARRER, LIST AND HÖRHAMMER, STAHL). Heptanal, *trans*-2-nonenal and *cis*-3-hexenyl butyrate are obtained by direct solvent extraction of the root, but are not in the root oil.[10]

The leaf oil contains $\alpha$-terpinyl acetate (29%), *cis*- and *trans*-ligustilides (18%), $\alpha$-phellandrene (17%), and $\alpha$-terpineol (5%).[10,11]

## PHARMACOLOGY OR BIOLOGICAL ACTIVITIES

Phthalides have been reported to have sedative activities on mice (see ***celery***).

Some of the coumarins are known to be phototoxic to humans as well as useful in treating psoriasis (see ***bergamot***).

Lovage extracts and oil have been reported to exhibit strong diuretic effects on rabbits and mice (LIST AND HÖRHAMMER); also spasmolytic.[12]

Available data from one source indicate lovage root oil to be nonirritating and nonsensitizing to human skin, though one case of sensitization has been reported from another source. The phototoxic effects of lovage oil on humans are not known.[13]

## USES

**Medicinal, Pharmaceutical, and Cosmetic.** Lovage oil is used as a fragrance component in soaps, creams, lotions, and perfumes, with maximum use level of 0.2% reported in perfumes.[13]

**Food.** Lovage oil and lovage extracts are used as flavor components in major food products, including alcoholic (liqueurs, etc.) and nonalcoholic beverages, frozen dairy desserts, candy, baked goods, gelatins and puddings, meat and meat products, and sweet sauces, among others. Average maximum use levels are generally below 0.005%, with the exceptions of 0.017% and about 0.013% (125 ppm) reported for lovage extract (type not indicated) in sweet sauces and in frozen dairy desserts, respectively.

Lovage (crude) is also reported used in alcoholic beverages, frozen dairy desserts, candy, and baked goods. Highest average maximum use level is 0.015% in alcoholic beverages.

**Health Food/Herb Teas.** Root occasionally used in digestive formulations in capsules, tablets, also tea ingredient (FOSTER).

**Traditional Medicine.** Used as a diuretic, stomachic, expectorant, and emmenagogue; conditions for which it is used include digestive problems, flatulence, gastric catarrh, skin problems, and menstrual difficulties.

During China's Cultural Revolution, the root was used as a substitute for wild-harvested *danggui* (*Angelica sinensis*), until cultivated supplies of the drug were developed (FOSTER AND YUE).

**COMMERCIAL PREPARATIONS**

Crude, extracts, and oil; lovage oil is official in F.C.C. Strengths (see ***glossary***) of extracts are expressed in flavor intensities.

*Regulatory Status.* Has been approved for food use (§172.510). The root is the subject of a German therapeutic monograph; allowed in irrigation therapy for inflammation of the lower urinary tract and to prevent kidney gravel.[12]

**REFERENCES**

See the General References for ARCTANDER; BAILEY 2; BLUMENTHAL; FEMA; FERNALD; FOSTER; FOSTER AND YUE; GRIEVE; GUENTHER; LUST; ROSE; STAHL.

1. M. Sekulic and M. Smodlaka, *Arhiv Farm. (Belgrade)*, **11**, 177 (1961); through *Chem. Abstr.*, **56**, 10301f (1962).
2. G. Tibori et al., *Rev. Med. (Tirgu-Mures, Rom.)*, **20**, 222 (1974); through *Chem. Abstr.*, **82**, 160093d (1975).
3. G. Pattenden in W. Herz et al., eds., *Fortschritte der Chemie Organischer Naturstoffe*, Vol. 35, Springer-Verlag, Vienna, 1978, p. 133.
4. M. J. M. Gijbels et al., *Planta Med.*, **44**, 207 (1982).
5. D. Albulescu et al., *Farmacia (Bucharest)*, **23**, 159 (1975); through *Chem. Abstr.*, **84**, 147632a (1976).
6. A. D. Dauksha, *Aktual. Vop. Farm.*, **23** (1968); through *Chem. Abstr.*, **76**, 70136s (1972).
7. J. Karlsen et al., *Medd. Nor. Farm. Selsk.*, **30**, 169 (1968); through *Chem. Abstr.*, **70**, 112379g (1969).
8. F. C. Fischer and A. B. Svendsen, *Phytochemistry*, **15**, 1079 (1976).
9. B. E. Nielsen and H. Kofod, *Acta Chem. Scand.*, **17**, 1167 (1963).
10. B. M. Lawrence, *Perfum. Flav.*, **15**(5), 57 (1990).
11. Toulemonde et al. in B. M. Lawrence et al., eds., *Flavors and Fragrances: A World Perspective*, Elsevier Science Publishers B.V., Amsterdam, 1988, p. 641.
12. Monograph *Levistici radix*, *Bundesanzeiger*, no. 101 (June 1, 1990).
13. D. L. J. Opdyke, *Food Cosmet. Toxicol.*, **16**(Suppl. 1), 813 (1978).

# LYCIUM FRUIT

**Source:** ***Lycium barbarum*** L. (*L. halimifolium* Mill.) and ***L. chinense*** Mill. (Family Solanaceae).

*Synonyms.* Duke of Argyll's tea tree, Ningxia *gouqi* (*L. barbarum*); Chinese matrimony vine (*L. chinense*); boxthorn and wolfberry; *gouqizi*.

**GENERAL DESCRIPTION**

*Lycium barbarum* is a deciduous shrub,

upright or spreading, reaching 2–3 m high; fruit (berry) subglobose to ovoid or short oblong, scarlet to orange red when ripe, 0.8–2 cm long and 0.5–1.0 cm in diameter; native to northern Asia, distributed and abundant in northern China, especially in the province of Ningxia; naturalized in the United States.

*Lycium chinense* is also a deciduous shrub, up to about 1 m high, with slightly smaller fruit; native to eastern Asia; now distributed throughout China; naturalized in the United States. Young shoots (leaves) are a popular vegetable in southern China and Hong Kong and are now also available in major Chinatowns in the United States.[1]

Part used is the ripe fruit collected in summer or fall from both wild and cultivated plants, rid of stalk, left in a shady and airy area until skin is wrinkled, and then sun dried or oven dried until skin is dried but the whole fruit is still soft to the touch. Lycium fruit comes in numerous grades; top grades consist of fruits that are large, bright red or purplish red, soft to the touch and taste sweet. Ningxia is the major producing province, which also produces the best grades (ZHU); most of the lycium fruit imported into the United States comes from Ningxia produced from *L. barbarum*.[2]

## CHEMICAL COMPOSITION

Most of the chemical and biological studies on lycium fruit have been performed on *L. barbarum*. Hence unless otherwise stated, information reported in the following is for fruits from this species.

According to a report from Ningxia, lycium fruit (Ningxia *gouqi*) contains large amounts of amino acids (8–10%), about half of which in the free form: aspartic acid (1.2%), proline (0.65%), glutamic acid (0.63%), alanine (0.37%), arginine (0.19%), serine (0.14%), and nine others. Even higher amounts were found in the leaves (10–17%); freeze-drying and conventional drying did not affect the concentration and relative distribution of these amino acids either in the fruit, fruit stalk or leaf.[3] The high concentrations of amino acids were confirmed by a more recent study on lycium fruits from four different sources, which were found to contain total amino acids ranging from 9.9 to 18.8%.[4] However, another study found a much lower concentration of free amino acids (ca. 0.47%) in lycium fruit of unspecified origin.[5] Also, lycium fruit (from *L. chinense*) produced in Shandong Province is reported to contain much lower amounts of total amino acids (ca. 5.3%).[6]

Other constituents reported include pigments (zeaxanthin, physalien (zeaxanthin dipalmitate) and cryptoxanthin);[6] betaine; $\beta$-carotene (7.38–8.88 mg/100 g); vitamins $B_1$, $B_2$, and C; nicotinic acid;[2,7] 58–64% carbohydrates (47–56% sugars, 5.4–8.2% polysaccharides);[2,8] scopoletin;[9] taurine and $\gamma$-aminobutyric acid;[4] cinnamic acid, fats (8–12%), proteins (11–20%); trace minerals; and others (HU).[2,10,11]

## PHARMACOLOGY OR BIOLOGICAL ACTIVITIES

A highly valued *yin* tonic in traditional Chinese medicine, lycium fruit is regarded as having antiaging, body-strengthening, and vision-brightening properties, all of which have at least some scientific basis as evidenced by the following modern findings: (*1*) Oral administration of 50 g/day of lycium fruit for 10 successive days to 43 patients age 60 years and over significantly raised the values of lysozyme, IgG, IgA, lymphocyte blastogenesis (Stimulation Index), cyclic AMP, and testosterone.[12] (*2*) Oral administration of 50 g/day of lycium fruit for 10 successive days to 25 healthy subjects aged 64–80 years significantly raised their serum levels of superoxide dismutase (SOD) and hemoglobin but lowered the level of lipid peroxides.[13] (*3*) In a double-blind, placebo controlled study, oral administration of lycium fruit extractives to normal subjects aged 56–81 years significantly improved their immune functions, reduced senility symptoms, and increased total peripheral white cell count and neutrophils.[14] (*4*) Oral administration of 50 g/day of lycium fruit (containing ca. 5 mg β-carotene) to healthy subjects aged 18–25 years significantly improved vision and dark adaptation, with serum vitamin A reaching saturation level at 1.16 ± 0.15 mol/L after 34 days of intake.[7]

Recent studies indicate that the polysaccharides are responsible for most of the biological activities of lycium fruit, including antiperoxidative effects on cell membrane of *Xenopus* oocytes, with activity comparable to that of SOD;[15] preventing lipid peroxidation in liver, spleen, and brain tissues of rats and mice induced by physical stress and by carbon tetrachloride;[16] increasing the interleukin-2 (IL-2) activity in adult mice and restoring the level of IL-2 activity of aged mice to that of adult mice;[17] markedly elevating the cellular immune response in mice, raising level of splenic plaque forming cells in aged mice to that of adult mice and enhancing T-lymphocyte proliferation and cytotoxicity of cytotoxic T-lymphocytes and natural killer cells in normal mice as well as in immunosuppressed mice treated with cyclophosphamide;[18,19] reducing the dosage of *Corynebacterium parvum* in a synergistic effect on the tumoristatic activity of mouse peritoneal macrophages against P815 and P388 cells;[20] and protecting genetic material from genetic damage (antimutagenic) by mitomycin both *in vitro* and in healthy subjects over 60 years old.[21,22]

Other biological effects include hypoglycemic, hypolipemic, preventing fatty liver, hypotensive, and increasing weight gain in mice (when used with ginseng) among others (WANG); markedly increasing tolerance to anoxia as well as increasing dermal hydroxyproline level (by 15.5%) in mice, indicating increased collagen synthesis.[23]

Juice expressed from fresh young leaves of *L. barbarum* or *L. chinense* collected in spring and summer was used in topically treating 1853 cases of mosquito and insect stings/bites with great success: 1703 cases cured with one to three applications, leaving no scars or pigmentation; only 10 cases did not respond.[24]

## USES

**Medicinal, Pharmaceutical, and Cosmetic.** Traditionally considered of benefit to the complexion and to prolong life (*mei rong yan nian*), lycium fruit has been consumed for 2000 years in China for these purposes; in recent years also used successfully in the topical treatment of burns, ulcers, bedsores, frostbite, canker sores, and furuncles.[25] Its high contents of free amino acids, β-carotene and bioactive polysaccharides as well recent evidence of its ability to increase skin hydroxyproline levels and at the same time with no known human toxicity make it a potentially useful cosmetic ingredient.

**Health Food/Herb Teas.** Powder and extracts (water and hydroalcoholic) are used in tonic formulas (especially for vision and male problems) in tablet, capsule, or liquid form; also sold whole for making tea (use 1–2 tablespoons/cup).

**Traditional Medicine.** First described in the *Ming Yi Bie Lu* (ca. A.D. 200), lycium fruit is one of the most commonly used Chinese *yin* tonics. Traditionally regarded as sweet tasting and neutral, liver- and kidney-nourishing, replenishing vital essence (*yi jing*), and vision improving. Used in treating general debility and deficient energy (*xu lao jing kui*), aching back and knee, tinnitus, dizziness, diabetes, blurred vision, cough, and nocturnal emission. It is said to grow muscles and to cause weight gain but is also used to reduce weight;[26] and in recent years also used in treating the damaging side effects of chemotherapy and radiotherapy.[27]

## COMMERCIAL PREPARATIONS

Crude (whole) and extracts (water, hydroalcoholic, and oil). Crude can be powdered after baked completely dry.

*Regulatory Status*. It is an ethnic food; U.S. regulatory status not known.

## REFERENCES

See the General References for BAILEY 1; CHP; FERNALD; FOSTER AND YUE; HU; JIANGSU; LU AND LI; WANG; ZHU.


1. S. Y. Zee and L. H. Hui, *Hong Kong Food Plants*, The Urban Council, Hong Kong, 1981, p. 27.
2. Z. S. Qi et al, *Zhongyao Tongbao*, **11**(3), 41 (1986).
3. X. Z. Meng et al., *Zhongyao Tongbao*, **12**(5), 42 (1987).
4. H. Q. Chen et al., *Zhongguo Yaoke Daxue Xuebao*, **22**(1), 53 (1991).
5. Y. X. Gong, *Zhongcaoyao*, **18**(11), 37 (1987).
6. J. Wang et al., *Shandong Zhongyi Zazhi*, **10**(1), 42 (1991).
7. Y. Shen et al., *Acta Nutrimenta Sinica*, **12**, 420 (1990).
8. Q. Wang et al., *Zhongcaoyao*, **22**, 67 (1991).
9. C. S. Li et al., *Zhongguo Zhongyao Zazhi*, **15**(3), 43 (1990).
10. Q. H. Yuan, *Zhongguo Zhongyao Zazhi*, **14**(4), 42 (1989).
11. Y. Q. Shao et al., *Shandong Zhongyi Zazhi*, **9**(2), 38 (1990).
12. P. G. Xiao and K. J. Chen, *Phytother. Res.*, **2**(2), 55 (1988).
13. W. Li et al., *Zhongcaoyao*, **22**, 251 (1991).
14. D. Y. Li et al., *Zhongcaoyao*, **20**(10), 26 (1989).
15. X. Zhang and X.C. Xie, *Zhongguo Zhongyao Zazhi*, **18**, 110 (1993).
16. H. Zhan et al., *Chin. J. Pharmacol. Toxicol.*, **3**, 163 (1989).
17. C. S. Geng et al., *Chin. J. Pharmacol. Toxicol.*, **3**, 175 (1989).
18. C. S. Geng et al., *Zhonghua Laonian Yixue Zazhi*, **8**, 236 (1989).
19. B. K. Wang et al., *Chin. J. Pharmacol. Toxicol.*, **4**, 39 (1990).
20. Y. X. Zhang et al., *Chin. J. Pharmacol. Toxicol.*, **3**, 169 (1989).
21. M. X. Tao and Z. L. Zhao, *Zhongcaoyao*, **23**, 474 (1992).
22. M. D. Wang et al., *Zhongcaoyao*, **23**, 251 (1992).
23. S. Q. Chen et al., *Zhongyao Yaoli Yu Linchuang*, **6**(3), 28 (1990).
24. P. X. Gao and X. L. Dang, *Shaanxi Zhongyi*, **9**, 320 (1988).
25. L. J. Jiang et al., *Zhongyi Zazhi*, **26**(5), 69 (1985).
26. H. X. Jing, *Xinzhongyi*, (7), 37 (1988).
27. S. N. Gu et al., *Zhongyao Yaoli Yu Linchuang*, **6**(6), 38 (1990).

# MAGNOLIA FLOWER

**Source:** ***Magnolia biondii*** Pamp. (syn. *M. fargesii* (Fin. et Gagnep.) Cheng), ***M. denudata*** Desr. (syn. *M. heptapeta*, (Buc'hoz) Dandy), ***M. sprengeri*** Pamp., ***M. sargentiana*** Rehd. et Wils. (syn. *M. emargenata* Cheng), ***M. wilsonii*** (Fin. et Gagnep.) Rehd. et Wils., ***M. salicifolia*** Maxim., and other ***Magnolia*** species (Family Magnoliaceae).

*Synonyms*. Magnolia flower bud, *xinyi*, *xinyihua*, *shin-i*, and flos magnoliae.

## GENERAL DESCRIPTION

Small to large trees, mostly deciduous, widely distributed in China; some extensively cultivated. At least 10 species of *Magnolia* serve as source of magnolia flower. However, *M. biondii*, *M. denudata*, and *M. sprengeri* are currently the major sources; and *M. liliflora* Desr., although still listed in many major works as the major source, is not a current source.[1,2] Produced mainly in eastern and southern provinces in China.

*Magnolia salicifolia* is source of Japanese magnolia flower.[3]

Part used is the flower bud collected before opening in early spring and carefully sun or oven dried. The dried bud measures 1–5 cm long and has a diameter of 0.5–2 cm at midsection, depending on the source.[4] It is hairy and its overall appearance resembles that of a pussy willow bud, but it emits a strong characteristic eucalyptus-like odor when crushed.

## CHEMICAL COMPOSITION

Contains 0.4–3.2% (mL/100 g) volatile oil, with *M. denudata* yielding the highest and *M. sargentiana* the lowest;[5] neolignans and lignans; alkaloids; flavonoids; tannins; and others (HU, WANG).[3,6–8]

More than 70 components of the volatile oil have been identified. Major ones include $\alpha$-pinene (0.63–5.67%), $\beta$-pinene (0.84–39.05%), camphene (0.39–5.54%), limonene (0.96–10.14%), 1,8-cineole (0.76–17.48%), *p*-cymene (0.50–25.69%), linalool (0.28–2.83%), camphor (0.14–44.20%), and $\alpha$-terpineol (0.49–4.00%), which are present in flower buds of all eight *Magnolia* (including *M. biondii*, *M. denudata*, *M. sprengeri*, *M. sargentiana*, and *M. salicifolia*) and one *Michelia* species tested; others include myrcene (0.40–2.70%), sabinene (1.15–13.42%), fenchone (0.06–9.62%), terpinen-4-ol (0.90–10.00%), borneol (0.40–3.18%), bornyl acetate (0.42–8.40%), methyleugenol (0.22–19.45%), caryophyllene oxide (0.76–3.30%), and eudesmol (0.70–4.41%), present in the majority of species tested. The proportions of these constituents vary considerably among the species. Thus $\beta$-pinene is the major component in *M. denudata* (39.05%) and camphor, the major component in *M. biondii* (44.20%) and *M. sargentiana* (40.00%). Although safrole is not present in the others tested, it is present in *M. salicifolia* (Japanese *xinyi*) in major concentration (29.87%) along with methyleugenol (19.45%) and fenchone (9.62%), which are absent or present only in minor amounts in the other species (HU).[5]

Neolignans and lignans include fargesone A, B, and C; denudatin B; pinoresinol dimethyl ether; lirioresinol B dimethyl ether; magnolin; and fargesin (from flower buds of *M. biondii*) as well as magnoshinin and magnosalin (from Japanese *xinyi*).[6,7]

Alkaloids include *d*-coclaurine, *d*-reticuline, 1-*N*-methylcoclaurine, and yuzirine (all benzylisoquinolines), which are present in minor to trace amounts in flower buds of *M. salicifolia* and *M. biondii*.[3]

## PHARMACOLOGY OR BIOLOGICAL ACTIVITIES

Magnolia flower has numerous biological activities, including antihistaminic in guinea pig trachea (due to volatile oil, alcoholic, and water extracts);[9,10] protective against allergic asthma in guinea pig (volatile oil);[10] antiinflammatory in mice (magnoshinin and magnosalin);[7] $Ca^{++}$-antagonistic on the taenia coli of guinea pig (neolignans);[6] central dopaminergic modulating in mice (*d*-coclaurine and *d*-reticuline);[8] neuromuscular blocking in isolated frog skeletal muscle and nerve–muscle preparations (alkaloids);[3] hypotensive in several species of experimental animals (alcoholic and water extracts); uterus stimulating in animals (decoction and fluid extract); antifungal, antibacterial, and antiviral (decoction); frog skeletal muscle contracting (decoction) and relaxant (alkaloids); local anesthetic in animals (decoction and infusion); and others (JIANGSU, WANG).

Magnolia flower in various forms (decoction, alcoholic extract, volatile oil, etc.) and in combination with other herbal drugs has been reported highly effective both locally and internally in treating allergic rhinitis (e.g., hay fever), chronic rhinitis, and paranasal sinusitis (WANG).[11,12]

Magnolia flower has very low toxicity: i.v. injection of decoction in dogs (1.0, g/kg) and rabbits (4.75 g/kg) produced no fatalities. The $LD_{50}$ (i.p.) of its tincture (after alcohol removal) in rats and mice were 22.5 ± 0.96 and 19.9 ± 0.25 g/kg, respectively, based on the crude drug (WANG).

## USES

**Medicinal, Pharmaceutical, and Cosmetic.** Extracts used in skin care products to minimize or counteract undesirable irritant effects of other cosmetic ingredients; also used for its traditional skin-whitening properties.

**Health Food/Herb Teas.** Used in allergy and cold preparations (JIANGSU).

**Traditional Medicine.** Traditionally considered acrid tasting and warming; disperses wind and cold (*san feng han*) and clears the nasal cavity. Chinese traditional use records date back 3000 years to the *Wu Shi Er Bing Fang* (*Prescriptions for 52 Diseases*; 1065–771 B.C.). Used both internally and externally to treat nasal congestion, running nose, the common cold, and headache as well as facial dark spots; also used topically to treat toothache.

## COMMERCIAL PREPARATIONS

Crude and extracts (mainly hydroalcoholic and hydroglycolic).

*Regulatory Status.* U.S. regulatory status not determined.

## REFERENCES

See General References for HU; JIANGSU; NATIONAL; WANG.

1. Z. F. Wu et al., *Zhongguo Zhongyao Zazhi*, **16**(1), 13 (1991).
2. W. Z. Song, *Zhongcaoyao*, **15**(3), 26 (1984).
3. I. Kimura et al., *Planta Med.*, **48**, 43 (1983).
4. Y. Y. Tong and W. Z. Song, *Yaoxue Xuebao*, **20**, 22 (1985).
5. Z. L. Xu et al., *Zhongguo Zhongyao Zazhi*, **14**(5), 38 (1989).
6. C. C. Chen et al., *Planta Med.*, **54**, 438 (1988).
7. M. Kimura et al., *Planta Med.*, **51**, 291 (1985).
8. H. Watanabe et al., *Planta Med.*, **42**, 213 (1981).
9. R. D. Xiang et al., *Zhongcaoyao*, **16**(2), 22 (1985).
10. D. Q. Zhou et al., *Zhongcaoyao*, **22**(2), 81 (1991).

11. T. C. Wang et al., *Chin. J. Integr. Trad. Western Med.*, **4**, 728 (1984).
12. Y. Ren, *Zhongyao Tongbao*, **10**(5), 45 (1985).

# MARJORAM (SWEET, POT AND WILD)

**Source:** ***Sweet marjoram*** *Origanum majorana* L. (syn. *Majorana hortensis* Moench); ***Pot marjoram*** *Origanum onites* L. (syn. *Majorana onites* (L.) Benth.); ***Wild marjoram*** *Origanum vulgare* L. (see ***oregano***) (Family Labiatae or Lamiaceae).

*Synonym*. Knotted marjoram (*O. majorana*).

## GENERAL DESCRIPTION

There is much confusion regarding the plant sources of marjoram (pot, sweet), oregano, and Spanish origanum (see ***origanum oil, Spanish***).

There do not seem to be any discrepancies with respect to sweet marjoram, as most authors agree to *Origanum majorana* L., also known as *Majorana hortensis* Moench, as its major botanical source (ARCTANDER, BAILEY 1, FURIA AND BELLANCA, ROSENGARTEN, TERRELL, UPHOF, §182.10). However, the sources of pot marjoram have been attributed by numerous authors to three plant names, *Origanum onites* L., *O. vulgare* L., and *Majorana onites* Benth. One source lists *M. onites* as a synonym of *O. vulgare*, both of which are entered under the heading of *pot marjoram*, and *O. vulgare* is also described as the source of origanum, which is a separate entry (FURIA AND BELLANCA). According to ROSENGARTEN, *O. vulgare* has been known as wild marjoram (in agreement with ARCTANDER, BAILEY 1, and TERRELL) and is one of the two major sources of ***oregano*** (European oregano). The other major type of oregano, Mexican oregano, is mostly derived from *Lippia* species, especially *L. graveolens* H. B. K. (ROSENGARTEN, TERRELL) (see ***oregano***).

The source of pot marjoram is attributed to a single species, *M. onites*, also called *O. onites* (BAILEY 1, ROSENGARTEN, TERRELL).

The major source of Spanish origanum is not an *Origanum* species but is *Thymus capitatus* (L.) Hoffmgg. et Link. Nevertheless, *Origanum* species are used and thus add to the confusion (see ***origanum oil, Spanish***).

Sweet marjoram is a tender, bushy perennial herb, with woolly hairy leaves, up to about 0.6 m high; native to the Mediterranean region; cultivated as an annual in colder climates. Part used is the dried flowering herb from which sweet marjoram oil is obtained by steam distillation in 0.2–0.8% yield. Major oil-producing countries include France, Tunisia, Morocco, and Bulgaria.

Pot marjoram is a sturdy perennial herb with coarse hairy, erect stems and sessile, hairy (long and soft) leaves; native to the Mediterranean region. Parts used are the dried leaves. It is considered of low quality and not much used.

## CHEMICAL COMPOSITION

Sweet marjoram contains up to 3% volatile oil (usually less than 1%), consisting primarily of $\alpha$-terpinene, $\gamma$-terpinene, *p*-cymene, 4-terpineol, sabinene, linalool, borneol, carvacrol, *cis*-sabinene hydrate, and *trans*-sabinene hydrate with linalyl acetate, ocimene, cadinene, geranyl acetate, citral, estragole, eugenol, and 3-carene also present, totaling more than 50 compounds. *cis*-Sabinene hydrate is reported to be responsible for the typical aroma of sweet marjoram oil (GUENTHER, LIST AND HÖRHAMMER).[1–4]

Other compounds present in sweet marjoram include flavonoid glycosides (luteolin-7-diglucoside, apigenin-7-glucoside, and diosmetin-7-glucuronide), arbutin, methylarbutin,[5] tannins, caffeic acid, labiatic acid, rosmarinic acid,[6] steroids (e.g., $\beta$-sitosterol), triterpenoids (oleanolic acid, ursolic acid, etc.), paraffins (e.g., *n*-triacontane), protein (ca. 13%), vitamins (especially A and C), and others (LIST AND HÖRHAMMER, MARSH)[7,8]

Pot marjoram is reported to contain 1.4–2.4% volatile oil, consisting mainly of carvacrol, linalool, and thymol (LIST AND HÖRHAMMER).

Extracts of sweet marjoram have antioxidative properties on lard, which are in part due to labiatic acid and the flavonoids present.[9–11]

## PHARMACOLOGY OR BIOLOGICAL ACTIVITIES

Sweet marjoram is considered to have carminative, antispasmodic, diaphoretic, and diuretic properties. Its aqueous extract has been reported to have antiviral activities against herpes simplex *in vitro*.[12]

Fresh sweet marjoram may cause inflammation of the skin and eyes (LIST AND HÖRHAMMER).

Sweet marjoram oil has been reported to be nonirritating and nonsensitizing to human skin.[13]

## USES

**Medicinal, Pharmaceutical, and Cosmetic.** Sweet marjoram oil is used as a fragrance component in soaps, detergents, creams, lotions, and perfumes, with maximum use level of 0.6% reported in perfumes.[13]

**Food.** Sweet marjoram oil and oleoresin are used as flavor ingredients in most food categories, including alcoholic (bitters, vermouths, etc.) and nonalcoholic beverages, frozen dairy desserts, candy, baked goods, gelatins and puddings, meat and meat products, condiments and relishes, and others. Average maximum use levels reported are generally below 0.004%.

Sweet marjoram is also used in baked goods, meat and meat products, condiments, and relishes, soups, snack foods, processed vegetables, and others, with highest average maximum use level of about 1% (9,946 ppm) reported in baked goods.

**Traditional Medicine.** Sweet marjoram has been used for treating similar types of conditions as *Origanum vulgare* (see ***oregano***). It has also been used in cancers.[14]

## COMMERCIAL PREPARATION

Sweet marjoram crude, oil, and oleoresin. Sweet marjoram oil is official in F.C.C.

*Regulatory Status*. GRAS (Sweet marjoram, §182.10 and §182.20; pot marjoram, §182.10).

## REFERENCES

See the General References for ARCTANDER; BAILEY 1; FEMA; GUENTHER; MARTINDALE; ROSE; ROSENGARTEN; TERRELL.

1. R. Granger et al., *Riv. Ital. Essenze, Profumi, Piante Office., Aromi, Saponi, Cosmet., Aerosol*, **57**, 446 (1975); through *Chem. Abstr.*, **84**, 79580e (1976).
2. J. Taskinen, *Acta Chem. Scand., Ser. B*, **28**, 1121 (1974).
3. G. Graner, *Präp. Pharm.*, **4**, 86 (1968).

4. E. Sarer et al., *Planta Med.*, **46**, 236 (1982).
5. M. H. Assaf et al., *Planta Med.*, **53**, 343 (1987).
6. U. Gerhardt and A. Schroeter, *Fleischwirtschaft*, **63**, 1628 (1983).
7. W. Olechnowicz-Stepien and E. Lamer-Zarawska, *Herba Pol.*, **21**, 347 (1975); through *Chem. Abstr.*, **85**, 74939q (1976).
8. G. Lossner, *Planta Med.*, **16**, 54 (1968).
9. Y. Saito et al., *Eiyo To Shokuryo*, **29**, 505 (1976); through *Chem. Abstr.*, **87**, 150314r (1977).
10. K. Herrmann, *Z. Lebensm. Unters. Forsch.*, **116**, 224 (1962).
11. S. Zalewski, *Przemysl Spozywczy*, **16**, 237 (1962); through *Chem. Abstr.*, **57**, 8970f (1962).
12. E. C. Herrmann Jr. and L. S. Kucera, *Proc. Soc. Exp. Biol. Med.*, **124**, 874 (1967).
13. D. L. J. Opdyke, *Food Cosmet. Toxicol.*, **14**, 469 (1976).
14. J. L. Hartwell, *Lloydia*, **32**, 247 (1969).

# MILK THISTLE

**Source:** ***Silybum marianum*** (L.) Gaertner. (syn. *Carduus marianum* L.) (Family Asteraceae or Compositae).

*Synonyms*. Mary thistle, St. Mary thistle, marian thistle, lady's thistle, and holy thistle (not to be confused with blessed thistle *Cnicus benedictus*).

## GENERAL DESCRIPTION

Stout, branching annual or biennial, 1–2 m high; leaves alternate, white mottled, glabrous, margins scalloped, with sharp spines on lobes; flowers purple; heads, solitary, to 6 cm in diameter; black shiny seeds (fruit) crowned with a spreading pappus; indigenous to Mediterranean region, southwest Europe; cultivated for centuries and naturalized in much of Europe; also naturalized in North America, especially California; South America from Uruguay to Chile, to Ecuador; weedy in Australia; common in abandoned fields, old pastures, and roadsides (FOSTER).

The part used is the seed (fruit).

## CHEMICAL COMPOSITION

A flavanolignan complex, silymarin, was first isolated from the seeds in 1968.[1,2] Silymarin (4–6% in ripe fruits) consists primarily of three flavanolignans, silybin (silibinin), silychristin (silichristin), and silidianin.[3] Other flavanolignans include dehydrosilybin, 3-desoxysilichristin, deoxysilydianin (silymonin), siliandrin, silybinome, silyhermin, and neosilyhermin.[4] Other constituents include apigenin, silybonol; a fixed oil (16–18%), consisting largely of linoleic and oleic acids, plus myristic, palmitic, and stearic acids; betaine hydrochloride, triamine, histamine, and others.[1–5]

## PHARMACOLOGY OR BIOLOGICAL ACTIVITIES

Few plant principles have been as extensively investigated in recent years as silymarin (WEISS). Primary activity is as a hepatoprotective and antioxidant.[1]

Hepatoprotective activity of silymarin has been demonstrated in numerous experimental models of toxic liver damage, including carbon tetrachloride, galactosamine, thioacetamide, hepatotoxic cold-blood frog virus ($FV_3$), lanthanides, and the toxins of *Amanita phalloides* (deathcap fungus) phalloidin and $\alpha$-amanitin.[1,6]

Efficacy is based on several separate mechanisms of action. Silymarin stimulates

RNA polymerase A, enhancing ribosome protein synthesis and resulting in activating the regenerative capacity of the liver through cell development.[1,4,7] Silymarin interacts with hepatic cell membranes, blocking binding cites and hindering the uptake of toxins as demonstrated in rabbit liver microsomes[8] and mononuclear lipid layers.[9] Strong antioxidant (free radical–scavenging activity 10-fold greater than vitamin E), blocking the release of malonyldialdehyde, and antiperoxidative activity has been reported.[10,11]

Clinical studies have suggested that pretreatment with silymarin inhibits alcohol-, industrial chemical-, and psychopharmaceutical-induced liver damage, accelerating normalization of impaired liver function.[1,7] Patients who received silymarin showed an accelerated improvement of increased serum levels of glutamic-oxaloacetic transaminase (GOT), glutamic-pyruvic transaminase (GPT) and $\gamma$-glutamyl-transpeptidase ($\gamma$-GT).[12]

Silibinin administered in i.v. infusion has shown protective and curative effect on liver damage resulting from the highly toxic compounds phalloidin and $\alpha$-amanitin (from *Amanita phalloides*). The antihepatotoxic effect of silymarin was found to depend on the time interval in which poisoning and therapy took place as well as the degree of liver damage. A multicenter trial involving 220 cases of *Amanita* poisoning treated in German, French, Swiss, and Austrian hospitals was carried out from 1979 to 1982, using silibinin in supportive treatment. Use of silibinin as an adjunct to current methods has lowered mortality rates below any levels that have previously been achieved.[13–15]

Silymarin products were used with success in toxic-metabolic liver damage, ranging from fatty liver through fatty liver hepatitis to actual hepatic cirrhosis, caused by toxic substances, drugs, or exposure to irradiation.[5]

Even in large doses silymarin is devoid of toxic effects and in particular has no harmful action on the embryo.[1,5,7] In isolated cases, a mild laxative effect has been observed.[6] A review of pharmacological and clinical studies, along with a comprehensive bibliography is available.[16]

## USES

**Medicinal, Pharmaceutical, and Cosmetic.** In Europe, clinical use is widespread for toxic liver damage in supportive treatment of chronic inflammatory liver disorders and cirrhosis, including chronic hepatitis and fatty infiltration of the liver by alcohol and other chemicals. In infusion therapy, silibinin preparations used for supportive treatment of *Amanita* mushroom poisoning.[6]

**Foods.** No commercial food use other than herb tea ingredient. Historically, grown in Europe as a vegetable, for edible peeled stalks; leaves a salad green or pot herb (with spines removed); roots eaten (soaked overnight in water to remove bitterness); flower receptacle eaten like artichoke; roasted seeds a coffee substitute.[1]

**Health Food/Herb Teas.** Seeds or seed extract in teas, capsules, tablets, tinctures or other preparations, primarily as a liver detoxicant (FOSTER).

**Traditional Medicine.** Milk thistle seeds used continuously for 2000 years for liver conditions (first mentioned by Pliny in the 1st century).[1] Historical references are particularly abundant in herbals of the Middle Ages, including the hepatoprotective activity.[5] Eclectic physicians in 19th-century America used seeds for liver congestion. Use of the seed for treatment of liver disease was revitalized by the German physician Rademacher in the mid-19th century. Reinvestigation of the value of milk thistle in modern practice began with H. Schulz in 1929 and G. Madaus in 1938.[5]

**COMMERCIAL PREPARATIONS**

Crude, ethanolic extracts, tablets, or capsules (35–70 mg) standardized to 70% silymarin (calculated as silibinin) in average daily dose of 200–400 mg.

*Regulatory Status*. Undetermined in the United States. Milk thistle is the subject of a positive German therapeutic monograph indicated for toxic liver damage, liver cirrhosis, and supportive treatment of chronic inflammatory liver disease.[6]

**REFERENCES**

See the General References for BLUMENTHAL; DER MARDEROSIAN AND LIBERTI; FELTER AND LLOYD; FOSTER; FOSTER AND DUKE; MARTINDALE; STEINMETZ; TYLER; UPHOF; WEISS, WREN.

1. S. Foster, *Milk Thistle—Silybum marianum,. Botanical Series, no. 305,* Austin, Tex., American Botanical Council, 1991.
2. H. Wagner et al., *Arzneim. Forsch.*, **18**(6), 688 (1968).
3. H. Wagner and O. Seligmann in H. M. Chang et al., eds. *Advances in Chinese Medicinal Materials Research*, World Scientific Publ. Co., Singapore, 1985, p. 247.
4. D. V. C. Awang, *Can. Pharm. J.*, 403 (Oct. 1993).
5. G. Hahn and A. Meyer, *Deut. Apoth.*, **40**(6–7), 2 (1988).
6. Monograph *Cardui Mariae Fructus, Bundesanzeiger*, no. 50 (Mar. 13, 1986).
7. H. Hikino and Y. Kiso in H. Wagner, H. Hikino, and N. R. Farnsworth, eds., *Economic and Medicinal Plant Research*, Vol. 2. Academic Press, New York, 1968, p. 39.
8. A. Wissemann in D. Lorenz et al. eds., *Aktuelle Hepatologie*, Hansiches Verlagskontor, Lübek, 1990, p. 55.
9. T. Parasassi et al., *Cell Biochem. Funct.*, **2**, 85 (1984).
10. A. Bindoli et al., *Biochem. Pharmacol.*, **26**, 2405 (1977).
11. A. Valenzuela and R. Guerra, *Experientia*, **42**, 139 (1986).
12. V. Fintelmann and A. Albert, *Therapiewoche*, **30**(35), 5589, (1980).
13. K. Hruby, *Forum* **8**(6), 23 (1984).
14. K. Hruby. *Intensivmed.*, **24**, 269 (1987).
15. K. Hruby et al., *Wien. Klin. Wochenschr.*, **95**(7), 225 (1983).
16. S. Foster, ed., *Milk Thistle Bibliography and Abstracts*, American Botanical Council, Austin, Tex., in press (1995).

# MINTS

**Source:** ***Peppermint*** *Mentha* × *piperita* L. (hybrid of *M. spicata* L. and *M. aquatica* L.); ***Spearmint*** *Mentha spicata* L. (syn. *M. viridis* L.); ***Cornmint*** *Mentha arvensis* L. var. *piperascens* Malinvaud (Family Labiatae or Lamiaceae).

*Synonyms*. Field mint and Japanese mint (*M. arvensis*).

**GENERAL DESCRIPTION**

Closely related perennial aromatic herbs with runners or stolons by which they are propagated; leaves of spearmint are sessile (no petioles), while those of peppermint and cornmint are petioled and short petioled, respectively; up to about 1 m high; cultivated worldwide. Each species has numerous varieties, strains, or chemotypes that produce essential oils with widely different chemical compositions.[1,2] There are 20 true species of *Mentha*, represented by as many as 2300 named variations, half of which are synonyms; half are legitimate infraspecific names.[3] Commercial varieties of mints produce oils that can be distinguished by their relative contents of menthol and carvone. Parts used are the dried leaves and the fresh or partially dried whole, aboveground flowering herb. The former furnishes the spice, while the latter is used for the production of the essential oil.

United States is the major producer of peppermint and spearmint and their essential oils, especially in Idaho, Indiana, Oregon, Washington, and Wisconsin (USDA). Major producers of cornmint and cornmint oil include Japan, Taiwan, and Brazil.

## CHEMICAL COMPOSITION

Peppermint yields 0.1–1.0% (usually 0.3–0.4%) of volatile oil that is composed mainly of menthol (29–48%), menthone (20–31%), and menthyl acetate (ca. 3–10%), with smaller amounts of menthofuran (1–7%) and limonene.[1,4–6] Other constituents present in peppermint oil include viridiflorol,[7,8] pulegone (1–11%),[9] 1,8-cineole (6–7.5%), piperitone, caryophyllene, bisabolene, isomenthone, isomenthol, α- and β-pinenes,[10] neomenthol,[5] ledol, *d-trans*-sabinene hydrate,[11] and bicycloelemene, among others (REMINGTON).[12]

Other constituents present in peppermint include flavonoids (e.g., menthoside, isorhoifolin, hesperetin, eriodictyol-7-*O*-rutinoside, luteolin-7-*O*-rutinoside, and rutin),[13] phytol, tocopherols (α and γ), carotenoids (e.g., α- and β-carotenes), betaine, choline, azulenes, rosmarinic acid,[14] and tannin.[15–20]

Spearmint yields normally about 0.7% volatile oil, consisting of 50–70% carvone, with lesser amounts of dihydrocarvone, phellandrene, and limonene.[21] Other compounds reported present include 6-hydroxycarvone,[22] menthone, menthol, pulegone, piperitenone, piperitenone oxide,[23] *cis*-carveyl acetate,[24] carveol, myrcene, α- and β-pinenes, cineole, linalool, α-terpineol, terpinen-4-ol, terpinolene,[1] dihydrocarveol, dihydrocarveol acetate; caryophyllene, 3-octyl acetate, 3-octanol; menthofuran,[25] and *cis*-hexenyl isovalerate, among others (JIANGSU, REMINGTON).[26]

Other constituents present in spearmint include flavonoids (e.g., diosmin and diosmetin)[27] and probably similar compounds as those found in peppermint.

Cornmint contains 1–2% volatile oil that consists of a high concentration of menthol (70–95%),[28–30] menthone (10–20%),[31] menthyl acetate, isomenthone, thujone, α-pinene, phellandrene, piperitone, menthofuran, and others (JIANGSU, LIST AND HÖRHAMMER, MORTON 3).[25]

A recent cultivar of *M. arvensis* has been reported to contain high concentrations of menthofuran (42%) and lesser amounts of menthol (21%), menthyl acetate (14%), and menthone (0.71%).[32]

Other compounds present in cornmint include oligosaccharides (raffinose and stachyose),[33] resin, tannin, and rosmarinic acid (JIANGSU).[20]

## PHARMACOLOGY OR BIOLOGICAL ACTIVITIES

Peppermint and cornmint oils have antimicrobial activities *in vitro*.[34–37]

Peppermint extracts have been reported to have antiviral activities against Newcastle disease, herpes simplex, vaccinia, Semliki Forest, and West Nile viruses in egg and cell-culture systems (see ***balm***).[38]

Peppermint oil has been demonstrated to exhibit spasmolytic activity on smooth muscles of experimental animals.[39]

The flavonoids in peppermint leaves reportedly have choleretic activity in dogs.[17]

Azulene isolated from peppermint had antiinflammatory and antiulcer effects in experimental animals.[40–42]

Peppermint and cornmint oils have been reported to have cytotoxic properties.[43]

Menthol, the major component of peppermint and cornmint oils, may cause allergic reactions (e.g., contact dermatitis, flushing, and headache) in certain individuals. Applying a menthol-containing ointment to the nostrils of infants for the treatment of cold symptoms may cause instant collapse (MARTINDALE).

## USES

**Medicinal, Pharmaceutical, and Cosmetic.** Peppermint oil is extensively used as a flavoring agent, carminative, antiseptic, and local anesthetic in cold, cough, and other preparations (lozenges, syrups, ointments, tablets, etc.).

Enteric coated peppermint oil capsules have been examined as a useful treatment for irritable bowel syndrome; enteric coating allows oil to reach colon in an unmetabolized state; treatment is contraindicated with meals (dosage recommended between meals), and in achlorhydria.[44–46]

Peppermint oil has been recommended as an adjunct to colonoscopy; a diluted suspension of the oil is sprayed on the endoscope to reduce colonic spasm.[47,48]

Spearmint oil is used primarily as a flavoring agent; it is also used as a carminative.

The mint oils (especially spearmint oil) are commonly used as fragrance components in toothpastes, mouthwashes, gargles, soaps, detergents, creams, lotions, and perfumes. Maximum use levels reported for spearmint and cornmint oils are 0.4%–0.8%, respectively, in perfumes.[26,49]

**Food.** Spearmint oil and peppermint oil (usually rectified) are extensively used in flavoring chewing gums, candies, and chocolates as well as in most other food products, including alcoholic (liqueurs, etc.) and nonalcoholic beverages, frozen dairy desserts, baked goods, gelatins and puddings, processed fruits, and sweet sauces. The highest average maximum use levels reported are 0.104% for peppermint oil in candy and about 0.132% (1318 ppm) for spearmint oil in baked goods.

Spearmint leaves and extracts are reported used in alcoholic and nonalcoholic beverages. Spearmint leaves are also used in baked goods, gelatins and puddings, and meat and meat products. Highest average maximum use level is about 0.597% (5967 ppm) for spearmint leaves in baked goods.

**Health Food/Herb Teas.** Leaves (or oil) of peppermint and spearmint, widely used as primary or adjunct flavoring for herb teas; capsules, tablets, tincture, etc., in formulations for digestion, colds, and fevers (FOSTER).

**Traditional Medicine.** Peppermint, spearmint, and their oils reportedly used in both western and eastern cultures as aromatic, stomachic, stimulant, antiseptic, local anesthetic, and antispasmodic in treating indigestion, nausea, sore throat, diarrhea, colds, headaches, toothaches, and cramps (LEUNG).

Cornmint is used in China for treating similar conditions; also used in relieving earache and treating tumors and sores.

All three mints have been reported used in cancers.[50]

**Others.** Peppermint oil and menthol are widely used in flavoring tobacco.

Cornmint oil is mainly used for the production of menthol. Due to the high concentrations of menthol in this oil, it will solidify at room temperature. Much of the menthol can be removed from the crude oil (which may contain about 90% menthol) by freezing. The "dementholized" oil still contains about 55% menthol; this oil is the commercial cornmint oil. It can be further used as a source of menthol.

## COMMERCIAL PREPARATIONS

Crudes (peppermint and spearmint) and oils. Peppermint, spearmint, and peppermint oil and spearmint oil are official in N.F. Peppermint oil, spearmint oil, and dementholized cornmint oil are official in F.C.C.

*Regulatory Status*. Peppermint, spearmint, and their derivatives (e.g., oils) are GRAS (§182.10 and §182.20).

The recent proposed European monograph suggests use of the oil, only under the direction of a physician, for bile duct and gallbladder inflammation and gallstones; also for spasms of the upper gastrointestinal tract, flatulence, symptomatic treatment of irritable bowel syndrome, and catarrh of the respiratory tract. External use includes oral mucosa inflammations, rheumatic conditions, and local muscle and nerve pain as well as skin conditions such as pruritus and urticaria (ESCOP 3).

Peppermint leaf and oil are subjects of German therapeutic monographs; leaves in infusion, or extract for spastic complaints of the gastrointestinal tracts as well as gallbladder and bile ducts, at average daily dose of 3–6 g of the leaves; 5–15 g tincture.[51,52]

## REFERENCES

See the General References for APhA; ARCTANDER; BAILEY 2; BLUMENTHAL; ESCOP 3; FARNSWORTH; FEMA; FOGARTY; FOSTER; GOSSELIN; GUENTHER; JIANGSU; LIST AND HÖRHAMMER; MARTINDALE; NANJING; REMINGTON; ROSENGARTEN; TERRELL; UPHOF; USD 26th.


1. F. W. Hefendehl and M. J. Murray, *Planta Med.*, **23**, 101 (1973).
2. B. M. Lawrence and J. K. Morton, *An. Acad. Bras. Cienc.*, **44**(Suppl.), 38 (1972); through *Chem. Abstr.*, **83**, 128676f (1975).
3. A. O. Tucker et al., *Taxon*. **29**(2–3), 233 (1980).
4. M. B. Embong et al., *Can. Inst. Food Sci. Technol. J.*, **10**, 247 (1977).
5. K. Belafi-Rethy et al., *Acta Chim. (Budapest)*, **76**, 167 (1973); through *Chem. Abstr.*, **80**, 19365w (1974).
6. A. K. Saxena et al., *Indian Perfum.*, **20**(1-B), 1 (1976); through *Chem. Abstr.* **88**, 141484q (1978).
7. W. Rojahn et al., *Dragoco Rep. (Ger. Ed.)*, **24**, 230 (1977); through *Chem. Abstr.*, **88**, 141478r (1978).
8. D. Karasawa and S. Shimizu, *Shinshu Daigaku Nogakubu Kiyo*, **13**, 89 (1976); through *Chem. Abstr.*, **87**, 28840u (1977).
9. A. Morkunas et al., *Polez. Rast. Priblat. Respub. Beloruss., Mater Nauch. Knof.*, **2**, 225 (1973); through *Chem. Abstr.*, **81**, 111398c (1974).
10. I. Calvarano, *Essenze Deriv. Agrum.*, **39**, 77 (1969); through *Chem. Abstr.*, **72**, 107806f (1970).
11. Y. Ono et al., *Bunseki Kagaku*, **24**, 589 (1975); through *Chem. Abstr.*, **83**, 204916b (1975).
12. B. M. Lawrence et al., *Flav. Ind.*, **3**, 467 (1972).
13. B. G. Hoffmann and L. T. Lunder, *Planta Med.*, **50**, 361 (1984).
14. U. Gerhardt and A. Schroeter, *Fleischwirtschaft*, **63**, 1628 (1983).
15. E. V. Gella et al., *Farmatsevt. Zh.*, **21**, 58 (1966); through *Chem. Abstr.*, **65**, 13810e (1966).
16. H. Wagner et al., *Chem. Ber.*, **102**, 2083 (1969).
17. I. K. Pasechnik, *Farmakol. Toksikol.*, **29**, 735 (1966); through *Chem. Abstr.*, **66**, 54111t (1967).
18. M. F. Shakhova and L. O. Shnaidman, *Rast. Resur.*, **4**, 53 (1968); through *Chem. Abstr.*, **69**, 25087w (1968).

19. N. M. Solodovnichenko and Y. G. Borisyuk, *Farmatsevt. Zh.*, **17**, 44 (1962); through *Chem. Abstr.*, **58**, 8846e (1963).
20. B. E. Ellis and G. H. N. Towers, *Biochem. J.*, **118**, 291 (1970).
21. M. J. Murray et al., *Crop Sci.*, **12**, 723 (1972).
22. T. Tsuneya et al., *Koryo*, **104**, 23 (1973); through *Chem. Abstr.*, **80**, 40925p (1974).
23. S. Shimizu et al., *Int. Congr. Essent. Oils (Pap.)*, **6**, 13 (1974).
24. T. Nagasawa et al., *Koryo*, **108**, 45 (1974); through *Chem. Abstr.*, **82**, 129171d (1975).
25. I. C. Nigam and L. Levi, *J. Pharm. Sci.*, **53**, 1008 (1964).
26. D. L. J. Opdyke, *Food Cosmet. Toxicol.*, **16**(Suppl. 1), 871 (1978).
27. S. S. Subramanian and A. G. R. Nair, *Phytochemistry*, **11**, 452 (1972).
28. M. L. Sharma et al., *Indian Perfum.*, **16**(Pt. 2), 27 (1972); through *Chem. Abstr.*, **80**, 100105h (1974).
29. M. M. Chopra and K. L. Handa, *Indian Perfum.*, **16**(Pt. 2), 15 (1972); through *Chem. Abstr.*, **80**, 100104g (1974).
30. A. K. Srivastava et al., *Indian Perfum.*, **20**(1-B), 61 (1976); through *Chem. Abstr.*, **88**, 141485r (1978).
31. M. J. Murray et al., *Crop Sci.*, **12**, 742 (1972).
32. N. G. R. Donalisio et al., *Int. Congr. Essent. Oils (Pap.)*, **6**, 10 (1974).
33. A. Lombard et al., *J. Chromatogr.*, **134**, 242 (1977).
34. A. Sanyal and K. C. Varma, *Indian J. Microbiol.*, **9**, 23 (1969).
35. A. C. Pizsolitto et al., *Rev. Fac. Farm. Odontol. Araraquara*, **9**, 55 (1975); through *Chem. Abstr.*, **86**, 12226s (1977).
36. K. K. Abdullin, *Uch. Zap. Kazansk. Vet. Inst.*, **84**, 75 (1962); through *Chem. Abstr.*, **60**, 11843b (1964).
37. F. M. Ramadan et al., *Chem. Mikrobiol. Technol. Lebensm.* **1**, 96 (1972).
38. E. C. Herrmann Jr. and L. S. Kucera, *Proc. Soc. Exp. Biol. Med.*, **124**, 874 (1967).
39. T. Shipochliev, *Vet. Med. Nauki*, **5**, 63 (1968); through *Chem. Abstr.*, **70**, 86144e (1969).
40. G. N. Maksimenko, *Farmakol. i. Toksikol.*, **27**, 571 (1964); through *Chem. Abstr.*, **62**, 3285c (1965).
41. B. D. Taylor et al., *Gut.* **24**, 992 (1983).
42. C. Briggs, *Can. Pharm. J.*, **89** (Mar. 1993).
43. K. Silyanovska et al., *Parfüm Kosmet.*, **50**, 293 (1969).
44. K. W. Somerville et al., *Br. J. Clin. Pharmac.*, **18**, 638 (1984).
45. W. D. Rees et al., *Brit. Med. J.*, 835 (Oct. 6, 1979).
46. S. Foster, *Peppermint—Mentha × piperita, Botanical Series, no. 301,* American Botanical Council, Austin, Tex., 1991.
47. R. J. Leicester and R. H. Hunt, *Lancet*, 989 (Oct. 30, 1982).
48. H. L. Duthie, *Br. J. Surg.*, **68**, 820 (1981).
49. D. L. J. Opdyke, *Food Cosmet. Toxicol.*, **13**, 771 (1975).
50. J. L. Hartwell, *Lloydia*, **32**, 247 (1969).
51. Monograph *Menthae piperitae aetheroleum*, *Bundesanzeiger*, no. 50 (Mar. 13, 1986).
52. Monograph *Menthae piperitae folium*, *Bundesanzeiger*, no. 223 (Nov. 30, 1985); corrected (Mar. 13, 1990).

# MISTLETOE

**Source:** ***Viscum album*** L. (Family Loranthaceae or Viscaceae).

## GENERAL DESCRIPTION

Parasitic shrub, stems yellow-green, up to 100 cm long; leaves opposite, obovate-oblong, 2–8 cm long; flowers in cymes, unisexual, four-merous; berry white; occurring on woody angiosperms and gymnosperms; most of Europe; naturalized in one California county.

The part used is the herb.

## CHEMICAL COMPOSITION

All plant parts contain β-phenylethylamine, tyramine, and related compounds; polypeptides, including viscotoxins I, II, III, IVb (II, III, IVb identical to viscotoxins B, A-2, and A-3, respectively);[1] glycoprotein lectins, including viscumin and lectins 1, II, and III;[2] phenylpropanoids, including syringin, syringenin-apiosylglucoside, and 4,4″-diglucoside (eleutheroside E);[3] caffeic and gentisic acids, polysaccharides; and others (WREN).

## PHARMACOLOGY OR BIOLOGICAL ACTIVITIES

Various pharmacological activities include hypotensive, cardiotonic, immunostimulant, antineoplastic, sedative and antispasmodic (WREN).

Lignans are believed responsible for cardiotonic activity, significantly inhibiting cAMP-phosphodiesterase activity.[4]

Antiinflammatory and immunostimulatory activity has been confirmed in a number of *in vitro* models, stimulating the production of lymphokines by lymphocytes.[5,6] However, a study on isolated polysaccharides from the stem (a galacturonan), and from the berries (an arabinogalactan), failed to increase phagocytosis of granulocytes and macrophages.[7]

Isolated polypeptides, viscotoxins II, III, and IVb have been associated with cardiotoxicity and have been found to exhibit cytotoxic activity against human tumor cells of the KB and HeLa lines in tissue culture.[1] A peptide with a molecular weight of 5000 was found to be cytotoxic to Dalton's lymphoma ascites tumor cells *in vitro* in mice, without affecting normal lymphocytes, indicating a cell-dependent specificity;[8] also cyctotoxic to Ehrlich ascites cells, both prophylactically and after tumor development.[9] Commercial mistletoe products have been used to treat various cancers in Europe with clinical success. A group of 50 cases of carcinomatous pleural effusions were treated with a topical preparation for an average of 3.3 application over 18 days; exudation disappeared in 92% of the patients.[10] In postoperative ovarian cancer patients a mistletoe preparation statistically increased survival.[11] Selective cytotoxicity and immunopotentiating activity of mistletoe lectins and their chains are considered primarily responsible for the controversial antineoplastic effects attributed to it.[12]

Often regarded as a poisonous plant, toxic effects are primarily associated with parenteral administration. Large quantities taken orally may be locally irritating and necrotising. Given potential toxicity, self-medication is not advised (FROHNE AND PFÄNDER). An analysis of more than 300 reported mistletoe ingestion cases in the United States found that a majority of patients were asymptomatic; no fatalities occurred; ingestion of fewer than three berries or two leaves will likely not produce serious toxicity.[2]

## USES

**Medicinal, Pharmaceutical, and Cosmetic.** In German phytomedicine fresh plant, cut, or powdered herb in injectable solutions are used for supportive therapy of nonspecific irritation in malignant tumors; degenerative inflammatory joint diseases; contraindicated in protein hypersensitivity, chronic progressive infections. Reported side effects include fever, headaches, angina pectoris symptoms, and allergic reactions.[13]

**Traditional Medicine.** Mistletoe has been employed for the treatment of various disorders, including sores and abscesses, and nervous disorders; an abortifacient and antineoplastic for over two millenia.[14]

**COMMERCIAL PREPARATIONS**

Five parenteral-administered product forms available in Germany.[14]

*Regulatory Status.* Leaves subject of a German therapeutic monograph.[13] Fruit indicated for numerous conditions in combination with other herbs is not recommended due to lack of documentation of efficacy.[15]

**REFERENCES**


See the General References for BLUMENTHAL; DER MARDEROSIAN AND LIBERTI; FOSTER AND DUKE; MARTINDALE; STEINMETZ; TUTIN 1; TYLER 1; WEISS, WREN.

1. J. Konopa et al., *Hoppe Seylers Z. Physiol Chem.*, **361**, 10 (1980).
2. B. Olin, Ed., *Lawrence Rev. Natl. Prod.* (Dec. 1992).
3. H. Wagner et al., *Oncology*, **43**(S1), 16 (1986).
4. H. Wagner et al., *Planta Med.*, **52**(2), 102 (1986).
5. N. Bloksma et al., *Immunobiology*, **156**(3), 309 (1979).
6. E. G. Coeugniet and E. Elek, *Onkologie*, **10**(S3), 27 (1987).
7. E. Jordan and H. Wagner, *Oncology*, **43**(S1), 8 (1986).
8. G. Kuttan et al., *Cancer Lett.*, **41**(3), 307 (1988).
9. G. Kuttan et al., *J. Ethnopharmacol*, **29**(1), 35 (1990).
10. G. Slazer and H. Muller, *Praxis und Klinik der Pneumologia*, **32**(11), 721 (1978).
11. W. Hassauer et al., *Onkolgie*, **2**(1), 28 (1979).
12. H. Franz, *Oncology*, **43**(S1), 23 (1984).
13. Monograph *Visci albi herba*, *Bundesanzeiger*, no. 228 (Dec. 5, 1984).
14. I. A. Bowman, *HerbalGram*, **26**, 16 (1992).
15. Monograph *Visci albi fructus*, *Bundesanzeiger*, no. 228 (July 14, 1993).


# MONOSODIUM GLUTAMATE (MSG)

**Source:** Wheat gluten and other natural sources of glutamic acid are used for the manufacture of MSG.

*Synonyms.* Sodium glutamate and monosodium L-glutamate.

**GENERAL DESCRIPTION**

Monosodium glutamate (MSG) is a monosodium salt of L-glutamic acid, which is a common amino acid present in high concentrations in brain tissues and is found in most proteins.[1] Certain proteins such as gluten (corn and wheat) are very rich in this amino acid. Wheat gluten contains about 38% glutamic acid and has been used as one of the major sources for the manufacture of MSG.[2–6]

Monosodium glutamate can generally be produced by three methods: (*1*) hydrolysis of proteins such as gluten or proteins present in sugar beet wastes, (*2*) synthesis, and (*3*) microbial fermentation.[5,6] In the hydrolysis method, the protein is hydrolized with a strong mineral acid to free amino acids, and the glutamic acid is then separated from the mixture, purified, and converted to its monosodium salt, MSG. This used to be the major method of MSG manufacture. Currently most of the world production of MSG is by bacterial fermentation. In this method bacteria (especially strains of *Micrococcus glutamicus*) are grown aerobically in a liquid nutrient medium containing a carbon source (e.g., dextrose or citrate), a nitrogen source such as ammonium ions or urea, and mineral ions and growth factors. The bacteria selected for this process have the ability to excrete glutamic acid they synthesize outside of their cell membrane into the medium and accumulate there. The glutamic acid is separated from the fermentation broth by filtration, concentration, acidification, and crystallization, followed by conversion to its monosodium salt, MSG.[5,6]

Monosodium glutamate is considered to be a flavor enhancer, which enhances or intensifies the flavor of other foods. Its flavor-enhancing property was discovered in the early 1900s by the Japanese who first started producing it in a commercial scale and have since been the major producer.[7] Monosodium glutamate has a sweetish meaty taste.

## PHARMACOLOGY OR BIOLOGICAL ACTIVITIES

Monosodium glutamate has been used in treating mental retardation and hepatic coma that is accompanied by a high blood level of ammonia. It has been reported effective in lowering the blood level of ammonia in many cases, though the mechanism of action is still not known (MARTINDALE, USD 26th).

It is presently embroiled in a controversy that appears unlikely to be resolved for years to come. Many scientific investigations have demonstrated MSG to cause brain damage in infant experimental animals (mice, rats, rhesus monkeys, etc.) and to produce the so-called Chinese restaurant syndrome (burning sensation, facial pressure, and chest pain) in humans.[8–21] However, as many research efforts have been spent by as many different groups of investigators resulting in negative findings.[22–33] It should be noted that only one laboratory has reported both positive and negative findings.[13,16,24] The other laboratories have observed consistently either one or the other but not both.[11,12,14,15,20–23,27]

Monosodium glutamate has also been reported to have emetic properties in experimental animals (dogs, cats, and monkeys),[34] to induce clonic movements and tonic-clonic seizures in rats,[35,36] to cause obesity in mice,[37] to cause sterility in female mice, and other activities.[17,26,38]

## USES

**Medicinal, Pharmaceutical, and Cosmetic.** It is used as flavor enhancer in certain liver and protein hydrolysate preparations.

**Food.** Monosodium glutamate is very popular in Japanese and Chinese cooking. It is liberally used in restaurants, especially in soups and vegetarian dishes.

Monosodium glutamate is used very extensively in processed foods, including nonalcoholic beverages, candy, baked goods, meat and meat products, condiments and relishes, breakfast cereals, milk products, cheeses, processed fruits, processed vegetables, soups, snack foods, nut products, gravies, seasonings and flavorings, and fish products, among others. Highest average maximum use levels are about 24.68% (246,785 ppm), 1.07%, and about 0.803% (8034 ppm) reported in seasonings and flavorings, breakfast cereals, and soups, respectively.

**COMMERCIAL PREPARATIONS**

Monosodium glutamate. It is official in N.F. and F.C.C.

*Regulatory Status.* GRAS.

**REFERENCES**

See the General References for FEMA; FURIA; MARTINDALE; MERCK; USD 26th.

1. A. White et al., *Principles of Biochemistry*, 2nd ed., McGraw-Hill, New York, 1959.
2. K. Hess and E. Hille, *Z. Lebensm. Unters. Forsch.*, **115**, 211 (1961).
3. H. Ludewig and T. Messing, Ger. (East) 37,074, (1965); through *Chem. Abstr.*, **63**, 18259c (1965).
4. E. Magyar et al., *Staerke*, **15**, 12 (1963); through *Chem. Abstr.*, **59**, 12913d (1963).
5. W. L. Faith et al., *Industrial Chemicals*, 3rd ed., John Wiley & Sons, New York, 1965, p. 521.
6. R. Powell, *Monosodium Glutamate and Glutamic Acid*, Noyes Development Corp., Park Ridge, N.J., 1968.
7. H. J. Sanders, *Chemistry*, **40**, 23 (1967).
8. T. Uehara et al., *Okinawa-ken Kogai Eisei Kenkyusho Ho*, **10**, 34 (1976); through *Chem. Abstr.*, **88**, 84335r (1978).
9. M. J. Kuhar, *Res. Commun. Chem. Pathol. Pharmacol.*, **2**, 95 (1971); through *Chem. Abstr.*, **74**, 109861k (1971).
10. T. Fujiwara et al., *Jutsugo Taisha Kenkyu Kaishi*, **10**, 385 (1976); through *Chem. Abstr.*, **88**, 69172s (1978).
11. E. A. Arees and J. Mayer, *Science*, **170**, 549 (1970).
12. P. E. Araujo and J. Mayer, *Am. J. Physiol.*, **225**, 764 (1973).
13. N. Lemkey-Johnston and W. A. Reynolds, *J. Neuropathol. Exp. Neurol.* **33**, 74 (1974).
14. N. Snapir et al., *Pathol. Eur.*, **8**, 265 (1973); through *Chem. Abstr.*, **81**, 58934w (1974).
15. B. Robinson et al., *Poult. Sci.*, **54**, 234 (1975).
16. N. Lemkey-Johnston et al., *J. Comp. Neurol.*, **167**, 481 (1976); through *Chem. Abstr.*, **85**, 187241z (1976).
17. J. L. Everly, *Diss. Abstr. Int.*, *B*, **33**, 1351 (1972).
18. S. Ungthavorn et al., *J. Fac. Med. Chulalongkorn Univ., Bangkok*, **16**, 265 (1971); through *Chem. Abstr.*, **78**, 24890b (1973).
19. H. H. Schaumburg et al., *Science*, **163**, 826 (1969).
20. J. W. Olney and L. G. Sharpe, *Science*, **167**, 1017 (1970).
21. J. W. Olney, *Science*, **165**, 1029 (1969).
22. R. Heywood et al., *Toxicol. Lett.*, **1**, 151 (1977).
23. G. Owen et al., *Toxicol. Lett.*, **1**, 217 (1978).
24. W. A. Reynolds et al., *Science*, **172**, 1342 (1971).
25. P. Morselli and S. Garattini, *Nature (London)*, **227**, 611 (1970).
26. G. Bazzano et al., *Science*, **169**, 1208 (1970).
27. A. J. Newman et al., *Toxicology*, **1**, 197 (1973).
28. M. E. Semprini et al., *Nutr. Metab.*, **16**, 276 (1974).
29. B. L. Oser et al., *Food Cosmet. Toxicol.*, **13**, 7 (1975).

30. N. J. Adamo and A. Ratner, *Science*, **169**, 673 (1970).
31. I. P. Barchenko and S. G. Vasiliu, *Vop. Ratsion. Pitan.*, **6**, 21 (1970); through *Chem. Abstr.*, **77**, 1345t (1972).
32. I. Rosenblum et al., *Toxicol. Appl. Pharmacol.*, **18**, 367 (1971).
33. S. Matsuyama et al., *Natl. Inst. Anim. Health Q.*, **13**, 91 (1973); through *Chem. Abstr.*, **79**, 144610c (1973).
34. W. P. Pi and M. T. Peng, *Taiwan I Hsueh Hui Tsa Chih*, **64**, 669 (1965); through *Chem. Abstr.*, **65**, 7863b (1966).
35. C. N. Stewart et al., *Toxicol. Appl. Pharmacol.*, **23**, 635 (1972).
36. C. B. Nemeroff and F. D. Crisley, *Pharmacol. Biochem. Behav.*, **3**, 927 (1975).
37. J. Bunyan et al., *Br. J. Nutr.*, **35**, 25 (1976).
38. P. Cooper, *Food Cosmet. Toxicol.*, **15**, 347 (1977).

# MUSK

**Source:** ***Moschus moschiferus*** L. (Family Moschidae).

*Synonyms*. Tonquin musk, musk Tonquin, and deer musk.

## GENERAL DESCRIPTION

The musk deer is a relatively small solitary animal standing about 0.5 m high at the shoulder and measuring 0.65–0.95 m long; both male and female are devoid of antlers. It is distributed in mountainous regions of Asia such as northern India, Tibet, and southern, western, northern, and northeastern China (JIANGSU).[1,2]

Musk is the highly odoriferous secretion derived from the musk gland present under the abdomen near the pubis of the male animal. There are two methods of obtaining musk. In the first method the male musk deer is trapped and killed in late winter or early spring and the whole musk gland is immediately removed from the abdomen. After drying, the whole gland is known as a pod and the secretion inside in the form of granules is called grained musk, or musk grains. This used to be the only method of collecting musk. Nowadays musk is collected more and more from male musk deer raised in captivity. In this method the male deer is tied to a special table on its back and the secretion is carefully removed from the musk gland with a special sterilized spoon. Healthy male deer at least 3 years old are used. Musk is collected once a year in late winter or early spring; sometimes it is also collected twice a year, in March or April and in July or August. The fresh secretion is a dark brown viscous semisolid, which turns into brownish yellow or purplish red granules when dried (JIANGSU, POUCHER).

Major musk producers include China and India.

It should be noted that the term *musk* is sometimes also used to describe substances or compounds that have an odor similar to that of true musk grains; these "musks" can be of synthetic or natural origins. (MERCK).[3]

## CHEMICAL COMPOSITION

Musk normally contains as its odorous and active principles 0.3–2% muscone together with small quantities of normuscone (JIANGSU, LIST AND HÖRHAMMER).[4–6]

Other constituents present in musk include steroids (cholesterol and its esters, cholest-4-en-3-one, etc.); paraffins (long chain and branched); triglycerides (oleodipalmitin, palmitodiolein, and triolein); methyl palmitate and methyl oleate; wax;[5,6]

muscopyridine, hydroxymuscopyridine A and B, and other nitrogenous substances (e.g., urea and ammonium carbonate);[7] mineral salts; and fatty acids (JIANGSU, LIST AND HÖRHAMMER).

## PHARMACOLOGY OR BIOLOGICAL ACTIVITIES

Musk has antiinflammatory and antihistaminic activities on experimental animals.[8-10] Its antiinflammatory activity was greater than that of phenylbutazone against arthritis in rats induced by injection of dead tubercle bacteria in liquid paraffin.[9] Its water-soluble fraction has the strongest antiinflammatory activity, being 36 times that of hydrocortisone in mouse ear edema induced by croton oil. The active principle is a polypeptide with a molecular weight of about 10,000 whose structure has not been determined.[11,12]

Musk has also been reported to have spasmolytic, CNS-depressant as well as stimulant, antibacterial, and other activities (JIANGSU).[10,13,14]

## USES

**Medicinal, Pharmaceutical, and Cosmetic.** The use of musk in cosmetics (e.g., fragrance and perfumes) dates back at least 1300 years when it was widely used during the Tang and Sung dynasties (NANJING).

It is used as a fragrance component and fixative in perfumes (especially Oriental and heavy floral types).

**Food.** Used for its "rounding off" effect in nut, caramel, and fruit-type flavors in major food products, including alcoholic and nonalcoholic beverages, frozen dairy desserts, candy, baked goods, and gelatins and puddings. Use levels are very low, generally much below 0.0001% (1 ppm).

**Traditional Medicine.** Musk has been used in Chinese medicine for thousands of years in treating stroke, coma, neurasthenia, convulsions, heart pains, ulcerous sores, and other conditions. It has been used in the clinical treatment of angina pectoris with results (ca. 74%) comparable or better than those of nitroglycerin (JIANGSU).

The musk used in Chinese medicine has been treated differently than the musk that is intended for export. For example, the former is dried only in the shade, while the latter is sun or heat dried (NANJING).

## COMMERCIAL PREPARATIONS

Crude and extracts (e.g., tincture and absolute). Musk was formerly official in U.S.P.

*Regulatory Status.* GRAS (§182.50).

## REFERENCES

See the General References for ARCTANDER; FEMA; JIANGSU; MARTINDALE; NANJING; POUCHER.

1. *The Larousse Encyclopedia of Animal Life*, McGraw-Hill, New York, 1967, p. 596.
2. M. Burton, Ed., *The World Encyclopedia of Animals*, Funk and Wagnalls, New York, 1972, p. 255.
3. W. E. Brugger and P. C. Jurs, *J. Agr. Food Chem.*, **25**, 1158 (1977).
4. H. Y. Hsu et al., *T'ai-wan Yao Hsueh Tsa Chih*, **25**, 26 (1973); through *Chem. Abstr.*, **84**, 95516u (1976).
5. J. C. Do, *Yongnam Taehakkyo Chonyonmul Hwahak Yonguso Yongu Pogo*, **3**, 19 (1976); through *Chem. Abstr.*, **88**, 110379e (1978).
6. J. C. Do et al., *Chem. Pharm. Bull.*, **23**, 629 (1975).

7. D. Q. Yu and B. C. Das, *Planta Med.*, **49**, 183 (1983).
8. R. K. Mishra et al., *J. Pharm. Pharmacol.*, **14**, 830 (1962).
9. H. H. Siddiqui, *Indian J. Pharm.*, **27**, 80 (1965).
10. S. D. S. Seth et al., *Jpn. J. Pharmacol.*, **23**, 673 (1973).
11. X. Y. Zhu et al., *Yaoxue Xuebao*, **23**, 406 (1988).
12. X. Y. Zhu et al., *Acta Acad. Med. Sin.*, **11**, 52 (1989).
13. M. Kimura et al., *Yakugaku Zasshi*, **88**, 130 (1968); through *Chem. Abstr.*, **69**, 1733c (1968).
14. A. Mukhopadhyay et al., *Indian J. Pharm.*, **35**, 169 (1973).

# MUSTARD

**Source:** ***Brown mustard*** *Brassica juncea* (L.) Czern. et Coss. (syn. *Sinapis juncea* L.); ***Black mustard*** *Brassica nigra* (L.) Koch; ***White mustard*** *Sinapis alba* L. (syn. *B. alba* (L.) Rabenh.; *B. alba* (L.) Boiss.; *B. hirta* Moench) (Family Cruciferae or Brassicaceae).

*Synonyms*. Indian mustard (*B. juncea*); yellow mustard (*S. alba*).

## GENERAL DESCRIPTION

Annual or biennial herbs; brown and white mustards up to about 1 m, black mustard up to 3 m high; white mustard and black mustard are native to Eurasia (probably the Mediterranean region), and brown mustard is native to Asia. All three are cultivated worldwide. Parts used are their dried ripe seeds.

Mustard flour (ground mustard) is powdered mustard seeds with their seed coats removed. It often consists of a mixture of brown (or black) and white seeds, especially certain British and Chinese types. The more pungent ("hot") mustard flours are those with the fixed oil removed. The fixed oil constitutes more than one-third of the flour and does not contribute to the pungency or aroma of the mustard. Ground mustard does not have any pungent aroma when dry. This aroma is produced when the mustard comes in contact with water whereby the enzyme systems present in the mustard hydrolyze sinigrin (a glucoside of brown and black mustard), setting free allyl isothiocyanate, which is mainly responsible for the aroma.

Mustard oil (volatile) is prepared by steam distillation from brown or black mustard after expressing the fixed oil (expressed mustard oil) and macerating in warm water to allow the hydrolysis of sinigrin by the enzyme myrosin. The volatile oil consists mainly of allyl isothiocyanate.

White mustard does not produce a volatile oil by water and steam distillation.

## CHEMICAL COMPOSITION

Brown mustard (*B. juncea*) contains the glucosinolate sinigrin (potassium myronate) and the enzyme myrosin (myrosinase); sinapic acid; sinapine (sinapic acid choline ester); fixed oils (25–37%), consisting mainly of glycerides of erucic, eicosenoic, arachidic, nonadecanoic, behenic, oleic, and palmitic acids, among others;[1] proteins (e.g., globulins);[2] and mucilage (JIANGSU, NANJING).

Sinigrin on hydrolysis by myrosin (myrosinase) yields allyl isothiocyanate, glucose, and potassium bisulfate. Allyl isothiocyanate is volatile; its yield from *B. juncea* is 0.25–1.4% (usually ca. 0.9%)[3–8] Other minor volatile components that are also set free by enzymatic hydrolysis include methyl, isopropyl, *sec*-butyl, butyl, 3 butenyl, 4-pentenyl, phenyl, 3-methylthiopropyl, benzyl, and $\beta$-phenylethyl isothiocyanates.[7–9]

Black mustard (*B. nigra*) contains similar constituents as *B. juncea* (LIST AND HÖRHAMMER).

White, or yellow, mustard (*S. alba*) contains the glucosinolate sinalbin, which on hydrolysis by enzymes present (myrosin or glucosinolases) yields *p*-hydroxybenzyl isothiocyanate (a relatively nonvolatile compound);[3,5,6,9,10] *p*-hydroxybenzylamine;[11] and other similar components (proteins, fixed oils, sinapine, mucilage, etc.) as brown mustard (JIANGSU, MARSH).[1–3,6,12,13]

**PHARMACOLOGY OR BIOLOGICAL ACTIVITIES**

Volatile mustard oil (or allyl isothiocyanate) is an extremely powerful irritant and produces blisters on the skin. It is also tear producing (lachrymatory) and has counterirritant properties when greatly diluted (e.g., 1 in 50). It should not be tasted or inhaled when undiluted. It is one of the most toxic essential oils (GOSSELIN, JIANGSU, MARTINDALE).[14]

*p*-Hydroxybenzyl isothiocyanate from white mustard does not have lachrymatory properties, but has a very pungent taste (ARCTANDER).

Isothiocyanates such as those present in mustard have been implicated in endemic goiter (hypothyroidism with thyroid enlargement). They have also been reported to produce goiter in experimental animals (LIENER)[15]

Volatile mustard oil has strong antimicrobial (bacteria and fungi) properties.[16,17]

Sinigrin has been reported to be toxic to certain insect larvae but harmless to others.[18]

Recent evidence suggests that glucosinolate products of the Brassicaceae may have protective effects against a variety of carcinogens.[19]

**USES**

**Medicinal, Pharmaceutical, and Cosmetic.** The volatile oil (allyl isothiocyanate) is used in certain rubefacient and counterirritant liniments.

**Food.** Mustard (esp. white) is extensively used in prepared mustards, where it is commonly used with vinegar and other spices; the bright yellow color is usually due to turmeric. Other food products in which mustard is used include baked goods, meat and meat products, processed vegetables, fats and oils, snack foods gravies, and nut products, among others. Highest average maximum use levels are about 12.4% (124,274 ppm) and 10.1% (101,503 ppm) reported for yellow and brown mustards, respectively, in condiments and relishes (e.g., prepared mustards).

Mustard oil (allyl isothiocyanate) is widely used as a flavor ingredient in nonalcoholic beverages, frozen dairy desserts, candy, baked goods, gelatins and puddings, meat and meat products, condiments and relishes (e.g., pickles), fats and oils, and gravies. Highest average maximum use level reported is about 0.02% (201 ppm) in gravies.

**Health Food/Herb Teas.** Sometimes used in ointments or liniments intended to relieve symptoms of colds. Mustard oil, absorbed through the skin, is eliminated via the lungs, hence use in liniment preparations for relief of lung congestion (WEISS).

**Traditional Medicine.** Both brown (black) and white mustards are used as an appetizer, stimulant, emetic, diuretic, and rubefacient. They are more commonly used externally for treating rheumatism, arthritis, and lumbago. Black mustard is also used in footbaths for sore and aching feet.

Brown and white mustards are used in

Chinese medicine for essentially the same types of illnesses (JIANGSU, NANJING).

**Others.** Mustard oil (allyl isothiocyanate) is used in cat and dog repellents.

Expressed mustard oil is used in soapmaking and as lubricant and illuminant.

**COMMERCIAL PREPARATIONS**

Crude (whole, ground) and volatile oil. Brown and black mustards were formerly official in N.F. and U.S.P. Mustard oil (volatile oil) formerly official in U.S.P. and currently official in F.C.C. where it is listed as equivalent to synthetic allyl isothiocyanate, except that its source (natural or synthetic) has to be indicated on the label. It is only required that volatile mustard oil or synthetic allyl isothiocyanate contain 93% (or more) allyl isothiocyanate ($C_3H_5NCS$); the remaining can be minor amounts of mostly known isothiocyanates (natural mustard oil) or impurities of unknown composition (synthetic allyl isothiocyanate) which can vary with the synthetic processes.

*Regulatory Status.* Has been affirmed as GRAS (§184.1527).[9] White mustard seed is the subject of a German therapeutic monograph; allowed externally in poultice for upper respiratory tract congestion and supportive treatment of joint and soft tissue diseases.[20]

**REFERENCES**

See the General References for ARCTANDER; BAILEY 1; BLUMENTHAL; CLAUS; FEMA; GOSSELIN; JIANGSU; LIST AND HÖRHAMMER; LUST; NANJING; ROSE; ROSENGARTEN; TERRELL; UPHOF, WEISS.

1. J. Pokorny and I. Zeman, *Nahrung*, **15**, 35 (1971).
2. S. L. MacKenzie and J. A. Blakely, *Can. J. Bot.*, **50**, 1825 (1972).
3. M. S. Karawya et al., *Egypt J. Pharm. Sci.*, **16**, 113 (1975).
4. H. Nakabayashi et al., *Utsunomiya Daigaku Nogakubu Gakutsu Hokoku*, **8**, 1 (1972); through *Chem. Abstr.*, **77**, 2865z (1972).
5. B. Reichert, *Dtsch. Apoth.*, **20**, 234 (1968).
6. A. Kjär in L. Zechmeister ed. *Fortschritte der Chemie organischer Naturstoffe*, Vol. 18, Springer-Verlag, Vienna, 1960, p. 122.
7. B. E. Wallbank and G. A. Wheatley, *Phytochemistry*, **15**, 763 (1976).
8. M. Kojima et al., *Yakugaku Zasshi*, **93**, 453 (1973); through *Chem. Abstr.*, **79**, 63527d (1973).
9. Anon., *Fed. Regist.*, **42**(166), 43092 (1977).
10. J. R. Vose, *Phytochemistry*, **11**, 1649 (1972).
11. P. O. Larsen, *Biochim. Biophys. Acta*, **107**, 134 (1965).
12. S. A. Taille and F. E. Weber, U.S. 3,878,195 (1975); through *Chem. Abstr.*, **83**, 75630c (1975).
13. V. K. Morozov, *Maslob. Zhir. Prom.*, **25**, 15 (1959); through *Chem. Abstr.*, **60**, 8239e (1964).
14. E. Skramlik, *Pharmazie*, **14**, 435 (1959).
15. P. Ahmad and A. J. Muztar, *Pak. J. Biochem.*, **4**, 72 (1971); through *Chem. Abstr.*, **77**, 160858d (1972).
16. K. K. Abdullin, *Zap. Kazansk. Vet. Inst.*, **84**, 75 (1962); through *Chem. Abstr.*, **60**, 11843b (1964).
17. I. Slavenas, *Tr. l-oi (Pervoi) Biokhim. Kon. Pribaltiisk. Resp. i Belorussii, Tartusk. Gos. Univ. Est. SSR, Estonsk, Biokhim. Obshchestvo, Tartu*, 258 (1960); through *Chem. Abstr.*, **58**, 8244c (1963).

18. P. A. Blau et al., *Science*, **200**, 1296 (1978).
19. R. McDannell et al., *Food. Chem. Toxicol.*, **26**, 59 (1988).
20. Monograph *Sinapis albae semen*, *Bundesanzeiger*, no. 22 (Feb. 1, 1990).

# MYRRH

**Source:** ***Commiphora myrrha*** (Nees) Engl.; ***C. molmol*** Engl. ex Tschirch; ***C. madagascariensis*** Jacq. (syn. *C. abyssinica* (Berg) Engl.); ***C. erythraea*** (Ehrenb.) Engl. and other ***Commiphora*** species (Family Burseraceae).

*Synonyms*. African myrrh and Somali myrrh (*C. myrrha*); Arabian and yemen myrrh (*C. madagascariensis*); myrrha and gum myrrh.

## GENERAL DESCRIPTION

*Commiphora* species yielding myrrh are shrubs to small trees up to about 10 m high; native to northeastern Africa and southwestern Asia, especially the Red Sea region (e.g., Somalia, Yemen, and Ethiopia). The whitish gray bark has schizogenous gum-oleo-resin cavities.[1] Part used is the exudation from the natural fissures in the bark or from man-made incisions. The exudation is a pale yellow liquid, which soon hardens to form yellowish red or reddish brown tears or masses that are then collected. Myrrh is an oleogum resin.

## CHEMICAL COMPOSITION

Myrrh contains 1.5–17% (usually ca. 8%) volatile oil composed of heerabolene, limonene, dipentene, pinene, eugenol, cinnamaldehyde, cuminaldehyde, cumic alcohol, *m*-cresol, cadinene, curzerene (11.9%), curzerenone (11.7%), dihydropyrocurzerenone (1.1%), furanoeudesma-1,3-diene (12.5%), 1,10(15)-furanodiene-6-one (1.2%), and lindestrene (3.5%), among others;[1,2] up to 40% (usually ca. 20%) resins consisting of $\alpha$-, $\beta$-, and $\gamma$-commiphoric acids; commiphorinic acid, $\alpha$- and $\beta$-heerabomyrrhols, heeraboresene, commiferin, campesterol, $\beta$-sitosterol, cholesterol, $\alpha$-amyrone, 3-epi-$\alpha$-amyrin, and others.[3,4] about 60% gum, which on hydrolysis yields arabinose, galactose, xylose, and 4-*O*-methylglucuronic acid; and others (GUENTHER; JIANGSU, LIST AND HÖRHAMMER).[2,4]

A mixture of furanoeudesma-1,3-diene and lindestrene possesses a typical myrrh fragrance, and the resinous myrrh fragrance is best represented by dihydropyrocurzerenone.[1]

Two furanosesquiterpenoids and a furanodienone have been isolated from a hexane extract of *C. erythraea*, traded as opopanax.[1]

## PHARMACOLOGY OR BIOLOGICAL ACTIVITIES

Myrrh is reported to have astringent properties on mucous membranes as well as antimicrobial activities in vitro (JIANGSU, MARTINDALE).

It has been reported to be nonirritating, nonsensitizing, and nonphototoxic to human and animal skins.[5]

*Z*-Guggulsterone (4,17(20)-*trans*-pregnanediene-3,16-dione) isolated from the oleogum resin of *Commiphora mukul* (Hook. ex Stocks) Engl. has thyroid-stimulating activity in rats.[6,7]

Furanosesquiterpenoids are ixodicidal (tick-killing) against the larvae of the African tick *Rhipicephalus appendiculatus*.[1]

## USES

**Medicinal, Pharmaceutical, and Cosmetic.** Myrrh is used mainly as an astringent in certain mouthwashes and gargles.

Myrrh oil is used as a fragrance component or fixative in soaps, detergents, creams, lotions, and perfumes, with maximum use level of 0.8% reported in perfumes.[5] The sweet, warm-balsamic notes of absolute, oil or resinoid used in Oriental spice fragrances, woody and forest notes; often blended with geranium, musk, patchouli and other heavy floral bases.[1]

**Food.** Myrrh is reported used in nonalcoholic beverages with an average maximum use level of 0.025%.

Myrrh oil is used as a flavor component in major food products, including alcoholic and nonalcoholic beverages, frozen dairy desserts, candy, baked goods, gelatins and puddings, and meat and meat products. Highest average maximum use level is about 0.002% reported in alcoholic beverages (25 ppm), baked goods (23.5 ppm), and gelatins and puddings (20 ppm).

**Health Food/Herb Teas.** Primarily in tincture, and other formulations as astringent for mucous membranes of the throat and mouth, indigestion, bronchial congestion, and emmenagogue; also in lip balms for chapped lips; externally for wounds, hemorrhoids, sores, etc. (LUST).

**Traditional Medicine.** Myrrh has been used since ancient times in Western culture as a stimulant, antiseptic, expectorant, antispasmodic, emmenagogue, and stomachic. Conditions for which it is used include cancers, leprosy, syphilitic ulcers, sores, sore throat, asthma, coughs, bad breath, weak gums, gingivitis, and loose teeth.[2,8]

Myrrh was introduced in Chinese medicine around the 7th century and has since been used mainly in treating conditions involving bleeding, pain, and wounds (e.g., bleeding hemorrhoids, menstrual difficulties, sores, tumors, and arthritic pain).[8]

## COMMERCIAL PREPARATIONS

Crude, extracts (tincture, fluid extract, etc.), and oil. Crude and tincture were formerly official in N.F. Myrrh oil is official in F.C.C.

*Regulatory Status.* Has been approved for food use (§172.510). Myrrh powder and tincture subjects of a positive German therapeutic monograph for treatment of mild inflammations of the oral and pharyngeal mucosa.[9]

## REFERENCES

See the General References for ARCTANDER; BLUMENTHAL; CLAUS; FEMA; JIANGSU; LIST AND HÖRHAMMER; LUST; NANJING; ROSE; TUCKER AND LAWRENCE; TYLER 1; UPHOF.

1. A. O. Tucker, *Econ. Bot.*, **40**(4), 425 (1986).
2. R. Pernet, *Lloydia*, **35**, 280 (1972).
3. E. Mincione and C. Iavarone, *Chim. Ind. (Milan)*, **54**, 525 (1972); through *Chem. Abstr.*, **77**, 111467s (1972).
4. E. Mincione and C. Iavarone, *Chim. Ind. (Milan)*, **54**, 424 (1972); through *Chem. Abstr.*, **78**, 13745g (1973).
5. D. L. J. Opdyke, *Food Cosmet. Toxicol.*, **14**, 621 (1976).
6. Y. B. Tripathi et al., *Planta Med.*, **50**, 78 (1984).
7. Y. B. Tripathi et al., *Planta Med.*, **54**, 271 (1988).
8. J. L. Hartwell, *Lloydia*, **31**, 71 (1968).
9. Monograph *Myrrha*, *Bundesanzeiger*, no. 193 (Oct. 15, 1987).

# NETTLE

**Source:** ***Urtica dioica*** L. (Family Urticaceae).

*Synonym.* Stinging nettle.

## GENERAL DESCRIPTION

Perennial to 30–150 dm, stems four angled; leaves and stems with stinging hairs; leaves opposite, oval; flowers green, in axillary panicles; June to September; found in waste places, moist thickets, fields, pastures; most of North America and Europe.

American material is designated *U. dioica* subsp. *gracilis* (Ait.) Seland., consisting of six varieties. The American material differs from the typical European *Urtica dioica* subsp. *dioica* primarily in that it has male and female flowers on the same plant. Some botanists treat the varieties of *U. dioica* subsp. *gracilis* as separate species.

The parts used are the herb, leaves, and root.

## CHEMICAL COMPOSITION

Herb contains amines, including histamine, serotonin (5-hydroxytryptamine), acetylcholine; flavonoids, including isoquercitrin, rutin, kaempferol, and isorhamnetin (in flowers);[1] a lectin, *Urtica dioica* agglutinin;[2] ubiquitous compounds, including carotenoids, vitamin C, triterpenes, $\beta$-sitosterol, formic and citric acids, and relatively high amounts of calcium and potassium salts, and silicic acids (BRADLEY, WREN).

## PHARMACOLOGY OR BIOLOGICAL ACTIVITIES

Described activities include diuretic, antispasmodic, antiallergenic, and hemostatic.

Diuretic activity has been suggested in animal models;[3] nettle juice produced a diuretic effect in a clinical study with patients suffering from myocardial or chronic venous insufficiency.[4]

Hemostatic and mild hypotensive activities have been reported (BRADLEY), along with antibacterial activity against *Staphylococcus aureus* and *Staphylacoccus albus*;[5] also CNS depressant and antispasmodic.[6]

A freeze-dried nettle extract produced positive, though limited results, in a double-blind clinical study in the treatment of allergic rhinitis.[7]

A recent study evaluated the effects of an ethanol extract of the roots of *U. dioica* and *U. urens* in 67 men over 60 years of age suffering from prostatic adenoma. Symptoms of nocturia were alleviated, especially in less severe cases.[8]

Fresh plants sting on contact, due to amines in bladderlike base of stinging hairs. Upon contact, the hair tip breaks off, injecting the amines and folic acid into the skin. Burning sensation may last for up to 1 h (FOSTER AND CARAS).

## USES

**Medicinal, Pharmaceutical, and Cosmetic.** In German phytomedicine an average daily dose of 8–10 g of the herb is used for supportive treatment of rheumatic complaints, inflammation of the lower urinary tract and for treatment of renal gravel; also for benign prostatic hyperplasia.[9,10]

Nettle extract is reported to be used as a biological additive in shampoos, permanent waves; hair conditioners; skin fresheners, and miscellaneous skin care products.

**Food.** Boiled fresh greens eaten as a pot herb.

**Health Food/Herb Teas.** Dried leaf (or roots) in teas, capsules, tablets, tinctures, primarily as a nutritional supplement for vitamin and mineral content; chlorophyll source (FOSTER).

**Traditional Medicine.** In European and American folk traditions, leaf tea, used as blood purifier, diuretic, astringent; for anemia, gout, glandular diseases, rheumatism, poor circulation, enlarged spleen, mucous discharges of lungs, internal bleeding, diarrhea, dysentery (FOSTER AND DUKE).

**Other.** *Urtica dioica* agglutinin has been suggested as a useful probe for analysis of T-cell activation by superantigens.[2]

**COMMERCIAL PREPARATIONS**

Crude dried or fresh, or freeze-dried; extract. Nettles are used as a commercial source of chlorophyll.

*Regulatory Status.* Undetermined in the United States. Nettle herb, leaves, and roots are subjects of positive German therapeutic monographs.[9,10] Herb and leaves indicated for supportive treatment of rheumatic complaints; internally for irrigation therapy in inflammatory conditions of the lower urinary tract, and prevention and treatment of gravel.[9] Root allowed for symptomatic relief of urinary difficulties associated with prostate enlargement (without decreasing enlargement).[10]

**REFERENCES**

See the General References for BLUMENTHAL; BRADLEY; FOSTER; FOSTER AND CARAS; FOSTER AND DUKE; MARTINDALE; NIKITAKIS; STEINMETZ; TYLER 1; WEISS, WREN.

1. N. Chaurasia and M. Wichtl, *Planta Med.*, **53**, 432 (1987).
2. A. Galelli and P. Truffa-Bachi, *J. Immunol*, **151**(4), 1821 (1993).
3. H. Schilcher, *Dtsch. Apoth. Ztg.*, **124**, 2429 (1984).
4. H. W. Kirchhoff, *Z. Phytotherapie*, **4**, 621 (1984).
5. L. Lezhneva et al., *Rastit Resur.*, **22**(2), 255 (1986).
6. E. Madrid et al., *Ann. R. Acad. Farm.*, **53**(2), 284 (1987).
7. P. Mittman et al., *Planta Med.*, **56**, 44 (1990).
8. P. Belachie and O. Lievoux, *Phytother. Res.*, **5**, 267 (1991).
9. Monograph *Urticae herba and Urticae folium*, *Bundesanzeiger*, no. 76 (Apr. 23, 1987).
10. Monograph *Urticae radix*, *Bundesanzeiger*, no. 173 (Sept. 18, 1986); revised (Mar. 6, 1990).

# NUTMEG (AND MACE)

**Source:** ***Myristica fragrans*** Houtt. (syn. *M. officinalis* L.f.) (Family Myristicaceae).

*Synonyms.* Myristica (nutmeg); macis (mace).

**GENERAL DESCRIPTION**

The nutmeg tree is an evergreen tree with spreading branches and dense foliage; up to about 20 m high; native to the Moluccas and nearby islands; cultivated in Indonesia (Java, Moluccas, etc.), Sri Lanka, and the West Indies (especially Grenada). Its fruit is fleshy like an apricot and up to 6 cm long; on ripening it splits in half, exposing a bright red netlike aril wrapped around a dark reddish brown and brittle shell within which lies a single seed. The netlike aril is mace, which on drying turns from red to yellowish or orange brown. The dried brown seed, after shell is broken and discarded, is nutmeg.

Major producers of mace and nutmeg include Indonesia, Sri Lanka, and Grenada. The first two produce the East Indian nutmegs while Grenada produces the West Indian nutmegs. East Indian nutmegs are considered superior in flavor to their West Indian counterparts. Nutmeg oil (myristica oil) is generally produced by steam distillation of worm-eaten nutmegs; these give a higher yield of essential oil, as the worms have eaten much of the starchy and fatty portions of the nutmegs, leaving behind portions that are rich in volatile oil (ARCTANDER, ROSENGARTEN).

## CHEMICAL COMPOSITION

Nutmeg contains 2–16% (usually ca. 10%) volatile oil;[1] 25–40% fixed oils consisting of free myristic acid and triglycerides of lauric, tridecanoic, palmitic, stearic, and myristic acids as well as branched isomers of myristic and stearic acids, with the West Indian nutmeg having larger amounts than East Indian nutmeg;[1–4] starch (ca. 30%); protein (ca. 6%); an oleanolic acid glycoside (saponin);[2] sclareol;[5] diarylpropanoids (dimeric phenylpropanoids);[6–10] catechins, proanthocyanidins;[11] and others (LIST AND HÖRHAMMER, MARSH, STAHL).[1,12,13]

Nutmeg oil contains large amounts of monoterpene hydrocarbons (ca. 88%) with camphene and pinene as the major components and dipentene, sabinene, cymene, $\alpha$-thujene, and $\gamma$-terpinene, among others in lesser amounts; monoterpene alcohols (geraniol, *d*-borneol, linalool, terpineol, etc.);[1,3,12] and 4–8% myristicin and smaller amounts of safrole and elemicin, with the East Indian oil higher in myristicin content than the West Indian oil.[14] Other constituents reported present include copaene, *trans*- and *cis*-sabinene hydrate, *cis*-piperitol, eugenol, isoeugenol, methyleugenol, dehydrodiisoeugenol, and *cis*-*p*-menth-2-enol, among others (LIST AND HÖRHAMMER).[3,4,9,13,14]

Mace and mace oil contain similar constituents as nutmeg and nutmeg oil, except with less fixed oil and more myristicin content (LIST AND HÖRHAMMER, MARSH, STAHL).[6,10,12,13,15]

Nutmeg, mace, and their extracts have strong antioxidant activities that do not seem to be due to their essential oils.[16–18]

## PHARMACOLOGY OR BIOLOGICAL ACTIVITIES

Nutmeg in sufficient dosage is reported to have psychotropic properties (hallucinations, feelings of unreality, euphoria, delusions, etc.); these are believed to be due to myristicin. Other toxic symptoms due to the ingestion of large doses of nutmeg (>5 g) or nutmeg oil include nausea, vomiting, stupor, disorientation, flushing, tachycardia, and dryness of mouth. Death may result (GOSSELIN, MARTINDALE, MERCK).[19]

Nutmeg oil and mace oil have been reported to have larvicidal properties.[20]

Nutmeg extracts have been demonstrated to inhibit the *in vitro* biosynthesis of prostaglandin by rat kidney tissue, and ground nutmeg fed orally to rats has decreased kidney prostaglandin levels, these effects being similar to those produced by indomethacin.[21]

Antiinflammatory activity of mace has been established in carrageenan-induced edema in rats. Vascular permeability in mice is attributed to myristicin.

East Indian nutmeg oil has been reported to be nonirritating and nonsensitizing to human skin, though it was moderately irritating to rabbit skin when applied undiluted for 24 h under occlusion.[19]

## USES

**Medicinal, Pharmaceutical, and Cosmetic.** Nutmeg oil is used in certain analgesic ointments and tonic preparations. It is also used as a flavoring agent in pharmaceuticals.

In European phytomedicine, nutmeg and/or mace have been used for gastrointestinal ailments, including diarrhea, gastric spasms, intestinal catarrh, and flatulence, though claimed efficacies are not sufficiently documented to warrant a positive regulatory status for claims.[22]

Nutmeg oil is used as a fragrance component in soaps, detergents, creams, lotions, and perfumes. Maximum use level reported is 0.3% for the East Indian oil in perfumes.[19]

**Food.** Nutmeg and mace are widely used as domestic spices in cooking, in cakes, drinks (e.g., eggnog and spiced hot wines). and other foods.

Nutmeg and mace are extensively used as flavor ingredients in many food products, including nonalcoholic beverages, baked goods, meat and meat products, condiments and relishes, processed vegetables, soups, snack foods, and gravies. Nutmeg is also used in alcoholic beverages, frozen dairy desserts, gelatins and puddings, breakfast cereals, and others. Highest average maximum use level is about 0.3% reported for nutmeg in gelatins and puddings (3125 ppm), sweet sauces (2600 ppm), and baked goods (2594 ppm).

Nutmeg oil, mace oil, and mace oleoresin are used in most above major food products, including alcoholic and nonalcoholic beverages, frozen dairy desserts, candy, baked goods, gelatin and puddings, meat and meat products, and condiments and relishes, among others. Highest average maximum use levels are about 0.078% (775 ppm) and 0.065% reported for mace oleoresin in alcoholic beverages and candy, respectively.

**Health Food/Herb Teas.** Nutmeg used as a tea flavoring.

**Traditional Medicine.** Nutmeg and mace have been used for centuries in both Western and Eastern cultures mainly as carminative and stimulants in treating flatulence, indigestion, nausea, and other stomach as well as kidney problems. They have also been reported used in cancers.[23] In the Peruvian Andes, nutmeg chopped in pork fat has been used externally in massages for paralysis, rheumatism, and as an antiparasitic (for mange).[24] Mace has been used externally as a treatment for rheumatism.[25]

In European tradition, nutmeg and mace reportedly used for gastric spasms, diarrhea, flatulence, and others.[22]

**Others.** The fixed oil of nutmeg (nutmeg butter) obtained by expression is used in soap making and in candles.

Due to the presence of sclareol in nutmeg oil distillation waste, it can be recovered as a byproduct of the nutmeg oil industry (see ***clary sage***).[5]

## COMMERCIAL PREPARATION

Crude, extracts (e.g., oleoresin), and oils (mace, East and West Indian nutmeg). Mace and nutmeg were formerly official in N.F. Nutmeg oil is official in N.F. and F.C.C., with specifications that sources (East Indian or West Indian) be indicated on the label.

*Regulatory Status.* GRAS (§182.20). Subject of a German therapeutic monograph; claimed efficacies not sufficiently documented; allowed as flavor or fragrance corrigent.[22]

## REFERENCES

See the General References for ARCTANDER; BLUMENTHAL; CLAUS; FEMA; GOSSELIN; GUENTHER; JIANGSU; MARTINDALE; ROSENGARTEN; TERRELL; UPHOF.

1. A. T. Shulgin et al., *Public Health Serv. Publ.*, **No. 1645**, 202 (1967).
2. I. P. Varshney and S. C. Sharma, *Indian J. Chem.*, **6**, 474 (1968).

3. K. J. Sanford and D. E. Heinz, *Phytochemistry*, **10**, 1245 (1971).
4. Z. Mobarak et al., *Chemosphere*, **6**, 633 (1977); through *Chem. Abstr.*, **88**, 20754u (1978).
5. N. F. Novotel'nova et al., U.S.S.R. 161,842 (1964); through *Chem. Abstr.*, **61**, 4145f (1964).
6. J. E. Forrest et al., *J. Chem. Soc., Perkin Trans. I* (2), 205 (1974).
7. A. Isogai et al., *Agr. Biol. Chem.*, **37**, 193 (1973).
8. A. Isogai et al., *Agr. Biol. Chem.*, **37**, 1479 (1973).
9. T. P. Forrest et al., *Naturwissenschaften*, **60**, 257 (1973).
10. D. J. Harvey, *J. Chromatogr.*, **110**, 91 (1975).
11. J. M. Schulz and K. Herrmann, *Z. Lebensm. Unters. Forsch.*, **171**, 278 (1980).
12. A. T. Weil, *Econ. Bot.*, **19**, 194 1965.
13. J. E. Forrest and R. A. Heacock, *Lloydia*, **35**, 440 (1972).
14. J. Baldry et al., *Int. Flav. Food Addit.*, **7**, 28 (1976).
15. J. E. Forrest et al., *Experientia*, **29**, 139 (1973).
16. A. Palitzsch et al., *Fleischwirtschaft*, **54**, 63 (1974); through *Chem. Abstr.*, **81**, 24345q (1974).
17. Y. Kihara and T. Inoue, *Nippon Shokuhin Kogyo Gakkaishi*, **9**, 290 (1962); through *Chem. Abstr.*, **59**, 13276a (1963).
18. Y. Saito et al., *Eiyo To Shokuryo*, **29**, 505 (1976); through *Chem. Abstr.*, **87**, 150314r (1977).
19. D. L. J. Opdyke, *Food Cosmet. Toxicol.*, **14**, 631 (1976).
20. K. Oishi et al., *Nippon Suisan Gakkaishi*, **40**, 1241 (1974); through *Chem. Abstr.*, **82**, 84722r (1975).
21. V. Misra et al., *Indian J. Med. Res.*, **67**, 482 (1978).
22. Monograph *Myristicae semen; Myristicae arillus*, *Bundesanzeiger*, no. 173, (Sept. 18, 1986).
23. J. L. Hartwell, *Lloydia*, **33**, 288 (1970).
24. V. De Feo, *Fitoterapia*, **63**(5), 417 (1992).
25. S. S. Handa et al., *Fitoterapia*, **63**(1), 3 (1992).

# OLIBANUM

**Source:** ***Boswellia carterii*** Birdw., ***B. frereana*** Birdw., ***B. serrata*** Roxb. ex Colebr., ***B. bhau-dajiana*** Birdw., and other ***Boswellia*** species (Family Burseraceae).

*Synonyms.* Frankincense, olibanum gum, Bible incense, (*B. carterii*); African elemi (*B. frereana*); Indian olibanum, and Indian frankincense (*B. serrata*).

## GENERAL DESCRIPTION

Olibanum-yielding *Boswellia* species are shrubs to small trees belonging to the same family as myrrh-producing trees (see ***myrrh***). They are native to the Red Sea region, growing wild throughout northeastern Africa. Olibanum is the exudation (an oleogum resin) from their bark. It is collected by making incisions in the bark; the milky liquid flows out and solidifies on the branches or on the ground as tears or masses. These are then sorted and graded. Major frankincense-producing countries include Somalia and Ethiopia.

## CHEMICAL COMPOSITION

Contains 3–10% volatile oil;[1,2] 60–70% resins composed mainly of 3-acetyl-β-boswellic acid and α-boswellic acid;[1,3,4] about 20% gum made up of galactose, arabinose, and 4-*O*-methylglucuronic acid; 5–8% bassorin; and bitter substances (JIANGSU; LIST AND HÖRHAMMER).[1]

The volatile oil contains predominantly terpene hydrocarbons (especially pinene, dipentene, limonene, α-thujene, α- and β-phellandrenes, *p*-cymene, myrcene, and terpinene; also α-copaene, β-bourbonene, β-ylangene, β-guaiene, *trans*-bergamotene, β-cadinene, humulene, β-caryophyllene, farnesene, etc.), with lesser amounts of oxygenated compounds (farnesol, borneol, carvone, *trans*-pinocarveol, terpinen-4-ol, menthadien-7-ol, verbenone, *trans*-verbenol, bornyl acetate, terpinyl acetate, ethyl laurate, etc.). Other important compounds reported present in olibanum oil include octyl acetate, octanol, incensole, isoincensole, incensole oxide, isoincensole oxide, and incensole acetate, among others.[3,5–12] The relative proportions of the volatile compounds vary considerably according to the types and sources of olibanum (GUENTHER; JIANGSU, LIST AND HÖRHAMMER).[1,7] A study of the essential oil from an olibanum sample of Somalian origin revealed a high content (62.1%) of esters comprised mostly of 1-octyl acetate; terpene hydrocarbons amounted to only 17% of the oil.[2]

Due to its high content of boswellic acid olibanum has been demonstrated to have antioxidant properties on seed oils (cottonseed and sunflower); 0.1% olibanum has activity comparable to 0.02% butylated hydroxyanisole (BHA).[4]

## PHARMACOLOGY OR BIOLOGICAL ACTIVITIES

The nonphenolic fraction of Indian olibanum (from *B. serrata*) has been reported to have strong analgesic effects on rats; it also had depressant effects.[13]

Antiinflammatory and antiarthritic activity has been established in the carrageenan-induced paw edema test and in arthritic rats. Clinical trials in India have been conducted on arthritis patients with positive results, and boswellic acid and other related pentacyclic triterpene acids are marketed as antiarthritic drugs in India.[14]

Olibanum and olibanum absolute have been reported to be nonirritating and nonsensitizing to human skin; olibanum absolute was also nonphototoxic. The phototoxicity of olibanum gum is not known. Olibanum gum was moderately irritating to rabbit skin when applied undiluted for 24 h under occlusion.[15,16]

## USES

**Medicinal, Pharmaceutical, and Cosmetic.** Olibanum, oil, and extracts (absolute and resinoid) are used as fixative and/or fragrance components in soaps, detergents, creams, lotions, and perfumes (e.g., Oriental types), with maximum use level of 0.8% reported for the gum and absolute in perfumes.[15,16] In perfumes the absolute, oil, and resinoid are used for a fresh balsamic, dry, resinous, somewhat green note in Oriental bases, ambers, florals, colognes, male fragrances, etc.[17]

**Food.** Only olibanum oil is used in food products, including alcoholic and nonalcoholic beverages, frozen dairy desserts, candy, baked goods, gelatins and puddings, and meat and meat products. Highest average maximum use level is about 0.001% (11.2 ppm) in meat and meat products.

**Health Food/Herb Teas.** A proprietary extract of *B. serrata* from India, containing a mixture of triterpene pentacyclic acid derivatives of boswellic acids has appeared on the market in recent years, with claims for antiinflammatory and antiarthritic activity.

**Traditional Medicine.** Olibanum (more popularly known as frankincense) has been used since antiquity as an incense in India, China, Egypt, and the Catholic Church. It was an ingredient of the embalming liquid ancient Egyptians used on their dead. It has been used as a stimulant, respiratory antiseptic, diuretic, and emmenagogue in both Western and Eastern cultures. Conditions for which it has been and still is used include syphilis, rheumatism, painful menstruation, pimples, sores, tumors, cancers, asthma, sore throat, abdominal pain, stomach troubles and nervous problems, among others (JIANGSU, LIST AND HÖRHAMMER, ROSE).[1,18] In Ayurvedic medicine the oleogum resin of *B. serrata*, known as "Salai guggal" has been used as a treatment for rheumatism, nervous diseases, and as a topical antiinflammatory.[14]

## COMMERCIAL PREPARATIONS

Crude, extracts (e.g., resinoid), and oil, Olibanum oil is official in F.C.C.

*Regulatory Status.* Has been approved for food use (§172.510).

## REFERENCES

See the General References for ARCTANDER; FEMA; GUENTHER; JIANGSU; LIST AND HÖRHAMMER; MERCK; TUCKER AND LAWRENCE; UPHOF.


1. R. Pernet, *Lloydia*, **35**, 280 (1972).
2. S. M. Abdel Wahab et al., *Planta Med.*, **53**, 382 (1987).
3. G. Snatzke and L. Vertesy, *Monatsh. Chem.*, **98**, 121 (1967).
4. A. H. Y. Abdel-Rahman, *Grasas Aceites (Seville)*, **27**,175 (1976); through *Chem. Abstr.*, **85**, 121943y (1976).
5. S. A. Higazy et al., *Egypt. J. Food Sci.*, **1**, 203 (1973).
6. S. A. Higazy et al., *Egypt. J. Food Sci.*, **2**, 29 (1974).
7. H. Obermann, *Dragoco Rep. (Ger. Ed.)*, **24**, 260 (1977); through *Chem. Abstr.*, **88**, 197432p (1978).
8. S. Corsano and R. Nicoletti, *Tetrahedron*, **23**, 1977 (1967).
9. R. Nicoletti and M. L. Forcellese, *Tetrahedron*, **24**, 6519 (1968).
10. M. L. Forcellese et al., *Tetrahedron*, **28**, 325 (1972).
11. L. Peyron et al., *Riv. Ital. Essenze, Profumi, Piante Offic., Aromi, Saponi, Cosmet., Aerosol*, **63**, 133 (1981).

12. P. Maupetit, *Perfum. Flav.*, **9**(6), 19 (1985).
13. A. Kar and M. K. Menon, *Life Sci.*, **8**, 1023 (1969).
14. S. S. Handa et al., *Fitoterapia*, **63**(1), 3 (1992).
15. D. L. J. Opdyke, *Food Cosmet. Toxicol.*, **16**(Suppl. 1), 837 (1978).
16. D. L. J. Opdyke, *Food Cosmet. Toxicol.*, **16**(Suppl. 1), 835 (1978).
17. A. O Tucker, *Econ. Bot.*, **40**(4), 425 (1986).
18. J. L. Hartwell, *Lloydia*, **31**, 71 (1968).

# ONION

**Source:** ***Allium cepa*** L. (Family Amaryllidaceae or Liliaceae).

## GENERAL DESCRIPTION

A perennial or biennial herb with hollow leaves and a taller and thicker scape, (flowering stem), also hollow; up to about 1.2 m high; generally considered to be native of western Asia; has long been cultivated worldwide and much varied. Part used is the fleshy bulb; onion oil is obtained by steam distillation.

There are numerous forms and varieties of onion, the most common ones being the white globe, yellow globe, and red globe onions. Due to the particularly high variability of onion as opposed to most other plants, results of chemical and biological studies from different countries should be evaluated with caution as they may be based on different varieties of *A. cepa*, which are often not specified in the reports.

## CHEMICAL COMPOSITION

Like garlic, onion contains many organic sulfur compounds, including *trans*-*S*-(1-propenyl) cysteine sulfoxide, *S*-methylcysteine sulfoxide, *S*-propylcysteine sulfoxide, and cycloalliin.[1,2] Except for cycloalliin, these sulfur compounds are converted to simpler sulfur compounds by the enzyme alliinase present when the onion is cut or crushed. These simpler compounds are unstable and undergo further decomposition to sulfides (di-, tri-, etc.) and other compounds that are responsible for the onion flavor (especially methylpropyl disulfide, methylpropyl trisulfide, and dipropyl trisulfide).[3,4]

The lachrymating (tear-producing) principle in crushed or cut onion is thiopropanal *S*-oxide (propanethial *S*-oxide) produced from its precursor, *trans*-*S*-(1-propenyl)-cysteine sulfoxide, by the action of alliinase.[5–8]

Onion also contains a trace of volatile oil composed mainly of sulfur compounds, with dipropyl disulfide as the main component (but not an important flavor contributor).[1,3] Others present include the three important flavor contributors methylpropyl disulfide, methylpropyl trisulfide, and dipropyl trisulfide as well as allylpropyl disulfide, dimethyl disulfide, 3,4-dimethylthiophene, methyl-*cis*-propenyl disulfide, and others.[1,3,9]

Other constituents present in onion include phenolic acids (caffeic, sinapic, *p*-coumaric, protocatechuic acids, etc.), flavonoids (e.g., quercetin and its glucosides),[10,11] sterols (cholesterol, stigmasterol, $\beta$-sitosterol, etc.),[12,13] saponins,[14] sugars, vitamins (A, C, $B_1$ and $B_2$), pectin, and anthocyanins, among others (JIANGSU, LIST AND HÖRHAMMER).[15]

## PHARMACOLOGY OR BIOLOGICAL ACTIVITIES

Onion is reported to have many similar

pharmacological properties as garlic, including antihypercholesterolemic (in experimental animals), hypoglycemic (in humans and experimental animals),[16–20] and antifungal (see ***garlic***).[21]

Other reported activities for onion include antiplatelet aggregation, antibiotic, and an increase in fibrinolytic activity (with conflicting results).[22,23]

A preliminary study suggests that increased consumption of *Allium* vegetables (including onion) correlated to a significant reduction in gastric cancer risk.[24]

## USES

**Food.** Onion oil is extensively used as a flavor ingredient in most food products, including alcoholic and nonalcoholic beverages, frozen dairy desserts, candy, baked goods, gelatins and puddings, meat and meat products, condiments and relishes, fats and oils (e.g., salad dressings and salad oils), soups, snack foods, and gravies. Highest average maximum use level reported is about 0.083% (826 ppm) in fats and oils.

**Health Food/Herb Teas.** Capsulated and tableted supplements of dehydrated onion; contradictory results have been reported for studies using fresh onions compared with supplements.[22]

**Traditional Medicine.** Although not as valued or widely used as garlic, onion is used as antispasmodic, carminative, diuretic, expectorant, stomachic, and anthelmintic for many of the same conditions as is garlic (see ***garlic***). Its use in Chinese medicine is of only recent origin and is limited (NANJING).

## COMMERCIAL PREPARATIONS

Oil; official in F.C.C.

*Regulatory Status*. GRAS (§182.20). Onion is the subject of a positive German therapeutic monograph, indicated for antibacterial, lipid lowering, blood pressure–lowering and antiplatelet aggregation activity.[25]

## REFERENCES

See the General References for ARCTANDER; BAILEY 2; BLUMENTHAL; FEMA; GUENTHER; JIANGSU; LIST AND HÖRHAMMER; LUST; TERRELL.

1. J. R. Whitaker, *Adv. Food Res.*, **22**, 73 (1976).
2. A. I. Virtanen and C. G. Spare, *Suomen Kemistilehti*, **34**, 72 (1961); through *Chem. Abstr.*, **56**, 10585h (1962).
3. W. G. Galetto and A. A. Bednarczyk, *J. Food Sci.*, **40**, 1165 (1975).
4. M. Boelens et al., *J. Agric. Food Chem.*, **19**, 984 (1971).
5. C. G. Spare and A. I. Virtanen, *Acta Chem. Scand.*, **17**, 641 (1963).
6. W. F. Wilkens, *Cornell Univ., Agr. Expt. Sta. Mem.*, **385** (1964); through *Chem. Abstr.*, **61**, 9771c (1964).
7. M. H. Brodnitz and J. V. Pascale, *J. Agr. Food Chem.*, **19**, 269 (1971).
8. T. H. Maugh, *Science*, **204**, 293 (1979).
9. M. H. Brodnitz et al., *J. Agr. Food Chem.*, **17**, 760 (1969).
10. J. P. Varshney and T. Ali, *Indian J. Appl. Chem.*, **34**, 142 (1971); through *Chem. Abstr.*, **77**, 85626w (1972).
11. M. Leonte and A. Leonte, *Lucr. Stiint., Inst. Politeh. Galati*, **4**, 195 (1970); through *Chem. Abstr.*, **77**, 163205z (1972).
12. L. A. R. Sallam et al., *Qual. Plant Plant Foods Hum. Nutr.*, **24**, 159

(1974); through *Chem. Abstr.*, **83**, 7464b (1975).
13. T. Itoh et al., *Phytochemistry*, **16**, 140 (1977).
14. D. Nitschke and A. Smoczkiewiczowa, *Zesz. Nauk Akad. Ekon. Poznaniu, Ser. 1*, **69**, 104 (1976); through *Chem. Abstr.*, **87**, 164252f (1977).
15. A. L. Feldman et al., *Konserv. Ovoshchesuch. Prom.*, **4**, 19 (1973); through *Chem. Abstr.*, **79**, 15877w (1973).
16. R. C. Jain and C. R. Vyas, *Br. Med. J.*, **2**, 730 (1974).
17. R. K. Gupta et al., *Indian J. Exp. Biol.*, **15**, 313 (1977).
18. K. K. Sharma et al., *Indian J. Med. Res.*, **65**, 422 (1977).
19. K. T. Augusti, *Indian J. Exp. Biol.*, **14**, 110 (1976).
20. K. T. Augusti and M. E. Benaim, *Clin. Chim. Acta*, **60**, 121 (1975).
21. P. S. Shekhawat and R. Prasada, *Sci. Cult.*, **37**, 40 (1971).
22. J. Kleijnen et al., *Br. J. Clin. Pharmac.*, **28**, 535 (1989).
23. B. S. Kendler, *Prev. Med.*, **16**, 670 (1987).
24. W. C. You et al., *J. Nat. Cancer Inst.* **81**(2), 162 (1989).
25. Monograph *Allii cepae bulbus, Bundesanzeiger*, no. 50 (Mar. 13, 1986).

# ORANGE (BITTER AND SWEET)

**Source:** ***Bitter orange*** *Citrus aurantium* L. (syn. *C. aurantium* L. ssp. *amara* (L.) Engl.; *C. vulgaris* Risso; *C. bigaradia* Risso); ***Sweet orange*** *Citrus sinensis* (L.) Osb. (syn. *C. aurantium* L. var. *sinensis* L.; *C. aurantium* L. var. *dulcis* Pers.; *C. aurantium* L. ssp. *dulcis* L.; *C. aurantium* Risso ssp. *sinensis* Engl.) (Family Rutaceae).

## GENERAL DESCRIPTION

Bitter orange is an evergreen glabrous (nonhairy) tree with long but not very sharp spines and very fragrant flowers; membranes and pulp of fruit very bitter and sour; up to about 10 m high; native to southern China and northeastern India; cultivated in China, southern Europe, and the United States. Its trunk is more resistant to plant diseases than those of other citrus trees and consequently serves as stock for the less resistant sweet orange. Parts used are the peel of the fruit, freshly picked flowers, and leaves and twigs.

Bitter orange oil is obtained from the fresh peel by cold expression in about 0.15% yield, by machine or hand. Its major producers include Guinea, southern European countries (especially Spain and Italy), Brazil, and the West Indies. The Guinean and Spanish oils are considered of the best quality.

Neroli oil (orange flower oil) is obtained from the freshly picked flowers of bitter orange by steam distillation in about 0.1% yield. Major producers include France, Italy, and Tunisia. The distillation water from which the oil layer has been removed is called orange flower water.

Petitgrain oil is obtained from the leaves (also twigs) of bitter orange by steam distillation in about 0.2% yield. Major producers include France, Haiti, Paraguay, and Guinea.

Sweet orange exists in numerous varieties (e.g., Navel, Jaffa, and Valencia). It is a smaller tree than the bitter orange tree, less hardy, and with few or no spines; fruits smaller, with sweet pulp and nonbitter membranes; generally believed to be native to China; extensively cultivated worldwide, especially in the United States (e.g., California and Florida) and Mediterranean countries. Part used is the peel of the partially or fully ripe fruit. The peel used for the preparation of Sweet Orange Peel Tincture N.F. is specified to be derived from the nonartificially colored ripe fruit and devoid of the albedo (inner white portion of the rind).

Sweet orange oil, commonly known simply as orange oil, is obtained by one of the three major methods: (*1*) cold expression (hand or machine) of the fresh peel, (*2*) steam distillation of the fresh peel (either already expressed or not yet processed), and (*3*) distillation of the essences recovered as a by-product in the manufacture of orange juice concentrates. Major producers of cold-expressed sweet orange oil include the United States, Cyprus, Guinea, Israel, and Brazil, while the United States is the major producer of the distilled sweet orange oil.

Terpeneless orange oils (both bitter and sweet) are obtained from the respective oils by vacuum distillation and/or extraction with dilute ethyl alcohol whereby all or most of the terpenes (e.g., limonene) are removed.

## CHEMICAL COMPOSITION

In addition to a volatile oil (1–2.5%), bitter orange peel contains appreciable quantities of neohesperidin (up to ca. 14% in unripe peel, usually 2.4–2.8% in ripe peel), naringin (0.9–4%), rhoifolin, lonicerin, hesperidin, and other flavonoids (tangeretin, nobiletin, sinensetin, auranetin, rutin, etc.);[1,2] vitamins (A, $B_1$, and C); coumarins (e.g., 6,7-dimethoxycoumarin and umbelliferone);[3] carotenoid pigments (citraurin, violaxanthin, and cryptoxanthin); pectin, citrantin; and others (FARNSWORTH, JIANGSU; LIST AND HÖRHAMMER).

The volatile oil (bitter orange oil) contains more than 90% monoterpenes (mainly *d*-limonene, also myrcene, camphene, pinene, ocimene, *p*-cymene, etc.); small amounts of alcohols (linalool, terpineol, nerol, farnesol, nerolidol, octanol, etc.); usually 0.5–1% aldehydes (mainly decanal, also nonanal, dodecanal, citronellal, neral, acetaldehyde, formaldehyde, etc.), and ketones (carvone, α-ionone, and jasmone); free acids (octadecadienoic, pelargonic, cinnamic, acetic, etc.);[4] about 2.4% esters (linalyl acetate, decyl pelargonate, octyl acetate, geranyl acetate, etc.); coumarins (osthole and auraptenol);[5,6] and others (JIANGSU, LIST AND HÖRHAMMER).

Neroli oil is reported to contain linalool (ca. 34%), linalyl acetate (6–17%), limonene (15%), β-pinene (11%), nerolidol (6%), geraniol, nerol, methyl anthranilate, indole, and jasmone as well as small amounts of other compounds including citral, nonanal, *cis*-8-heptadecene, 2,5-dimethyl-2-vinyl-4-hexenal, neryl acetate, and valeric acid, among others (GUENTHER, JIANGSU, LIST AND HÖRHAMMER; MERCK, REMINGTON).[7–9]

Petitgrain oil contains large amounts of esters (40–80%, depending on sources; composed mainly of linalyl acetate, with lesser quantity of geranyl acetate. Other compounds present include linalool, nerol, α-terpineol, geraniol, nerolidol, farnesol, and limonene (GUENTHER, MERCK).[8,9]

Sweet orange peel contains 1.5–2% volatile oil, numerous flavonoids (neohesperidin, hesperidin, naringin, tangeretin, auranetin, nobiletin, etc.),[10,11] vitamins (e.g., C and E), limonin,[12] coumarins (e.g., 6,7-dimethoxycoumarin),[3] carotenoids, pectin, citrantin, and other similar constituents present in bitter orange peel (JIANGSU, LIST AND HÖRHAMMER).

Sweet orange oil is very similar to bitter orange oil in chemical composition, containing about 90% or more of *d*-limonene and normally 1.2–2.5% aldehydes (mostly decanal, octanal, etc.).[13–15] The cold-expressed oil also contains coumarins (e.g., bergapten and auraptenol), acids (octadecadienoic, etc.), valencene, $\alpha$-ylangene,[16] and other compounds present in bitter orange oil (JIANGSU, LIST AND HÖRHAMMER, MERCK, REMINGTON).[4,15] A recent study on the use of absorbents for cold-pressed orange oil found that silica gel produced the best results, reducing monoterpene hydrocarbons, while maintaining high oxygenated component content, especially decanal.[17,18]

Sweet orange oil does not taste bitter as bitter orange oil. Important flavor contributors reported in sweet orange essence include octanol, ethyl butyrate, and acetaldehyde, while $\alpha$-terpineol and *trans*-2-hexenal contribute to the off flavor of the oil.[14,15]

Certain flavonoid glycosides (e.g., naringin and neohesperidin) are intensely bitter; the specific linkages in the sugars present are reported to be the determining factors for their bitterness.[19]

## PHARMACOLOGY OR BIOLOGICAL ACTIVITIES

Both bitter and sweet orange peels exhibit numerous distinct pharmacological activities, including antiinflammatory (due to flavonoids, especially naringin), antibacterial and antifungal (flavonoids and pectin), antihypercholesterolemic in humans and laboratory animals due to the contained pectin (see ***pectin***), and choleretic, among others (FARNSWORTH, JIANGSU, LIST AND HÖRHAMMER).

Ingestion of large amounts of orange peel (bitter or sweet) by children has been reported to cause intestinal colic, convulsions, and even death (JIANGSU).

Citrantin is reported to have antifertility properties in rabbits when fed orally at a dose of 0.75 mg/kg/day (FARNSWORTH, JIANGSU).

Bitter orange, sweet orange, and neroli oils have been reported to exhibit antifungal and antibacterial activities *in vitro* (see ***lemon oil*** and ***pine needle oil***).[19–23]

Expressed sweet orange oil, bitter orange oil, and neroli oil are generally reported to be nonirritating and nonsensitizing to humans.[23–25] However limonene present in citrus oils has been known to cause contact dermatitis in humans (LEWIS AND ELVIN-LEWIS).

Bitter orange oil is reported to have distinct phototoxic activity, while none is reported for expressed sweet orange oil even though both oils contain coumarins (see ***lime oil***).[24,25]

3-(4-Hydroxy-3-(3-methyl-2-butenyl)-phen-yl)-2(*E*)-propenal, a phenolic cinnamaldehyde found in wound gum from injured peels of *C. sinensis*, is antifungal against *Penicillium digitatum*, and *Cladosporium cucumerinum*.[26]

Sweet orange oil has been reported to promote tumor formation on mouse skin treated with a primary carcinogen (see ***lemon oil*** and ***lime oil***).[24,27] Its major component, *d*-limonene, is also reported to have anticarcinogenic activity.[24]

## USES

**Medicinal, Pharmaceutical, and Cosmetic.** Bitter orange peel is occasionally used as an ingredient in certain stomachic, carminative, and laxative preparations.

Neroli oil (orange flower oil), orange flower water, and sweet orange peel tincture are used in flavoring pharmaceuticals.

Bitter and sweet orange, neroli, and petitgrain oils are extensively used as fragrance components in soaps, detergents, creams, lotions, and cosmetics. Highest maximum use level is 1.0% reported for bitter orange oil in perfumes.[23–25]

**Food.** Bitter orange oil is extensively used as a flavor component in most major food products, including alcoholic (e.g., liqueurs, especially triple sec) and nonalcoholic beverages, frozen dairy desserts, candy, baked goods, gelatins and puddings, meat and meat products, and condiments and relishes. Highest average maximum use level reported is about 0.043% (424 ppm).

Neroli and petitgrain oils are also widely used in major food products, including alcoholic and nonalcoholic beverages, frozen dairy desserts, candy, baked goods, and gelatins and puddings, among others, with highest average maximum use levels of about 0.001% reported for neroli oil in alcoholic beverages (11.5 ppm) and baked goods (11.3 ppm) and about 0.004% (37.7 ppm) for petitgrain oil in condiments and relishes.

Sweet orange oil (produced by various methods) is by far the most widely used in food products, including all above categories in addition to jams and jellies, gravies, sweet sauces, breakfast cereals, and processed vegetables, among others. Highest average maximum use level is 0.75% reported for the distilled oil in sweet sauces. Highest average maximum use level reported for its terpeneless type is 0.01% in breakfast cereals.

**Health Food/Herb Teas.** Dried bitter and sweet orange peels used in tea formulations and in digestive formulas.

**Traditional Medicine.** Dried bitter orange peel and, to a lesser extent, sweet orange peel are used as tonic and carminative in treating dyspepsia. In Chinese medicine, dried sweet orange peel is used to reduce phlegm and in treating coughs, colds, anorexia (lack of appetite), and malignant breast sores. Dried bitter orange and less commonly the dried peel are used in treating prolapse of the uterus and of the anus (also rectum), diarrhea, blood in feces, and others, in addition to above conditions (JIANGSU).

They have also been reported used in cancers.[28]

Bitter orange flowers and oil have reportedly been used in European tradition as prophylactics for gastrointestinal complaints, nervous conditions, gout, sore throat, as a sedative, and for sleeplessness.

**Others.** Due to the high contents of *d*-limonene, in both bitter and sweet orange oils, the oils are also used as major sources of *d*-limonene, which in turn serves as a starting material for the synthesis of carvone, an important flavor chemical (see ***caraway*** and ***spearmint***).

Neohesperidin and naringin from bitter orange peel could serve as starting materials for the production of neohesperidin dihydrochalcone (NHDHC) and naringin dihydrochalcone (NDHC); both are sweeteners, with NDHC being as sweet as saccharin and NHDHC 20 times more so.[2]

## COMMERCIAL PREPARATIONS

Dried peels (bitter and sweet), their extracts (solid, fluid, and tincture), and oils. Bitter and sweet orange peels were formerly official in N.F. and U.S.P., respectively. Bitter orange oil was formerly official in N.F., and sweet orange oil formerly in U.S.P. Both oils as well as petitgrain oil Paraguay are official in F.C.C., while neroli oil, orange flower water, and sweet orange peel tincture are official in N.F. Qualities of oils vary according to sources. Fluid extracts (see ***glossary***) prepared by the direct extraction method have flavor qualities superior to those produced by the dilution method.

*Regulatory Status.* All GRAS (§182.20). Bitter orange flowers subject of a German therapeutic monograph; traditional uses are not recommended as efficacy has not been demonstrated; use is allowed as a flavor corrigent.[29] Sweet orange peels subject of a positive monograph for treatment of loss of appetite.[30]

## REFERENCES

See the General References for ARCTANDER; BAILEY 1; BAILEY 2; BIANCHINI AND CORBETTA; BLUMENTHAL; FARNSWORTH; FEMA; GOSSELIN; GUENTHER; JIANGSU; LIST AND HÖRHAMMER; LUST; MARTINDALE; TERRELL; TUCKER AND LAWRENCE; UPHOF.


1. T. Nakabayashi, *Nippon Nogei Kagaku Kaishi*, **35**, 945 (1961); through *Chem. Abstr.*, **60**, 11044b (1964).
2. S. Natarajan et al., *Econ. Bot.*, **30**, 38 (1976).
3. J. H. Tatum and R E. Berry, *Phytochemistry*, **16**, 1091 (1977).
4. G. Lamonica et al., *Essenze Deriv. Agrum.*, **43**,115 (1973); through *Chem. Abstr.*, **80**, 131772c (1974).
5. W. L. Stanley et al., *Tetrahedron*, **21**, 89 (1965).
6. W. L. Stanley and L. Jurd, *J. Agr. Food Chem.*, **19**, 1106 (1971).
7. B. Corbier and P. Teisseire, *Recherches*, **19**, 289 (1974).
8. W. D. Fordham in L. W. Codd et al., eds., *Chemical Technology: An Encyclopedic Treatment*, Vol. 5, Barnes & Noble, New York, 1972, p. 1.
9. M. Stoll in A. Standen, ed., *Kirk-Othmer Encyclopedia of Chemical Technology*, Vol. 14, 2nd ed., Wiley-Interscience, New York, 1967, p. 178.
10. C. W. Wilson and P. E. Shaw, *J. Agr. Food Chem.*, **25**, 221 (1977).
11. H. Wagner et al., *Chem. Ber.*, **102**, 2089 (1969).
12. W. C. Scott, *Proc. Fla. State Hort. Soc.*, **83**, 270 (1970); through *Chem. Abstr.*, **75**, 75067j (1971).
13. M. G. Moshonas and E. D. Lund, *J. Food Sci.*, **34**, 502 (1969).
14. E. D. Lund and W. L. Bryan, *J. Food Sci.*, **42**, 385 (1977).
15. P. E. Shaw and R. L. Coleman, *J. Agr. Food Chem.*, **19**, 1276 (1971).
16. S. K. Ramaswami et al. in B. M. Lawrence et al., eds., *Flavors and Fragrances: A World Perspective*, Elsevier Science Publishers B.V., Amsterdam, 1988, p. 951.
17. O. J. Ferrer and R. F. Matthews, *J. Food Sci*, **52**, 801 (1987).
18. B. M. Lawrence, *Perfume. Flav.*, **15**, 6, 45 (1990).
19. S. Kamiya et al., *Agr. Biol. Chem.*, **38**, 1785 (1974).
20. M. S. Subba et al., *J. Food Sci.*, **32**, 225 (1967).
21. B. G. V. N. Rao and P. L. Joseph, *Riechst., Aromen, Körperpflegem.*, **21**, 405 (1971).
22. D. I. Murdock and W. E. Allen, *Food Technol.*, **14**, 441 (1960).
23. D. L. J. Opdyke, *Food Cosmet. Toxicol.*, **14**(Suppl.), 813 (1976).
24. D. L. J. Opdyke, *Food Cosmet. Toxicol.*, **12**, 733 (1974).
25. D. L. J. Opdyke, *Food Cosmet. Toxicol.*, **12**, 735 (1974).
26. R. R. Stange Jr. et al., *J. Nat. Prod*, **56**(9), 1627 (1993).
27. F. M. Nacino et al., *Kalikasan*, **4**, 240 (1975); through *Chem. Abstr.*, **84**, 39528n (1976).
28. J. L. Hartwell, *Lloydia*, **34**, 103 (1971).
29. Monograph *Aurantii flos*, *Bundesanzeiger* (July 14, 1993).
30. Monograph *Citri sinensis pericarpium*, *Bundesanzeiger* (Feb. 1, 1990).

# OREGANO

**Source:** ***Origanum vulgare*** L. (Family Labiatae or Lamiaceae), ***Lippia graveolens*** H. B. K. (syn. *L. berlandieri* Schauer), ***L. palmeri*** S. Wats. (Family Verbenaceae), and other ***Origanum*** and ***Lippia*** species.

*Synonyms*. Wild marjoram, origanum, and European oregano (*O. vulgare*); Mexican oregano, Mexican marjoram, and Mexican wild sage (*Lippia* species).

## GENERAL DESCRIPTION

Oregano is derived from several genera of plants from mainly two families (e.g., *Origanum*, *Lippia*, *Coleus*, *Lantana*, and *Hyptis*). It is not one or two well-defined species but rather any one of more than two dozen known species that yield leaves or flowering tops having the flavor recognized as being oregano. The most commonly used plants are *Origanum vulgare*, *O. onites* (as Turkish oregano) *Lippia graveolens*, and *L. palmeri* (see ***marjoram***).[1,2]

European oregano (*O. vulgare*) is a hardy perennial herb with erect, more or less hairy, branching stems, and hairy leaves; up to about 0.9 m high; native to Europe. Parts used are the dried herb and leaves.

Mexican oregano is usually derived from *Lippia graveolens* and occasionally from *L. palmeri*. Both are aromatic shrubs native to Mexico; *L. graveolens* is larger and is sometimes a tree, up to 9 m high. Parts used are the dried leaves.[1]

## CHEMICAL COMPOSITION

*Origanum vulgare* (also known as wild marjoram) contains 0.1–1.0% volatile oil composed of thymol, carvacrol, β-bisabolene, caryophyllene, *p*-cymene, borneol, linalool, linalyl acetate, geranyl acetate, α-pinene, β-pinene, α-terpinene, and others, with highly variable relative proportions, depending on sources; the phenols (thymol and carvacrol) content has been reported to vary from 0 to 90% in the oil (JIANGSU, LIST AND HÖRHAMMER, MARSH).[3,5] Other constituents include luteolin, apigenin, diosmetin, caffeic acid, ursolic acid, oleanolic acid, rosmarinic acid, protein, and vitamins (LIST AND HÖRHAMMER, MARSH)[6,7]

The essential oil of Greek oregano (*O. vulgare* subsp. *viride* (Boiss) Hayek; syn. *O. heracleoticum* L.) has been reported to contain carvacrol, thymol, linalool, borneol, and other similar constituents found in *O. vulgare* as well as carvone, camphor, amyl alcohol, and 4,5-epoxy-*p*-menth-1-ene.[8,9]

The essential oil of *O. onites* has been found to be highly variable in wild populations in Greece, ranging from 1.85 to 4.37%, with carvacrol as the major component (51–84.5%), plus borneol (1.3–8.2%), *p*-cymene (5.1–12.2%), γ-terpinene (2.3–13.6%), and recently isolated minor components including β-phellandrene, *trans*-thujanol, *cis*-thujanol, linalyl acetate, α-muurolene, and others.[2]

The essential oils of numerous *Lippia* species have been reported to have highly variable compositions.[10–15] Nevertheless *Lippia* species (e.g., *L. graveolens* and *L. origanoides* H. B. K.) known to be used as sources of oregano contain *p*-cymene, 1,8-cineole, thymol and/or carvacrol as major components.[14–16]

Recently, the flavanones pinocembrin and naringenin along with lapachenole have been isolated from *L. graveolens* (aerial parts and root).[17]

## PHARMACOLOGY OR BIOLOGICAL ACTIVITIES

Thymol and carvacrol, usually the major phenols present in oregano, have strong fungicidal, anthelmintic, irritant, and other properties (see ***thyme***).

The essential oil of European oregano is reported to have spasmolytic, expectorant, diuretic, and choleretic properties (JIANGSU, LIST AND HÖRHAMMER).

Lapachenole has antifertility as well as carcinogenic activities.[17]

## USES

**Food.** European oregano is extensively used as a major flavor ingredient in pizza. The more spicy Mexican oregano is widely used in Mexican dishes (chili, chili con carne, etc.); it is less preferred for use in pizza than the milder European type. Oregano is also widely used in other foods, including alcoholic beverages, baked goods, meat and meat products, condiments and relishes, milk products, processed vegetables, snack foods, fats and oils, and others. Highest average maximum use level reported is about 0.3% in condiments and relishes (3168 ppm) and milk products (2900 ppm).

**Traditional Medicine.** *Lippia graveolens* is used in Mexico as a stimulant emmenagogue, and demulcent.[18]

*Origanum vulgare* has been used as a stimulant, carminative, diaphoretic, and nerve tonic and as a cure for asthma, coughs, indigestion, rheumatism, toothaches, headaches, spider bites, and coronary conditions (ROSENGARTEN).

In European phytomedicine, oregano and its preparations have been used for the treatment of respiratory ailments, coughing, bronchitis; antispasmodic and expectorant. Also used as an appetite stimulant, diuretic, and mild sedative.[19]

In China, in addition to some of the above uses, *O. vulgare* is used to treat fevers, vomiting, diarrhea, jaundice, and itchy skin conditions (JIANGSU).

## COMMERCIAL PREPARATIONS

Mainly the spice.

*Regulatory Status.* GRAS (Mexican oregano, §182.20; Origanum, §182.20). *O. vulgare* is the subject of a German therapeutic monograph; use not recommended, since efficacy has not been confirmed.[19]

## REFERENCES

See the General References for BAILEY 1; BLUMENTHAL; FEMA; GOSSELIN; JIANGSU; LIST AND HÖRHAMMER; LUST; MARTINDALE; ROSE; ROSENGARTEN; TERRELL; TUCKER.


1. L. Calpouzos, *Econ. Bot.*, **8**, 222 (1954).
2. D. Vokou et al., *Econ. Bot.*, **42**(3), 407 (1988).
3. U. Asllani, *Bull. Shkencave Nat., Univ. Shteteror Tiranes*, **28**, 61 (1974); through *Chem. Abstr.*, **82**, 28499m (1975).
4. H. Maarse and F. H. L. Van Os, *Flav. Ind.*, **4**, 481 (1973).
5. C. H. Brieskorn and H. Brunner, *Planta Med.* (Suppl.), 96 (1967).
6. W. Olechnowicz-Stepien and E. Lamer-Zarawska, *Herba Pol.*, **21**, 347 (1975); through *Chem. Abstr.*, **85**, 74939q (1976).
7. U. Gerhardt and A. Schroetu, *Fleischwirtschaft*, **63**, 1628 (1983).
8. B. M. Lawrence et al., *Phytochemistry*, **13**, 1012 (1974).
9. V. Staikov et al., *Soap, Perfum. Cosmet.*, **41**, 327 (1968).
10. P. Rovesti, *Riv. Ital. Essenze, Profumi, Piante Offic., Aromi, Saponi, Cosmet., Aerosol*, **54**, 254 (1972); through *Chem. Abstr.*, **77**, 105509r (1972).
11. C. A. N. Catalan et al., *Riv. Ital. Essenze, Profumi, Piante Offic.*,

*Aromi, Saponi, Cosmet., Aerosol*, **59**, 513 (1977); through *Chem. Abstr.*, **88**, 78947a (1978).
12. M. Leao da Silva et al., *Acta Amazonica*, **3**(3), 41 (1973); through *Chem. Abstr.*, **82**, 135732d (1975).
13. J. A. Retamar et al., *Essenze Deriv. Agrum.*, **45**, 31 (1975); through *Chem. Abstr.*, **83**, 197678v (1975).
14. A. Alpande de Morais et al., *Ann. Acad. Bras. Cienc.*, **44**(Suppl.), 315 (1972); through *Chem. Abstr.*, **83**, 120651y (1975).
15. W. H. Stahl et al., *J. Assoc. Anal. Chem.*, **52**, 1184 (1969).
16. C. M. Compadre et al., *Planta Med.*, **53**, 495 (1987).
17. X. A. Dominguez et al., *Planta Med.*, **55**, 208 (1989).
18. P. C. Standley, *Trees and Shrubs of Mexico*, Vol. 23, Part 3 of *Contributions from the United States National Herbarium*, Smithsonian Press, Washington, D.C., 1923, p. 1242.
19. Monograph *Origani vulgaris herba*, *Bundesanzeiger*, no. 122 (July 6, 1988).

# ORIGANUM OIL, SPANISH

**Source:** ***Thymus capitatus*** (L.) Hoffmanns. et Link (syn. *Coridothymus capitatus* Reichb. f.; *Satureja capitata* L.) and carvacrol-rich ***Origanum*** species (Family Labiatae or Lamiaceae).

*Synonym.* Origanum oil.

## GENERAL DESCRIPTION

*Thymus capitatus* is a subshrub with stout, erect branches and very short, stiff linear leaves; up to about 30 cm high; native to the Mediterranean region. Part used is the flowering top from which Spanish origanum oil is obtained by steam distillation. Spanish origanum oil is characterized by its content of carvacrol, which is its major phenolic component. *Origanum* species that are used for the production of Spanish origanum oil are the ones that yield essential oils with carvacrol as their major phenolic component (see ***marjoram***). Major producing countries include Spain, Israel, Lebanon, and Turkey.

## CHEMICAL COMPOSITION

Contains 60–75% phenols consisting mostly of carvacrol (13–74%) and thymol (1–39%),[1–3] plus $\alpha$- and $\beta$-pinene, thujone, camphene, carene, myrcene, terpinene, limonene, $\alpha$-phellandrene, *cis*-*O*-ocimene, $\gamma$-terpinene, and others.[1–6]

## PHARMACOLOGY OR BIOLOGICAL ACTIVITIES

Carvacrol has antifungal and anthelmintic activities, though they are weaker than those of thymol (see ***thyme***).

Origanum oil has been reported to be nonirritating, nonsensitizing, and nonphototoxic to human skin but is moderately to severely irritating to rabbit and mouse skin when applied undiluted.[7]

## USES

**Medicinal, Pharmaceutical, and Cosmetic.** Origanum oil is used as a fragrance component in soaps, detergents, creams, lotions, and perfumes, with maximum use level of 0.2% reported in perfumes.[7]

**Food.** Used as a flavor component in most food products, including alcoholic and nonalcoholic beverages, frozen dairy desserts, candy, baked goods, gelatins and puddings, meat and meat products, condiments and relishes, soups, and gravies. Average maximum use levels reported are usually below 0.004%, except in condiments and relishes, soups, and gravies, which are about 0.007 (74.3 ppm), 0.062, and 0.01% (99 ppm), respectively.

## COMMERCIAL PREPARATIONS

Oil; it is official in F.C.C.

*Regulatory Status.* Has been approved for food use (§172.510); listed as *Thymus capitatus* (Spanish "origanum").

## REFERENCES

See the General References for ARCTANDER; BAILEY 2; FEMA; GUENTHER; MARTINDALE; MERCK; POLUNIN AND SMYTHIES; TERRELL; TUCKER.

1. D. V. Zaitschek and S. Levontin, *Harokeach Haivri*, **14**, 284 (1971); through *Chem. Abstr.*, **77**, 9523v (1972).
2. B. G. Skrubis, *Flav. Ind.*, **3**, 566 (1972).
3. J. M. Hagemann et al., *Lipids*, **2**, 371 (1967).
4. M. De Vincenzi and M. R. Dessi, *Fitoterapia*, **62**(1), 39 (1991).
5. B. M. Lawrence, *Perfum. Flav.*, **9**(5), 41 (1984).
6. B. M. Lawrence, *Perfum. Flav.*, **13**(4), 69 (1988).
7. D. L. J. Opdyke, *Food Cosmet. Toxicol.*, **12**(Suppl.), 945 (1974).

# PAPAIN

**Source:** ***Carica papaya*** L. (Family Caricaceae).

*Synonym.* Vegetable pepsin.

## GENERAL DESCRIPTION

A nonwoody tree with large palmately and deeply lobed leaves; up to about 10 m high; native to tropical America; extensively cultivated in the tropics, including the West Indies, Central America, Africa (e.g., Tanzania), Hawaii, India, Sri Lanka, China, and Indonesia. Part used is the latex from the full-grown but unripe fruit; it is also present in the leaves and trunk.

Commercial crude papain is collected from the green fruit by making shallow cuts in the fruit; the latex is collected, allowed to coagulate, and sun dried or artificially dried (MORTON 3).

The qualities and activities of crude papain vary considerably, depending on the drying process; fast artificial drying at low temperatures (<50°C) yields a cleaner and more active product. Crude papain-producing countries include Tanzania, South Africa, Mozambique, Sri Lanka, and India.

The imported crude papain is purified in the United States by numerous methods, which generally involve filtration, solvent or chemical precipitation, and/or spray drying to yield commercial purified papain of various grades. These grades are usually standardized by adding lactose, sucrose, or starch as diluents. These purified grades of papain are not pure papain, which is not commercially available on a large scale and is not used commercially. Pure papain is currently used mainly for research purposes only.

Like bromelain and ficin, commercial papain is a buff to light brown powder that may be odorless or strongly putrid, depending on the quality. It is relatively soluble in water and glycerol but insoluble in most organic solvents.

In addition to papain itself and appropriate diluents, commercial papain contains chymopapain, lysozyme, and small amounts of papaya peptidase A, lipase, and other constituents.[1–3]

Commercial papain is considered the most stable among the commercial plant proteases. It is active over a wide pH range (3–11) and is normally quite stable in solution at neutral pH. It can withstand temperatures up to 70°C at pH 7 for up to 30 min without much loss in activity. In its powder form, papain can resist dry heat at 100° for 3 h.[2,4]

The optimal pH for papain activity varies with the substrate and is usually between 5 (for gelatin) and 7 (for casein, hemoglobin, and egg albumin). Its optimal temperature range is normally 60–70°C, depending on the substrates.

## CHEMICAL COMPOSITION AND PROPERTIES

Like bromelain and ficin, papain is a sulfhydryl protease, but unlike these enzymes, it does not contain a carbohydrate moiety. It has 212 amino acid residues with cysteine-25 bearing the essential active thiol (-SH) group. It has an isoelectric point at pH 8.75. Its molecular weight has been reported to be 21,000–23,700. Its three-dimensional structure has been shown by x-ray crystallography to be composed of two parts each containing about 100 amino acid residues separated by a big cleft, with the essential cysteine-25 residue lying to one side of the cleft.[1,2,5]

Papain is activated by reducing agents (e.g., thioglycolate, glutathione, cysteine, sulfide, bisulfite, thiosulfate, and cyanide) but is inactivated or inhibited by oxidizing agents and thiol reagents (e.g., iodoacetate, iodoacetamide, methyl bromide, and *N*-ethylmaleimide) as well as heavy metallic ions (e.g., $Zn^{2+}$, $Fe^{2+}$, $Cu^{2+}$, $Pb^{2+}$, and

$Hg^{2+}$) and organic mercurial salts. The lost activity due to inactivation by heavy metals can often be completely restored by EDTA and reducing agents (e.g., cysteine).[1–4] Other compounds reported to activate papain or increase its activity include ammonium sulfate, soil humic acids (especially those with high molecular weight), fumaric acid, malic acid, dihydroxyfumaric acid, and propyl alcohol.[6–10] Other inhibitors include ascorbic, tartaric, dihydroxytartaric, maleic, dihydroxymaleic, and succinic acids as well as oligopeptides that have phenylalanine as the second amino acid from the carboxyl terminus.[2,7,11]

Papain has very broad specificity; it hydrolyzes proteins, small peptides, amides, and esters.[2,4] Its hydrolysis of most protein substrates is more extensive than other proteases (e.g., trypsin and pepsin). Papain also has synthetic activity, producing proteinlike substances (plasteins) from proteolytic hydrolysates of proteins; this property may be useful in industrial applications.[2]

Chymopapain is very similar to papain in its proteolytic activities and its behavior toward activating and inhibiting agents. However, even though its milk-clotting activity is the same as that of papain, its proteolytic activity is only one-half that of papain; it is also more water soluble, more stable, and more active at low pH (down to pH 1.8), and more heat resistant than papain (MERCK).[12]

## PHARMACOLOGY OR BIOLOGICAL ACTIVITIES

Papain has been reported to have numerous biological activities, including anti-inflammatory and antiedemic properties (USD 26th), which are disputed by later findings.[13,14] Its ability to digest dead tissue without affecting the surrounding live tissue has gained itself a reputation as a "biological scalpel." Papain is also known to cause contact allergic reactions in certain individuals and can be a severe irritant and vesicant. Internally, it may cause severe gastritis (USD 26th).[15]

Crude papain has been reported to be teratogenic and embryotoxic when administered orally or intraperitoneally to rats.[16] However, these results should be viewed with caution as crude papain or any crude material usually contains many unknown extraneous substances that might have been responsible for the deleterious effects.

In the early 1980s chymopapain was approved as an intradural injection for herniated lumbar intervertebral discs in patients not responding to conventional therapy. Anaphylactic shock has been reported in 0.5–1% of patients receiving the drug.[17]

In a reassessment of a preliminary report suggesting $\beta$-carotene may be associated with increased risk of prostate cancer, it was found that the increased prostate cancer incidence among men age 70 years and older was instead related to papaya consumption.[18]

Efficacy of the seeds as an effective indigenous treatment for ascariasis, thread worms, and other organisms has been associated with seed content of benzylisothiocyanate, both by inhibiting energy metabolism and in affecting motor activity of *Ascaridia galli in vitro*.[19]

## USES

**Medicinal, Pharmaceutical, and Cosmetic.** Formerly used in treating a wide variety of illnesses or conditions, including infected wounds, sores, ulcers, chronic diarrhea, tumors, hay fever, esophageal obstruction due to meat impaction, catarrh, and psoriasis. Major current uses are as digestive aids and in preparations to control edema and inflammation associated with surgical or accidental trauma, infections, or allergies. It is also used in certain face creams, cleansers, "face lift" formulations, and dentifrices, and as the active ingredient in enzyme cleaners for soft lenses, among others.

Chymopapain is used in the treatment of degenerative intervertabral disk disorders (low back pain, sciatica, etc.).

**Food.** Widely used in tenderizing commercial meat and meat products, and as the active ingredient (sometimes in combinations with ***bromelain*** and/or ***ficin***) in instant meat tenderizers for domestic use. As with bromelain and ficin, much of the tenderizing (hydrolysis of meat fibers and connective tissues) occurs during cooking, but due to the higher temperature optima of papain, overtenderizing resulting in mushiness of the meat often occurs if the meat is cooked at a temperature below the inactivation temperature of papain for an extended period of time.

Papain is extensively used in stabilizing and chillproofing beer (at ca. pH 4.5), hydrolyzing proteins that normally are soluble at room temperature but likely to precipitate from the beer when chilled. It is also used in the preparation of protein hydrolysates.

Other uses and potential uses include modifying dough in baking, clarifying fruit juices, in fish oil extraction, in manufacture of collagen heat-sealable films, and manufacture of sausage casings, among others (see ***bromelain*** and ***ficin***).

**Health Food/Herb Teas.** Papain is used as a digestive aid ingredient in capsules, tablets, and others (WREN).

**Traditional Medicine.** In the Peruvian Andes chopped fruits reportedly used externally as a vulnerary, topical antiinfective, and escharotic; internally, macerated fruits used as a vermifuge and in the treatment of children's enteritis; leaves in infusion as a hypotensive and digestive.[20] In northeastern India juice of the root bark used as a galactogogue.[21] Leaf latex used for warts, cancers, tumors, corns, and various skin ailments. All plant parts reportedly used for a wide variety of conditions throughout the tropics (DUKE 2). The seeds are used in various indigenous medicine systems as a dependable anthelmintic.[19]

In Europe, leaf preparations reportedly used for treatment of gastrointestinal discomfort due to intestinal parasites (nematodes); sedative, diuretic.[22]

**Others.** Used in bating hide and in desizing fabric.

## COMMERCIAL PREPARATIONS

Available as purified papain in numerous grades; formerly official in N.F. and currently official in U.S.P. and F.C.C. Activities are expressed in many ways, depending on the suppliers (N.F. or F.C.C. units; milk clotting units, MCU; gelatin digestion units, GDU; casein digestion units, CDU; etc.). The minimum grade specified in U.S.P. and F.C.C. represents one of the lower grades commercially available.

*Regulatory Status.* GRAS (§182.1585); also has been approved by the USDA Meat Inspection Division. Leaves subject of a German therapeutic monograph; however, use not recommended since effectiveness is not well documented.[22]

## REFERENCES

See the General References for BAILEY 1; BLUMENTHAL; DUKE 2; MARTINDALE; MERCK; MORTON 3; UPHOF; USD 26th; WREN.

1. I. E. Liener, *Adv. Chem. Ser.*, **136**, 202 (1974).
2. A. Yamamoto in G. Reed, ed., *Enzymes in Food Processing*, 2nd ed., Academic Press, New York, 1975, p. 123.

3. A. N. Glazer and E. L. Smith in P. D. Boyer, ed., *The Enzymes*, Vol. 3, Academic Press, New York, 1971, p. 501.
4. R. Arnon in G. E. Perlmann and L. Lorand, eds., *Methods in Enzymology*, Vol. 19, Academic Press, New York, 1970, p. 226.
5. B. G. Wolthers et al. in P. Desnuelle et al., eds., *Structure-Function Relationships of Proteolytic Enzymes*, Academic Press, New York, 1970, p. 272.
6. R. C. Sinha and K. Bahadur, *Proc. Natl. Acad. Sci. India*, **26A**(Pt.1), 53 (1957).
7. A. Ohara et al., *Kyoto Yakka Daigaku Gakuho*, **12**, 32 (1964); through *Chem. Abstr.*, **63**, 16685f (1965).
8. A. Maekawa et al., *J. Vitaminol. (Osaka)*, **6**, 124 (1960); through *Chem. Abstr.*, **54**, 24919 (1960).
9. J. N. Ladd and J. H. A. Butler, *Aust. J. Soil Sci.*, **7**, 253 (1969); through *Chem. Abstr.*, **73**, 13549t (1970).
10. E. Van den Eeckhout and R. Ruyssen, *Farm. Tijdschr. Belg.*, **47**, 73 (1970); through *Chem. Abstr.*, **74**, 576v (1971).
11. I. G. S. Skelton, *Enzymologia*, **25**, 201 (1963).
12. D. K. Kunimatsu and K. T. Yasunobu, in G. E. Perlmann and L. Lorand, eds., *Methods in Enzymology*, Vol. 19, Academic Press, New York, 1970, p. 244.
13. C. Netti et al., *Farmaco, Ed. Prat.*, **27**, 453 (1972); through *Chem. Abstr.*, **77**, 109523p (1972).
14. Anon., *Fed. Regist.*, **40**, 30995 (1975).
15. K. Hwang and A. C. Ivy, *Ann. N. Y. Acad. Sci.*, **54**, 161 (1951).
16. S. Singh and S. Devi, *Indian J. Med. Res.*, **67**, 499 (1978).
17. B. R. Olin, ed., *Lawrence Rev. Nat. Prod.* (Aug. 1991).
18. L. Le Marchand et al., *Am. J. Epidemiol.*, **133**(3), 215 (1991).
19. D. Kumar et al., *Fitoterapia*, **62**(5), 403 (1992).
20. V. De Feo, *Fitoterapia*, **63**(5), 417 (1992).
21. S. K. Borthaku, *Fitoterapia*, **63**(6), 483 (1992).
22. Monograph *Caricae papayae folium*, *Bundesanzeiger*, no. 193 (Oct. 15, 1987).

# PARSLEY

**Source:** ***Petroselinum crispum*** (Mill.) Nym. ex A. W. Hill (syn. *P. sativum* Hoffm.; P. *hortense* Hoffm.; *Apium petroselinum* L.; *Carum petroselinum* Benth. et Hook. f.) (Family Umbelliferae or Apiaceae).

*Synonyms*. Common parsley and garden parsley.

### GENERAL DESCRIPTION

A nonhairy biennial or short-lived perennial with a much-branched stem; up to about 0.7 m high; often cultivated as an annual for its foliage; native to the Mediterranean region; extensively cultivated (especially in California, Germany, France, Belgium, and Hungary), with numerous varieties. Parts used are the ripe fruits (commonly called seeds), the aboveground herb, and the leaves. Parsley seed oil and parsley herb oil are obtained from the fruits and the aboveground herb, respectively, by steam distillation. The leaves (devoid of stems), after drying, furnish the familiar dehydrated parsley flakes; when fresh they are also a familiar culinary herb carried in most supermarkets.

The major producers of parsley oils (seed and herb) include France, Hungary, Germany, and The Netherlands. Domestically used dehydrated parsley flakes are mostly produced in California.

Parsley herb oil has a flavor that resembles the fresh herb, but parsley seed oil has a distinctly different flavor.

Parsley oleoresin is obtained from the fruits by solvent extraction (see ***glossary***).

## CHEMICAL COMPOSITION

Parsley seed contains 2–7% volatile oil;[1] 13–22% fixed oil consisting mainly of petroselinic acid (*cis*-6-octadecenoic acid), with lesser amounts of palmitic, myristic, stearic, oleic, linoleic, and myristolic acids as well as 7-octadecenoic acid;[2–5] flavonoids (apiin (apigenin-7-apiosylglucoside) and luteolin-7-apiosylglucoside);[6] and traces of bergapten (LIST AND HÖRHAMMER).

Parsley seed oil contains mainly apiole, myristicin, tetramethoxyallybenzene, and $\alpha$-pinene.[1,7–9] It also contains petroselinic acid and other volatile fatty acids (see also ***celery seed oil***).[4]

Parsley oleoresin contains mainly apiole (CLAUS).

Parsley herb (leaf) contains 0.05–0.3% volatile oil;[1] furocoumarins, consisting mostly of bergapten (up to 0.02%) with smaller amounts of xanthotoxin (up to 0.003%) and isopimpinellin;[10,11] flavonoids (apiin, luteolin-7-apiosylglucoside, apigenin-7-glucoside, and luteolin-7-diglucoside); 2–22% protein; fats (ca. 4%); vitamins (especially A and C); sugars; and others (LIST AND HÖRHAMMER, MARSH).[12] Oxypeucedanin has been reported as a major furocoumarin (though absent in some varieties).[13]

Parsley herb (leaf) oil contains myristicin (up to 85% in the curly moss variety), $\beta$-phellandrene, 1,3,8-*p*-menthatriene, myrcene, apiole, terpinolene, and 1-methyl-4-isopropenylbenzene as major components.[1,8,9,14] Other compounds present include $\alpha$- and $\beta$-pinenes, *trans*-$\beta$-ocimene, $\gamma$-terpinene, methyl disulfide, $\alpha$-terpineol, $\alpha$-copaene, caryophyllene, and carotol, among others.[14–16] A recent assessment of 104 accessions found leaf oil content to range from 0.00 to 0.16% (fresh weight). Major constituents were 1,3,8-*p*-menthatriene (68%), myristicin (60%), $\beta$-phellandrene (33%), apiol (22%), myrcene (16%), plus terpinolene and 1-methyl-4-isopropenylbenzene at 13%. Thymol, first reported for the leaf oil, was found at 2% or less in seven samples. More than 45 components have been reported from parsley leaf oil.[17]

1,3,8-*p*-Menthatriene is reported to be one of the compounds that contribute to the aroma of parsley.[14]

$\beta$-Elemene, myristicin, and 1,3,8-*p*-menthatriene have been found to accumulate in detectable amounts in young undifferentiated cultures of *P. crispum* cv. 'Paramount.'[18]

## PHARMACOLOGY OR BIOLOGICAL ACTIVITIES

Parsley (herb, seed, extracts, and oils) has been reported to have numerous pharmacological and biological properties, including laxative, hypotensive, antimicrobial, and tonic (on uterine muscles).[19–23]

Myristicin is reported to have psychedelic activities as well as toxic properties (see ***nutmeg***), and apiole is reported to have antipyretic properties (MERCK).

Furocoumarins are phototoxic and may cause skin inflammations or contact dermatitis (see ***bergamot*** and ***lime***).[11]

Apiole is reported to be a spasmolytic, vasodilator, and emmenagogue. Parsley seed oil, due to high concentrations of apiole, may cause vascular congestion and increase smooth muscle contractibility in the bladder, intestine, and uterus; damage to kidney epithelia and heart arrhythmia reported; may cause fatty liver, emaciation, and bleeding of mucous membranes of the gastrointestinal tract.[24]

## USES

**Medicinal, Pharmaceutical, and Cosmetic.** In German phytomedicine cut or ground herb used; daily doses of 6.0 g in infusions or other galenic preparations for systemic irrigation for ailments of the lower urinary tract; also irrigation therapy for the prevention of renal gravel.[25] In France, topical preparations used for dermatological conditions, such as cracks, grazes, chapped skin, and insect bites (BRADLEY).

Parsley seed oil is used as a fragrance component in soaps, detergents, creams, lotions, and perfumes (especially Oriental types, men's fragrances and colognes), with maximum use level of 0.2% reported in perfumes.[19]

**Food.** Parsley (both fresh and dehydrated) is widely used in home cooking.

Chinese parsley is not a variety of parsley but is the young leaf of coriander and has a chemical composition and aroma quite different from parsley (see ***coriander***).

Parsley, parsley oils (herb and seed), and oleoresin are extensively used in flavoring meat sauces, sausages, canned meats, pickles, and spice blends.

Parsley is also used in baked goods, fats and oils, processed vegetables, soups, snack foods, gravies, and others. Highest average maximum use level reported is about 1.5% (14,963 ppm) in processed vegetables.

Other food products in which parsley oils and oleoresin are used include alcoholic and nonalcoholic (e.g., soft drinks) beverages, frozen dairy desserts, candy, baked goods, gelatins and puddings, and soups, among others. Highest average maximum use level reported is about 0.039% (392 ppm) for the oleoresin in condiments and relishes (e.g., pickles).

**Health Foods/Herb Teas.** Parsley herb or root sometimes used as a tea ingredient; also in diuretic formulations (FOSTER).

**Traditional Medicine.** Parsley herb and seed are reportedly used to treat jaundice, menstrual difficulties, asthma, coughs, indigestion, and dropsy, usually in the form of a tea. Parsley herb tea is used to treat gallstones, dyspepsia, dysuria, and rheumatic conditions. In combination with garlic and rue, the leaves are used as a vulnerary in Italy.[26] The herb is also used as a breath freshener when chewed (LUST, ROSENGARTEN). Parsley seed traditionally used as an abortifacient.

## COMMERCIAL PREPARATIONS

Flakes, seed, oils (herb and seed), and extracts (e.g., oleoresin). Seed and oleoresin were formerly official in U.S.P. Both oils are official in F.C.C.

*Regulatory Status.* GRAS (§182.10 and §182.20) with botanical source listed as *Petroselinum crispum* (Mill.) Mansf. Parsley herb is the subject of a positive German therapeutic monograph. Use is contraindicated in pregnancy and inflammatory kidney diseases.[25] Parsley fruits are covered by a German monograph, however use is not recommended because efficacy is not well documented and potential risks outweigh benefits.[24]

## REFERENCES

See the General References for ARCTANDER; BAILEY 1; BLUMENTHAL; BRADLEY; FEMA; FOSTER; GUENTHER; ROSENGARTEN; TERRELL.

1. C. Franz and H. Glasl, *Ind. Obst. Gemueseverwert.*, **59**, 176 (1974); through *Chem. Abstr.*, **81**, 62325r (1974).
2. S. I. Balbaa et al., *Egypt J. Pharm. Sci.*, **16**, 383 (1976).
3. E. Constantinescu et al., *Riv. Ital. Essenze, Profumi, Piante Offic.*,

*Aromi, Saponi, Cosmet., Aerosol*, **54**, 419 (1972); through *Chem. Abstr.*, **78**, 20100w (1973).
4. A. R. S. Kartha and R. A. Khan, *Chem. Ind. (London)*, (52), 1869 (1969).
5. O. S. Privett et al., *J. Am. Oil Chem. Soc.*, **40**, 28 (1963).
6. J. B. Harborne and C. A Williams, *Phytochemistry*, **11**, 1741 (1972).
7. H. Wagner and J. Hölzl., *Deut. Apoth. Ztg.*, **108**, 1620 (1968).
8. J. B. Harborne et al., *Phytochemistry*, **8**, 1729 (1969).
9. C. Franz and H. Glasl, *Qual. Plant. Plant Foods Hum. Nutr.*, **25**, 253 (1976); through *Chem. Abstr.*, **85**, 174323c (1976).
10. G. Innocenti et al., *Planta Med.*, **29**, 165 (1976).
11. J. Palicska and B. Lengyel, *Borgyogy. Venerol. Szemle*, **45**, 118 (1969); through *Chem. Abstr.*, **71**, 110986z (1969).
12. R. A. Komarova, *Sb. Tr. Aspir. Molodykh Nauch. Sotrudnikov, Vses. Nauch. Issled Inst. Rastenievod.*, **8**, 276 (1967); through *Chem. Abstr.*, **69**, 25090s (1972).
13. S. K. Chaudhary et al., *Planta Med.*, **52**, 462 (1986).
14. R. Kasting et al., *Phytochemistry*, **11**, 2277 (1972).
15. G. G. Freeman et al., *J. Sci. Food Agr.*, **26**, 465 (1975).
16. J. Garnero and Y. Chretien-Bessiere, *Fr. Ses Parfums*, **11**, 332 (1968); through *Chem. Abstr.*, **70**, 80785t (1969).
17. J. E. Simon and J. Quinn, *J. Agr. Food Chem.*, **36**, 467 (1988).
18. A. A. Gbolade and G. B. Lockwood, *Fitoterapia*, **62**(3), 237 (1991).
19. D. L. J. Opdyke, *Food Cosmet. Toxicol.*, **13**(Suppl.), 897 (1975).
20. F. Kaczmarek et al., *Biul. Inst. Roslin Leczniczych*, **8**, 111 (1962); through *Chem. Abstr.*, **58**, 14589g (1963).
21. I. Tsonev et al., *Farmatsiya (Sofia)*, **17**, 39 (1967); through *Chem. Abstr.*, **68**, 33128e (1968).
22. J. Kresanek and J. Vittek, *Farm. Obzor.*, **31**, 202 (1962); through *Chem. Abstr.*, **61**, 8625c (1964).
23. A. Sharaf et al., *Qual. Plant. Mater. Veg.*, **17**, 337 (1969); through *Chem. Abstr.*, **72**, 93299d (1970).
24. Monograph *Petroselini fructus, Bundesanzeiger*, no. 43 (Mar. 2, 1989).
25. Monograph *Petroselini herba/radix, Bundesanzeiger*, no. 43 (Mar. 2, 1989).
26. V. De Feo et al., *Fitoterapia*, **63**(4), 337, (1992).

# PASSION FLOWER

**Source:** ***Passiflora incarnata*** L. (Family Passifloraceae).

*Synonyms*. Maypop, maypop passion flower, passiflora, apricot vine, wild passion flower, and passion vine.

## GENERAL DESCRIPTION

A perennial vine with deeply three- to five-lobed leaves and large showy flowers; climbing by axillary tendrils to about 9 m; native to the United States; distributed in southern states from Virginia to Florida and west to Texas; commercial cultivation in

Florida and Guatemala. Part used is the dried flowering and fruiting top (FOSTER).

## CHEMICAL COMPOSITION

Contains small and highly variable amounts (<0.01–0.09%) of indole alkaloids, consisting mainly of harman, with lesser amounts of harmol, harmaline, harmine, and harmalol.[1–6] Presence of the last four alkaloids have been disputed.[1,7]

Other constituents present include flavonoids (isovitexin 2″-β-D-glucoside, isoorientin 2″-β-D-glucoside, apigenin, luteolin, quercetin, kaempferol, schaftoside, isoschaftoside, saponaretin, saponarin, vitexin, orientin, and rutin);[8–13] a cyanogenic glucoside, gynocardin (0.01%);[14,15] sugars (raffinose and sucrose predominant);[16] sterols (stigmasterol and sitosterol); *n*-nonacosane, and gum, among others (LIST AND HÖRHAMMER).

Maltol and ethyl maltol have been isolated from the plant.[17] The coumarins umbelliferone and scopoletin have been detected in the root.[18]

## PHARMACOLOGY OR BIOLOGICAL ACTIVITIES

Passion flower has been reported to have analgesic and sedative properties. It depressed motor activity, increased rate of respiration, and produced a transient reduction in blood pressure. It appears to be nontoxic (GOSSELIN, MERCK).

The harman alkaloids and flavonoids reportedly have tranquilizing effects.[19]

Antispasmodic, anxiolytic, and hypotensive activity reported;[20–22] a synergistic association of sedative activity at high dosage and anxiolytic activity at low dosage, have been observed.[22] Animal studies have repeatedly reported a depression of motor reflexes (ESCOP 2).

Oral administration and peritoneal injections in rats decreased brain stimulus in a number of pharmacological models; the latter significantly prolonged sleep time and produced protective anticonvulsive activity; locomotor activity also affected. Active components ascribed to both hydrophilic and lipophilic fractions; however activity cannot be attributed to specific alkaloids or flavonoids; neuropharmacological activity cannot be clearly attributed to a single chemical fraction. Fractions of a fluid extract have been shown to raise the nociceptive threshold of rats in the tail-flick and hot-plate tests and to prolong sleeping time as well as protect the animals from the convulsive effect of pentylenetetrazole. The active compounds have not yet been characterized but are neither alkaloids nor flavonoids.[23] Future research should evaluate possible relationships with CNS neurotransmitters.[23]

## USES

**Medicinal, Pharmaceutical, and Cosmetic.** Fresh or dried herb and extracts are used as ingredients in some sedative preparations for nervous anxiety. Preparations include tisanes (tea), tinctures, fluid extracts, solid extracts, and sedative chewing gums available in Europe; used in European proprietary tranquilizing phytomedicines in combination with valerian and hawthorn. Passionflower and hawthorn combinations are employed as antispasmodics for digestive spasms, such as in gastritis and colitis.[24]

A recent proposed European monograph indicates use for nervous tension, especially in cases of sleep disturbance or exaggerated awareness of heart palpitations at doses ranging from 0.5 to 2 g of the herb and 2.5 g of the herb in infusion, etc. (ESCOP 2).

**Food.** Extract (type not specified) of passion flower is reported used as a flavor component in alcoholic beverages, nonalcoholic beverages, and frozen dairy desserts with average maximum use levels of 0.05, 0.32, and 0.05%, respectively.

**Health Food/Herb Teas.** Herb in capsules, tablets, teas, and other product forms; primarily in sleep aid formulations (FOSTER).

**Traditional Medicine.** Used as a sedative in treating neuralgia, insomnia, restlessness, headache, hysteria, epilepsy, and other nervous conditions. Also used in bath mixtures for its allegedly calming and soothing effects.

Fruits cultivated or managed for fruit production before arrival of Europeans in Algonkian settlements in Virginia.[25]

**COMMERCIAL PREPARATIONS**

Crude and extracts; crude was formerly official in N.F. Strengths (see ***glossary***) of extracts are expressed in weight-to-weight ratios. European monograph specifications comply with French, German, and Swiss pharmacopoeias, with crude drug to contain not less than 0.8% total flavonoids, calculated as vitexin. A German monograph specifies harmala alkaloids not to exceed 0.01%. Other *Passiflora* species are often distributed in commerce as passion flower (*P. incarnata*).

*Regulatory Status.* Has been approved for food use (§172.510). The dried flowering and fruiting tops of *P. incarnata* in the fourth (1916) and fifth (1926) N.F.; dropped in sixth N.F. (1936). Formerly an approved OTC sedative and sleep aid; removed in 1978. Herb subject of a positive German therapeutic monograph for the treatment of nervous anxiety, in daily dosages equivalent to 4–8 g.[26]

**REFERENCES**

See the General References for BAILEY 1; BLUMENTHAL; BRADLY; CLAUS; ESCOP 2; FEMA; FERNALD; FOSTER; GOSSELIN; KROCHMAL AND KROCHMAL; LEWIS AND ELVIN-LEWIS; LIST AND HÖRHAMMER; LUST; ROSE; TERRELL; UPHOF.

1. W. Poethke et al., *Planta Med.*, **18**, 303 (1970).
2. E. Bennati and E. Fedeli, *Boll. Chim. Farm.*, **107**, 716 (1968); through *Chem. Abstr.*, **70**, 93911f (1969).
3. E. Bernnati, *Boll. Chim. Farm.*, **110**, 664 (1971); through *Chem. Abstr.*, **77**, 24849z (1972).
4. J. Lutomski et al., *Herba Pol.*, **14**, 139 (1968); through *Chem. Abstr.*, **71**, 6564z (1969).
5. N. Svanidze et al., *Rev. Cubana Farm.*, **8**, 309 (1974); through *Chem. Abstr.*, **82**, 90005a (1975).
6. J. Lutomski and B. Malek, *Planta Med.*, **27**, 381 (1975).
7. J. Loehdefink and H. Kating, *Planta Med.*, **25**, 101 (1974).
8. N. M. Gavasheli et al., *Khim. Prir. Soedin.*, **10**, 95 (1974); through *Chem. Abstr.*, **81**, 60811 (1974).
9. H. Schilcher, *Z. Naturforsch.*, *B*, **23**, 1393 (1968).
10. N. M. Gavasheli, *Soobhch. Akad. Nauk Gruz. SSR*, **60**, 353 (1970); through *Chem. Abstr.*, **74**, 72818y (1971).
11. L. Qimin et al., *J. Chromatogr.*, **562**, 435 (1991).
12. H. Geiger and K. R. Markham. *Z. Natuforsch.*, **41c**, 949 (1986).
13. C. Congora et al., *Helv. Chim. Acta*, **69**, 251 (1986).
14. K. C. Spencer and D. S. Seigler, *Planta Med.*, **50**, 356 (1984).
15. K. C. Spencer and D. S. Seigler, *Phytochemistry*, **24**, 2615 (1985).
16. N. M. Gavasheli et al., *Khim. Prir. Soedin.*, **11**, 84 (1975); through *Chem. Abstr.*, **83**, 75359w (1975).
17. N. Aoyagi et al., *Chem. Pharm. Bull.*, **22**, 1008 (1974).

18. N. M, Gavasheli et al., *Khim. Prir. Soedin.*, **9**, 552 (1973); through *Chem. Abstr.*, **80**, 118198a (1974).
19. J. Lutomski et al., *Planta Med.*, **27**, 112 (1975).
20. B. R. Olin, ed., *Lawrence Rev. Nat. Prod.* (May, 1989).
21. R. Kimura et al., *Chem. Pharm. Bull.* **28**(9), 2570 (1980)
22. R. Della Loggia et al., *Rivista di Neurologia*, **51**(5), 297, (1981).
23. E. Speroni and A. Minghetti. *Planta Medica*, **54**, 488 (1988).
24. Brasseur and L. Angenot. *J. Pharm. Belg.*, **38**(1), 15 (1984).
25. K. J. Gremillion, *J. Ethnobiol.*, **9**(2), 135 (1989).
26. Monograph *Passiflorae herba*, *Bundesanzeiger*, no. 223, (Nov. 30, 1985).

# PATCHOULY OIL

**Source:** ***Pogostemon cablin*** (Blanco) Benth. (syn. *P. heyneanus* Benth.; *P. patchouly* Pellet.) (Family Labiatae or Lamiaceae).

*Synonym.* Patchouli oil.

## GENERAL DESCRIPTION

A perennial herb with a sturdy, hairy stem, much branched at the top; leaves opposite, with fragrant odor when rubbed; up to about 1 m high; native to tropical Asia (especially Indonesia and the Philippines) and extensively cultivated in the tropics (Indonesia, Philippines, Malaysia, India, southern China, Seychelles, Brazil, etc.). Parts used are the dried leaves from which patchouly oil is obtained by steam distillation. The leaves are usually subjected to some sort of fermentation or curing process before distillation so as to increase oil yield. Major oil-producing countries include Malaysia, Indonesia, the Seychelles, China, India, and the Philippines. The oil is also distilled in Europe and the United States.

## CHEMICAL COMPOSITION

The leaves contain 1.5–4% volatile oil that is composed mainly of patchouly alcohol (ca. 32–40%) and other sesquiterpenes such as pogostol, bulnesol, norpatchoulenol (ca. 2.2%), α-guaiene, α-bulnesene, and β-patchoulene.[1–5] Other compounds found in the oil include cycloseychellene (a tetracyclic sesquiterpene);[6,7] patchoulipyridine, epiguaipyridine, and guaipyridine (sesquiterpene alkaloids);[8,9] eugenol, cinnamaldehyde, and benzaldehyde;[10] pogostone or dhelwangine (a lactone);[11] and oxygenated sesquiterpenes (e.g., 1α,5α-epoxy-α-guaiene and epoxycaryophyllene);[12] among others (LIST AND HÖRHAMMER).[13–15]

Patchouly alcohol and norpatchoulenol are mostly responsible for the odor of patchouly oil.[4,12]

## PHARMACOLOGY OR BIOLOGICAL ACTIVITIES

Results from one short-term feeding study indicate patchouly oil to be nontoxic to rats.[16]

Dhelwangine (pogostone) is reported to have antimicrobial activities (fungi and bacteria) and is responsible for the bactericidal properties of patchouly oil (LIST AND HÖRHAMMER).[7]

Eugenol, cinnamaldehyde, and benzaldehyde isolated from patchouly (*P. heyneanus*) reportedly have insecticidal activity against insects in stored grain.[10]

## USES

**Medicinal, Pharmaceutical, and Cosmetic.** Patchouly oil is extensively used as a fragrance component in cosmetic preparations. It is one of the most used ingredients in perfumes (especially Oriental types) and is also widely used in soaps and in depilatory creams (to mask the undesirable odor of the active hair-removal ingredients).

Patchouly resinoid (concrète) is used as a fixative.

**Food.** Patchouly oil is extensively used as a flavor ingredient in most major food products, including alcoholic and nonalcoholic beverages, frozen dairy desserts, candy, baked goods, gelatins and puddings, and meat and meat products. Use levels reported are generally very low, mostly below 0.0002% (2.21 ppm).

**Health Food/Herb Teas.** Patchouly oil, widely available in health food stores, primarily used topically as a fragrance (ROSE).

**Traditional Medicine.** The herb is used in Chinese medicine to treat colds, headaches, nausea, vomiting, diarrhea, and abdominal pain, usually decocted with other drugs (JIANGSU).

It is also used to treat bad breath either used alone as a decoction for gargling or combined with cardamom and another herb (*Eupatorium japonicum* Thunb.), the latter especially for bad breath due to drinking of alcohol.

The leaf, fruit, and flower of *P. heyneanus* have been used in tumors.[17]

## COMMERCIAL PREPARATIONS

Mainly the oil. A number of adulterants have been reported in commercial supplies, including gurjun balsam oil, copaiba balsam oil, cedarwood oil, and others.[18]

*Regulatory Status.* Has been approved for food use (§172.510).

## REFERENCES

See the General References for ARCTANDER; BAILEY 2; FEMA; GUENTHER; JIANGSU; NANJING; TERRELL; UPHOF.

1. W. D. Fordham in L. W. Codd et al., eds., *Chemical Technology: An Encyclopedic Treatment*, Vol. 5, Barnes & Noble, New York, 1972, p. 1.
2. P. Teisseire et al., *Recherches*, **19**, 8 (1974).
3. H. Hikino et al., *Chem. Pharm. Bull.*, **16**, 1608 (1968).
4. F. W. Hefendehl, *Seifen, Öle, Fette, Wachse*, **103**, 159 (1977).
5. A. Akhila and M. C. Nigam, *Fitoterapia*, **55**, 363 (1984).
6. V. V. Dhekne and S. K. Paknikar, *Indian J. Chem.*, **12**, 1016 (1974).
7. S. J. Terhune et al., *Int. Congr. Essent. Oils, (Pap.)*, **6**, 153 (1974).
8. G. Buechi et al., *J. Am. Chem. Soc.*, **88**, 3109 (1966).
9. A Van der Gen et al., *Recl. Trav. Chim. Pays Bas*, **91**, 1433 (1972); through *Chem. Abstr.*, **78**, 58643p (1973).
10. R. S. Deshpande et al., *Bull. Grain Technol.*, **12**, 232 (1974); through *Chem. Abstr.*, **84**, 55270c (1976).
11. Anon., *K'o Hsueh T'ung Pao*, **22**, 318 (1977); through *Chem. Abstr.*, **87**, 172771q (1977).
12. P. Teisseire, *Riv. Ital. Essenze, Profumi, Piante Offic., Aromi, Saponi, Cosmet., Aerosol*, **55**, 572 (1973); though *Chem. Abstr.*, **80**, 124571h (1974).

13. S. Nakahara et al., *Phytochemistry*, **14**, 2712 (1975).
14. B. M. Lawrence, *Perfum. Flav.*, **6**(4), 73 (1981).
15. B. M. Lawrence, *Perfum. Flav.*, **15**(2), 75 (1990).
16. B. L. Oser et al., *Food Cosmet. Toxicol.*, **3**, 563 (1965).
17. J. L. Hartwell, *Lloydia*, **32**, 247 (1969).
18. A. Akhila and R. Tewari, *Curr. Res. Aromat. Plants*, **6**(1), 38 (1984).

# PECTIN

**Source:** Primarily from Lemon peel *Citrus limon* (L.) Burm. f. (syn. *C. limonum* Risso) (Family Rutaceae).

## GENERAL DESCRIPTION

Pectin or pectic substances are complex polysaccharides universally present in the cell walls of plants, especially in the spaces between cell walls (middle lamellae) where they act as binders to hold adjacent cell walls together. The peel (especially albedo, the white portion) of citrus fruits (especially lemon, lime, orange, and grapefruit) and apple pomace are rich in pectin. Other pectin-rich plant materials are sugar beet pulp, sunflower heads, carrot, potato, and tomato.[1–5]

Commercial pectin is isolated mainly from lemon peel and to a lesser degree from apple pomace. Lemon peel contains 2–4% pectin when fresh and 20–40% when dried, while dried apple pomace contains 10–20% pectin. The manufacturing process generally involves extraction of the raw materials with water containing a mineral acid (e.g., nitric acid, hydrochloric acid, or sulfur dioxide) at pH 1.5–3.0 and at an elevated temperature of 60°–100°C, centrifugation, and filtration. The filtrate can be concentrated and spray dried or roller dried to yield a low-grade pectin. However most often the filtrate is treated either with alcohols (especially isopropyl) or with soluble salts of aluminum or less often copper to precipitate the pectin, which is then washed, dried, and milled (WHISTLER AND BEMILLER).[6]

Two types of pectins are produced: low-ester (low-methoxy) and high-ester (high-methoxy) pectins. To obtain low-ester pectin, extended hydrolysis or deesterification is allowed to take place some time (step) during the manufacturing process. They are usually standardized to uniform grades; thus "100 gel power" for low-ester pectin and "150 jelly grade" for high-ester pectin. Diluents used are dextrose or other sugars. The "150 jelly grade" or "150 grade" means 1 part of the pectin will set 100 parts of sugar in solution to a jelly of standard strength and firmness containing 65% of sugar (GLICKSMAN, MARTINDALE, USD 26th).

The major pectin producer is the United States (especially California). Other producing countries include the UK, France, and Germany.

Pectin dissolves in water to form a viscous sol with an acidic pH. Its solutions (sols) are most stable at a pH range of 3–4. Outside this range, the viscosity and gel strength of these solutions decrease due to decomposition of pectin. They are especially unstable under alkaline conditions. Pectin is also incompatible with heavy metals, tannin, and salicylic acid. It is insoluble in alcohol and other organic solvents.

The gelling time of pectin varies with the types of pectin and ranges from 20–70 s (rapid set) to 180–250 s (slow set). The gel strength and gelling time are the two most important factors determining the quality of pectin (GLICKSMAN, WHISTLER AND BEMILLER).

## CHEMICAL COMPOSITION

In addition to pectin itself, commercial pectin normally contains sugar (e.g., dextrose) and sodium citrate or other buffer salts (sodium and potassium carbonates, lactates, etc.).

The molecular weight of pectin has been reported to range from 150,000 to 400,000. It is a complex polysaccharide, with D-galacturonic acid as its major sugar component. Other sugars present include D-galactose, L-arabinose, and L-rhamnose. Its molecule consists mainly of a linear galacturonoglycan of $\alpha(1\rightarrow4)$-linked D-galactopyranosyluronic acid. It is partially methylated. Pectins from different sources may vary widely in their methoxy content (0.2–12%) and in their relative proportions of sugars (GLICKSMAN, MERCK, WHISTLER, AND BEMILLER).[1]

## PHARMACOLOGY OR BIOLOGICAL ACTIVITIES

Pectin does not appear to be digested by animals or humans. However on passage through their gastrointestinal tracts, much of the pectin is degraded by bacteria present in their large intestine.[4,7] There is evidence indicating that a small amount of the products of degradation is absorbed, resulting in a hemostatic effect manifested in a shorter coagulation time of drawn blood samples (WHISTLER AND BEMILLER).

As with other gums (see ***algin***, ***guar***, etc.), the cholesterol-lowering property of pectin has been extensively studied. Many studies have demonstrated pectin to have antihypercholesterolemic activities in laboratory animals (rats, fowl, dogs, etc.) and in humans, with the high molecular weight and high-methoxy pectin being the most active.[4,7–16] Protopectin and low-methoxy pectin however do not appear to be active.[4] One study on humans did not produce antihypercholesterolemic effects in patients with hyperlipoproteinemia.[17]

Pectin (from apple) has been reported to exhibit bactericidal effects on Gram-negative and nonspore-forming Gram-positive bacteria, including *Salmonella typhi* and *Escherichia coli* from the former group and *Staphylococcus aureus* from the latter. It had little or no activity on spore-forming bacilli, yeast, and fungi.[18,19]

Other biological and pharmacological properties of pectin include increasing the excretion of heavy metals (e.g., lead and mercury) in experimental animals (see also ***algin***),[20,21] lowering fat absorption in rats,[15] and others (MARTINDALE).[4]

## USES

**Medicinal, Pharmaceutical, and Cosmetic.** Pectin (150 grade) is used as an ingredient in numerous antidiarrheal preparations, often in combination with kaolin. It is used as an emulsifier and/or thickener in creams and lotions.

**Food.** By far the largest use of pectin is in jams, jellies, and preserves, with low-methoxy pectin being primarily used in low-sugar or sugar-free products.

Other food products in which pectin (both or either type) is used include candy, frozen dairy desserts, bakers' jellies, and nonalcoholic beverages (GLICKSMAN, WHISTLER AND BEMILLER).

## COMMERCIAL PREPARATIONS

Several types, including 100 and 150 grades. Pectin (150 grade) is official in U.S.P. and both 100 and 150 grades are official in F.C.C.

*Regulatory Status.* GRAS (§182.1775).

## REFERENCES

See the General References for APhA; GLICKSMAN; LAWRENCE; MARTINDALE; USD 26th; WHISTLER AND BEMILLER.


1. R. L. Whistler in H. W. Schultz et al., eds., *Symposium on Foods: Carbohydrates and Their Roles*, AVI, Westport, Conn., 1969, p. 73.
2. R. M. McCready and H. S. Owens, *Econ. Bot.*, **8**, 29 (1954).
3. J. W. Kesterson and R. Hendrickson, *Econ. Bot.*, **12**, 164 (1958).
4. W. L. Chenoweth and G. A. Leveille, *ACS Symp. Ser.*, **15**, 312 (1975).
5. D. B. Nelson et al. in H. D. Graham, ed., *Food Colloids*, AVI, Westport, Conn., 1977, p. 418.
6. A. H. Rouse and P. G. Crandall, *J. Food Sci.*, **43**, 72 (1978).
7. T. A. Miettinen and S. Tarpila, *Clin. Chim. Acta*, **79**, 471 (1977).
8. P. A. Judd et al., *Nutr. Metab.*, **21**(Suppl. 1), 84 (1977).
9. S. Kiriyama et al., *J. Nutr.*, **97**, 382 (1969).
10. D. J. A. Jenkins et al., *Ann. Int. Med.*, **86**, 20 (1977).
11. T. A. Anderson and R. D. Bowman, *Proc. Soc. Exp. Biol. Med.*, **130**, 665 (1969).
12. H. Fisher et al., *J. Atherosclerosis Res.*, **6**, 292 (1966).
13. H. Fisher et al., Science, **146**, 1063 (1964).
14. K. Tsuji et al., *Eiyogaku Zasshi*, **26**, 113 (1968); through *Chem. Abstr.*, **70**, 65756r (1969).
15. M. L. W. Chang and M. A. Johnson, *J. Nutr.*, **106**, 1562 (1976).
16. D. Mathe et al., *J. Nutr.*, **107**, 466 (1977).
17. F. Delbarre et al., *Am. J. Clin. Nutr.*, **30**, 463 (1977).
18. M. A. El-Nakeeb and R. T. Yousef, *Planta Med.*, **18**, 201 (1970).
19. M. A. El-Nakeeb and R. T. Yousef, *Planta Med.*, **18**, 295 (1970).
20. O. D. Livshits, *Vop. Pitan.*, **28**, 76 (1969); through *Chem. Abstr.*, **71**, 99967w (1969).
21. O. G. Arkhipova and L. A. Zorina, *Prof. Zabolevaniya v Khim. Prom.*, 210 (1965); through *Chem. Abstr.*, **65**, 1286a (1966).


# PEONY (PEONY BARK AND PEONY ROOTS)

Peony yields three commonly used herbs (see individual entries): **PEONY BARK (*MUDANPI*)** and **PEONY ROOTS, RED** and **WHITE (*CHISHAOYAO*** and ***BAI-SHAOYAO*)**.

# PEPPER (BLACK AND WHITE)

**Source:** ***Piper nigrum*** L. (Family Piperaceae).

### GENERAL DESCRIPTION

A perennial woody vine with many nodes, climbing to about 5 m; native to southwestern India and widely cultivated in tropical countries.

Black pepper is the dried full-grown but unripe fruit.

White pepper is the dried ripe fruit with the outer part of the pericarp removed by soaking in water, followed by rubbing. It is less aromatic than black pepper but has a more delicate flavor.

Pepper oil is the volatile oil obtained from black pepper by steam distillation; yield is usually 2–4%.

The oleoresin is obtained from black pepper by solvent extraction followed by removal of the solvent.

Major producers of black and white peppers include India, Indonesia, Malaysia, and China.

Black and white peppers should not be confused with red pepper or cayenne pepper, which is obtained from *Capsicum* species (see ***capsicum***).

## CHEMICAL COMPOSITION

Black pepper contains 2–4% volatile oil and 5–9% piperine, piperidine, piperettine, and a few other minor alkaloids (piperyline, piperolein A, piperolein B, piperanine, etc.).[1–3] Piperine and piperanine are the known pungent principles. Fruit maturity has been found to have little effect on piperine content.[4] Chavicine, formerly believed to be *cis,cis*-piperine, is reported to be a mixture of piperine and minor alkaloids.[5]

White pepper contains little volatile oil but has the same pungent principles and alkaloids as black pepper. Both also contain about 11% protein, 65% carbohydrates, lipids, crude fiber, and others (MARSH).

Black pepper is also reported to contain flavonol glycosides (especially those of kaempferol, rhamnetin, and quercetin) in considerable concentration.[6]

Pepper oil contains a complex mixture of monoterpenes (70–80%), sesquiterpenes (20–30%), and small amounts of oxygenated compounds, with no pungent principles present. Concentration and composition vary, depending on sources. Major monoterpenes include $\alpha$-thujene, $\alpha$-pinene, camphene, sabinene, $\beta$-pinene, myrcene, 3-carene, limonene, and $\beta$-phellandrene. Sesquiterpenes include $\beta$-caryophyllene (major component), $\beta$-bisabolene, $\beta$-farnesene, *ar*-curcumene, humulene, $\beta$-selinene, $\alpha$-selinene, $\beta$-elemene, $\alpha$-cubebene, $\alpha$-copaene, and sesquisabinene. Oxygenated components include linalool, *l*-terpinen-4-ol, myristicin, nerolidol, safrole, $\beta$-pinone, and *N*-formypiperidine, among others (JIANGSU).[3,7–14]

The oleoresin contains pungent principles and volatile oil.

## PHARMACOLOGY OR BIOLOGICAL ACTIVITIES

Black pepper has diaphoretic, carminative, and diuretic properties as well as stimulating activities on the taste buds, producing a reflex increase in gastric secretion. It has also been reported to have strong lipolytic activity that resides in the outer layer of the fruit.[15]

A 0.1-g dose of pepper (type not specified) when held in the mouth without swallowing caused a temporary increase in blood pressure of all 24 subjects tested; pulse rate was not significantly affected (JIANGSU).

Data from one source indicate black pepper oil to be nonirritating and nonsensitizing to human skin but moderately irritating to rabbit skin when applied undiluted under occlusion for 24 h. Its phototoxicity on humans is not known.[16]

## USES

**Medicinal, Pharmaceutical, and Cosmetic.** Used in certain tonic and rubefacient preparations.

**Food.** Both black and white peppers are used extensively as domestic spices. They are also widely used as flavor ingredients in most major food products, including nonalcoholic beverages, candy, baked goods, meat and meat products, cheese, and condiments and relishes. Their oleoresins and oils are used in major categories of foods, including alcoholic beverages, frozen dairy desserts, and gelatins and puddings. Highest average maximum use level reported is 0.42% for white pepper in nut products and about 0.2% for oleoresins in baked goods.

**Traditional Medicine.** Used as a stimulant, carminative, and tonic. Both types of pepper have been reportedly used in cancers.[17]

In Chinese medicine, white pepper is used to treat stomachache, malaria, and cholera in addition to above uses.

**Others.** Along with clove, allspice and ginger, black pepper has been reported to have antioxidant properties and can be a potential source of natural antioxidants.[18]

## COMMERCIAL PREPARATIONS

Crude, oleoresin (black pepper), and oil (black pepper). Black pepper was formerly official in N.F. Black pepper oil is official in F.C.C.

*Regulatory Status.* Both are GRAS (§182.10 and §182.20).

## REFERENCES

See the General References for ARCTANDER; FEMA; GOSSELIN; GUENTHER; JIANGSU; MARTINDALE; MASADA; MERCK; ROSENGARTEN; YOUNGKEN.

1. J. T. Traxler, *J. Agr. Food Chem.*, **19**, 1135 (1971).
2. M. L. Raina et al., *Planta Med.*, **30**, 198 (1976).
3. C. K. Atal et al., *Lloydia*, **38**, 256 (1975).
4. E. R. Jansz et al, *J. Sci. Food Agr.*, **35**, 41 (1984).
5. R. Grewe et al., *Chem. Ber.*, **103**, 3752 (1970).
6. B. Voesgen and K. Herrmann, *Z. Lebensm. Unters. Forsch.*, **170**, 204 (1980).
7. J. Debrauwere and M. Verzele, *Bull. Soc. Chim. Belg.*, **84**, 167 (1975); through *Chem. Abstr.*, **83**, 25073c (1975).
8. C. J. Muller and W. G. Jennings, *J. Agr. Food Chem.*, **15**, 762 (1967).
9. H. M. Richard and W. G. Jennings, *J. Food Sci.*, **36**, 584 (1971).
10. G. F. Russell and W. G. Jennings, *J. Agr. Food Chem.*, **17**, 1107 (1969).
11. S. J. Terhune et al., *Can. J. Chem.*, **53**, 3285 (1975).
12. V. S. Govindarajan in T. E. Furia, ed., *CRC Critical Reviews in Food Science and Nutrition*, Vol. 9, CRC Press, Cleveland, 1977, p. 115.
13. B. M. Lawrence, *Perfum. Flav.*, **10**(2), 51 (1985).
14. B. M. Lawrence, *Major Tropical Spices—Pepper (Piper nigrum* L.*) Essential Oils 1979–1980*, Allured Publishing, Corp., Weaton, Ill., 1981.
15. E. Halbert and D. G. Weeden, *Nature (London)*, **212**, 1603 (1966).
16. D. L. J. Opdyke, *Food Cosmet. Toxicol.*, **16**(Suppl.1), 651 (1978).
17. J. L. Hartwell, *Lloydia*, **33**, 288 (1970).
18. B. Al-Jalay et al., *J. Food Prod.* **50**, 25 (1987).

# PINE BARK, WHITE

**Source:** ***Pinus strobus*** L. (Family Pinaceae).

*Synonym*. Eastern white pine.

## GENERAL DESCRIPTION

An evergreen tree much used for its timber, with leaves in five-leaved fascicles (clusters); branches horizontal and in regular whorls; bark on old trunks deeply fissured; up to about 45 m high, occasionally reaching 67 m; native to northeastern North America and distributed throughout eastern United States and Canada, from Newfoundland west to Iowa and south to Georgia; also cultivated in Europe. Part used is the dried inner bark after having removed the outer layer (cork).

## CHEMICAL COMPOSITION

Reported to contain mucilage, coniferin, coniferyl alcohol, diterpenoids (strobol, strobal, *cis-* and *trans*-abienol, manoyl oxide, etc.), a triterpenoid (3β-methoxyserrat-14-en-21-one), a volatile oil, and others (LIST AND HÖRHAMMER).[1,2]

The bark of Scotch pine (*P. sylvestris*), a related species, has been reported to contain similar but more terpenic compounds, including monoterpenes, diterpenes, and triterpenes as well as sterols, many of which are present in rosin or turpentine (see ***turpentine*** and ***rosin***).[3]

## PHARMACOLOGY OR BIOLOGICAL ACTIVITIES

White pine bark has been reported to have expectorant, demulcent and diuretic properties (CLAUS, WREN).

## USES

**Medicinal, Pharmaceutical, and Cosmetic.** Used as a constituent in some cough syrups that are based on White Pine Compound or related formulations. More commonly used in Europe than the United States.

**Health Food/Herb Teas.** Bark rarely used in tea formulations.

**Traditional Medicine.** Has been used by American Indians for centuries to treat coughs, colds, and congestion; also as a poultice to treat wounds, sores, abscesses, boils, rheumatism, bruises, felons, and inflammation (FOSTER AND DUKE). It is still used as a home remedy.

## COMMERCIAL PREPARATIONS

White pine bark, White Pine Compound, and their extracts; both were formerly official in N.F. Strengths (see ***glossary***) of extracts are expressed in weight-to-weight ratios.

*Regulatory Status*. White pine bark has been approved for use in alcoholic beverages only (§172.510). Oil of pine (*P. sylvestris* and other species) subject of a German therapeutic monograph; allowed internally and externally for congestions of the respiratory tract; externally for rheumatic and neuralgic complaints.[4]

## REFERENCES

See the General References for APhA; BAILEY 1; FOSTER AND DUKE; GOSSELIN; LIST AND HÖRHAMMER; LUST; SARGENT; TERRELL; WREN.

1. D. F. Zinkel et al., *Phytochemistry*, **11**, 425 (1972).
2. D. F. Zinkel et al., *Phytochemistry*, **11**, 3387 (1972).
3. T. Norin and B. Winell, *Acta Chem. Scand.*, **26**, 2297 (1972).
4. Monograph *Pini aetheroleum*, *Bundesanzeiger*, no. 154 (Aug. 21, 1985); revised (Mar. 13, 1990).

# PINE NEEDLE OIL (DWARF AND SCOTCH)

**Source:** ***Dwarf pine*** *Pinus mugo* Turra (syn. *P. montana* Mill.) and *P. mugo* var. *pumilio* (Haenke) Zenari (syn. *P. pumilio* Haenke); ***Scotch pine*** *Pinus sylvestris* L. (Family Pinaceae).

*Synonym*. Swiss mountain pine oil (*P. mugo*).

## GENERAL DESCRIPTION

Dwarf pine is a prostrate shrub or pyramidal tree (up to ca. 12 m high) with leaves in two-leaved fascicles (clusters), stiff and twisted; native to mountains of central and southern Europe.

Scotch pine is a tree also with stiff and twisted leaves in two-leaved fascicles; bark deeply fissured; up to about 40 m high; native to Eurasia and cultivated in eastern United States.

Parts used are the leaves (needles) and twigs from which the essential oils are obtained by steam distillation.

Dwarf pine needle oil (dwarf pine oil, pine needle oil N.F., Pinus pumilio oil, or pumilio pine oil) is produced mainly in Austria (Tirol), and Italy.

Scotch pine needle oil (Scotch pine oil or Pinus sylvestris oil) is mostly produced in Austria (Tirol), Russia, and Scandinavia.

## CHEMICAL COMPOSITION

Dwarf pine needle oil has been reported to contain mostly monoterpene hydrocarbons (ca. 70%), including *d*-limonene, 3-carene, $\alpha$- and $\beta$-pinenes, $\beta$-phellandrene, dipentene, camphene, and myrcene; 4–10% bornyl acetate and other esters; aldehydes (e.g., hexanal, cuminaldehyde, and anisaldehyde); *d*-cryptone; small amounts of sesquiterpenes (e.g., cadinene); and alcohols; among others (LIST AND HÖRHAMMER, MARTINDALE, MERCK, REMINGTON).[1–3]

The presence of the simple aldehydes is believed to be responsible for the characteristic sweet balsamic odor of dwarf pine oil.[1]

Scotch pine needle oil contains 50–97% monoterpene hydrocarbons composed mostly of $\alpha$-pinene, with lesser amounts of 3-carene, dipentene, $\beta$-pinene, *d*-limonene, $\alpha$-terpinene, $\gamma$-terpinene, *cis*-$\beta$-ocimene, myrcene, camphene, sabinene, terpinolene, and others.[2,4–9] Other compounds reported present include bornyl acetate (3 to 3.5%), borneol, 1,8-cineole, citral, terpineol, T-cadinol, T-muurolol, $\alpha$-cadinol, caryophyllene, chamazulene, butyric acid, valeric acid, caproic acid, and isocaproic acid (MERCK, LIST AND HÖRHAMMER).[2,6,8,10,11]

## PHARMACOLOGY OR BIOLOGICAL ACTIVITIES

Both dwarf pine oil and Scotch pine oil have varying degrees of antimicrobial activities. Limonene, dipentene, and bornyl acetate have been reported as the active principles responsible for the antiviral and antibacterial activities of some essential oils (also see ***turpentine***)[3,4,12]

Dwarf pine oil (but not Scotch pine oil) has been reported to be irritating to human skin. Both have also been demonstrated to be sensitizing to certain individuals; both were nonphototoxic.[3,4]

## USES

**Medicinal, Pharmaceutical, and Cosmetic.** Dwarf pine needle oil is used as a fragrance and flavor component in pharmaceutical preparations, including cough and cold medicines, vaporizer fluids, nasal decongestants, and analgesic ointments.

Both dwarf pine oil and Scotch pine oil are used as fragrance ingredients in soaps, detergents, creams, lotions, and perfumes, with maximum use level of 1.2% reported for both oils in perfumes.[3,4]

**Food.** Dwarf pine needle oil and Scotch pine oil are used as flavor components in major food products, including alcoholic and nonalcoholic beverages, frozen dairy desserts, candy, baked goods, and gelatins and puddings. Use levels are generally low, with average maximum level usually below 0.001%.

## COMMERCIAL PREPARATION

Both oils. Dwarf pine needle oil is official in N.F. Both oils are official in F.C.C. Only leaves are specified in N.F. for the production of dwarf pine oil.

*Regulatory Status.* Both oils have been approved for food use (§172.510).

## REFERENCES

See the General References for ARCTANDER; BAILEY 1; FEMA; FERNALD; GOSSELIN; GUENTHER; LIST AND HÖRHAMMER; TERRELL; UPHOF.

1. W. D. Fordham in L. D. Codd et al., eds., *Chemical Technology: An Encyclopedic Treatment*, Vol. 5, Barnes & Noble, New York, 1972, p. 1.
2. R. M. Ikeda et al., *J. Food Sci.*, **27**, 455 (1962).
3. D. L. J. Opdyke, *Food Cosmet. Toxicol.*, **14**, 843 (1976).
4. D. L. J. Opdyke, *Food Cosmet. Toxicol.*, **14**, 845 (1976).
5. R. Hiltunen, *Ann. Acad. Sci. Fenn., Ser. A4*, **208**, 1 (1976); through *Chem. Abstr.*, **88**, 86184q (1978).
6. Y. A. Poltavchenko and G. A. Rudakov, *Biol. Nauki*, **15**, 95 (1972); through *Chem. Abstr.*, **77**, 123879r (1972).
7. J.-C. Chalchát et al., *Phytochemistry.*, **24**, 2443 (1985).
8. J.-C. Chalchat et al., *Planta Med.*, **51**, 285 (1985).
9. B. M. Lawrence, *Perfume Flav.*, **16**(2), 59 (1991).
10. S. Z. Ivanova et al., *Khim. Drev.*, **1**, 103 (1978); through *Chem. Abstr.*, **88**, 117788w (1978).
11. R. D. Kolesnikova et al., *Rastit. Resur.*, **13**, 351 (1977); through *Chem. Abstr.*, **87**, 35863g (1977).
12. L. Joubert and M. Gattefosse, *Mezhdunar. Kongr. Efirnym Maslam (Mater.), 4th*, **1**, 99 (1968); through *Chem. Abstr.*, **78**, 119653r (1973).

# PIPSISSEWA

**Source:** ***Chimaphila umbellata*** Nutt. (syn. *C. corymbosa* Pursh) (Family Pyrolaceae or Ericaceae).

*Synonyms*. Chimaphila, prince's pine, bitter wintergreen, spotted wintergreen, and holly.

## GENERAL DESCRIPTION

Perennial evergreen herb; leaves lanceolate, leathery, prominently toothed; with long creeping subterranean shoots; up to about 30 cm high; flowers five-merous, nodding, petals waxy, native to Eurasia and northern North America (Quebec to Georgia and west to British Columbia, California); also in Central America. Part used is the dried leaf. Commercial supplies in the United States largely sourced from the Pacific Northwest in recent years.

## CHEMICAL COMPOSITION

Contains arbutin (ca. 7.5% in aerial parts)[1] and isohomoarbutin as well as other glycosides (e.g., reinfolin and glucosides of homogentisic acid and toluquinol) and flavonoids (hyperoside, avicularin, kaempferol, etc.);[1–6] about 0.2% chimaphilin (2,7-dimethyl-1,4-naphthoquinone);[6,7] ursolic acid; epicatechin gallate;[1] $\beta$-sitosterol; taraxasterol; nonacosane and hentriacontane; methyl salicylate; resins; tannins; gums; starch; sugar; and others (LIST AND HÖRHAMMER, MERCK).

## PHARMACOLOGY OR BIOLOGICAL ACTIVITIES

Pipsissewa has been reported as one of numerous plants that have elicited hypoglycemic activity in experimental animals.[8]

Arbutin and its derivatives have urinary antiseptic properties due to formation of the hydroquinone aglycone after hydrolysis (see ***uva ursi***).

Chimaphilin is reported to have urinary antiseptic as well as tonic and astringent activities; it also had bacteriostatic properties (LIST AND HÖRHAMMER, MERCK).

## USES

**Medicinal, Pharmaceutical, and Cosmetic.** Has been used in urinary antiseptic preparations.

**Food.** Pipsissewa extracts (especially solid) are used as flavor components mainly in beverages (e.g., root beer, sarsaparilla, and other soft drinks) and candy. Other food products in which they are used include frozen dairy desserts, baked goods, gelatins and puddings, and sweet sauces. Highest average maximum use level is about 0.03% reported for the extract (type not specified) in sweet sauces (365 ppm) and baked goods (290 ppm).

**Health Food/Herb Teas.** Herb used in teas as flavor ingredient; also in capsules and tablet formulations for traditional uses (below).

**Traditional Medicine.** Used as diuretic (also as antidiuretic), astringent, diaphoretic, and mild disinfectant and in treating bladder stones, usually in the form of a tea; also reported to be an antispasmodic for epilepsy, nervous disorders (FOSTER AND DUKE). Used externally in treating ulcerous sores, blisters, and others. It is also reported used in cancers.[9]

**Others.** Herb reportedly used in poison baits for rodents.[10]

## COMMERCIAL PREPARATIONS

Crude and extracts (e.g., solid). Crude was

formerly official in N.F. Strengths (see ***glossary***) of extracts are either expressed in weight-to-weight ratios or in flavor intensities.

*Regulatory Status.* GRAS (§182.20).

## REFERENCES


See the General References for BAILEY 1; FEMA; FOSTER AND DUKE; LEWIS AND ELVIN-LEWIS; LUST; UPHOF; YOUNGKEN.

1. A. A. Trubachev, *Tr. Leningrad, Khim. Farm. Inst.*, **21**, 176 (1967); through *Chem. Abstr.*, **69**, 74507s (1968).
2. E. Walewska and H. Thieme, *Pharmazie*, **24**, 423 (1969).
3. E. Walewska, *Herba Pol.*, **17**, 242 (1971); through *Chem. Abstr.*, **76**, 158253p (1972).
4. A. A. Trubachev and V. S. Batyuk, *Farmatsiya (Moscow)*, **18**, 48 (1969); through *Chem. Abstr.*, **71**, 46684v (1969).
5. A. A. Trubachev and V. S. Batyuk, *Khim. Prir. Soedin.*, **4**, 320 (1968); through *Chem. Abstr.*, **70**, 65170v (1969).
6. K. H. Bolkhart and M. H. Zenk, *Z. Pflanzenphysiol.*, **61**, 356 (1969).
7. K. H. Bolkhart et al., *Naturwissenschaften*, **55**, 445 (1968).
8. N. R. Farnsworth and S. B. Segelman, *Tile Till*, **57**, 52 (1971).
9. J. L. Hartwell, *Lloydia*, **34**, 103 (1971).
10. D. M. Secoy and A. E. Smith, *Econ. Bot.*, **37**(1), 28 (1983).


# PODOPHYLLUM (PODOPHYLLIN)

**Source: *Podophyllum peltatum*** L. (Family Berberidaceae).

*Synonyms.* Mayapple, mandrake, American mandrake, devil's apple, wild lemon, and vegetable mercury.

### GENERAL DESCRIPTION

Perennial herb with an erect stem bearing at its apex one or two large peltate (shield-like) leaves that are five to nine lobed and measure up to about 33 cm across; a solitary flower borne on fork between the two leaves; up to about 45 cm high; native to eastern North America from Quebec to Florida and west to Minnesota and Texas. Parts used are the died rhizome and roots collected early in the spring or in the fall after the aboveground parts have died down. The resin (podophyllin) is prepared from the root and rhizome by alcohol extraction followed by partial removal of the solvent and precipitating the resin with acidified water.

Indian podophyllin obtained similarly from Indian podophyllum (rhizome and roots of *Podophyllum hexandrum* Royle, also known as *P. emodi* Wall.) is now more widely used. In recent years drastic declines from commercial harvest of wild populations in India and Pakistan has prompted listing of the species under the provisions of Appendix II of the Convention on International Trade of Endangered Species of Wild Fauna and Flora (CITES).[1]

### CHEMICAL COMPOSITION

Podophyllum contains lignans (mainly podophyllotoxin, $\beta$-peltatin, and $\alpha$-peltatin, with minor amounts of 4′-demethylpodophyllotoxin, dehydropodophyllotoxin, desoxypodophyllotoxin, etc.), lignan glucosides (e.g., those of podophyllotoxin, β-peltatin, α-peltatin, 4′-demethylpodophyllotoxin, and podorhizol), flavonoids (e.g., quercetin, quercetin-3-galactoside, kaempferol, kaempferol-3-glucoside or astragalin, and isorhamnetin), starch, gum, and others (LIST AND HÖRHAMMER, MORTON 3).[2–4]

Podophyllin or podophyllum resin is the alcohol-soluble and water-insoluble resinous material present in podophyllum in 3–6%. Due to the method of preparation, podophyllin generally contains little or no lignan glucosides. Most of the active principles present are the lignan aglycones consisting of about 20% podophyllotoxin, 13% β-peltatin, 7% α-peltatin, and small amounts of 4′-demethylpodophyllotoxin and dehydropodophyllotoxin. Quercetin (ca. 5%) is also reported present.[5]

Indian podophyllin contains no peltatins; it contains much higher amounts of podophyllotoxin (ca. 40%), 4′-demethylpodophyllotoxin (ca. 2%), and dehydropodophyllotoxin (2 to 3%).[5]

## PHARMACOLOGY OR BIOLOGICAL ACTIVITIES

Podophyllin has strong cathartic and antineoplastic properties. These properties are due to the lignans and their glucosides present, with the former being more potent.[6] The cathartic properties are reported to be due mainly to the peltatins while the antitumor properties are attributed to podophyllotoxin and 4′-demethylpodophyllotoxin as well as the peltatins.[5–8]

Podophyllin and podophyllotoxin have been demonstrated to be strongly embryocidal and growth retarding but not teratogenic in animals (e.g., mice and rats).[9–13]

Podophyllin is very irritating to mucous membranes (especially those of the eye) and to the skin. It is also highly toxic, and fatalities as a result of its use have been reported (MARTINDALE).

Immunotherapeutic features of podophyllotoxin have been investigated. It inhibits mitogen-induced human lymphocyte proliferation and macrophage growth factor-stimulated macrophage proliferation. Alone, it induces lymphocyte activating factor/interleukin 1, and with mitogen, induced T-cell growth factor/interleukin II. The compound directly stimulates macrophage proliferation and potentiates the effects of low doses of macrophage growth factor.[14]

Podophyllotoxin has been found to be a microtubule-disrupting agent with strong antimitotic activity, blocking cells in mitosis. Semisynthetic podophyllotoxin derivatives, VP 16213 (etoposide) and VM26 (teniposide) have no antimitotic activity but are highly active in causing breakage of DNA strands, thus blocking cancer cells in a premitotic stage of the cell cycle.[14–17]

## USES

**Medicinal, Pharmaceutical, and Cosmetic.** Podophyllin is used in treating venereal warts (condyloma acuminatum) and other papillomas, most commonly as a dispersion in Compound Benzoin Tincture or as a solution in alcohol. It is also rather extensively used in laxative preparations, usually along with other cathartic agents.

A semisynthetic podophyllotoxin derivative, VP 16213 (etoposide) has been approved in the United States as a chemotherapeutic agent (often in combination regimes with cisplatin, bleomycin, and others) for the treatment of refractory testicular tumors and for the treatment of small cell lung cancer. VM26 (teniposide) is being investigated for the treatment of acute lymphoblastic leukemia, various lymphomas, and other carcinomas. Semisynthetic derivatives are being used in Europe for treatment of psoriasis and rheumatoid arthritis.[14–17]

**Traditional Medicine.** Used as a cathartic and in treating jaundice, fever, liver ailments, and syphilis; also used in cancers.[18]

## COMMERCIAL PREPARATIONS

Crude (podophyllum) and resin (podophyllin); both are official in U.S.P. Current U.S.P. standards do not specify contents or identities of active principles, and strengths (see ***glossary***) of podophyllin usually vary, depending on suppliers.

An adulterant of Indian podophyllum has recently been identified as *Ainsliaea latifolia* (D. Don) Sch. Bip. (Family Asteraceae). It appears that its rhizome has been used as adulterant for a long time.[19]

*Regulatory status.* Once widespread use of podophyllum resins in OTC laxative preparations has been disallowed.[16] Subject of a German therapeutic monograph; external preparations for condyloma acuminatum.[20]

## REFERENCES

See the General References for BAILEY 1; GOSSELIN; KROCHMAL AND KROCHMAL; LIST AND HÖRHAMMER; LUST; MARTINDALE; MORTON 3; STAHL; TYLER 3.

1. S. Foster, *HerbalGram*, **23**, 19 (1990).
2. J. L. Hartwell and A. W. Schrecker in L. Zechmeister, ed., *Fortschritte der Chemie Organischer Naturstoffe*, Vol. 15, Springer-Verlag, Vienna, 1958, p. 83.
3. A. Stoll et al., *J. Am. Chem. Soc.*, **76**, 6413 (1954).
4. A. Wartburg, *Helv. Chim. Acta*, **40**, 1331 (1957).
5. H. Auterhoff and O. May, *Planta Med.*, **6**, 240 (1958).
6. I. H. Emmenegger et al., *Arzneim. Forsch.*, **11**, 327 (1961).
7. H. Staehelin, *Planta Med.*, **22**, 337 (1972).
8. S. M. Kupchan et al., *J. Pharm. Sci.*, **54**, 659 (1965).
9. J. B. Thiersch, *Int. Congr. Chemotherapy, Proc., 3rd, Stuttgart*, **2**, 1741 (1963); through *Chem. Abstr.*, **65**, 4484a (1966).
10. M. G. Joneja and W. C. LeLiever, *Toxicol. Appl. Pharmacol.*, **27**, 408 (1974).
11. J. B. Thiersch, *Proc. Soc. Exptl. Biol. Med.*, **113**, 124 (1963).
12. M. G. Joneja and W. C. LeLiever, *Can. J. Genet. Cytol.*, **15**, 491 (1973).
13. Anon., *Fed. Regist.*, **40**(56), 12902 (1975).
14. Q. Y. Zheng et al., *Int. J. Immunopharm.* **9**(5), 539 (1987).
15. B. F. Issell et al., eds., *Etoposide (VP-16) Current Status and New Development*, Academic Press, Orlando, Fl., 1984.
16. B. R. Olin, ed. *Lawrence Rev. Nat. Prod.* (Jan. 1992).
17. Anon., *Am. Pharm.*, **NS24**(3), 31 (1984).
18. J. L. Hartwell, *Lloydia*, **31**, 71 (1968).
19. H. S. Puri and S. P. Jain, *Planta Med.*, **54**, 269 (1988).
20. Monograph *Podophylli peltati rhizoma* and *Podophylli peltati resina*, *Bundesanzeiger*, no. 50 (Mar. 13, 1986); revised (Mar. 13, 1990).

# PORIA

**Source:** ***Poria cocos*** (Schw.) Wolf. (syn. *Pachyma cocos* Franch.; *Pachyma hoelen* Sacc.; *Sclerotium cocos* Schw.) (Family Polyporaceae).

*Synonyms*. *Yunling* (Yunnan *fuling*), *fulingge* (whole poria), *baifuling* (white poria), *chifuling* (red poria), *fulingpi* (poria skin), *fuling*, hoelen, Indian bread, and tuckahoe.

## GENERAL DESCRIPTION

*Poria cocos* is a polypore fungus parasitic on roots or saprophytic on rotten stumps of certain pine trees growing in loose sandy soil, the most common including *Pinus massoniana* Lamb, *P. densiflora* Sieb. et Zucc., *P. yunnanensis* Franch., *P. taiwanensis* Hayata, and *P. thunbergii* Parl. Part used is its sclerotium (fungal tissue mass), which is irregularly shaped, ranging from fist-size to large masses with diameter of 10–30 cm or larger, growing 20–30 cm deep in the ground. Its presence in the native state ready for harvesting is indicated by several signs, including cracks on ground around the tree trunk, which when struck with a shovel produce a hollow sound; white mycelia (fungal tissue strands) or grayish white powder on surface above root system around the tree; and little or no vegetative growth around the tree where water also drains quickly after rain. Cultivated *fuling* is harvested after 2 years, with best quality after 3–4 years. The sclerotia from wild plants are dug up between July and March while those of cultivated plants are dug in autumn, rid of dirt and allowed to "sweat" (ferment) a few times (5–8 days each) until most moisture has evaporated and then dried completely in an airy, shady place. The resulting product is wrinkled and brown or dark brown called *fulingge* (whole *fuling*). *Fulingge* is white in the middle (with brown root embedded) and light brown to pink toward the outside. From different parts of *fulingge* after sweating and before final drying, numerous products of various grades are produced: *fulingpi* (*fuling* skin), *baifuling* (white *fuling*), *chifuling* (red *fuling*), *fushen* (slices of peeled *fulingge* containing pieces of pine root), *fushenmu* (pine root removed from *fuling*), and others (CMH, JIANGSU, ZHU).[1,2]

## CHEMICAL COMPOSITION

Contains ca. 93% of a polysaccharide named $\beta$-pachyman composed of pachymaran in $\beta$-(1$\rightarrow$3) linkages and side chains of $\beta$-(1$\rightarrow$6) linkages; triterpene acids, including pachymic acid, eburicoic acid, tumulosic acid, pinicolic acid, and poriatin (mixture); adenine, ergosterol, choline, lecithins, histidine, sucrose, fructose, proteases, trace minerals, and others (ETIC, HU, JIANGSU, NATIONAL).[3–8]

## PHARMACOLOGY OR BIOLOGICAL ACTIVITIES

Poriatin has been shown to potentiate the effects of antitumor agents against mouse sarcoma $S_{180}$, with tumor inhibition of 38.9% for dactinomycin (vs. 19.6% when used without poriatin), 48% for mitomycin (vs. 35% for mitomycin alone), 69.0% for cyclophosphamide (vs. 32.3% for cyclophosphamide alone) and 59.1% for fluorouracil (vs. 38.6% for fluorouracil alone); against murine leukemia (L1210), increasing survival time by 168.1% when cyclophosphamide was used with poriatin (vs. 70% when cyclophosphamide was used alone), with the optimal dose for poriatin at 1/40 of $LD_{50}$.[5] Poriatin also has immunoregulatory effects in mice.[4,9]

Polysaccharides (especially pachymaran) have exhibited antitumor (mouse sarcoma, $S_{180}$) and immunomodulating effects in mice, enhancing phagocytosis of macrophages and activating T and B lymphocytes (HU).[3,10]

Decoctions of both *fuling* and *fushen* had sedative effects in mice, with those of *fushen* stronger. The diuretic effects of *fuling* have been reported but are so far equivocal; however, at doses of 30–100 g/day, it had a strong diuretic effect in patients with cardiac edema.[11] Acetone extractives of *fuling* had marked inhibitory effects on mutagenesis induced by aflatoxin $B_1$ per the Ames test.[12] Other biological effects include hypoglycemic in rabbits, relaxation of isolated rabbit intestine, inhibiting isolated frog heart, and antibacterial *in vitro* (alcohol but not water extract) (WANG).

## USES

**Medicinal, Pharmaceutical, and Cosmetic.** *Fuling* (especially *baifuling*) is an ingredient in several well-known skin (facial) treatment formulas used by women of the Chinese imperial courts to maintain clean and "radiant" skin and to prevent pimples, dark spots, and wrinkles. Powder used in facial scrubs and extracts (hydroalcoholic) in moisturizing and nourishing creams and lotions for its traditional cleansing, whitening, soothing and moisturizing properties (ETIC).[13]

**Food.** Used as an ingredient in *fuling* breads and *fuling* cakes that are traditionally eaten for their tonic properties.

**Health Food/Herb Teas.** Used as an ingredient in various herbal formulas for its traditional diuretic, tonic, and calming properties, especially in weight-control and sedative products.

**Traditional Medicine.** First mentioned in the *Shi Jing* (ca. 400–300 B.C.) and its medicinal properties later described for the first time in the *Shen Nong Ben Cao Jing* (ca. 200 B.C.–A.D. 100), *fuling* is a widely used tonic food in China. Traditionally considered as sweet and bland tasting, neutral, diuretic (*li shui shen shi*), kidney invigorating and nerve calming, it is used in treating dysuria (urination difficulties), edema, cough due to phlegm retention (*tan yin ke sou*), diarrhea, nervousness, insomnia, spermatorrhea, and forgetfulness, etc. It is extensively used in diet therapy, especially for older people, in soups, cakes, breads, and wines, etc.[14–16]

*Fushen* is considered superior to *fuling* in treating nervousness and insomnia.

## COMMERCIAL PREPARATIONS

Crude and extracts (aqueous and hydroalcoholic). Crude normally comes as *baifuling* (white), *chifuling* (red), and *fushen* in thick slices and as a powder; identity of powder can be verified by microscopy (XU AND XU); extracts in liquid and powdered forms with no established standards.

*Regulatory Status*. *Fuling* is an ethnic food; its U.S. regulatory status not determined.

## REFERENCES

See the General References for CHP; CMH; HU; IMM-CAMS; JIANGSU; LU AND LI; NATIONAL; WANG; XU AND XU; ZHU.

1. B. Liu, *Medicinal Fungi of China*, People's Press, Taiyuan, Shanxi, 1984, p. 89.
2. Z. G. Song, *Sichuan Zhongyi* **9**, 51 (1990).
3. C. X. Chen, *Zhongcaoyao*, **16**(4), 40 (1985).
4. J. Xu et al., *Faming Zhuanli Gongbao*, **5**(35), 8 (1989); through *Chin. Pharm. Abstr.*, 9104649.
5. D. D. Li et al., *Zhongguo Kangshengsu Zazhi*, **15**(1), 63 (1990).
6. C. Goro, *Nature*, **225**, 943 (1970).
7. J. Hamuro et al., *Nature* **233**, 486 (1971).

8. H. Saito et al., *Agric. Biol. Chem.*, **32**, 1261 (1968).
9. G. J. Wang et al., *Zhongguo Kangshengsu Zazhi*, **17**(1), 42 (1992).
10. S. C. Lu et al., *Diyi Junyi Daxue Xuebao*, **10**, 267 (1990).
11. A. Q. Kang and Z. X. Zhang, *Tianjin Zhongyi* (1), 14 (1989).
12. C. C. Ruan et al., *Chin. J. Cancer*, **8**(1), 29 (1989).
13. X. F. Zhang, *Dazhong Zhongyiyao*, (3), 37 (1991).
14. S. Y. Chen, *Edible Fungi China*, **9**(3), 24, 25 (1990).
15. S. Y. Chen, *Edible Fungi China*, **9**(5), 42 (1990).
16. S. Y. Chen, *Edible Fungi China*, **9**(6), 37 (1990).

# PSYLLIUM

**Source:** ***Plantago psyllium*** L., ***P. indica*** L. (syn. *P. arenaria* Waldst. et Kit.), and ***P. ovata*** Forsk. (syn. *P. decumbens* Forsk. and *P. ispaghula* Roxb.) (Family Plantaginaceae).

*Synonyms*. Black psyllium (*P. indica*); blond psyllium, ispaghula, ispagol, spogel, and Indian plantago (*P. ovata*); brown psyllium, French psyllium, and Spanish psyllium (*P. psyllium*, *P. indica*); plantago; and plantain. *P. arenaria* is accepted in taxonomic literature as the correct name for *P. psyllium* (ambiguous name) and *P. indica* (illegitimate name); however, both names are still used in commercial trade over *P. arenaria*.

## GENERAL DESCRIPTION

All three are annual herbs, less than 0.5 m high. *Plantago ovata* is stemless (acaulescent) or nearly so; it is native to the Mediterranean, North Africa, and western Asia; extensively cultivated in India and Pakistan. *Plantago psyllium* and *P. indica* have erect and branched stems; native to the Mediterranean region; cultivated in Spain and southern France.

Parts used are the dried ripe seed and its husk, a thin membranous layer on the seed coat. When soaked in water, the seed increases in volume many fold but contracts to its original volume when excess alcohol is added.[1]

## CHEMICAL COMPOSITION

Contains 10–30% mucilage; present mainly in the husk; it is composed of a mixture of polysaccharides with D-xylose as the major residue, and L-arabinose and an aldobiouronic acid also present.[2]

Other constituents present include monoterpene alkaloids such as (+)-boschniakine (indicaine), (+)-boschniakinic acid (plantagonine), and indicainine;[3–5] aucubin (a glucoside);[3,4] planteose (a trisaccharide), sucrose, glucose, and fructose;[3,6,7] sterols ($\beta$-sitosterol, stigmasterol, and campesterol) and triterpenes ($\alpha$- and $\beta$-amyrins);[4] a fixed oil and fatty acids (e.g., linoleic, oleic, stearic, and palmitic acids); tannins; and others (JIANGSU, MORTON 3).

## PHARMACOLOGY OR BIOLOGICAL ACTIVITIES

Psyllium seed has laxative properties due to the swelling of its husk in water to form a gelatinous mass, thus keeping the feces hydrated and soft. The resulting bulk promotes peristalsis and laxation (GOODMAN AND GILMAN).

An alcoholic extract of blond psyllium seed (*P. ovata*) has been reported to lower the blood pressure of anesthetized dogs and cats, to inhibit isolated rabbit, rat, and guinea pig ileum; it also has cholinergic activity (JIANGSU).[8]

The husk has been reported to depress the growth of chickens by 15% when 2% of it was included in an otherwise balanced diet.[9]

Blond psyllium seed powder has been reported to counteract strongly the deleterious effects of feeding a supplement of 2% sodium cyclamate, 2% FD & C red no. 2, and 4% polyoxyethylene (20) sorbitan monostearate to rats.[10]

In a recent clinical study hydrophilic muciloid psyllium preparations were found to lower serum cholesterol, LDL cholesterol, and LDL:HDL ratios,[11] and triglycerides,[12] by binding bile acids, hence increasing fecal excretion while increasing bile salt synthesis from cholesterol (ESCOP 2).

Blond psyllium seeds are a partly fermentable dietary fiber supplement that increases stool bulk and has mucosa-protective effects. Rats fed 100–200 g of blond psyllium seeds/kg in a fiber-free elemental diet for 4 weeks were found to have an increase of fecal fresh weight up to 100%, fecal dry weight up to 50%, and fecal water content up to 50%. Length and weight of the large intestine (but not small intestines) increased significantly. The seeds and husks also increase total fecal bile secretion while reducing activity of $\beta$-glucuronidase.[13]

## USES

**Medicinal, Pharmaceutical, and Cosmetic.** The husks are widely used as an ingredient in bulk laxatives; seeds are also used, but less so.

When seeds are used, they should not be ground or chewed, as it has been reported that they release a pigment that deposits in renal tubules. (USD 26th).[2]

A recent European monograph indicates use for habitual constipation, in case where soft stool is desired as in cases of anal fissures, hemorrhoids, and post-rectal surgery; irritable bowel syndrome, diverticulosis; as a dietary supplement where increases in dietary fiber are required; and adjuvant therapy for diarrhea (ESCOP 2).

**Food.** The husk mucilage is used as a thickener or stabilizer in certain frozen dairy desserts.

**Health Food/Herb Teas.** Ground seeds or husks used in various dietary supplement formulations for increased fiber, cholesterol reduction, laxative activity; e.g., weight loss products.

In 1989, psyllium was introduced into processed foods, including General Foods' Benefits cereal and Kellogg's Heartwise cereal, with claims for cholesterol-reducing benefits and soluble fiber health benefits. Shortly after introduction, the FDA suggested the products were "misbranded drugs" due to insufficient evidence to support the labeling claims. Benefits was removed from the market, while Heartwise is still marketed.

The introduction of these products largely stimulated the current regulatory and legislative debate on "acceptable health claims" for foods, indirectly leading to passage of the Nutrition and Labeling Act of 1990.

A recent report links a case of anaphylactic reaction to Heartwise; asthma and anaphylaxis had previously been associated with rare allergies often resulting from industrial or occupational sensitivity.[14] In response, Kellogg added a label, reading: "New Users: A very small percentage of individuals, particularly some nurses and health care providers who have been occupationally exposed to psyllium dust, may develop a sensitivity to psyllium. This sensitivity may result in an allergic reaction."[15]

**Traditional Medicine.** Used in the United States and Europe primarily as a bulk laxative to treat chronic constipation; also used as an emollient and demulcent. Fresh leaves of *Plantago* species applied topically for poison ivy, insect bites and stings; an uncontrolled study reported that use of the fresh leaves prevented itching and spread of dermatitis in poison ivy–induced dermatitis.[16]

Used in India as a diuretic and to treat diarrhea, gonorrhea, urethritis, hemorrhoids, kidney, and bladder problems, among others (MORTON 3).

In China both the seeds and whole herbs of related *Plantago* species (e.g., *P. asiatica* L. and *P. depressa* Willd.) are used for similar purposes. In addition, the seeds are used to treat hematuria (bloody urine), coughing, high blood pressure (with ca. 50% success in clinical trials), and other ailments. The seeds used in Chinese medicine are often specially treated by frying with saltwater before drying; these cannot be directly compared with the psyllium seeds used in the United States (FARNSWORTH, JIANGSU, NANJING).

Seeds of psyllium and other *Plantago* species as well as their roots, juice, leaves, and whole herbs have been reported used in treating cancer.[17]

## COMMERCIAL PREPARATIONS

Husks (blond) and whole seeds (black and blond); whole seeds contain both husk and kernels; husks are official in U.S.P. Blond psyllium husk subject of a positive German therapeutic monograph used for habitual constipation, and supportive treatment of irritable bowel syndrome.[18]

## REFERENCES


See the General References for BLUMENTHAL; ESCOP 2; MARTINDALE; MERCK; MORTON 3; TERRELL; TUCKER; UPHOF; YOUNGKEN.

1. R. Wasicky, *Planta Med.*, **9**, 232 (1961).
2. J. N. BeMiller in R. L. Whistler, Ed., *Industrial Gums*, Academic Press, New York, 1973, p. 339.
3. M. S. Karawya et al., *U. A. R. J. Pharm. Sci.*, **12**, 53 (1971); through *Chem. Abstr.*, **78**, 13729e (1973).
4. S. I. Balbaa et al., *U. A. R. J. Pharm. Sci.*, **12**, 35 (1971); through *Chem. Abstr.*, **77**, 156311c (1972).
5. G. A. Cordell, in R. H. F. Manske, Ed., *The Alkaloids*, Vol. 16, Academic Press, New York, 1977, p. 431.
6. M. S. Karawya, *Planta Med.*, **20**, 14 (1971).
7. D. French et al., *J. Am. Chem. Soc.*, **75**, 709 (1953).
8. M. L. Khorana et al., *Indian J. Pharm.*, **20**, 3 (1958).
9. P. Vohra and F. H. Kratzer, *Poultry Sci.*, **43**, 1164 (1964).
10. B. H. Ershoff, *J. Food Sci.*, **41**, 949 (1976).
11. J. W. Anderson et al., *Arch. Intl. Med.*, **148**, 292 (1988).
12. A. Danielsson et al., *Acta Hepatogastroenterol.*, **26**, 148 (1979).
13. E. Leng-Peschlow, *Br. J. Nutr.*, **66**, 331 (1991).
14. M. J. Kaplin, *N. Engl. J. Med*, **323**, 1072 (1990).
15. D. R. Schaller, *N. Engl. J. Med*, **323**, 1073 (1990).
16. S. Duckeet, *N. Engl. J. Med*, **303**, 583 (1980).
17. J. L. Hartwell, *Lloydia*, **33**, 288 (1970).
18. Monograph *Plantaginis ovatae testa and semen*, *Bundesanzeiger*, no. 22 (Feb. 1, 1990).

# QUASSIA

**Source:** ***Picrasma excelsa*** (Sw.) Planch. and ***Quassia amara*** L. (Family Simaroubaceae).

*Synonyms*. Jamaican quassia (*P. excelsa*); Surinam quassia (*Q. amara*); quassia wood and bitterwood.

## GENERAL DESCRIPTION

*Picrasma excelsa* is a tree with a trunk diameter of 0.5–1 m; up to about 25 m high; native to the West Indies and growing in Jamaica and other Caribbean Islands. *Quassia amara* is a shrub or small tree up to about 3 m high; native to northern South America and growing in Surinam, Brazil, Colombia, Venezuela, and other tropical American countries. Part used is the wood.

## CHEMICAL COMPOSITION

Surinam quassia (*Q. amara*) contains quassin, quassinol, 18-hydroxyquassin, and neoquassin, while Jamaican quassia (*P. excelsa*) contains isoquassin (picrasmin), neoquasin and 18-hydroxyquassin as their bitter principles (LIST AND HÖRHAMMER).[1,2] These bitter principles are reported to be about 50 times more bitter than quinine (STAHL).

Other constituents include two additional quassinoids, quassimarin and simalikalactone D, isolated from *Q. amara*[3], $\beta$-sitosterol and $\beta$-sitostenone;[4] 1.8% thiamine (in *P. excelsa*); and alkaloids of the $\beta$-carboline type including canthin-6-one, 5-methoxycanthin-6-one, 4-methoxy-5-hydroxycanthin-6-one, and *N*-methyl-1-vinyl-$\beta$-carboline (LIST AND HÖRHAMMER, WILLAMAN AND SCHUBERT).[5]

Jamaican quassia is reported to contain no tannin (MARTINDALE).

The Asian *P. quassioides* D. Don. Benn is reported to contain the carboline alkaloids 1-carboxy-$\beta$-carboline and picrasidines A, B, L, M, and P. Quassinoids include kumijians A, B, C, and G; picrasins A, B, C, D, E, F, G, N, O and Q; and kusulactone (GLASBY 2).

## PHARMACOLOGY OR BIOLOGICAL ACTIVITIES

Has bitter tonic properties but in large doses is reported to cause stomach irritation and to produce vomiting (LEWIS AND ELVIN-LEWIS).

Quassimarin has been reported to have antileukemic properties.[3]

Quassinoids reported to be amebicidal *in vitro* and *in vivo* (WREN).

Reported traditional antimalarial activity has not been confirmed in quassinoids.[6]

## USES

**Medicinal, Pharmaceutical, and Cosmetic.** Occasionally used in certain laxative preparations.

**Food.** Extracts and purified mixtures of bitter principles (commercially known simply as "quassin") are used to impart a bitter flavor to various food products, especially alcoholic (e.g., liqueurs and bitters) and nonalcoholic beverages. Food products in which the extracts are used include alcoholic and nonalcoholic beverages, frozen dairy desserts, candy, baked goods, and gelatins and puddings. Highest average maximum use level reported is about 0.007% (71.8 ppm) in nonalcoholic beverages. Used as a flavoring substitute for quinine.

**Traditional Medicine.** Used as a bitter and anthelmintic (via enema). Also used to treat fevers.

**Others.** Bark reported used as an insecticide.

## COMMERCIAL PREPARATIONS

Crude, extracts, and quassin; crude and fluid extract were formerly official in N.F.

Strengths (see ***glossary***) of extracts are usually expressed in weight-to-weight ratios.

*Regulatory Status.* Has been approved for food use (§172.510).

## REFERENCES

See the General References for BIANCHINI AND CORBETTA; CLAUS; GLASBY 2; GOSSELIN; LIST AND HÖRHAMMER; MERCK; TERRELL; UPHOF.

1. J. Polonsky in W. Herz et al., eds., *Fortschritte der Chemie Organischer Naturstoffe*, Vol. 30, Springer-Verlag, Vienna, 1973, p. 101.
2. H. Wagner et al., *Planta Med.* **38**, 204 (1980).
3. S. M. Kupchan and D. R Streelman, *J. Org. Chem.*, **41**, 3481 (1976).
4. D. Lavie and I. A. Kaye, *J. Chem. Soc.* (10), 5001 (1963).
5. H. Wagner et al., *Planta Med.* **36**, 113 (1979).
6. D. H. Bray et al., *Phytother. Res.*, **1**(1), 22 (1987).

# QUEBRACHO

**Source:** ***Aspidosperma quebracho-blanco*** Schlecht. (Family Apocynaceae) and ***Schinopsis quebracho-colorado*** (Schlecht.) Barkl. et T. Meyer (syn. *S. lorentzii* (Griseb.) Engl. and *Quebrachia lorentzii* Griseb.) (Family Anacardiaceae).

*Synonyms.* Quebracho blanco and white quebracho (*A. quebracho-blanco*); quebracho colorado and red quebracho (*S. quebracho-colorado*).

## GENERAL DESCRIPTION

Large trees native to Argentina and neighboring countries. Part used is their dried bark.

Although red quebracho has been approved for food use, it is doubtful that it is actually used in food products. Its wood extracts, containing large amounts of tannins and an alkaloid, have been primarily used in tanning leather and in dying. There has been relatively little scientific work (chemical or pharmacology) done on red quebracho relating to its food or drug use (MERCK).

## CHEMICAL COMPOSITION

White quebracho contains 0.3–1.5% indole alkaloids, including aspidospermine, aspidospermatine, aspidosamine, yohimbine (quebrachine), *l*-quebrachamine, eburnamenine, aspidospermidine, *l*-pyrifolidine, deacetylpyrifolidine, rhazidine, and akuammidine, among others (GLASBY 1, LIST AND HÖRHAMMER, RAFFAUF, WILLAMAN AND SCHUBERT).[1–3]

Numerous alkaloids, including rhazinilam (a lactam), have also been isolated from the leaves of *A. quebracho-blanco*.[1,4–8]

Other constituents reported present in white quebracho include 3–4% tannin, sugars, β-sitosterol (quebrachol), and the triterpenic alcohols lupeol and α-amyrin (LIST AND HÖRHAMMER).[9]

## PHARMACOLOGY OR BIOLOGICAL ACTIVITIES

White quebracho alkaloids (especially aspidospermine, aspidosamine, yohimbine, quebrachamine, and akuammidine) have been reported to have numerous pharmacological properties, including hypotensive, spasmolytic, diuretic, peripheral vasoconstrictor, arterial hypertensive, respiratory stimulant, uterine sedative, and local anesthetic, among others (LIST AND HÖRHAMMER, MARTINDALE).[1]

Large doses of white quebracho may cause nausea and vomiting (MARTINDALE).

## USES

**Medicinal, Pharmaceutical, and Cosmetic.** Now rarely used in pharmaceutical preparations.

**Food.** Quebracho is primarily used in foods. Its extract (type not specified) is used as a flavor ingredient in major categories of foods, including alcoholic and nonalcoholic beverages, frozen dairy desserts, candy, baked goods, and gelatins and puddings. Highest average maximum use level is about 0.003% reported in candy (29.8 ppm) and baked goods (34.5 ppm).

**Health Food/Herb Teas.** Sometimes used as a tea ingredient; seldom used in the United States; liquid extracts used in respiratory preparations in the UK (WREN).

**Traditional Medicine.** White quebracho is used as a febrifuge, antispasmodic, and respiratory stimulant. Also used as an aphrodisiac. The latex of *A. nitidum* is used in Columbia for controlling leprosy. *A. schultesii* latex is used topically to control sores of probable fungal origin.[10]

## COMMERCIAL PREPARATIONS

Mainly crude (white quebracho); white quebracho was formerly official in U.S.P.

*Regulatory Status.* Has been approved for food use (§172.510).

## REFERENCES

See the General References for CLAUS; FEMA; GLASBY 1; GOSSELIN; LEWIS AND ELVIN-LEWIS; LIST AND HÖRHAMMER; MARTINDALE; TERRELL; UPHOF; WREN; YOUNGKEN.

1. R. L. Lyon et al., *J. Pharm. Sci.*, **62**, 218 (1973).
2. P. Tunmann and D. Wolf, *Z. Naturforsch. B*, **24**, 1665 (1969).
3. S. Markey et al., *Tetrahedron Lett.* (2), 157 (1967).
4. P. S. Benoit et al., *J. Pharm. Sci.*, **62**, 1889 (1973).
5. R. L. Lyon et al., *J. Pharm. Sci.*, **62**, 833 (1973).
6. D. J. Abraham et al., *Tetrahedron Lett.* (10), 909 (1972).
7. H. K. Schnoes et al., *Tetrahedron Lett.* (22), 993 (1962).
8. R. L. Lyon, *Diss. Abstr. Int., B.* **35**, 2673 (1974).
9. P. Tunmann and G. Hermann, *Pharmazie*, **25**, 361 (1970).
10. R. E. Schultes and R. F. Raffauf, *The Healing Forest*, Dioscorides Press, Portland, Ore., 1990.

# QUILLAIA

**Source:** ***Quillaja saponaria*** Mol. (Family Rosaceae).

*Synonyms.* Soapbark, soap tree bark, murillo bark, quillaja, Panama bark, Panama wood, and China bark.

## GENERAL DESCRIPTION

A large evergreen tree with shiny coriaceous (leathery) leaves and thick bark; native to Chile and Peru; cultivated in southern California. Part used is the dried inner bark deprived of cork.

## CHEMICAL COMPOSITION

Quillaia contains 9–10% triterpenoid saponins consisting of glycosides of quillaic acid (quillaja sapogenin);[1–3] tannin; 11% calcium oxalate; sugars; starch; and others (LIST AND HÖRHAMMER).

## PHARMACOLOGY OR BIOLOGICAL ACTIVITIES

Saponins are generally reported to have widely different (both in kind and in intensity) pharmacological and biological activities. Some of the more important activities include hemolytic (strong *in vitro*, much weaker *in vivo*),[4] local irritant, inflammatory (e.g., on intestine),[5–7] anti-inflammatory,[7] antimicrobial,[8] cytotoxic,[9] and antihypercholesterolemic in laboratory animals (also see ***alfalfa***).[10–13]

Powdered quillaja bark or saponin concentrate (saponin) has highly local irritant and sternutatory (causing sneezing) properties. It also has expectorant properties as well as depressant activities on the heart and respiration. Saponin is reported to be too strongly hemolytic and irritating in the gastrointestinal tract to be used internally. Severe toxic effects due to large doses include liver damage, gastric pain, diarrhea, hemolysis of red blood corpuscles, respiratory failure, convulsions, and coma (LIST AND HÖRHAMMER, MARTINDALE).

An immunostimulating complex formed from a semipurified quillaia saponin fraction by a protein antigen has proven useful as a protective vaccine for equine influenza virus; also the subject of HIV research in humans.[14]

The chronic effects in humans due to the ingestion of low levels of saponin (esp. in root beer) are not known. However two studies (one short-term in rats and the other long-term in mice) from one laboratory have indicated quillaia saponins to be nontoxic.[2,15]

## USES

**Medicinal, Phamaceutical, and Cosmetic.** Quillaia extracts are used in some dermatological creams and in hair tonic preparations and shampoos for treating dandruff.

**Food.** Quillaia extracts (especially saponins concentrate) are used quite extensively as a foaming agent in root beer and cocktail mixes (see ***yucca***). Other food products in which quillaia (product form not specified) is also reported used include frozen dairy desserts, candy, baked goods, and gelatins and puddings. Highest average maximum use level is about 0.01% for alcoholic and nonalcoholic beverages.

**Traditional Medicine.** Used for the relief of coughs, chronic bronchitis, and other pulmonary ailments; a component in herbal vaginal douches. Also used to relieve itchy scalp or dandruff, skin sores, and athlete's foot.

**Others.** Saponin used as a foaming agent in fire extinguishers (MABBERLY).

## COMMERCIAL PREPARATIONS

Crude, extracts, and saponins concentrates; crude was formerly official in N.F.

*Regulatory Status.* Has been approved for food use (§172.510).

## REFERENCES

See the General References for BAILEY 2; CLAUS; FEMA; GOSSELIN; LIST AND HÖRHAMMER; LUST; MABBERLY; ROSE; STAHL; TERRELL; UPHOF; WREN; YOUNGKEN.

1. R. A. Labriola and V. Deulofeu, *Experientia*, **25**, 124 (1969).
2. I. F. Gaunt et al., *Food Cosmet. Toxicol.*, **12**, 641 (1974).
3. R. Higuchi et al., *Phytochemistry*, **26**(1), 229 (1987).
4. C. D. Thron, *J. Pharmacol. Exp. Ther.*, **145**, 194 (1964).
5. R. Richou et al., *Rev. Immunol. Ther. Antimicrob.*, **33**, 155 (1969); through *Chem. Abstr.*, **72**, 11053g (1970).
6. R. Richou et al., *Compt. Rend.*, **260**, 3791 (1965); through *Chem. Abstr.*, **63**, 3503c (1965).
7. P. Lallouette et al., *C. R. Acad. Sci., Paris, Ser. D*, **265**, 582 (1967); through *Chem. Abstr.*, **67**, 107231m (1967).
8. B. Wolters, *Planta Med.*, **14**, 392 (1966).
9. C. D. Thron et al., *Toxicol. Appl. Pharmacol.*, **6**, 182 (1964).
10. G. Wulff, *Deut. Apoth. Ztg.*, **108**, 797 (1968).
11. G. Vogel, *Planta Med.*, **11**, 362 (1963).
12. A. J. George, *Food Cosmet. Toxicol.*, **3**, 85 (1965).
13. E. Heftmann, *Phytochemistry*, **14**, 891 (1975).
14. P. Newmark, *Biotechnology*, **6**, 23 (1988).
15. J. C. Phillips et al., *Food Cosmet. Toxicol.*, **17**, 23 (1979).

# REHMANNIA

**Source:** ***Rehmannia glutinosa*** (Gaertn.) Libosch. ex Fisch. et Mey. (syn. *Rehmannia glutinosa* Libosch.; *R. glutinosa* Libosch. form *hueichingensis* (Chao et Schih) Hsiao; *R. glutinosa* (Gaertn.) Libosch.; *R. chinensis* Libosch. ex Fisch. et Mey.; and *Digitalis glutinosa* Gaertn.) (Family Scrophulariaceae or Gesneriaceae).

*Synonyms.* Radix Rehmanniae, Rhizoma Rehmanniae, dihuang ("earth yellow," general name for rehmannia), *xian dihuang* (fresh rehmannia), *sheng dihuang* (raw rehmannia), *gan dihuang* (dried rehmannia), *shu dihuang* (cooked or cured rehmannia), *huaiqing dihuang* (*R. glutinosa* f. *hueichingensis*), and Chinese foxglove.

## GENERAL DESCRIPTION

Hardy herbaceous perennial, 25–40 cm high; whole plant covered with long soft hairs (pubescent-hirsute); root/rhizome thick and fleshy, tuberous, cylindrical or spindle-shaped, which is the part used, being referred to as root by some authors (CHP, CMH, FOSTER AND YUE, HU, IMM-2, MA, ZHU) and as rhizome by others (IMM-CAMS, JIANGSU, NATIONAL). The plant is distributed throughout most of China, especially northern, northeastern, eastern and central provinces; extensively cultivated (IMM-CAMS).

Three major types of rehmannia are generally used, namely, fresh rehmannia (*xian dihuang*), raw or dried rehmannia (*sheng dihuang* or *gan dihuang*), and cured or cooked rehmannia (*shu dihuang*). However, only the latter two are available in the United States. Most rehmannia is produced from cultivated plants. Henan province is the largest producer, which also produces the best grades (ZHU).

The tubers are carefully dug up in autumn (sometimes also in spring) to avoid bruises that would promote rotting and rid of rootlets and dirt; these constitute fresh rehmannia, which can be used immediately or laid on the ground and covered with dry sand and dirt for later use as needed for up to 3 months.

Methods for producing dried or raw rehmannia from fresh rehmannia vary but basically consist of baking the fresh tubers at carefully controlled temperatures until the inside turns black, followed by kneading to round masses that are further dried to completion or first sliced and then sun dried to completion.

Cured or cooked rehmannia is generally produced by two methods: wine curing and steaming. To produce wine-cured rehmannia, dried rehmannia is first partially rehydrated, followed by mixing with wine (30–50% w/w) in a sealed container and steaming with the container sealed until all wine is absorbed. The tubers are then removed, partially dried, cut into thick slices and further dried to completion. To produce steam-cured rehmannia, dried rehmannia is first partially rehydrated followed by steaming for hours until the tuber turns black and moist, which is then dried or first sliced into thick pieces and then dried (CMH, FOSTER AND YUE, JIANGSU, MA).

## CHEMICAL COMPOSITION

Raw and cured rehmannia contain similar types of chemical components. The main difference is their relative proportions as a result of processing. The chemical constituents present include at least 25 iridoid, naphthopyrone, and phenethyl alcohol glycosides (catalpol (0.08–0.5%, with highest concentration in fresh and lowest in cured rehmannia); dihydrocatalpol; leonuride; aucubin; monomelittoside; melittoside; rehmanniosides A, B, C, and D; acteoside; isoacteoside; purpureaside C; echinacoside; cistanosides A and F; jionosides $A_1$ and $B_1$; etc.);[1–9] a water-soluble polysaccharide fraction (RPS-b) composed of galactose, glucose, xylose, mannose and arabinose in a 12:6:2:2:1 ratio, with molecular weights of 162,000 (70%), 66,000 (15%), 37,000 (8%), and 3,000 (7%);[10] at least 15 free amino acids (ca. 0.16–6.15%), with higher concentrations in raw than in cured rehmannia;[4,5,11,12] fatty acid esters (ca. 0.01%) composed mainly of methyl linoleate, methyl palmitate and methyl *n*-octadecanoate;[13] β-sitosterol, daucosterol, palmitic acid, succinic acid and ($S_8$) cyclic compounds;[14] also sugars, including stachyose (a tetrasaccharide, 32.1–48.3%), sucrose, and monosaccharides (>3 times more in cured than in raw rehmannia);[15] trace minerals;[16] 1-ethyl-β-D-galactoside; mannitol; campesterol; α-aminobutyric acid; etc. (HU, IMM-2).[4,5,17]

## PHARMACOLOGY OR BIOLOGICAL ACTIVITIES

Practically all the pharmacologic studies on rehmannia have been conducted in China and Japan and most of the earlier reports did not specify the types of rehmannia used.

More recent specific studies have shown that cured rehmannia has tranquilizing, hypotensive and diuretic activities and that the methods of curing (wine or steam) do not significantly affect these activities or its chemistry and clinical efficacy.[18] While dried rehmannia has exhibited both immunoenhancing and immunosuppressive effects in mice, the immunoenhancing effects are reduced while its immunosuppressive activities still remain after curing; it is suggested that the ether-soluble β-sitosterol, daucosterol and 1-ethyl-β-D-galactoside play a role in the immunosuppression.[19] Cured rehmannia has "blood tonic" effects, normalizing red cell counts and hemoglobin values in experimental posthemorrhagic anemia in mice as well as markedly enhancing hematopoiesis (production and differentiation of CFU-S and CFU-E);[20] it also has shown "*yin*-nourishing" effects, being able to normalize serum levels of aldosterone, $T_3$, $T_4$ and other parameters in a hyperthyroid *yin*-deficiency rat model.[21]

The polysaccharide fraction, PRS-b, has exhibited immunomodulating/antitumor effects. Thus when administered i.p., it was found to inhibit the growth of various transplanted tumors (sarcoma S180, Lewis lung carcinoma, melanoma B16, hepatoma H22, etc.) in mice. It was also effective when administered p.o. in experiments with S180 but was ineffective *in vitro* against S180 and HL22 cells.[10]

Other effects of *dihuang* (type unclear) include hypoglycemic, hypotensive, hypertensive, vasoconstricting, vasodilating, shortening of rabbit blood coagulation time, cardiotonic, liver protective, antifungal, antiinflammatory, etc.; also found effective in treating rheumatoid arthritis, eczema, urticaria and neurodermatitis (raw rehmannia, p.o.) (IMM-2, JIANGSU, WANG).

Toxicity of *dihuang* (type not specified) is low: decoction and alcoholic extract administered to mice p.o. 60 g/kg for 3 days caused no deaths or adverse reactions after observation for 1 week (WANG).

## USES

**Medicinal, Pharmaceutical, and Cosmetic.** Extracts of *dihuang* (especially raw rehmannia) are used in skin-care (e.g., toilet water, cleanser and bath preparations) and hair-care products (e.g., shampoos) for its antiinflammatory and antimicrobial effects as well as its traditional detoxicant and nourishing properties (ZHOU).

**Health Food/Herb Teas.** Used in soup mixes where both raw and cured rehmannia are frequently used together; powder and extracts used in general and blood tonic formulas (FOSTER AND YUE).

**Traditional Medicine.** Earliest records of *dihuang* date back to the *Shen Nong Ben Cao Jing* (ca. 200 B.C.–A.D. 100) for dried rehmannia and to the *Ben Cao Tu Jing* (A.D. 1061) for cured rehmannia. Both are traditionally considered sweet tasting; the former cold while the latter slightly warming. Although both are *yin* nourishing, dried rehmannia is normally used for its heat-dispersing and blood-cooling properties in treating conditions related to febrile diseases (e.g., restlessness and thirst, skin eruptions, measles, vomiting blood, nosebleed, diabetes, restless fetus, etc.), and cured rehmannia is commonly used for its blood-tonifying properties and its beneficial effects on the vital essence (*jing*) in such conditions as lumbago, weak knees, night sweat, spermatorrhea (nocturnal emission), hectic fever, diabetes, palpitation, irregular menses, metrorrhagia (uterine bleeding), vertigo, tinnitus and premature graying of hair. The two types of rehmannia are frequently used together in formulations and are among the most commonly used Chinese herbs. *Dihuang* is also highly valued as a tonic food for disease prevention and for prolonging life.[22,23]

## COMMERCIAL PREPARATIONS

Crude and extracts of *sheng dihuang* or *gan dihuang* (raw or dried rehmannia) and *shu dihuang* (cured rehmannia).

*Regulatory Status.* Rehmannia is an ethnic food; its regulatory status not determined.

## REFERENCES

See the General References for CHP; CMH; FOSTER AND YUE; HU; IMM-2; IMM-CAMS; JIANGSU; LU AND LI; MA; NATIONAL; WANG; XIAO; ZHOU.

1. L. S. Yu and S. Y. Sha, *Analytical Methods for Active Ingredients of Chinese Herbal Drugs*, People's Health Publications, Beijing, 1984, p. 138.
2. S. M. Wong et al., *Planta Med.*, **54**, 566 (1988).
3. H. Sasaki et al., *Planta Med.*, **55**, 458 (1989).
4. M. Tomoda et al., *Chem. Pharm. Bull.*, **19**, 1455 (1971).
5. M. Tomoda et al., *Chem. Pharm. Bull.*, **19**, 2411 (1971).
6. M. Yoshikawa et al., *Chem. Pharm. Bull.*, **34**, 1403 (1986).
7. M. Yoshikawa et al., *Chem. Pharm. Bull.*, **34**, 2294 (1986).
8. M. Y. Ni and B. L. Bian, *Zhongguo Zhongyao Zazhi*, **14**(7), 40 (1989).
9. Y. Shoyama et al., *Phytochemistry*, **25**, 1633 (1986).
10. L. Z. Chen et al., *Chin. J. Pharmacol. Toxicol.*, **7**, 153 (1993).
11. Y. X. Gong, *Zhongcaoyao*, **18**(11), 37 (1987).
12. M. Y. Ni et al., *Zhongguo Zhongyao Zazhi*, **14**(3), 21 (1989).
13. B. L. Bian et al., *Zhongguo Zhongyao Zazhi*, **16**(6), 339 (1991).
14. M. Y. Ni et al., *Zhongguo Zhongyao Zazhi*, **17**, 297 (1992).
15. Z. Y. Liu, *Zhongyao Tongbao*, **9**(1), 17 (1984).
16. M. Y. Ni and B. L. Bian, *Zhongyao Tongbao*, **13**(4), 18 (1988).
17. S. J. Wu et al., *Zhongcaoyao*, **15**(7), 6 (1984).
18. Z. L. Cao et al., *Henan Zhongyi* (3), 36 (1989).
19. Z. L. Cao et al., *Zhongyao Tongbao*, **13**(10), 22 (1988).
20. Y. Yuan et al., *Zhongguo Zhongyao Zazhi*, **17**, 366 (1992).
21. S. L. Hou and J. W. Sheng, *Zhongguo Zhongyao Zazhi*, **17**, 301 (1992).
22. M. Y. Chang, *Zhongguo Shipin* (9), 25 (1987).
23. X. P. Li, *Zhongchengyao*, **15**(4), 47 (1993).

# RHUBARB

**Source:** ***Chinese rhubarb*** *Rheum officinale* Baill., *R. palmatum* L., *R. tanguticum* Maxim. ex Reg., and other *Rheum* species or hybrids grown in China; ***Indian rhubarb*** *Rheum australe* D. Don. (syn. *R. emodi* Wall.); ***Garden rhubarb*** *Rheum* × *cultorum* Hort. (syn. *R. rhabarbarum* L. erroneously attributed to *R. rhaponticum* L.) (Family Polygonaceae).

*Synonyms*. Medicinal rhubarb (*R. officinale*); Himalayan rhubarb (*R. australe*); common rhubarb and pie plant (*R.* × *cultorum*).

## GENERAL DESCRIPTION

Large and sturdy, perennial herbs with large leaves borne on thick petioles; stem up to 2–3 m high; native to Asia (e.g., China, India, and southern Siberia); widely cultivated. Parts used are the dried rhizome and roots deprived of periderm (corky layer). Only plants 3 years or older are used. Chinese rhubarb, especially those from R. *officinale* and *R. palmatum*, are considered to be of the best quality. Those species with palmate rather than undulate leaves are generally considered the official drug source species or substitutes in China.

The species cultivated as ornamental plants in the United States are generally *R. palmatum* and *R.* × *cultorum*; the latter is also grown for its edible stalks (petioles). *R.* × *cultorum*, a hybrid that evolved in the 18th century, probably involved the rare eastern European species *R. rhaponticum*, a binomial commonly associated with the garden rhubarb (FOSTER AND YUE, MABBERLY).[1,2]

## CHEMICAL COMPOSITION

Chinese rhubarbs contain two major classes of active constituents, anthraglycosides and tannins.

The amounts of anthraglycosides and related free anthraquinones range from about 1% (in *R. tanguticum*) to more than 5% (in *R. palmatum* and *R. officinale*).[3,4] Anthraglycosides constitute 75% or more of the mixture, with the rest being free anthraquinones that are mainly chrysophanic acid (chrysophanol), emodin (rheum emodin), aloe-emodin, rhein, and physcion (parietin). The major anthraglycosides are *O*-glucosides (chrysophanol-1-monoglucoside, emodin-6-monoglucoside, aloe-emodin-8-monoglucoside, rhein-8-monoglucoside, physcione monoglucoside, etc.), with lesser amounts of the dianthrone glycosides sennosides A, B, C, D, E, and F (see ***cascara***, ***frangula***, and ***senna***).[3–7] Although the total anthracene derivatives remain relatively constant year-round, the relative concentrations of anthrone and anthraquinone forms of the glycosides vary with time of harvest.[8] A recent study found physcion-8-*O*-*β*-D-gentiobioside to be a major constituent of the hydroxyanthracene complex.[9] Recently isolated constituents with laxative activity from Chinese rhubarb include rheinosides A, B, C, and D.[10] Nonanthraquinone glycosides have also been isolated from Chinese rhubarb; they are stilbene glycosides (e.g., 3,5,4′-trihydroxystilbene-4′-*O*-*β*-D-glucopyranoside) related to rhaponticin (rhapontin) found in *R. rhaponticum* (JIANGSU).[11]

The tannins in Chinese rhubarb are of both the catechin and gallic acid types, including *d*-catechin, *d*-epicatechin gallate, glucogallin (galloylglucose), and others (LIST AND HÖRHAMMER).[11,12]

Chinese rhubarb also contains a trace amount of volatile oil consisting of 100 constituents, including chrysophanic acid and other anthraquinones, diisobutyl phthalate, cinnamic acid, phenylpropionic acid, and ferulic acid, but no *p*-hydroxycinnamic, caffeic, or quinic acid.[13,14]

Other constituents present in Chinese rhubarbs include calcium oxalate (ca. 6%), fatty acids (oleic acid, palmitic acid, etc.), sugars (glucose, fructose, etc.), rutin and other flavonoids, starch (ca. 16%), resins, and others (BRADLEY, JIANGSU, NANJING, STAHL).

Indian rhubarb and garden rhubarb have been much less extensively investigated than Chinese rhubarb. Indian rhubarb is reported to contain tannins and anthraglycosides many of which are probably the same as those of Chinese rhubarb. Garden rhubarb has been reported to contain chrysophanic acid (probably also its glucosides) but not the other anthraquinones or sennosides found in Chinese rhubarb;[3] it contains rhaponticin not present in Chinese or Indian rhubarb (MARTINDALE, NANJING, STAHL).[4]

## PHARMACOLOGY OR BIOLOGICAL ACTIVITIES

Chinese rhubarb has both astringent and cathartic properties. In small doses or when used under certain conditions it is astringent and is used to treat diarrhea, while in larger doses or under other conditions it is cathartic (JIANGSU, MARTINDALE, STAHL).[15]

The cathartic properties of Chinese rhubarb are due mainly to the sennosides and the other anthraglycosides (see ***cascara***, ***frangula***, and ***senna***). The tannins are responsible for its astringent properties (see ***tannic acid***). Laxative action occurs as the result of inhibition of water and electrolyte absorption from the large intestine, influencing intestinal motility. Anthraquinone glycosides are bound to emodin aglycones by microbes, partially absorbed, then reduced to anthranols and anthrones, which are responsible for the laxative effect.[2,16]

Leaf blades of rhubarb are poisonous, and accidental ingestion will cause severe vomiting and may also cause liver and kidney damage. The toxic effect was formerly attributed to oxalates present in the leaves, but more recent data indicate it to be due to monoanthrones (see ***cascara***).[7]

Other pharmacological properties of rhubarb and its extracts include antitumor,[17] antimicrobial, hypotensive, and antiinflammatory effects, among numerous others (JIANGSU).[18]

## USES

**Medicinal, Pharmaceutical, and Cosmetic.** Rhubarb and its extracts are used in certain laxative preparations.

**Food.** Chinese rhubarb is used as a flavor component (bitter note) in major food products, including alcoholic (e.g., bitters) and nonalcoholic beverages, frozen dairy desserts, candy, baked goods, and gelatins and puddings. Highest average maximum average use level is 0.05% in alcoholic beverages and baked goods.

**Health Food/Herb Teas.** Chinese rhubarb used in various laxative formulations; included as an ingredient in the unconventional cancer treatment formula, Essiac; for which the garden rhubarb is sometimes erroneously substituted. Tonic rhubarb alcoholic beverages are made in China and Italy.[2]

**Traditional Medicine.** Chinese rhubarb is considered a very valuable and versatile drug in Chinese medicine having been used for thousands of years not only for treating constipation but also other conditions such as chronic diarrhea, thermal burns, jaundice, sores, and cancers (WANG).[19] In recent years, it has also been used in treating upper GI bleeding with considerable success.[20–22]

## COMMERCIAL PREPARATIONS

Crudes (Chinese, numerous grades; Indian) and extracts (e.g., solid and fluid). Crude, tincture, and fluid extract were formerly official in N.F. Strengths (see ***glossary***) of extracts are expressed in weight-to-weight ratios.

*Regulatory Status.* Chinese rhubarb and garden rhubarb have been approved for food use, the latter in alcoholic beverages only (§172.510); Indian rhubarb is not listed. Chinese rhubarb (*R. palmatum*) is the subject of a positive German therapeutic monograph.[16]

## REFERENCES

See the General References for BAILEY 1; BAILEY 2; BLUMENTHAL; BRADLY; CLAUS; FEMA; FOSTER AND YUE; JIANGSU; MARTINDALE; NANJING; STAHL; TERRELL; UPHOF.


1. D. E. Marshall, *A Bibliography of Rhubarb and Rheum Species*, National Agricultural Library, Bibliography and Literature of Agriculture, no. 62, Beltsville, Md., 1988.
2. C. M. Foust, *Rhubarb: The Wondrous Drug,* Princeton University Press, Princeton, N.J., 1992.
3. J. H. Zwaving, *Planta Med.*, **21**, 254 (1972).
4. B. Klimek, *Ann. Acad. Med. Lodz.*, **14**, 133 (1973); through *Chem. Abstr.*, **85**, 189212q (1976).
5. J. H. Zwaving, *Pharm. Weekbl.*, **109**, 1169 (1974).
6. H. Oshio et. al., *Chem. Pharm. Bull.*, **22**, 823 (1974).
7. F. H. L. van Os, *Pharmacology*, **14**(Suppl. 1), 7 (1976).
8. E. H. C. Verhaeren et al., *Planta Med.*, **45**, 15 (1982).
9. L. Holzschuh et al., *Planta Med.*, **46**, 159 (1982).
10. T. Yamagishi et al., *Chem. Pharm. Bull.*, **35**, 3132 (1987).
11. G. Nonaka et al., *Chem. Pharm. Bull.*, **25**, 2300 (1977).
12. H. Friedrich and J. Höhle, *Arch. Pharm.*, **299**, 857 (1966).
13. C. Frattini et al., *Riv. Ital. Essenze, Profumi, Piante Offic., Aromi, Saponi, Cosmet., Aerosol*, **58**, 132 (1976); through *Chem. Abstr.*, **85**, 25261d (1976).
14. C. Frattini et al., *Riv. Ital. Essenze, Profumi, Piante Offic., Aromi, Saponi, Cosmet., Aerosol*, **56**, 597 (1974); through *Chem. Abstr.*, **83**, 84693c (1975).
15. J. W. Fairbairn, *Pharmacology*, **14**(Suppl. 1), 48 (1976).
16. Monograph *Rhei radix*, *Bundesanzeiger*, no. 228 (Dec. 5, 1984); revised (Apr. 27, 1989).
17. O. K. Kabiev and S. M. Vermenichev, *Vopr. Onkol.*, **12**, 61 (1966); through *Chem. Abstr.*, **65**, 2875c (1966).
18. X. S. Gao et al., eds., *Abstracts of the First International Symposium on Rhubarb, Chengdu, 29–31 May 1990,* Institute of Chinese Materia Medica, China Academy of Traditional Chinese Medicine, Beijing 1990.
19. J. L. Hartwell, *Lloydia*, **33**, 288 (1970).
20. D. H. Jiao et al., *Pharmacology*, **20**(Suppl. 1), 128 (1980).
21. D. A. Sun et al., *Chin. J. Integr. Trad. Western Med.*, **6**, 4589 (1986).
22. D. H. Jiao et al., *Zhejiang Zhongyi Zazhi*, **23**, 179 (1988).

# ROSE OIL (AND ABSOLUTE)

**Source:** ***Rosa alba*** L., ***R. centifolia*** L., ***R. damascena*** Mill., ***R. gallica*** L., and their varieties (Family Rosaceae).

*Synonyms*. Bulgarian otto of rose, Bulgarian rose oil, and Bulgarian attar of rose (*R. damascena* var. *alba*); Moroccan otto of rose, Moroccan rose oil (*R. centifolia*); Turkish otto of rose, Turkish rose oil, and Turkish attar of rose (*R. damascena*); French rose absolute and rose de mai absolute (*R. centifolia*).

## GENERAL DESCRIPTION

Small prickly shrubs up to about 1.2 to 2.4-m high; generally considered to be natives of Europe and western Asia; widely cultivated. Parts used are the fresh flowers, from which rose oil is obtained normally in 0.02 to 0.03% yield by steam distillation and rose absolute by petroleum ether extraction followed by alcohol solubilization (see ***concrète*** and ***absolute*** in ***glossary***).[1–3] Rose water (more specifically called stronger rose water) is the aqueous portion of the steam distillation after rose oil is removed.

Flowers are usually harvested early in the morning when their essential oil content is the highest.[1,4] Major rose oil and absolute producing countries include France, Bulgaria, Morocco, Turkey, Italy, and China.

It has been proposed that *R. damascena* evolved as an eastern Mediterranean hybrid between *R. gallica* and *R. phoenicia*.[5]

Major commercial cultivars of *R. damascena* include 'Trigintipetala' ("Kananlik rose"), which is confused in the horticultural trade with 'Prof. Emile Perrot,' 'Alika,' and 'Bella Donna' (TUCKER).

Major commercial cultivars of *R. gallica* include 'Conditorum' (Hungarian rose) and 'Officinalis' (apothecary rose) (TUCKER).

## CHEMICAL COMPOSITION

Rose oil contains usually as its major components geraniol, citronellol, nerol, $\beta$-phenethyl alcohol, geranic acid, and eugenol, which together make up 55 to 75% of the oil, with citronellol up to 60%.[4,6,7] These components are found in free and bound forms in ratios specific to individual species or cultivar.[8] Other components present include terpene hydrocarbons (especially $\alpha$- and $\beta$-pinenes, myrcene, etc.), esters, $C_{14}$ to $C_{23}$ *n*-paraffins making up the so-called stearoptenes (accounting for 15 to 23% of the oil), nerol oxide, and others (GUENTHER, JIANGSU).[4,9–11]

Important fragrance components of rose oil include (4*R*)-*cis*-rose oxide, (4*R*)-*trans*-rose oxide, (±)-nerol oxide, 3-(4-methyl-3-pentenyl)-2-buten-4-olide, 3-methyl-4(3-methyl-2-butenyl)-2-buten-4-olide, *cis*- and *trans*-2-(3-methyl-2-butenyl)-3-methyl tetrahydrofuran, $\beta$-damascenone, 3-hydroxy-$\beta$-damascenone, and $\beta$-damascone.[8]

Rose absolute contains mainly phenethyl alcohol, with lesser amounts of citronellol, geraniol, and nerol.[12] Other compounds present include eugenol esters, and others.[13]

A major review of rose oil and extracts has been published.[8]

## PHARMACOLOGY OR BIOLOGICAL ACTIVITIES

Rose oil when added to the food has been reported to have choleretic effects on cats.[14]

Bulgarian rose oil has been reported to decrease urinary corticosteroids and serum ceruloplasmin as well as elicit other effects on laboratory animals when administered intraperitoneally or intravenously.[15]

Available data from one source indicate rose oil (Moroccan, Bulgarian, and Turkish) to be nonirritating, nonsensitizing, and nonphototoxic to human skin, but slightly to moderately irritating to rabbit skin when applied undiluted.[16–18] Rose absolute French was nonirritating and nonphototoxic, but it produced one sensitization reaction in 25 subjects tested.[19]

## USES

**Medicinal, Pharmaceutical, and Cosmetic.** Rose oil and stronger rose water are used primarily as fragrance components and as astringent in pharmaceutical preparations (e.g., cold creams, ointments, and lotions).

Rose oil and rose absolute are extensively used as fragrance ingredients in perfumes, creams, lotions, soaps, and sometimes detergents. Maximum use level is 0.2% reported for French rose absolute and rose oils Moroccan, Bulgarian, and Turkish in perfumes.[15–18]

**Food.** Rose oil and absolute are used extensively as flavor ingredients (usually in very low use levels) in fruit-type flavors. Food products in which they are used include alcoholic and nonalcoholic beverages, frozen dairy desserts, candy, baked goods, and gelatins and puddings. Reported average maximum use levels are generally below 0.0002% (2 ppm).

**Health Food/Herb Teas.** Rose oil used in aromatherapy for aphrodisiac and rejuvenating claims (ROSE).

**Others.** Rose oil and to a lesser extent rose absolute are used in flavoring tobacco.

## COMMERCIAL PREPARATIONS

Rose oil (Moroccan, French, Turkish, etc.) and absolute (French).

Rose oil is official in N.F. and F.C.C. Stronger rose water is official in N.F.

*Regulatory Status.* GRAS (§182.20). *Rosa gallica* flowers are the subject of a German therapeutic monograph; allowed for mild inflammations of the oral and pharyngeal mucosa.[20]

## REFERENCES

See the General References for ARCTANDER; BAILEY 1; BLUMENTHAL; FEMA; GUENTHER; JIANGSU; MARTINDALE; ROSE; TUCKER; YOUNGKEN.

1. S. Kapetanovic, *Kem. Ind.*, **21**, 355 (1972); through *Chem. Abstr.*, **78**, 47636n (1973).
2. D. Ivanov, *Parfums, Cosmet., Savons Fr.*, **2**, 153 (1972); through *Chem. Abstr.*, **77**, 39035e (1972).
3. S. Kapetanovic, *Kem. Ind.*, **23**, 629 (1974); through *Chem. Abstr.*, **83**, 168304q (1975).
4. M. R. Narayana, *Indian Perfum.*, **13**(P.2), 46 (1969); through *Chem. Abstr.*, **76**, 158202w (1972).
5. M. P. Widrlechner, *Econ. Bot.*, **35**(1), 42 (1981).
6. Y. Ohno and S. Tanaka, *Agr. Biol. Chem.*, **41**, 399 (1977).
7. Y. Mikhailova et al., *Dokl. Bolg. Akad. Nauk*, **30**, 89 (1977); through *Chem. Abstr.*, **86**, 177158j (1977).
8. B. M. Lawrence, *Perfum. Flav.*, **16**(6), 43, (1991).
9. V. Staikov, *Dokl. Akad. Sel'skokhoz. Nauk Bolg.*, **4**, 199 (1971); through *Chem. Abstr.*, **77**, 39033c (1972).
10. G. Igolen et al., *Riv. Ital. Essenze Profumi, Piante Offic., Aromi, Saponi, Cosmet. Aerosol*, **50**, 352 (1968); through *Chem. Abstr.*, **70**, 31623e (1969).
11. C. Ehret and P. Teisseire, *Recherches*, **19**, 287 (1974).
12. M. S. Karawya et al., *Bull. Fac. Pharm., Cairo Univ.*, **13**, 183 (1974); through *Chem. Abstr.*, **86**, 60396u (1977).

13. Y. Ohno et al., *Kanzei Chuo Bunsekishoho*, **15**, 47 (1975); through *Chem. Abstr.*, **87**, 44126w (1977).
14. E. N. Vasil'eva and V. P. Gruncharov, *Farmakol. Toksikol. (Moscow)*, **35**, 312 (1972); through *Chem. Abstr.*, **77**, 56647y (1972).
15. A. Maleev et al., *Eksp. Med. Morfol.*, **10**, 149 (1971); through *Chem. Abstr.*, **76**, 135927x (1972).
16. D. L. J. Opdyke, *Food Cosmet. Toxicol.*, **12**(Suppl.), 981 (1974).
17. D. L. J. Opdyke, *Food Cosmet. Toxicol.*, **12**(Suppl.), 979 (1974).
18. D. L. J. Opdyke, *Food Cosmet. Toxicol.*, **13**(Suppl.), 913 (1975).
19. D. L. J. Opdyke, *Food Cosmet. Toxicol.*, **13**(Suppl.), 911 (1975).
20. Monograph *Rosae flos*, *Bundesanzeiger*, no. 164 (Sept. 1, 1990).

# ROSE HIPS

**Source:** ***Rosa canina*** L., ***R. gallica*** L., ***R. rugosa*** Thunb., ***R. villosa*** L. (syn. *R. pomifera* Herrm.), and other ***Rosa*** species (Family Rosaceae).

*Synonym*. Hipberries.

## GENERAL DESCRIPTION

Prickly bushes or shrubs; native to Europe and Asia; extensively cultivated. Part used is the ripe fruit. *R. canina* is the major commercial source of rose hips; with ellipsoid; globose, or ovoid fruits. The *R. canina* group is separated into 17 or more species; found in most of Europe (TUTIN 2).

## CHEMICAL COMPOSITION

Rose hips contain high concentrations of vitamin C (ascorbic acid), ranging from 0.24 to 1.25% depending on sources, climate, degrees of ripeness, and other factors[1–4] (compare with ***acerola***).

Other constituents present in rose hips include carotenoids (0.01 to 0.05%),[3,4] flavonoids (0.01 to 0.35%),[4,5] pectic substances (3.4 to 4.6%), polyphenols (2.02 to 2.64%), leucoanthocyanins (1.35 to 1.75%), catechins (0.8 to 0.91%),[4] riboflavin,[3] sugars (glucose, fructose, sucrose, etc.), and plant acids (e.g., citric and malic), among others (JIANGSU).

Depending on the species, rose hips may also contain purgative glycosides (e.g., multiflorin A and multiflorin B in *Rosa multiflora* Thunb.),[6] saponins (ca. 17% in *R. laevigata* Michx.), and other compounds that have widely different pharmacological properties (JIANGSU) (see ***Cherokee rosehip***).

## PHARMACOLOGY OR BIOLOGICAL ACTIVITIES

Vitamin C has antiscorbutic properties.

Mild laxative and diuretic effects reported, probably attributable to malic and citric acid content (TYLER 1).

## USES

**Food.** Rose hips are mainly used as a source of natural vitamin C. However, as with acerola, much of the vitamin C in rose hips is destroyed during ordinary drying or extraction (MERCK).[7] For this reason most rose hips products are supplemented with synthetic vitamin C. They are usually in the form of tablets or capsules.

**Health Food/Herb Teas.** Rose hips are widely used as an herb tea ingredient; also in capsules, tablets, etc., as a questionable source of vitamin C.[8]

## COMMERCIAL PREPARATIONS

Crude and extracts, usually adjusted with vitamin C to a specified amount.

*Regulatory Status*. GRAS (§182.20); with only *Rosa alba* L., *R. centifolia* L., *R. damascena* Mill., *R. gallica* L., and their varieties listed. Subject of a German therapeutic monograph; claimed efficacy to prevent colds and flu (due to vitamin C) content are not sufficiently documented; corrigent for tea mixtures.[8]

## REFERENCES

See the General References for BAILEY 1; BIANCHINI AND CORBETTA; BLUMENTHAL; JIANGSU; NANJING; TUTIN 2; TYLER 1; UPHOF.

1. Z. Butkiene, *Liet. TSR Mokslu Akad. Darb., Ser. C*, **1**, 51 (1969); through *Chem. Abstr.*, **71**, 78136j (1969).
2. A. N. Nizharadze et al., *Konservn. Ovoshchesush. Prom-st.*, **4**, 36 (1977); through *Chem. Abstr.*, **87**, 20827z (1977).
3. G. P. Shnyakina and E. P. Malygina, *Rastit. Resur.*, **11**, 390 (1975); through *Chem. Abstr.*, **83**, 128706r (1975).
4. S. G. Mel'yantseva, *Konservn. Ovoshchesush. Prom-st.* **2**, 13 (1978); through *Chem. Abstr.*, **88**, 168634a (1978).
5. M. Retezeanu et al., *Farmacia (Bucharest)*, **20**, 167 (1972); through *Chem. Abstr.*, **77**, 58790p (1972).
6. S. Takagi et al., *Yakugaku Zasshi*, **96**, 1217 (1976); through *Chem. Abstr.*, **86**, 13815v (1977).
7. S. Mrozewski, *Przem. Spozyw.*, **22**, 294 (1968); through *Chem. Abstr.*, **71**, 11911q (1969).
8. Monograph *Rosae fructus*, *Bundesanzeiger*, no. 164 (Sept. 1, 1990).

# ROSELLE

**Source:** ***Hibiscus sabdariffa*** L. (Family Malvaceae).

*Synonyms*. Hibiscus, Jamaica sorrel, and Guinea sorrel.

## GENERAL DESCRIPTION

Large strong annual, often shrublike; up to about 2.4 m high; native to the Old World tropics; widely cultivated in the tropics and subtropics (e.g., Sudan, China, Thailand, Egypt, Mexico, and the West Indies). Parts used are the dried red, fleshy sepals (calyx) together with the bracts. Major producing countries include Sudan, China, and Thailand.

## CHEMICAL COMPOSITION

Roselle contains about 1.5% anthocyanins (see ***grape skin extract***) consisting mainly of delphinidin-3-sambubioside (hibiscin or daphniphylline) and cyanidin-3-sambubioside, with cyanidin-3-glucoside, delphinidin-3-glucoside, delphinidin, and other pigments in lesser amounts;[1–6] large amounts (23%) of hibiscic acid;[6,7] fruit acids consisting mostly of citric acid (12 to 17%), with small amounts of malic, tartaric, and other acids also present;[7] resins; sugars; trace of an unidentified alkaloid; and others (LIST AND HÖRHAMMER).[8]

The pigments have been tested as food colors (red) and found to be stable in fruit jellies and jams but unstable (hence unsuitable) in carbonated beverages.[4,5]

## PHARMACOLOGY OR BIOLOGICAL ACTIVITIES

Roselle extracts have been reported to have antispasmodic activity on intestinal and uteral muscles as well as anthelmintic (tapeworm) properties. Its decoction or infusion reportedly has hypotensive properties with no side effects.[7,8]

Extracts have been shown to both stimulate and inhibit isolated rabbit intestine. An aqueous extract of the calyces has been found to decrease intestinal motility in intact rats, as well as the oral-caecal transit time in dogs.[9]

It also reportedly has bactericidal properties *in vitro* (LIST AND HÖRHAMMER).[7,8,10]

## USES

**Food.** Roselle is reported used as a flavor ingredient in alcoholic and nonalcoholic beverages, frozen dairy desserts, candy, baked goods, and gelatins and puddings. Average maximum use level reported is 0.02% for all above categories.

Roselle in coarse ground form is used in herb teas, primarily to impart a deep reddish-brown color and a pleasant slightly acid taste.

Fresh roselle is used in jams, juices, and jellies with a cranberrylike flavor. Also used in wines.

**Health Food/Herb Teas.** Dried flowers one of the most widely used herbal tea ingredients, imparting red color and tart flavor.

**Traditional Medicine.** Leaves are used in Egypt for treating heart and nerve diseases.[11]

Roselle (calyx) is used as a refrigerant. It has also been used in cancer.[12]

In European tradition, flowers reportedly used for loss of appetite, colds, upper respiratory tract congestion, circulatory impairment; diuretic, mild laxative.[12]

## COMMERCIAL PREPARATIONS

Mainly crude.

*Regulatory Status.* Has been approved for use in alcoholic beverages only (§172.510). Flowers subject of a German therapeutic monograph; claimed efficacies are not substantiated.

## REFERENCES


See the General References for BAILEY 1; BAILEY 2; BLUMENTHAL; FEMA; LIST AND HÖRHAMMER; MORTON 2; TERRELL; TYLER 1; UPHOF.

1. C. T. Du and F. J. Francis, *J. Food Sci.*, **38**, 810 (1974).
2. M. Shibata and M. Furukawa, *Shokubutsugaku Zasshi*, **82**, 341 (1969); through *Chem. Abstr.*, **72**, 96801x (1970).
3. M. S. Karawya et al., *Egypt J. Pharm. Sci.*, **16**, 345 (1976).
4. W. B. Esselen and G. M. Sammy, *Food Prod. Dev.*, **9**, 37 (1975).
5. W. B. Esselen and G. M. Sammy, *Food Prod. Dev.*, **7**, 80 (1973).
6. H. Schilcher, *Dtsch. Apoth. Ztg.*, **116**, 1155 (1976).
7. J. Kerharo, *Plant. Med. Phytother.*, **5**, 277 (1971).
8. A. Sharaf, *Planta Med.*, **10**, 48 (1962).
9. M. B. Ali et al., *Fitoterapia*, **62**(6), 475 (1991).

10. M. B. Ali, et al., *J. Ethnopharmacol.*, **31**, 249 (1991).
11. A. M. Osman et al., *Phytochemistry*, **14**, 829 (1975).
12. J. L. Hartwell, *Lloydia*, **33**, 97 (1970).
13. Monograph *Hibisci flos*, *Bundesanzeiger*, no. 22 (Feb. 1, 1990).

# ROSEMARY

**Source:** ***Rosmarinus officinalis*** L. (Family Labiatae or Lamiaceae).

## GENERAL DESCRIPTION

A small evergreen shrub with thick aromatic, linear leaves; up to about 2 m high; native to the Mediterranean region, cultivated worldwide (California, England, France, Spain, Portugal, Morocco, China, etc.). Part used is the dried leaf, which supplies the spice. Rosemary oil is prepared by steam distillation of the fresh flowering tops. Major oil-producing countries include Spain, France, and Tunisia.

## CHEMICAL COMPOSITION

Contains about 0.5% volatile oil; flavonoids (diosmetin, diosmin, genkwanin, genkwanin-4′-methyl ether, 6-methoxygenkwanin, luteolin, 6-methoxyluteolin, 6-methoxyluteolin-7-glucoside, 6-methyoxyluteolin-7-methyl ether, hispidulin, apigenin, etc.);[1–4] phenolic acids (rosmarinic, labiatic, chlorogenic, neochlorogenic, and caffeic acids);[2] carnosic acid;[5–7] rosmaricine and isorosmaricine (reaction products of carnosic acid);[5–6] triterpenic acids (mainly ursolic and oleanolic acids, with traces of 19$\alpha$-hydroxyursolic, 2$\beta$-hydroxyoleanolic, and 3$\beta$-hydroxyurea-12,20(30)-dien-17-oic acids);[8] rosmanol, 7-ethoxyrosmanol, betulinic acid, and carnosol;[9] and others (JIANGSU, MARSH).

The essential oil contains mainly monoterpene hydrocarbons ($\alpha$- and $\beta$-pinenes, camphene, limonene, etc.), cineole (eucalyptol), and borneol, with camphor, linalool, verbenol, terpineol, 3-octanone, and isobornyl acetate also present (GUENTHER, JIANGSU).[10–12]

Rosemary extracts have antioxidative properties comparable to those of butylated hydroxyanisole (BHA) and butylated hydroxytoluene (BHT); carnosic acid and labiatic acid are reported as active components (see ***sage***).[13–16]

## PHARMACOLOGY OR BIOLOGICAL ACTIVITIES

Rosemary oil has antimicrobial (bacteria and fungi) activities.[17–19] It also stimulates locomotor activity of mice when administered orally or by inhalation. Cineole is believed to be the active principle.[20]

Diosmin is reported to be stronger than rutin in decreasing capillary permeability and fragility as well as less toxic (JIANGSU).

A rosmaricine derivative (*O,O,N*-trimethylrosmaricine) has been demonstrated to exhibit significant smooth-muscle stimulant effects *in vitro* as well as moderate analgesic activity.[6]

Rosemary extracts have exhibited inhibitory activities against urease, and rosmanol was found to be the active agent.[9]

Rosemary galenicals are reported to be spasmolytic on gall passages and small intestines, and have a positive inotropic effect, increasing coronary flow-through.[21]

Rosemary oil has been reported to be nonirritating and nonsensitizing to human skin but moderately irritating to rabbit skin when applied undiluted.[19]

## USES

**Medicinal, Pharmaceutical, and Cosmetic.** Rosemary leaves are used in European phytomedicine for dyspeptic complaints and as supportive therapy for rheumatic diseases; externally for circulatory problems; in baths, the herb is used as an external stimulant for increased blood supply to the skin.[21]

Rosemary oil is extensively used in cosmetics as a fragrance component and/or a masking agent. Products in which it is used include soaps, detergents, creams, lotions, and perfumes (especially colognes and toilet waters). Maximum use level reported is 1% in the last category.[19]

**Food.** Both the spice and oil are extensively used in foods.

The spice is reported used in alcoholic beverages, baked goods, meat and meat products, condiments and relishes, processed vegetables, snack foods, gravies and others, with highest average maximum use level of about 0.41% (4,098 ppm) in baked goods.

The oil is used in alcoholic and nonalcoholic beverages, frozen dairy desserts, candy, baked goods, gelatins and puddings, meat and meat products, and condiments and relishes, among others, with highest average maximum use level of about 0.003% (26.2 ppm) reported in meat and meat products.

**Health Food/Herb Teas.** More widely used in Europe than the U.S. in infusions, powder, dry extracts or other galenic preparations for internal and external use, primarily as a stomachic (FOSTER).[21]

**Traditional Medicine.** Has been used since ancient times in Europe as a tonic, stimulant, and carminative and in treating dyspepsia (indigestion), stomach pains, headaches, head colds, and nervous tension.

In China, rosemary herb (leaves and branches) has been used for centuries in treating singular conditions, particularly headaches. An infusion of a mixture of rosemary herb and borax is used for the prevention of baldness. A similar formula is also used in European folk medicine for the prevention of scurf and dandruff (GRIEVE).

Rosemary has been reported used in cancers.[22]

**Others.** Rosemary can serve as a source of natural antioxidants.[23,24]

## COMMERCIAL PREPARATIONS

Crude and oil. Crude was formerly official in U.S.P., and oil was formerly in N.F. Oil is currently official in F.C.C.

*Regulatory Status.* GRAS (§182.10 and §182.20). Rosemary leaves are the subject of a positive German therapeutic monograph; allowed internally for dyspeptic complaints; externally for supportive therapy for rheumatic diseases and circulatory problems.[21]

## REFERENCES

See the General References for ARCTANDER; BAILEY 1; BIANCHINI AND CORBETTA; BLUMENTHAL; CLAUS; FEMA; FOSTER; GRIEVE; JIANGSU; LUST; ROSENGARTEN; TERRELL; UPHOF.

1. C. H. Brieskorn et al., *Dtsch. Lebensm. Rundsch.*, **69**, 245 (1973); through *Chem. Abstr.*, **79**, 102803n (1973).
2. V. I. Litvinenko et al., *Planta Med.*, **18**, 243 (1970).
3. V. Plouvier, *C. R. Acad. Sci. Ser. D*,

**269**, 646 (1969); through *Chem. Abstr.*, **72**, 39742e (1970).

4. C. H. Brieskorn and H. Michel, *Tetrahedron Lett.*, (30), 3447 (1968).
5. L. D. Yakhontova et al., *Khim. Prir. Soedin.*, **7**, 416 (1971); through *Chem. Abstr.*, **75**, 141010p (1971).
6. A. Boido et al., *Studi Sassar.*, *Sez. 2*, **53**, 383 (1975); through *Chem. Abstr.*, **88**, 69046d (1978).
7. C. H. Brieskorn and H. J. Dömling, *Z. Lebensm. Unters. Forsch.*, **14**, 10 (1969).
8. C. H. Brieskorn and G. Zweyrohn, *Pharmazie*, **25**, 488 (1970).
9. T. Hayashi et al., *Planta Med.*, **53**, 394 (1987).
10. J. Cabo Torres et al., *Boll. Chim. Farm.*, 111, 573 (1972); through *Chem. Abstr.*, **78**, 62049s (1973).
11. A. Koedam and M. J. M. Gijbels, *Z. Naturforsch., C. Biosci.*, **33C**, 144 (1978); through *Chem. Abstr.*, **88**, 141489v (1978).
12. B. G. Skrubis, *Flavour Ind.*, **3**, 566, (1972).
13. H. Schulze et al., *Fleischwirtschaft*, **51**, 303 (1971); through *Chem. Abstr.*, **75**, 18717x (1971).
14. A. Palitzsch et al., *Fleischwirtschaft*, **54**, 63 (1974); through *Chem. Abstr.*, **81**, 24345q (1974).
15. Y. Watanabe and Y. Ayano, *Eiyo To Shokuryo*, **27**, 181 (1974); through *Chem. Abstr.*, **81**, 150686q (1974).
16. Y. Saito et al., *Eiyo To Shokuryo*, **29**, 505 (1976); through *Chem. Abstr.*, **87**, 150314r (1977).
17. B. G. V. N. Rao and S. S. Nigam, *Indian J. Med. Res.*, **58**, 627 (1970).
18. B. G. V. N. Rao and P. L. Joseph, *Riechst., Aromen, Körperpflegem.*, **21**, 405 (1971).
19. D. L. J. Opdyke, *Food Cosmet. Toxicol.*, **12**(Suppl.), 977 (1974).
20. K. A. Kovar et al., *Planta Med.*, **53**, 315 (1987).
21. Monograph *Rosmarini folium*, *Bundesanzeiger*, no. 223 (Nov. 30, 1985); revised (Nov. 28, 1986; Mar. 13, 1990).
22. J. L. Hartwell, *Lloydia*, **32**, 247 (1969).
23. S. S. Chang et al., *J. Food Sci.*, **42**, 1102 (1977).
24. A. G. Maggi, Neth. Appl. 6,600,754 (1966); through *Chem. Abstr.*, **66**, 1740c (1967).

# ROYAL JELLY

**Source:** Pharyngeal glands of the worker bee (honeybee).

*Synonyms*. Queen bee jelly, apilak, Weiselfuttersaft, Gelee royale, and *feng wang jiang*.

## GENERAL DESCRIPTION

Royal jelly is a milky white, viscous substance secreted by the pharyngeal glands of the worker bee (honeybee), *Apis mellifera* L., an insect of the family Apidae. It is the food for all bee larvae for the first three days of life but is reserved as food for queen bees for the rest of their lives, hence the name "royal jelly." It is collected from bee hives by manual scraping using nonmetallic utensils or by suction through a glass tube, at the same time removing impurities such as wax and insect fragments. The crude royal jelly is placed in sterilized glass jars and refrigerated until enough is collected for processing or shipment to central processing facilities. Processing involves filtration to further remove impurities of smaller particle size; the resulting liquid is either frozen in special plastic containers or freeze-dried (lyophilized). Normally the fresher materials are used for direct freezing while the less fresh materials are freeze-dried. Most of the royal jelly used in the United States is imported from China; the best grades are produced in northeastern provinces.

## CHEMICAL COMPOSITION

The chemistry of royal jelly has been extensively studied. Chemical constituents reported to be present: hydroxycarboxylic acids, including 10-hydroxy-*trans*-2-decenoic acid (royal jelly acid; 10-HDA) which ranges between 1.64 and 4.24% (usually 2.0–2.5%) in fresh royal jelly and up to 7.0% in the freeze-dried product (usual range: 4.5–5.5%),[1–5] 10-hydroxydecanoic acid (0.60–1.25%),[2,4,6] gluconic acid,[2] sebacic acid and 2-decenedioic acid,[3] *p*-hydroxybenzoic acid,[6] 3-hydroxydecanoic, 8-hydroxydecanoic and 3,10-dihydroxydecanoic acids;[7] sterols including methylenecholesterol (ca. 0.3%), cholesterol, stigmastanol, stigmasterol and testosterone;[6,8–11] acetylcholine (467–1113 $\mu$g/g);[12–14] free amino acids (total 1.59% according to one report),[15] including proline, lysine, glutamic acid, serine, alanine, arginine, aspartic acid, glycine, isoleucine, leucine, methionine, tyrosine, valine, glutamine and taurine, with proline accounting for 54–60% of all free amino acids present, followed by lysine and glutamic acid;[15–17] and biopterin, among others (JIANGSU, MERCK).[18]

The following is its general composition: 24 to 70% water (usually ca. 65%); 11 to 31% crude protein (usually ca. 12%); 4.9 to 23% total lipids, including phospholipids, glycerides, waxes and fatty acids; 8.5 to 16% carbohydrates; vitamins, including thiamine 1.2 to 7.4 mg%, riboflavine 5.2 to 10 mg%, niacin 60 to 150 mg%, ascorbic acid 12 mg%, pyridoxine 2.2 to 10.2 mg%, $B_{12}$ 0.15 mg%, pantothenic acid 65 to 200 mg%, biotin 0.9 to 3.7 mg%, inositol 80 to 150 mg%, folic acid 0.2 mg%, and vitamin E <0.2 mg%; and trace elements (MERCK).[10,18,19]

Generally recognized as the major active principle of royal jelly, 10-HDA has been reported to be very stable to heat, remaining chemically intact despite deterioration of other components in the royal jelly; royal jelly itself is also quite stable when refrigerated or in the frozen or dried state.[20–23] However, one recent study found honey accelerated the decomposition of 10-HDA when the two were dissolved in water along with royal jelly.[4]

## PHARMACOLOGY OR BIOLOGICAL ACTIVITIES

Royal jelly exhibited a transient vasodilating effect on dog femoral artery, which was due to the acetylcholine present.[12]

Royal jelly has exhibited weak to strong antibacterial activities against several bacterial species (*Bacillus subtilis*, *Staphylococcus aureus*, *Escherichia coli*, *Streptococcus hemolyticus*, *Enterococcus*, etc.) *in vitro* and *in vivo* (experimentally infected rats), with 10-HDA being the major active agent.[24–27]

Other activities of 10-HDA and royal jelly include: strongly inhibiting growth of transplantable tumor and leukemia in mice, with enhanced phagocytosis of peritoneal macrophages;[28,29] protecting DNA of peripheral leukocytes from damage by 4-nitrosoquinoline-*N*-oxide (4-NQO);[30] immunopotentiating, liver-protectant, antiinflammatory, radiation-protectant, etc.[18,31]

Although rare, royal jelly can cause allergic reactions in some individuals (JIANGSU).[32]

## USES

**Medicinal, Pharmaceutical, and Cosmetic.** Royal jelly (in liquid, powder, or extract form) is used in various types of skin-care products (creams, lotions, soap, etc.) for its alleged antiwrinkle, skin-nourishing and whitening properties; also used in hair-care and oral (e.g., toothpaste) products.

**Health Food/Herb Teas.** Extensively used as a food supplement in tablet, capsule or liquid (honey syrup) form, often in combination with Asian ginseng and other herbs for their "energizing" effects.

**Traditional Medicine.** Used as a general nutrient and tonic by different cultures. Although adopted only recently (20th century) into traditional Chinese medical practice, it is considered sweet and sour tasting, neutral and to have invigorating, nourishing and strengthening properties; used in treating malnutrition in children, general debility in the elderly, chronic hepatitis, diabetes, rheumatism and arthritis, and hypertension (JIANGSU).

## COMMERCIAL PREPARATIONS

Royal jelly normally is available in frozen or freeze-dried form (strength = 3.5X). Quality is determined by the amount of 10-HDA present: high-quality frozen royal jelly contains about 2% while the freeze-dried powder contains about 5% 10-HDA.

*Regulatory Status.* U.S. regulatory status not determined.

## REFERENCES

See the General References for JIANGSU; MERCK; NATIONAL, TYLER.

1. V. E. Tyler, *The New Honest Herbal*, George F. Stickley Co., Philadelphia, 1987, p. 197.
2. T. Echigo et al., *Tamagawa Daigaku Nogakubu Kenkyu Hokoku*, (22), 67 (1982); through *Chem. Abstr.*, **99**, 50582f.
3. W. H. Brown and R. J. Freure, *Can. J. Chem.*, **37**, 2042 (1959).
4. M. Matsui, *Shokuhin Eiseigaku Zasshi*, **29**(5), 297 (1988); through *Chem. Abstr.*, **110**, 37984q.
5. L. P. Zhang, *Zhongguo Zhongyao Zazhi*, **15**(6), 30 (1990).
6. W. H. Brown et al., *Can. J. Chem.*, **39**, 1086 (1961).
7. N. Weaver et al., *Lipids*, **3**, 535 (1968).
8. M. Barbier and D. Bogdanovsky, *Compt, Rend.*, **252**, 3407 (1961); through *Chem. Abstr.*, **55**, 23853d.
9. J. Matsuyama et al., *Tamagawa Daigaku Nogakubu Kenkyu Hokoku*, (**13**), 46 (1973); through *Chem. Abstr.*, **82**, 70417h.
10. Y. F. Yang, *Zhongcaoyao*, **19**(11), 33 (1988).
11. J. Vittek and B. L. Slomiany, *Experientia*, **40**(1), 104 (1984).
12. M. Shinoda et al., *Yakugaku Zasshi*, **98**(2), 139 (1978); through *Chem. Abstr.*, **88**, 183546s.
13. S. M. Abdel-Wahab et al., *Egypt. J. Pharm. Sci.*, **20**(1–4), 353 (1982); through *Chem. Abstr.*, **97**, 214454c.
14. W. B. Rice and F. C. Lu, *Can. Pharm. J.*, **97**(3), 34 (1964).
15. K. Baek and B. Y. Cho, *Kangwon Tachak Yon'gu Nonmunjip*, **6**, 7 (1972); through *Chem. Abstr.*, **79**, 29574g.
16. J. J. Pratt Jr. and H. L. House, *Science*, **110**, 9 (1949).
17. T. Takenaka and T. Echigo, *Honeybee Science*, **5**(1), 7 (1984); through *Apic. Abstr.*, **36**, 640/85.
18. B. X. Wang, *Therapeutic Efficacy of Honeybee Products*, Jilin People's Publishers, Changchun, PRC, 1981.
19. S. R. Howe et al., *J. Apic. Res.*, **24**(1), 52 (1985).
20. Saitama Yoho Co., Ltd., *Honeybee Science*, **2**(3),123 (1981); through *Apic. Abstr.*, **33**, 995/82.

21. T. Takenaka et al., *Nippon Shokuhin Kogyo Gakkaishi*, **33**(1), 1 (1986); *through Chem. Abstr.*, **105**, 23337r.
22. E. L. Lee et al., *Shih Pin K'o Hsueh (Taipei)*, **15**(1), 81 (1988); through *Chem. Abstr.*, **109**, 148242v.
23. A. Dietz and M. H. Haydak, *J. Georgia Entomol. Soc.*, **5**(4), 203 (1970); through *Apic. Abstr.*, **24**, 208/73.
24. M. S. Blum et al., *Science*, **130**, 452 (1959).
25. K. Yatsunami and T. Echigo, *Bull. Faculty Agric., Tamagawa University*, **25**, 13 (1985); through *Apic. Abstr.*, **40**, 997/89.
26. Kh. N. Muratova et al., *Eksp. Khir. Anesteziol.*, **12**(2), 52 (1967); through *Chem. Abstr.*, **67**, 31309v.
27. Y. D. Xu, *Zhongguo Yangfeng*, (6), 28 (1989).
28. G. F. Townsend et al., *Nature*, **183**, 1270 (1959).
29. J. Z. Dai et al., *Yiyao Gongye*, **16**(5), 219 (1985).
30. X. L. Yang et al., *Beijing Yike Daxue Xuebao*, **22**(1) 75 (1990).
31. Z. Q. Xie et al., *Zhongguo Yike Daxue Xuebao*, **21**(3) 167 (1990).
32. M. Takahashi et al., *Contact Dermatitis*, **9**(6), 452 (1983); through *Apic. Abstr.*, **37**, 680/86.

# RUE

**Source:** ***Ruta graveolens*** L. (Family Rutaceae).

*Synonyms*. Common rue and garden rue.

## GENERAL DESCRIPTION

An erect glaucous and nonhairy perennial herb, with a strong disagreeable odor; up to about 1 m high; native to the Mediterranean region; cultivated worldwide (Europe, Africa, Asia, America, etc.). Part used is the dried herb. Rue oil is obtained by steam distillation of the fresh flowering plant. *Ruta montana* L. and *R. bracteosa* L. are also reported used; Algerian rue oil is derived from these species.

## CHEMICAL COMPOSITION

Common rue (*R. graveolens*) contains a volatile oil (ca. 0.1%), rutin (ca. 2%),[1] numerous alkaloids ($\gamma$-fagarine, arborinine, kokusaginine, skimmianine, graveoline, graveolinine, 6-methoxydictamnine, rutacridone, etc.)[2–7] coumarin derivatives (bergapten, xanthotoxin, rutamarin, psoralen, isoimperatorin, pangelin, rutarin, chalepensin, etc.),[2,6,8–12] and others (KARRER, JIANGSU). An unusual coumarin, naphthoherniarin, has recently been isolated from root of common rue.[13]

Rue oil contains varying amounts of 2-nonanone, 2-decanone, and 2-undecanone (methyl-*n*-nonyl ketone) as major components, with 2-undecanone about 90% in Algerian rue oil.[14–16] Other components include 2-heptanone, 2-octanone, 2-nonanol, 2-undecanol, undecyl-2-acetate, anisic acid, phenol, guaiacol, small amounts of coumarins (bergapten, herniarin, and xanthotoxin), monoterpenes (cineole, $\alpha$- and $\beta$-pinenes, limonene),[15,17–19] and others (KARRER, JIANGSU, MARTINDALE).

## PHARMACOLOGY OR BIOLOGICAL ACTIVITIES

Rue alkaloids (especially arborinine and $\gamma$-fagarine) as well as furocoumarins (e.g., bergapten and xanthotoxin) and rue oil are reported to have spasmolytic effects on smooth muscles (e.g., isolated rabbit ileum). Arborinine and furocoumarins also have antiinflammatory and antihistaminic properties (JIANGSU).[2,3,8]

Furocoumarins have phototoxic properties and are useful in treating psoriasis (see ***bergamot***).[16]

Chalepensin has been shown to have antifertility (anti-implantation) effects in rats.[11]

Rue oil has been reported to have anthelmintic (worm, leech, and nematode) activities *in vitro* that were attributed to its major component, 2-undecanone.[16]

When applied to human skin, rue oil may produce a burning sensation, erythema (redness), and vesication (blisters). Taken internally, it causes severe stomach pain, vomiting, exhaustion, confusion, and convulsion. Large doses may be fatal (JIANGSU, MARTINDALE).

Rue oil has been reported to cause abortion in pregnant guinea pigs and in pregnant women.[19]

A single oral dose of 400 mg/kg given to guinea pigs has been reported to be fatal due to internal hemorrhages, particularly of the adrenal gland, liver and kidney. However, an oral daily dose of 30 mg given to human subjects for three months did not cause abnormal liver functions (JIANGSU).

Rutin, first isolated from common rue, has numerous pharmacological activities (see ***rutin***).

## USES

**Medicinal, Pharmaceutical, and Cosmetic.** Rue oil is used as a fragrance ingredient in soaps, detergents, creams, lotions, and perfumes, with maximum use level of 0.15% reported in perfumes.[16]

**Food.** Rue oil is used as a flavor component (e.g., coconut type) in most major food products, including alcoholic (vermouths, bitters, etc.) and nonalcoholic beverages, frozen dairy desserts, candy, baked goods, gelatins and puddings, among others. Average maximum use levels reported are below 0.001%.

Rue is also used in certain foods, including nonalcoholic beverages, frozen dairy desserts, candy, and baked goods. Average maximum use levels are below 0.0002%, (2 ppm).

**Traditional Medicine.** Rue is used as an emmenagogue, intestinal antispasmodic, uterine stimulant, hemostatic, and vermifuge.

In Chinese medicine, rue is also used for essentially the same purposes. In addition, it is used to treat colds, fevers, toothache, and especially snake and insect bites.

**Others.** Rue oil can serve as a source of natural 2-undecanone, which is a starting material for the synthesis of methylnonyl acetaldehyde, a valuable perfume chemical (ARCTANDER).

## COMMERCIAL PREPARATIONS

Crude and oil. Crude was formerly official in U.S.P., and oil in N.F. Oil is official in F.C.C.

*Regulatory Status.* Has been affirmed as GRAS (rue, §184.1698; oil of rue, §184.1699).[20,21] Subject of a German therapeutic monograph; effectiveness of claimed application is not verified.[22]

## REFERENCES

See the General References for ARCTANDER; BAILEY 1; BIANCHINI AND CORBETTA; BLUMENTHAL; KARRER; JIANGSU; LUST; MARTINDALE; MERCK; TERRELL; UPHOF.

1. F. R. Humphreys, *Econ. Bot.*, **18**, 195 (1964).
2. I. Novak et al., *Acta Pharm. Hung.*, **37**, 131 (1967); through *Chem. Abstr.*, **67**, 25374w (1967).

3. I. Novak et al., *Planta Med.*, **15**, 132 (1967).
4. J. Reisch et al., *Phytochemistry*, **15**, 240 (1976).
5. O. Nieschulz, *Sci. Pharm., Proc., 25th,* **2**, 559 (1965); through *Chem. Abstr.*, **70**, 18805w (1969).
6. A. Gonzalez et al., *An. Quim.*, **70**, 60 (1974); through *Chem. Abstr.*, **80**, 130487b (1974).
7. K. Szendrei et al., *Herba Hung.*, **10**,131 (1971); through *Chem. Abstr.*, **79**, 15853k (1973).
8. I. Novak et al., *Pharmazie*, **20**, 738 (1965).
9. E. Varga et al., *Fitoterapia*, **47**, 107 (1976).
10. J. Reisch et al., *Magy. Kem. Foly.*, **78**, 6 (1972); through *Chem. Abstr.*, **76**, 138149f (1972).
11. Y. C. Kong et al., *Planta Med.*, **55**, 176 (1989).
12. A. Nahrstedt et al., *Planta Med.*, **51**, 517 (1985).
13. Z. Rozsa et al., *Planta Med.*, **55**, 68 (1989).
14. T. M. Andon and N. V. Belova, *Rastit. Resur.*, **11**, 539 (1975); through *Chem. Abstr.*, **84**, 21982w (1976).
15. D. H. E. Tattje, *Pharm. Weekbl.*, **105**, 1241 (1970).
16. D. L. J. Opdyke, *Food Cosmet. Toxicol.*, **13**, 455 (1975).
17. D. H. E. Tattje et al., *Pharm. Weekbl.*, **109**, 881 (1974).
18. K. H. Kubeczka, *Phytochemistry*, **13**, 2017 (1974).
19. Anon., *Fed. Regist.*, **39** (185), 34215 (1974).
20. Anon., *Fed. Regist.*, **41**(236), 53620 (1976).
21. Anon., *Fed. Regist.*, **43**(19), 3704 (1978).
22. Monograph *Ruta graveolens*, *Bundesanzeiger*, no. 43 (Mar. 2, 1989).

# RUTIN

**Source:** Currently, *Sophora japonica* L., *Eucalyptus macrorhyncha* F. v. M., and *Fagopyrum esculentum* Moench are the major sources of rutin.

*Synonyms*. Quercetin-3-rutinoside, rutoside, eldrin, and sophorin.

## GENERAL DESCRIPTION

Rutin is a glycoside containing quercetin as its aglycone and rutinose (rhamnose and glucose) as its sugar portion. It is widely distributed in the plant kingdom, being found in many families of higher plants as well as in ferns.[1–4]

It is present in high concentrations in leaves of *Eucalyptus macrorhyncha* (10 to 24%), flowers of *Viola tricolor* L. var. *maxima* (18 to 21%), flower buds of *Sophora japonica* (13 to 30%), and in buckwheat, *Fagopyrum esculentum* (0.1 to 6.4%).[1,4]

Currently, *S. japonica*, *E. macrorhyncha*, and *F. esculentum* are the major sources of rutin. Its production is relatively simple to rather complex, depending on the raw material and the process used.

## PHARMACOLOGY OR BIOLOGICAL ACTIVITIES

Rutin has been reported to have many pharmacological properties. The most well-known one is its ability to decrease capillary permeability and fragility, though evidence is inconclusive (CLAUS, MARTINDALE, REMINGTON, USD 26th).[1,5] It is considered a "vitamin P" or "permeability" vitamin.

Other pharmacological activities include antiedemic,[6] antiatherogenic in chicks fed cholesterol,[7] increasing survival time of rats fed a thrombogenic diet,[8] inhibiting tumor formation on mouse skin by the carcinogen benzo(*a*)pyrene,[9] antiinflammatory, antispasmodic, hypotensive, and protective against X-ray irradiation in mice, among others[10–11] (JIANGSU).

Rutin is a noncompetitive inhibitor of angiotensin II and prostaglandin $E_2$ on the guinea-pig ileum.[12] Antispasmodic activity has been observed in isolated guinea-pig colon and rat duodenum.[13,14]

Rutin has been found to induce a release of endogenous histamine and 5-hydroxytryptamine 2 hours after administration, however a predicted antispasmodic effect, as observed *in vitro*, was absent *in vivo*, suggesting a physiological antagonism mechanism contrary to previously reported *in vitro* results.[14]

Rutin has been generally considered to lack toxicity (USD 26th). However, it has been reported by one group of investigators to cause concretion formation in human as well as in laboratory animals.[15,16]

## USES

Rutin was formerly an official drug in the United States and has been used in treating capillary hemorrhage due to increased capillary fragility in degenerative vascular diseases (e.g., arteriosclerosis and hypertension), diabetes, and allergic manifestations.

Although no longer official in N.F., rutin still is quite widely used, both as a prescription drug and as a vitamin supplement.[17] The latter use is by far the more widespread. It is usually used in formulations with vitamin C or together with other bioflavonoids as well as rose hips, especially in the health food industry.

Flower buds of *Sophora japonica* containing high concentrations (usually ca. 20%) of rutin have been used for centuries in Chinese medicine for the treatment of internal bleeding (e.g., bloody urine, spitting blood, and intestinal bleeding) and bleeding hemorrhoids, It is also used for the prevention of strokes (FARNSWORTH, FOGARTY, JIANGSU).

## COMMERCIAL PREPARATIONS

Pure rutin; often adulterated; it was formerly official in N.F.

## REFERENCES

See the General References for APhA; BAILEY 2; JIANGSU.

1. F. R. Humphreys, *Econ. Bot.*, **18**, 195 (1964).
2. G. Berti and F. Bottari, in *Progress in Phytochemistry*, Vol. 1, L. Reinhold and Y. Liwschitz, Eds., Interscience, New York, 1968, p. 589.
3. J. Davidek, *Veda Vyzkum Prumyslu Potravinarskem*, **12**, 179 (1963); through *Chem. Abstr.*, **62**, 7027c (1965).
4. V. A. Bandyukova and N. V. Sergeeva, *Khim. Prir. Soedin.*, **4**, 524 (1974); through *Chem. Abstr.*, **82**, 54179b (1975).
5. K. Venkataraman, in *Fortschritte der Chemie organischer Naturstoffe*, Vol. 17, L. Zechmeister, Ed., Springer-Verlag, Vienna, Austria, 1959, p. 1.
6. V. M. Samvelyan and L. G. Khlgayan, *Farmakol. Toksikol. (Moscow)*, **32**, 447 (1969); through *Chem. Abstr.*, **71**, 89851q (1969).
7. A. N. Chernov et al., *Lipidy Tkanei Eksp. Giperkholesterinemin*, **115** (1967); through *Chem. Abstr.*, **71**, 79554f (1969).
8. R. C. Robbins, *J. Atheroscler. Res.*, **7**, 3 (1967).

9. B. L. Van Duuren et al., *J. Natl. Cancer Inst.*, **46**, 1039 (1971).
10. V. A. Baraboi, *Fenol'nye Soedin, Ikh Biol. Funkts., Mater Vses. Simp.*, **1st**, 353 (1966); through *Chem. Abstr.*, **71**, 1957v (1969).
11. G. Vogel and H. Ströcker, *Arzneim.-Forsch.*, **16**, 1630 (1966).
12. O. Altinkurt and N. Abacioglu, *Arzneim.-Forsch.*, **30**, 610 (1980).
13. O. Altinkurt and Y. Öztürk, *J. Fac. Pharm. Ankara*, **17**, 49 (1987).
14. N. Yildizoglu et al., *Phytother. Res.*, **5**(1), 19 (1991).
15. K. Pfeifer et al., *Dtsch. Gesundheilsw.*, **24**, 260 (1969); through *Chem. Abstr.*, **71**, 69165r (1969).
16. K. Pfeifer et al., *Dtsch. Gesundheitsw.*, **25**, 386 (1970); through *Chem. Abstr.*, **73**, 43640h (1970).
17. N. R. Farnsworth in L. P. Miller, ed., *Phytochemistry*, Vol. 3, Van Nostrand Rheinhold, New York, 1973, p. 351.

# SAFFRON

**Source:** ***Crocus sativus*** L. (Family Iridaceae).

*Synonyms.* Saffron crocus

## GENERAL DESCRIPTION

A perennial herb with a large fleshy corm from which leaves and flowers are produced in the fall; native to the eastern Mediterranean region; cultivated as an annual or perennial worldwide (Spain, France, Italy, India, etc.). Part used is the dried stigma; 100,000 to 140,000 flowers are reportedly required to yield 1 kg saffron. Hence it is the most expensive spice. Major producing countries include Spain, France, Turkey, and India.

## CHEMICAL COMPOSITION

Contains about 2% crocin-1 (ester of crocetin with 2 molecules gentiobiose); about 2% picrocrocin (bitter principle); small amounts of crocin-2 (ester of crocetin with 1 molecule each of gentiobiose and glucose), crocin-3 (monogentiobiose ester of crocetin), and crocin-4 (ester of monomethylcrocetin with 1 molecule glucose); small amounts of free crocetin ($\alpha$-crocetin), methylcrocetin ($\beta$-crocetin), *trans*-dimethylcrocetin ($\gamma$-crocetin), and *cis*-dimethylcrocetin; starch (ca. 13%); vitamins $B_1$ and $B_2$; fixed oils (8 to 13%); other carotenoids; and 0.4 to 1.3% of a volatile oil consisting of safranal, oxysafranal, pinene, cineole, isophorone, naphthalene, 2-butenoic acid lactone, 2-phenylethanol, 3,5,5-trimethyl-4-hydroxy-1-cyclohexanone-2-ene, and others (JIANGSU, LIST AND HÖRHAMMER).[1–3]

Crocin is mainly responsible for the color of saffron, while picrocrocin is responsible for its bitter taste and aroma (after hydrolysis to yield safranal).

## PHARMACOLOGY OR BIOLOGICAL ACTIVITIES

Extracts of saffron have been demonstrated to have various pharmacological properties, including stimulation of the uteri of experimental animals, lowering the blood pressure of anesthetized dogs and cats as well as stimulating respiration, and strongly inhibiting the isolated toad and rat hearts, among others (JIANGSU).[4]

Crocetin has been found to increase oxygen diffusion in plasma by 80%. Crocetin binds to serum albumin, and has been found to lower serum cholesterol levels in rabbits.[5,6]

Recent studies suggest that saffron extracts have *in vitro* anticancer activity, limiting the growth of experimentally induced cancers by inhibiting cellular nucleic acid synthesis.[7,8]

No risk associated with food or therapeutic use at 1.5 g or less. Toxic reactions have been reported at dose of 5 g including necrosis of the nose, thrombocytopenia and uremia collapse. Associated symptoms include vomiting, bleeding of the uterus, bloody diarrhea, nose bleed, vertigo, dizziness, and others.[9]

## USES

**Medicinal, Pharmaceutical, and Cosmetic.** Extracts (e.g., tincture) are used as fragrance components in perfumes (especially oriental types). An Australian patent has been issued for use of an aqueous extract of the corm (in combination with other ingredients for treatment of baldness).[6]

**Food.** Saffron is used both as a coloring (yellow) and as a flavoring agent. It is often used as a domestic spice, especially in Spanish and French cooking (e.g., arroz con pollo and bouillabaisse).

Saffron and saffron extract (type not specified) are used in alcoholic (e.g., bitters and vermouths) and nonalcoholic beverages, candy, baked goods, and other food products. Highest average maximum use level is about 0.1% (969 ppm) reported for the crude in baked goods.

**Traditional Medicine.** Reportedly used as a sedative, diaphoretic, antispasmodic, emmenagogue, anodyne, and aphrodisiac. Conditions for which it is used include coughs, whooping cough, stomach gas, and insomnia. A tincture formerly used as a sedative in Germany (WEISS).

In Chinese medicine, it is traditionally used to treat conditions resulting from depression, fright, or shock; spitting blood; pain and difficulties in menstruation and after childbirth; and others (JIANGSU).

Saffron has been widely used in cancers, which has been experimentally confirmed *in vitro*.[7,8,10]

### COMMERCIAL PREPARATIONS

Mainly crude (Spanish); it was formerly official in N.F.

*Regulatory Status.* GRAS (§182.10 and §182.20); also has been approved as a food color additive exempt from certification (§73.500). Subject of a German therapeutic monograph, indicated as a sedative for spasms and asthma, with the caveat that claimed efficacy is not documented.[9]

### REFERENCES

See the General References for ARCTANDER; BLUMENTHAL; CLAUS; FEMA; JIANGSU; LIST AND HÖRHAMMER; LUST; ROSENGARTEN; TERRELL; UPHOF, WEISS.

1. V. K. Dhingra et al., *Indian J. Chem.*, **13**, 339 (1975).
2. P. Duquenois, *Bull. Soc. Pharm. Strasbourg*, **15**, 149 (1972); through *Chem. Abstr.*, **79**, 45885q (1973).
3. N. S. Zarghami, *Diss. Abstr. Int. B*, **31**, 5235 (1971).
4. P. Y. Chang et al., *Yao Hsueh Hsueh Pao*, **11**, 94 (1964); through *Chem. Abstr.*, **61**, 2348b (1964).
5. B. R. Olin, ed., *Lawrence Rev. Nat. Prod.*, April 1993.
6. T. L. Miller et al., *J. Pharm. Sci.*, **71**, 173 (1982).
7. F. I. Abdullaev et al., *Biofactors*, **3**, 201 (1992).
8. S. C. Nair et al., *Cancer Lett*, **57**, 109 (1991).
9. Monograph *Croci stigma*, *Bundesanzeiger*, no. 76 (Apr. 23, 1987).
10. J. L. Hartwell, *Lloydia*, **32**, 247 (1969).

# SAGE

**Source:** ***Sage*** *Salvia officinalis* L.; ***Spanish sage*** *Salvia lavandulaefolia* Vahl (Family Labiatae or Lamiaceae).

*Synonyms.* Garden sage, true sage, and Dalmation sage (*S. officinalis*).

### GENERAL DESCRIPTION

*Salvia officinalis* is a small, evergreen shrubby perennial with woody stems near the base and herbaceous ones above, much branched; up to about 0.8 m high; native to the Mediterranean region; cultivated worldwide (Albania, Turkey, Greece, Italy, United States, etc.). Part used is the dried leaf from which sage oil is obtained by steam distillation.

*Salvia lavandulaefolia* is closely related to *S. officinalis*. It grows wild in Spain and southwestern France. Spanish sage oil is obtained by steam distillation of its leaves.

A recent study found that most commercial sage sold in the United States (from 50 to 95%) was represented by *S. fruticosa* Mill. (*S. triloba* L. f.), characterized by compound or simple leaves with 1–2 pairs of lateral segments and a large terminal segment, rather than *S. officinalis* as purported (TUCKER).[1]

## CHEMICAL COMPOSITION

Sage (*S. officinalis*) contains 1.0 to 2.8% volatile oil;[2] picrosalvin and carnosol (bitter principles);[3] salvin, salvin monomethyl ether, and carnosic acid;[4,5] flavonoids including genkwanin, 6-methoxygenkwanin, luteolin, luteolin-7-methyl ether, 6-methoxyluteolin, 6-methoxyluteolin-7-methyl ether, hispidulin, and salvigenin;[6,7] phenolic acids (rosmarinic, labiatic, caffeic, and trace of chlorogenic);[8,9] salviatannin (a tannin of the condensed catechin type which on storage undergoes degradation to phlobaphenes);[10] and others (MARSH, MARTINDALE, STAHL).[9]

Sage oil contains $\alpha$- and $\beta$-thujones (normally ca. 50%) as the main components. Other compounds present include cineole, borneol, viridiflorol, 2-methyl-3-methylene-5-heptene, and sesquiterpenes (STAHL).[2,11–13]

Spanish sage contains a volatile oil composed of highly variable amounts of camphor (11 to 34%), cineole (18 to 35%), limonene (1 to 41%), camphene (5 to 30%), $\alpha$-pinene (4 to 20%), $\beta$-pinene (6 to 19%), linalool, linalyl acetate, borneol, and others (GUENTHER, STAHL).[14–17] It also contains numerous polyphenolic compounds, including luteolin-4′-*O*-glucuronide, rosmarinic acid, salvigenin, eupatorin, nepetin and apigenin, among others.[18]

Sage extracts, like those of rosemary, have strong antioxidative activities; labiatic acid and carnosic acid are reported to be the active compounds (see ***rosemary***).[8] Sage is also reported to have fish odor-suppressant properties.[19]

## PHARMACOLOGY OR BIOLOGICAL ACTIVITIES

Sage reportedly has antibacterial, fungistatic, virustatic, astringent, secretion-stimulating, and perspiration-inhibiting effects.[20]

Phenolic acids (e.g., salvin and salvin monomethyl ether) isolated from sage have antimicrobial activities, especially against *Staphylococcus aureus*.[4,9]

Spanish sage oil has also been reported to have antimicrobial properties.[15]

Sage oil has been reported to have neurotropic antispasmodic effects against acetylcholine spasms in laboratory animals.[21]

Although sage oil contains more thujone than absinthium oil, it has not been reported to be toxic (see ***absinthium***).

Dalmation sage oil has been reported to be nonirritating and nonsensitizing to human skin when tested in a diluted form. When applied undiluted it produced one irritation reaction in 20 subjects and was moderately irritating to rabbits.[22]

Spanish sage oil was nonirritating and nonsensitizing to human skin and skin of laboratory animals; it was also nonphototoxic on mice and swine.[15]

Infusions and suspensions of Spanish sage have hypoglycemic activity in rabbits.[23]

## USES

**Medicinal, Pharmaceutical, and Cosmetic.** The dried leaves, the essential oil, tincture and fluid extract, are used in European phytomedicine for dyspeptic symptoms and diaphoretic effects; external use (gargles and rinses) for inflamed mucous membranes of the oral mucosa and throat.[20]

Both sage oil and Spanish sage oil are

used (the former much more extensively) as fragrance components in soaps, detergents, creams, lotions, and perfumes (e.g., colognes and after-shave lotions), with maximum use level of 0.8% reported for both oils in perfumes.[15,22] Spanish sage oil is generally more commonly used in soaps, detergents, and industrial fragrances.

**Food.** Sage is widely used as a flavor ingredient in baked goods, meat and meat products, condiments and relishes, processed vegetables, soups, gravies, fats and oils, and others. Highest average maximum use level reported is 0.477% in baked goods.

Sage oleoresin is also widely used in baked goods, meat and meat products, and condiments and relishes. Highest average maximum use level reported is about 0.014% (139 ppm) in meat and meat products.

Sage oil and Spanish sage oil are extensively used in most categories of food products, including alcoholic (e.g., vermouths and bitters) and nonalcoholic beverages, frozen dairy desserts, candy, baked goods, gelatins and puddings, meat and meat products, and condiments and relishes. Highest average maximum use levels reported are about 0.013% (126 ppm) and 0.004% (40.5 ppm) for sage oil and Spanish sage oil, respectively, in meat and meat products.

**Health Food/Herb Teas.** Dried leaves used as a tea ingredient; occasionally in tablets, capsules, tincture, etc., for traditional indications (FOSTER).

**Traditional Medicine.** Sage is used as a tonic, digestive, antiseptic, astringent, and antispasmodic. It is used to reduce perspiration (e.g., night sweats), to stop the flow of milk, to treat nervous conditions (e.g., trembling, depression, and vertigo), dysmenorrhea, diarrhea, gastritis, sore throat, insect bites, and others, usually in the form of a tea or infusion.

Sage has been reported used in cancers.[24]

**Others.** Like rosemary, sage can serve as source of natural antioxidants (see ***rosemary***).

## COMMERCIAL PREPARATIONS

Sage, sage oleoresin, and Dalmation sage oil; Spanish sage oil. Sage was formerly official in N.F. Dalmation sage oil and Spanish sage oil are official in F.C.C.

*Regulatory Status.* GRAS (sage, §182.10 and §182.20; Spanish sage, §182.20); no thujone limit for sage oil is specified (see ***absinthium***). Subject of a positive German therapeutic monograph; allowed for internal use in dyspeptic complaints, and as a diaphoretic; internally for inflammed oral mucous membranes.[20] A separate monograph on *S. fruticosa* (*S. triloba*) is pending.

## REFERENCES

See the General References for ARCTANDER; BAILEY 1; BIANCHINI AND CORBETTA; BLUMENTHAL; FEMA; FOSTER; GUENTHER; KROCHMAL AND KROCHMAL; LUST; POLUNIN AND SMYTHIES; ROSENGARTEN; TUCKER.


1. A. O. Tucker et al., *Econ. Bot*, **34**, 1, 16, (1980).
2. A. Ceylan, *Ege Univ. Ziraat Fak., Derg., Seri A*, **13**, 283 (1976); through *Chem. Abstr.*, **87**, 141106x (1977).
3. E. Ghigi et al., *Ann. Chim. (Rome)*, **59**, 510 (1969); through *Chem. Abstr.*, **71**, 113102u (1969).
4. V. N. Dobrynin et al., *Khim. Prir. Soedin.*, **5**, 686 (1976); through *Chem. Abstr.*, **86**, 117603r (1977).
5. C. H. Brieskorn and H. J. Dömling, *Z.*

*Lebensm.-Unters. Forsch.*, **141**, 10 (1969).
6. C. H. Brieskorn and W. Biechele, *Arch. Pharm. (Weinheim)*, **304**, 557 (1971).
7. C. H. Brieskorn and W. Biechele, *Dtsch. Apoth.-Ztg.*, **111**, 141 (1971).
8. K. Herrmann, *Z. Lebensm.-Unters. Forsch.*, **116**, 224 (1962).
9. N. Z. Alimkhodzhaeva and R. L. Khazanovich, *Mater. Yubileinoi Resp. Nauchn. Konf. Farm., Posvyashch. 50-Letiyu Obraz. SSSR*, **37** (1972); through *Chem. Abstr.*, **82**, 167491r (1975).
10. D. Murko et al., *Planta Med.*, **25**, 295 (1974).
11. C. H. Brieskorn and S. Dalferth, *Ann.*, **676**, 171 (1964); through *Chem. Abstr.*, **61**, 14463d (1964).
12. M. B. Embong et al., *Can. Inst. Food Sci. Technol. J.*, **10**, 201 (1977).
13. C. Karl et al., *Planta Med.*, **44**, 188 (1982).
14. B. M. Lawrence et al., *Chromatogr.*, **50**, 59 (1970).
15. D. L. J. Opdyke, *Food Cosmet. Toxicol.*, **14**(Suppl.), 857 (1976).
16. C. H. Brieskorn and S. Dalferth, *Dtsch. Apoth. Ztg.*, **104**, 1388 (1964).
17. M. E. Crespo et al., *Planta Med.*, **52**, 366 (1986).
18. S. Cañigueral et al., *Planta Med.*, **55**, 92 (1989).
19. T. Kikuchi et al., *Eiyo To Shokuryo*, **21**, 253 (1968); through *Chem. Abstr.*, **70**, 95560q (1969).
20. Monograph *Salviae folium*, *Bundesanzeiger*, no. 90 (May 15, 1985); revised (Mar. 13, 1990).
21. A. M. Debelmas and J. Rochat, *Plant. Med. Phytother.*, **1**, 23 (1967).
22. D. L. J. Opdyke, *Food Cosmet. Toxicol.*, **12**(Suppl.), 987 (1974).
23. J. Jimenez et al., *Planta Med.*, **52**, 260 (1986).
24. J. L. Hartwell, *Lloydia*, **32**, 247 (1969).

# SANDALWOOD OIL

**Source: *Santalum album*** L. (Family Santalaceae).

*Synonyms*. Santal oil, East Indian sandalwood oil, white sandalwood oil, yellow sandalwood oil, and white saunders oil.

### GENERAL DESCRIPTION

A small evergreen tree, much-branched, with opposite leathery leaves; up to about 9 m high; native to and cultivated in tropical Asia (especially India, Sri Lanka, Malaysia, Indonesia, and Taiwan). Part used is the heartwood. Sandalwood oil is obtained in 3 to 5% yield from the coarsely powdered dried heartwood by steam or water distillation.

India is the major producer of sandalwood oil (East Indian sandalwood oil). A closely related oil, Australian sandalwood oil, is derived from the wood of *Eucarya spicata* Sprag. et Summ. (syn. *Santalum spicatum* DC.); it has a different topnote than the East Indian oil but is similar to the East Indian oil in overall odor.[1]

### CHEMICAL COMPOSITION

Sandalwood oil contains up to 90% or more of $\alpha$- and $\beta$-santalols. Minor constituents present include 6% sesquiterpene hydrocarbons (mostly $\alpha$- and $\beta$-santalenes and *epi*-$\beta$-santalene with small amounts of $\alpha$- and $\beta$-curcumenes, possibly $\beta$-farnesene, and dendrolasin), dihydro-$\beta$-agarofuran, santene, teresantol, borneol, teresantalic acid, tricycloekasantalal, santalone, and santanol, among others (JIANGSU).[1–6]

$\alpha$-Santalol (ca. 46%) and $\beta$-santalol (ca. 20%) account for most of the odor of sandalwood oil.[2,8]

## PHARMACOLOGY OR BIOLOGICAL ACTIVITIES

Sandalwood oil is reported to have diuretic and urinary antiseptic properties. Santalol can reportedly cause contact dermatitis in sensitive individuals (CLAUS, LEWIS, AND ELVIN-LEWIS).

Sandalwood oil has been reported to be nonirritating, nonsensitizing, and nonphototoxic to human skin, though it was slightly irritating to mouse skin and irritating to rabbit skin when applied undiluted.[7]

It also has antifungal activities.[8]

## USES

**Medicinal, Pharmaceutical, and Cosmetic.** Extensively used as a fragrance ingredient in soaps, detergents, creams, lotions, and perfumes (especially oriental types), with maximum use level of 1% reported in perfumes.[7] It is also commonly used in incenses.

**Food.** Used as a flavor component in major categories of food products, including alcoholic and nonalcoholic beverages, frozen dairy desserts, candy, baked goods, and gelatins and puddings, with reported average maximum use levels generally below 0.001%.

**Health Food/Herb Teas.** Sandalwood oil is used in aromatherapy formulations (ROSE).

**Traditional Medicine.** In Chinese medicine, sandalwood oil is reportedly used to treat stomachache, vomiting, and gonorrhea (JIANGSU). Oil formerly used in Europe for pains, fevers, and "strengthening the heart."

## COMMERCIAL PREPARATIONS

Oil. It was formerly official in N.F. and is official in F.C.C.

*Regulatory Status.* Has been approved for food use (§172.510). Wood is the subject of a German therapeutic monograph; allowed as an antibacterial and spasmolytic for supportive therapy for lower urinary tract infections.[9]

## REFERENCES

See the General References for ARCTANDER; CLAUS; FEMA; GUENTHER; JIANGSU; TERRELL; UPHOF.

1. W. D. Forham in L. W. Codd et al., eds., *Chemical Technology: An Encyclopedic Treatment*, Vol. 5, Barnes & Noble, New York, 1972, p. 1.
2. E. Demole et al., *Helv. Chim. Acta*, **59**, 73 (1976).
3. D. R. Adams et al., *Phytochemistry*, **14**, 1459 (1975).
4. S. K. Paknikar and C. G. Naik, *Tetrahedron Lett.*, (15), 1293 (1975).
5. H. C. Kretschmar et al., *Tetrahedron Lett.*, (1), 37 (1970).
6. B. M. Lawrence, *Perfum. Flav.*, **6**(5), 27 (1981).
7. D. L. J. Opdyke, *Food Cosmet. Toxicol.*, **12**(Suppl.) 989 (1974).
8. A. Dikshit and A. Husain, *Fitoterapia*, **55**, 171 (1984).
9. Monograph *Santali albi lignum*, *Bundesanzeiger*, no. 43 (Mar. 2, 1989).

# SARSAPARILLA

**Source:** ***Smilax medica*** Schlecht. (syn. *S. aristolochiifolia* Mill.), ***S. regelii*** Killip et Morton, ***S. officinalis*** Kunth, ***S. febrifuga*** Kunth, ***S. ornata*** Lem. and other ***Smilax*** species (Family Liliaceae).

*Synonyms*. Mexican sarsaparilla (*S. medica*), Honduras sarsaparilla (*S. regelii* and *S. officinalis*), Ecuadorian sarsaparilla (*S. febrifuga* and other *Smilax* species), and Jamaican sarsaparilla (*S. regelii*).

## GENERAL DESCRIPTION

Mostly climbing or trailing perennial vines with prickly stems, short and thick rhizomes, and very long slender roots; native to tropical America and the West Indies. Part used is the dried root.

## CHEMICAL COMPOSITION

Contains steroids (sarsasapogenin, smilagenin, sitosterol, stigmasterol, and pollinastanol) and their glycosides (saponins) including sarsasaponin (parillin), smilasaponin (smilacin), sarsaparilloside, and sitosterol glucoside, among others (KARRER, REMINGTON).[1,2]

Other constituents present include starch, resin, cetyl alcohol, and a trace of a volatile oil (KARRER, MERCK).

## PHARMACOLOGY OR BIOLOGICAL ACTIVITIES

Sarsaparilla saponins have been reported to facilitate the absorption by the body of other drugs when used with sarsaparilla (also see ***quillaia*** and ***yucca***).

Sarsaparilla has been used in the United States in treating syphilis and rheumatism, but its effectiveness has not been substantiated.

Sarsaparilla products, along with other sterol containing plants, have in recent years been touted as performance-enhancing or body-building substitutes for anabolic steroids sold primarily to athletes. No human or animal studies substantiate these claims. Plant sterols cannot be biochemically transformed *in vivo* into steroidal compounds, and have not been shown to promote anabolic effects in humans.[3]

Sarsaparilla preparations may cause gastric irritation or temporary kidney impairment; adverse drug interaction includes increased absorption of *Digitalis* glycosides, while accelerating elimination of hypnotic drugs.[4]

Sarsaparilla is reported to have hepatoprotective,[5] diuretic, and antiinflammatory activity.[6]

## USES

**Medicinal, Pharmaceutical, and Cosmetic.** It is used in certain tonic preparations. In European tradition the root used for skin disease, particularly psoriasis, as well as rheumatic complaints, and kidney disease; diuretic and diaphoretic. Claims for efficacy have not been substantiated.[4]

**Food.** Sarsaparilla extracts are used extensively as flavor components in root beer even though they are essentially odorless and have hardly any taste; and it is doubtful that the reported average maximum use level of approximately 0.001% (12.9 ppm), even for the strongest commercial extract (solid extract), will contribute to the foaming properties of the root beer (see ***quillaia*** and ***yucca***). Other food products in which they are used include frozen dairy desserts, candy, and baked goods, with the highest average maximum use level of 0.2% reported in baked goods.

**Health Food/Herb Teas**. Root used as a flavoring ingredient in teas; numerous product forms, including tablets, capsules, and nutritional powders touted as an anabolic enhancing dietary supplement for athletes and bodybuilders. Claims that the ingredient contains testosterone are not substantiated (TYLER 1).

**Traditional Medicine.** Used generally as a tonic. Rhizome of Mexican sarsaparilla is reportedly used in Mexico in treating gonorrhea, skin diseases, rheumatism, fevers, and digestive disorders, usually as a decoction. Rhizome and root of Honduras sarsaparilla are used for similar purposes.

In Chinese medicine, the roots and/or rhizomes of several related *Smilax* species native to China are used. They include *S. sieboldi* Miq., *S. stans* Maxim., *S. scobinicaulis* C. H. Wright, and *S. glabra* Roxb., among others. As with sarsaparilla, they are used mainly in treating rheumatism, arthritis, sores, and skin problems. The rhizome of *S. glabra* is also used in treating mercury poisoning, syphilis, and acute bacterial dysentery, among others. In clinical observations, its effectiveness on primary syphilis has been reported to be about 90% (negative blood test). It is sometimes decocted with other plant drugs (JIANGSU).

Numerous *Smilax* species have been used in cancers both in the Old World and the New World.[7]

## COMMERCIAL PREPARATIONS

Crude and extracts; crude and fluid extract were formerly official in N.F. Strengths (see **glossary**) of extracts are expressed in weight-to-weight ratios.

*Hemidesmus indicus* R. Br. (Asclepidaceae) is reported to be a widespread adulterant to commercial supplies of sarsaparilla, which is easily distinguished organoleptically by its strong vanilla fragrance.[8]

*Regulatory Status*. Has been approved for food use (§172.510). Subject of a German therapeutic monograph; not recommended since claims for skin diseases and psoriasis have not been substantiated.[4]

## REFERENCES

See the General References for ARCTANDER; BLUMENTHAL; BRADLEY; CLAUS; FEMA; GOSSELIN; JIANGSU; LUST; STAHL; TERRELL; TUCKER AND LAWRENCE; TUCKER; TYLER 1; UPHOF; YOUNGKEN.

1. M. Devys et al., *C. R. Acad. Sci., Ser. D*, **269**, 2033 (1969); through *Chem. Abstr.*, **72**, 75627e (1970).
2. R. Tschesche et al., *Chem. Ber.*, **102**, 1253 (1969).
3. R. L. Barron and G. J. Vanscoy, *Annals of Pharmacotherapy*, **27**, 607 (1993).
4. Monograph *Sarsaparillae radix*, *Bundesanzeiger*, no. 164 (Sept. 1, 1990).
5. S. Rafatullah et al., *Int. J. Pharmacognosy*, **29**, 296 (1991).
6. A. M. Ageel et al., *Drugs Exptl. Clin. Res.*, **15**, 369 (1989).
7. J. L. Hartwell, *Lloydia*, **33**, 97 (1970).
8. C. Hobbs, *HerbalGram*, **17**(1), 10 (1988).

# SASSAFRAS

**Source:** ***Sassafras albidum*** (Nutt.) Nees [syn. *S. officinale* Nees et Eberm.; *S. variifolium* (Salisb.) Kuntzel] (Family Lauraceae).

*Synonym*. Common sassafras.

## GENERAL DESCRIPTION

An aromatic deciduous tree with leaves ranging in shape from three-lobed to two-lobed to unlobed; up to about 40 m high; native to eastern United States from Maine to Florida and west to Michigan and Texas. Part used is the dried root bark.

Safrole-free sassafras extract is obtained by dilute alcoholic extraction of the bark followed by concentrating under vacuum, diluting the concentrate with water, and separating and discarding the oily fraction.[1]

## CHEMICAL COMPOSITION

Contains 5 to 9% volatile oil, about 0.02% alkaloids (boldine, norboldine, isoboldine, norcinnamolaurine, cinnamolaurine, and reticuline), two lignans (sesamin and desmethoxyaschantin), sitosterol, tannins, resin, and starch (CLAUS).[2]

The volatile oil contains safrole (80 to 90%) as the major component. Other compounds present include $\alpha$-pinene, $\alpha$- and $\beta$-phellandrenes, 5-methoxyeugenol, asarone, piperonylacrolein, apiole, coniferaldehyde, camphor, myristicin, thujone, *l*-menthone, caryophyllene, elemicin, copaene, anethole, and eugenol, among others.[3,4]

## PHARMACOLOGY OR BIOLOGICAL ACTIVITIES

Sassafras and its oil have been reported to have carminative and diaphoretic properties. The oil also reportedly has antiinfective and pediculicide (lice destroying) activities (MERCK).

Safrole has produced hepatomas (liver tumors) in laboratory animals (GOSSELIN, LEWIS AND ELVIN-LEWIS).[3–6]

Sassafras tea has been associated with clinical diaphoresis with hot flashes.[7]

## USES

**Food.** Sassafras, its extracts, and oil were formerly extensively used in flavoring root beer; this use has been discontinued. Only safrole-free bark extract is reported used in nonalcoholic beverages and in candy, with average maximum use levels of 0.022 and 0.015%, respectively. As most of the flavor is removed along with safrole, these uses of the safrole-free extract are rather limited.

**Health Food/Herb Teas.** Bulk sassafras is readily available; usually labeled "not for internal use." Popularity as a "spring tonic," continues in the Ozarks and Appalachians, where fresh root is available in the produce section of supermarkets in spring (FOSTER).

**Traditional Medicine.** Sassafras is traditionally used in treating bronchitis, high blood pressure of elderly people, rheumatism, gout, arthritis, skin problems, and kidney problems, among others, usually as a tea or infusion, used both internally and externally.

Sassafras has also been used in cancers.[8]

**Others.** Safrole present in sassafras oil is used as a starting material for the synthesis of heliotropin (piperonal), an important fragrance and flavor chemical.

## COMMERCIAL PREPARATIONS

Crude, oil, and safrole-free extract; crude and oil were formerly official in N.F.

*Regulatory Status.* Only safrole-free sassafras extract (§172.580) and safrole-free sassafras leaves and extracts (§172.510) have been approved for food use. Safrole, sassafras, and sassafras oil are prohibited from use in foods (§189.180).[9]

## REFERENCES

See the General References for BAILEY 1; CLAUS; FEMA; FOSTER; FOSTER AND DUKE; KROCHMAL AND KROCHMAL; LUST; MORTON 1; SARGENT; TERRELL; TYLER; UPHOF.

1. Anon., *Fed. Regist.*, **27**, 9449 (1962).
2. B. K. Chowdhury et al., *Phytochemistry*, **15**, 1803 (1976).
3. M. L. Sethi et al., *Phytochemistry*, **15**, 1773 (1976).
4. A. B. Segelman et al., *J. Am. Med. Assoc.*, **236**, 477 (1976).
5. F. Homburger et al., *Arch. Pathol.*, **73**, 118 (1962).
6. P. Borchert et al., *Cancer Res.*, **33**, 575 (1973).
7. J. D. Haines, *Postgraduate Medicine*, **90**, 4, 75 (1991).
8. J. L. Hartwell, *Lloydia*, **32**, 247 (1969).
9. Anon., *Fed. Regist.*, **41**(92), 19207 (1976).

# SAVORY (SUMMER AND WINTER)

**Source:** ***Summer Savory*** *Satureja hortensis* L. (syn. *Calamintha hortensis* Hort.); ***Winter savory*** *Satureja montana* L. (syn. *S. obovata* Lag.; *Calamintha montana* Lam.) (Family Labiatae or Lamiaceae).

## GENERAL DESCRIPTION

Summer savory (*S. hortensis*) is an annual herb with oblong-linear leaves and hairy, erect branching stems; up to about 45 cm high; native to Europe and widely escaped from cultivation elsewhere (e.g., United States). Parts used are the dried leaves and tender stems; summer savory oil is obtained by steam distillation of the whole dried herb. Major producing countries include Spain, France, and the United States.

Winter savory (*S. montana*) is a bristly perennial subshrub with a woody base and oblong-linear leaves; up to about 38 cm high; native to the Mediterranean region and widely cultivated. Parts used are the leaves and tender stems.

The savory used in American households is normally summer savory.

## CHEMICAL COMPOSITION

Summer savory (*S. hortensis*) contains a volatile oil (ca. 1%) consisting mostly of carvacrol and monoterpene hydrocarbons ($\beta$-pinene, *p*-cymene, $\beta$-phellandrene, limonene, camphene, etc.), with borneol, cineole, camphor, and others also present (GUENTHER, ROSENGARTEN).[1–4]

Other constituents present include labiatic acid, proteins (ca. 7%), vitamins (especially A), and minerals (especially Ca and K), among others (MARSH).[5]

Winter savory (*S. montana*) contains a volatile oil (1.6%) composed mainly of carvacrol, *p*-cymene, and thymol (total phenols ca. 50%), with lesser amounts of $\alpha$- and $\beta$-pinenes, limonene, cineole, borneol, and $\alpha$-terpineol.[6,7] It also contains triterpenic acids (ursolic and oleanolic acids).[8] Flavonoids include apigenin, apigenin-4′-methyl ether, scutellarein-6,7-dimethyl ether, and others.[9]

## PHARMACOLOGY OR BIOLOGICAL ACTIVITIES

Summer savory oil has antimicrobial (fungi and bacteria) activities. It is also reported to have spasmolytic effects on isolated smooth muscles. When applied undiluted to the backs of hairless mice it was lethal to half of the animals in 48 hours; it was also strongly irritating to rabbit and guinea pig skin, though it was nonphototoxic to hairless mice and swine. In diluted form, it was nonirritating and nonsensitizing to human skin.[2]

Winter savory has diuretic activity in rats.[10]

## USES

**Food.** Summer savory is quite extensively used as a flavor component in baked goods, meat and meat products (e.g., canned meats), condiments and relishes (e.g., spice blends), processed vegetables, soups, and gravies, among others, with highest average maximum use level of 0.519% reported in condiments and relishes.

Summer savory oil and oleoresin are used in candy, baked goods, meat and meat products (e.g., canned meats), condiments and relishes, and others, with highest average maximum use level of about 0.036% reported for the oil in meat and meat products (358 ppm) and in condiments and relishes (373 ppm).

Winter savory oil and oleoresin are not widely used. Food products in which they are used include candy, baked goods, meat and meat products, and condiments and relishes. Highest average maximum use level reported is about 0.013% (127 ppm for the oleoresin in condiments and relishes.

**Traditional Medicine.** Both summer savory and winter savory are used as tonic, carminative, astringent, and expectorant in treating stomach and intestinal disorders (e.g., cramps, nausea, and indigestion), diarrhea, and sore throat, generally in the form of a tea. Fresh summer savory is also used for insect bites (e.g., bee sting); for this purpose it is rubbed on the affected area.

Savory has been used as an aphrodisiac.

## COMMERCIAL PREPARATIONS

Crude, oleoresin, and oil (mostly summer savory). Summer savory oil is official in F.C.C.

*Regulatory Status.* GRAS (§182.10 and §182.20).

## REFERENCES


See the General References for ARCTANDER; BAILEY 1; FEMA; FERNALD; GUENTHER; LUST; ROSE; ROSENGARTEN; UPHOF.

1. M. S. Karawya et al., *Amer. Perfum. Cosmet.*, **85**, 23 (1970).
2. D. L. J. Opdyke, *Food Cosmet. Toxicol.*, **14**(Suppl.), 859 (1976).
3. A. Herisset et al., *Plant Med. Phytother.*, **8**, 287 (1974).
4. B. M. Lawrence, *Perfum. Flav.*, **13**, 1, 46 (1988).
5. K. Herrmann, *Z. Lebensm.-Unters. Forsch.*, **116**, 224 (1962).
6. B. Srepel, *Acta Pharm. Jugosl.*, **4**, 167 (1974); through *Chem. Abstr.*, **82**, 47673e (1975).
7. M. Paulet and D. Felisaz, *Riv. Ital. Essenze, Profumi, Piante Offic., Aromi, Saponi, Cosmet, Aerosol*, **53**, 618 (1971); through *Chem. Abstr.*, **76**, 131366r (1972).
8. J. Susplugas et al., *Trav. Soc. Pharm. Montpellier*, **29**, 129 (1969); through *Chem Abstr.*, **72**, 63643u (1970).
9. E. Wollenweber and K. M. Valant-Vetschera, *Fitoterapia*, **62**, 5, 462 (1991).
10. G. Stanic and I. Samarzija, *Phytother. Res.*, **7**(5), 363 (1993).

# SAW PALMETTO

**Source:** ***Serenoa repens*** (Bart.) Small. (syn. *Sabal serrulata* Schultes and Schultes) (Family Palmaceae).

*Synonym*. Sabal.

## GENERAL DESCRIPTION

Shrub, 3–4 m tall; leaves palmate, without continuing rib, divided into lance-shaped linear-lanceolate leaflets, to 2.5 dm long; petioles armed with spiny teeth; inflorescence many-branched, less than 1 m, substending the leaves, with white flowers; fruit a prominent olivelike mesocarp, 16–25 mm long, yellowish turning blue-black when ripe, with single large oblong seed to 18 mm long (17.7% of dry fruit weight);[1] ripening September through December; forming large colonies in southeastern U.S. in coastal plain from South Carolina, to south Mississippi and throughout Florida (FOSTER AND DUKE).

The part used is the fruit.

## CHEMICAL COMPOSITION

Fruits reported to contain 1–2% essential oil (formed by reaction of alcohols and acids during distillation); fixed oil with 75% free fatty acids and 25% neutral substances including free or esterified sterols or esters of fatty acids with alcohols; acids including caproic, caprylic (ca. 1.3%), caprinic (ca. 1.8%), lauric (ca. 24%), linoleic (ca. 3.6%), linolenic, myristic (ca. 11.6%), oleic (ca. 33.2%), palmitic (ca. 8.7%), and stearic acids;[2] anthranilic acid in fruit alcohol extracts;[3,4] sterols including $\beta$-sitosterol,[5] $\beta$-sitosterol 3-*O*-$\beta$-D-glucoside, campesterol, stigmasterol, lupeol, 24-methylene-cycloartanol, hexacosanol, and 1-octacosanol;[2] alcohols including farnesol, and phytol;[6] polysaccharides S1, S2, S3, and S4 with varying ratios of glucose, galactose, mannose, fucose, arabinose, rhamnose, and glucuronic acid.[7–11]

## PHARMACOLOGY OR BIOLOGICAL ACTIVITIES

Fruits are considered to have diuretic, sedative, antiandrogenic, antiinflammatory, antiexudative, and both antiestrogenic and estrogenic effects.

In mice experiments intraperitoneal administration of partially purified $\beta$-sitosterol fractions of the fruit extract showed significant estrogenic activity;[5] an antiestrogenic effect has also been demonstrated.[12]

A fruit extract has been shown to selectively antagonize and prevent binding to 52% of dihydrotestosterone receptors in the prostate.[13] Oral administration of a fruit extract in mice and rats has demonstrated inhibition of up to 90% of the activity of prostate 5-$\alpha$-reductase (which transforms testosterone into metabolites increasing the size of the prostate).[14]

A number of double-blind, placebo controlled clinical studies published in recent years (involving over 600 patients) suggest that lipoid hexane extracts or supercritical $CO_2$ extracts improve objective and subjective symptoms, including dysuria, nocturia, and frequent and poor urinary flow in benign prostatic hypertrophy.[15–21]

## USES

**Medicinal, Pharmaceutical, and Cosmetic.** In European phytomedicine cut sifted, or powdered crude fruit in galenical preparations; hexane, or ethanol extracts are used in antitestosterone and antiexudative preparations for treatment of difficulty in micturition in benign prostatic hypertrophy, stages I and II. The drug only relieves symptoms associated with prostrate enlargement, without reducing enlargement.[22]

**Food.** Historically used as a food source by indigenous groups of Florida; survival food of early European settlers (FOSTER AND DUKE).

**Health Food/Herb Teas.** Fruits in tea, capsules, tablets, tinctures, and other product forms as a dietary supplement, primarily as an endocrine and anabolic agent; also for prostate enlargement (FOSTER AND DUKE).

**Traditional Medicine.** Fruit traditionally considered expectorant, sedative, diuretic; used to treat prostate enlargement and inflammation; also used for colds, coughs, irritated mucous membranes, sore throat, asthma, chronic bronchitis, head colds, and migraine; suppository of powdered fruits used as a uterine and vaginal tonic; also a folk cancer remedy (FOSTER AND DUKE).

**Others.** Leaf wax investigated as a potential valuable material for wax-consuming industries, but was found to be of less value than the principal hard vegetable waxes currently available.[23]

## COMMERCIAL PREPARATIONS

Crude, ethanol, hexane, and supercritical $CO_2$ extracts.

*Regulatory Status.* Undetermined in the U.S. The fruit is the subject of a positive German therapeutic monograph, indicated for micturition trouble in benign prostatic hypertrophy stages I and II.[22]

## REFERENCES

See the General References for FOSTER AND DUKE; MARTINDALE; STEINMETZ; TYLER; WEISS; WREN.

1. J. B. Hilmon, Autecology of Saw Palmetto, Ph.D. Dissertation, Duke University, 1968.
2. W. Breu, *Arzneim. Forsch.*, **42**, 547 (1992).
3. R. Hansel et al., *Planta Med.*, **12**, 169 (1964).
4. R. Hansel et al., *Planta Med.*, **14**, 261 (1966).
5. M. I. Elghamry and R. Hänsel, *Experientia*, **25**, 8, 828 (1969).
6. G. Jommi et al., *Gazzetta Chim. Ital.*, **118**, 823 (1988).
7. G. Harnischfeger, *Z. Phyotherapie*, **10**, 71 (1989).
8. P. Hatinguais et al., *Traveux Soc. Pharmacie Montpellier*, **41**, 253 (1981).
9. H. Wagner et al., *Planta Med.*, **41**, 244 (1981).
10. H. Wagner et al., *Arzneim. Forsch.*, **34**, 659 (1984).
11. H. Wagner et al., *Arzneim. Forsch.*, **35**, 1069 (1985).
12. F. Di Silverio et al., *Eur. Urol.*, **21**, 309 (1992).
13. C. Sultan et al., *J. Steroid Biochem.*, **20**, 515 (1984).
14. A. Stenger et al., *Gaz. Med. De France*, **89**, 2041 (1982).
15. E. Emili et al., *Urologia*, **50**, 1040 (1983).
16. C. Boccafoschi et al., *Urologia*, **50**, 1257 (1983).
17. G. Champault et al., *Br. J. Clin. Pharm.*, **18**, 461 (1984).
18. A. Tasca et al., *Minerva Urol., Nefrol.*, **37**, 87 (1985).
19. P. Cukier et al., *C. R. Ther. Pharmacol. Clin.*, **4**, 15 (1985).
20. C. Casarosa et al., *Clin. Ther.*, **10**, 5, 585 (1988).
21. S. M. Adriazola et al., *Arch. Esp. Urol.*, **45**, 3, 211 (1992).

22. Monograph *Sabal Fructus*, *Bundesanzeiger*, no. 43 (Mar. 2, 1989); revised (Feb. 1, 1990; Jan. 17, 1991).

23. E. A. Wilder and E. D. Kitzke, *Science*, **120**, 108 (1954).

# SCHISANDRA

**Source:** ***Schisandra chinensis*** (Turcz.) Baill., ***S. sphenanthera*** Reid. et Wils., and other ***Schisandra*** species (Family Schisandraceae).

*Synonyms*. *Wuweizi*, meaning "five-flavor seed" (general term for all varieties); northern schisandra or *beiwuweizi* (*S. chinensis*); southern schisandra or *nanwuweizi* and western schisandra or *xiwuweizi* (*S. sphenanthera*); and gomishi.

## GENERAL DESCRIPTION

*Schisandra chinensis* is a deciduous woody vine, up to 8 m long; berries bright red when mature; native to northern and northeastern China and adjacent regions of Russia and Korea. Part used is the fully ripe, sun-dried fruit which yields northern schisandra; it is oval and wrinkled, with a diameter of 5 to 8 mm, ranging from bright red, dull red, to purplish red; flesh is soft, with a weak characteristic odor and tastes primarily sour and sweet, with a salty note; its 1 to 2 yellowish brown, kidney-shaped seeds are fragrant when crushed and taste simultaneously pungent, bitter and salty; these five flavor elements give schisandra its name, "five-flavor seed." Northern schisandra is mainly produced in northern and northeastern provinces, including Inner Mongolia, Ningxia, Shanxi, Hebei, Liaoning, Jilin, and Heilongjiang.

*Schisandra sphenanthera* is a climbing shrub similar to *S. chinensis*, up to about 5 m long; native to western, central and southern China. Its fully ripe, sun-dried fruit yields southern or western schisandra, which is similar in properties to northern schisandra fruit, but is smaller, with thinner flesh and is reddish brown to dull brown. Southern schisandra is mainly produced in western, central and southern provinces, including Gansu, Shaanxi, Henan, Hunan, Hubei, and Sichuan.

Although several other *Schisandra* species also serve as commercial or potential commercial sources of schisandra fruit, its current major sources are *S. chinensis* and *S. sphenanthera*.[1,2]

## CHEMICAL COMPOSITION

*Schisandra chinensis* fruit contains roughly 1–3% volatile oil (composed mainly of citral, sesquicarene, $\beta$-2-bisabolene, $\alpha$-ylangene, $\beta$-chamigrene, $\alpha$-chamigrene and chamigrenal), 12% citric acid, 10% malic acid, small amounts of tartaric acid, monosaccharides, resins, pectin, vitamins A, C and E, phospholipids, sterols, tannins, etc.[3,4]

Seeds of *S. chinensis* contain about 19% lignans (those of *S. sphenanthera* about 10%) which are the major active principles.[1] Extensive chemical studies have been performed on these schisandra lignans during the past 30 years, primarily by Russian, Japanese and Chinese researchers.[3,5] However, their nomenclature is extremely confusing: Russian researchers named them schizandrins and schizandrols while Japanese researchers called them gomisins and the Chinese called them wuweizisus and wuweizi esters, among others.[5–10] Thus far, over 40 such lignans have been isolated from the nonsaponifiable fraction of the seed oil. The more important ones include: schizandrin (= schizandrol A), schizandrin A (= deoxyschizandrin), schizandrin B (= gamma-schizandrin), schizandrin C, schizandrol B, schisantherin D, schizandrer A, schizandrer B (= schisantherin), pseudo-γ-schizandrin, wuweizisu A (= schizandrin A), wuweizisu B, wuweizisu C, gomisins A to R and derivatives, wuweizi ester A (= gomisin C, schisantherin A, schizandrer A), wuweizi ester B (= gomisin B, schisantherin B, schizandrer B), wuweichun A (= schizandrin), wuweichun B, and others.[1,3,5–11]

Other lignans isolated from seeds of other *Schisandra* species (e.g., *S. sphenanthera*, *S. henryi* Clarke, and/or *S. rubriflora* Rhed. et Wils.) include: schisanhenol, schisandrone, epischisandrone, enshicine, epienshicine, wulignan A1, wulignan A2, epiwulignan A1, (−)-rubschisandrin, rubschisantherin, schisanhenol acetate, schisanhenol B, and others.[5,11–14]

## PHARMACOLOGY OR BIOLOGICAL ACTIVITIES

Schisandra fruit exhibits a wide variety of pharmacological activities in humans and in laboratory animals, including: "adaptogenic" properties (increasing nonspecific resistance) similar to those of ginseng and Siberian ginseng, but with weaker effects and lower toxicities;[15] stimulating the central nervous system and improving the reflexes, endurance and work performance of healthy individuals;[16] tranquilizing and anticonvulsive effects in rodents, with schizandrol A as an active agent;[17] antidepressant effect in mice;[18] antifatigue effects in rodents and horses and improving markedly the performance of race horses;[19] heart stimulating, vasodilating and blood-pressure normalizing, especially in circulatory failure;[20] promoting thymidine incorporation in DNA synthesis in human lymphocytes, and enhancement as well as suppression of immune functions;[20,21] stimulating rabbit uterus *in vivo* and *in vitro*;[20,22] stimulating respiration in animals, with lignans as active agents; antitussive and expectorant effects in mice; antibacterial *in vitro*;[22,23] strongly anti-oxidant;[24,25] and others (JIANGSU; JILIN; WANG).

The most extensively studied effect of schisandra fruit is its ability to protect human and animal liver from toxins and diseases. Studies over the past 20 years have shown it to have the following antihepatotoxic effects: lowering serum alanine aminotransferase (SGPT) levels in humans and in experimental animals, and improving symptoms in patients with chronic hepatitis; protecting the liver from damage caused by carbon tetrachloride and other toxins; promoting the biosynthesis of serum and liver proteins; stimulating the formation of liver glycogen; induction of cytochrome P-450 of animal liver microsomes; inhibiting microsomal lipid peroxidation by liver toxins; inhibiting the covalent binding between liver toxins and microsomal lipids; lowering the mortality rate of animals poisoned by acetaminophen, digitoxin and indomethacin; increasing serum and liver cyclic AMP levels in mice; and others.[5,6,8,9,23,24,26–30] These antihepatotoxic effects are due mainly to the lignans present in the seed, especially wuweizisu C, schisantherin D, gomisin A, gomisin C, gomisin N, deoxygomisin A,[5,8] schizandrin A, schizandrin B, schizandrin C, schizandrol A, schizandrol B, schizandrer A, schizandrer B, and schisanhenol, among others.[6,9,27,29,30]

The acetone extract of schisandra fruit exhibited marked inhibition against the mutagenic effect of aflatoxin $B_1$ per the Ames test.[31]

Adverse side-effects of schisandra fruit are rare; they include stomach upset, decreased appetite, and urticaria (WANG).[32]

## USES

**Medicinal, Pharmaceutical, and Cosmetic.** Biphenyldimethyldicarboxylate (BDD), an intermediate of schizandrin C synthesis, is now used in China to treat viral hepatitis with much higher efficacy than silymarin.[6]

**Health Food/Herb Teas.** Extracts and crude powder are used in preparations marketed mainly for their antioxidant and energy-enhancing as well as adaptogenic effects. Crude is also used in soup mixes (FOSTER AND YUE).

**Traditional Medicine.** In traditional Chinese medicine, schisandra fruit is considered a lung astringent, and kidney and male tonic. It is used in treating cough, asthma, insomnia, neurasthenia, chronic diarrhea, night sweat, spontaneous sweating, involuntary seminal discharge, thirst, impotence, physical exhaustion and excessive urination. The official *Chinese Pharmacopoeia* lists its daily oral dose as 1.5 to 6 g.

The vines and roots of various *Schisandra* species (e.g., *S. sphenanthera* and *S. henryi*) are used in China to treat painful joints, rheumatism and traumatic injuries.

## COMMERCIAL PREPARATIONS

Crude and extracts; seeds must be crushed before extraction. Most extracts are expressed in weight to weight ratios with no uniform quality standards; some extracts are standardized to schizandrins content.

*Regulatory Status.* Schisandra is an ethnic food; U.S. regulatory status not determined.

## REFERENCES

See the General References for CHP; FOSTER AND YUE; HU; IMM-3; JIANGSU; JILIN; LU AND LI; NATIONAL; WANG; ZHU.

1. W. Z. Song and Y. Y. Tong, *Yaoxue Xuebao*, **18**, 138 (1983).
2. W. Z. Song, *Zhongyao Tongbao*, **13**(8), 3 (1988).
3. Y. S. Huang et al., *Chin. J. Pharmacol. Toxicol.*, **4**, 275 (1990).
4. Y. Ohta and Y. Hirose, *Tetrahedron Lett.*, **20**, 2483 (1968).
5. H. Hikino et at., *Planta Med.*, **50**, 213 (1984).
6. P. G. Xiao and K. J. Chen, *Phytother. Res.*, **2**, 55 (1988).
7. R. Tan et al., *Planta Med.*, **52**, 49 (1986).
8. Y. Kiso et al., *Planta Med.*, **51**, 331 (1985).
9. G. T. Liu in H. M. Chang et al., eds., *Advances in Chinese Medicinal Materials Research*, World Scientific Publishing Co., Singapore, 1985, p. 257.
10. Y. Ikeya et al., *Chem. Pharm. Bull.*, **30**, 3207 (1982).
11. H. B. Zhai and P. Z. Cong, *Yaoxue Xuebao*, **25**, 110 (1990).
12. L. N. Li and H. Xue, *Planta Med.*, **51**, 217 (1985).
13. J. S. Liu et al., *Huaxue Xuebao*, **46**, 483 (1988).
14. H. J. Wang and Y. Y. Chen, *Yaoxue Xuebao*, **20**, 832 (1985).
15. I. I. Brekhman and I. V. Dardymov, *Ann. Rev. Pharmacol.*, **419** (1969).
16. B. X. Wang, *Tianjin Yiyao Zazhi*, **7**, 338 (1965).
17. X. Y. Niu et al., *Yaoxue Xuebao*, **18**, 491 (1983).
18. J. L. Hancke et al., *Planta Med.*, **53**, Poster P85 (1987).
19. F. Ahumada et al., *Phytother. Res.*, **3**, 175 (1989).
20. K. Sun, *Yaoxue Xuebao*, **7**, 277 (1959).
21. M. Li and H. Liu, *Zhongcaoyao*, **14**(4), 31 (1983).
22. S. S. Liu, Ed., *Zhongyao Yanjiu Wenxian Zhaiyao* (1820–1961), Science Press, Beijing, 1975, p. 85.
23. S. S. Liu, Ed., *Zhongyao Yanjiu Wenxian Zhaiyao* (1962–1974), Science Press, Beijing, 1979, p. 108.
24. T. M. Zhang et al., *Acta Pharmacol. Sinica*, **10**, 353 (1989).

25. T. J. Lin et al., *Chinese J. Pharmacol. Toxicol.*, **3**, 153 (1989).
26. M. Li, *Zhongyao Tongbao*, **9**(4), 41 (1984).
27. G. T. Liu et al., *Chinese J. Integr. Trad. Western Med.*, **3**, 182 (1983).
28. H. Nagai et al., *Planta Med.*, **55**, 13 (1989).
29. G. T. Liu and H. L. Wei, *Acta Pharmacol. Sinica*, **6**, 41 (1985).
30. G. T. Liu et al., *Yaoxue Xuebao*, **15**, 206 (1980).
31. C. C. Ruan et al., *Chin. J. Cancer*, **8**, 29 (1989).
32. H. W. Song and F. X. Wang, *Zhongguo Zhongyao Zazhi*, **15**(4), 51 (1990).

# SENNA

**Source:** ***Cassia senna*** L. (syn. *C. acutifolia* Del.; *Senna alexandrina* Mill.) and ***C. angustifolia*** Vahl (syn. *Senna alexandrina* Mill.) (Family Leguminosae or Fabaceae).

*Synonyms*. Alexandrian senna and Khartoum senna (*C. senna*); Tinnevelly senna and Indian senna (*C. angustifolia*).

## GENERAL DESCRIPTION

Alexandrian senna (*C. senna*) is an herbaceous subshrub up to about 1 m high; native to the Nile region of northern Africa (Egypt, Sudan, etc.) and cultivated there as well as in tropical Asia (e.g., southern China and India).

Indian or Tinnevelly senna (*C. angustifolia*) is also an herbaceous subshrub up to about 1 m high; native to India and northeastern Africa; cultivated mainly in southern and northwestern India and in Pakistan.

Parts used are the dried leaflets (commonly called leaves) and pods. Alexandrian senna leaves and pods are reportedly derived mostly from wild plants, while Indian senna leaves are collected at three to five months after planting and pods one to two months hence.

The two species of senna are very closely related and have formerly been recognized as a single species, though later research data have established them as distinct species.[1,2] Recent taxonomic treatment once again merges *C. senna*, *C. acutifolia*, and *C. angustifolia*, into one taxonomic entity, *Senna alexandrina* Mill., which is not generally recognized in the trade.

Most of the senna currently used in the United States is Indian senna (Tinnevelly senna).

## CHEMICAL COMPOSITION

Alexandrian and Indian senna leaves have similar chemical compositions, especially in their anthracene derivatives. They contain as their active constituents dianthrone glucosides (usually 1.5–3.0%), consisting mostly of sennosides A and B (rhein-dianthrone glucosides), with minor amounts of sennosides C and D (rhein-aloe-emodin-heterodianthrone glucosides) and aloe-emodin-dianthrone glucoside also present.[3–9] These dianthrone glycosides are reportedly absent in fresh leaves, and it appears that they are formed during the drying process through enzymatic oxidation of monoanthrone glycosides that are present in fresh leaves but normally absent in dried leaves (also see ***cascara***).[4,10]

There is also evidence of the existence of primary glycosides of the sennosides (with additional sugar molecules) that are more active than the sennosides.[5]

Senna leaves also contain small amounts of free anthraquinones (rhein, aloe-emodin, chrysophanol, etc.) and their *O*-glycosides and *C*-glycosides.[1,4–7,10,11]

Alexandrian senna leaves generally contain higher sennosides content than Indian senna leaves.

Other constituents present in senna leaves include free sugars (glucose, fructose, sucrose, and pinitol), a mucilage (consisting of galactose, arabinose, rhamnose, and galacturonic acid), and polysaccharides (in *C. angustifolia*), whose chemical structures have recently been determined;[12,13] flavonoids (isorhamnetin, kaempferol, etc.); a trace of volatile oil; and resins; among others (LIST AND HÖRHAMMER).

Senna pods contain normally 2 to 5% sennosides, with Alexandrian pods having higher values than Indian pods. In addition to sennosides A and B, a closely related glucoside, named sennoside $A_1$, has been isolated from Alexandrian senna pods.[14]

A galactomannon consisting of D-glucose and D-mannose in a 3:2 molar ratio has recently been isolated from the seeds of Indian senna.[15]

## PHARMACOLOGY OR BIOLOGICAL ACTIVITIES

Sennosides are cathartic, with a similar mode of action as cascarosides (see ***cascara***).

Sennosides A and C have equal purgative potency in mice. But sennoside C has potentiating effects on the activity of sennoside A, exerting a potentiating effect of about 1.6 when 20% of the dose of sennoside A is replaced by sennoside C.[16]

Senna products along with cascara products are generally considered the drugs of choice among anthraquinone cathartics and are also generally considered safe (APhA).[17,18] Excessive or prolonged use of senna, as with laxatives in general, may lead to colon damage and other problems (APhA, MARTINDALE).[19] However, given in doses sufficient to produce a motion of physiological water content, senna can be safely administered, even over a long period of time. Stimulating laxatives should not be used more than 1–2 weeks without medical advice.[20] Daily treatment, as with any laxative is not recommended. Senna does not induce specific lesions in the nerve plexus of the intestinal wall, and when used rationally, does not lead to electrolyte losses or habituation.[21] Chronic abuse can disturb electrolyte balance, leading to potassium deficiency, heart dysfunction, and muscular weakness, especially under concomitant use of heart-affecting glycosides, thiazide diuretics, corticoadrenal steroids, and licorice root.[20]

One of the polysaccharides exhibited a significant inhibitory effect against solid Sarcoma-180 in CD1 mice.[13]

## USES

**Medicinal, Pharmaceutical, and Cosmetic.** Senna leaves, pods, their extracts, and sennosides are extensively used as active ingredients in laxative preparations (syrups, tablets, etc.). Often termed a "bowel irritant" or "stimulant," recent scientific consensus suggests such terms should be avoided, in favor of a more specific characterization as a prokinetic agent with a secretory component.[21]

**Health Food/Herb Teas.** Senna leaves or pods used in laxative formulations, usually in tablet, capsule or tea-bag form.

**Traditional Medicine.** Senna leaves have been used for centuries in both western and eastern cultures as a laxative, usually taken as a tea or swallowed in powdered form.

## COMMERCIAL PREPARATIONS

Crudes (leaves and pods), extracts (solid, fluid, etc.), and sennosides A and B (15 to 100%). Senna leaf (both Indian and Alexandrian), its fluid extract, and sennosides A and B are official in U.S.P. Commercial grades of leaves are graded according to their size and shape (entire, broken, etc.), with whole leaves being the most expensive, yet they may often contain less sennosides than the lower grades (e.g., broken leaves). Strengths (see ***glossary***) of extracts are based on weight-to-weight ratios, though more and more manufacturers are using sennosides contents as inhouse standards or guides.

*Regulatory Status.* Only Alexandrian senna (leaf and pod) has been approved for food used (§172.510). Senna leaves, pods, and their preparations are subjects of a positive German therapeutic monograph.[9,20,22]

## REFERENCES


See the General References for BLUMENTHAL; CLAUS; JIANGSU; LIST AND HÖRHAMMER; LUST; MORTON 3; NANJING; TUCKER; WREN; YOUNGKEN.

1. J. W. Fairbairn and A. B. Shrestha, *Lloydia*, **30**, 67 (1967).
2. J. Lemli et al., *Planta Med.*, **49**, 36 (1983).
3. A. Stoll and B. Becker in L. Zechmeister, ed., *Fortschritte der Chemie organischer Naturstoffe*, Vol. 7, Springer-Verlag, Vienna, Austria, 1950, p. 248.
4. J. W. Fairbairn, *Lloydia*, **27**, 79 (1964).
5. F. H. L. van Os, *Pharmacology*, **14**(Suppl. 1), 7 (1976).
6. H. Friedrich and S. Baier, *Planta Med.*, **23**, 74 (1973).
7. W. D. Brendel and D. Schneider, *Planta Med.*, **25**, 342 (1974).
8. R. Atzorn et al., *Planta Med.*, **41**, 1 (1981).
9. Monograph *Sennae*, *Bundesanzeiger*, no. 228 (Dec. 5, 1984).
10. J. Lemli and J. Cuveele, *Phytochemistry*, **14**, 1397 (1975).
11. S. C. Y. Su and N. M. Ferguson, *J. Pharm. Sci.*, **62**, 899 (1973).
12. J. Lemli and J. Cuveele, *Plant. Med. Phytother.*, **10**, 175 (1976).
13. B. M. Müller et al., *Planta Med.*, **55**, 536 (1989).
14. B. Christ et al., *Arzneim.-Forsch.*, **28**, 225 (1978).
15. N. Alam and C. Gupta, *Planta Med.*, **52**, 308 (1986).
16. K. Kisa et al., *Planta Med.*, **42**, 302 (1981).
17. Anon., *Fed. Regist.*, **40**(56), 10902 (1975).
18. J. W. Fairbairn, ed., *The Anthraquinone Laxatives*, S. Karger, Basel, Switzerland, 1976, in *Pharmacology*, **14**(Suppl. 1), 1 (1976).
19. B. Smith, *The Neuropathology of the Alimentary Tract*, Arnold, London, 1972.
20. Monograph *Sennae folium*, *Bundesanzeiger* (July 21, 1993).
21. E. Leng-Peschlow, Ed., *Intl. J. of Exp. Clin. Pharmacol.*, **44**, S1, (1992).
22. Monograph *Sennae fructus*, *Bundesanzeiger* (July 21, 1993).


# SOUR JUJUBE KERNEL (see also JUJUBE, COMMON)

**Source:** ***Ziziphus spinosa*** Hu and ***Z. jujuba*** Mill. var. ***spinosa*** (Bge.) Hu ex H.F. Chow (Family Rhamnaceae). Also listed as sources/synonyms: *Z. jujuba* Mill.; *Z. jujuba* Mill. var. *spinosus* Bge. [*Z. spinosa* (Bge.) Hu].

*Synonyms.* Sour date kernel, spiny jujube kernel, *suan zao ren*, *zao ren* and *shan zao ren*.

## GENERAL DESCRIPTION

Deciduous shrub, 1 to 3 m high, bearing two types of spines, one sturdy and erect and the other small and curved; fruit (drupe) almost round, 1 to 1.4 cm in diameter, dull reddish brown when ripe, with thin flesh, and tastes sour; grows in northern and central China. Part used is the seed. Fruits are wildcrafted when they ripen and turn red in autumn, soaked overnight and rid of flesh; the pits are then crushed open and the seeds collected and dried. The better grades are full and deep red or purplish brown in appearance and contain less than 5% impurities (e.g., husks). Major producing provinces include Hebei, Shaanxi, Liaoning and Henan.

## CHEMICAL COMPOSITION

Contains flavonoids that include swertisin, spinosin (2″-β-*O*-glucopyranosyl swertisin) and zivulgarin (4″-β-*O*-glucopyranosyl swertisin); triterpenes and triterpene saponin glycosides, including betulin, betulinic acid, jujubosides A, B, $B_1$ and jujubogenin, ceanothic acid, alphitolic acid; numerous alkaloids, including aporphines, benzylisoquinolines and 14-member cyclopeptides (e.g., zizyphusine, nuciferine, nornuciferine, coclaurine, norisocorydine, caaverine, *N*-methylasimilobine, sanjoinines A, B, D, F, G1, sanjoinenine, etc.) (STEINER); ferulic acid; β-carotene; sterols (daucosterol); fatty oil (ca. 32%, composed of 42% oleic and 46% linoleic acids and lesser amount of linolenic acid);[1] and cyclic AMP and cyclic GMP, among others (IMM-3).[2-6]

## PHARMACOLOGY OR BIOLOGICAL ACTIVITIES

*Suan zao ren* is probably the best known and most frequently used Chinese herbal sedative. Modern studies have shown it to have strong sedative and hypnotic effects in humans and in experimental animals (mice, rats, guinea pigs, cats, rabbits, and dogs),[3] with the flavonoid glycosides (spinosin, swertisin and zivulgarin), alkaloids and saponins (jujubosides A and B) being the active principles (STEINER).[7-10]

An ethanolic extract of *suan zao ren* (per os) markedly increased feeding and weight gain as well as immune response in mice. Its polysaccharides (per os) also exhibited immunopotentiating and radiation-protective effects in mice.[11,12]

Its water extractives had antiarrhythmic activities in experimental animals; and its total saponins had protective effects on cultured rat myocardial cells.[13,14]

Oral administration of the fatty oil to quails (2.5 mL/kg/day) for 53 days not only markedly reduced their serum lipids but also significantly inhibited blood platelet aggregation.[1]

Other pharmacologic effects of *suan zao ren* include: analgesic, antipyretic, antispasmodic, hypotensive, uterine stimulant, and others.[3]

Toxicities are very low: 150 g/kg crude drug or 50 g/kg decoction fed to mice did not produce any toxic reactions; and no adverse side effects have been reported from clinical use of this drug. However, since it is known to be a uterine stimulant, caution is advised in pregnant women (JIANGSU, WANG).[3]

## USES

**Medicinal, Pharmaceutical, and Cosmetic.** Its water and/or hydro-alcoholic extracts are used along with Dahurian angelica in freckle-removal creams and lotions; also used in other skin-care products (sunscreen, acne, deodorant, antiperspirant etc.) for its traditional skin-protectant, antiperspirant, calming and nourishing properties (ZHOU).

**Health Food/Herb Teas.** Powder and extract used in herbal sedative formulas in tea, capsule or tablet form.

**Traditional Medicine.** Sour jujube was first recorded over 2000 years ago and the fruit was listed in the *Shen Nong Ben Cao Jing* (ca. 200 B.C.–A.D. 100) as a tonic in the superior category. Use of the seed, *suan zao ren*, however, was not described until the 3rd century A.D.[15] Traditionally regarded as sweet and sour tasting; neutral; heart nourishing and tranquilizing (*yang xin an shen*); promoting the secretion of body fluids; and stopping excessive perspiration. It is one of the major traditional Chinese brain tonics for treating neurasthenia, insomnia, excessive dreams (nightmares), night sweat, forgetfulness and palpitations. Traditional medical writings distinguish between raw and stir-fried *suan zao ren*: the former for treating sleepiness while the latter for treating insomnia (Li Shi-Zhen's *Ben Cao Gang Mu*, ca. 1590 A.D.). However, modern chemical studies have not confirmed the validity of this distinction.[9,16,17]

*Suan zao ren* is normally used in the crushed or powdered form.

## COMMERCIAL PREPARATIONS

Crude (raw and stir-fried) in whole or powdered form; whole seeds must be crushed or powdered before extraction.

*Regulatory Status*. Not determined.

## REFERENCES

See the General References for CHP; FOSTER AND YUE; IMM-3; JIANGSU; LU AND LI; NATIONAL; STEINER; XIAO 4,6; ZHOU; ZHU.

1. S. X. Wu et al., *Zhongguo Zhongyao Zazhi*, **16**, 435 (1991).
2. K. H. Shin et al., *Planta Med.*, **44**, 94 (1982).
3. G. X. Hong and B. Cao, *Zhongyao Tongbao*, **12**(8), 51 (1987).
4. L. Zeng et al., *Acta Botanica Sinica*, **28**, 517 (1986).
5. S. Y. Li and R.Y. Zhang, *Zhongyao Tongbao*, **11**(11), 43 (1986).
6. L. Zeng et al., *Yaoxue Xuebao*, **22**, 114 (1987).
7. W. S. Woo et al., *Phytochemistry*, **18**, 353 (1979).
8. I. Watanabe et al., *Jpn. J. Pharmacol.*, **23**, 563 (1973).
9. J. Wang, *Zhongchengyao*, **11**(1), 19 (1989).
10. C. L. Yuan et al., *Zhongyao Tongbao*, **12**(9), 34 (1987).
11. X. C. Lang et al., *Zhongyao Tongbao*, **13**(11), 43 (1988).
12. X. C. Lang et al., *Zhongguo Zhongyao Zazhi*, **16**, 366 (1991).
13. S. Y. Xu et al., *Zhongcaoyao*, **18**(12), 18 (1987).
14. X. J. Chen et al., *Acta Pharmacol. Sinica*, **11**, 153 (1990).
15. F. X. Liu et al., *Zhongguo Zhongyao Zazhi*, **16**, 444 (1991).
16. F. X. Liu and J. F. Gao, *Zhongguo Zhongyao Zazhi*,**15**(5), 28 (1990).
17. H. Jin and Y. M. Xu, *Liaoning Zhongyi Zazhi*, **20**(2), 40 (1993).

# SQUILL

**Source:** ***Urginea maritima*** (L.) Baker (syn. *U. Scilla* Steinh., *Scilla maritima* L., *Drimia maritima* (L.) Stearn) and ***U. indica*** (Roxb.) Kunth. (syn. *Scilla indica* L., *Drimia indica* (L.) Stearn) (Family Liliaceae).

*Synonyms*. Scilla, sea onion, sea squill, Mediterranean squill, red squill, white squill (*U. maritima*); Indian squill, white squill (*U. indica*).

## GENERAL DESCRIPTION

*Urginea maritima* is a bulbous perennial

herb belonging to the lily family; native to the Mediterranean region. Parts used are the dried fleshy inner scales of the bulb. White squill is derived from the white variety, while red squill is from its red variety. *U. indica* is used as an acceptable substitute for *U. maritima*. Recent taxonomic treatments placed the taxa in the genus *Drimia*, however, the genus name *Urginea* persists in the trade and scientific literature (BRADLY).

*Urginea maritima* is a species complex, now consisting of six or more species, with both genetic and phytochemical differences. *U. maritima sensu stricto*, native to the Iberian peninsula, a hexaploid ($2n = 60$) has the highest bufadienolide content (BRADLY).[1]

## CHEMICAL COMPOSITION

White squill contains as active constituents several steroid glycosides (bufadienolides), including scillaren A (scillarenin + rhamnose + glucose), glucoscillaren A (scillaren A + glucose), proscillaridin A (scillarenin + rhamnose), scillaridin A, scillicyanoside, scilliglucoside, scilliphaeoside (12β-hydroxyproscillaridin A), and glucoscilliphaeoside (12β-hydroxyscillaren),[2–5] with the most important ones being scillaren A and proscillaridin A.[3,5] Scillaren B has been used to describe a mixture of squill glycosides as opposed to pure scillaren A.

Other constituents present in white squill include flavonoids (vitexin, isovitexin, orientin, isoorientin, scoparin, vicenin-2, quercetin, dihydroquercetin or taxifolin, dihydroquercetin-4′-monoglucoside, etc.), stigmasterol, scilliglaucosidin, mucilage (glucogalactans) and others (KARRER, MERCK).[4–8]

Red squill contains scilliroside; it also contains cardiac glycosides as white squill (KARRER, MARTINDALE).[9]

Indian squill has same types of active bufadienolides as the Mediterranean squill, but with contents of proscillaridin A and scillaren A in some cytotypes (diploid and tetraploid races) considerably higher than the latter.[10]

## PHARMACOLOGY OR BIOLOGICAL ACTIVITIES

The glycosides present in squill have digitalis-like cardiotonic properties which are due to their aglycones (GOODMAN AND GILMAN).[11] Action is more rapid, but shorter than that of digitalis glycosides.[12,13]

White squill also reportedly has expectorant, emetic, and diuretic properties (MARTINDALE, MERCK).

Methanol extracts of red squill have been claimed to be effective as hair tonics in treating chronic seborrhea and dandruff, with the active principle being attributed to scilliroside.[14]

Red squill is very toxic to rats, the active principle being scilliroside. It is also extremely irritating to the skin, and handling with rubber gloves is strongly recommended (MARTINDALE, REMINGTON).

## USES

**Medicinal, Pharmaceutical, and Cosmetic.** White squill is used mainly as an expectorant in some cough preparations. In German phytomedicine used for mild cardiac insufficiency and impaired kidney function.[12]

**Traditional Medicine.** Has been used for centuries in Europe as a diuretic, emetic, expectorant, and cardiac stimulant. It has also been used in cancers.[15]

**Others.** The primary use of red squill is as an effective rat poison. However it is generally not effective with mice. It is quite safe with humans and animals such as pets (cats and dogs) and domestic animals (e.g., hogs) because it causes vomiting in these animals promptly, thus not allowing enough poison to be ingested (REMINGTON).

**COMMERCIAL PREPARATIONS**

Crude and extracts; crude, tincture, and fluid extract were formerly official in N.F. Strengths (see ***glossary***) of extracts are expressed in weight-to-weight ratios.

*Regulatory Status*. Subject of a German therapeutic monograph, indicated for diminished kidney capacity and mild cardiac insufficiency (contraindicated with digitalis glycosides or potassium deficiency).[12]

**REFERENCES**

See the General References for BLUMENTHAL; BRADLY; CLAUS; GOSSELIN; LEWIS AND ELVIN-LEWIS; MARTINDALE; TERRELL; TUCKER; UPHOF; YOUNGKEN.

1. L. Krenn et al., *Planta Med.*, **57**, 560 (1991).
2. H. Lichti et al., *Helv. Chim. Acta*, **56**, 2088 (1973).
3. M. S. Karawya et al., *Planta Med.*, **23**, 213 (1973).
4. A. von Wartburg et al., *Helv. Chim. Acta*, **51**, 1317 (1968).
5. P. Garcia Casado et al., *Pharm. Acta Helv.*, **52**, 218 (1977).
6. F. S. Hakim and F. J. Evans, *Pharm. Acta Helv.*, **51**, 117 (1976).
7. M. Fernandez et al., *Phytochemistry*, **14**, 586 (1975).
8. M. Fernandez et al., *Galencia Acta*, **24**, 45 (1971); through *Chem. Abstr.*, **78**, 26440d (1973).
9. A. von Wartburg, *Helv. Chim. Acta*, **49**, 30 (1966).
10. S. Jha and S. Sen, *Planta Med.*, **47**, 43 (1983).
11. C. L. Gemmill, *Bull. N.Y. Acad. Med.*, **50**, 747 (1974).
12. Monograph *Scillae bulbus*, *Bundesanzeiger*, no. 154 (Aug. 21, 1985); corrected (Mar. 2, 1989).
13. W. E. Court, *Pharm. J.*, **235**, 194 (1985).
14. S. Giannopoulos, Ger. Offen., 2,715,214 (1977); through *Chem. Abstr.*, **88**, 55079e (1978).
15. J. L. Hartwell, *Lloydia*, **33**, 97 (1970).

# STEVIA

**Source:** ***Stevia rebaudiana*** (Bertoni) Bertoni (syn. *Eupatorium rebaudianum* Bertoni) (Family Compositae or Asteraceae).

*Synonym*. Sweet herb.

**GENERAL DESCRIPTION**

Stevia is an herbaceous perennial to 80 cm high, leaves opposite, toothed; indigenous to highlands in the Amambay and Iguaçu districts of the border area of Brazil and Paraguay;[1] commercially produced in Paraguay, Brazil, Japan, Korea, Thailand and China. Much of the Asian production is for export to Japan, the largest consumer.

**CHEMICAL COMPOSITION**

The main constituents of the leaves responsible for sweetness are the ent-kaurene diterpene glycosides stevioside (2.2–18.5%), rebaudiosides A and C, and dulcoside A; rebaudiosides B, D and E and steviolbioside are of much less importance. Labdane diterpenes include jhanol, austroinulin, 6-*O*-acetylaustroinulin; triterpenes include $\beta$-amyrin acetate and lupeol; other constituents include $\beta$-sitosterol, stigmasterol, tannins, and a volatile oil (0.12–

0.43%) from which at least 25 compounds have been identified.[2]

## PHARMACOLOGY OR BIOLOGICAL ACTIVITIES

Stevioside is about 300 times sweeter than sucrose at 0.4% sucrose concentration, 150 times sweeter at 4% sucrose, and 100 times sweeter at 10% sucrose concentration.[2]

Given interest in the plant and its glycosides as sweeteners, most biological testing has been to assess toxicity. Stevioside has been found to be nontoxic in acute toxicity tests with rabbits, guinea pigs and fowl, being excreted without structural modification. A leaf extract (with 50% stevioside) administered i.p. to rats has an $LD_{50}$ of 3.4 g/kg. Separate 2.0 g/kg doses of stevioside, rebaudiosides A–C, steviolbioside, or dulcoside A, administered to mice by oral intubation showed no acute toxicity. Two weeks after administration no significant difference in body or organ weights were reported.[2]

Subacute toxicity studies on rats over a 50-day period up to 7.0% concentration of stevioside in feed produced no remarkable toxic effects.[2]

Studies on the effect of stevioside on fertility found that it produced no abnormal mating performance or fertility, and produced no teratogenic effects. Various tests have not shown mutagenic or genotoxic activity, however mutagenic activity of metabolized steviol has been reported.[2]

Other reported activities include claims of hypoglycemic activity of leaf extracts and stevioside in animals or humans; however, follow-up studies have failed to support a hypoglycemic effect. Weak antimicrobial activity of unidentified fractions have been reported against *Pseudomonas aeruginosa* and *Proteus vulgaris*.[2]

## USES

**Food.** Stevia leaf is reported to have been used by Indian groups in Paraguay as a sweetener for maté (*Ilex paraguayensis*) as early as the sixteenth century.[2] Use as a sweetener by the Guaraní Paraguayan Indians, Mestizos, and others is documented back to 1887.[3] It was planted in England during World War II as a possible sugar substitute.[3] Stevioside and stevia leaves are used as non-fermentive, noncaloric commercial sweetening agent primarily in Japan, as well as Paraguay and Brazil. As much as 1700 metric tons dry leaf were used in Japan in 1987.[4] Given high stability to heat and acid, a sweetness similar to sucrose (with only a mild after taste), it has been found to be of use as a sweetener in many food categories in Japan. Reported Japanese use includes flavoring pickles, dried seafood, fish, and meat products; vegetables; confectionery; desserts; soy sauce; miso (and other products with high levels of sodium chloride); soft drinks and others.[2]

**Health Food/Herb Teas.** Used since the early 1980s; dried leaf a sweetening agent in herbal teas, as well as leaf powder or extracts as a table or cooking sweetener.[5]

**Traditional Medicine.** Stevia has been traditionally used in Paraguay to treat diabetes, and as a contraceptive.[1]

## COMMERCIAL PREPARATIONS

Crude leaf, extracts, stevioside (Japan).

*Regulatory Status.* Since May 17, 1991, the FDA has treated stevia leaf as an "unsafe food additive," banning importation into the United States.[6] A petition by the American Herbal Products Association to allow the herb to be sold as a food flavor, arguing that the ingredient is a food, commonly used before 1958, (rather than a food additive), hence exempt from the import alert, has been rejected by FDA.[4,5,7] The agency claims there is not sufficient data to establish common food use prior to 1958, and raised questions about possible toxic effects related to reduced fertility rates in mice.[7,8]

## REFERENCES

See the General References for GLASBY 2; UPHOFF.

1. D. D. Soejarto et al., *Econ. Bot.*, **37**(1), 71 (1983).
2. A. D. Kinghorn and D. D. Soejarto, in H. Wagner, H. Hikino and N. R. Farnsworth, eds., *Economic and Medicinal Plant Research*. Vol. 1. Academic Press, Orlando, Fla; 1985, p. 1.
3. W. H. Lewis, *Econ. Bot.*, **46**(3), 336 (1992).
4. M. Blumenthal, *Whole Foods*, 29, Feb. (1992).
5. M. Blumenthal, *HerbalGram*, **26**, 22 (1992).
6. FDA Import Alert No. 45-06 (May 17, 1991).
7. M. Blumenthal, *HerbalGram*, [in edit] (1994).
8. P. Nunes et al., *Brazilian Review of Pharmacy*, **69**, 1, 46 (1988) (in M. Blumenthal, *HerbalGram*, **31**, 1994).

# STORAX

**Source:** ***American storax*** *Liquidambar styraciflua* L.; ***Levant (Asiatic) storax*** *Liquidambar orientalis* Mill. (Family Hamamelidaceae or Altingiaceae).

*Synonyms*. Liquid storax (*L. orientalis*) and styrax.

## GENERAL DESCRIPTION

*Liquidambar styraciflua* (sweet gum or red gum) is a deciduous tree with five to seven-lobed leaves that resemble stars and are subtruncate at their base; trunk tall and straight, with furrowed bark; up to about 46 m high; native to North and Central America (Connecticut to Florida and west to Oklahoma and Texas; central Mexico to Nicaragua); cultivated in eastern United States as ornamental tree. Part used is the exudation (balsam) collected in natural pockets between the wood and the bark and may be located by excrescences on the trunk. The balsam flows readily into containers when the pockets are tapped, yielding the crude American storax, a solid or semisolid. Major producing countries include Honduras and Guatemala.

*Liquidambar orientalis* (oriental sweet gum or Turkish sweet gum) is a deciduous tree smaller than American sweet gum; leaves usually deeply five-lobed with irregular secondary lobes; up to about 15-m high; native to Asia Minor. Levant storax is obtained as a pathological secretion by pounding the tree bark, inducing the sapwood to secrete the balsam which then accumulates in the bark. The balsam is collected by stripping the bark, followed by pressing it and boiling it with water. Levant storax is often packed in tin cans with a layer of water to prevent the evaporation of volatile ingredients; it is a viscous semiliquid.

## CHEMICAL COMPOSITION

American storax and Levant storax are reported to contain similar constituents in highly variable concentrations, including free cinnamic acid (5–15%), 5–10% styracin (cinnamyl cinnamate), about 10% phenylpropyl cinnamate, a resin (storesin) consisting of triterpenic acids (oleanolic and

3-epioleanolic acids) and their cinnamic acid esters, and a volatile oil.[1]

The volatile oil present in Levant storax is usually less than 1%, but that in American storax has been reported to range from less than 7% to over 20% (ARCTANDER, LIST AND HÖRHAMMER, MERCK, NANJING, REMINGTON).

Compounds present in storax oil include styrene, phenylpropyl alcohol, cinnamic alcohol, benzyl alcohol, ethyl alcohol, and vanillin (KARRER, LIST AND HÖRHAMMER).

## PHARMACOLOGY OR BIOLOGICAL ACTIVITIES

Storax has antiseptic and expectorant properties. It is also reported to have antimicrobial and antiinflammatory properties (JIANGSU).

## USES

**Medicinal, Pharmaceutical, and Cosmetic.** Storax is mainly used as an ingredient of Compound Benzoin Tincture (see ***aloe***, ***balsam tolu***, and ***benzoin***).

Storax oil and resinoid are used as fragrance components and/or fixatives in soaps and perfumes (especially oriental and floral types).

**Food.** Storax and storax extracts (e.g., resinoid and absolute) are used as flavor components or fixatives in major food products, including alcoholic and nonalcoholic beverages, frozen dairy desserts, candy, baked goods, and gelatins and puddings, among others. Use levels are usually quite low (0.001%), with the highest average maximum levels being about 0.002% reported for storax in candy (15.3 ppm) and baked goods (25 ppm).

**Traditional Medicine.** American storax is used as antiseptic and expectorant in treating wounds and skin problems (e.g., scabies) as well as coughs and colds.

Levant storax is used in Chinese medicine for similar purposes as is American storax in the West. In addition, it is used in treating epilepsy.

Both American storax and Levant storax have been used in cancers.[2]

## COMMERCIAL PREPARATIONS

American and Levant storax, purified storax, oil, and extracts (e.g., resinoid). Storax is official in U.S.P.; purified storax was formerly official in U.S.P. According to ARCTANDER, storax oil and storax resinoid have been frequently adulterated.

*Regulatory Status*. Has been approved for food use (§172.510).

## REFERENCES

See the General References for ARCTANDER; BAILEY 2; FEMA; GRIEVE; GUENTHER; JIANGSU; KROCHMAL AND KROCHMAL; LIST AND HÖRHAMMER; LUST; MORTON 3; NANJING; REMINGTON; SARGENT; TERRELL; TYLER 3; UPHOF.

1. S. Huneck, *Tetrahedron*, **19**, 479 (1963).
2. J. L. Hartwell, *Lloydia*, **32**, 247 (1969).

# TAGETES

**Source:** ***Tagetes erecta*** L., ***T. patula*** L., and ***T. minuta*** L. (syn. *T. glandulifera* Schrank) (Family Compositae or Asteraceae).

*Synonyms*. African marigold, Aztec marigold, and big marigold (*T. erecta*); French marigold (*T. patula*); Mexican marigold (*T. minuta*); and marigold.

## GENERAL DESCRIPTION

Strong-scented annual herbs, usually 0.3 to 1 m high; *T. erecta* bears the largest flower heads (ca. 5 to 10 cm across) among the three species; generally considered to be natives of Mexico (*T. erecta* and *T. patula*) and South America (*T. minuta*); cultivated or found growing wild worldwide, including Ethiopia, Kenya, Nigeria, and Australia (*T. minuta*) and Europe, India, and China (*T. erecta* and *T. patula*).

Tagetes oil is obtained by steam distillation of the aboveground parts of all three species (especially *T. minuta*).

Tagetes meal is the dried ground flower petals of Aztec marigold (*T. erecta*); tagetes extract is the hexane extract of the flower petals of Aztec marigold.[1]

## CHEMICAL COMPOSITION

Tagetes oil from *T. minuta* contains tagetones, ocimene, β-myrcene, linalool, limonene, α- and β-pinenes, carvone, citral, camphene, and salicylaldehyde as major components, with phenylethanol, valeric acid, ocimenones [e.g., (5*E*)-ocimenone], geraniol, *p*-cymene, sabinene, cineole, linalyl acetate, linalool monoxide, aromadendrene, and α-terpineol, among others also present.[2–7]

The volatile oils from *T. erecta* and *T. patula* have been reported to have compositions similar to that from *T. minuta*.[8,9]

The flower petals of Aztec marigold (*T. erecta*) is reported to contain mainly carotenoids, especially lutein and its esters (dipalmitate, dimyristate, and monomyristate), as the major pigments (JIANGSU); also α-terthienyl.[10,11]

## PHARMACOLOGY OR BIOLOGICAL ACTIVITIES

Tagetes oil (from *T. minuta*) has been reported to have tranquilizing, hypotensive, bronchodilatory, spasmolytic, and antiinflammatory properties in experimental animals.[11,12]

*Tagetes* species have been reported to cause contact dermatitis.[13,14]

(5*E*)-Ocimenone reportedly has larvicidal activity against mosquito larvae.[3]

α-Terthienyl from *T. erecta* is reportedly a nematocide and larvicide.[11]

Patulin, derived from *T. patula* has been shown to reduce capillary permeability; is antispasmodic and hypertensive.[11]

## USES

**Medicinal, Pharmaceutical, and Cosmetic.** Tagetes oil (from *T. minuta*) is used as a fragrance component in perfumes.

**Food.** Tagetes oil is used as a flavor component in most major food products, including alcoholic and nonalcoholic beverages, frozen dairy desserts, candy, baked goods, gelatins and puddings, and condiments and relishes. Highest average maximum use level reported is 0.003% in condiments and relishes.

**Traditional Medicine.** Flower heads and foliage of *T. erecta* are used as anthelmintic and emmenagogue and in treating colic.

The herb of *T. minuta* is used as stomachic, carminative, diuretic, and diaphoretic.

In China, the flowerheads of *T. erecta* are used in treating whooping cough, coughs, colds, mumps, mastitis, and sore eyes, usually as a decoction. The leaves are used in treating sores and ulcers (JIANGSU). The whole herb of *T. patula* is used in coughs and dysentery, taken internally in the form of a powder or a decoction (JIANGSU).

In India the juice of the leaves of *T. erecta* is used as a treatment for eczema.[15]

In Peru the aerial parts of *T. minuta* are used in decoction as a digestive, vermifuge, cholagogue, sedative in gastric pain, and antiabortifacient.[16]

**Others.** Tagetes meal and tagetes extract are extensively used in chicken feed to give the characteristic yellow color to chicken skin and egg yolk.

*T. erecta* flowers are used as a mosquito repellent, *T. minuta* oil for treatment of wound maggots, *T. lucida* Cav. in fumigants and repellents for mosquitoes, and *T. ternifloa* H. B. & K as an insecticide in South America.[17]

## COMMERCIAL PREPARATIONS

Tagetes oil and tagetes (Aztec marigold) meal and extract.

*Regulatory Status.* Tagetes oil has been approved for food use (§172.510); tagetes (Aztec marigold) meal and extract have been approved for use in chicken feed to enhance the yellow color of chicken skin and eggs, exempt from certification (§73.295).

## REFERENCES

See the General References for ARCTANDER; BAILEY 1; FEMA; FERNALD; JIANGSU; TERRELL; UPHOF.

1. Anon., *Fed. Regist.*, **31**, 5069 (1966).
2. P. C. Carro de la Torre and J. Retamar, *Arch. Bioquim., Quim. Farm.*, **18**, 39 (1973); through *Chem. Abstr.*, **83**, 183286d (1975).
3. A. Maradufu et al., *Lloydia*, **41**, 181 (1978).
4. S. H. L. De Mucciarelli and A. L. Montes, *An. Soc. Cient. Argent.*, **190**, 145 (1970); through *Chem. Abstr.*, **75**, 67381s (1971).
5. N. A. Kekelidze et al., *Aktual. Probl. Izuch. Efirnomaslich. Rast. Efirn. Masel*, 135 (1970); through *Chem. Abstr.*, **76**, 89943z (1972).
6. Y. N. Gupta and K. S. Bhandari, *Indian Perfum.*, **19**(1), 29 (1975).
7. M. Koketsu et al., *An. Acad. Bras. Cienc.*, **48**, 743 (1976); through *Chem. Abstr.*,**88**, 141480k (1978).
8. Y. N. Gupta and K. S. Bhandari, *Indian Perfum.*, **18**(2), 29 (1974).
9. Y. N. Gupta et al., *Indian Perfum.*, **17**(2), 24 (1973).
10. T. Phillip and J. W. Berry, *J. Food Sci.*, **40**, 1089 (1975).
11. R. A. Bye, Jr., *Econ. Bot.* **40**(1), 103 (1986).
12. N. Chandhoke and B. J. R. Ghatak, *Indian J. Med. Res.*, **57**, 864 (1969).
13. J. C. Mitchell in V. C. Runeckles, ed., *Recent Advances in Phytochemistry*, Vol. 9, Plenum Press, New York, 1975, p. 119.
14. E. Rodriguez et al., *Phytochemistry*, **15**, 1573 (1976).
15. M. B. Siddiqui et al., *Econ. Bot.*, **43**(4), 480 (1989).
16. V. De Feo, *Fitoterapia*, **63**, 5, 417 (1992).
17. D. M. Secoy and A. E. Smith, *Econ. Bot.*, **37**(1), 28 (1983).

# TAMARIND

**Source:** ***Tamarindus indica*** L. (Family Leguminosae or Fabaceae).

*Synonym.* Tamarindo.

## GENERAL DESCRIPTION

Evergreen tree with large trunk and dark-gray bark; up to about 20 m high; native to tropical Asia and Africa; cultivated worldwide (e.g., China, India, South Florida, Africa, and the West Indies). Part used is the partially dried ripe fruit (pod). The fruit is a legume 6 to 15 cm long, with a thin brittle shell and a brown, sweet-sour, stringy pulp, enclosing up to 12 seeds. The brittle shell is removed and the fruit is preserved in syrup, or the whole fruit (shell, pulp, and seeds) is mixed with salt and pressed or made into cakes or balls. Producing countries include tropical American countries (e.g., Jamaica and the Antilles) and India. The West Indian countries produce the syrup-preserved product, while East Indian countries produce the salt-preserved product.

## CHEMICAL COMPOSITION

Contains plant acids (16 to 18%) composed mainly of *d*-tartaric acid (up to ca. 18%), with minor amounts of *l*-malic acid.[1–3] Citric acid has also been reported as a major component in the old literature, though it has not been detected more recently in Indian tamarind (YOUNGKEN).[1] Other constituents include sugars (20 to 40%),[2,4] pectin, protein (2.8%), fat, vitamins (e.g., $B_1$ and C), minerals (Ca, K, P, etc.) and tartrate (MERCK, WATT AND MERRILL).[2,4,5]

It also contains a volatile fraction that consists of over 60 identified compounds, including limonene, terpinen-4-ol, neral, $\alpha$-terpineol, geranial, and geraniol, which are responsible for its citrus note; methyl salicylate, safrole, ionones ($\beta$- and $\gamma$-), cinnamaldehyde; and ethyl cinnamate which contribute to its warm spicy notes; piperitone; and several pyrazines and alkylthiazoles that normally occur only in roasted or fried foods such as roasted peanuts, coffee, and potato chips.[4]

$\beta$-Sitosterol and a bitter principle tamarindienal (5-hydroxy-2-oxo-hexa-3,5-dienal are reported from the dried fruit pulp (ca. 0.67%).[3]

## PHARMACOLOGY OR BIOLOGICAL ACTIVITIES

It has mild laxative properties, which are reportedly destroyed on cooking (JIANGSU).

Aqueous extract of the fruit pulp has been found to be highly toxic to all life cycle stages of the parasitic blood trematode, *Schistosoma mansoni*; molluscicidal against *Bulinus truncatus* vector snails.[3]

Tamarindienal is fungicidal against *Aspergillus niger* and *Candida albicans*; antibacterial against *Bacillus subtilis*, *Staphylococcus aureus*, *Escherichia coli*, and *Pseudomonas aeruginosa*.[3]

## USES

**Food.** Widely used in Asia as an ingredient in chutneys and curries and in pickling fish; also extensively used in making a refreshing drink in the tropics where tamarind grows.

Extracts are widely used as a flavor ingredient in Worcestershire and other steak sauces. Other food categories in which tamarind extracts are used include alcoholic and nonalcoholic beverages, frozen dairy desserts, candy, baked goods, gelatins and puddings, fats and oils, and gravies. Highest average maximum use level is about 0.81% (8,072 ppm) reported in gravies.

**Traditional Medicine.** Used mainly as a refrigerant (in a refreshing drink to cool down fever) and a laxative. In China it is also used to treat nausea in pregnancy and as an anthelmintic for children. In Saudi Arabia the fruit pulp juice is used for stomach problems, colds, and fevers; a thick paste of ground seeds has been used as a cast for broken bones.[6]

**COMMERCIAL PREPARATIONS**

Crude (in cake form or preserved in syrup) and extracts; crude was formerly official in N.F. Extracts come in varying flavor strengths based primarily on total acids contents; tartaric and citric acids are most commonly used.

*Regulatory Status.* GRAS (§182.20).

**REFERENCES**

See the General References for ARCTANDER; BAILEY 2; FEMA; JIANGSU; LUST; MARTINDALE; MORTON 2; TERRELL; UPHOF; WHISTLER AND BeMILLER; YOUNGKEN.

1. Y. S. Lewis and S. Neelakantan, *Food Sci.* (Mysore), **9**, 405 (1960); through *Chem. Abstr.*, **61**, 3622g (1964).
2. Y. S. Lewis et al., *Food Sci.* (Mysore), **10**, 49 (1961); through *Chem. Abstr.*, **55**, 22639f (1961).
3. E. S. Imbabi et al., *Fitoterapia*, **63**, 6, 537 (1992).
4. P. L. Lee et al., *J. Agr. Food Chem.*, **23**, 1195 (1975).
5. A. Vialard-Goudou et al., *Qualitas Plant. et Materiae Vegetabiles*, **3–4**, 426 (1958); through *Chem. Abstr.*, **53**, 6476i (1959).
6. H. A. Abulafatih, *Econ. Bot.*, **41**(4), 354 (1987).

# TANNIC ACID

**Source:** Nutgalls from oaks (***Quercus*** spp.).

*Synonyms.* Tannin, gallotannic acid, and gallotannin.

**GENERAL DESCRIPTION**

Tannins are complex polyphenolic substances isolated from nutgalls formed on young twigs of certain Middle Eastern oak trees *Quercus infectoria* Olivier and other related *Quercus* species (Family Fagaceae) by insects (e.g., *Cynips gallae-tinctoriae* Olivier). The nutgalls are collected preferably before the insect matures and leaves the gall through a hole bored in the wall, as galls with holes contain less tannin. Other sources of tannin include the seed pods of Tara [*Caesalpinia spinosa* (Mol.) O. Kuntze] of South America and Chinese nutgalls formed by insects on certain *Rhus* species growing in China [e.g., *R. chinensis* Mill. and *R. potaninii* Maxim. (Family Anacardiaceae)].

Tannic acid is generally extracted from the galls by a mixture of solvents involving water, alcohol, ether, and acetone. The complex tannic acid is separated from the simple acids such as gallic acid (3,4,5-trihydroxybenzoic acid) and is then purified.

Tannic acid is very soluble in water, alcohol, acetone, and glycerol; it is practically insoluble in carbon disulfide, chloroform, benzene, ether, hexane, and fixed oils. Its aqueous solution is acidic and decomposes on standing.

**CHEMICAL COMPOSITION**

Commercial tannic acid is not the flavanol (catechin) type. It is a mixture of glycosides of phenolic acids (mainly gallic acid). Its structure varies in complexity from a few to many molecules of gallic acid per molecule of sugar (especially glucose). Pharmaceutical-grade tannic acid is generally considered to be pentadigalloylglucose (consisting of ten gallic acid molecules per glucose molecule) or its higher molecular weight derivatives. However, in reality, the pharmaceutical-grade (or food-grade) tannic acid in commerce can differ widely in molecular weight and structure. There are no official tests that distinguish these differences. It may also contain gallic and digallic acid as well as other impurities.

## PHARMACOLOGY OR BIOLOGICAL ACTIVITIES

The most well-known property of tannins is their astringency, due to their ability of precipitating proteins.[1]

Tannins (both gallic acid and catechin types) have both carcinogenic and anticancer properties in experimental animals or systems.[2–6] They have also been implicated in human cancers (LEWIS AND ELVIN-LEWIS).[5]

Tannins have been reported to have numerous other biological or pharmacological properties, including antiviral (see ***balm***), antimicrobial,[8] growth depressant (in rats),[2,9] non-specific CNS-depressant (in mice),[10] and cariostatic (in hamsters),[11] among others (see ***catechu***).[12,13]

Toxic effects of tannic acid in humans include fatal liver damage which may result from use of tannic acid on burns or as an ingredient of enemas. There is evidence that the toxic agent may not be tannic acid but rather digallic acid, which is present as an impurity.[2] Ingestion of large doses of tannic acid may cause gastric irritation, nausea, and vomiting (MARTINDALE, USD 26th).

## USES

**Medicinal, Pharmaceutical, and Cosmetic.** Mainly used as an ingredient in some ointments and suppositories for treating hemorrhoids.

**Food.** Tannic acid is used in clarifying alcoholic beverages (especially beer and wines). It is also used as a flavor ingredient in most major categories of foods, including alcoholic and nonalcoholic beverages, frozen dairy desserts, candy, baked goods, gelatins and puddings, and meat and meat products. Highest average maximum use level is about 0.018% (182 ppm) reported in frozen dairy desserts.

**Traditional Medicine.** Nutgalls (especially Chinese nutgalls) containing high concentrations of tannic acid (50 to 80%) have been used for centuries in China in treating numerous conditions, including bleeding, chronic diarrhea or dysentery, bloody urine, hard-to-heal sores, painful joints, and persistent cough, both internally and externally in the form of a powder or decoction.

Nutgalls have also been used in cancers.[14]

**Others.** Tannic acid is used extensively in tanning hide and in the manufacture of inks. For these purposes, industrial grades are usually used.

## COMMERCIAL PREPARATIONS

Numerous grades (pharmaceutical, food, industrial, etc.). Tannic acid is official in U.S.P. and in F.C.C. Grades differ in their molecular complexities.

*Regulatory Status*. GRAS (§182.20); source not limited to *Quercus* species.

## REFERENCES

See the General References for APhA; CLAUS; FEMA; GOSSELIN; HORTUS 3rd; JIANGSU; MARTINDALE; NANJING; SAX; YOUNGKEN.

1. E. C. Bate-Smith, *Phytochemistry*, **12**, 907 (1973).
2. Z. Glick, *Diss. Abstr., B*, **29**, 1416 (1968).
3. A. Oler et al., *Food Cosmet. Toxicol.*, **14**, 565 (1976).
4. G. J. Kapadia et al., *J. Natl. Cancer Inst.*, **57**, 207 (1976).
5. J. F. Morton, *Quart. J. Crude Drug Res.*, **12**, 1829 (1972).
6. H. H. S. Fong et al., *J. Pharm. Sci.*, **61**, 1818 (1972).
7. B. E. Aizenman et al., *Mikrobiol. Zh. (Kiev)*, **30**, 403 (1968); through *Chem. Abstr.*, **71**, 1067e (1969).
8. V. E. Sokolova et al., *Prikl. Biokhim. Mikrobiol.*, **5**, 694 (1969); through *Chem. Abstr.*, **72**, 63944t (1970).
9. M. A. Joslyn and Z. Glick, *J. Nutr.*, **98**, 119 (1969).
10. R. N. Takahashi et al., *Planta Med.*, **4**, 272 (1986).
11. A. Stralfors, *Arch. Oral Biol.*, **12**, 321 (1967).
12. B. Toth, *Kolor. Ert.*, **9**, 77 (1967); through *Chem. Abstr.*, **67**, 65451a (1967).
13. G. V. N. Rayudu et al., *Poultry Sci.*, **49**, 957 (1970).
14. J. L. Hartwell, *Lloydia*, **32**, 153 (1969).

# TARRAGON

**Source:** ***Artemisia dracunculus*** L. (Family Compositae or Asteraceae).

*Synonym*. Estragon.

## GENERAL DESCRIPTION

A green, nonhairy perennial herb with an erect branched stem; up to about 1.2 m high; native to Europe (southern Russia) and western Asia; cultivated in Europe (France, Germany, Italy, etc.), United States (California), Argentina, and other countries. Parts used are the leaves or aboveground herb; from the latter tarragon oil is obtained by steam distillation. Two varieties traded; *A. dracunculus* cv. 'Sativa', French tarragon, and *A. dracunculus*, Russian tarragon (syn. *A. redowskii*). Russian tarragon does not have the fine flavor of French tarragon. Major oil producing countries include France and the United States.

## CHEMICAL COMPOSITION

The above ground herb usually contains 0.25–1% volatile oil; coumarins (coumarin, esculetin dimethyl ether, herniarin, scopoletin, etc.);[1,2] isocoumarins (e.g., artemidin and artemidinal)[3,4]; flavonoids (e.g., rutin and quercetin);[5] sterols ($\beta$-sitosterol, stigmasterol, etc.);[1,2] a saturated hydrocarbon, $C_{29}H_{60}$ (m.p. 63 to 64° C);[1] tannin; protein; and others (LIST AND HÖRHAMMER, MARSH).

Tarragon oil consists mainly of estragole (methyl chavicol) at 70%–81%.[6–9] Other components present include capillene, ocimene, nerol, thujone, 1,8-cineole, 4-methoxycinnamaldehyde, $\alpha$-pinene (0.89%), $\beta$-phellandrine (1.07%), limonene (2.68%) and $\gamma$-terpinene (10.40%), among others;[6,7,10,11] nerol has been reported to be the major component of an oil of British origin.[10] Elemicin, *trans*-isoelemicin, eugenol, methyl eugenol and *trans*-methyl isoeugenol were found in the oil of Russian tarragon.[12]

The root has been reported to contain several oligosaccharides (including inulobiose), polyacetylenes, artemidiol [3-(1,2-dihydroxybutyl)-isocoumarin], and other isocoumarins.[13–16]

Another isocoumarin, artemidinol, has also been reported isolated from tarragon.[17]

## PHARMACOLOGY OR BIOLOGICAL ACTIVITIES

Estragole, the main constituent of tarragon oil, has been reported to produce tumors in mice (see ***sweet basil*** and ***avocado***).

Essential oil strongly antibacterial (though estragole at 81% of oil content not responsible for activity).[8,9]

Although undiluted tarragon oil has been reported to be irritating to rabbit skin and the backs of hairless mice, it was found to be nonirritating and nonsensitizing to humans at a concentration of 4% in petrolatum; it was also not phototoxic.[18]

## USES

**Medicinal, Pharmaceutical, and Cosmetic.** The oil is used as a fragrance component in soaps, detergents, creams, lotions, and perfumes, with maximum use level of 0.4% reported in perfumes.[18]

**Food.** The leaf is used commonly as a domestic herb. It is also used extensively as a flavor component in numerous food products, including nonalcoholic beverages, candy, meat and meat products, condiments and relishes (e.g., vinegar), fats and oils, and gravies. Highest average maximum used level reported is about 0.27% (2,731 ppm) in condiments and relishes.

In addition to all above food categories, the oil is used in alcoholic beverages (e.g., liqueurs), frozen dairy desserts, baked goods, and gelatins and puddings. Highest average maximum use level reported is about 0.04% (414 ppm) in baked goods.

**Traditional Medicine.** The herb is used as a stomachic, diuretic, hypnotic, emmenagogue, and in treating toothache. Also reported used for treating tumors.[19]

## COMMERCIAL PREPARATIONS

Mainly the leaves and oil; oil is official in F.C.C. According to ARCTANDER, adulteration of tarragon oil has been common.

*Regulatory Status.* GRAS (§182.10 and §182.20).

## REFERENCES

See the General References for ARCTANDER; BAILEY 1; FEMA; FOSTER; GUENTHER; LIST AND HÖRHAMMER; LUST; ROSENGARTEN; TERRELL; UPHOF.


1. A. Mallabaev et al., *Khim. Prir. Soedin.*, **5**, 320 (1969); through *Chem. Abstr.*, **72**, 51802g (1970).
2. P. Tunmann and E. Mann, *Z. Lebensm.-Unters. Forsch.*, **138**, 146 (1968).
3. A. Mallabaev et al., *Khim. Prir. Soedin.*, **6**, 531 (1970); through *Chem. Abstr.*, **74**, 76270m (1971).
4. A. Mallabaev et al., *Khim. Prir. Soedin.*, **7**, 120871 (1971); through *Chem. Abstr.*, **75**, 115871a (1971).
5. G. A. Lukovnikova, *Prikl. Biokhim. i Mikrobiol.*, **1**, 594 (1965); through *Chem. Abstr.*, **64**, 4170e (1966).
6. H. Thieme and N. T. Tam, *Pharmazie*, **23**, 339 (1968).
7. G. M. Nano et al., *Riv. Ital. Essenze-Profumi, Piante Offic., Aromi-Saponi,*

*Cosmet. Aerosol*, **98**, 409 (1966); through *Chem. Abstr.*, **66**, 31930v (1967).
8. S. G. Deans and K. P. Svoboda, *J. Hort Sci.*, **63**, 503 (1988).
9. B. M. Lawrence, *Perfum. Flav.*, **15**(2), 75 (1990).
10. D. V. Banthorpe et al., *Planta Med.*, **20**, 147 (1971).
11. S. R. Srinivas, *Atlas of Essential Oils*, S. R. Srinivas, Bronx, N.Y. 1986.
12. B. M. Lawrence, *Perfum. Flav.*, **13**(1), 44 (1988).
13. A. Lombard et al., *Atti Accad. Sci. Torino, Cl. Sci. Fis. Mat. Nat.*, **109**, 439 (1975); through *Chem. Abstr.*, **85**, 74938p (1976).
14. A. Mallabaev and G. P. Sidyakin, *Khim. Prir. Soedin.*, **6**, 720 (1974); through *Chem. Abstr.*, **82**, 121665x (1975).
15. F. Bohlmann and K. M. Kleine, *Chem. Ber.*, **95**, 39 (1961).
16. H. Greger et al., *Phytochemistry*, **16**, 795 (1977).
17. A. Mallabaev and G. P. Sidyakin, *Khim. Prir. Soedin.*, **6**, 811 (1976); through *Chem. Abstr.*, **86**, 136302h (1977).
18. D. L. J. Opdyke, *Food Cosmet. Toxicol.*, **12**, 709 (1974).
19. J. L. Hartwell, *Lloydia*, **31**, 71 (1968).

# TEA

**Source:** ***Camellia sinensis*** (L.) Kuntze (syn. *C. thea* Link; *C. theifera* Griff.; *Thea sinensis* L.; *T. bohea* L.; *T. viridis* L.) and its varieties (Family Theaceae).

## GENERAL DESCRIPTION

Tea originated in China; its use dates back several thousand years. It is an evergreen shrub to occasionally a tree, much branched; young leaves hairy; up to about 9-m high if free growing, but usually maintained at 1 to 1.5 m high by regular pruning; native to the mountainous regions of southern China, Japan and India; extensively cultivated in China, India, Japan, Sri Lanka, Indonesia, and other tropical and subtropical countries (Kenya, Uganda, Turkey, Argentina, etc.); cultivated in the United States in the Carolinas. Parts used are the dried, cured leaf bud and the two adjacent young leaves together with the stem, broken between the second and third leaf; older leaves are also used but are considered of inferior quality. The young leaves and leaf bud together are called "tea flush" and are collected from spring to fall (JIANGSU).[1,2]

There are two major kinds of tea, black tea and green tea. The main difference between black tea and green tea is that during tea manufacture the former undergoes a fermentation step whereby enzymes (polyphenol oxidase, peptidase, alcohol dehydrogenase, etc.) present in tea flush convert certain constituents present (proteins, amino acids, fatty acids, polyphenols, etc.) to compounds that are responsible for the characteristic aroma and flavor of black tea.[1–3] In green tea manufacture this fermentation step is eliminated by initially subjecting the flush to steaming (Japanese process) or dry heating (Chinese process), whereby the enzymes are inactivated.[1,2]

There are many grades of black tea and green tea which differ enormously in price. India and Sri Lanka are the major black tea producers, and China and Japan are the major green tea producers.

Tea drinking has evolved into a very delicate art in some Asian countries (especially China, Japan, and India), where apart from the tea used, types of water (spring, well, etc.), brewing utensils, and brewing conditions, among others, are carefully controlled by some connoisseurs. Certain types of tea in China are served in thimble-sized cups and are so strong (astringent) that they would be unpalatable to most Americans.

## CHEMICAL COMPOSITION

The chemistry of tea is extremely complicated.[1,2] Both black tea and green tea contain caffeine (1 to 5%), with small amounts of other xanthine alkaloids (theobromine, theophylline, dimethylxanthine, xanthine, adenine, etc.) also present.[4] Part of the caffeine is in bound form (see ***kola nut***). They also contain large amounts of tannins or phenolic substances (5–27%) consisting of both catechin (flavanol) and gallic acid units, with those in green tea being higher than those in black tea (JIANGSU, LIST AND HÖRHAMMER, STAHL).[1–3,5–8]

Other components present in tea include 4–16.5% fats; flavonoids (quercetin, quercitrin, rutin, etc.); amino acids (higher in green tea);[2] sterols; vitamin C; flavor and aroma chemicals including theaflavin, thearubigin, *l*-epicatechin gallate, theogallin, theaspirone, dihydroactinidiolide, dimethyl sulfide, ionones ($\alpha$- and $\beta$-), damascones ($\alpha$- and $\beta$-), jasmone, furfuryl alcohol, geranial, *trans*-hexen-2-al, and others, totaling over 300 compounds; proteins; triterpenoids; and others (JIANGSU, LIST AND HÖRHAMMER).[1–3]

Theaflavin, thearubigin, and *l*-epicatechin gallate are reported to be important taste components of black tea, while its important aroma compounds include ionones, damascones, theaspirone, *trans*-hexen-2-al, and dihydroactinidiolide.[1–3,9]

Dimethyl sulfide has been reported to be an important aroma compound of green tea,[2] along with benzylaldehyde, benzyl alcohol, cyclohexanones, dihyhroactinodiolide, *cis*-hexen-3-ol, hexenyl hexanoate, *cis*-jasmone, linalool, linalool oxides, nerolidol, and phenylethanol.[10]

Certain constituents found in tea, especially the tannic substances catechin, epigallocatechin, and epigallocatechin gallate have been reported to have antioxidative properties, with epigallocatechin being the strongest.[8,11,12]

A cup of tea contains comparable amounts of caffeine (ca. 100 mg) as a cup of coffee (see ***coffee***), but it has a much higher tannins content.

## PHARMACOLOGY OR BIOLOGICAL ACTIVITIES

The pharmacological properties of tea are due primarily to its alkaloids (mostly caffeine) and tannins (polyphenolic compounds).

Caffeine has diuretic and CNS-stimulant activities as well as numerous other kinds of pharmacological properties (see ***coffee***).

Tannins have anticancer, carcinogenic, and other activities (see ***tannic acid***).[13]

In addition to its stimulant activities and diuretic properties, tea (usually in the form of a decoction) has various pharmacological effects on humans. It has been reported effective in clinically treating bacterial dysentery (>95% effective in acute cases; 85% chronic), amebic dysentery (100%, 12 cases), acute gastroenteritis (100%, 20 cases), acute and chronic enteritis (>90% acute; >83% chronic), acute infectious hepatitis, and dermatitis due to handling of rice plants, among others (JIANGSU).

Tea has also been reported to have antiatherosclerotic effects, "vitamin P" activities, and others.[2]

Cancer-inducing N-nitrosation by-products have been shown to be inhibited by green tea polyphenols. Expression of experimentally-induced chromosome damage in bone marrow cells (*in vitro*) have been found to be suppressed by pretreatment with green tea polyphenols.[16] A recent study found that green tea polyphenols and epigallocatechin gallate arrested skin tumor growth, size, and number in mice.[15]

## USES

**Food.** The major use of tea domestically is as a beverage usually in the form of tea bags (pure ground tea) or as instant teas (tea extracts mixed with other ingredients). The tea used is mostly black tea.

Tea extract (type not specified) is also reported used as a flavor component in most major food products, including alcoholic beverages, frozen dairy desserts, candy, baked goods, and gelatins and puddings. Highest average maximum use level is 3% in baked goods.

**Health Food/Herb Teas.** Various products containing green tea extracts are appearing in this market, presumably in response to reports of antioxidant activity.

**Traditional Medicine.** The common tea bag is used as a wash for sunburn, as a poultice for baggy eyes, and as a compress for headache or tired eyes (ROSE); also used to stop bleeding of a tooth socket.

Tea has been used for millennia in Chinese medicine as a stimulant, diuretic, stomachic, expectorant, and antitoxic. Conditions for which it is traditionally used include headache, dysentery, and excess phlegm.

In India the leaf juice is used as a topical hemostatic for cuts and injuries.[16]

**Others.** Tea (especially low grade) is a potential source of food colors (green, yellow, orange, black, etc.).[17,18]

## COMMERCIAL PREPARATIONS

Crude and extracts.

*Regulatory Status.* GRAS (§182.20).

## REFERENCES

See the General References for GOODMAN AND GILMAN; JIANGSU; LEWIS AND ELVIN-LEWIS; LIST AND HÖRHAMMER; MARTINDALE; TERRELL.

1. G. W. Sanderson in V. C. Runeckles and T. C. Tso, eds., *Recent Advances in Phytochemistry*, Vol. 5, Academic Press, New York, 1972, p. 247.
2. R. L. Wiskremasinghe in C. O. Chichester et al., eds., *Advances in Food Research*, Vol. 24, Academic Press, New York, 1978, p. 229.
3. D. Reymond, *Chemtech*, **7**(11), 664 (1977).
4. M. Mironescu, *Rev. Fiz. Chim., Ser. A*, **11**, 218 (1974); through *Chem. Abstr.*, **82**, 84684e (1975).
5. C. T. Wu et al., *Chung Kuo Nung Yeh Hua Hsueh Hui Chih*, **13**, 159 (1975); through *Chem. Abstr.*, **85**, 45106k (1976).
6. D. J. Cattell and H. E. Nursten, *Phytochemistry*, **15**, 1967 (1976).
7. T. Bryce et al., *Tetrahedron Lett.*, (6), 463 (1972).
8. G. Kajimoto et al., *Eiyo To Shokuryo*, **22**, 473 (1969); through *Chem. Abstr.*, **73**, 33942u (1970).
9. P. Coggon et al., *J. Agric. Food Chem.*, **25**, 278 (1977).
10. H. N. Graham, *Preventive Medicine*, **21**, 334 (1992).
11. M. A. Bokuchava et al., *Maslo-Zhir. Prom-st.*, **2**, 15 (1975); through *Chem. Abstr.*, **82**, 153982q (1975).
12. S. Uchida et al., *Life Sci.*, **50**(2), 147 (1992).
13. H. E. Kaiser, *Cancer*, **20**, 614 (1967).
14. B. R. Olin, ed., *Lawrence Rev. Nat. Prod.* (May 1993).

15. Z. Y. Wang et al., *Cancer Res.*, **52**, 6657, (1992).
16. R. R. Rao and N. S. Jamir, *Econ. Bot.*, **36**(2), 176 (1982).
17. M. A. Bokuchava and G. N. Pruidze, Br. 1,181,079 (1970); through *Chem. Abstr.*, **72**, 120263w (1970).
18. M. A. Bokuchava et al., *Biokhim. Progr. Tekhnol. Chai. Proizvod., Akad. Nauk SSSR, Inst. Biokhim.*, 335 (1966); through *Chem. Abstr.*, **66**, 10045c (1967).

# THYME

**Source:** ***Thymus vulgaris*** L. (Family Labiatae or Lamiaceae).

*Synonyms*. Common thyme, garden thyme, and French thyme.

## GENERAL DESCRIPTION

There are many species and varieties of thyme whose classification is complicated. Estimates of legitimate species range from 100 to 400. The most commonly used species is *Thymus vulgaris*. It is an erect evergreen subshrub with numerous white, hairy stems and a woody fibrous root; up to about 45 cm high; native to the Mediterranean region (Greece, Italy, Spain, etc.); extensively cultivated in France, Spain, Portugal, Greece, and the United States (California). Parts used are the dried or partially dried leaves and flowering tops from which thyme oil is produced by water and steam distillation.

Other thyme species used include: *T.* × *citriodorus* (Pers.) Schreb. (syn. *T. serpyllum* L. var. *vulgaris* Benth.), which is known as lemon thyme; *T. zygis* L.; and *T. serpyllum* L., known as creeping thyme, wild thyme, or mother of thyme.

Thyme oil is derived from *T. vulgaris* and *T. zygis* and its var. *gracilis* Boiss. Two types of thyme oil are produced, red thyme oil and white thyme oil. White thyme oil is obtained from red thyme oil by redistillation; it has been reported to be much adulterated (ARCTANDER). The major oil producing country is Spain (see also ***origanum*, *Spanish***).

## CHEMICAL COMPOSITION

Common thyme contains 0.8–2.6% (usually ca. 1%) volatile oil consisting of highly variable amounts of phenols (20–80%);[1–4] monoterpene hydrocarbons (e.g., 51% according to one report) such as *p*-cymene and $\gamma$-terpinene;[3] and alcohols (e.g., linalool, $\alpha$-terpineol, and thujan-4-ol each of which can be the major component and constitute up to 80% or more of the volatile oil).[1,2,4–6] Thymol is normally the major phenolic component in common thyme with carvacrol being only a minor component.[1–7] Thymol and carvacrol occurring as glucosides and galactosides have been reported.[8]

Other constituents present include tannin, flavonoids, caffeic acid, labiatic acid, ursolic acid, and oleanolic acid (STAHL).

Wild thyme (*T. serpyllum*) contains 0.4–2.3% volatile oil consisting of highly variable amounts of phenols (e.g., 47–74%);[9,10] alcohols;[9] and monoterpene hydrocarbons. Either thymol or carvacrol can be the major phenol in wild thyme, depending on sources (see ***origanum*, *Spanish***).[1,2,7]

Thyme oil and thymol have been demonstrated to have antioxidative activities on dehydrated pork.[11] The labiatic acid present in thyme as well as in marjoram, oregano, sage, and other plants of the mint family also has antioxidative properties (see ***marjoram***, ***rosemary***, and ***sage***).

## PHARMACOLOGY OR BIOLOGICAL ACTIVITIES

Thyme oil is reported to have antispasmodic, expectorant, and carminative properties; it also has antimicrobial (bacteria and fungi) activities. These activities are due to thymol and carvacrol, with the former more potent.[4,12–14]

Thyme oil has been reported to be lethal to mosquito larvae.[15] When administered to rabbits orally or per intramuscular injection, it caused arterial hypotension accompanied by increased rhythmic contraction of the heart, and in higher dosage also increased respiratory frequency.[16] When given intravenously to cats as a 5% emulsion in saline solution, thyme oil also increased respiratory volume and lowered blood pressure.[17]

Red thyme oil has been reported to be nonirritating, nonsensitizing, and nonphototoxic to human skin but severely irritating to mouse and rabbit skin when applied undiluted.[18]

Thymol, a major constituent of thyme oil, is reported to have strongly fungicidal and anthelmintic (especially hookworms) as well as mildly local irritant properties. It is considered to be very toxic. Toxic symptoms include nausea, vomiting, gastric pain, headache, dizziness, convulsions, coma, cardiac, and respiratory collapse (GOSSELIN, MERCK, USD 26th).

Fluid extracts of *T. vulgaris* and *T. serpyllum* have been shown to have spasmolytic activity on the smooth muscle of guinea-pig trachea and ileum *in vitro*. However, this activity was not due to thymol or carvacrol but to as yet unidentified nonphenolic components.[19]

*Thymus vulgaris* has been shown to have an antithyrotropic effect in rats.[20]

## USES

**Medicinal, Pharmaceutical, and Cosmetic.** Thyme oil is used as a flavor component, antispasmodic, carminative, counterirritant, or rubefacient in certain cough drops, antiseptic mouthwashes, and liniments. Thymol is similarly used; in addition, it is used in antifungal preparations (for fungal skin infections), dental formulations, and others.

In German phytomedicine, preparations (tea) prescribed at 1 to 2 g of dried herb (calculated to contain at least 0.5% phenols, calculated as thymol), are used for symptoms of bronchitis, whooping cough, and catarrhs of the upper respiratory tract.[21]

Thyme oil is also used in toothpastes, soaps, detergents, creams, lotions, and perfumes, with maximum use level of 0.8% of the red type reported in perfumes.[18]

**Food.** Lemon thyme is used primarily in spice blends (especially for salads).

Thyme is widely used in baked goods, meat and meat products, condiments and relishes, processed vegetables, soups, gravies, and fats and oils; among others. Highest average maximum use level is about 0.172% (1,716 ppm) reported in meat and meat products.

Thyme oil, white thyme oil, tincture, and fluid extract are used as flavor components in most major food products, including alcoholic (e.g., liqueurs) and nonalcoholic beverages, frozen dairy desserts, candy, baked goods, gelatins and puddings, meat and meat products, and condiments and relishes. Average maximum use levels are usually below 0.003%.

**Health Food/Herb Teas.** Thyme is sometimes used as a flavoring ingredient in teas.

**Traditional Medicine.** Common thyme (both fresh and dried) is reportedly used as an anthelmintic, antispasmodic, bronchiospasmolytic, carminative, sedative, diaphoretic, and expectorant, usually in the form of an infusion or tincture. Conditions for which it is used include acute bronchitis, laryngitis, whooping cough, chronic gastritis, diarrhea, and lack of appetite. Also used externally in baths to help rheumatic and skin problems (bruises, sprains, etc.). In Chinese medicine it is used to treat similar conditions (JIANGSU).

Creeping thyme is used for similar purposes as common thyme. Its infusion is also reputed to be useful in breaking the alcohol habit (LUST).

Several species of thyme (e.g., *T. vulgaris*, *T. serpyllum*, and *T. zygis*) have been used in cancers.[22]

**COMMERCIAL PREPARATIONS**

Crude and oils. Thyme and thyme oil were formerly official in N.F. Thyme oil is official in F.C.C.

*Regulatory Status.* GRAS: Common thyme and creeping thyme (§182.10 and §182.20); derivatives (e.g., essential oils) of *T. vulgaris*, *T. zygis* var. *gracilis*, and *T. serpyllum* (§182.20). *T. vulgaris* is the subject of a positive German therapeutic monograph, indicated for symptoms of bronchitis and congestion of the upper respiratory tract.[21]

**REFERENCES**

See the General References for ADA; APhA; ARCTANDER; BAILEY 2; BIANCHINI AND CORBETTA; BLUMENTHAL; CLAUS; FEMA; FOSTER; GOSSELIN; GRIEVE; GUENTHER; JIANGSU; LUST; MARTINDALE; MORTON 3; ROSE; ROSENGARTEN; STAHL; UPHOF; USD 26th.

1. B. M. Lawrence, *Perfum. Flav.*, **2**, 3 (1977).
2. J. D. Miguel et al., *J. Agr. Food Chem.*, **24**, 833 (1976).
3. A. J. Poulose and R. Croteau, *Arch. Biochem. Biophys.*, **187**, 307 (1978).
4. M. Simeon de Bouchberg et al., *Riv. Ital. Essenze, Profumi, Piante Offic., Aromi, Saponi, Cosmet., Aerosol*, **58**, 527 (1976); through *Chem. Abstr.*, **86**, 84201c (1977).
5. A. B. Svendsen and J. Karlsen, *Planta Med.*, **14**, 376 (1966).
6. M. S. Karawya and M. S. Hifnawy, *J. Assoc. Off. Anal. Chem.*, **57**, 997 (1974).
7. M. Grims and R. Senjkovic, *Acta Pharm. Jugoslav.*, **17**, 3 (1967); through *Chem. Abstr.*, 67, 57279k (1967).
8. K. Skopp and H. Hoerster, *Planta Med.*, **29**, 208 (1976).
9. U. Asllani, *Bul. Shkencave Natyr., Univ. Shteteror Tiranes*, **27**, 111 (1973); through *Chem. Abstr.*, **80**, 149025w (1974).
10. M. Mihajlov and J. Tucakov, *Bull., Acad. Serbe Sci. Arts, Cl. Sci. Med.*, **44**, 57 (1969); through *Chem. Abstr.*, **73**, 80399g (1970).
11. H. Fujio et al., *Nippon Shokuhin Kogyo Gakkai-Shi*, **16**, 241 (1969); through *Chem. Abstr.*, **74**, 2846g (1971).
12. A. C. Pizsolitto et al., *Rev. Fac. Farm. Odontol. Araraquara*, **9**, 55 (1975); through *Chem. Abstr.*, **86**, 12226s (1977).
13. M. De Vincenzi and M. R. Dessi, *Fitoterapia*, **62**, 1, 39 (1991).
14. D. Patakova and M. Chladek, *Pharmazie*, **29**, 140 (1974).
15. D. Novak, *Arch. Roum. Pathol. Exp. Microbiol.*, **27**, 721 (1968); through *Chem. Abstr.*, **71**, 58264w (1969).
16. K. M. Kagramanov et al., *Azerb. Med. Zh.*, **54**, 49 (1977); through *Chem. Abstr.*, **87**, 145903c (1977).

17. T. Shipochliev, *Vet.-Med. Nauki*, **5**, 63 (1968); through *Chem. Abstr.*, **70**, 86144e (1969).
18. D. L. J. Opdyke, *Food Cosmet. Toxicol.*, **12**(Suppl.), 1003 (1974).
19. C. O. Van Den Broucke and J. A. Lemli, *Planta Med.*, **41**, 129 (1981).
20. H. Sourgens et al., *Planta Med*, **45**, 78 (1982).
21. Monograph *Thymi herba (T. vulgaris), Bundesanzeiger,* no. 228 (Dec. 5, 1984).
22. J. L. Hartwell, *Lloydia*, **32**, 247 (1969).

# TIENCHI GINSENG

**Source:** ***Panax notoginseng*** (Burk.) F. H. Chen [syn. *P. pseudo-ginseng* Wall. var. *notoginseng* (Burk.) Hoo et Tseng; *P. sanchi* Hoo] (Family Araliaceae). *P. pseudo-ginseng* Wall (also a synonym for ginseng) is sometimes listed as source; this has resulted in much confusion whenever herbal formulas are presented in English with no Chinese name given.

*Synonyms*. *Sanqi*, *sanchi ginseng*, *tianqi*, *tianchi*, *shen sanqi*, *tian sanqi* and *renshen sanqi*.

## GENERAL DESCRIPTION

Perennial herb, 30 to 60 cm high, with spindle-shaped, fleshy main root; distributed in southern China; now mostly cultivated in the provinces of Yunnan, Guangxi, Guangdong, Jiangxi and Hubei, especially at altitudes between 800 and 1000 m.

Part used is the tuberous root collected from 3- to 4-year-old plants before flowering or after fruits have ripened; spring harvest is of better quality and preferred. After being rid of root crown, lateral roots, rootlets, sand and dirt, the root is exposed to the hot sun for a day followed by gentle hand kneading and again sun-drying. This process is repeated until root is hard and completely dried, which is then placed in a vessel along with pieces of wax and shaken back and forth until its surface turns shiny and dark brown (MA). Yunnan, Guangdong and Guangxi are the major producers.

## CHEMICAL COMPOSITION

*Sanqi* contains: 4.42–12% saponin glycosides (WANG);[1–3] amino acids, including dencichine ($\beta$-$N$-oxalo-L-$\alpha$,$\beta$-diaminopropionic acid); volatile oil; flavonoids; phytosterols (e.g., $\beta$-sitosterol, stigmasterol and daucosterol); polysaccharides (e.g., an arabinogalactan named sanchinan-A);[4] trace minerals, and others (HU, WANG).[1]

Saponin glycosides are similar to those of ginseng (dammarane type) and include: ginsenosides $R_{b1}$, $R_d$, $R_e$, $R_{g1}$, $R_{g2}$, and $R_{h1}$, with $R_{b1}$ and $R_{g1}$ in predominant amounts; notoginsenosides $R_1$, $R_2$, $R_3$, $R_4$, and $R_6$ in minor amounts; and gypenoside XVII in trace amount. The sapogenins so far identified are 20($S$)-protopanaxadiol and 20($S$)-protopanaxatriol; no oleanolic acid is present, hence no ginsenoside $R_0$ (see ***ginseng***).[1–5] Saponin glycoside contents in raw and, cured *sanqi* are similar, except that there is an increase of monodesmosides (ginsenosides $R_{g2}$, $R_{g3}$ and notoginsenoside $R_2$) and a decrease of bidesmosides (ginsenosides $R_{b1}$, $R_d$, $R_e$, $R_{g1}$ and notoginsenosides $R_1$ and $R_4$) in the cured drug.[5] Also, compared to raw *sanqi*, steam-curing markedly increased amounts of total extractives and saponin glycosides while deep frying reduced amounts of both.[6]

## PHARMACOLOGY OR BIOLOGICAL ACTIVITIES

*Sanqi* has broad biologic activities, including: hematologic, cardiovascular, immunomodulating, antiinflammatory, and effects on the nervous system (central stimulant, due to panaxatriol saponins), digestive system (anti-ulcer in rats), metabolism and endocrine system.[7–9]

*Sanqi* is best known for its traditional use as a hemostatic and modern studies have verified the rationale for this use by demonstrating its shortening of coagulation time as well as its hemostatic effects in experimental animals (dencichine being an active component).[7] On the other hand, *sanqi* has also exhibited anticoagulant and antiplatelet aggregation activities under other conditions. Its total saponins have hemolytic activity *in vitro*, though some of the saponins only hemolyzed red cells of certain animals (guinea pig and monkey) but not others (rabbit, sheep and pigeon) (WANG). *Sanqi* has been reported to have both hypolipemic and hyperlipemic effects and a recent study found raw *sanqi* to be hypolipemic but cured *sanqi* to be hyperlipemic.[8] Oral administration of total saponins in rabbits inhibited aortic atherosclerotic plaque formation; their oral administration in rats increased the contents of prostacyclin in carotid artery and decreased amount of thromboxane A in blood platelets.[10]

Most of the cardiovascular studies on *sanqi* have been performed in China using the total saponins. Their effects include: anti-arrhythmic in mice, rats and rabbits, with the panaxatriol saponins being most active;[11,12] vasodilating and hypotensive in experimental animals;[13–15] protective against experimental hemorrhagic shock in rabbits;[16] protective against experimental myocardial injury in rats, inhibiting lipid peroxidation and preventing reduction of superoxide dismutase activity, with ginsenosides $R_{b1}$ and $R_{g1}$ being the most active;[17] selective blocking of calcium channels;[18] and increasing coronary blood flow volume and reducing coronary arterial resistance as well as peripheral vascular resistance, among others (WANG).[15]

*Sanqi* decoction stimulated the activities of natural killer cells, macrophages and plaque-forming cells while crude polysaccharides stimulated the activities of only macrophages and plaque-forming cells in mice;[19] sanchinan-A (a polysaccharide) has activating effects on the reticuloendothelial system per the carbon clearance test in mice.[4]

*Sanqi* total saponins and those from the rootlets all exhibited strong antiinflammatory activities in several experimental models, with rootlet saponins (100 mg) stronger than cortisone (50 mg) in the ear edema ($1.74 \pm 0.41$ vs $5.54 \pm 0.83$ mg; control: $9.35 \pm 0.72$ mg) induced by croton seed oil in mice.[20,21]

Oral administration of *sanqi* powder to rats markedly reduced lipid peroxide formation and greatly increased superoxide dismutase activity in brain tissue but not in other tissues (heart, liver and lung).[22]

Total *sanqi* saponins (i.p.) markedly decreased adrenal ascorbic acid in rats and increased plasma corticosteroids in guinea pigs;[23] rootlet total saponins also greatly increased plasma corticosterone in mice.[21]

Rootlet total saponins have been shown to promote growth in mice and to have androgenic activities in rats.[21]

Oral toxicity of *sanqi* powder is low: stomach feeding of 15 g/kg to mice did not cause any fatalities and examination of tissues (heart, liver, kidney, spleen and gastrointestinal tract) did not reveal any abnormalities.[7]

## USES

**Medicinal, Pharmaceutical, and Cosmetic.** Powder and extracts used primarily in Asia as an ingredient in shampoos and in skin-care products (e.g., acne creams and lotions) for its vasodilating effects and its traditional ability to remove dark spots (ZHOU).[24]

**Health Food/Herb Teas.** Powder used in tonic formulas, usually in tablet or capsule form.

**Traditional Medicine.** First described in Li Shi-Zhen's *Ben Cao Gang Mu* (ca. 1590 A.D.), *sanqi* is traditionally regarded to be sweet and slightly bitter tasting, warming, and to have stasis-dispersing, hemostatic, antiswelling and analgesic properties. Used in hemorrhages of various kinds (e.g., coughing blood, vomiting blood, nosebleeds, hematochezia and metrorrhagia), traumatic injuries with bleeding and pain, stabbing pain in chest and abdomen, coronary heart disease and other conditions (CHP, JIANGSU, WANG).

Its most well-known use is as a major ingredient in *Yunnan Bao Yao* which was carried by both Chinese and American airmen (the Flying Tigers) during the Second World War for stopping bleeding resulting from wounds and injuries.

## COMMERCIAL PREPARATIONS

Crude comes in over 20 grades with top grades being heavy and hard, 3 to 6 cm long, and with a grayish brown or grayish green fracture; powdered crude is most likely produced from the crown, lateral roots, rootlets or other traditionally considered low-grade materials.

*Regulatory Status.* U.S. regulatory status not determined.

## REFERENCES

See the General References for CHP; IMM-1; IMM-CAMS; JIANGSU; MA; NATIONAL; WANG; ZHOU; ZHU.

1. F. Y. Gan and G. Z. Zheng, *Zhongguo Yaoxue Zazhi*, **27**, 138 (1992).
2. H. Matsuura et al., *Chem. Pharm. Bull.*, **31**, 2281 (1983).
3. C. R. Yang et al., *Yaoxue Tongbao*, **20**, 337 (1985).
4. K. Ohtani et al., *Planta Med.*, **53**, 166 (1987).
5. C. R. Yang et al., *Zhongyao Tongbao*, **10**(9), 33 (1985).
6. S. Zhang, *Zhongchengyao*, **11**(11), 20 (1989).
7. B. H. Zhang, *Zhongcaoyao*, **15**(11), 34 (1984).
8. G. Z. Chen et al., *Chin. J. Integr. Trad. Western Med.*, **4**, 540 (1984).
9. H. X. Quan et al., *Henan Zhongyi*, **10**(1), 41 (1990).
10. L. Shi et al., *Acta Pharmacol. Sinica*, **11**, 29 (1990).
11. S. Liu and J. X. Chen, *Acta Pharmacol. Sinica*, **5**, 100 (1984).
12. B. Y. Gao et al., *Yaoxue Xuebao*, **27**, 641 (1992).
13. J. D. Wang and J. X. Chen, *Acta Pharmacol. Sinica*, **5**, 181 (1984).
14. J. X. Wu and J. X. Chen, *Acta Pharmacol. Sinica*, **9**, 147 (1988).
15. Y. P. Zhou and W. H. Liu, *Zhongcaoyao*, **19**(4), 25 (1988).
16. L. X. Li et al., *Acta Pharmacol. Sinica*, **9**, 52 (1988).
17. X. Li et al., *Acta Pharmacol. Sinica*, **11**, 26 (1990).
18. Z. G. Xiong et al., *Acta Pharmacol. Sinica*, **10**, 122 (1989).
19. J. M. Wang et al., *Zhongguo Yiyao Xuebao*, **4**(4), 29 (1989).
20. C. Q. Hao and F. Yang, *Acta Pharmacol. Sinica*, **7**, 252 (1986).
21. Q. S. Chen et al., *Zhongyao Tongbao*, **12**(3), 45 (1987).
22. E. B. Dong et al., *Zhongcaoyao*, **21**(4), 26 (1990).

23. J. D. Wang and J. X. Chen, *Acta Pharmacol. Sinica*, **5**, 50 (1984).

24. Y. X. Ling, *Shanghai Zhongyiyao Zazhi*, (5), 22 (1991).

# TRAGACANTH

**Source:** ***Astragalus*** spp. (especially *A. gummifer* Labill.) (Family Leguminosae or Fabaceae).

*Synonym.* Gum tragacanth.

## GENERAL DESCRIPTION

Tragacanth is the dried gummy exudate from various species of *Astragalus* (e.g., *A. gummifer*). They are low thorny shrubs, up to about 1 m high, native to the mountainous regions of the Middle East (especially Iran, Turkey, and Syria). Tragacanth is obtained by tapping the branches and tap roots (near the ground surface) whereby the gum exudes, which after drying becomes horny and is collected. The word tragacanth is derived from Greek, meaning goat's horn, probably describing the appearance and texture of the gum. Iran is the major producer; it also produces the best quality product. Tragacanth occurs in two main forms, ribbons and flakes; ribbons are considered to be of superior quality (FURIA, GLICKSMAN).[1]

Tragacanth partly dissolves and partly swells in water to give a viscous colloidal solution (sol); the maximum viscosity is only attained after 24 hours at room temperature or after heating for 8 hours at 40°C or 2 hours at 50°C. The viscosity of its solutions is generally regarded as the highest among all the plant gums (see ***guar***, ***karaya***, and ***locust bean***). A 1% solution has a slightly acidic pH (5.1 to 5.9). Its solutions are stable under acidic conditions (down to pH 2), with maximum viscosity at pH 5; lowering or increasing its pH beyond 5 would decrease its viscosity.[2] They are also heat stable.

Like karaya gum, tragacanth forms heavy, thick gel-like pastes at high concentrations (e.g., 2 to 4%).

Solutions or gels of tragacanth are especially susceptible to microbial degradation, sometimes even in the presence of a preservative. The most effective preservative system for tragacanth gels or sols at neutral and acidic pHs has been reported to be a mixture of parabens (MARTINDALE, WHISTLER AND BEMILLER).

Tragacanth is compatible with other polysaccharides and with proteins.

## CHEMICAL COMPOSITION

Tragacanth is reported to contain 20–30% of a water-soluble fraction called tragacanthin (consisting of tragacanthic acid and an arabinogalactan) and 60–70% of a water-insoluble fraction called bassorin as well as 1–3% starch, 1–4% cellulose; 3% ash, small amounts of invert sugar, 2–3% of a volatile acid (probably acetic acid), and about 15% water. Tragacanthic acid is composed of 40% D-galacturonic acid, 40% D-xylose, 10% L-fucose, 4% D-galactose, and three aldobiuronic acids. Arabinogalactan consists of 75% L-arabinose, 12% D-galactose, 3% D-galacturonic acid, and small amounts of L-rhamnose. Bassorin is reported to contain about 5% methoxyl groups (LIST AND HÖRHAMMER).[3,4]

The structures and molecular weights of the major polysaccharides in tragacanth (bassorin, tragacanthic acid, and arabinogalactan) have not been reported, but the molecular weight of tragacanth itself has been reported to be 310,000 and 840,000 (FURIA, GLICKSMAN, WHISTLER AND BEMILLER).

Tragacanthin dissolves in water to form a sol, while bassorin swells to form a thick gel.

## PHARMACOLOGY OR BIOLOGICAL ACTIVITIES

Tragacanth preparations contaminated with enterobacteria have been reported to cause death of fetuses when administered intraperitoneally to pregnant mice.[5]

Tragacanth has also been reported to have strong inhibitory activity on cancer cells (MORTON 3).

## USES

Tragacanth has been in use since ancient times. Most of its uses are based on its emulsifying, thickening, and suspending abilities as well as its stability to acid and heat.

**Medicinal, Pharmaceutical, and Cosmetic.** Extensively used in vaginal jellies and creams, in emulsions (e.g., cod liver oil), low-calorie syrups and elixirs, and as binding agent or demulcent in tablets and lozenges.

It is also used in toothpastes and hand lotions, among others.

**Food.** Tragacanth is used extensively in salad dressings (due to its acid stability), flavor formulations (emulsions), ice cream, and in confectionery. Other food products in which it is used include nonalcoholic beverages, baked goods, gelatins and puddings, meat and meat products, condiments and relishes, processed fruits, processed vegetables, gravies, and others. Highest average maximum use level reported is about 1.3% (13,015 ppm) in fats and oils (e.g., salad dressings).

## COMMERCIAL PREPARATIONS

Ribbon, flake, and powder in numerous grades; tragacanth is official in N.F. and F.C.C. Grades meeting N.F. or F.C.C. standards vary considerably in the viscosity of their sols.

*Regulatory Status.* Has been affirmed as GRAS (§184.1351).

## REFERENCES

See the General References for FURIA; GLICKSMAN; LAWRENCE; LIST AND HÖRHAMMER; MARTINDALE; MORTON 3; UPHOF; WHISTLER AND BEMILLER.

1. H. S. Gentry, *Econ. Bot.*, **11**, 40 (1957).
2. T. W. Schwarz et al., *J. Am. Pharm. Assoc., Sci. Ed.*, **47**, 695 (1958).
3. G. O. Aspinall and J. Baillie, *J. Chem. Soc.*, 1702 (1963).
4. G. O. Aspinall and J. Baillie, *J. Chem. Soc.*, 1086 (1967).
5. H. Frohberg et al., *Arch. Toxikol.*, **25**, 268 (1969).

# TURMERIC

**Source:** ***Curcuma longa*** L. (syn. *C. domestica* Val. and *C. domestica* Loir.) (Family Zingiberaceae).

*Synonyms.* Curcuma and Indian saffron.

## GENERAL DESCRIPTION

A perennial herb of the ginger family with a thick rhizome from which arise large, oblong, and long-petioled leaves; up to about 1 m high; native to southern Asia; extensively cultivated in India, China, Indonesia, and other tropical countries (e.g.,

Jamaica and Haiti). Part used is the cured (boiled, cleaned, and sun-dried) and polished rhizome. India is the major producer of turmeric (up to 94% of annual world production).[1,2]

## CHEMICAL COMPOSITION

Contains 0.3–7.2% (usually 4–5%) of an orange-yellow volatile oil that is composed mainly of turmerone (ca. 60%), ar-turmerone, $\alpha$-atlantone, $\gamma$-atlantone, and zingiberene (25%), with minor amounts of 1,8-cineole, $\alpha$-phellandrene, *d*-sabinene, borneol, and dehydroturmerone, among others;[3–5] yellow coloring matter including 0.3–5.4% curcumin, monodesmethoxycurcumin, and didesmethoxycurcumin;[3,5] *p*-coumaroylferuloylmethane and di-*p*-coumaroylmethane;[6] sugars (28% glucose, 12% fructose, and ca. 1% arabinose); fixed oil; protein (ca. 8%); minerals (especially high in potassium); vitamins (especially C); resin; and others (JIANGSU, LIST AND HÖRHAMMER, MARSH).[3]

Turmeric and its water-, alcohol-, and ether-soluble fractions have been reported to have antioxidative activities;[7] curcumin is mostly responsible for these activities.[8–10]

## PHARMACOLOGY OR BIOLOGICAL ACTIVITIES

A fraction of curcuma oil (b.p. 80–110°C) has been demonstrated to have antiinflammatory and antiarthritic activities in rats.[11] Curcumin is also reported to exhibit antiedemic effects in rats (MARTINDALE).

Other activities of turmeric and its derivatives include choleretic in dogs (aqueous extracts); hypotensive in dogs (alcohol extract); antibacterial (curcumin, volatile oil, etc.);[12] protective of mouse liver and cultured rat hepatocytes from injury induced by carbon tetrachloride and galactosamine (curcumin, *p*-coumaroylferuloylmethane, and di-*p*-coumaroylmethane);[6] insecticidal against houseflies (petroleum ether extracts);[13] and others (JIANGSU, LIST AND HÖRHAMMER).

Choleretic action of the essential oil is attributed to tolmethyl carbinol.[2] Antiinflammatory activity is attributed, as least in part, to curcumin and the volatile oil;[14] inhibition of leukotriene synthesis may be in part responsible for antiinflammatory activity.[15] Extracts have antispasmodic activity on isolated guinea pig ileum; lower serum cholesterol and triglyceride levels in mice; anticancer activity against Dalton's lymphoma cells in the Chinese hamster; plus anticoagulant, antifungal, antioxidant, and antimutagenic, among others.[14]

A recent study found no visible signs of acute toxicity with extracts, though at 3 g/kg CNS stimulation was observed; in chronic toxicity feeding, an increase in weights of the heart and lung were observed, though general visceral condition was normal; increased sperm motility (without increase in sperm count) suggested a possible androgenic effect.[16]

## USES

**Medicinal, Pharmaceutical, and Cosmetic.** Cut or ground drug (1.5–3 g), or other preparations are used in European phytomedicine for dyspeptic conditions. Use is contraindicated in obstruction of gall passages; used only under medical advice for gallstones.[17]

The essential oil (called curcuma oil) is used to a limited extent in certain perfumes (especially oriental types).

**Food.** Turmeric is a major ingredient of curry powder and is also used in prepared mustard.

Turmeric and turmeric oleoresin are used extensively both for its color and flavor in many food products, including baked goods, meat and meat products, condiments and relishes (especially pickles), fats and oils, egg products, soups, and gravies, among others. Highest average maximum use levels are 22% and about 0.883% (8,834 ppm) reported for turmeric in seasonings and flavorings and in condiments and relishes, respectively.

**Health Food/Herb Teas.** Turmeric is used as an antioxidant in capsules, tablets; flavoring in tea (LEUNG).

**Traditional Medicine.** Reportedly used in Chinese medicine to treat numerous conditions including flatulence, liver problems, menstrual difficulties, bloody urine, hemorrhage, toothache, bruises and sores, chest pain, and colic, usually decocted with other drugs. It is also used as a poultice to relieve pain and itching of sores and ringworms (JIANGSU).

Root tuber of *Curcuma* spp. is used in China for treating epilepsy and unconsciousness due to febrile diseases (*re bing shen hun*) (CHP, JIANGSU).

In Polynesia, the root used for asthma, skin diseases, constipation, and religious rituals.[18]

## COMMERCIAL PREPARATIONS

Crude and oleoresin; crude was formerly official in U.S.P. Strength (see Glossary) of oleoresin is often expressed in terms of curcumin content.

*Regulatory Status.* GRAS (§182.10 and §182.20); turmeric and turmeric oleoresin have also been approved as food colorants exempt from certification (§73.600 and §73.615). The rhizome is subject of a positive German therapeutic monograph for treatment of dyspeptic conditions.[17]

## REFERENCES

See the General References for ARCTANDER; BAILEY 1; BLUMENTHAL; CHP; FEMA; JIANGSU; LEUNG; LIST AND HÖRHAMMER; MARTINDALE; ROSENGARTEN; STAHL; TERRELL; UPHOF.

1. M. Ilays, *Econ. Bot.*, **32**(3), 238 (1988).
2. G. S. Randhawa and R. K. Mahey in L. E. Craker and J. E. Simon, eds., *Herbs, Spices, and Medicinal Plants: Recent Advances in Botany, Horticulture, and Pharmacology*, Vol. 3, Oryx Press, Phoenix, 1988, p. 71.
3. A. Khalique and M. N. Amin, *Sci. Res. (Dacca, Pakistan)*, **4**, 193 (1967); through *Chem. Abstr.*, **69**, 701d (1968).
4. C. R. Mitra, *Riechst., Aromen, Körperpflegem.*, **25**, 15 (1975).
5. N. Krishnamurthy et al., *Trop. Sci.*, **18**, 37 (1976).
6. Y. Kiso et al., *Planta Med.*, **49**, 185 (1983).
7. F. Hirahara et al., *Eiyogaku Zasshi*, **32**, 1 (1974); through *Chem. Abstr.*, **82**, 2764u (1975).
8. S. B. Xu et al., *Zhongcaoyao*, **22**, 140 (1991).
9. S. B. Xu et al., *Zhongcaoyao*, **22**, 264 (1991).
10. B. L. Zhao et al., *Cell Biophys.*, **14**, 175 (1989).
11. D. Chandra and S. S. Gupta, *Indian J. Med. Res.*, **60**, 138 (1972).
12. J. Lutomski et al., *Planta Med.*, **26**, 9 (1974).
13. R. S. Dixit and S. L. Perti, *Bull. Reg. Res. Lab. Jammu, India*, **1**, 169 (1963); through *Chem. Abstr.*, **60**, 11312c (1964).
14. H. P. T. Ammon and M. A. Wahl, *Planta Med.*, **57**, 1, (1991).
15. H. P. T. Ammon et al., *Planta Med.*, **58**, 226 (1992).
16. S. Qureshi et al., *Planta Med.*, **58**, 124 (1992).
17. Monograph *Curcumae longae rhizoma*, *Bundesanzeiger*, no. 223 (Nov. 30, 1985); revised (Sept. 1, 1990).
18. W. McClatchey, *Econ. Bot.*, **47**(3), 291 (1993).

# TURPENTINE (AND ROSIN)

**Source:** ***Longleaf pine*** *Pinus palustris* Mill.; ***slash pine*** *P. elliottii* Engelm.; and other *Pinus* spp.

*Synonyms.* Gum terpentine, gum thus, turpentine oil, and turpentine balsam.

## GENERAL DESCRIPTION

The term "turpentine" is rather loosely used to describe either the oleoresin obtained from the longleaf pine (*Pinus palustris*), slash pine (*P. elliottii*), and other *Pinus* species that yield exclusively terpene oils, or the essential oil obtained from the above oleoresin. The oleoresin is commonly called gum turpentine or turpentine balsam while the essential oil is called turpentine oil; both the oleoresin and the essential oil are also called simply turpentine. To avoid confusion, gum turpentine (though not a true gum; see ***glossary***) is here reserved only for the oleoresin, while turpentine or turpentine oil (spirits of turpentine) is used for the essential oil. Thus, gum turpentine on steam distillation yields turpentine (turpentine oil) and rosin (a terpenic resin), also known as colophony.

Turpentine and rosin are also produced by solvent extraction of heartwood chips of pine stumps, which are by-products of the lumber industry, and as by-products of the paper (sulfate or kraft pulping) industry. The last source is reported to account for the largest volumes of turpentine and rosin produced in the United States. These products derived from pines and other resinous conifers are commonly called naval stores.

Turpentine is the largest (in volume) essential oil in the world. Its current biggest producer is the United States; other major producing countries include New Zealand, China, Mexico, Portugal, the former U.S.S.R., and the Scandinavian countries.

## CHEMICAL COMPOSITION

On account of their widely different sources (plant, geographic, method of manufacture, etc.) turpentine and rosin often vary considerably in their relative composition.

Turpentine contains mostly monoterpene hydrocarbons, the major ones being $\alpha$-pinene (45–95%) $\beta$-pinene (0–35%), and 3-carene (20–60%).[1–4] Others present in lesser amounts include camphene, dipentene, terpinolene, $\beta$-myrcene, $\beta$-phellandrene, and *p*-cymene (LIST AND HÖRHAMMER).[1,5,6]

Rosin contains mainly diterpene resin acids including abietic acid (22–50%), dehydroabietic acid (6–30%), palustric acid (10–25%), neoabietic acid (4–20%), isopimaric acid (10–17%), and pimaric acid (4–6%).[1] Others present include dehydropimaric acid, levopimaric acid, and sandaracopimaric acid.[1,5,6] It also contains small amounts of diterpene alcohols and aldehydes, sterols (mainly sitosterol), and phenolic compounds.[1]

## PHARMACOLOGY OR BIOLOGICAL ACTIVITIES

Turpentine has rubefacient and counterirritant properties.

Turpentine has contact allergenic activities that are mainly due to the pinenes and 3-carene as well as dipentene (MARTINDALE, MORTON 3).[7–11] It also has antimicrobial activities that are reportedly due to $\alpha$-pinene, 3-carene, and dipentene (LEWIS AND ELVIN-LEWIS).[12–14]

Turpentine oil has been reported to promote tumor development on rabbit but not mouse skin.[15,16]

Toxic effects of turpentine include erythema, urticaria, headache, insomnia, coughing, vomiting, hematuria, albuminuria, and coma (MARTINDALE).[7] Turpentine has caused fatalities in children ingesting as little as 15 mL.[17]

Turpentine is used in an experimental animal model to induce systemic inflammatory immune response.[18]

Rosin also has irritant properties on some individuals (MARTINDALE).

## USES

**Medicinal, Pharmaceutical, and Cosmetic.** Turpentine is used as an ingredient in many ointments, liniments, and lotions for treating minor aches and pains as well as colds. Rosin is an ingredient in some soaps and ointments; it is also used as a fixative in perfumes.

**Food.** Steam-distilled turpentine oil is reported used as a flavor component in most major food products, including alcoholic and nonalcoholic beverages, frozen dairy desserts, candy, baked goods, gelatins and puddings, meat and meat products, and condiments and relishes. Highest average maximum use level is about 0.002% (20.6 ppm) in baked goods.

**Traditional Medicine.** Turpentine is mainly used as a counterirritant and rubefacient in treating rheumatism and aching muscles.

In Chinese medicine gum turpentine and rosin (mainly from *Pinus tabulaeformis* Carr., *P. massoniana* Lamb., and *P. yunnanensis* Franch.) have been used for centuries in treating rheumatism, stiff joints, toothache, boils, and sores. Furthermore rosin is used in treating ringworms, chronic bronchitis, and neurogenic dermatitis, among others. They are used both internally and externally.

**Others.** The major use of turpentine oil is as a solvent (e.g., paints) and as a starting material for the synthesis of useful chemicals such as camphor, menthol, terpin hydrate (an expectorant), $\alpha$-terpineol and other fragrance compounds, and resins (adhesives, chewing gum, etc.), among others.[1]

## COMMERCIAL PREPARATIONS

Rosin, turpentine oil, and rectified turpentine oil; all were formerly official in N.F.

*Regulatory Status.* Turpentine oil has been approved for food use, and rosin has been approved for use in alcoholic beverages only (§172.510).

## REFERENCES

See the General References for ARCTANDER; CLAUS; FEMA; GUENTHER; JIANGSU; LIST AND HÖRHAMMER; MORTON 3; NANJING; USDA.

1. D. F. Zinkel, *Chemtech*, **5**(4), 235 (1975).
2. Y. G. Drochnev, *Lesokhim. Podsochka*, **3**, 7 (1977); through *Chem. Abstr.*, **87**, 70053q (1977).
3. W. D. Fordham in L. W. Codd et al., eds., *Chemical Technology: An Encyclopedic Treatment,* Vol. 5, Barnes & Noble, New York, 1972, p. 1.
4. J. Fousseteau et al., *Bull. Soc. Fr. Dermatol. Syphiligr.*, **77**, 415 (1970); through *Chem. Abstr.*, **74**, 109687h (1971).
5. I. I. Bardyshev et al., *Isv. Vyssh. Ucheb. Zaved., Les Zh.*, **12**, 161 (1969); through *Chem. Abstr.*, **71**, 40493k (1969).
6. I. I. Bardyshev et al., *Vestsi Akad.*

*Navuk Belarus. SSR, Ser Khim. Navuk*, **5**, 123 (1971); through *Chem. Abstr.*, **76**, 87417g (1972).

7. P. Mikhailov et al., *Allerg. Asthma*, **16**, 201 (1970); through *Chem. Abstr.*, **76**, 95369h (1972).
8. P. Mikhailov and N. Berova, *Dermatol. Veenerol. (Sofia),* **19**, 20 (1970); through *Chem. Abstr.*, **73**, 43266r (1970).
9. M. Pambor, *Dermatol. Monats.*, **162**., 992 (1976); through *Chem. Abstr.*, **86**, 153782g (1977).
10. D. L. J. Opdyke, *Food Cosmet. Toxicol.*, **12**, 703 (1974).
11. E. Rudzki et al., *Contact Dermatitis*, **24**, 317 (1991).
12. B. N. Uzdennikov, *Nauch. Tr. Tyumen. Sel.-Khoz. Inst.*, **7**, 116 (1970); through *Chem. Abstr.*, **77**, 84292x (1972).
13. S. V. Iliev and G. Y. Papanov, *Khim.-Farm. Zh.*, **3**, 35 (1969); through *Chem. Abstr.*, **71**, 10636s (1969).
14. B. N. Uzdennikov, *Tr. Tyumenskogo Sel'skokhoz. Inst.*, **7**, 120 (1970); through *Chem. Abstr.*, **77**, 160481a (1972).
15. F. Homburger and E. Boger, *Cancer Res.*, **28**, 2372 (1968).
16. F. J. C. Roe and W. E. H. Field, *Food Cosmet. Toxicol.*, **3**, 311 (1965).
17. B. R. Olin, *Lawrence Rev. Nat. Prod.*, (April 1993).
18. C. Pous et al., *Inflammation*, **156**, 197 (1992).

# UVA URSI

**Source:** ***Arctostaphylos uva-ursi*** (L.) Spreng. (syn. *Arbutus uva-ursi* L.) and its varieties ***coactitis*** and ***adenotricha*** Fern. et Macbr. (Family Ericaceae).

*Synonyms.* Bearberry, common bearberry, beargrape, hogberry, and rockberry.

## GENERAL DESCRIPTION

Trailing evergreen shrub with small leathery obovate to spatulate leaves; up to about 15 cm high; native to the temperate regions of the northern hemisphere (e.g., Europe, northern United States, Canada, and Asia). Part used is the dried leaf. The major producing country is Spain.

## CHEMICAL COMPOSITION

Contains as its active principles 5–18% (usually 7–9%) arbutin (hydroquinone $\beta$-glucoside) and lesser amounts of methylarbutin, with concentrations varying according to ages of leaves, season, localities, and other factors;[1–3] a third glucoside, piceoside, has also recently been isolated.[4] Other constituents include flavonoids (quercetin mono- and diglucosides, myricetin),[5] allantoin,[6] tannins (6–27.5%) of the gallic and ellagic acid types,[2] ursolic acid (0.4–0.75%), phenolic acids (e.g., gallic, ellagic, and quinic acids), an iridoid glycoside (monotropein), uvaol, trace of volatile oil, resin, and others (ESCOP 3; LIST AND HÖRHAMMER STAHL).[7]

## PHARMACOLOGY OR BIOLOGICAL ACTIVITIES

Uva ursi is reported to have diuretic and astringent as well as urinary antiseptic properties (MARTINDALE).

Arbutin undergoes hydrolysis to yield hydroquinone (in the intestines) which has urinary disinfectant activities. Arbutin is reported to be an effective urinary disinfectant if taken in large doses and if the urine is alkaline (LIST AND HÖRHAMMER).[8] However caution should be exercised when using large doses of arbutin as hydroquinone is toxic. Toxic symptoms include tinnitus (ringing in the ear), vomiting, delirium, convulsions, and collapse; death may result (ESCOP 3; GOSSELIN, MERCK). Use is contraindicated in kidney disorders and irritated digestive conditions. Crude drug preparations may induce nausea and vomiting (ESCOP 3). *In vitro* results suggest that antibacterial activity is due either to hydroquinone sulfate ester or to the free hydroquinone.

## USES

**Medicinal, Pharmaceutical, and Cosmetic.** Crude and extracts are used quite extensively as components in certain diuretic as well as laxative preparations. Also used as a urinary disinfectant, especially in Europe.[8] Reported dosage is 1.5–2.5 g of crude drug, infusion, or cold aqueous extract, containing not less than 6.0% of hydroquinone derivatives calculated as anhydrous arbutin. Treatment is limited to seven days or less (ESCOP 3).

**Health Food/Herb Teas.** Crude and extracts used in various capsule, tablet, and tea formulations with an intended diuretic or urinary antiseptic effect; also as tea flavoring due to tannin content (FOSTER AND DUKE).

**Traditional Medicine.** Reportedly used as a diuretic, astringent, and urinary antiseptic. Conditions for which it is used included chronic cystitis, nephritis, kidney stones, and bronchitis, usually in the form of a tea or tincture (LUST).

## COMMERCIAL PREPARATIONS

Crude and extracts; crude and fluid extract were formerly official in N.F. and U.S.P.

Strengths (see ***glossary***) of extracts are generally expressed in weight-to-weight ratios and sometimes in arbutin contents.

*Regulatory Status*. Undetermined in the U.S. Subject of a positive German therapeutic monograph for mild urinary tract infections.[9]

## REFERENCES

See the General References for APhA; BAILEY 1; BIANCHINI AND CORBETTA; BLUMENTHAL; CLAUS; ESCOP 3; FOSTER AND DUKE; LEWIS AND ELVIN-LEWIS; LIST AND HÖRHAMMER; LUST; STAHL; TERRELL; UPHOF; YOUNGKEN.

1. V. Moretti, *Boll. Soc. Ital. Farm. Osp.*, **23**, 207 (1977); through *Chem. Abstr.*, **88**, 60105q (1978).
2. A. A. Makarov, *Uch. Zap. Yakutsk. Gos. Univ.*, **18**, 41 (1971); through *Chem. Abstr.*, **79**, 2724u (1973).
3. E. Rubine, *Nauch. Tr., Irkutsk. Gos. Med. Inst.*, **113**, 9 (1971); through *Chem. Abstr.*, **81**, 60901b (1974).
4. G. A. Karikas et al., *Planta Med.*, **53**, 307 (1987).
5. K. E. Denford, *Experientia*, **29**, 939 (1973).
6. E. Constantinescu et al., *Herba Hung.*, **8**, 101 (1969); through *Chem. Abstr.*, **73**, 101980g (1970).
7. T. Kawai et al., *Osaka Kogyo Daigaku Kiyo, Rikohen*, **19**, 1 (1974); through *Chem. Abstr.*, **88**, 34499b (1978).
8. D. Frohne, *Planta Med.*, **18**, 1 (1970).
9. Monograph *Uvae ursi folium*, *Bundesanzeiger*, no. 228 (Dec.5, 1984).

# VALERIAN ROOT

**Source:** ***Valeriana officinalis*** L. and ***V. jatamansii*** Jones (syn. *V. wallichii* DC.) (Family Valerianaceae).

*Synonyms*. Common valerian, Belgian valerian, all heal, fragrant valerian, and garden valerian (*V. officinalis*); Indian valerian (*V. jatamansii*).

## GENERAL DESCRIPTION

Common valerian (*V. officinalis*) is a perennial herb with deeply dissected leaves each bearing seven to ten pairs of lance-shaped leaflets; stems erect, longitudinally grooved and hollow; up to about 1.5 m high; native to Eurasia and naturalized in North America.

Indian valerian (*V. jatamansii*) is a perennial herb with basal (lower) leaves long-petioled, large and heart-shaped; stems green or with a purplish tint and fine hairy; up to 0.7 m high; native to Asia, especially the Himalayan region (e.g., India and southwestern China).

The genus *Valeriana* includes about 250 northern temperate species; also from South Africa and the Andes; 20 species indigenous to Europe; 16 species are found in the United States and Canada.[1] Parts used are the dried rhizomes and roots, commonly called valerian roots. Major producers of common valerian root include Belgium, France, the former U.S.S.R, and China, while India is the major producer of Indian valerian.

## CHEMICAL COMPOSITION

Common valerian contains as its primary active constituents several iridoid compounds called valepotriates including valtrates (valtrate, valtrate isovaleroxhydrin, acevaltrate, valechlorine, etc.), didrovaltrates (didrovaltrate, homodidrovaltrate, deoxydodidrovaltrate, homodeoxydodidrovaltrate, isovaleroxyhydroxydidrovaltrate, etc.), and isovaltrates (isovaltrate, 7-epideacetylisovaltrate, etc.);[2–7] valtrate and didrovaltrate are the major valepotriates.[2] In addition, it contains valerosidatum (an iridoid ester glycoside)[8] and a volatile oil (0.5–2%) consisting of many components including bornyl acetate and isovalerate (major compounds), caryophyllene, $\alpha$- and $\beta$-pinenes, valerenal, valerenic acid, valeranone, $\beta$-ionone, eugenyl isovalerate, isoeugenyl isovalerate, patchouli alcohol, valerianol, borneol, camphene, $\beta$-bisabolene, ledol, isovaleric acid, and terpinolene, among others (JIANGSU).[9–16]

Common valerian also contains several alkaloids including actinidine valerianine, valerine, and chatinine.[17–25]

Other constituents present in common valerian include choline (ca. 3%), methyl 2-pyrrolyl ketone, chlorogenic acid, and caffeic acid;[25] $\beta$-sitosterol;[6] tannins; gums; and others (JIANGSU).

Indian valerian is generally reported to contain similar constituents as common valerian including valepotriates, valerosidatum, and a volatile oil.[8,26,27] In addition, it contain 2‴-*O*- and 3‴-*O*-2-methylbutyryl esters of acetylated linarin (JIANGSU).[28]

Valepotriates are also present in aerial parts of Indian valerian, but they are absent in those of common valerian.[3,29,30]

## PHARMACOLOGY OR BIOLOGICAL ACTIVITIES

As with the chemical studies, most biological work on valerian has been performed on common valerian.

Valerian has CNS-depressant activities, and it is reported to have antispasmodic and equalizing (sedative in states of agitation and stimulant in fatigue) activities (JIANGSU, STAHL).[31]

In a double-blind study, an aqueous extract of common valerian root was found to decrease sleep latency in eight human subjects who had problem falling asleep.[32]

The valepotriates have been reported to be mainly responsible for the CNS-depressant and antispasmodic effects in laboratory animals.[6,33,34] Valtrate and didrovaltrate as well as baldrinal (a valtrate degradation product) also have antitumor activities against experimental tumors.[27]

A recent study reports that valepotriates, valerenic acid, valeranone and the volatile oil of valerian were ineffective as CNS depressant agents when tested individually. An as yet uncharacterized chemical fraction was proposed as the active substance.[35]

Other activities of valerian include hypotensive in experimental animals; antibacterial, especially against Gram-positive bacteria (due to the alkaloids); antidiuretic; protective against experimental liver necrosis;[36] and others (JIANGSU).

Valerian oil has been reported to be the least toxic among numerous common volatile oils tested orally in rats.[37] Valerenal, valerenic acid, and, to a lesser extent, valeranone, have central depressive effect in mice.[16,38]

Common valerian preparations are considered safe despite the known *in vitro* cytotoxic activity of valepotriates, as these compounds degrade easily, are absent from most products, and are poorly absorbed via oral administration. Acute side effects have not been reported (ESCOP 1).

An ethanol extract of valerian is said to have antidandruff properties.[39]

## USES

**Medicinal, Pharmaceutical, and Cosmetic.** Crude, extracts, tinctures are used in certain sedative preparations, especially in Europe, at doses equivalent to 2–3 g of the drug, 1–3 times per day.[35,40]

**Food.** Extracts and the essential oil (produced by steam distillation) of common valerian are used as flavor components in most major food products, including alcoholic (liqueur, beer, etc.) and nonalcoholic (e.g., root beer) beverages, frozen dairy desserts, candy, baked goods, gelatins and puddings, and meat and meat products, among others. Highest average maximum use levels are about 0.01% of the extract (no type given) reported in alcoholic beverages (96.1 ppm) and baked goods (94.3 ppm), and about 0.002% (16.7%) reported for the oil in baked goods.

**Health Food/Herb Teas.** Widely used in sleep aid and sedative formulations in various dosage forms (teas, tincture, capsule, tablet, etc.).

**Traditional Medicine.** Common valerian root (fresh or dried) is used as antispasmodic, carminative, stomachic, and sedative. Conditions for which it is used include migraine, insomnia, hysteria, neurasthenia, fatigue, stomach cramps that cause vomiting, and other nervous conditions, usually as a tea or an infusion. It is also used externally to treat sores and pimples.

In Chinese medicine, both common and Indian valerian roots as well as those of *V. coreana* Briq., *V. stubendorfi* Kreyer ex Kom., *V. amurensis* P. Smirn. ex Kom., and *V. hardwickii* Wall. are similarly used. In addition, they are used in treating chronic backache, numbness due to rheumatic conditions, colds, menstrual difficulties, and bruises and sores, among others, generally as a decoction or alcoholic infusion.

## COMMERCIAL PREPARATIONS

Crude, extracts, and oil; crude and fluid extract were formerly official in N.F. Strengths (see ***glossary***) of extracts are expressed in weight-to-weight ratios, although certain manufacturers have their own in-house bioassays.

*Regulatory Status*. Has been approved for food use (§172.510); only *V. officinalis* is listed. Subject of a positive German therapeutic monograph.

## REFERENCES

See the General References for ARCTANDER; BAILEY 1; BIANCHINI AND CORBETTA; BLUMENTHAL; ESCOP 1; FEMA; FOSTER; FOSTER AND DUKE; GOSSELIN; JIANGSU; LUST; MARTINDALE; STAHL; TYLER 1; UPHOF.

1. S. Foster, *Valerian, Botanical Series, No. 312,* American Botanical Council Austin, Texas, 1991.
2. S. Popov et al., *Phytochemistry*, **13**, 2815 (1974).
3. E. D. Funke and H. Friedrich, *Planta Med.*, **28**, 215 (1975).
4. N. Marekov et al., *Izv. Khim.*, **8**, 672 (1975); through *Chem. Abstr.*, **85**, 74960q (1976).
5. J. H. Van Meer et al., *Pharm. Weekbl.*, **112**, 20, (1977).
6. V. D. Petkov et al., *Dokl. Bolg. Akad. Nauk*, **27**, 1007 (1974); through *Chem. Abstr.*, **82**, 51636n (1975).
7. G. Verzarne-Petri et al., *Herba Hung.*, **15**, 79 (1976); through *Chem. Abstr.*, **86**, 78610u (1977).
8. P. W. Thies, *Tetrahedron Lett.*, (28), 2471 (1970).
9. H. Wagner et al., *Arzneim.-Forsch.*, **22**, 1204 (1972).
10. E. Lemberkovics et al., *Sci. Pharm.*, **45**, 281 (1977); through *Chem. Abstr.*, **88**, 197425p (1978).
11. H. Hendricks et al., *Phytochemistry*, **16**, 1853 (1977).
12. H. Hoerster et al., *Phytochemistry*, **16**, 1070 (1977).
13. G. Ruecker and J. Tautges, *Phytochemistry*, **15**, 824 (1976).
14. G. Jommi et al., *Collect. Czech. Chem. Commun.*, **34**, 593 (1969).
15. E. Pethes et al., *Sci. Pharm.*, **43**, 173 (1975); through *Chem. Abstr.*, **84**, 21981v (1976).
16. H. Hendricks et al., *Planta Med.*, **42**, 62 (1981).
17. K. Torssell and K. Wahlberg, *Tetrahedron Lett.*, (4), 445 (1966).
18. E. Cionga et al., *Adv. Mass Spectrom. Biochem Med.*, **1**, 299 (1976); through *Chem. Abstr.*, **85**, 74964u (1976).
19. A. Buckova et al., *Cesk. Farm.*, **26**, 308 (1977); through *Chem. Abstr.*, **88**, 86063z (1978).
20. B. Franck et al., *Angew, Chem., Int. Ed. Engl.*, **9**, 891 (1970); through *Chem. Abstr.*, **74**, 54058w (1971).
21. D. Gross et al., *Arch. Pharm. (Weinheim),* **304**, 19 (1971).
22. C. G. Casinovi, *Fitoterapia*, **39**, 45 (1968).
23. K. Torssell and K. Wahlberg, *Acta Chem. Scand.*, **21**, 53 (1967).
24. R. D. Johnson and G. R. Waller, *Phytochemistry*, **10**, 3334 (1971).
25. G. R. Szentpetery et al., *Pharmazie*, **18**, 816 (1963).
26. G. D. Joshi et al., *Perfum. Essent. Oil Records*, **59**, 187 (1968).
27. C. Bounthanh et al., *Planta Med.*, **41**, 21 (1981).
28. V. M. Chari et al., *Phytochemistry*, **16**, 1110 (1977).
29. J. Hoelzl and K. Jurcic, *Planta Med.*, **27**, 133 (1975).
30. E. D. Funke and H. Friedrich, *Phytochemistry*, **13**, 2023 (1974).
31. E. Cionga, *Pharmazie*, **16**, 43 (1961).
32. P. D. Leathwood and F. Chauffard, *Planta Med.*, **51**, 144 (1985).
33. P. Manolov and V. Petkov, *Farmatsiya (Sofia),* **26**, 29 (1976); through *Chem. Abstr.*, **85**, 137513u (1976).

34. K. W. von Eickstedt and S. Rahman, *Arzneim.-Forsch.*, **19**, 316 (1969).
35. J. Krieglstein and D. Grusla, *Deutsche Apotheker Zeitung.*, **40**, 2041 (1988).
36. M. A. Farooki, *Pakistan Med. Forum*, **1**, 19 (1966); through *Chem. Abstr.*, **67**, 20399y (1967).
37. E. Skramlik, *Pharmazie*, **14**, 435 (1959).
38. H. Hendricks et al., *Planta Med.*, **51**, 28 (1985).
39. T. Abe, Jpn Kokai 72 47, 664 (1972); through *Chem. Abstr.*, **80**, 19389g (1974).
40. C. Hobbs, *HerbalGram*, **21**, 19 (1989).

# VANILLA

**Source:** ***Vanilla planifolia*** Andr. [syn. *V. fragrans* (Salisb.) Ames] and ***V. tahitensis*** J. W. Moore (Family Orchidaceae).

*Synonyms*. Bourbon vanilla, Réunion vanilla, and Mexican vanilla (*V. planifolia*); Tahiti vanilla (*V. tahitensis*); common vanilla.

## GENERAL DESCRIPTION

Large green-stemmed perennial herbaceous vines, reaching a length of about 25 m or more in their wild state; native to tropical America (especially Mexico); cultivated in the tropics (Madagascar, Comoros Islands, French Polynesia, Tahiti, Indonesia, Réunion, Seychelles, Mexico, Tanzania, Uganda, etc.). Part used is the fully grown but unripe fruit (a capsule) commonly called pod or bean, collected four to nine months after pollination; pollination is all done artificially, except in Mexico where it is partly performed artificially and partly by certain indigenous hummingbirds and butterflies not found elsewhere. The pods are then subject to a complicated labor-intensive curing (fermentation) and drying process which requires five to six months to complete. During this period, vanillin is enzymatically produced and may accumulate as white crystals on the surface of the beans; the pods also turn brown and lose 80% of their weight. A much faster process which takes only a few days is reportedly used in Uganda where the beans are processed mechanically. By far the largest vanilla producer is Madagascar. The Mexican beans and the Bourbon beans, the latter grown in Madagascar, are considered of the best quality; they are larger than the Tahiti beans. The United States is the leading importer and consumer.[1]

The quality of vanilla beans does not depend on the vanillin content, even though vanillin is generally recognized as having the "vanilla" odor. Other constituents present together with vanillin as a whole are responsible for the flavor and quality of vanilla and its extracts. Value is determined by fragrance rather than vanillin content.[1]

The so-called "single-strength vanilla extract" is comparable to a 1:0.1 or 10% tincture (see ***glossary***) containing at least 35% alcohol. The beans used are required by federal regulations to contain no more than 25% moisture, otherwise proportionately more beans will have to be used. The so-called "tenfold extract" is comparable to a 1:1 fluidextract (see ***glossary***) in terms of extract to beans (crude) ratio and is ten times stronger than the single-strength extract.

Due to the high price of vanilla and the low cost of vanillin, vanilla extracts have been extensively adulterated (ARCTANDER). There is still no simple method to detect with certainty whether or not a vanilla extract is authentic.[2,3]

## CHEMICAL COMPOSITION

Vanilla contains vanillin (1.3–3.0%) as the major flavor component, with over 150 other aroma chemicals also present, most of which are present in traces, including *p*-hydroxybenzaldehyde, acetic acid, isobutyric acid, caproic acid, eugenol, furfural, *p*-hydroxybenzyl methyl ether, vanillyl ethyl ether, anisyl ethyl ether, and acetaldehyde.[2,4–8]

The vanillin content differs in different varieties of vanilla, with Bourbon beans containing generally higher amounts than Mexican and Tahiti beans (MARTINDALE, ROSENGARTEN, STAHL).

Other constituents present include resins, sugars, and fixed oil.

## PHARMACOLOGY OR BIOLOGICAL ACTIVITIES

Vanilla has been reported to have allergenic properties in humans, and vanillin was found not to be the principal active agent.[9]

## USES

Many of the uses of vanilla have been replaced by vanillin. However vanillin cannot replace vanilla in many applications where a delicate natural vanilla flavor or fragrance is called for.

**Medicinal, Pharmaceutical, and Cosmetic.** Vanilla extracts (especially tincture N.F.) are used in pharmaceutical preparations such as syrups, primarily as a flavoring agent.

Vanilla extracts (tincture, absolute, etc.) are used as fragrance ingredients in perfumes.

**Food.** Vanilla, vanilla extract, and vanilla oleoresin are widely used as flavor ingredients in most food products, including alcoholic (e.g., liqueurs) and nonalcoholic beverages, frozen dairy desserts (especially ice cream and yogurt), candy, baked goods, gelatins and puddings, and others. Highest average maximum use level is about 0.964% (9,642 ppm) reported for vanilla in baked goods.

## COMMERCIAL PREPARATIONS

Crude and extracts. Crude and tincture are official in N.F.

*Regulatory Status.* GRAS (§182.10, §182.20, and §169.3).

## REFERENCES

See the General References for ARCTANDER; BAILEY 2; CLAUS: FEMA; GUENTHER; KARRER; MARTINDALE; MERCK; REMINGTON; ROSENGARTEN; STAHL; TERRELL; UPHOF.

1. E. Westpahl and P. C. M. Jansen, eds., *Plant Resources of South-East Asia: A Selection*, Pudoc, Wageningen, Netherlands, 1989.
2. G. E. Martin et al., *Food Technol.*, **29**, 54 (1975).
3. G. E. Martin et al., *Food Sci.*, **42**, 1580 (1977).
4. B. M. Lawrence, *Perfum. Flav.*, **2**, 3 (1977).
5. H. Shiota and K. Itoga, *Koryo*, **113**, 65 (1975); through *Chem. Abstr.*, **84**, 163018k (1976).
6. I. Klimes and D. Lamparsky, *Int. Flavours Food Addit.*, **7**, 272 (1976).
7. H. Bohnsack, *Riechst. Aromen, Körperpflegem.*, **17**, 133 (1967).
8. H. Bohnsack, *Riechst. Aromen, Körperpflegem.*, **15**, 284 (1965).
9. D. L. J. Opdyke, *Food Cosmet. Toxicol.*, **14**, 633 (1976).

# WINTERGREEN OIL

**Source:** ***Gaultheria procumbens*** L. and other ***Gaultheria*** species (Family Ericaceae).

*Synonyms.* Checkerberry, teaberry, or gaultheria oil.

## GENERAL DESCRIPTION

An evergreen shrub with slender and extensively creeping stems from which arise erect branches bearing at the top oval, leathery leaves with toothed (often bristly), margins; up to about 15 cm high; native to North America, growing from Newfoundland to Manitoba and south to Georgia and Alabama. Part used is the leaf. Wintergreen oil is obtained by steam distillation of the warm water-macerated leaves whereby gaultherin present in the leaves is enzymatically hydrolyzed to yield methyl salicylate (which is subsequently distilled with steam), D-glucose, and D-xylose (see ***sweet birch oil***). The yield of oil is normally 0.5–0.8% from the leaves (LIST AND HÖRHAMMER).[1]

## CHEMICAL COMPOSITION

Wintergreen oil contains almost exclusively methyl salicylate (≥98%).

## PHARMACOLOGY OR BIOLOGICAL ACTIVITIES

See ***sweet birch oil***.

## USES

**Medicinal, Pharmaceutical, and Cosmetic.** Wintergreen oil is used interchangeably with sweet birch oil or methyl salicylate (see ***sweet birch oil***).

**Food.** Wintergreen oil is used for similar flavoring purposes as sweet birch oil, though its use levels reported are generally lower than those of sweet birch oil; the highest average maximum use level is about 0.04% (405 ppm) in candy (see ***sweet birch oil***).

**Health Food/Herb Teas.** The dried leaves are used as a tea flavoring ingredient.

**Traditional Medicine.** Leaf tea reportedly for colds, headache, stomachache, fevers, kidney ailments; externally, wash for rheumatism, sore muscles, and lumbago (FOSTER AND DUKE).

## COMMERCIAL PREPARATIONS

Volatile oil; formerly official in U.S.P. and currently official in F.C.C. (see ***sweet birch oil***). Wintergreen oil is slightly levorotatory, while sweet birch oil and synthetic methyl salicylate are optically inactive.

*Regulatory Status.* Not listed under §172.510, §182.10, or §182.20.

## REFERENCES

See the General References for APhA; ARCTANDER; BAILEY 1; FERNALD; FEMA; FOSTER AND DUKE; GUENTHER; LIST AND HÖRHAMMER; MARTINDALE; TERRELL; UPHOF; USD 26th.

1. C. C. Mu and I. C. Yang, *Yao Hsueh Hsueh Pao*, **13**, 451 (1966); through *Chem. Abstr.*, **66**, 8821e (1967).
2. B. R. Olin, ed., *Lawrence Rev. Nat. Prod.* (Aug. 1992).

# WITCH HAZEL

**Source:** ***Hamamelis virginiana*** L. (Family Hamamelidaceae).

*Synonym.* Hamamelis.

## GENERAL DESCRIPTION

A deciduous shrub or small tree flowering in the fall; up to about 7.5 m high; native to North America; distributed from Quebec to Georgia and west to Minnesota. Parts used are the dried leaves, bark, and partially dried dormant twigs. Witch hazel water (also known as hamamelis water and distilled witch hazel extract) is obtained from the recently cut and partially dried dormant twigs. *Hamamelis vernalis* Sarg. leaves and twigs vicariously enter commercial supplies from the Ozark plateau.

The twigs are macerated for about 24 hours in twice their weight of warm water followed by distilling and adding the required amount of alcohol to the distillate and thoroughly mixing, such that 1000 volumes witch hazel water are derived from 1000 parts crude.

## CHEMICAL COMPOSITION

Witch hazel leaf contains 8–10% tannin that is composed of hamamelitannin or digallyhamamelose, gallotannins, and/or proanthocyanidins (LIST AND HÖRHAMMER).[1–3] Other constituents present include free gallic acid, free hamamelose, saponins, choline, resins, flavonoids [quercetin, kaempferol, astragalin (kaempferol-3-glucoside), quercitrin, afzelin, myricitrin, etc.], 0.5% volatile oil (*n*-hexen-2-al, hexenol, $\alpha$- and $\beta$-ionones, safrole, sesquiterpenes, etc.), and others (LIST AND HÖRHAMMER).[4–7]

The bark contains 1 to 7% hamamelitannin and smaller amounts of condensed tannins (e.g., *d*-gallocatechin, *l*-epicatechin gallate, and *l*-epigallocatechin); saponins; fixed oil (0.6%); wax; 0.5% volatile oil (sesquiterpenes, a phenol, etc.); and a resin (LIST AND HÖRHAMMER).[3,8,9]

Witch hazel water contains a trace of volatile oil consisting of eugenol, carvacrol, and probably similar compounds as the volatile oils of leaf and bark.[7] As it is a steam distillate, it does not contain tannins.

## PHARMACOLOGY OR BIOLOGICAL ACTIVITIES

Witch hazel leaf, witch hazel bark, and witch hazel water have all been reported to have astringent and hemostatic properties. These properties can be attributed to the tannins contained in the leaves and bark, but it is not known what is responsible for these activities in hamamelis water (LIST AND HÖRHAMMER, MARTINDALE).

## USES

**Medicinal, Pharmaceutical, and Cosmetic.** Witch hazel leaf extract, witch hazel bark extract, and witch hazel water are all used as astringent and hemostatic in preparations (suppositories, ointments, lotions, cloth wipes, etc.) for use in treating hemorrhoids, itching, irritations, and minor pains, with witch hazel water the most commonly used. They are also used in eye drops, shaving lotions, and others.

In European phytomedicine preparations are used as an astringent, antiinflammatory and local hemostyptic for mild skin injuries, hemorrhoids, varicose veins, and local inflammations of the skin and mucous membranes.[10]

The bottled "witch hazel" in most domestic medicine cabinets is simply witch hazel water.

**Health Food/Herb Teas.** Leaves sometimes used as tea ingredient.

**Traditional Medicine.** The leaf and bark are reportedly used internally to treat diarrhea and externally to treat mouth and throat irritations, hemorrhoids, eye inflammation, insect bites, minor burns, and other skin irritations, usually as a decoction, poultice, or ointment. They have also been used in cancers.[11]

**COMMERCIAL PREPARATIONS**

Crudes (leaf and bark), extracts (solid, fluid, etc.), and witch hazel water. Witch hazel leaf and its fluid extracts were formerly official in N.F.; witch hazel bark was formerly official in U.S.P.; and witch hazel water was formerly official in N.F. Strengths (see ***glossary***) of extracts and distillate are expressed in weight-to-weight or volume to weight ratios.

*Regulatory Status.* Used in OTC preparations in the U.S. The leaves and branches are the subject of a positive German therapeutic monograph, indicated for minor skin injuries, local skin and mucous membrane irritation, hemorrhoids, and varicose veins.[10]

**REFERENCES**


See the General References for BAILEY 1; BLUMENTHAL; CLAUS; FERNALD; FOSTER; KROCHMAL AND KROCHMAL; LIST AND HÖRHAMMER; LUST; MARTINDALE; TERRELL; USD 26th.

1. G. Netien and R. Rochan, *Bull. Trav. Soc. Pharm. Lyon*, **12**, 121 (1968); through *Chem. Abstr.*, **71**, 94709d (1969).
2. H. Friedrich and N. Krueger, *Planta Med.*, **26**, 327 (1974).
3. B. Vennat et al., *Planta Med.*, **54**, 454 (1988).
4. P. Bernard et al., *J. Pharm. Belg.*, **26**, 661 (1971); through *Chem. Abstr.*, **76**, 90099k (1972).
5. W. Messerschmidt, *Arch. Pharm. (Weinheim)*, **300**, 550 (1967); through *Chem. Abstr.*, **67**, 97610t (1967).
6. W. Messerschmidt, *Arzneim.-Forsch.*, **18**, 1618 (1968).
7. H. Janistyn, *Parfüm Kosmet.*, **45**, 335 (1964).
8. H. Glick et al., *Carbohydr. Res.*, **39**, 160 (1975).
9. H. Friedrich and N. Krueger, *Planta Med.*, **25**, 138 (1974).
10. Monograph *Hamamelidis folium et cortex*, *Bundesanzeiger*, no. 154 (Aug. 21, 1985); revised (Mar. 13, 1990).
11. J. L. Hartwell, *Lloydia*, **32**, 247 (1969).


# WOODRUFF, SWEET

**Source:** ***Galium odoratum*** (L.) Scop. (syn. *Asperula odorata* L.) (Family Rubiaceae).

*Synonyms.* Woodruff, master of the wood, and woodward.

**GENERAL DESCRIPTION**

A small perennial herb with a creeping rhizome from which smooth erect stems arise (up to 30-cm high) bearing lance-shaped leaves in whorls of six to eight (usually eight); native to Eurasia and northern Africa; naturalized in North America. Part used is the dried whole flowering herb which does not have any odor when fresh but develops a new-mown hay (coumarinlike) odor on drying. European countries (especially Germany) are the major producers.

## CHEMICAL COMPOSITION

Generally reported to contain coumarin in bound form (glycoside) that is set free by enzymatic action during wilting or drying. However one study did not detect any coumarins in woodruff.[1]

Other constituents reported present include asperuloside (0.05%), monotropein, tannins, anthracene and naphthalene derivatives, traces of nicotinic acid, fixed oil, and bitter principle, among others (LIST AND HÖRHAMMER, MERCK).[2,3]

## PHARMACOLOGY OR BIOLOGICAL ACTIVITIES

Asperuloside and the leaves are reported to have antiinflammatory activity (LIST AND HÖRHAMMER).[4]

Asperuloside has been suggested as a starting material for prostaglandins.[5]

Coumarin is reported to be toxic (see ***deertongue***).

## USES

**Medicinal, Pharmaceutical, and Cosmetic.** Used in sachets. Extracts (e.g., concrète and absolute) are used as fragrance components in perfumes, mostly in Europe.[6]

**Food.** It is reported used as a flavor component in various major food products, including alcoholic (May wines, vermouths, bitters, etc.) and nonalcoholic beverages, frozen dairy desserts, candy, baked goods, and gelatins and puddings. Highest average maximum use level is 0.04% in baked goods.

**Traditional Medicine.** The herb is traditionally used as a diaphoretic, antispasmodic, sedative (particularly for children and elderly people), and diuretic. Conditions for which it is used include restlessness, insomnia, stomachache, migraine, neuralgia, and bladder stones, usually as a tea (FOSTER).

In European tradition, reportedly used for prophylaxis and therapy of respiratory conditions, as well as gallbladder, kidney, and circulatory disorders; topically for venous conditions, hemorrhoids; antiinflammatory; claimed efficacy not documented.[7]

## COMMERCIAL PREPARATIONS

Mainly crude.

*Regulatory Status.* Has been approved for use in alcoholic beverages only (§172.510). Subject of a German therapeutic monograph; use not recommended as efficacy is not documented.[7]

## REFERENCES

See the General References for BAILEY 1; BIANCHINI AND CORBETTA; FEMA; FERNALD; FOSTER; LIST AND HÖRHAMMER; LUST; TUCKER; UPHOF.

1. M. I. Borisov, *Khim. Prir. Soedin.*, **10**, 82 (1974); through *Chem. Abstr.*, **81**, 60805y (1974).
2. A. Buckova et al., *Acta Fac. Pharm., Univ. Comeniana*, **19**, 7 (1970); through *Chem. Abstr.*, **75**, 59839g (1971).
3. A. R Burnett and R. H. Thomson, *J. Chem. Soc., C*, (7), 854 (1968).
4. N. Mascolo et al., *Phytother. Res.*, **1**(1), 28 (1987).
5. W. F. Berkowitz et al., *J. Org. Chem.*, **47**, 824 (1982).
6. L. Trabaud, *Perfum. Essent. Oil Records*, **54**, 382 (1963).
7. Monograph *Galii odorati herba*, *Bundesanzeiger*, no. 193 (Oct. 15, 1987).

# XANTHAN GUM

**Source:** Derived from a pure-culture fermentation of glucose using the bacterium *Xanthomonas campestris*.

## GENERAL DESCRIPTION

Xanthan gum is a polysaccharide secreted by certain species of bacteria. It is produced by a pure-culture fermentation of glucose using the bacterium *Xanthomonas campestris*, reported to be a minor plant pathogen, though the strain used is reportedly nonpathogenic and nontoxic to humans.[1-5]

During fermentation the bacteria utilize the glucose and other nutrients in the culture medium to produce a high molecular weight polysaccharide which they excrete into the culture broth (see also ***monosodium glutamate***). The polysaccharide is recovered by precipitation and purification with isopropyl alcohol, followed by drying and milling to yield commercial xanthan gum.

Xanthan gum is a cream-colored powder that is readily soluble in cold or hot water to form neutral, viscous, and nonthixotropic (thinning when disturbed) solutions. They have relatively high viscosity which is unusually stable toward changes in temperatures, acidity, alkalinity, and salt content. The also have good freeze-thaw stability (GLICKSMAN).[1,3]

## CHEMICAL COMPOSITION

Xanthan gum is a very long, linear polymer. Its molecular weight has been reported to range from 1 million to 10 million. The molecule consists of a chain composed of D-glucose, D-mannose, and D-glucuronic acid with short side chains. Pyruvic acid, present as side chains, accounts for 2.5–4.8% of the molecule (GLICKSMAN, WHISTLER, AND BEMILLER).[5]

Xanthan gum samples with high pyruvate contents have been reported to yield solutions with higher viscosity than those of low pyruvate samples.[5]

## PHARMACOLOGY OR BIOLOGICAL ACTIVITIES

Xanthan gum has been reported to be nontoxic to three animal species (rat, cat, and dog) studied.[6] However no toxicity or safety data on humans are reported.

Xanthan gum has been reported to be active against Ehrlich ascites tumor and S-180 in mice, being synergistic with 5-fluorouracil or bleomycin in S-180.[7]

## USES

**Medicinal, Pharmaceutical, and Cosmetic.** Xanthan gum is used as a stabilizer, thickener, and an emulsifying agent in water-based pharmaceutical and cosmetic preparations.

**Food.** Xanthan gum was approved for food use in 1969 as a stabilizer, emulsifier, thickener, suspending agent, bodying agent, and foam enhancer.[2] It was also listed as an optional emulsifier for French dressing in 1971 under the Standard of Identity for French dressing.[8] Since then, it has been extensively used in many types of food products, including salad dressings (e.g., French), dairy products (chocolate milk drinks, puddings, cheese spread, etc.) canned products (meat, fish, and poultry), and others.

## COMMERCIAL PREPARATIONS

Available in powdered forms. Xanthan gum is official in N.F. and F.C.C.

*Regulatory Status*. Has been approved for food use (§172.695).

## REFERENCES

See the General References for FURIA; GLICKSMAN; LAWRENCE; MARTINDALE; WHISTLER AND BEMILLER.

1. C. T. Blood in L. W. Codd et al., eds., *Chemical Technology: An Encyclopedic Treatment,* Vol. 5, Barnes & Noble, New York, 1972, p. 27.
2. *Fed. Regist.*, **34** (53), 5376 (1969).
3. J. K. Rocks, *Food Technol.*, **25**, 476 (1971).
4. R. W. Silman and P. Rogovin, *Biotechnol. Bioeng.*, **14**, 23 (1972).
5. P. A. Sandford et al., *ACS Symp. Ser.*, **45**, 192 (1977).
6. W. H. McNeely and P. Kovacs, *ACS Symp. Ser.*, **15**, 269 (1975).
7. M. Oda, *Yakuri to Chiryo*, **13**, 5743 (1985); through *Chem. Abstr.*, **106**, 131313j (1987).
8. *Fed. Regist.*, **36**, 9010 (1971).

# YARROW

**Source:** ***Achillea millefolium*** L. (Family Compositae or Asteraceae).

*Synonyms*. Milfoil, common yarrow, nosebleed, and thousand leaf.

## GENERAL DESCRIPTION

A perennial herb with a simple stem bearing aromatic bipinnately parted and dissected leaves, giving a lacy appearance; up to about 1 m high; native to Eurasia and naturalized in North America; found in most temperate zones of the world (e.g., United States, Canada, throughout Europe, and northern China). There are numerous varieties or forms. The taxon *A. millefolium* is a species complex represented by a number of other species.[1] Part used is the entire flowering aboveground herb.

## CHEMICAL COMPOSITION

Much work has been performed on yarrow especially regarding its volatile oil composition, which is highly variable.

Yarrow contains about 0.1–1.4% volatile oil that is composed of azulene (0–51%), $\alpha$- and $\beta$-pinenes (mainly $\beta$-), caryophyllene, borneol, terpineol, cineole, bornyl acetate, camphor, sabinene, isoartemisia ketone, and other compounds (including a trace of thujone).[2–9] The relative composition varies considerably (especially its azulene content) depending on sources (JIANGSU, LIST AND HÖRHAMMER).[3–6,9] *A. millefolium* L. sensu stricto (a hexaploid) contains no azeulene.[1] Its oil primarily contains monoterpenes including linalool (26%), camphor (18%), borneol, 1,8-cineole, and others.[10]

Other constituents reported include lactones (achillin, millefin, deacetylmatricarine, etc.);[11,12] flavonoids (e.g., rutin);[13] tannins; resin; coumarins; saponins;[14] sterols (e.g., $\beta$-sitosterol and its acetate); alkanes (mainly tricosane, heptadecane, and pentacosane); fatty acids (linoleic, palmitic, oleic acids, etc.);[15] sugars (glucose, galactose, sucrose, arabinose, inositol, dulcitol, mannitol, etc.); alkaloids or bases (betaine, choline, trigonelline, betonicine, and stachydrine); amino acids (alanine, leucine, lysine, histidine, glutamic acid; etc.); and acids (succinic acid and salicylic acid);[16] among others (JIANGSU, LIST AND HÖRHAMMER).

## PHARMACOLOGY OR BIOLOGICAL ACTIVITIES

An aqueous extract of yarrow flower heads has been reported to have antiinflammatory activity in laboratory animals; this activity was due to a mixture of protein–carbohydrate complexes that had very low toxicity.[17,18]

Yarrow alkaloids (bases) are reported to have hypotensive and weakly antipyretic, astringent, antibacterial, and choleretic properties. Yarrow extracts also exhibited hemostatic properties, among others (JIANGSU).[19,20]

Yarrow may cause contact dermatitis in certain individuals.[21]

## USES

**Medicinal, Pharmaceutical, and Cosmetic.** In European phytomedicine herb (at a daily dose of 4.5 g) is used for appetite loss, dyspeptic complaints, spasmodic gastrointestinal disturbances, etc. Use is contraindicated in known hypersensitivity to other Compositae members.[20]

Extracts used in preparations (e.g., baths) for their alleged soothing and quieting effects on the skin (de NAVARRE),[22] in hair tonic and antidandruff preparations.[23]

**Health Food/Herb Teas.** Used in fever and cold formulation, mostly in combinations, as minor ingredient. Tincture used as topical styptic (FOSTER).

**Food.** Used in bitters and vermouths, with average maximum use level of less than 0.001% (5 ppm) reported.

Yarrow flower is used in herb teas.

**Traditional Medicine.** Reportedly used as a tonic, carminative, febrifuge, antispasmodic, astringent, hemostatic, and others, usually in the form of an infusion or decoction, or the fresh juice. Conditions for which it is used include lack of appetite, stomach cramps, flatulence, gastritis, enteritis, internal and external bleeding of all kinds (coughing blood, nosebleed, hemorrhoidal bleeding, bloody urine, etc.), wounds, sores, and skin rash.

In China both the fresh herb and the dried herb are used to treat similar conditions. The fresh herb (mashed as a poultice) is especially recommended for all sorts of sores and wounds as well as dog and snake bites, while the dried herb is recommended for internal bleeding (especially menstrual and hemorrhoidal).

It has been used in cancers.[24]

## COMMERCIAL PREPARATIONS

Mainly crude; it was formerly official in U.S.P.

*Regulatory Status.* Has been approved for use in alcoholic beverages only; finished beverage must be thujone free (§172.510). Yarrow oil normally contains little or no thujone whereas sage oil contains normally about 50% thujone (see ***sage***). The herb is the subject of a positive German therapeutic monograph.[20]

## REFERENCES

See the General References for BAILEY 1; BIANCHINI AND CORBETTA; BLUMENTHAL; FEMA; FERNALD; FOSTER; FOSTER AND DUKE; JIANGSU; KROCHMAL AND KROCHMAL; LIST AND HÖRHAMMER; LUST.

1. R. F. Chandler et al., *Econ. Bot.*, **36**, 203 (1982).
2. G. Verzar-Petri and A. S. Shalaby, *Acta Agron. Acad. Sci. Hung.*, **26**, 337 (1977); through *Chem. Abstr.*, **88**, 65858x (1978).
3. J. Kozlowski and J. Lutomski, *Planta Med.*, **17**, 226 (1969).
4. H. Popescu and H. Winand, *Clujul Med.*, **50**, 78 (1977); through *Chem. Abstr.*, **87**, 172694s (1977).
5. A. Ruminska, *Acta Agrobot.*, **23**, 53 (1970); through *Chem. Abstr.*, **75**, 16226n (1971).
6. M. Y. Haggag et al., *Planta Med.*, **27**, 361 (1975).
7. R. B. Chelishvili and A. I. Tavberidze, *Maslo-Zhir. Prom.*, **2**, 24 (1974); through *Chem. Abstr.*, **81**, 54273g (1974).
8. G. Verzar-Petri and Hanh N. Cuong, *Acta Pharm. Hung.*, **47**, 134 (1977); through *Chem. Abstr.*, **87**, 73219j (1977).
9. A. J. Falk et al., *Lloydia*, **37**, 598 (1974).
10. L. Hofmann et al., *Phytochemistry*, **31**, 537 (1992).
11. S. J. Smolenski et al., *Lloydia*, **30**, 144 (1967).
12. S. Z. Kasymov and G. P. Sidyakin, *Khim. Prir. Soedin.*, **2**, 246 (1972); through *Chem. Abstr.*, **77**, 85556y (1972).
13. I. D. Neshta et al., *Khim. Prir. Soedin.*, **5**, 676 (1972); through *Chem. Abstr.*, **78**, 94818h (1973).
14. K. S. Tillyaev et al., *Rast. Resur.*, **9**, 58 (1973); through *Chem. Abstr.*, **78**, 121284q (1973).
15. C. Ivanov and L. Yankov, *God. Vissh. Khimikotekhnol. Inst. Sofia*, **14**, 61 and

73 (1970); through *Chem. Abstr.*, **77**, 111473r and 111474s (1972).
16. C. Ivanov and L. Yankov, *God. Vissh. Khimikotekhnol. Inst. Sofia*, **14**, 195 and 223 (1971); through *Chem. Abstr.*, **77**, 111471p and 111472q (1972).
17. A. S. Goldberg and E. C. Mueller, *J. Pharm. Sci.*, **58**, 938 (1969).
18. A. Goldberg et al., *U.S. Pat.* 3,552,350 (1970); through *Chem. Abstr.*, **73**, 102048w (1970).
19. F. W. Kudrzycka-Bieloszabska and K. Glowniak, *Diss. Pharm. Pharmacol.*, **18**, 449 (1966); through *Chem. Abstr.*, **67**, 62837v (1967).
20. Monograph *Achillea millefolium*, *Bundesanzeiger*, no. 22 (Feb. 1, 1990).
21. J. C. Mitchell in V. C. Runeckles, ed., *Recent Advances in Phytochemistry*, Vol. 9, Plenum Press, New York, 1975, p. 119.
22. P. Alexander, *Cosmet. Perfum.*, **88**, 35 (1973).
23. H. Greger and O. Hofer, *Planta Med.*, **55**, 216 (1989).
24. J. L. Hartwell, *Lloydia*, **31**, 71 (1968).

# YERBA SANTA

**Source:** ***Eriodictyon californicum*** (Hook. et Arn.) Torr. (syn. *E. glutinosum* Benth. and *Wigandia californicum* Hook. et Arn.) (Family Hydrophyllaceae).

*Synonyms*. Eriodictyon, bear's weed, consumptives' weed, mountain balm, and tarweed.

## GENERAL DESCRIPTION

An evergreen aromatic shrub with woody rhizomes from which arise glutinous nonhairy stems bearing lance-shaped leaves (5–15 cm long) that are glutinous and nonhairy above but hairy beneath; up to 2.2 m high; native to California, extending north to Oregon and south to northern Mexico. Part used is the dried leaf.

## CHEMICAL COMPOSITION

Contains 3–6% eriodictyonine (homoeriodictyol); 0.23–0.6% eriodictyol (5,7,4′,3′-tetrahydroxyflavanone); other flavonoids including 3′-methyl-4′-isobutyryleriodictyol, pinocembrin, sakuranetin, cirsimaritin, chrysoeriol, hispidulin, chrysin, among others; eriodictyonic acid (probably impure eriodictyonine); a resin consisting of triacontane, pentatriacontane, cerotic acid, chrysoeriodictyol, xanthoeriodictyol (xanthoeriodol), and eriodonol (eriodonal), among others; tannin; gum; sugars; fats; formic and acetic acids; a trace of volatile oil; and others (LIST AND HÖRHAMMER).[1]

## PHARMACOLOGY OR BIOLOGICAL ACTIVITIES

Eriodictyol is reported to have expectorant properties (MERCK).

## USES

**Medicinal, Pharmaceutical, and Cosmetic.** Yerba santa extracts are used mainly in flavoring pharmaceutical preparations, particularly those containing bitter drugs such as quinine whose bitter taste can be masked by the yerba santa extracts.

**Food.** The fluid extract is used as a flavor component in alcoholic and nonalcoholic beverages, frozen dairy desserts, and baked goods, with highest average maximum use level of 0.05% in baked goods.

**Health Food/Herb Teas.** Tincture available, primarily from local California manufacturers, used for traditional indications.

**Traditional Medicine.** Traditionally used as an expectorant, antispasmodic, febrifuge, and tonic. Conditions for which it is used include asthma and chronic bronchitis. Also used as a poultice to treat wounds, bruises, sprains, and insect bites (LUST; UPHOF).

**COMMERCIAL PREPARATIONS**

Crude and extracts; crude, fluid extract, and aromatic syrup are official in N.F. Strengths (see ***glossary***) of extracts are expressed in weight-to-weight ratios.

*Regulatory Status.* Has been approved for food use (§172.510).

**REFERENCES**

See the General References for FEMA; LIST AND HÖRHAMMER; LUST; MARTINDALE; MUNZ AND KECK; REMINGTON; UPHOF.

1. Y. L. Liu et al., *J. Nat. Prod.*, **55**, 357 (1992).

# YLANG YLANG OIL

**Source:** ***Cananga odorata*** (Lam.) Hook. f. et Thoms. forma ***genuina*** (Family Annonaceae).

**GENERAL DESCRIPTION**

Both ylang ylang oil and cananga oil are derived from the same species, *Cananga odorata*, a tree native to tropical Asia (especially Indonesia and the Philippines). The species is reported to exist in different forms; ylang ylang is a different form than cananga (see ***cananga oil***). Part used is the fresh flower picked early in the morning. Ylang ylang oil is obtained by steam distillation or water and steam distillation of the freshly picked flowers. The first distillate (usually ca. 40% of the total distillate) constitutes ylang ylang extra, which is considered to be the best grade. As distillation continues, other inferior grades (first, second, and third) are obtained in decreasing degrees of quality. Major producers include the Comoro Islands, Madagascar, and Réunion Island (ARCTANDER, GUENTHER).[1,2]

**CHEMICAL COMPOSITION**

Ylang ylang oil contains *d*-$\alpha$-pinene, linalool, geraniol, sesquiterpenes (caryophyllene, $\gamma$-, $\sigma$- and $\varepsilon$-cadinenes, ylangene, farnesol, farnesyl acetate, $\gamma$-muurolene, etc.),[3] acids (acetic, valeric, isovaleric, benzoic, 2-methylbutyric, hexanoic, heptanoic, octanoic, nonanoic, and *trans*-geranic acids), phenols (eugenol and isoeugenol), benzyl acetate, methyl benzoate, *p*-cresol methyl ether, eugenol methyl ether (methyl eugenol), geranyl acetate, *p*-tolyl methyl ether, safrole, and isosafrole, among others (LIST AND HÖRHAMMER, MASADA).[4–6]

**PHARMACOLOGY OR BIOLOGICAL ACTIVITIES**

Data from one source indicate ylang ylang oil to be nonirritating to mouse skin, slightly irritating to rabbit skin, and nonirritating and nonsensitizing to human skin; no phototoxic effects were reported.[7]

## USES

**Medicinal, Pharmaceutical, and Cosmetic.** Ylang ylang oil is extensively used as fragrance components in soaps, detergents, creams, lotions, and perfumes (especially floral and heavy oriental types), with maximum use level of 1% reported in perfumes. The higher grades are generally used in perfumes, while the lower grades are used in scenting soaps and detergents.

**Food.** Used as a flavor component (e.g., fruit flavors) in major food products, including alcoholic and nonalcoholic beverages, frozen dairy desserts, candy, baked goods, gelatins and puddings, and others. Use levels reported are generally below 0.001% (5.03 ppm).

## COMMERCIAL PREPARATIONS

Available in several grades, especially extra and third. Ylang ylang oil and canaga oil have been extensively adulterated (ARCTANDER).

*Regulatory Status.* GRAS (§182.20).

## REFERENCES

See the General References for ARCTANDER; FEMA; FURIA AND BELLANCA; GUENTHER; MASADA; MERCK; UPHOF.

1. W. D. Fordham in L. W. Codd et al., eds., *Chemical Technology: An Encyclopedic Treatment,* Vol. 5, Barnes & Noble, New York, 1972, p. 1.
2. M. Stoll in A. Standen, ed., *Kirk-Othmer Encyclopedia of Chemical Technology*, Vol. 14, 2nd ed., Interscience, New York, 1967, p. 178.
3. E. M. Gaydou et al., *J. Agric. Food Chem.*, **34**, 481 (1986).
4. D. B. Katague and E. R. Kirch, *J. Pharm. Sci.*, **52**, 252 (1963).
5. J. A. Wenninger et al., *Proc. Sci. Sect. Toilet Goods Assoc.*, **46**, 44 (1966); through *Chem. Abstr.*, **66**, 79494h (1967).
6. R. Timmer et al., *Int. Flavours Food Addit.*, **6**, 189 (1975).
7. D. L. J. Opdyke, *Food Cosmet. Toxicol.*, **12**(Suppl.), 1015 (1974).

# YOHIMBE

**Source:** ***Pausinystalia yohimba*** Pierre ex Beille. (*P. johimbe* (Schumann) Beille, *Corynanthe johimbi* Schumann) (Family Rubiaceae).

## GENERAL DESCRIPTION

Tall evergreen forest tree with large, glabrous, leathery leaves with upcurving lateral nerves fading out at margins; flowers white arranged in umbel-like clusters at the ends of the shoots. The range of this West African tree extends from southwestern Nigeria to Gabon.[1]

The part used is the bark, entering commerce in flattened or slightly quilled pieces.

## CHEMICAL COMPOSITION

Bark contains a number of alkaloids, the most important of which is yohimbine (reportedly up to 6% in crude bark) (17 $\alpha$-hydroxy-yohimban-16$\alpha$-carboxylic acid methyl ester); also yohimbinine, $\alpha$-yohimbane, yohimbenine, isoyohimbine (mesoyohimbine), dihydroyohimbine, corynantheine,[2] and others.

## PHARMACOLOGY OR BIOLOGICAL ACTIVITIES

Biological activity mainly attributed to the alkaloid yohimbine (rather than crude bark) includes aphrodisiac and CNS stimulant activity. Early studies suggest that the alkaloid stimulated the respiratory center in small doses, while depressed respiration in large doses (USD). It readily penetrates the CNS causing a complex response including antidiuresis (due to release of antidiuretic hormone), general excitation, with elevated blood pressure, heart rate, and increased motor activity in both humans and animals. It is generally considered an $\alpha_2$-adrenergic blocking agent. According to Goodman and Gilman, the activity of yohimbine may result from its activity as a relatively selective inhibitor of $\alpha_2$-adrenergic receptors, enhancing neural release of norepinephrine at concentrations less than those required to block postsynaptic $\alpha_1$ receptor.[3] Yohimbine also blocks peripheral 5-HT receptors (GOODMAN AND GILMAN). Aphrodisiac activity has also been attributed to the enlargement of blood vessels in the genitalia, transmission of nerve impulses to genital tissue, and an increased reflex excitability in the sacral region of the spinal cord. Yohimbine is reportedly a monoamine oxidase inhibitor (DUKE 2; TYLER 1); also a weak calcium channel blocking effect.[4]

A widely reported clinical study at Queens University (Kingston, Ontario) found that in 23 impotent men, yohimbine improved measurements of nocturnal penile tumescence in 43% of those studied. Yohimbine was determined to be most effective in patients with vascular dysfunction.[5]

Yohimbine is reported to produce nausea, salivation, irritability, elevated heart rate and blood pressure (USD).

## USES

**Medicinal, Pharmaceutical, and Cosmetic.** Yohimbine has been applied in the form of prescription drugs for impotence of vascular, diabetic or psychogenic origin; contraindicated when chronic inflammation of the prostate gland or related organs is present; as an MAO inhibitor it is contraindicated with tyramine-containing foods, antidepressants, and other mood-changing drugs. Risks associated with use of the crude drug have included excitation, tremor, sleeplessness, anxiety, increase in blood pressure, tachycardia, nausea and vomiting. Observations of interactions with psychopharmacological drugs have been reported but are not well documented.[6] Cardiovascular and behavioral effects of clonidine are antagonized by yohimbe.[7]

**Health Food/Herb Teas.** Bark in teas, capsules, tablets, tinctures, and other product forms has been promoted as an aphrodisiac and to increase athletic performance (as an alternative to anabolic steroids) (TYLER 1).

**Traditional Medicine.** Bark reportedly used as an aphrodisiac for centuries, though little evidence support claims that it has an effect on sexual desire or performance. The majority of pharmacological data is on the alkaloid yohimbine, rather than the crude drug.

## COMMERCIAL PREPARATIONS

Crude drug; yohimbine hydrochloride (prescription), often with other drugs such as strychnine, thyroid and/or methyltestosterone (TYLER 1).

*Regulatory Status.* Crude bark undetermined in the U.S., though of substantial regulatory agency interest. The bark is the subject of a negative German monograph, as the efficacy of the bark and its preparations for the claimed applications (male impotence, aphrodisiac activity, stimulant) are not well-documented.[6]

## REFERENCES


See the General References for BLUMENTHAL; DUKE 2; GOODMAN AND GILMAN; LEWIS AND ELVIN-LEWIS; MARTINDALE; TYLER 1; TYLER 3; USD; WREN.

1. R. W. J. Keay, *Trees of Nigeria*, Clarendon Press, Oxford, 1989.
2. J. W. Schermerhorn and M. W. Quimby, eds., *The Lynn Index*, Vol. V, Massachusetts College of Pharmacy, Boston, 1962.
3. B. B. Hoffman and R. J. Lefkowitz, *New Engl. J. Med.*, **302**, 1390 (1980).
4. K. Watanabe et al., *J. Pharm. Pharmacol.*, **39**, 439 (1987).
5. A. Moreles et al., *J. Urol.*, **128**, 45, (1982).
6. Monograph *Yohimbe Cortex*, *Bundesanzeiger*, no. 193 (Oct. 15, 1987).
7. R. D. Robson et al., *Eur. J. Pharmacol.*, **47**, 431 (1978).


# YUCCA

**Source:** ***Mohave yucca*** *Yucca schidigera* Roezl ex Ortgies (syn. *Y. mohavensis* Sarg.); ***Joshua tree*** *Yucca brevifolia* Engelm. (syn. *Y. arborescens* Trel.), and other *Yucca* spp. (Family Liliaceae or Agavaceae).

## GENERAL DESCRIPTION

Mohave yucca (*Y. schidigera*) is a tree seldom exceeding 4.5 m in height with a simple or branched trunk that is 15–20 cm in diameter; leaves narrow, up to 1.5 m long, with few coarse marginal fibers; native to southwestern United States (especially southern Nevada, northwestern Arizona, the Mojave Desert, and southern California into northern Baja California).

Joshua tree (*Y. brevifolia*) is a tree up to about 20 m high, branched mostly at 1–3 m aboveground with lance-shaped leaves (blades 20–35 cm long) crowded in dense clusters near ends of branches; native to southwestern United States (especially the Mojave Desert in California, southwestern Utah, and western Arizona).

Parts used are the leaves from which a solid extract is prepared by hot-water extraction.

## CHEMICAL COMPOSITION

Yucca extracts contain steroidal saponins, with the major sapogenins being sarsasapogenin and tigogenin in Mohave yucca and Joshua tree, respectively (see ***sarsaparilla*** and ***fenugreek***).[1] The concentration of saponins in these extracts have not been reported, except that Mohave yucca extracts contain about 60% solids.[2]

## PHARMACOLOGY OR BIOLOGICAL ACTIVITIES

Saponins generally have hemolytic properties *in vitro*, but when given orally or intravenously these activities are much weaker. They also have numerous other pharmacological activities (see ***quillaia***).

Water extracts of *Y. glauca* Nutt. in mice experiments have shown antitumor activity against B16 melanoma (FOSTER AND DUKE).

The effects of yucca extracts (or saponins) in humans due to long-term ingestion of small amounts of these extracts (especially in root beer) are not known. However, one 12-week study has indicated Mohave yucca extract (ca. 60% solids) to be nontoxic to rats; it also demonstrated that the extract was about half as hemolytic *in vitro* as commercial soap bark saponin.[2]

## USES

**Food.** Extract (especially Mohave yucca) is extensively used as the foaming agent in root beer and other frothy drinks, with average maximum use level of about 0.062% (618 ppm) reported.

**Health Food/Herb Teas.** Used in various preparations to treat arthritic conditions (TYLER 1).

**Traditional Medicine.** American Indian groups used root of various *Yucca* spp. in salves or poulticed for sores, skin diseases, inflammations, to stop bleeding; steam bath for sprains and broken limbs; hair wash for dandruff and baldness (FOSTER AND DUKE).

## COMMERCIAL PREPARATIONS

Extracts (mainly Mohave yucca).

*Regulatory Status.* Both have been approved for food use (§172.510).

## REFERENCES

See the General References for FEMA; FOSTER AND DUKE; HOCKING; MUNZ AND KECK; SARGENT; TYLER 1; UPHOF.

1. M. E. Wall and C. S. Fenske, *Econ. Bot.*, **15**, 131 (1961).
2. B. L. Oser, *Food Cosmet. Toxicol.*, **4**, 57 (1966).
3. L. Liberti, ed., *Lawrence. Rev. Nat. Prod.* (Jan. 1988).

# Chinese Cosmetic Ingredients

As in other ancient cultures, use of herbs by the Chinese to modify and improve physical appearance dates back thousands of years. During the course of using herbs for treating illnesses and undesirable physical conditions, they have accumulated considerable experience in cosmetic treatments and much of this experience has been documented, and this documentation dates back at least two thousand years. Common modern-day cosmetic problems such as facial dark spots, wrinkles, pimples, and dry skin were already dealt with in ancient medical treatises such as the *Huang Di Nei Jing* which was written two thousand years ago. However, unlike their Western counterparts, the Chinese have always considered external beauty as an extension of one's whole self. Thus, cosmetic conditions are often simply considered as manifestations of an unbalanced whole. Hence, treatments are often a combination of both internal and external applications.

Among the more than 6,000 documented natural drugs used in traditional Chinese medicine, one can find a sizable number that are used in treating skin and oral problems which can be considered as "cosmetic" conditions.[1,2]

To take advantage of the vast resource of Chinese herbal medicine as a source of cosmetic ingredients, it is necessary to define in a broad sense the scope of "cosmetic" properties and "cosmetic" usages.

For practical purposes, any herb or herbal formula reported to have the following properties, either traditional or modern, can be considered a potential source of natural cosmetic ingredient(s).

**TRADITIONAL PROPERTIES:**

alleviates depression,
benefits/improves complexion,
detoxifying,
invigorates/nourishes blood,
lightens skin,
moistens skin/removes dryness,
prevents scar formation,
promotes flesh growth,
promotes hair growth/prevents hair loss,
removes heat,
removes swelling,
etc.

**MODERN PROPERTIES:**

analgesic,
anti-allergic,
antiinflammatory,
antihistaminic,
antimicrobial,
antipruritc,
astringent,
local anesthetic,
vasodilator,
whitening,
wound healing,
etc.

Alternatively, any herb or herbal formula that has been recorded as beneficial in external application in one or more of the following conditions can be considered within the scope of "cosmetic" usage:

dark spots on skin,
lacquer sores,
acne/pimples,

chapped skin,
dandruff,
burns,
dry skin,
itching,
hand/facial wrinkles,
skin rash,
wounds,
premature graying of hair,
hair loss,
insect bites,
snake bites,
urticaria,
vitiligo,
ringworm,
sore gums,
toothache,
eczema,
skin sores,
etc.

In fact, among the more than 100,000 Chinese herbal formulas recorded over the past three thousand years, at least 10% of them are for the treatment of above conditions.

Thus, in the *Wu Shi Er Bing Fang* (*Prescriptions for Fifty-two Diseases*) compiled sometime between 1065 B.C. and 771 B.C., 283 known prescriptions from 247 drugs were recorded. Among them, over one-third are for treating diseases/conditions that are relevant to cosmetics:

| *Disease* | *No. of Prescriptions* |
|---|---|
| Skin ulcers/carbuncles | 42 |
| Wounds/injuries | 17 |
| Frostbite | 14 |
| Snake bite | 13 |
| External hemorrhoids | 4 |
| Poison arrow wounds | 7 |
| Lacquer sores (dermatitis) | 7 |

The most extensive collection of formulas ever published is the *Pu Ji Fang* (*Prescriptions for General Relief*), a formulary describing 61,739 prescriptions for treating diseases and human conditions of practically every kind. It was compiled during the later part of the 14th century A.D. by Zhu Xiao and others of the Ming court. In this compilation, there are at least 5,000 formulas for treating skin sores and ulcers, wounds and injuries, insect and snake bites, hemorrhoids, and ringworm; and at least 2,000 for hair, facial, and oral problems, including 350 for facial dark spots, pimples, rash, chapped skin, scars and "lack of luster," 400 for hair, beard and eyebrow conditions (e.g., premature graying and hair loss), 80 for dryness in mouth, bad breath, and mouth sores, and 1,000 for tooth and gum problems (e.g., toothache, swollen gums, tooth decay, yellow and stained teeth, etc.). Most of these formulas are for topical application and thus are potential source for cosmetic ingredients.

The most extensive work on natural products ever published is the *Zhong Yao Da Ci Dian* (*Encyclopedia of Chinese Materia Medica*). Compiled by the Jiangsu Institute of New Medicine and published in 1977, it describes 5,767 natural drugs, many in great detail, with 4,500 drawings. It contains both traditional (properties, uses, use history, etc.) and modern (pharmacology, chemistry, clinical reports, etc.) information, with a comprehensive index/appendix. The index/appendix lists hundreds of substances and chemical compounds that exhibit certain pharmacological properties, such as antiinflammatory, antimicrobial, (including antifungal), analgesic, wound-healing, anti-allergic, antihistaminic, vasodilating, and others, which can provide useful cosmetic ingredients.

Among the hundreds to thousands of single herbs and herbal formulas recorded to have "cosmetic" properties, numerous natural drugs have been repeatedly used. A few of them are already being used in Western cosmetics and may be familiar to some readers. And a few dozen more are now being used in certain Asian cosmetics, which are bound to find their way into cosmetic products in the United States in the foreseeable future.

Chinese herbal medicine as a potential source of modern cosmetic ingredients can be appreciated from a recent analysis of 85 beautifying formulas in the *Qian Jin Fang* (*Precious Formulas*), a formulary compiled by the well-known physician, Sun Si-Miao, during the 7th century A.D., which contains 8,200 formulas. Among the 160 natural substances used in these beauty formulas, 105 were herbs, 32 were animal by-products and 23 were minerals. At least two dozen of these substances repeatedly appeared in the formulas, some over twenty times. Most of them are still being used in Chinese cosmetic products.

Although most of the cosmetic usage of Chinese herbal drugs and formulas is based on traditional rationale, some of the uses do appear to have a scientific basis.

Thus, among the herbs used to remove dark spots on the skin, some (especially *chuanxiong*, *fangfeng*, *gaoben*, and *danggui*) have been shown to exhibit marked tyrosinase inhibitory effects *in vitro*.[3,4]

Among more than 100 Chinese herbal drugs screened for *in vitro* antihistaminic activities, numerous were demonstrated to exhibit strong effects. They include *wumei*, *xinyi*, *baizhi*, ginger and star anise; all have traditionally been used to treat cosmetic problems.[5–7]

Also, many of the herbal drugs traditionally used for their healing and antiswelling properties (e.g., in wounds and burns) have been scientifically demonstrated to have antiinflammatory effects. They include *xinyi*, *gaoben*, *huzhang*, *lianqiao*, *wumei*, *fangfeng*, *chishao*, *sanqi* and *rehmannia* (see corresponding entries).

Based on above criteria, the following herbs and/or their derivatives (extracts, essential oils, special chemical fractions, etc.) can serve as potential sources of cosmetic ingredients.

ATRACTYLODES (*BAIZHU* AND *CANGZHU*)
BAIZHU (ATRACTYLODES)
BLETILLA TUBER (*BAIJI*)
*CANGZHU* (ATRACTYLODES)
DAHURIAN ANGELICA (*BAIZHI*)
DITTANY BARK (*BAIXIANPI*)
FORSYTHIA FRUIT (*LIANQIAO*)
GARDEN BURNET (*DIYU*)
HONEYSUCKLE FLOWER (*JINYINHUA*)
KNOTWEED, GIANT (*HUZHANG*)
LIGUSTICUM (*GAOBEN*)
LUFFA (*SIGUALUO*)
MUME (SMOKED PLUM OR *WUMEI*)
PEARL (*ZHENZHU* OR MARGARITA) AND MOTHER-OF-PEARL (*ZHENZHUMU*)
PEONY (PEONY BARK AND PEONY ROOT)
PEONY BARK (*MUDANPI*)
PEONY ROOT, RED AND WHITE (*SHAOYAO*: *CHISHAOYAO* AND *BAISHAOYAO*)
PHELLODENDRON BARK (*HUANGBAI*)
PURSLANE, COMMON (*MACHIXIAN*)
RED SAGE (*DANSHEN*)
SAFFLOWER (FALSE SAFFRON; *HONGHUA*)
SICHUAN LOVAGE (*CHUANXIONG*)
SKULLCAP, BAIKAL (*HUANGQIN*)

## REFERENCES

1. A. Y. Leung, *D & C I*, April, 34 (1989).
2. A. Y. Leung, "Chinese Medicinals," in *Advances in New Crops*, J. Janick and J. E. Simon, eds., Timber Press, Portland, Ore., 1990, p. 499; reprinted in *HerbalGram*, **23**, 21 (1990).
3. X. T. Liu, *Zhongchengyao*, **13**(3), 9 (1991).
4. Y. Masamoto et al., *Planta Medica*, **40**, 361 (1980).
5. R. D. Xiang et al., *Zhongcaoyao*, **15**(2), 22 (1985).
6. Z. X. Zhang and L. S. Liu, *Zhongchengyao*, **14**(11), 30 (1992).
7. D. Q. Zhou et al., *Zhongcaoyao*, **22**(2), 81 (1991).

# ATRACTYLODES (*BAIZHU* AND *CANGZHU*)

*Atractylodes* yields two widely used herbs, *baizhu* and *cangzhu* (see individual entries).

## BAIZHU

Rhizome of *Atractylodes macrocephala* Koidz. (Family Compositae or Asteraceae), an erect perennial herb, up to ca. 80 cm high. Thick fist-shaped rhizome is collected from 2- to 3-year-old plants in early winter after lower leaves have withered and turned yellow, rid of sand, dirt and rootlets, and oven or sun dried. Produced mainly in eastern and central provinces.

First recorded use dates back at least 2,000 years when it was simply called *zhu*, not distinguishable from ***cangzhu*** (see that entry); listed in the *Shen Nong Ben Cao Jing* (ca. 200 B.C.–100 A.D.) under the first category of drugs to which most tonics belong. It was not until around 600 A.D. that *baizhu* became a distinct separate entity. Traditionally regarded as bitter and sweet tasting; warming; and to have spleen- and *qi*-invigorating, as well as wetness-drying (*zao shi*), diuretic and fetus-calming properties. Used in treating spleen-deficient (*pi xu*) conditions, including dyspepsia (indigestion), flatulence, diarrhea and fluid retention; also used for spontaneous perspiration, restless fetus and as an ingredient in numerous well-known tonic foods, including soups, cakes, and specialty rices (PENG).

In recent years also used in treating constipation, leukopenia, and toxic effects due to chemotherapy and radiotherapy (WANG).[1,2]

Contains 0.25–1.42% volatile oil (depending on geographic sources and processing methods), with atractylon (a furan derivative) as its major component (9.59–27.4%) (IMM-1); others include sesquiterpene lactones [e.g., selina-4(15),7(11)-dien-8-one; atractylenolide I; atractylenolide II; atractylenolide III; and 8-β-ethoxyasterolid], scopoletin, and acetylenes.[3]

Also contains Atractylodes polysaccharides, AM-1 (β-D-mannan), AM-2 (β-D-fructan), and AM-3 (β-D-mannan), with mol. wt. of $3.1 \times 10^4$, $1.1 \times 10^4$ and $1.2 \times 10^4$, respectively.[4,5]

Although high concentrations of vitamin A substances (0.2599%) were earlier reported to be present in *baizhu*, later findings could not substantiate them (WANG).

The atractylenolides have been reported to have antiinflammatory activities while the polysaccharides were shown to have immunopotentiating effects.[4–6]

In endurance tests, decoctions of cured *baizhu* markedly prolonged the swimming time of mice while those of raw *baizhu* were not active. On the other hand, decoction of raw *baizhu* enhanced the phagocytosis activity of mouse phagocytes while that of cured *baizhu* had no such effect.[7]

Other documented pharmacologic effects of *baizhu* decoctions in animals (p.o.) include: diuretic; hypoglycemic; improving stamina; liver protectant (preventing liver glycogen reduction caused by carbon tetrachloride); and anticoagulant. The volatile oil was active against experimental tumors (JIANGSU; ZHOU AND WANG).[8]

Toxicity of *baizhu* is low: $LD_{50}$ of its decoction in mice was 13.3 g/kg (i.p.); and a daily oral dose of 0.5 g/kg decoction in rats moderately lowered white blood cells in 14 days, which was contrary to clinical results (WANG).

Used in many traditional topical formulas for wrinkles and dark spots on the skin, especially the face and hand.

**REFERENCES**

See the General References for CHP; IMM-1; JIANGSU; PENG; WANG; ZHOU AND WANG.

1. C. X. Zou et al., *Zhongyiyao Xinxi*, (3), 34 (1992).
2. C. Q. Ling, *Henan Zhongyi*, **13**, 94 (1993).
3. Z. L. Chen, *Planta Med.*, **53**, 493 (1987).
4. Y. C. Gu et al., *Zhongcaoyao*, **23**, 507 (1992).
5. Y. C. Gu et al., *Zhongguo Yaoxue Zazhi*, **28**, 275 (1993).
6. K. Endo et al., *Chem. Pharm. Bull.*, **27**, 2954 (1979).
7. B. Sun et al., *Shandong Zhongyi Xueyuan Xuebao*, **17**, 51 (1993).
8. Z. H. Li et al., *Zhongcaoyao*, **17**(10), 37 (1986).

# BLETILLA TUBER (*BAIJI*)

Rhizome of *Bletilla striata* (Thunb.) Reichb. f. (Family Orchidaceae), a perennial herb, 30 to 70 cm high, with a thick and fleshy rhizome; widely distributed in central, eastern and southern China; also cultivated. Rhizome is collected in autumn and early winter and cured by steaming or boiling in water until thoroughly cooked and then sun- or oven-dried.

Has been traditionally used for over two thousand years to treat and heal sores, burns, wounds, and chapped skin as well as to stop bleeding (both internal and external); also has astringent and anti-swelling properties.

Bletilla tuber contains large amounts of starch (30.48% in fresh tuber) and mucilage (Bletilla mannan) which is a polysaccharide consisting of mannose and glucose in a 4:1 ratio.

Its methanol extractives (phenolic fraction) have been shown to have antimicrobial activities (esp. against *Staphylococcus aureus*).[1] Decoction has exhibited protective effects on HCl-induced gastric mucosa damage in rats.[2]

Used in skin creams and lotions for its healing and whitening effects (per Li Shi-Zhen's *Ben Cao Gang Mu*). Its mucilage powder (water extracted, alcohol precipitated) is used as an adhesive and molding material, especially in dentistry.[3]

**REFERENCES**

See the General References for FOSTER AND YUE; HU; IMM-1; JIANGSU.

1. S. Takagi et al., *Phytochemistry*, **22**, 1011 (1983).
2. Z. G. Geng et al., *Zhongcaoyao*, **21**(2), 24 (1990).
3. J. D. Zhang and F. H. Tao, *Zhongguo Zhongyao Zazhi*, **14**(4), 34 (1989).

# CANGZHU

Rhizomes of *Atractylodes lancea* (Thunb.) DC., *Atractylodes chinensis* Koidz., and *Atractylodes japonica* Koidz. ex Kitam. (Family Compositae or Asteraceae), all perennial herbs, 30 to 80 cm high, with thick rhizomes. They are also known as *nan cangzhu* (southern), *bei cangzhu* (northern) and *dong cangzhu* (eastern), respectively. Rhizomes normally collected in spring or autumn (preferably autumn) from 1- to 2-year-old plants (if from cultivation), rid of residual stems, rootlets and dirt, and sun dried. Southern *cangzhu* is produced mainly in Jiangsu, Hubei, Hunan and other relatively southern provinces; northern *cangzhu* is produced primarily in northern provinces, including Inner Mongolia, Hebei, Shanxi and Shaanxi; and eastern *cangzhu* is produced mainly in northeastern provinces, including Heilongjiang, Jilin and Liaoning (IMM-1, JIANGSU). Eastern *cangzhu* (*A. japonica*) is considered equivalent to ***baizhu*** (see that entry) in Japan (HU).[1] The *cangzhu* available in the United States is mostly southern or northern *cangzhu*.

First recorded use dates back at least 2,000 years when it was simply referred to as *zhu* in the *Shen Nong Ben Cao Jing* (ca. 200 B.C.–100 A.D.) without differentiation from ***baizhu***. Traditionally considered as acrid and bitter tasting and warming; spleen-invigorating; wetness-drying (*zao shi*); antiinflammatory (*qu feng*); alleviating depression (*jie yu*); cold-dispersing; and eye-brightening. Used in treating various conditions, including: abdominal distention; diarrhea; edema; lack of appetite; indigestion; tiredness and sleepiness; rheumatism and arthritic pain; common cold; eczema; night blindness; and warts, among others.[2]

Also, a traditional practice during an epidemic was to burn *cangzhu* to drive off the "evil" that was believed to cause the epidemic. Recently, this practice has been adapted for the routine sterilization of operating rooms, using 1 g/m$^3$ of *cangzhou* with considerable success and safety. The fumes from burning *cangzhu* not only killed bacteria (esp. *Staphylococcus aureus*) but also viruses and fungi;[3] however, they had no effect on mosquitoes and other insects (WANG).

Contains 1.1–9.0% (w/w) volatile oil, depending on botanical, geographic and reporting sources, with southern *cangzhu* in the general range of 5–9%, northern *cangzhu* 3–5%, and eastern *cangzhu* 1–3% (IMM-1, JIANGSU, NATIONAL). The major volatile components present in southern *cangzhu* include atractylol (mixture of β-eudesmol and hinesol), elemol, and atractylodin (an acetylenic furan);[4] minor components present include atractylon, vitamin A substances, vitamin B and inulin. Major compounds present in northern *cangzhu* include atractylodin (12.50–20.93%), atractylon, and atractylol. The volatile oil in eastern *cangzhu* contains little or no atractylodin; its major components being atractylol and atractylon (IMM-1, JIANGSU); also contains hypoglycemic glycans (atractans A, B and C).[1]

Although fumes from burning *cangzhu* have strong disinfectant effects, its decoction does not. The fumes are reported to be nonirritating and to have very low toxicity: mice and rats exposed to burning *cangzhu* and *aiye* (*Artemisia argyi* leaves) for 0.5–2 hours exhibited no abnormalities on external and pathological examination; and over 4,000 normal subjects exposed nightly to one round of burning *cangzhu-aiye* incense for 30 days did not show any adverse reactions (WANG).

Used as a component in hair tonic liniments and in skin care formulas (e.g., acne creams) for its antimicrobial activity as well as its whitening properties.[5]

## REFERENCES

See the General References for CHP; HU; IMM-1; IMM-CAMS; JIANGSU; NATIONAL; WANG; ZHOU.

1. C. Konno et al., *Planta Med.*, **51**, 102 (1985).
2. C. F. Zhang, *Zhongyiyao Xuebao*, (4), 32 (1992).
3. L. E. Jiang and Z. X. Zhu, *Chin. J. Integr. Trad. Western Med.*, **9**, 245 (1989).
4. H. Y. Sun and A. W. Wang, *Zhongcaoyao*, **23**, 298 (1992).
5. Li Shi-Zhen's *Ben Cao Gang Mu*, 16th century, reprinted.

# DAHURIAN ANGELICA (*BAIZHI*)

Root of *Angelica dahurica* (Fisch. ex Hoffm.) Benth. et Hook. f. and *Angelica dahurica* (Fisch. ex Hoffm.) Benth. et Hook. f. var. *formosana* (Boiss.) Shan et Yuan. (Family Umbelliferae or Apiaceae), both perennial herbs, 1–2.5 m high; the former grows throughout northeastern China while the latter is distributed in southeastern China and Taiwan. Both are now extensively cultivated; and Dahurian angelica produced from different regions provides the various commercial types available, especially *Chuan* (Sichuan) *baizhi* (from *A. dahurica* var. *formosana* grown in Sichuan), *Yu baizhi* (from *A. dahurica* grown in Yu County in Henan), and *Hang baizhi* (from *A. dahurica* var. *formosana* grown in Hangzhou, Zhejiang Province) (IMM-1).[1] The types imported into the United States are *Chuan baizhi* and *Hang baizhi*. *Dian* (Yunnan) *baizhi* (from *Heracleum scabridum* Franch. grown in Yunnan) is used as "*baizhi*" in Yunnan, and has also been the subject of studies reported in the Chinese literature.

The root is harvested in late summer to early autumn from plants planted the year before, rid of stem remnants and leaves, rootlets and dirt and sun-dried; sulfur fumigation is often used. In the production of *Hang baizhi*, the cleaned root is mixed with lime and kept for a week before drying. Thick, starchy, hard and heavy roots with strong aroma are considered of best quality (HU, ZHU).

Traditionally considered acrid tasting and warming; removing rheumatic and arthritic pain; sinus-clearing, pain relieving, antiswelling and pus expelling (*pai nong*); and used to treat the common cold and headache, nasal congestion, sinusitis, toothache, leucorrhea, and others (CHP, JIANGSU).

Used externally since ancient times in treating various skin conditions such as acne, dark spots, freckles, carbuncles, ringworm, scabies, and itching, among others. Now also used in treating psoriasis, eczema and vitiligo. First known use was recorded in the *Wu Shi Er Bing Fang* (*Prescriptions for Fifty-two Diseases*, ca. 1065–771 B.C.) in a formula together with ginger and magnolia flower (see ***magnolia***) for the topical treatment of carbuncles. *Baizhi* has since been extensively used in many beauty formulas, including numerous skin and hair formulas of the imperial court, such as Empress Dowager Cixi's "Fragrant Hair Powder" for reportedly keeping her hair full and dark.[2] It is the single most frequently used herbal ingreedient in traditional beauty formulas.[3]

*Baizhi* contains numerous furocoumarins (see also ***angelica***), including: byakangelicin (ca. 2%), anhydrobyakangelicin, byakangelicol (ca. 0.2%), neobyakangelicol, oxypeucedanin (2.56%), oxypeucedanin hydrate, imperatorin (1.86%), isoimperatorin (1.96%), alloisoimperatorin, phellopterin, xanthotoxin, scopoletin, 5-(2-hydroxy-3-methoxy-3-methylbutoxy) psoralen, knidilin, demethylsuberosin, cedrelopsin, bergapten, and others (HU, IMM-1).[4–10] Also reported present are sterols (sitosterol and daucosterol), stearic acid and (*S*)-2-hydroxy-3,4-dimethyl-2-buten-4-olide which gives *baizhi* its characteristic odor.[6,7,11] Sulfur fumigation during processing was shown to markedly reduce the concentrations of furocoumarins in *baizhi*.[12]

Decoction has antibacterial and antifungal activities *in vitro* as well as antiinflammatory, analgesic and antipyretic effects in experimental animals, with effects comparable among the four types of *baizhi* tested (*Yu*, *Chuan*, *Hang and Dian*).[1] Alcohol extractives exhibited antihistaminic effects on isolated guinea pig trachea.[13]

Among the numerous furocoumarins tested, several (oxypeucedanin hydrate, imperatorin, phellopterin, etc.) activated adrenaline- and ACTH-induced lipolysis while others (byakangelicin, neobyakangelicol, isopimpinellin, etc.) strongly inhibited insulin-stimulated lipogenesis in fat cells.[14]

Although furocoumarins have photosensitizing effects, their presence in *baizhi* has not prevented its extensive use in China for over three thousand years, both internally and externally, without any major toxic side effects.

Powder and extracts used in skin-care products (especially acne creams, freckle-removing and antifungal creams) and in hair-care products (anti-dandruff shampoos, hair-growth tonics, etc.) for its antimicrobial and antiinflammatory effects and its traditional anti-itching, wound-healing, and skin-whitening (tyrosinase inhibitory) properties (ZHOU).[15]

## REFERENCES

See the General References for CHP; HU; IMM-1; JIANGSU; ZHOU; ZHU.

1. H. Y. Li et al., *Zhongguo Zhongyao Zazhi*, **16**, 560 (1991).
2. A. Y. Leung, *Drug Cosmet. Ind.*, (4), 34 (1989).
3. W. X. Hong, *Fujian Zhongyiyao*, **18**(1), 39 (1987).
4. J. Han and Q. Yang, *Beijing Yike Daxue Xuebao*, **21**, 186 (1989).
5. L. R. Wang et al., *Yaoxue Xuebao*, **131** (1990).
6. R. Y. Zhang et al., *Beijing Yixueyuan Xuebao*, **17**, 104 (1985).
7. H. Q. Zhang et al., *Yaoxue Tongbao*, **15**(9), 2 (1980).
8. J. Han and H. L. Zhang, *Zhongcaoyao*, **17**(8), 13 (1986).
9. M. Kozawa et al., *Shoyakugaku Zasshi*, **35**(2), 90 (1981); through *Chem. Abstr.*, **95**, 209449j (1981).
10. J. M. Zhou et al., *Zhongcaoyao*, **18**(6), 2 (1987).
11. K. Baba et al., *Planta Med.*, **51**, 64 (1985).
12. H. Y. Li et al., *Zhongguo Zhongyao Zazhi*, **16**, 27 (1991).
13. R. D. Xiang et al., *Zhongcaoyao*, **16**(2), 22 (1985).
14. Y. Kimura et al., *Planta Med.*, **45**, 183 (1982).
15. Y. Masamoto et al., *Planta Med.*, **40**, 361 (1980).

# DITTANY BARK (*BAIXIANPI*)

Root bark of *Dictamnus albus* L. var. *dasycarpus* (Turcz.) T. N. Liou et Y. H. Chang (syn. *Dictamnus dasycarpus* Turcz.) (Family Rutaceae), a strong scented perennial herb, up to 1 m high, with fleshy root; widely distributed in northern, northeastern and central China as well as Mongolia and Siberia. Root is dug up in spring and autumn, rid of rootlets and washed free of dirt; bark is then removed, cut in sections and normally sun-dried.

First described in the *Shennong Ben Cao Jing* (ca. 200 B.C.–100 A.D.) as cold-natured and bitter-tasting. Traditionally used to treat pruritus (itching), eczema, urticaria, ringworm, arthritis, and rheumatism; also used to treat acute and chronic hepatitis as well as traumatic injuries and external bleeding. Its use in pruritus is common.

Root contains 0.19–0.39% alkaloids, mainly quinoline type (dictamnine, skimmianine, preskimmianine, isodictamnine, dasycarpamine, $\gamma$-fagarine (8-methoxydictamnine), isomaculosindine, trigonelline, choline, etc.); lactones (dictamnolactone, obaculactone or limonin, rutaevin,[1] fraxinellone, etc.); sterols (sitosterol and campesterol); saponins and volatile oil. Fraxinellone has antifertility (antiimplantation) activities.[2]

Its decoction exhibited antifungal activities *in vitro* and lowered experimental fever in rabbits.

Extracts used in cosmetic creams and bath preparations for their antifungal and allegedly whitening effects.

### REFERENCES

See the General References for ETIC; JIANGSU; JILIN; NATIONAL.

1. Z. C. Wang et.al., *Zhongguo Zhongyao Zazhi*, **17**, 551 (1992).
2. W. S. Woo et al., *Planta Med.*, **53**, 399 (1987).

## FORSYTHIA FRUIT (*LIANQIAO*)

Fruit of *Forsythia suspensa* (Thunb.) Vahl. (Family Oleaceae), a deciduous shrub, 2–4 m high, bearing bright golden yellow flowers early in spring; native to China, distributed mostly in northern and northeastern provinces; cultivated in China; also cultivated as an ornamental plant in the United States. Fruit is a dehiscent capsule; both the newly ripened (not yet opened) and the fully ripened (opened) fruits are used, with the latter more common, which is also the one imported into the United States. This is mainly wildcrafted in October when the capsule has turned yellow and opened, rid of impurities and sun dried. This *lianqiao* does not have seeds; major producing provinces include Shanxi, Henan, Shaanxi and Shandong (ZHU).

One of the most commonly used herbal drugs in China, *lianqiao* has enjoyed a continuous documentation of at least 3,000 years. Traditionally considered bitter tasting and cooling and to have the following properties: heat dissipating, detoxifying, breaking up nodules, and removing swelling. Widely used in treating "toxic" and "hot" conditions that correlate to inflammatory and infectious diseases, such as erysipelas, carbuncles, inflammation of the lymph nodes (*luo li*), influenza, the common cold with its associated symptoms, fevers, skin eruption (rash) and urination difficulties (e.g., frequent, slow, or painful urination). Now widely used in commercial herbal formulas for treating the common cold and influenza as well as allergies.

*Lianqiao* contains triterpenoids, including oleanolic acid (0.73–2.28%),[1–3] ursolic acid, betulinic acid, amyrin acetate, *iso*-bauerenyl acetate and 20(*S*)-dammar-24-ene-3$\beta$,20-diol-3-acetate;[4] forsythol (a phenol with chemical structure not yet determined),[5] forsythosides A, C and D (1.46–1.63%), forsythin (phillyrin, phillyroside), forsythigenin (phillygenol), caffeic acid, sterols, and rutin (IMM-3).[5–7] High concentrations (ca. 4%) of an antiviral and antibacterial volatile oil has also been reported in the seeds,[5] which is probably also present in minor amounts in the dried ripe fruit.

Decoctions have exhibited broad biological activities: strong antibacterial *in vitro* (especially against *Staphylococcus aureus* and *Shigella dysenteriae*), with forsythol being one of the active components; antiviral *in vitro*; antipyretic in rabbits; and protective against liver damage by carbon tetrachloride in rats. Other activities include diuretic, antiinflammatory, hypotensive, antioxidant, and others (WANG).[8]

Oleanolic acid has anti-allergic, liver-protective and weak cardiotonic activities, among others (see also ***ligustrum***).

Acute toxicity is low: $LD_{50}$ of decoction (1:1) in mice was 29.37 g/kg (s.c.) (WANG).

Used in hair-care (hair-growth liniments, anti-dandruff shampoos, etc.), foot-care (e.g. athlete's foot) and skin-care products (moisturizing, nourishing, and acne creams) for its antibacterial and antifungal activities and for its traditional skin-protective, wrinkle-removal and moisture-preserving properties (ETIC, ZHOU).

**REFERENCES**

See the General References for BAILEY 2; CHP; HU; IMM-3; JIANGSU; WANG; ZHOU; ZHU.

1. H. B. Li et al., *Zhongchengyao*, **11**(6), 33 (1989).
2. N. J. Wu et al., *Zhongcaoyao*, **23**, 467 (1992).
3. B. Wang and C. H. Jiang, *Zhongguo Yaoxue Zazhi*, **27**, 393 (1992).
4. W. Y. Hu and S. D. Luo, *Zhongcaoyao*, **22**, 147 (1991).
5. L. Q. Fang and J. X. Guo, *Zhongchengyao*, **12**(9), 35 (1990).
6. Y. Y. Cui et al., *Yaoxue Xuebao*, **27**, 603 (1992).
7. J. X. Guo et al., *Zhongchengyao*, **12**(8), 29 (1990).
8. Y. L. Zhou and R. X. Xu, *Zhongguo Zhongyao Zazhi*, **17**, 368 (1992).

# GARDEN BURNET (*DIYU*)

Root and rhizome of *Sanguisorba officinalis* L. or its variety, *S. officinalis* L. var. *longifolia* (Bert.) Yu et Li (Family Rosaceae), a sturdy perennial with deep-red or purplish spikes, also known as salad burnet or burnet-bloodwort, native to Eurasia, widely distributed in China and can now also be found in the United States from Maine to Minnesota. Root with rhizome is collected in spring before budding or in autumn after aboveground parts have withered, rid of rootlets, washed and dried; or first sliced and then dried. Major production in China is along eastern coastal and adjacent provinces.

First recorded use dates back 2,000 years in the *Shen Nong Ben Cao Jing*. Traditionally regarded as bitter- and sour-tasting, cold-natured, and to have blood-cooling, hemostatic, heat-clearing and detoxifying (*qing re jie du*) properties. Used in treating nosebleeds, vomiting blood, hematochezia (bloody stools), bleeding hemorrhoids, bloody diarrhea, metrorrhagia (*beng lou*), burns and scalds, eczema, skin sores, and swelling; also in dog and snake bites; among others (CHP, JIANGSU). Some of the uses have been well documented in recent years, especially for burns,[1–4] metrorrhagia,[5,6] acne,[7,8] and eczema (JIANGSU, WANG).

Garden burnet contains 14.0–40.4% phenolic substances and 2.5–4.0% triterpene glycosides as its major constituents. Phenolic substances include: (+)-catechin; (+)-gallo-catechin; (+)-catechol; (+)-gallo-catechol; ellagic acid; and leucoanthocyanins. The triterpene glycosides include: Ziyu-glycoside I [aglycone = pomolic acid (19$\alpha$-hydroxyursolic acid); glycone = arabinose and glucose]; Ziyu glycoside II (pomolic acid and arabinose); sanguisorbins A, B, C, D and E (aglycone = ursolic acid; glycone = arabinose and glucose); sauvissimoside $R_1$ and pomolic acid 28-*O*-$\beta$-D-glucopyranoside (JIANGSU, JILIN).[9–11]

Decoction had antimicrobial activities *in vitro* and anti-emetic effects in pigeons. Crude powder fed to mice and rabbits markedly shortened bleeding and coagulation times; effects were only partially due to tannic substances present; when applied to experimental burns in rabbits and dogs, it reduced inflammation and promoted healing, with effects superior to those of tannins.

Toxicity of garden burnet is low: daily feeding of water extract (1:3) 20 mL/kg to rats for 10 days did not produce obvious toxic symptoms (JIANGSU).

Extracts used in acne creams, bath preparations and toilet waters for their antimicrobial effects as well as their alleged properties in preventing eczema and contact dermatitis, especially in children and infants (ZHOU).

## REFERENCES

See the General References for CHP; FERNALD; FOSTER AND DUKE; IMM-2; JIANGSU; JILIN; WANG; ZHOU.

1. P. J. Li and C. Z. Li, *Hebei Zhongyi*, **11**(6), 16 (1989).
2. W. B. Lu and Y. Q. Zhao, *Henan Zhongyi*, **8**(5), 30 (1988).
3. Y. L. Chen and H. Y. Zhou, *Zhongguo Yiyuan Yaoxue Zazhi*, **9**, 277 (1989).
4. T. S. Wang and B. X. Jiang, *Zhongyi Yanjiu*, **2**(1), 41 (1989).
5. G. S. Xu, *Shaanxi Zhongyi*, **10**, 204 (1989).
6. Z. Y. Zhang, *Xinzhongyi*, **23**(4), 18 (1991).
7. G. P. Zhou et al., *Chin. J. Dermatol.*, **24**, 192 (1991).
8. M. X. Liu, *Faming Zhuanli Gongbao*, **7**(21), 11 (1991).
9. G. W. Qin et al., *Zhongcaoyao*, **22**, 483 (1991).
10. F. Abe et al., *Chem. Pharm. Bull*, **35**, 1148 (1987).
11. I. Yosioka et al., *Chem. Pharm. Bull*, **19**, 1700 (1971).

# HONEYSUCKLE FLOWER (*JINYINHUA*)

Flower buds of *Lonicera japonica* Thunb. and numerous other *Lonicera* species (*L. confusa* DC., *L. hypoglauca* Miq., *L. dasystyla* Rehd., etc.) collectively known as honeysuckle (Family Caprifoliaceae). They are erect or climbing shrubs with opposite leaves and mildly fragrant to very fragrant flowers, with *L. japonica* (Japanese honeysuckle) being the major source. Native to Asia, Japanese honeysuckle now runs wild in many parts of North America, especially eastern United States; its climbing or twining stem reaches 9 m long and its flowers are very fragrant.

The flower buds are collected from both wild and cultivated plants in late spring to early summer. Traditionally, they are picked in the morning after the dew has evaporated and sun-dried or air-dried in the shade, avoiding harsh midday and early afternoon sun. The resulting dried flower buds have a characteristic aroma but not that of fresh honeysuckle flower. Produced mainly in Henan and Shandong from cultivated plants and in other provinces (Guangxi, Zhejiang, Sichuan, etc.) from wild plants. Honeysuckle flower comes in numerous grades; Henan produces the best grade while Shandong produces the largest quantity. Top grades consist of minimal amounts of opened flowers, leaves and twigs (ZHU).

Also known in Chinese as *rendong* ("winter-resistant") and *shuanghua* ("double flower"), honeysuckle has a long recorded history, dating back to the *Ming Yi Bie Lu* (ca. 200 A.D.) according to some records (JIANGSU PROVINCIAL 3).[1] Traditionally considered sweet tasting, cold, fever-relieving and detoxicant (*qing re jie du*), it is extensively used in treating "heat" related conditions such as fevers, inflammations and infections (especially viral and bacterial). Also, together with other detoxifying herbs (especially dandelion, licorice, chrysanthemum, mung bean, soybean, etc.), honeysuckle flower is often used for what Chinese medicine calls "toxic" conditions, such as swellings, sores and boils as well as in food, drug and industrial (pesticide, heavy metal, etc.) poisoning.

Flower (*jinyinhua*) contains 0.089 to 12.00% chlorogenic and isochlorogenic acids (amounts varying greatly depending on botanical sources and methods of processing),[2–5] a volatile oil composed predominantly of linalool, aromadendrene and geraniol;[6] saponins (ca. 1% in *L. fulvotomentosa* Hsu et S. C. Cheng);[7,8] flavonoids, including luteolin and luteolin-7-glucoside; inositol (ca. 1%); and tannins. Leaves and stems contain flavonoids, loganin and secologanin.[9]

Honeysuckle flower is active against various bacteria (*Staphylococcus aureus*, *Salmonella typhi*, *Mycobacterium tuberculosis*, dysentery bacilli, etc.) and viruses (e.g., HIV and influenza viruses) *in vitro* and/or *in vivo*; also active against dermatophytes, though less so. It has antiinflammatory effects on several experimental inflammation models. Other effects include: strengthening body resistance and activating phagocytosis of leukocytes in mice; lowering absorption of cholesterol in the intestinal tract of rabbits; and others (NATIONAL, WANG).[10,11]

Some of the biological activities are due to chlorogenic acid and the saponins. In addition, chlorogenic acid inhibited nitrosation *in vitro* and *in vivo* in rats;[12] inhibited tumor formation in experimental animals;[13] and also has central stimulant effects in mice and rats when administered per os, effect being 1/6 that of caffeine (WANG) (see also ***coffee***).

Toxicity of *jinyinhua* is low: $LD_{50}$ in mice was 53 g/kg (s.c.).

*Jinyinhua* is a major ingredient in some well-known Chinese cold remedies such as *Yinqiao Jiedu San*, which is also sometimes used externally to treat itching and inflammatory conditions. Aqueous and hydro-alcoholic extracts of *jinyinhua* are used in skin-care products (creams, lotions, cleansers) for its antimicrobial and astringent properties (ETIC).

## REFERENCES

See the General References for CHP; FERNALD; FOSTER AND YUE; HU; JIANGSU; JIANGSU PROVINCIAL 3; LEUNG; NATIONAL; WANG; ZHU.

1. G. M. Ding and L. C. Sun, *Jiangxi Zhongyiyao*, (5), 45 (1988).
2. X. G. Dong et al., *Zhongyao Tongbao*, **10**(5), 31 (1985).
3. H. R. Li et al., *Jilin Zhongyiyao*, (2), 39 (1989).
4. H. B. Lin et al., *Shandong Zhongyi Zazhi*, **9**(4), 34 (1990).
5. B. Q. Liu et al., *Jilin Zhongyiyao*, (3), 41 (1992).
6. G. L. Wang et al., *Zhongguo Zhongyao Zazhi*, **17**, 268 (1992).
7. J. Liu et al., *Acta Pharmacol. Sin.*, **9**, 395 (1988).
8. Q. Mao and X. S. Jia, *Yaoxue Xuebao*, **24**, 269 (1989).
9. A. Hermans-Lokkerbol and R. Verpoorte, *Planta Med.*, **53**, 546 (1987).
10. G. Y. Song et al., *Zhongcaoyao*, **16**(5), 37 (1985).
11. Y. S. Nan et al., *Zhongchengyao*, **11**(8), 17 (1989).
12. B. Pignatelli et al., *Carcinogensis*, **3**, 1045 (1982).
13. Anonymous, *C&EN*, **69**(37), 27 (1991).

# KNOTWEED, GIANT (*HUZHANG*)

Rhizome and root of *Polygonum cuspidatum* Sieb. et Zucc. (Family Polygonaceae), a stout perennial with mottled stems, 1–2.5 m high, also known as Japanese knotweed and "*huzhang* = tiger cane" in Chinese; the plant is native to eastern Asia; has escaped in North America and is now a weed throughout New England and neighboring states and Canada; young shoots edible.[1] Rhizome and root are dug up in spring or autumn, cut into sections, and sun dried.

Recorded use in China dates back at least 2,000 years. Traditionally considered slightly bitter and cold; used to treat arthritic pain, jaundice (*shi re huang dan*), amenorrhea, abdominal mass (*zheng jia*), cough with excessive phlegm, traumatic injuries, skin sores and boils, and burns and scalds (CHP); also used in Japan to treat suppurative dermatitis, gonorrhea, favus, athlete's foot and hyperlipemia.[2]

Recently used in treating burns, acute viral hepatitis, leukocytopenia due to chemotherapy and radiotherapy, and acute infections (lung infection, appendicitis, etc.) with considerable success (WANG).[3]

Contains anthraquinones and their glycosides (chrysophanol, physcione, emodin, emodin-8-*O*-D-glucoside (polygonin), physcione-8-*O*-D-glucoside, etc.); stilbenes (resveratrol (3,5,4′-trihydroxy stilbene) and piceid (polydatin; resveratrol-3-*O*-D-glucoside)); 2-methoxy-6-acetyl-7-methyljuglone (a naphthoquinone); and others, including fallacinol, citreorosein, questin, questinal, protocatechuic acid, (+)-catechin, 2,5-dimethyl-7-hydroxychromone, torachrysone-8-*O*-D-glucose, 7-hydroxy-4-methoxy-5-methylcoumarin, condensed tannin and polysaccharides (IMM-1, JIANGSU).[2,4–7]

Aqueous extracts have antibacterial (polygonin and polydatin active principles) and strong antiviral activities *in vitro*, especially against influenza virus Asian strain 68-1, $ECHO_{11}$, herpes simplex virus, and Coxsackie enteroviruses A and B as well as hepatitis B virus (WANG).[8]

*Huzhang* decoction exhibited antitussive (polydatin an active principle) and antihistaminic effects in experimental animals; decoction and polydatin also had hypotensive and vasodilating actions; polydatin (not decoction; per os) markedly lowered serum lipid levels in rats; an alcoholic extract was antioxidant *in vitro*;[9] resveratrol and polydatin (i.p. or per os) reduced triglyceride synthesis from palmitate in mouse liver;[4] polydatin also had liver-protectant and antioxidant (vs lipid peroxidation) effects as well as inhibited platelet aggregation of rabbits both *in vitro* and *in vivo*.[4,10]

The anthraquinones and their glycosides (esp. emodin) exhibited cytotoxic effects on HL-60 cells.[6] Emodin and the stilbenes have recently been shown to be inhibitors of a protein-tyrosine kinase partially purified from bovine thymus.[11]

Extracts used in skin lotions and creams (e.g., antifatigue, massage, and cleansing creams) for their antimicrobial and astringent properties (ZHOU); their intensely yellow color may limit their scope of application (see ***phellodendron bark***).

## REFERENCES

See the General References for CHP; FERNALD; JIANGSU; WANG; ZHOU; ZHU.

1. J. Richardson, *Wild Edible Plants of New England*, Delorme Publishing Co., Yarmouth, Maine, 1981.
2. Y. Kimura et al., *Planta Med.*, **48**, 164 (1983).
3. W. Z. Huang, *Fujian Zhongyiyao*, **18**(4), 27 (1987).
4. C.W. Shan, *Yaoxue Xuebao*, **23**, 394 (1988).
5. H. Arichi et al., *Chem. Pharm. Bull.*, **30**, 1766 (1982).

6. S. F. Yeh et al., *Planta Med.*, 413 (1988).
7. C. G. Ouyang, *Zhongcaoyao*, **18**(8), 45 (1987).
8. J. Y. Yang et al., *Chin. J. Integr. Trad. Western Med.*, **9**, 494 (1989).
9. Y. L. Zhou and R. X. Xu, *Zhongguo Zhongyao Zazhi*, **17**, 368 (1992).
10. P. Tong and Z. T. Zhang, *Zhongguo Yaoxue Zazhi*, **26**, 363 (1991).
11. G. S. Jayatilake et al., *J. Nat. Prod.*, **56**, 1805 (1993).

# LIGUSTICUM (*GAOBEN*)

Roots and rhizomes of several *Ligusticum* species, especially *Ligusticum sinense* Oliv. and *L. jeholense* Nakai et Kitag. (Family Umbelliferae or Apiaceae), which are aromatic perennial herbs, erect, up to 1 m high, the former with irregular cylindrical to round rhizome bearing many thin roots while the latter has a short rhizome. *Ligusticum sinense* is distributed and produced in central China, including the provinces of Hubei, Shaanxi and Sichuan while *L. jeholense* (syn. Liaoning *gaoben*) is distributed and produced in northeastern provinces, including Liaoning, Jilin, Inner Mongolia, Hebei and Shandong (HU, JIANGSU). Roots and rhizomes are mostly wildcrafted, in spring and autumn, and after being rid of dirt, stem and shoots, they are sun-dried or dried by artificial heat (ZHU).

Although *Gaoben* is closely related to Sichuan lovage (chuanxiong) and are used interchangeably in some areas, they are distinctly different drugs, with some distinctly different uses. They can be differentiated by their microscopic features (XU AND XU) as well as chemical compositions.[1]

Earliest written record dates back to the *Shan Hai Jing* (ca. 800 B.C.); and later also described in the *Shen Nong Ben Cao Jing* (ca. 200 B.C.–100 A.D.) Traditionally regarded as acrid tasting and warming; removing rheumatic and arthritic pain (*qu feng*); dispersing cold; eliminating wetness; and stopping pain. Used to treat the common cold and headache associated with it, headache on top of head (*dian ding tong*), migraine, rheumatic and arthritic pain, acne, acne rosacea, skin blemishes (freckles and dark spots) and abdominal pain and diarrhea; also used in treating ringworm, scabies and dandruff (CHP, JIANGSU, NATIONAL). In Jiangxi, *gaoben* (*chaxiong*, *L. sinense*) is often brewed with tea for the prevention of diseases.[2] Like Sichuan lovage, it is one of the most commonly used ingredients in traditional Chinese beauty formulas (see ***sichuan lovage***).[3]

Contains 0.3–1.8% volatile oil;[1,4] $\beta$-sitosterol, ferulic acid, and others. Amounts of volatile oil and its individual components vary with the geographic and botanical sources; major components present include: neocnidilide (0–25.57%), cnidilide (2.93–10.78%), myristicin (1.63–9.08%), ligustilide (0–6.23%), butylidene phthalide (0–2.01%), $\beta$-phellandrene (0–33.32%), 4-terpinyl acetate (3.59–13.82%), limonene (0–14.44%), terpineol-4 (2.7–8.0%), and terpinolene (2.67–3.24%).[4] Other compounds present include methyleugenol, butyl phthalide, 3-butylidene-4,5-dihydrophthalide, and senkyunolides A, G, H, and I (HU, IMM-2).[2,5]

Decoction (15–30%) has exhibited antifungal activities against dermatophytes *in vitro* (JIANGSU). Fat- and water-soluble extractives reduced inflammation (croton seed oil induced otitis in mice) by 75.3% and 72.9% respectively; water extract also active against experimental edema (egg white induced).[5]

The neutral fraction of the volatile oil (*L. sinense*) has been shown to have numerous biological activities, including:

sedative, analgesic, antipyretic and antiinflammatory in experimental animals;[6,7] antispasmodic and antihistaminic;[8] markedly decreasing oxygen consumption, prolonging survival time, increasing ability of tissue to tolerate anoxia and extending survival time in mice under cerebral ischemic anoxia.[9]

Ferulic acid and ligustilides are some of the active principles of *gaoben* (see also ***Sichuan lovage***).

Powder and extracts used in hair care and skin care products, (especially acne and whitening creams), often together with *Dahurian angelica*, for many of the same functions (e.g., anti-allergic, antiinflammatory, and tyrosinase inhibitory) as Sichuan lovage;[3,10,11] also imparts special aroma to products.

**REFERENCES**

See the General References for CHP; HU; IMM-2; JIANGSU; XU AND XU; ZCYX.

1. B. C. Zhang et al., *Zhongcaoyao*, **17**(8), 34 (1986).
2. Q. S. Li et al., *Zhongcaoyao*, **24**(4), 180 (1993).
3. W. X. Hong, *Fujian Zhongyiyao*, **18**(1), 39 (1987).
4. B. Dai, *Yaoxue Xuebao*, **23**, 361 (1988).
5. Y. G. Xi et al., *Zhongcaoyao*, **18**(2), 6 (1987).
6. Y. Q. Shen et al., *Chin. J. Integr. Trad. Western Med.*, **7**, 738 (1987).
7. Y. Q. Shen et al., *Zhongcaoyao*, **20**(6), 22 (1989).
8. G. J. Chen et al., *Zhongyao Tongbao*, **12**(4), 48 (1987).
9. C. K. Tang and Q. Y. Xu, *Zhongguo Zhongyao Zazhi*, **17**, 745 (1992).
10. X.T. Liu, *Zhongchengyao*, **13**(3), 9 (1991).
11. Y. Masamoto et al., *Planta Med.*, **40**, 361 (1980).

# LUFFA (*SIGUALUO*)

Luffa is also called loofah, vegetable sponge, or dishcloth gourd. It is the fibrous remains of the old mature fruit of either *Luffa cylindrica* (L.) Roem. (water gourd, smooth loofah) or *Luffa acutangula* Roxb. (silky gourd, angled loofah) (Family Cucurbitaceae). Both are annual vines, native to tropical Asia. The former is cultivated throughout China while the latter mainly in the southern provinces of Guangdong and Guangxi. The fruit of *L. cylindrica* is elongated and smooth while that of *L. acutangula* is elongated but bears ten prominent longitudinal ridges. Young fruits are eaten as vegetables.[1] For medicinal and cosmetic uses, the fruit of *L. cylindrica* is allowed to grow old and is harvested in autumn, usually after the first frost. The pulp, skin, and seeds are then removed by rubbing or the fruit is soaked in water until the skin and pulp disintegrate, which are then washed off along with the seeds; the resulting spongelike luffa is then sun dried. This is the form most familiar to Westerners. The mature fruit of *L. acutangula* is also collected in the fall but is dried without removing skin and seeds.

Luffa has been used in Chinese medicine since the 16th century. Traditionally considered sweet tasting and neutral and to promote blood circulation and facilitate energy flow in the body (*huoxue tongluo*) as well as having antiinflammatory, fever-reducing and detoxifying properties, among others (LEUNG). Used in treating numerous conditions, especially rheumatism, arthritic pain, muscle pain, chest pain, amenorrhea and lack of milk flow in nursing mothers. Luffa charcoal has recently been reported effective in the topical treatment of shingles (herpes zoster) in the face and eye region (LEUNG).[2]

Apart from presence of polysaccharides (including cellulose, xylan and mannan-galactan) the chemistry of luffa sponge is basically unknown.

Decoctions of luffa sponge (i.p. or s.c.) exhibited marked antiinflammatory, analgesic and tranquilizing effects in mice.[3,4] Toxicity of luffa is very low: $LD_{50}$ (i.p.) of decoction in mice was 137.40 ± 16.71 g/kg.[4]

Luffa sponge is used to remove dead skin tissue and to stimulate the skin; powdered luffa and extracts used in facial scrubs, skin cleansers, and other skin care products for their antiinflammatory and traditional detoxicant properties.

## REFERENCES

See the General References for CHEUNG AND LI; CHP; IMM-3; JIANGSU; LEUNG; NATIONAL.

1. S. Y. Zee and L. H. Hui, *Hong Kong Food Plants*, The Urban Council of Hong Kong, 1981, p. 42.
2. X. L. Gu, *Zhejiang Zhongyi Zazhi*, **23**, 88 (1988).
3. B. Kang et al., *Shiyong Zhongxiyi Zazhi*, **6**, 227 (1993).
4. B. Kang et al., *Zhongcaoyao*, **24**, 248 (1993).

# MUME (SMOKED PLUM OR *WUMEI*)

It is the dried unripe fruit of *Prunus mume* (Sieb.) Sieb. et Zucc., a deciduous tree up to 10 m high, also known as Japanese apricot (Family Rosaceae). The green, about-to-ripen fruit is collected in May and oven-dried at ca. 40°C for 2–3 days followed by leaving in the closed oven for 2 to 3 more days until it turns black. Produced mostly in southern provinces, especially Sichuan which is the largest producer; Zhejiang produces the best *wumei* which is large and jet black, with thick meat and small pit, and tastes sour (ZHU).

Its earliest record dates back to the *Shen Nong Ben Cao Jing* (ca. 200 B.C.–100 A.D.). Traditionally regarded as sour tasting, astringent, neutral and to promote secretion of body fluids as well as expel parasites; used to treat chronic cough, chronic diarrhea, diabetes, ascariasis (roundworm infection) and hookworm infection, neurodermatitis, eczema and hard-to-heal sores, among other conditions (CHP, IMM-3, JIANGSU). In recent years, often used as an ingredient with other detoxicant herbs (e.g., schisandra, licorice, and *fangfeng*) both internally and externally in the treatment of allergic conditions such as asthma, urticaria, allergic rhinitis and pruritus;[1,2] also to treat polyps, tumors, and capillary hemangioma (ZHOU AND WANG).[3] It is also boiled in water and sweetened with sugar to make *suan mei tang* (sour plum decoction), a refreshing drink very popular in southern China and Taiwan.

*Wumei* contains various plant acids, especially citric acid (19%) and malic acid (15%); oleanolic acid; β-sitosterol; amino acids; carbohydrates; wax; and others (HU, JIANGSU). Seed contains amygdalin (see ***almond***).

Decoctions and alcoholic extracts of *wumei* have exhibited marked *in vitro* antibacterial effects against numerous bacteria (both Gram positive and Gram negative);

and its decoction strongly active against pathogenic fungi *in vitro*. Its alcoholic extract also exhibited strong antihistaminic effects on isolated guinea pig trachea;[4] and acetone-soluble extractives were strongly active against aflatoxin $B_1$ mutagenesis.[5] Oleanolic acid has been reported to have various biological activities, including antiallergic (see ***ligustrum***). It appears that the use of *wumei* in treating allergic conditions has some scientific basis.

Extracts (water and hydro-alcoholic) are used in anti-allergic ointments and in skin creams and lotions for their antimicrobial and anti-allergic effects; also used in haircare products for their traditional hair-darkening and growth-stimulating properties.[6]

## REFERENCES

See the General References for BAILEY 1; CHP; HU; IMM-3; JIANGSU; WANG; ZHOU AND WANG; ZHU.

1. Z. H. Dong, *Zhongchengyao*, **11**(11), 37 (1989).
2. L. H. Liu and Z. X. Tang, *Zhongyi Zazhi*, (1), 15 (1989).
3. J. P. Yang, *Jiangsu Zhongyi*, **14**(1), 28 (1993).
4. R. D. Xiang et al., *Zhongcaoyao*, **15**(2), 22 (1985).
5. C. C. Ruan et al., *Chin. J. Cancer*, **8**(1), 29 (1989).
6. J. H. Chen, *Shandong Zhongyi Xueyuan Xuebao*, **15**(1), 58 (1991).

# PEARL (*ZHENZHU* OR MARGARITA) AND MOTHER-OF-PEARL (*ZHENZHUMU*)

Pearl is found in certain mollusks, including oysters, clams and mussels. It is composed of concentric layers of nacre secreted by these animals in response to irritation, especially that caused by a foreign substance. Nacre is also the substance that makes up mother-of-pearl, the shiny layer that forms the inner lining of the shells. The most common source species of Chinese pearl and/or mother of pearl are *Pteria martensii* (Dunker) (syn. *Pinctada martensii*), *Pteria margaritifera* (L.) (syn. *Pinctada margaritifera*) (Family Pteriidae); *Hyriopsis cumingii* (Lea) and *Cristaria plicata* (Leach) (Family Unionidae). The first two are marine while the latter two are freshwater species.

Pearl is collected from both natural and artificially implanted animals, washed free of mucilage and towel dried. The better grades are large, round, white and lustrous and, when broken, show distinct layers of nacre deposits and no hard nucleus. Major producers are coastal provinces in China, especially Guangxi, Guangdong, Hainan and Zhejiang (CMH, ZCYX, ZHU).

Mother-of-pearl is produced primarily from the shells of the freshwater species, which are boiled in alkaline water followed by soaking in fresh water. The dark surface layer is then scraped off and the shells are baked until crisp. Better grades are white, come in large pieces and crisp. Much of the commercial mother-of-pearl is produced from shells which are by-products of buttons production. Major producers are the same as those of pearl (NATIONAL, ZCYX).

For medicinal or cosmetic use, pearl must be ground to an extremely fine powder while mother-of-pearl can be used as a coarse powder or simply broken up in pieces. A typical traditional method of producing pearl powder is to wrap the clean pearls in cheesecloth or muslin, place them between two pieces of tofu (bean curd) so that the pearls are completely embedded in the tofu and cook them for 2 hours. After cooking, the pearls are washed with clean water, placed in a mortar with a suitable amount of water, and ground to an extremely fine powder (when no more sound is produced by the pearl particles rubbing against the mortar) and dried. The tofu treatment makes the pearl easier to grind and at the same time preserving many of its active components that would otherwise be destroyed if it were to be subjected to high baking temperatures to render it crisp (JIANGSU, MA).

Earliest recorded use of pearl in traditional Chinese medicine dates back to the *Ben Cao Jing Ji Zhu* (ca. 500 A.D.). Traditionally regarded as sweet and salty tasting, cold, and to have tranquilizing, vision-brightening, detoxifying and healing properties, it is used in treating anxiety, infantile convulsions, insomnia, epilepsy, nebula or opacity of the cornea (*yunyi*), sore throat, mouth sores, and difficult-to-heal sores and ulcers (CHP, JIANGSU, NATIONAL).

Pearl powder has long been regarded as of special benefit to the skin, having been recorded in the *Hai Yao Ben Cao* (ca. 907–925 A.D.) and the *Ben Cao Gang Mu* (1593 A.D.) as good for removing facial dark spots and making one's skin smooth and young looking. For this, it can be applied directly to the skin or taken internally once every 10–15 days, up to 3 g each time (CHP, JIANGSU).[1]

Mother-of-pearl is traditionally considered salty tasting, cold and to have some of the properties and uses of pearl (e.g., vision brightening, calming, and treating insomnia). However, it is traditionally not known to have healing properties nor special benefits to the skin (CHP, JIANGSU, NATIONAL).

Pearl contains mostly calcium carbonate, ranging from ca. 81–95%, depending on source species, natural or cultured. It also contains ca. 6–13% organic substances which include 16 amino acids (leucine, methionine, alanine, phenylalanine, glycine, aspartic acid, glutamic acid, proline, serine, etc.) and a small amount of taurine; numerous trace minerals etc. (HU, JIANGSU, NATIONAL).[2]

Mother-of-pearl contains similar constituents as pearl but with lesser amount of organic substances (JIANGSU) that include conchiolin (a protein).

At 3% level in an ointment, pearl has exhibited strong wound-healing activities, ranking second among 39 Chinese traditional drugs tested.[3] Treatment of 20 cases of canker sore due to adverse reactions to chemotherapy by topical application of pearl powder (4 × daily; 3 g/time) resulted in complete resolution of the condition in 3 to 7 days.[4]

Recent studies on mother-of-pearl have shown it to have numerous biological effects, including: anti-anoxia effects in mice (water extract; not conchiolin);[5] cardiovascular effects in patients with coronary heart disease, markedly decreasing their serum lipid peroxide levels;[6] antiulcer effects in humans;[7] antihistaminic; inhibiting contraction of isolated uterus and intestine of guinea pig, etc. (JIANGSU).[8] Its acute toxicity is low: $LD_{50}$ in rats were >21,500 mg/kg (p.o.) and >31,600 mg/kg (s.c.).[8]

It appears that both pearl and mother-of-pearl are used in cosmetics, even though the latter lacks prior use documentation.

Pearl powder is extensively used in skincare products (e.g., acne and freckle creams and lotions, nourishing creams, etc.) for its healing and traditional skin-lightening, smoothing and anti-wrinkle properties. However, as pearl is a very expensive ingredient and there is no meaningful assays to determine its identity and quality, it is prone to adulteration, especially with mother-of-pearl and other shell products.

**REFERENCES**

See the General References for CHP; CMH; HU; JIANGSU; MA; NATIONAL; ZCYX; ZHU.

1. Y. B. Xie, *Dazhong Zhongyiyao*, (4), 39 (1990).
2. L. F. Wen, *Zhongguo Yaoxue Zazhi*, **24**, 276 (1989).
3. S. X. Sun, *Chin. J. Integr, Trad. Western Med.*, **6**, 408 (1986).
4. H. J. Yang et al., *Sichuan Zhongyi*, (4), 57 (1988).
5. W. Y. Yang et al., *Haiyang Yaowu Zazhi*, (1), 29 (1986).
6. Y. W. Huang et at., *Chin. J. Integr, Trad. Western Med.*, **7**, 596 (1987).
7. Z. H. Zhu et al., *Zhonghua Xiaohua Zazhi*, **2**, 167 (1982).
8. Y. N. Yang et al., *Haiyang Yaowu Zazhi*, (1), 16 (1986).

# PEONY (PEONY BARK AND PEONY ROOT)

Peony yields three commonly used herbs (see individual entries): **peony bark (*mudanpi*)** and **peony root, red** and **white (*chishaoyao* and *baishaoyao*)**.

# PEONY BARK (*MUDANPI*)

It is the root bark of tree peony, *Paeonia suffruticosa* Andr. (syn. *P. moutan* Sims and *P. arborea* Donn) (Family Paeoniaceae or Ranunculaceae), a small perennial deciduous shrub, 1–1.5 m high, with short robust stems and thick roots; native to China, now extensively cultivated, also as ornamental in the United States (HAY AND SYNGE). The root from 3- to 5-year-old plants are dug up in autumn or early spring, rid of dirt and rootlets; the bark is removed and sun dried (*yuan danpi* or original bark) or the outer bark is first scraped off with a bamboo knife or broken porcelain and then dried (*gua danpi* or scraped bark). Produced mainly in central and eastern provinces. There are numerous grades between the two types of peony bark.

Earliest recorded medicinal use of peony bark in China dates back at least 2,000 years to the *Shen Nong Ben Cao Jing* (ca. 200 B.C.–100 A.D.). Traditionally considered pungent and bitter tasting, cooling and to have heat-dispersing, blood-cooling and blood-activating as well as stasis-removing properties; it is used in treating skin rashes, nosebleed and vomiting blood due to "heat and toxins" (*wen du fa ban*) (e.g., viral or bacterial infections such as flu, measles, acute appendicitis, etc.), neurodermatitis, carbuncles, amenorrhea and dysmenorrhea, abdominal pain, hypertension, allergic sinusitis, urticaria, and traumatic injuries and contusions, etc.

Peony bark contains paeonol, paeonoside (paeonol glucoside), paeonolide (paeonoside–arabinoside), and the monoterpene glycosides paeoniflorin, benzoylpaeoniflorin and oxypaeoniflorin (JIANGSU PROVINCIAL 1);[1] 1,2,3,4,6-pentagalloylglucose;[2] a volatile oil (0.15–0.4%) and phytosterols (NATIONAL). The amounts of paeonol vary greatly depending on geographic and reporting sources, ranging from a low of 0.19–0.54%;[1,3] to a high of 3.5%,[4] with a more common range of 1.08–2.51% (HU).[5]

Many of the biological activities of peony bark can be attributed to paeonol, which include: antiinflammatory, analgesic, antipyretic, central depressant, antibacterial and antifungal, diurectic, and antiatherosclerotic and antiplatelet aggregation, among others.[3,6–10] The toxicities of paeonol are low, its $LD_{50}$ in mice being 196 mg/kg (i.v.), 781 mg/kg (i.p.) and 3430 mg/k (p.o.).[8] 1,2,3,4,6-Pentagalloylglucose had antiviral activities.[2]

Extracts (water and hydro-alcoholic) and paeonol are used in dental products (e.g., toothpaste for inflamed and sore gums), hair-care and skin-care products (e.g., antiallergy creams and lotions) for their antibacterial, antiinflammatory and traditional skin-soothing and skin-protectant properties (ETIC).

**REFERENCES**

See the General References for BAILEY 1; CHP; FOSTER AND YUE; HU; JIANGSU; JIANGSU PROVINCIAL 1; NATIONAL; ZHU.

1. J. Yu et al., *Yaoxue Xuebao*, **20**, 229 (1985).
2. M. Takechi and Y. Tanaka, *Planta Med.*, **45**, 252 (1982).
3. K. Kawashima et al., *Planta Med.*, **187** (1985).
4. T. Tani et al., *J. Ethnopharmacology*, **21**, 37 (1987).
5. Y. S. Zhou et al., *Zhongchengyao*, **14**(7), 23 (1992).
6. A. B. Wang and X. C. Tang, *Zhongcaoyao*, **14**(10), 26 (1983).
7. L. Shi et al., *Acta Pharmacol. Sin.*, **9**, 555 (1988).
8. Q. A. Li, *Zhongcaoyao*, **19**(6), 36 (1988).
9. T. Ohta et al., *Yakugaku Zasshi*, **81**, 100 (1961).
10. M. Harada et al., *Yakugaku Zasshi*, **92**, 750 (1972).

# PEONY ROOT, RED AND WHITE (*SHAOYAO*: *CHISHAOYAO* AND *BAISHAOYAO*)

Peony root is known as *shaoyao* in Chinese; there are two types: *chishaoyao* or *chishao* (red peony root) and *baishaoyao* or *baishao* (white peony root). Both are the root of *Paeonia lactiflora* Pall. (syn. *P. albiflora* Pall., *P. edulis* Salisb. and *P. fragrans* Redoute) (Family Paeoniaceae or Ranunculaceae), a perennial herb, 60 to 80 cm high, with glabrous erect stems and thick cylindrical to spindle-shaped root; native to Siberia and China and distributed throughout northern, northeastern, eastern and central China; extensively cultivated, also as ornamental in the United States (BAILEY 1, HAY AND SYNGE).

Red peony root is collected from wild plants of *P. lactiflora* (*P. obvata* Maxim. and *P. veitchii* Lynch are also used) in spring and autumn (autumn is preferred as better grades are obtained), rid of rhizomes and rootlets, washed free of dirt and sun dried or dried in the shade; no cooking is involved. Inner Mongolia produces the best grade which is thick and long, fracture white and starchy.

White peony root is collected from 3- to 4-year-old cultivated plants of *P. lactiflora* in summer and autumn, rid of rhizomes and rootlets, washed free of dirt, scraped off outer bark and boiled in water for 5–15 minutes until soft; it is then sun dried. Zhejiang produces the best grade which is thick, tough, starchy and with no white center or cracks.

Topical use of red peony root in treating carbuncles (*ju*) was first mentioned in the *Wu Shi Er Bing Fang* (1065–771 B.C.) and its medicinal properties and uses were first described in the *Shen Nong Ben Cao Jing* (ca. 200 B.C.–100 A.D.) while white peony root was not described medicinally until the mid-11th century in the *Tu Jing Ben Cao*.[1] However, according to another report, a sauce made with white peony root is said to be a favorite of Confucius' (ca. 500 B.C.); also, it was cooked with animal organs to prevent food poisoning.[2] Hence, there is still much confusion regarding the history and identity of the two herbal food/drugs. The major difference between the two is that red peony root, like peony bark, is heat-dispersing and blood-cooling while white peony root is a liver and blood tonic. The following properties and uses are described in the current Chinese Pharmacopeia: Red peony root is bitter tasting and slightly cold, with analgesic and stasis-dispersing properties; it is used in treating tight chest and abdominal pain due to stagnation of Liver *qi* (*gan yu xie tong*), abdominal mass, and "heat and toxins" conditions (see ***peony bark***). White peony root is bitter and sour tasting, slightly cold, with liver-calming (*ping gan*), analgesic, blood-nourishing and menstrual regulating properties. It is used in treating headache, dizziness, abdominal pain, tight chest, stiff and painful joints (limbs), pale complexion due to blood deficiency, irregular menses, spontaneous perspiration and night sweat, etc. (CHP). Whenever peony root (*shaoyao*) is prescribed without specifics, white peony root is normally used. White peony root is also more frequently used for general tonic purposes and there are now numerous *baishao* based food and drink products commercially available in China, including wines, fruit juices, and soft drinks.[2]

Both red and white peony roots contain very similar chemical components, especially the monoterpene glycosides paeoniflorin (3–5%), benzoylpaeoniflorin and oxypaeoniflorin but little or no paeonol, paeonoside or paeonolide, the latter only present in shrubby peonies (see peony bark).[3–5] Other constituents present include: albiflorin, lactiflorin, and (*Z*)-(1*S*,5*R*)-β-pinen-10-yl-vicianoside (monoterpene glycosides), β-sitosterol, β-sitosterol-α-glucoside, benzoic acid (ca. 1%), palmitic acid, *cis*-9,12-octadecadienoic acid, alkanes ($C_{24}$–$C_{26}$), daucosterol, gallic acid, methyl gallate, *d*-catechin, myo-inositol, sucrose, and glucogallin, among others (JIANGSU).[5–9]

The glycosides (especially paeoniflorin) of peony root (red and white) are responsible for many of their biological activities, which include antibacterial, antifungal and antiviral;[10,11] antiinflammatory and immunomodulating;[10,12–14] analgesic; sedative; antispasmodic; antiplatelet aggregation; antifatigue, prolonging survival and improving memory;[15] antitumor, enhancing phagocytosis of macrophages and elevating cyclic AMP levels;[16] antimutagenic;[17] and others in humans and experimental animals (HU, JIANGSU).

Acute toxicities of peony root and paeoniflorin are low: $LD_{50}$ (p.o.) of white peony root in rats was 81 g/kg; $LD_{50}$ (i.v.) and $LD_{50}$ (i.p.) of paeoniflorin in mice were 3530 mg/kg and 9530 mg/kg respectively (HU).

Aqueous and hydro-alcoholic extracts of both red and white peony roots are used in skin-care products for their antimicrobial (acne creams, etc.), antiinflammatory and astringent properties; used with Dahurian angelica (see that entry) in freckle-removal creams and lotions for their traditional ability to remove blood stasis and activate blood circulation; also used as part of a natural preservative system in cosmetic products due to its relatively high content of benzoic acid (ETIC, ZHOU).

## REFERENCES

See the General References for BAILEY 1; CHP; ETIC; FOSTER AND YUE; HU; JIANGSU; NATIONAL; ZHOU; ZHU.

1. R. J. Chai, *Beijing Zhongyi Xueyuan Xuebao*, **14**(1), 49 (1991).
2. J. C. Li, *Zhongyao Tongbao*, **12**(8), 54 (1987).
3. J. Yu et al., *Yaoxue Xuebao*, **20**, 229 (1985).
4. C. D. Jin et al., *Zhongcaoyao*, **24**, 183 (1993).
5. C. F. Wu, *Zhongyao Tongbao*, **10**(6), 43 (1985).
6. H. Y. Lang et al., *Planta Med.*, **50**, 501 (1984).
7. H. S. Chen et al., *Zhongguo Yaoxue Zazhi*, **28**, 137 (1993).
8. M. Kaneda et al., *Tetrahedron*, **28**, 4309 (1972).
9. H. Y. Lang et al., *Yaoxue Xuebao*, **18**, 551 (1983).
10. M. Y. Wang et al., *Liaoning Zhongyi Zazhi*, (9), 43 (1992).
11. B. Y. Liang et al., *Shanghai Zhongyiyao Zazhi*, (6), 4 (1989).
12. J. S. Liang et al., *Chin. J Pharmacol. Toxicol.*, **4**, 258 (1990).
13. M. R. Liang et al., *Xinzhongyi*, (3), 51 (1989).
14. H. Zhang et al., *Chin. J Pharmacol. Toxicol.*, **4**, 190 (1990).
15. D. Zhou et al., *Jilin Zhongyiyao*, (2), 38 (1993).
16. K. W. Huang et al., *Chin. J. Oncol.*, **6**, 319 (1984).
17. X. B. Ni, *Zhongcaoyao*, **22**, 429 (1991).

# PHELLODENDRON BARK (*HUANGBAI*)

Stem bark of *Phellodendron amurense* Rupr. (Amur corktree) or *P. chinense* Schneid. (Chinese corktree) (Family Rutaceae); the former (called *huangbai*) from northern and northeastern China while the latter (*chuan huangbai* or *huangpishu*) from central and southern China. Both deciduous trees, with *P. amurense* up to 25 m and *P. chinense* up to 12 m high. Bark is collected between March and June from trees at least 10 years old, rid of outer cork layer, cut into small sections and sun dried.

Recorded use of *huangbai* dates back at least 2,000 years. Traditionally considered to taste bitter and cold, with heat-clearing, wetness-drying (*zao shi*), fire-purging (*xie huo*) and detoxifying properties. Used in treating numerous heat (*re*) conditions, including: acute bacterial dysentery, acute enteritis, acute icterohepatitis, jaundice, urinary infections, night sweating, wet dreams, leukorrhagia, and oral sores. Externally used in treating eczema, pruritus, and skin sores, among others.

In recent years, it has been extensively and successfully used in China as an ingredient in numerous formulas (in extract or powder form) for treating burns, often together with giant knotweed, garden burnet, baikal scullcap (see individual entries), and *zicao*;[1–7] also used in wounds and injuries and to treat acne and facial dark spots.[8,9]

Amur corktree bark contains alkaloids composed mainly of berberine (0.6–2.5%), phellodendrine, magnoflorine, jatrorrhizine, candicine, and palmatine; limonin (bitter principle); obakunone, dictamnolide, $\gamma$-sitosterol, $\beta$-sitosterol, 7-dehydrostigmasterol, stigmasterol, and mucilage, etc.

Chinese corktree bark contains similar constitutents, but with higher berberine content (4–8%).

Aqueous extracts of Amur corktree bark have strong antioxidant[10] as well as strong and broad antibacterial and antifungal activities (berberine is one of the active components); the alkaloids (esp. berberine, phellodendrine, and palmatine) also have hypotensive action in animals; other activities include hypoglycemic, hypocholesterolemic, and blood platelet protective effects (JILIN, WANG).

Extracts used in baby powder and acne creams and as natural preservatives in products that are not incompatible with a yellow tone imparted by these extracts (ETIC, ZHOU).

## REFERENCES

See the General References for CHP; ETIC; HU; JIANGSU; JILIN; NATIONAL; WANG; ZHOU.

1. Y. D. Ge, *Xinzhongyi*, (1), 25 (1986).
2. H. T. Mao et al., *Chin. J. Integr. Trad. Western Med.*, **7**, 532 (1987).
3. Y. H. Dong, *Zhejiang Zhongyi Zazhi*, **23**, 495 (1988).
4. Z. L. Li and Z. P. Xie, *Sichuan Zhongyi*, (2), 43 (1990).
5. H. Q. Song and F. X. Wang, *Yunnan Zhongyi Zazhi*, **12**(3), 34 (1991).
6. D. C. Chen, *Shiyong Zhongxiyi Jiehe Zazhi (PJCM)*, **4**, 414 (1991).
7. X. W. Sun et al., *Xinzhongyi*, (12), 30 (1992).
8. X. D. Zhou, *Chin. J. Integr. Trad. Western Med.*, **5**, 726 (1985).
9. L. T. Song and H. J. Jiang, *Jilin Zhongyiyao*, (3), 36 (1992).
10. L. C. Song et al., *Shaanxi Zhongyi*, **14**, 185 (1993).

# PURSLANE, COMMON (*MACHIXIAN*)

Aboveground parts of *Portulaca oleracea* L. (Family Portulacaceae), a prostrate, smooth herbaceous annual with succulent spatula-shaped leaves and tiny yellow flowers, up to 30 cm high; probably native to Eurasia but is now found worldwide and considered a weed throughout most of the United States and southern Canada; distributed in most of China; used both fresh and dried. To prepare the dried herb, aerial parts are collected in summer and early autumn when stems and leaves are in their fullest, washed free of dirt, briefly treated with boiling water, and sun dried. Produced throughout China.

Also known as garden purslane, green purslane and pigweed, it is eaten as a salad and vegetable by peoples around the world; and is used medicinally for various conditions, including headache, stomachache, painful urination, dysentery, enteritis, mastitis, lack of milk flow in nursing mothers and in postpartum bleeding; and externally in treating burns, earache, insect stings, inflammations, skin sores, ulcers, pruritus, eczema and abscesses, for which the fresh herb is normally used as poultice or expressed juice (FOSTER AND DUKE, GRIEVE).[1]

Its earliest recorded use in China dates back to about 500 A.D. in the *Ben Cao Jing Ji Zhu*. Traditionally considered sour tasting and cold, with heat-relieving and detoxicant (*qing re jie du*) as well as blood-cooling and hemostatic properties; used internally in treating bacillary dysentery, hematochezia (bloody stool), bleeding hemorrhoids and metrorrhagia; and externally to treat the same conditions listed above except in addition to using the fresh herb, the Chinese also use decoctions and powder of the dried herb for topical application. In recent years, it has also been used to treat colitis, acute appendicitis, diabetes, dermatitis and shingles (IMM-4, JIANGSU).

Purslane contains large amounts of *l*-norepinephrine (*l*-noradrenaline; 0.25% in fresh herb), a neurohormone that has vasopressor and antihypotensive activities and reduces hemorrhage at the tissue level (JIANGSU, MARTINDALE).

It also contains numerous common nutrients (varying from low to high concentrations depending on report), including: vitamins (A, $B_1$, $B_2$, C, niacinamide, nicotinic acid, $\alpha$-tocopherol, $\beta$-carotene, etc.); minerals (especially potassium); fatty acids, especially omega-3 acids whose concentration in purslane is the highest found in leafy vegetables;[2] glutathione; glutamic acid; and aspartic acid. Other constituents include a mucilage composed of an acidic and a neutral fraction with structure determined,[3] calcium oxalate, malic and citric acids, dopamine and dopa, coumarins, flavonoids, alkaloids, saponins, and urea, among others (JIANGSU, WATT AND MERRILL).[1–7]

An aqueous extract of purslane has exhibited skeletal muscle relaxant effects both *in vitro* and *in vivo*; it also relaxed guinea pig gastric fundus, taenia coli, and rabbit jejunum as well as contracted the rabbit aorta and raised blood pressure.[8–10] Topical application of the aqueous extract onto the skin was effective in relieving muscle spasms.[9]

Other biological effects include: antibacterial and antifungal; wound healing; antiinflammatory; uterine stimulant; and diuretic in rabbits (JIANGSU, NATIONAL).[11,12]

Although norepinephrine may account for some pharmacologic activities, the active principles for most of the biological activities and medicinal properties of purslane are still unidentified.

Due to its high content of nutrients, especially antioxidants (vitamins A and C, $\alpha$-tocopherol, $\beta$-carotene, glutathione) and omega-3 fatty acids, and its wound-healing and antimicrobial effects as well as its traditional use in the topical treatment of inflammatory conditions, purslane is a highly likely candidate as a useful cosmetic ingredient. Since most of the reported effects of purslane are due to its fresh juice or to its decoction, water extractives would be most suitable.

## REFERENCES

See the General References for CHP; FERNALD; FOSTER AND DUKE; FUJIAN; GRIEVE; IMM-4; JIANGSU; MARTINDALE; NATIONAL.

1. C. Whiteman, *Aust. J. Med. Herbalism*, **5**(2), 29 (1993).
2. A. P. Simopoulos et al., *J. Am. Coll. Nutr.*, **11**, 374 (1992).
3. E. S. Amin and S. M. El-Deeb, *Carbohydr. Res.*, **56**(1), 123 (1977); through *Chem. Abstr.*, **87**, 35876p.
4. F. R. Bharucha and G. V. Joshi, *Naturwissenschaften*, **44**, 263 (1957).
5. Zh. Stefanov et al., *Farmatsiya (Sofia)*, **16**(3), 27 (1966)(Bulg.); through *Chem. Abstr.*, 17557f.
6. J. Gillaspy, *The UWPT Newsletter*, **1**(2), 2 (1993).
7. T. M. Zennie and C. D. Ogzewalla, *Econ. Bot.*, **31**(1), 76 (1977).
8. F. Okwuasaba et al., *J. Ethnopharmacology*, **17**, 139 (1986).
9. O. Parry et al., *J. Ethnopharmacology*, **19**, 247 (1987).
10. O. Parry et al., *J. Ethnopharmacology*, **22**, 33 (1988).
11. Q. G. Hao and C. L. Wang, *Shandong Zhongyi Zazhi*, **10**(3), 39 (1991).
12. S. J. Xu and L. N. Liu, *Shandong Zhongyi Zazhi*, **10**(3), 52 (1991).

# RED SAGE (*DANSHEN*)

Root and rhizome of *Salvia miltiorrhiza* Bge. (Family Labiatae or Lamiaceae), a hairy perennial herb, 30–80 cm high; native to China and widely distributed there. Root (with rhizome) is dug up in spring and autumn, rid of rootlets, sand and dirt, and sun dried. Other species also used as source of *danshen* include *Salvia przewalskii* Maxim. (*gansu danshen*), *S. przewalskii* Maxim. var. *mandarinorum* (Diels) Stib., *S. bowleyana* Dunn (southern *danshen*) and *S. yunnanensis* C. H. Wright; they all have similar chemistry.

Recorded use dates back 2,000 years. Considered one of the major *huo xue hua yu* (activating blood circulation to dissipate stasis) herbs, *danshen* is traditionally used in blood and blood circulation problems, including angina pectoris (chest pain), irregular menses, menstrual pain, amenorrhea, metrorrhagia (*xue beng*), leukorrhagia (*dai xia*), abdominal masses (*zheng jia ji ju*), abdominal pain, and insomnia due to palpitations and tight chest, among others.

Contains several phenanthrene diketones and derivatives: tanshinones I, IIA, IIB, V, and VI, isotanshinones I and II, cryptotanshinone, isocryptotanshinone, and dihydrotanshinone; hydroxytanshinone IIA and methyltanshinonate; tanshinol I and tanshinol II;[1] tanshindiols A, B and C, nortanshinone and 3-$\beta$-hydroxytanshinone IIA.[2] Also contains miltirone, Ro-090680 and salvinone (diterpenoids);[2,3] salvianolic acids A and B, rosmarinic acid;[4,5] danshensu [D(+)-$\beta$-(3,4-dihydroxyphenyl)-lactic acid);[6,7] protocatechuic aldehyde, protocatechuic acid, oleanolic acid, ferruginol, dehydromiltirone, $\beta$-sitosterol, vitamin E and others.[8–10]

Modern scientific studies have confirmed many of its traditional properties and uses. Thus, its extracts and chemical components have been shown to have the following activities: anticoagulant or antiplatelet aggregation (decoction, injection, tanshinones, miltirone, ferruginol, Ro-090680, danshensu, protocatechuic aldehyde, salvinone, salvianolic acid A and rosmarinic acid);[3,4,6,8,11–13] antibacterial (decoction, alcoholic extract, tanshinones, methyltanshinonate); antiinflammatory (tanshinones);[14] estrogenic and anti-androgenic (tanshinones); retardation of cholesterol biosynthesis in cells and inhibition of lipoprotein oxidation (danshensu);[7] antioxidant *in vitro* (strong: salvianolic acids A & B, rosmarinic acid; fair to weak: danshensu, alcoholic extract);[5,7,15] antimutagenic (acetone extractives);[14] as well as positive cardiovascular and other effects (WANG).[1,17,18] Toxicity is low: 43 g/kg of decoction (i.p.) in mice caused no fatality within 48 hrs while 64 g/kg only resulted in 2 deaths in 10 mice; gastric feeding of 2% tanshinones emulsion to mice (0.5 ml) for 14 days and to rats (2.5 mL) for 10 days produced no obvious toxic reactions (WANG).

Tanshinone IIA sulfonate (a major active principle) did not promote growth or metastasis of Lewis carcinoma transplated in mice.[19]

In addition to cardiovascular diseases, *danshen* extracts (per os) have been successfully used in treating acne, psoriasis, eczema and other skin diseases (WANG).[12,20]

Tanshinones can be extracted with ether, acetone, or related solvents while danshensu is extracted with water.

Extracts used in hair liniments and shampoos for their alleged ability to prevent hair loss and maintain hair color (see ***safflower***); also used in skin creams and lotions for their alleged whitening effects (ZHOU).

## REFERENCES

See the General References for CHP; FOSTER AND YU; HU; JIANGSU; WANG; ZHOU.

1. A. Yagi et al., *Planta Med.*, **55**, 51 (1989).
2. H. W. Luo et al., *Phytochemistry*, **24**, 815 (1985).
3. N. Wang et al., *Planta Med.*, **55**, 390 (1989).
4. L. N. Li et al., *Planta Med.*, 227 (1984).
5. Y. S. Huang and J. T. Zhang, *Yaoxue Xuebao*, **27**, 96 (1992).
6. C. Z. Li et al., *Chin. J. Integr. Trad. Western Med.*, **3**, 297 (1983).
7. X. M. Sun et al., *Zhongcaoyao*, **22**, 20 (1991).
8. H. W. Luo et al., *Yaoxue Xuebao*, **23**, 830 (1988).
9. X. M. Xu and Z. Y. Xiao, *Zhongcaoyao*, **15**(1), 1 (1984).
10. N. Wang and H. W. Luo, *Zhongcaoyao*, **20**(4), 7 (1989).
11. Y. D. Shi et al., *Zhongyao Tongbao*, **11**(7), 48 (1986).
12. D. B. Wang, *J. Trad. Chin. Med.*, **3**, 227 (1983).
13. Z. W. Zou et al., *Yaoxue Xuebao*, **28**, 241 (1993).
14. Y. G. Gao et al., *Chin. J. Integr. Trad. Western Med.*, **3**, 300 (1983).
15. Y. L. Zhou and R. X. Xu, *Zhongguo Zhongyao Zazhi*, **17**, 368 (1992).
16. C. C. Ruan et al., *Chin. J. Cancer*, **8**, 29 (1989).
17. P. G. Xiao and K. J. Chen, *Phytother. Res.*, **1**(2), 53 (1987).
18. H. J. Deng et al., *Zhongguo Zhongyao Zazhi*, **17**, 233 (1992).
19. M. Z. Liu et al., *Acta Pharmacol. Sin.*, **12**, 534 (1991).
20. D. B. Wang et al., *Chin. J. Dermatol.*, **21**, 167 (1988).


## SAFFLOWER (FALSE SAFFRON; *HONGHUA*)

Flowers of *Carthamus tinctorius* L. (Family Compositae), a glabrous annual herb, up to about 1 m high, flowering from May to July; extensively cultivated worldwide, especially for its seeds. When the tubular florets turn from yellow to red, they are collected early in the morning and dried (sun, oven, or in the shade). Xinjiang Province is the largest producer.

First described in the *Shan Han Lun* (circa 3rd century, A.D.), safflower is one of the commonly used *huo xue hua yu* (activating circulation to dissipate blood stasis) herbs. Traditionally used to invigorate blood, break up stasis, facilitate menstruation, and relieve pain; used in cardiovascular conditions (e.g., amenorrhea, menorrhalgia coronary heart disease, chest pain, and traumatic injuries). Although also used to "calm" live fetus and abort dead fetus, caution in pregnancy is normally advised.

Contains a complex mixture of red and yellow pigments including 20–30% safflower yellow (safflor yellow; SY) which is composed of safflomin A (75%), SY-2 (15%), SY-3 and SY-4, all chalcones;[1] glycosides of chalcone (e.g., carthamin; yellow) and quinone (e.g., carthamone; red), with the latter predominant in the commercial product; also colorless flavonoids and flavonoid glycosides (carthamidin, isocarthamidin, neocarthamin); safflower polysaccharide (glucose, xylose, arabinose and galactose in $\beta$-linkages);[2] lignans, fatty acids; and others.[3]

Safflower yellow has immunosuppressive and strong anticoagulant activities;[1,4] and safflower polysaccharide has immunopotentiating effects.[2] Other activities of safflower extracts include: cardiac stimulant, vasodilating, hypolipemic, hypotensive, uterine stimulant, etc.[5] Toxicities are low: i.v. $LD_{50}$ of carthamin in mice is $2.35 \pm 0.14$ g/kg while safe oral dose is >8 g/kg; i.p. MLD of decoction in mice is 1.2 g/kg; and a 50% injection when dropped in rabbit eye produced no irritation to its conjunctiva.

Extract used in most Chinese hair growth liniments along with *danshen* and tonic herbs for its blood-invigorating and vasodilating effects that are thought to facilitate transportion of other tonic ingredients to nourish the hair follicles; in facial and body massage preparations and bath preparations for the same effects; also can be used as a coloring agent.

**REFERENCES**

See the General References for HU; JIANGSU; WANG; ZHOU.

1. Z. W. Lu et al., *Acta Pharmacol. Sin.*, **12**, 537 (1991).
2. H. Huang et al., *Zhongcaoyao*, **15**(5), 21 (1984).
3. X. Q. An et al., *Zhongcaoyao*, **21**(4), 44 (1990).
4. Z. L. Huang et al., *Zhongcaoyao*, **18**(4), 22 (1987).
5. C. Z. Li et al., *Zhongcaoyao*, **14**(7), 27 (1983).

# SICHUAN LOVAGE (*CHUANXIONG*)

Rhizome of *Ligusticum chuanxiong* Hort. (syn. *L. wallichii* auct. sin. non Franch.) (Family Umbelliferae or Apiaceae), an aromatic perennial herb, up to 1 m. high, with erect stem and irregularly knobby fist-like rhizome; all cultivated, mainly in southern China, including the provinces of Sichuan, Guizhou and Yunnan, with Sichuan as the primary producer. Rhizome is collected in late May to early June from plants planted the year before; after being rid of stem, leaves, rootlets and dirt, it is sun-dried or dried by artificial heat, followed by further removing of all rootlets and dirt by stirring in bamboo baskets (IMM-CAMS).

First recorded use dates back to the *Shen Nong Ben Cao Jing* (circa 200 B.C.–100 A.D.). Traditionally considered to be acrid tasting and warming; promoting blood circulation and activating vital energy (*huo xue xing qi*); alleviating mental depression (*kai yu*); and removing rheumatic and arthritic pain (*qu feng*). Traditionally used in treating irregular menses, amenorrhea, dysmenorrhea, abdominal mass (*zheng jia*), abdominal pain, chest pain, swelling and pain due to traumatic injuries, headache, rheumatism and arthritic pain. Extensively used over the centuries as an ingredient in many famous beauty formulas (both internal and topical), ranking third or fourth among the most frequently used herbs in traditional Chinese beauty formulas (cosmetics).[1–3]

Contains alkaloids (0.15% in raw and 0.20% after wine-curing),[4] including tetramethylpyrazine (ligustrazine, chuanxiongzine), L-isobutyl-L-valine anhydride and perlolyrine;[5] phthalides, including butylphthalide, 4-hydroxy-3-butylphthalide (chuanxiongol), butylidene phthalide, hydroxybutylidene phthalide, dihydroxybutylidene phthalide, ligustilide, sedanenolide, cnidilide, neocnidilide, etc.;[5–10] phenols, including ferulic acid, caffeic acid, chrysophanol and vanillic acid;[5,11,12] volatile oil (1.0–1.6%, depending on sources) containing the phthalides;[13] adenine and adenosine;[14] spathulenol;[15] sedanonic acid, and others (HU, IMM-2).[11]

The chemical composition of *chuanxiong* is similar but distinctly different from that of ***gaoben*** (Chinese ligusticum).[13]

Chuanxiong has exhibited various biological activities, including: cardiovascular (coronary dilatation, increasing coronary, cerebral and renal blood flow, reducing vascular resistance, hypotensive, calcium antagonistic, and others, with tetramethylpyrazine being a major active component and perlolyrine a minor active component);[9,16–20] antispasmodic (ligustilide; cnidilide, ferulic acid and alkaloids);[9,11] antiplatelet aggregation (tetramethylpyrazine and ferulic acid);[9,11] Sedative (volatile oil); antibacterial, antifungal and antiviral; radiation protective and other effects (WANG, ZHOU AND WANG).

In addition to its cardiovascular and hematologic activities, tetramethylpyrazine also has antimetastatic effects in animals.[21] It is now used in China to treat ischemic cerebrovascular diseases, coronary heart disease and angina pectoris (WANG, ZHOU);[11,22] with one case of drug-induced skin rash reported.[23]

Adenosine has exhibited antiplatelet aggregation and central inhibitory activities (analgesic, reducing spontaneous motor activity and prolonging death time induced by caffeine); also has weak muscle relaxant effects (see ***ganoderma***).[24]

Ferulic acid has anti-allergic, broad cardiovascular and hematologic effects (antiplatelet aggregation, antithrombic, etc.) as well as inhibitory effects on lipid peroxidation (see ***angelica***).[25,26]

Toxicities of ferulic acid and tetramethylpyrazine were both fairly low [$LD_{50}$(i.v.) = 866 ± 29 mg/kg and 416 ± 17 mg/kg, respectively, in mice], but their toxicities were much lower when used together, indicating the traditional rationale of using both *danggui* (ferulic acid) and *chuanxiong* (tetramethylpyrazine) in the same formula to obtain the desired effects with lower toxicity.[27]

Powder and extracts (aqueous, oil, and hydroalcoholic) are used in hair care products for their vasodilating and traditional hair-nourishing properties to prevent hair loss and premature graying; and in skin-care products (e.g., cleansing creams, nourishing creams, acne creams and lotions) for their vasodilating, antimicrobial, anti-allergic effects as well as traditional skin-whitening (tyrosinase inhibitory),[28,29] anti-swelling and anti-wrinkle properties (ETIC, ZHOU).[1–3] Tetramethylpyrazine hydrochloride eye-drops are used in the prevention and treatment of near-sightedness.[30]

## REFERENCES

See the General References for CHP; ETIC; IMM-2; IMM-CAMS; JIANGSU; WANG; ZHOU; ZHOU AND WANG; ZHU.


1. W. X. Hong, *Fujian Zhongyiyao*, **18**(1), 39 (1987).
2. Y. B. Xie, *Jiangsu Zhongyi*, (5), 38 (1989).
3. J. H. Chen, *Shandong Zhongyi Xueyuan Xuebao*, **15**(1), 58 (1991).
4. Y. Q. Ou, *Zhongchengyao*, **11**(9), 18 (1989).
5. F. Y. Cao et al., *Zhongcaoyao*, **14**(6), 1 (1983).
6. M. Puech-Baronnat et al., *Planta Med.*, **50**, 105 (1984)
7. M. Kaouadji, *Plantes Med. Phytother.*, **17**, 147 (1983).
8. P. S. Wang et al., *Zhongcaoyao*, **16**(3), 41 (1985).
9. J. M. Xu, *Zhongchengyao*, **11**(1), 37 (1989).
10. Y. S. Wen et al., *Zhongcaoyao*, **17**(3), 26 (1986).
11. Y. J. Ma and S. S. Zhu, *Chin. J. Integr. Trad. Western Med.*, **4**, 574 (1984).
12. P. S. Wang et al., *Zhongcaoyao*, **16**(5), 45 (1985).
13. B. C. Zhang et al., *Zhongcaoyao*, **17**(8), 34 (1986).
14. Y. X. Wang et al., *Zhongcaoyao*, **16**(11), 17 (1985).

15. P. S. Wang et al., *Zhongcaoyao*, **16**(4), 30 (1985).
16. L. B. Hou et al., *Zhongcaoyao*, **23**(11), 583 (1992).
17. P. G. Xiao and K. J. Chen, *Phytother. Res.*, **1**(2), 53 (1987).
18. J. A. O. Ojewole, *Planta Med.*, **42**, 223 (1981).
19. M. G. Feng et al., *Acta Pharmacol. Sin.*, **9**, 548 (1988).
20. Y. L. Wang and Y. K. Ba, *Chin. J. Integr. Trad. Western Med.*, **5**, 291 (1985).
21. J. R. Liu and S. B. Ye, *Chin. J. Pharmacol. Toxicol.*, **7**, 149 (1993).
22. J. W. Yang, *Liaoning Zhongyi Zazhi*, **12**(5), 26 (1988).
23. K. L. Zhang et al., *Chin. J. Integr. Trad. Western Med.*, **6**, 375 (1986).
24. Y. Kasahara and H. Hikino, *Phytother. Res.*, **1**, 173 (1987).
25. J. L. Wu and D. Y. Wang, *Zhongguo Yaoxue Zazhi*, **28**, 267 (1993).
26. H. J. Hu and B. Q. Hang, *Acta Pharmacol. Sin.*, **12**, 426 (1991).
27. J. Xu et al., *Zhongguo Zhongyao Zazhi*, **17**, 680 (1992).
28. X. T. Liu, *Zhongchengyao*, **13**(3), 9 (1991).
29. Y. Masamoto et al., *Planta Med.*, **40**, 361 (1980).
30. X. L. Hu and X.Q. Hu, *Zhongcaoyao*, **14**(5), 16 (1983).

# SKULLCAP, BAIKAL (*HUANGQIN*)

Root of *Scutellaria baicalensis* Georgi (Family Labiatae or Lamiaceae), a perennial herb up to 60 cm high, with a large and long taproot; plant native to eastern Asia. Root is collected in spring and autumn, usually from 3- to 4-year-old plants. After having rootlets and dirt removed, the root is partially dried, rid of root bark, and then further dried to completion. Produced mainly in northern China. Several other species of *Scutellaria* from other regions of China are also used, with specific names indicating their geographical or botanical origins (JIANGSU).[1]

Earliest use (in ointment form for treating wounds and cramps) was described in the *Wu Shi Er Bing Fang* (ca. 1065 B.C. to 771 B.C.).[2] It was later listed under the middle category of drugs in the *Shen Nong Ben Cao Jing* (ca. 200 B.C.–100 A.D.). Traditionally considered to taste bitter and cold and to have heat-clearing, wetness-drying (*zao shi*), fire-purging (*xie huo*), detoxifying, hemostatic and fetus-calming properties; used to treat heat (*re*) related conditions, including restlessness and thirst (*fan ke*), cough, diarrhea, tight chest and abdominal distention (*pi man*), jaundice, fever, vomiting of blood, sores, carbuncles, furuncles, red eye with swelling and pain, and threatened abortion (restless fetus).

In recent years, it is also used in treating burns, often combined with phellodendron, giant knotweed and garden burnet (see individual entries), as well as different kinds of infections (bacterial, viral, etc.) and hypertension, among others (JIANGSU).

Baikal skullcap contains flavonoids and their glycosides, including baicalein (5,6,7-trihydroxyflavone), wogonin (5,7-dihydroxy-8-methoxyflavone), skullcapflavone I (5,6′-dihydroxy-7,8-dimethoxyflavone), skullcapflavone II (2′,5-dihydroxy-6,6′,7,8-tetramethoxyflavone), oroxylin A (5,7-dihydroxy-6-methoxyflavone), koganebanain (5,7-dihydroxy-6,8,2′,3′-tetramethoxyflavone), (2*S*), 2′, 5, 6′, 7-tetrahydroxyflavanone, (2*R*,3*R*), 2′,3,5,6′,7-pentahydroxyflavanone, 2′,5,5′,7-tetrahydroxy-6′,8-dimethoxyflavone, baicalin (5,6,7-trihydroxyflavone-7-*O*-D-glucuronide) wogonoside (5,7-dihydroxy-8-methoxyflavone-7-*O*-glucuronide), oroxylin A glucuronide and others, with baicalin in major concentration (3.6–6.2%); benzoic acid, $\beta$-sitosterol, etc, (IMM-1; JIANGSU).[1] Roots of other *Scutellaria* species have similar chemistry as Baikal skullcap.[1]

Decoction and alcohol extractives have exhibited broad antibacterial and antifungal effects *in vitro*; also active against several viruses, including influenza strains $PR_8$ and Asian A; baicalein being one of the major active principles. (2*S*),2′,5,6′,7-Tetrahydroxyflavanone is active against Gram-negative bacteria (WANG).[3]

Baicalein, baicalin, wogonin, skullcapflavone II and 2′,5,5′,7-tetrahydroxy-6′,8-dimethoxyflavone have antiinflammatory and antiallergic effects in animals; skullcapflavone II also exhibited cytotoxic effects on L1210 cells (JIANGSU).[4–6]

Other biological effects of *huangqin* include: sedative; antipyretic; hypotensive, diuretic; antiarthritic; hypolipemic; cholagogic; antispasmodic; and detoxicant (WANG);[7] also strongly antioxidant (due to flavonoids).[8–10]

Baikal skullcap has very low toxicity: 10 g/kg of decoction p.o. and 2g/kg i.v. of alcohol extractives produced only sedation but no deaths in rabbits; and $LD_{50}$ of baicalin in mice is 3.081 g/kg (i.v.). However, 15 mg/kg of baicalin administered i.v. to rabbits was fatal within 48 hours (WANG).

Extracts used in skin care products (esp. skin freshener, acne creams and lotions, etc.) for their astringent, antiinflammatory and antimicrobial effects (ETIC, ZHOU); also used in toothpaste.[11]

## REFERENCES

See the General References for BAILEY 1; CHP; ETIC; FOSTER AND YU; HU; IMM-1; JIANGSU; WANG; ZHOU.

1. W. Z. Song, *Yaoxue Xuebao*, **16**, 139 (1981).
2. Q. M. Sun, *Yaoxue Tongbao*, **17**(5), 33 (1982).
3. M. Kubo et al., *Planta Med.*, **43**, 194 (1981).
4. Y. Kimura et al., *Planta Med.*, **51**, 132 (1985).
5. H. Otsuka et al., *J. Nat. Prod.*, **51**, 74 (1988).
6. S. H. Ryu et al., *Planta Med.*, **51**, 462 (1985).
7. Z. Li and X. Y. Guo, *Chin. J. Integr. Trad. Western Med.*, **9**, 698 (1989).
8. Y. Kimura et al., *Planta Med.*, **50**, 290 (1984).
9. Y. Kimura et al., *Chem. Pharm. Bull.*, **29**, 2610 (1981).
10. L. C. Song et al., *Shaanxi Zhongyi*, **14**, 185 (1993).
11. Z. Lin, *Dazhong Zhongyiyao*, (3), 47 (1990).

# General References


American Dental Association (ADA). 1973. *Accepted Dental Therapeutics*. 35th ed. A.D.A. Chicago, Ill.

American Herbal Products Association (AHPA). 1992. *Herbs of Commerce*. American Herbal Products Association. Austin, Texas.

American Pharmaceutical Association (APhA). 1990. *Handbook of Nonprescription Drugs*. 10th ed. A.Ph.A. Washington, D.C.

Arctander, S. 1960. *Perfumes and Flavor Materials of Natural Origin*. Published by the author. Elizabeth, N.J.

Ayensu, E. S. 1981. *Medicinal Plants of the West Indies*. Reference Publications. Algonac, Mich.

Bailey, L. H. 1949. *Manual of Cultivated Plants*. MacMillan. New York.

Bailey, L, H. 1942. *The Standard Cyclopedia of Horticulture*, 3 vols. MacMillan. New York,

Balsam, M. S. and E. Sagarin, Eds. 1972. *Cosmetics Science and Technology*. 2nd ed. 2 vols. Wiley-Interscience. New York.

Bauer, K., D. Garbe, and H. Surburg. 1990. *Common Fragrance and Flavor Materials*. 2nd ed. VCH Publishers. New York.

Bensky, D. and A. Gamble. 1986. *Chinese Herbal Medicine: Materia Medica*. Eastland Press, Inc. Seattle.

Bianchini, F. and F. Corbetta. 1977. *Health Plants of the World-Atlas of Medicinal Plants*. Newsweek Books. New York.

Blumenthal, M. Ed., S. Klein, trans. 1995. German Bundesgesuntheitsamt (BGA) Commission E Therapeutic Monographs on Medicinal Products for Human Use. (English translation). American Botanical Council. Austin, Texas (in edit).

Bradly, P. R., Ed. 1992. *British Herbal Compendium*. Vol. 1., British Herbal Medicine Association. Dorset, England.

Brouk, B. 1975. *Plants Consumed by Man*. Academic. London.

Burkill, I. H. 1966. *A Dictionary of the Economic Products of the Malay Peninsula*. 2 vols. Ministry of Agriculture and Cooperatives. Kuala Lumpur.

Cheung, S. C. and N. H. Li, Eds. 1978, 1981, 1983, 1985, 1986. *Chinese Medicinal Herbs of Hong Kong*. 5 vols. Commercial Press. Hong Kong. (in Chinese and English.)

Chinese Ministry of Health Bureau of Drug Administration (CMH). 1990. *Manual of Chinese Drugs*. 2nd ed. People's Health Publications. Beijing. (in Chinese.)

Chinese Pharmacopeia Committee (CHP), Chinese Ministry of Health. *Chinese Pharmacopeia* (1990). People's Health Publications. Beijing. (in Chinese.)

Chittendon, F. J. 1956. *Dictionary of Gardening*. 4 vols. Oxford University Press. London.

Claus, E. P. 1961. *Pharmacognosy*. 4th ed. Lea & Febiger. Philadelphia, Pa.

Coon, N. 1974. *The Dictionary of Useful Plants*. Rodale. Emmaus, Pa.

Council of Scientific and Industrial Research. (CSIR). 1948–1985. *The Wealth of India*. 11 vols. Publications & Information Directorate, Council of Scientific & Industrial Research. New Delhi.

Crellin, J. K. and J. Philpott. 1989. *Herbal Medicine Past and Present*. 2 vols. Duke University Press. Durham, N.C.

Davidson, R. L., Ed. 1980. *Handbook of Water Soluble Gums and Resins*. McGraw-Hill. New York.

Deng, W. L., Ed. 1990. *Pharmacology and Applications of Traditional Chinese Formulas*. Chongqing Publications. Chongqing, Sichuan. (in Chinese.)

Der Marderosian, A. and L. Liberti. 1988. *Natural Product Medicine: A Scientific Guide to Foods, Drugs, Cosmetics*. George F. Stickley Co. Philadelphia.

Devon, T. K. and A. I. Scott. 1975. *Handbook of Naturally Occurring Compounds. Vol. 1. Acetogenins, Shikimates, and Carbohydrates*. Academic. New York.

Devon, T. K. and A. I. Scott. 1972. *Handbook of Naturally Occurring Compounds. Vol. 2. Terpenes*. Academic. New York.

Duke, James A. 1981. *Handbook of Legumes of World Economic Importance*. Plenum Press. New York.

Duke, James A. 1985. *CRC Handbook of Medicinal Herbs*. CRC Press. Boca Raton, Florida.

Duke, James A. 1989. *CRC Handbook of Nuts*. CRC Press. Boca Raton, Florida.

Economic and Technical Information Center (ETIC), Ministry of Light Industries. 1987. *Utilization of Natural Additives in Cosmetics*. ETIC. People's Republic of China. (in Chinese.)

Erichsen-Brown, C. 1989. *Medicinal and Other Uses of North American Plants*. Dover Publications. New York.

ESCOP, 1990. *Proposal for European Monographs*. Vol. 1, ESCOP Secretariat. Bevrijdingslaan, The Netherlands.

ESCOP, 1992. *Proposal for European Monographs*. Vol. 2. ESCOP Secretariat. Bevrijdingslaan, The Netherlands.

ESCOP, 1992. *Proposal for European Monographs*. Vol. 3, ESCOP Secretariat. Bevrijdingslaan, The Netherlands.

Farnsworth, N. R. 1969. *The Lynn Index: A Bibliography of Phytochemistry*. Vol. 6. Norman R. Farnsworth. Pittsburgh.

Farnsworth, N. R. 1974. *The Lynn Index: A Bibliography of Phytochemistry*. Vol. 8. Norman R. Farnsworth. Chicago.

Farnsworth, N. R. 1975. *An Evaluation of "Atlas of Common Chinese Drugs, a compilation by the Chinese College of Medical Sciences. Peking*." University of Illinois Medical Center. Chicago, Ill. See also *Herbal Pharmacology in the People's Republic of China*.

Farnsworth, N. R. et al. 1971. *The Lynn Index: A Bibliography of Phytochemistry*. Vol. 7. Norman R. Farnsworth. Chicago.

Felter, H. W. and J. U. Lloyd. 1906. *King's American Dispensatory*. 2 vols. 18th ed. Reprinted. 1983. Eclectic Medical Publications. Portland, Ore.

Fernald, M. L. 1950. *Gray's Manual of Botany*. American Book. New York.

Flavor and Extract Manufacturer's Association of the United States (FEMA). 1975. *Results of Second FEMA Survey of Flavoring Ingredients – Average Maximum Use Levels*. FEMA. Washington, D.C.

J. E. Fogarty International Center for Advanced Study in the Health Sciences (FOGARTY). 1974. *A Barefoot Doctor's Manual*. Department of Health Education, and Welfare Publication No. (NIH) 75-695. National Institutes of Health. Washington, D.C. (Translation of Chinese text).

*Food Chemicals Codex*. 2nd ed. 1972. National Academy of Sciences. Washington, D.C.

Foster, S. 1993. *Herbal Renaissance*. Gibbs Smith Publisher. Layton, Utah.

Foster, S. and R. Caras. 1994. *A Field Guide to Venomous Animals and Poisonous Plants of North America (North of Mexico)*. Houghton Mifflin Co. Boston.

Foster, S. and J. A. Duke. 1990. *A Field Guide to Medicinal Plants: Eastern and Central North America*. Houghton Mifflin Co. Boston.

Foster, S, and C. X. Yue. 1992. *Herbal*

*Emissaries: Bringing Chinese Herbs to the West*. Healing Arts Press. Rochester, Vt.

Frohne, D. and H. J. Pfänder. 1983. *A Colour Atlas of Poisonous Plants*. Wolfe Publishing, Ltd. London.

Fujian Provincial Institute of Medical and Pharmaceutical Research. 1979, 1982. *Records of Fujian Materia Medica*. 2 vols. Fujian People's Press and Fujian Scientific and Technical Publications. Fuzhou. (in Chinese.)

Furia, T. E., Ed. 1975. *Handbook of Food Additives*. 2nd ed. CRC Press. Cleveland, Ohio.

Furia, T. E. and N. Bellanca, Eds. 1975. *Fenaroli's Handbook of Flavor Ingredients*. 2nd ed. Vol. 1. CRC Press. Cleveland, Ohio.

Glasby, J. S. 1976. *Encyclopedia of Antibiotics*. Wiley-Interscience. London.

Glasby, J. S. 1991. *Dictionary of Plants Containing Secondary Metabolites*. Taylor & Francis. New York.

Gleason, H. A., and A. Cronquist. 1991. *Manual of Vascular Plants*. 2nd. ed. New York Botanical Garden. New York.

Glicksman, M. 1969. *Gum Technology in the Food Industry*. Academic. New York.

Goodman, L. S. and A. Gilman, Eds. 1985. *The Pharmacological Basis of Therapeutics*. 7th ed. MacMillan. New York.

Gosselin, R. E. et al. 1976. *Clinical Toxicology of Commercial Products: Acute Poisoning*. 4th ed. Williams & Wilkins. Baltimore, Md.

Grieve, M. 1967. *A Modern Herbal*. 2 vols. Hafner. New York.

Guenther, E. 1948. *The Essential Oils*. 6 vols. Van Nostrand. New York.

Harborne, J. B. 1973. *Phytochemical Methods*. Chapman & Hall. London.

Harbourne, J. B. and H. Baxter. 1993. *Phytochemical Dictionary: A Handbook of Bioactive Compounds from Plants,* Taylor & Francis. London.

Hardin, J. W. and J. M. Arena. 1974. *Human Poisoning from Native and Cultivated Plants*. 2nd ed. Duke University Press. Durham, N.C.

Harris, R. S. and E. Karmas, Eds. 1975. *Nutritional Evaluation of Food Processing*. 2nd ed. AVI. Westport, Conn.

Hay, R. and P. M. Synge. 1975. *The Color Dictionary of Flowers and Plants for Home and Garden*. Crown. New York.

*Herbal Pharmacology in the People's Republic of China – A Trip Report of the American Herbal Pharmacology Delegation*. 1975. National Academy of Sciences. Washington, D.C.

Hickman, J. C., Ed. 1993. *The Jepson Manual – Higher Plants of California*, University of California Press. Berkeley.

Hocking, G. M. 1955. *A Dictionary of Terms in Pharmacognosy and Economic Botany*. Thomas. Springfield, Ill.

Hoffman, D. 1987. *The Herbal Handbook: A User's Guide to Medical Herbalism*. Healing Arts Press. Rochester, Vt.

Hortus Third: *A Concise Dictionary of Plants Cultivated in the United States & Canada*. L. H. Bailey Hortorium Staff. Cornell University. 1976. MacMillan. New York.

Hsu, H. H. et al. 1989. *Oriental Materia Medica – A Concise Guide*. Oriental Healing Arts Institute. Long Beach, Calif.

Hu, S. L. 1989. *Indigenous Chinese Drugs*. Heilongjiang Scientific and Technical Publishers. Harbin. (in Chinese.)

Institute of Materia Medica (IMM-1), Chinese Academy of Medical Sciences. 1982. *Records of Chinese Materia Medica. Vol. 1.* People's Health Publications. Beijing. (in Chinese.)

Institute of Materia Medica, (IMM-2) Chinese Academy of Medical Sciences. 1982. *Records of Chinese Materia Medica. Vol. 2.* People's Health Publications. Beijing. (in Chinese.)

Institute of Materia Medica (IMM-3), Chinese Academy of Medical Sciences. 1981. *Records of Chinese Materia Medica. Vol. 3.* People's Health Publications. Beijing. (in Chinese.)

Institute of Materia Medica (IMM-4) and Institute of Medicinal Plant Development, Chinese Academy of Medical Sciences. 1988. *Records of Chinese Materia Medica. Vol. 4.* People's Health Publications. Beijing. (in Chinese.)

Institute of Materia Medica, Chinese Academy of Medical Sciences (IMM-CAMS). 1979. *Techniques for Cultivating Chinese Herbs.* People's Health Publications. Beijing. (in Chinese.)

Isler, O. et al., Eds. 1971. *Carotenoids.* Halsted (Wiley). New York.

Jiangsu Institute of Modern Medicine. 1977. *Encyclopedia of Chinese Drugs.* 3 vols. Shanghai Scientific and Technical Publications. Shanghai, People's Republic of China. (in Chinese).

Jiangsu Provincial Institute of Botany (JIANGSU PROVINCIAL 1). 1988. *Essentials of Medicinal Plants of New China. Vol. 1.* Shanghai Scientific and Technical Publications. Shanghai. (in Chinese.)

Jiangsu Provincial Institute of Botany (JIANGSU PROVINCIAL 3). 1990. *Essentials of Medicinal Plants of New China.* Vol. 3. Shanghai Scientific and Technical Publications. Shanghai. (in Chinese.)

Jilin Provincial Institute of Traditional Chinese Medicine and Materia Medica. 1982. *Records of Plant Drugs of Changbai Mountain.* Jilin People's Press. Changchun. (in Chinese.)

Johnson, A. H. and M. S. Peterson. 1974. *Encyclopedia of Food Technology.* AVI. Westport, Conn.

Karrer, W. 1958. *Konstitution und Vorkommen der organischen Pflanzenstoffe (exclusive Alkaloide).* Birkhäuser Verlag. Basel, Switzerland. (In German).

Kartesz, J. T. and R. Kartesz. 1980. *A Synonymized Checklist of the Vascular Flora of the United States, Canada, and Greenland.* University of North Carolina Press. Chapel Hill.

Kartesz, J. T. 1994. *A Synonymized Checklist of the Vascular Flora of the United States, Canada, and Greenland.* Second Edition, 2 vols. Timber Press, Portland, Ore.

Keay, R. W. J. 1989. *Trees of Nigeria.* Clarendon Press Oxford. Oxford, England.

Kennedy, J. F. Ed. 1988. *Carbohydrate Chemistry.* Oxford University Press. London.

Kindscher, K. 1992. *Medicinal Wild Plants of the Prairie.* University Press of Kansas. Lawrence.

Kreig, M. B. 1964. *Green Medicine. The Search for Plants that Heal.* Rand McNally. Skokie, Ill.

Krochmal, A. and C. Krochmal. 1975. *A Guide to the Medicinal Plants of the United States.* Quadrangle/The New York Times Book Co. New York.

Kutsky, R. J. 1973. *Handbook of Vitamins and Hormones.* Van Nostrand Reinhold. New York.

Lawrence, A. A. 1976. *Natural Gums for Edible Purposes.* Noyes. Park Ridge, N.J.

Leung, A. Y. 1984. *Chinese Herbal Remedies.* Universe Books. New York. [Republished as: *Chinese Healing Foods and Herbs.* 1993. AYSL Corp. Glen Rock, N.J.]

Lewis, W. H. and M. P. H. Elvin-Lewis. 1977. *Medical Botany. Plants Affecting Man's Health.* Wiley-Interscience. New York.

Liener, I. E., Ed. 1969. *Toxic Constituents of Plant Foodstuffs.* Academic. New York.

List, P. H. and L. Hörhammer. 1969–1976. *Hagers Handbuch der Pharmazeutischen Praxis.* Vols. 2–5. Springer-Verlag. Berlin. (In German).

Lloyd, J. U., 1921. *Origin and History of all the Pharmacopeial Vegetable Drugs,*

*Chemicals, and Preparations, with Bibliography*. Vol. 1, The Caxton Press. Cincinnati.

Lu, Q. Y. and M. Li. 1987. *Chinese Herbs for Life Extension*. People's Health Publications. Beijing. (in Chinese.)

Lucas, R. 1966. *Nature's Medicines – The Folklore, Romance and Value of Herbal Remedies*. Parker. West Nyack, N.Y.

Lust, J. B. 1974. *The Herb Book*. Benedict Lust. Simi Valley, Calif.

Ma, X. M. 1980. *Processing Methods for Chinese Materia Medica. Revised*. Shaanxi Scientific and Technical Publishers. Xi'an. (in Chinese.)

Mabberly, D. J. 1987. *The Plant Book: A Portable Dictionary of the Higher Plants*. Cambridge University Press. New York.

Marsh, A. C. et al. 1977. *Composition of Foods, Spices and Herbs. Raw, Processed, Prepared*. Agriculture Handbook No. 8-2. Agricultural Research Service, U.S. Department of Agriculture. Washington, D.C.

*Martindale: The Extra Pharmacopoeia*. 1982. The Pharmaceutical Press. London.

Masada, Y. 1976. *Analysis of Essential Oils by Gas Chromatography and Mass Spectrometry*. Halsted (Wiley). New York.

*The Merck Index. An Encyclopedia of Chemicals and Drugs*. 10th ed. 1983. Merck. Rahway, N.J.

Merory, J. 1968. *Food Flavorings, Composition, Manufacture and Use*. AVI. Westport, Conn.

Moerman, D. E. 1986. *Medicinal Plants of Native America*. 2 vols. Technical Reports, No. 19, Research Reports in Ethnobotany, Contribution 2. Univ. of Michigan Museum of Anthropology. Ann Arbor.

Moore, M. 1989. *Medicinal Plants of the Desert and Canyon West*. Museum of New Mexico Press. Santa Fe, N.M.

Moore, M. 1993. *Medicinal Plants of the Pacific West*. Red Crane Books, Santa Fe, N.M.

Morton, J. F. 1974. *Folk Remedies of the Low Countries*. Seemann. Miami, Fla.

Morton, J. F. 1976. *500 Plants of South Florida*. Seemann. Miami, Fla.

Morton, J. F. 1977. *Major Medicinal Plants: Botany, Culture, and Uses*. Thomas. Springfield, Ill.

Morton, J. F. 1981. *Atlas of Medicinal Plants of Middle America*. Charles C. Thomas. Springfield, Illinois.

Morton, J. F. 1987. *Fruits of Warm Climates*. Creative Resources Systems, Inc. Winterville, S.C.

Munz, P. A. and D. D. Keck. 1968. *A California Flora*. University of California Press. Berkeley, Calif.

Nakanishi, K. et al., Eds. 1975. *Natural Products Chemistry*. 2 vols. Kodansha, Tokyo and Academic. New York.

Nanjing Pharmaceutical Institute. 1960. *Materia Medica*. Shao Hwa Society for Cultural Services. Hong Kong. (In Chinese.)

National Collection of Chinese Herbal Drugs Editorial Committee. 1983. *National Collection of Chinese Herbal Drugs*. 2 vols. People's Health Publications. Beijing. (in Chinese.)

*National Formulary*. 16th ed. (NF XVI). 1985. American Pharmaceutical Association. Washington, D.C.

de Navarre, M. G. 1975. *The Chemistry and Manufacture of Cosmetics*. 2nd. ed. Vol. 3 and 4. Continental. Orlando, Fla.

Nikitakis, J. M. 1988. *CTFA Cosmetic Ingredient Handbook*. The Cosmetic, Toiletry and Fragrance Association, Inc. Washington, D.C.

Peng, M. Q. 1987. *Comprehensive Treatise on Chinese Medicinal Foods*. Sichuan Scientific and Technical Publications. Chengdu. (in Chinese.)

Peterson, M. S. and A. H. Johnson. 1978. *Encyclopedia of Food Science*. AVI. Westport, Conn.

Petrides, G. A. 1988. *Peterson Field Guide*

*to Eastern Trees*. Houghton Mifflin Co. Boston.

Phillips, R. W. 1973. *Science of Dental Materials*. 7th ed. Saunders. Philadelphia, Pa.

Polunin, O. and B. E. Smythies. 1973. *Flowers of Southwest Europe*. Oxford University Press. London.

Poucher, W. A. 1974. *Perfumes, Cosmetics, and Soaps, Vol. 1 The Law Materials of Perfumery*. 7th ed. Halsted (Wiley). New York.

Raffauf, R. F. 1970. *A Handbook of Alkaloids and Alkaloid-Containing Plants*. Wiley-Interscience. New York.

*Remington's Pharmaceutical Sciences*. 15th ed. 1975. Mack. Easton, Pa.

Rose, J. 1976. *The Herbal Body Book*. Grosset & Dunlap. New York.

Rosengarten, F., Jr. 1969. *The Book of Spices*. Livingston. Wynnewood, Pa.

Sargent, C. S. 1965. *Manual of Trees of North America*. 2 vols. Dover. New York.

Sax, N. I. 1975. *Dangerous Properties of Industrial Materials*. 4th ed. Van Nostrand Reinhold. New York.

Simon, J. E., A. F. Chadwick, and L. E. Craker. 1984. *Herbs: An Indexed Bibliography, 1971–1980*. Archon Books. Hamden, Conn.

Stahl, E., Ed. 1973. *Drug Analysis by Chromatography and Microscopy*. Ann Arbor Science. Ann Arbor, Mich.

Steiner, R. P., Ed. 1986. *Folk Medicine: The Art and the Science*. American Chemical Society. Washington, D.C.

Steinmetz, E. F. 1957. *Codex Vegetabilis*. E. F. Steinmetz. Amsterdam.

Tang, W. and G. Eisenbrand, Eds. 1992. *Chinese Drugs of Plant Origin*. Springer-Verlag. Berlin.

Terrell, E. E. 1977. *A Checklist of Names for 3,000 Vascular Plants of Economic Importance*. Agriculture Handbook No. 505. Agricultural Research Service. U.S. Department of Agriculture. Washington, D.C.

Trease, G. E. and W. C. Evans. 1978. *Pharmacognosy*. 11th ed. Bailliere Tindall. London.

Tu, G. S. 1988. *Pharmacopoeia of the People's Republic of China* (English Translation of 1985 *Pharmacopeia of the People's Republic of China*). China Pharmaceutical Books Co. Hong Kong.

Tucker, A. O. 1986. *Botanical Nomenclature of Culinary Herbs and Potherbs*. In L. E. Craker and J. E. Simon, Eds. *Herbs, Spices, and Medicinal Plants: Recent Advances in Botany, Horticulture, and Pharmacology*. Vol. 1. Oryx Press. Phoenix, Ariz., pp. 33–80.

Tucker, A. O., J. A. Duke and S. Foster. 1989. *Botanical Nomenclature of Medicinal Plants*. In L. E. Craker and J. E. Simon, Eds. *Herbs, Spices, and Medicinal Plants: Recent Advances in Botany, Horticulture, and Pharmacology*. Vol. 4. Oryx Press. Phoenix, Ariz., pp. 169–242.

Tucker, A. O. and B. M. Lawrence. 1987. *Botanical Nomenclature of Commercial Sources of Essential Oils, Concretes, and Absolutes*. In L. E. Craker and J. E. Simon, Eds. *Herbs, Spices, and Medicinal Plants: Recent Advances in Botany, Horticulture, and Pharmacology*. Vol. 2. Oryx Press. Phoenix, Ariz., pp. 183–220.

Tutin, T. G. et al., Eds. 1964. *Flora Europaea*. Vol. 1, Cambridge University Press. Cambridge.

Tutin, T. G. et al., Eds. 1968. *Flora Europaea*. Vol. 2, Cambridge University Press. Cambridge.

Tutin, T. G. et al., Eds. 1972. *Flora Europaea*. Vol. 3, Cambridge University Press. Cambridge.

Tutin, T. G. et al., Eds. 1976. *Flora Europaea*. Vol. 4, Cambridge University Press. Cambridge.

Tutin, T. G. et al., Eds. 1980. *Flora*

*Europaea*. Vol. 5, Cambridge University Press. Cambridge.

Tyler, V. E. 1993. *The Honest Herbal*. 3d ed. Pharmaceutical Products Press. Binghamton, N.Y.

Tyler, V. E. 1994. *Herbs of Choice – The Therapeutic Use of Phytomedicinals*, Pharmaceutical Products Press. Binghamton, N.Y.

Tyler, V. E., L. R. Brady, and J. E. Robbers. 1988. *Pharmacognosy*. 9th ed. Lea & Febiger. Philadelphia, Pa.

United States Department of Agriculture (USDA). 1978. *Agricultural Statistics*. U.S. Government Printing Office. Washington, D.C.

*The Dispensatory of the United States of America*. 23rd ed. (USD 23rd). 1943. Lippincott. Philadelphia, Pa.

*The United States Dispensatory and Physicians Pharmacology*. 26th ed: (USD 26th). 1967. Lippincott. Philadelphia, Pa.

*The United States Pharmacopeia*. 21th rev. (USP XXI). 1985. U.S.P. Convention. Rockville, Md.

Uphof, J. C. T. 1968. *Dictionary of Economic Plants*. J. Cramer. Stechert-Hafner. New York.

Wallis, T. E. 1967. *Textbook of Pharmacognosy*. 5th ed. Churchill. London.

Wang, Y. S. et al., Eds. 1983. *Pharmacology and Applications of Chinese Materia Medica*. People's Health Publications. Beijing. (in Chinese.). [English translation: Chang, H. M. and P. P. But. 1986. *Pharmacology and Applications of Chinese Materia Medica*. World Scientific. Hong Kong.]

Watt, B. K. and A. L. Merrill. 1975. *Composition of Foods, Raw, Processed, Prepared*. Agriculture Handbook No. 8. Agricultural Research Service, U.S. Department of Agriculture. Washington, D.C.

Watt, J. M. and M. G. Breyer-Brandwijk. 1962. *The Medicinal and Poisonous Plants of Southern and Eastern Africa*. 2d. ed. E. & S. Livington Ltd. Edinburgh and London.

Weiss, R. F. 1988. *Herbal Medicine* (translated from German by A. R. Meuss). Beaconsfield Publishers Ltd. Beaconsfield, England.

Weniger, B. and L. Robineau. 1988. *Elements for a Caribbean Pharmacopeia*. TRAMIL 3 Workshop. Havana, Cuba. Nov. 1988.

Whistler, R. L. and J. N. BeMiller, Eds. 1973. *Industrial Gums*. Academic. New York.

Willaman, J. J. and B. G. Schubert. 1961. *Alkaloid-Bearing Plants and Their Contained Alkaloids*. U.S. Department of Agriculture Technical Bulletin 1234. Washington, D.C.

Wren, R. C., Revised by E. M. Williamson, and F. J. Evans. 1988. *Potter's New Cyclopedia of Botanical Drugs and Preparations*. 8th ed. C. W. Daniel Co. Essex, England.

Xiao, P .G., Ed. 1988, 1988, 1989, 1989, 1989, 1989, 1990, 1990, 1990, 1990, 1991. *Pictorial Records of Chinese Herbal Drugs*. 10 vols. plus index. People's Health Publications and Commercial Press. Beijing and Hong Kong.

Xu, G. J. and L. S. Xu, Eds. 1986. *Microscopy of Powdered Chinese Drugs*. People‘s Health Publications. Beijing. (in Chinese.)

Youngken, H. W. 1943. *Textbook of Pharmacognosy*. 5th ed. Blakiston. Philadelphia, Pa.

*Zhong Cao Yao Xue* Editorial Committee (ZCYX), Nanjing College of Pharmacy. 1987, 1976, 1980. *Chinese Herbal Drugs*. 3 vols. Jiangsu People's Press and Jiangsu Scientific and Technical Publications. Nanjing. (in Chinese.)

Zhou, J. H. and J. M. Wang, Eds. 1986. *Pharmacology of Chinese Materia Medica*. Shanghai Scientific and Technical Publications. Shanghai. (in Chinese.)

Zhou, X. C. 1989. *Chinese Herbs and Cosmetics*. Vacation Publishers, Ltd. Taipei. (in Chinese.)

Zhu, S. H. 1990. *Chinese Materia Medica of Commerce*. People's Health Publications. Beijing. (in Chinese.)

# General Index



The general and chemical indices for *Encyclopedia of Common Natural Ingredients* have been prepared by a professional indexer under commission from John Wiley & Sons, Inc. The authors assume no credit or responsibility for the accuracy, comprehensiveness, or quality of such indexing.

# Chemical Index



The general and chemical indices for *Encyclopedia of Common Natural Ingredients* have been prepared by a professional indexer under commission from John Wiley & Sons, Inc. The authors assume no credit or responsibility for the accuracy, comprehensiveness, or quality of such indexing.